The Chemistry of
Aqua Ions

The Chemistry of Aqua Ions

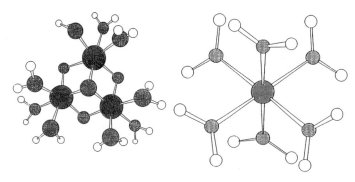

Synthesis, structure and reactivity

A tour through the periodic table of the elements

David T. Richens, C. Chem M.R.S.C.
School of Chemistry,
University of St Andrews, UK

JOHN WILEY & SONS

Chichester · New York · Weinheim · Brisbane · Singapore · Toronto

Copyright © 1997 by John Wiley & Sons Ltd,
Baffins Lane, Chichester,
West Sussex PO19 IUD, England

National   01243 779777
International (+ 44) 1243 779777
e-mail (for orders and customer service enquiries): cs-books@
wiley.co.uk
Visit our Home Page on http://www.wiley.co.uk
or http ://www.wiley.com

Other Wiley Editrorial Offices

John Wiley & Sons, Inc., 605 Third Avenue,
New York, NY 10158-0012, USA

VCH Verlagsgessellschaft mbH,
Pappelallee 3, D-69469 Weinheim, Germany

Jacaranda Wiley Ltd, 33 Park Road, Milton,
Queensland 4064, Australia

John Wiley & Sons (Asia) Pte Ltd, 2 Clementi Loop #02-01,
Jin Xing Distripark, Singapore 129809

John Wiley & Sons (Canada) Ltd, 22 Worcester Road,
Rexdale, Ontario M9W lLl, Canada

Library of Congress Cataloging-in-Publication Data

Richens, David T.
The chemistry of aqua ions/by David T. Richens.
      p.   cm.
   Includes bibliographical references (p.   —   ) and index.
   ISBN 0-471-97058-1 (alk. paper)
   1. Aqua ions. 2. Solution (Chemistry) I. Title.
QD562. A75R53   1997   96-30341
541.3'72—dc20   CIP

British Library Cataloguing in Publication Data

A catalogue record for this book is available from the British Library

ISBN 0 471 97058 1

Typeset in 10/12 pt Times from the author's disks by Thomson Press (India) Ltd.,
New Delhi
Printed and bound in Great Britain by Biddles Ltd, Guildford, Surrey
This book is printed on acid-free paper responsibly manufactured from sustainable
forestation, for which at least two trees are planted for each one used for paper
production.

To Caroline and Eleanor

Contents

Preface

ἀριστον μεν ύδωρ
'the noblest of elements is water'—Pindar

Water is the most abundant compound on the surface of the earth and is the principal constituent of all living organisms. It is the basis for life on this planet earth. Even before its molecular composition was established by Cavendish in the 1780's man, as far back as the time of the Thales, had long since recognised the vital importance of the substance water for sustaining life on earth. We depend on water, we are largely made up of water (65% of the human body by weight) and yet too much water can kill us. Water is the cheapest and most abundant solvent on this earth and by far our most precious resource. A human can survive for one month without food but only one week without water. We have developed, in a relatively short time, a reliance on a chemical economy based upon the use of 'non-aqueous' solvents whose prime source is fossil fuels. Although this reliance will continue well into the next century, it is not a limitless resource. However, water could be and indeed would have to be because we depend upon it for our own existence.

The earliest inorganic chemistry and indeed much of the earliest organic chemistry (based largely upon research into carbohydrates) was based upon water. This is apparent in the vast early literature of Gmelin and Beilstein. We may have to return to water as the solvent of the future. This book is produced at a time when environmental issues and the general move towards 'clean technology' are high on the world's political and economic agenda in the 1990s. Water is the ultimate environmentally friendly solvent. Understanding the equilibria and dynamic process within species dissolved and indeed largely based upon water is thus fundamental and surprisingly still evolving some 200 years after Cavendish's discovery. This book is concerned with an up to date review of the chemistry of what remains the simplest and most fundamental species present in aqueous solution for a chemical element—an *'aqua'* ion.

The motivation for writing this book was the fact that although aqua ions invariably get a mention in most descriptive inorganic textbooks covering the chemistry of the elements it struck me that this subject has now been developed to the extent of demanding independent and comprehensive coverage in a text of its own. Furthermore, no previous texts have sought to highlight preparative aspects along with properties and the salient reaction chemistry of the representative species for each element. If one expands the discussion of true 'aqua ions', i.e. the species $[M(OH_2)_n]^{m+}$, to include hydrolytic polymers and cluster species, and even hydroxy and oxyanions as I have done, then it is clear that a rich and diverse chemistry is at one's disposal with an array of simple but beautiful structural motifs, (see Section 1.1) enticing the chemist to get deeply involved in explaining the structures and reactivity patterns emerging.

I became familiar with the area of aqua ions when I worked as a postdoctoral fellow in the early 1980s in the laboratory of Professor Geoffrey Sykes at the University of Newcastle, UK to whom I owe much for the knowledge gained during the four years spent there and the motivation to continue research in the area. I learnt quickly to appreciate the simple beauty of studying these species. Next to the free ions themselves the aqua ions must represent the simplest species characteristic of a metal ion. It struck me very early on just how fundamentally important was the study of these species towards an understanding of not only a given element's aqueous solution chemistry but the chemical properties of that element in general. What could be more straightforward and satisfying than studying the kinetics and thermodynamics of simple replacement of a coordinated water ligand with another ligand or the simple water exchange process itself. In understanding the remarkable range of rates for these processes on different metal ions (spanning 20 orders of magnitude!) one gets an appreciation of the vastly differing reactivity throughout regions of the periodic table from the most inert of the metal ions ($Ir^{3+}, t_{1/2} \sim 10^{10}$s) to the most labile ($Cs^+$, $t_{1/2} \sim 10^{-10}$s). It is now appreciated that the major factors defining the particular kinetic lability of a coordinated water ligand are the size of the cationic charge density on the central atom (electrostatic) and the presence (or absence) of ligand field effects (electronic).

This unique comparison of 'like' species throughout the periodic table is possible since $[M(OH_2)_n]^{m+}$ ions (or hydrolytic polymeric derivatives thereof) are formed by the majority of the chemical elements, except those that are highly electronegative. Even then the absence of an aqua ion and the nature of the aqueous solution species formed instead tells us much about the chemistry of that particular element. In aqueous solution all ions are hydrated to some extent. For example good quality neutron diffraction data is now available to show that there is a well defined first hydration shell of water molecules around the halide anions in aqueous solution even if the bonding (now via the δ^+ H atoms) is somewhat weak. Therefore even here one can talk in terms of an '*aqua ion*' such as $[Cl(OH_2)_n]^-$ ($n \sim 4$–6). In the very highest oxidation states the positive charge on the central atom increases to such an extent as to polarise the coordinated waters resulting in spontaneous deprotonation by free water acting as a base and the formation of oxo and/or hydroxo species. Thus I^{7+} (aq) and Cl^{7+} (aq) exist not as $[I(OH_2)_n]^{7+}$ and $[Cl(OH_2)_n]^{7+}$ but as the oxo/hydroxo anions $[IO_4(OH)_2]^{3-}$ and $[ClO_4]^-$, explained in Section 1.3. The increased strength of the M–O bonding that parallels the formation of such species is exemplified in the half life for exchange of the oxygen groups on ClO_4^- with free water at 25°C (> 100 years) as opposed to the diffusion controlled rate $(t_{1/2} < 10^{-10}s)$ for water exchange on the hydrated Cl^- ion, $[Cl(OH_2)_n]^-$. Here lies a taste of the chemistry to be described.

This book is the first to treat the 'chemistry of aqua ions' as a self-contained topic. The book also focuses on molecular and electronic structure/reactivity relationships in highlighting specific examples as well as trends throughout the periodic table. Although this book is a comprehensive research review of the topic it is hoped that the compilation of structural and reactivity parameters contained within, together with an easy to read style, will provide the reader/student/ teacher with an enjoyable trip through the periodic table providing countless examples of how one can use the properties of '*aqua ions*' to illustrate so much of the chemistry of the elements as well as many of the fundamental principles of inorganic chemistry.

The material in the book assumes a basic understanding of inorganic chemistry but since many of the concepts are explained from a fundamental point of view (e.g. in Section 1.1 there is a basic introduction to the concepts of ionic hydration and of crystal field splitting) it should be suitable as a text for both undergraduate and postgraduate chemistry students. There are a number of appendices at the back of the book to help students with some of the background (e.g. the assignment of electronic spectral bands for transition metal ions), to provide help with nomenclature and for supplying useful supplementary data. The book should also have wide appeal to those studying, researching and teaching in topics such as environmental, biological and geochemistry.

I would like to thank Dr. J. Burgess, and Professors P. Moore, T. W. Swaddle and A. E. Merbach for help with proof reading and for useful discussions in the preparation of the manuscript. Finally I wish to thank my parents, close family and friends for their constant support throughout.

D. T. Richens
June 1996

Chapter 1

Introduction

The physical chemistry of aqueous solutions and ion hydration in particular has received extensive coverage in most inorganic and physical chemistry textbooks. Several texts in particular, e.g. the two *Ions in Solution* books by Burgess [1] and the books by Marcus and Magini [2] have given an excellent overview of metal ion hydration and the methods employed. In addition, Baes and Mesmer [3] have provided a comprehensive coverage and critical treatment of the data concerning metal ion hydrolysis up to the mid 1970s in their excellent book *The Hydrolysis of Metal Cations*. Solution X-ray diffraction data on aqua ions and complexes has been compiled and discussed in an excellent series of comprehensive reviews by Marcus [4], Ohtaki and Radnai [5] and Johansson [6]. Finally, Enderby and coworkers [7] have published an excellent series of articles on the use of neutron scattering and diffraction for the study of metal ion hydration.

We will start with a basic introduction to the concept of ion hydration and a general survey of aqua ions throughout the periodic table in Section 1.1. The methods used for the study of the solution structure of hydrated ions are then surveyed in Section 1.2 Section 1.3 covers metal ion hydrolysis and Section 1.4 deals with our current understanding as to the mechanisms of water exchange and replacement reactions on aqua metal ions. Finally, Chapter 1 finishes with a survey of techniques and methodology for the synthesis, handling, manipulation and purification of representative aqua ion species in Section 1.5.

We hope to show how far our understanding of this area of chemistry has come in the past 10–15 years. It is apparent that there is clearly much we still need to learn about water as a substance and its behaviour as a ligand.

1.1 ION HYDRATION

1.1.1 Basic Concepts

All ions, whether they be positively or negatively charged, are hydrated to varying extents in water. We should start by attempting to define hydration. Ionic hydration is an exothermic process viewed from the energy released when a hydrated ion forms from the gaseous free ion, e.g.

$$M^{n+}(g) + nH_2O(1) \xrightarrow{\Delta H^{0}_{hyd}} M^{n+}(aq)(1) \tag{1.1}$$

This energy can be thought of simplistically as deriving from the ionic 'electrostatic' forces that constitute the ion–water interactions. The same situation holds for a negative ion also. What are the nature of these interactions? In a simplistic way one can imagine the ion as assembling firstly a sheath of water molecules that will be in direct contact with the ion, termed the primary hydration shell. As a result the properties of this primary hydration shell can be said to become somewhat removed from those of free bulk water. Outside this primary shell one can envisage a further more diffuse assemblage of water molecules having somewhat different properties from the bulk water stemming from a transmitted effect through hydrogen bonding with the primary shell. This can be termed the secondary hydration shell. The existence of a secondary shell has become clearly apparent from earliest measurements of ion mobility, e.g. stemming from measurements of conductance and diffusion coefficients. Direct evidence for the secondary shell in a number of trivalent metal ions is now available from X-ray diffraction and X-ray scattering studies as well as from vibrational spectra. As will be emphasised later, both the primary and secondary shells have an intimate role to play in water ligand exchange dynamics. A further highly diffuse region, the tertiary hydration shell, may well be present before distances are reached that are so far removed from the ion that the water molecules become essentially indistinguishable from those of bulk water. The result is of a picture of the ion diffusing from one region of the condensed phase to another while appearing to drag with it an assemblage of water molecules having a greater degree of order than that present in the normal bulk region. Figure 1.1 provides a schematic picture of the hydration sphere of a metal cation having a primary shell of 6 waters and a secondary shell of ~ 13 waters. Such a situation is probably the case for trivalent metal ions such as Cr^{3+} and Rh^{3+}.

The degree of hydration depends upon a number of factors, not least ionic size and charge density. This is nicely reflected in the trend in hydration enthalpies showing a direct correlation of increasing exothermicity with decreasing size and increasing charge density, (Fig. 1.2).

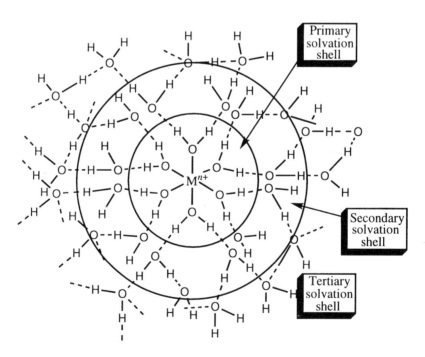

Figure 1.1. The localised structure of a hydrated metal cation in aqueous solution (the metal ion assumed to have a primary hydration number of six)

The hydration enthalpy is a parameter that one can use to correlate with the degree of hydration. The equation used in Fig. 1.2 is

$$\Delta H^{\theta}_{hyd} = -69\,500\,Z^2/r_{eff}\ (\text{kJ}\,\text{mol}^{-1}) \tag{1.2}$$

where r_{eff} (pm) is equal to the effective ionic radius of the cation plus a constant, 85 pm, the radius of the O atom of water in the primary shell. Thus r_{eff} is the interatomic distance in the hydrate and the simple coulombic relationship (1.2) applies [8]. Cations are more strongly hydrated in general than anions due to

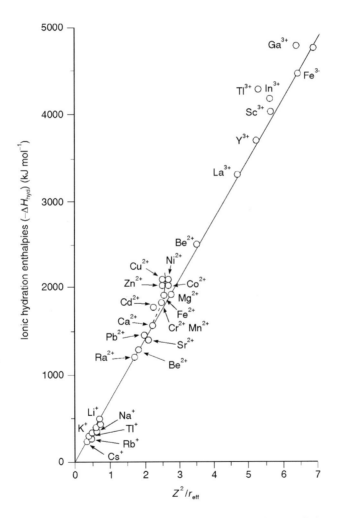

Figure 1.2. Hydration enthalpies as a function of size and charge of the cation (see Appendix 3)

a combination of high positive charge density and a particularly strong interaction with the negatively polarised oxygen atom of water. How far the ion–water interaction can be termed 'ionic' rather than 'covalent' or a combination of both is a point of continuing debate. As will become apparent, this will largely depend upon the particular ion in question.

The modern picture of the structure of liquid water is as a network of clusters of water molecules held together by hydrogen bonds continually bereaking and reforming. The dynamic behaviour of water has been likened to that of a 'flickering cluster', wherein in one instance a given water molecule can be in the centre of a cluster and in another instance on the edge of the cluster and then free (Fig. 1.3). This process, despite the apparent constraints of the hydrogen-bonding network, is normally going on at the diffusion control limit of $\sim 10^{10}-10^{11}\,\text{s}^{-1}$. It is of course this clustering together that gives rise to water behaving as if it were a compound of much higher Mwt (molecular weight), such as the anomously high boiling point (for such a small molecule) and the highest latent heat of vaporisation of any substance. The characteristically high surface tension is believed to be the result of greater clustering at the surface due to the reduction in the number of directional degrees of freedom available for the hydrogen bond breaking and reforming process. Water clusters are even believed to persist in the vapour phase.

It should be therefore readily apparent that both ions and water molecules are highly mobile in an aqueous solution and thus a given water molecule might in one instance be far removed from the influence of the ion, as in pure water, and in another instance be located within the secondary or even the primary solvation

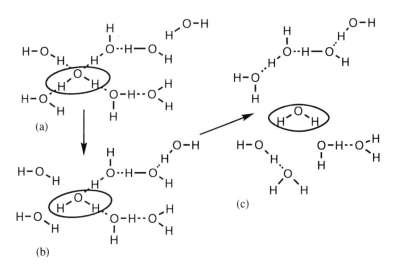

Figure 1.3. The 'flickering cluster' model of liquid water. A given water molecule (circled) is (a) in the middle of a cluster at time t_1, (b) at the edge of a different cluster at time t_2 and (c) free at time t_3

shell. Metal cations that induce a rigidly defined arrangement of water molecules within the primary shell, e.g. the octahedral six coordination (Section 1.2) found for many d-transition metal ions, (see Fig. 1.2) or others with a high charge density, may influence significantly the diffusion of molecules within the secondary shell, but on the whole this is not the case and the secondary shell probably differs little in properties from one metal ion to the next. This has important consequences, as will be seen later.

Recent good-quality neutron diffraction data have shown that around even the halide anions there is a well-defined first hydration shell of water melecules even if the bonding (via the H atoms) is somewhat weaker. For example, the Cl^- ion has been shown to have a primary hydration shell consisting of four to six waters, with the precise number acutely sensitive to concentration and also to the nature of the accompanying cation [5,9]. Thus anions cannot and should not be considered as 'naked' species within the water assemblage, a situation that is frequently overlooked in discussions of anion ligand substitution reactions on aqua metal ions (see later). Even the nobel gases are hydrated to some extent in aqueous solution, their solubility in water increasing down the group paralleling the increase in size and polarisability of their outer filled valence shell and correspondingly the extent of hydration. For example, the formation of a hydrate of argon is undoubtedly the principal reason why the gas is 10 times more soluble than both dioxygen and dinitrogen in water and as a result it is more efficient than dinitrogen in acting as a blanketing inert gas.

1.1.2 The Representative Nature of Aqua Ions

1.1.2.1 GENERAL PERIODIC TRENDS

Much of the chemistry of aqua ions is somewhat representative of the general chemical properties of the metal ion and the oxidation state in question, e.g. the influence of electronegativity, i.e. the formation of simple aqua compounds rather than hydroxo species or oxo anions, stabilities of oxidation states and the tendency towards disproportionation or comproportionation. The aqua ligand coordination geometry one sees reflects obvious factors such as size as well as electronic factors, e.g. crystal field d-orbital splitting (see below) and the tendency towards forming metal–metal bonded and polynuclear oxo(hydroxo)-bridged structures.

For example, the extremely small Be^{2+} ion has four tetrahedrally coordinated primary shell water molecules and indeed tetrahedral hydrates survive in solid Be^{2+} salts. However, for its larger partner Mg^{2+} and also Al^{3+}, Ga^{3+}, In^{3+} and all of the simple transition metal M^{n+} aqua ions, the primary hydration number is six in the familiar octahedral arrangement. Six coordination is very common in aqua ions as can be judged from Table 1.1. It possesses high symmetry and the arrangement is appropriate in balancing the size of the available cavity (radius of M^{n+}) and that of the adjacent water ligands while providing an appropriate

Table 1.1. Selected properties of some aqua ions

Metal ion	Configuration	LFSE	$M-OH_2$[a] (pm)	pK_{11} (hydrolysis)	k_{H_2Oex}[i] (s^{-1})	$\Delta V^{\ddagger}_{H_2Oex}$ ($cm^3\ mol^{-1}$)
Octahedral M³⁺(aq)						
$[Ti(OH_2)_6]^{3+}$	t_{2g}^1	$-0.4\Delta_o$	203^b	2.20	1.8×10^5	-12.1
$[V(OH_2)_6]^{3+}$	t_{2g}^2	$-0.8\Delta_o$	199^b	2.26	5.0×10^2	-8.9
$[Cr(OH_2)_6]^{3+}$	t_{2g}^3	$-1.2\Delta_o$	$198, 196^b$	4.00	2.4×10^{-6}	-9.3
$[Mo(OH_2)_6]^{3+}$	t_{2g}^3	$-1.2\Delta_o$	209^b	?	$\sim 0.1 - 1.0?^f$	$-ve?^g$
$[Mn(OH_2)_6]^{3+}$	$t_{2g}^3 e_g^1$	$-0.6\Delta_o$	199^b	0.70	$\sim 10^3?$	?
$[Fe(OH_2)_6]^{3+}$	$t_{2g}^3 e_g^2$	$0.0\Delta_o$	200	2.16	1.6×10^2	-5.4
$[Ru(OH_2)_6]^{3+}$	t_{2g}^5	$-2.0\Delta_o$	$203, 201^b$	2.40	3.5×10^{-6}	-8.3
$[Co(OH_2)_6]^{3+}$	t_{2g}^6	$-2.4\Delta_o$	187^b	~ 1.8	?	?
$[Rh(OH_2)_6]^{3+}$	t_{2g}^6	$-2.4\Delta_o$	203	3.45	2.2×10^{-9}	-4.2
$[Ir(OH_2)_6]^{3+}$	t_{2g}^6	$-2.4\Delta_o$	204^b	4.37	1.1×10^{-10}	?
$[Al(OH_2)_6]^{3+}$	—	$0.0\Delta_o$	188	4.97	1.29	$+5.7$
$[Ga(OH_2)_6]^{3+}$	$t_{2g}^6 e_g^4$	$0.0\Delta_o$	194^b	~ 3.8	4.0×10^2	$+5.0$
$[In(OH_2)_6]^{3+}$	$t_{2g}^6 e_g^4$	$0.0\Delta_o$	211^b	~ 3.0	$\sim 10^5$	$+ve?$
Octahedral M²⁺(aq)						
$[V(OH_2)_6]^{2+}$	t_{2g}^3	$-1.2\Delta_o$	$212, 212^c,$ $213^d, 213^e$	?	8.7×10^1	-4.1
$[Ru(OH_2)_6]^{2+}$	t_{2g}^6	$-2.4\Delta_o$	212	?	1.8×10^{-2}	-0.9
$[Mn(OH_2)_6]^{2+}$	$t_{2g}^3 e_g^2$	$0.0\Delta_o$	$220, 215^c,$ $218^d, 214^e$	10.60	2.1×10^7	-5.4
$[Fe(OH_2)_6]^{2+}$	$t_{2g}^4 e_g^2$	$-0.4\Delta_o$	$211, 213^c,$ $213^d, 210^e$	9.50	4.4×10^6	$+3.8$
$[Co(OH_2)_6]^{2+}$	$t_{2g}^5 e_g^2$	$-0.8\Delta_o$	$209, 209^d,$ 208^e	9.65	3.2×10^6	$+6.1$

$[Ni(OH_2)_6]^{2+}$	$t_{2g}^6 e_g^2$	$-1.2\Delta_o$	$207, 205^c,$ $206^d, 204^e$	9.86	3.4×10^4	$+7.2$
$[Zn(OH_2)_6]^{2+}$	$t_{2g}^6 e_g^4$	$0.0\Delta_o$	$209, 209^c,$ $210^d, 208^e$	8.96	$\sim 10^7$	$\sim +5?$
$[Cd(OH_2)_6]^{2+}$	$t_{2g}^6 e_g^4$	$0.0\Delta_o$	230	10.08	$\sim 10^8$	$\sim -7?$
$[Hg(OH_2)_6]^{2+}$	$t_{2g}^6 e_g^4$	$0.0\Delta_o$	233	3.4	$\sim 10^9$	$-ve?$
Tetragonally distorted M^{2+}(aq)						
$[Cr(OH_2^{eq})_4(OH_2^{ax})_2]^{2+h}$			$203^{eq}, 227^{ax}$ $207^{eq}, 239^{axc}$ $209^{ea}, 233^{axd}$	?	$\sim 10^9$	$+ve?$
$[Cu(OH_2^{eq})_4(OH_2^{ax})_2]^{2+}$			$196^{eq}, 230^{ax}$ $199^{eq}, 230^{axd}$	7.97^{eq}	4.4×10^9	$+ve?$
Tetrahedral M^{2+}(aq)						
$[Be(OH_2)_4]^{2+}$			167	5.80	7.3×10^2	-13.6
Square-planar M^{2+}(aq)						
$[Pd(OH_2)_4]^{2+}$				2.3	5.6×10^2	-2.2
$[Pt(OH_2)_4]^{2+}$				~ 2.5	3.9×10^{-4}	-4.6

[a] Average value determined by XRD or EXAFS (see Section 1.3).
[b] Average value determined from XRD on the caesium alum.
[c] Average value determined by neutron diffraction at low T (liquid He) on the deuterated ammonium Tutton salt.
[d] Average value determined from XRD on the ammonium Tutton salt.
[e] Average value determined by XRD on the hexafluorosilicate salt.
[f] Estimated value based upon comparisons of complex formation rate constants with those of $[Cr(OH_2)_6]^{3+}$.
[g] $\Delta V'_{obs}$ for 1:1 complexation by NCS^- anation on $[Mo(OH_2)_6]^{3+}$ is -11.4 cm^3 mol^{-1}, suggestive of a highly negative value (< -11) for $\Delta V^{\ddagger}_{H_2oex}$.
[h] eq = equatorial, ax = axial.
[i] At 25°C.

M—OH_2 distance to facilitate a significant bonding interaction. For the transition metal aqua ions, crystal field splitting and the accompanying stabilisation energy (CFSE or LFSE) is an important influencing factor. The dependence of ΔH_{hyd}^{θ} on $Z^2/r_{M—OH_2}$ (1.2) and not on hydration number is nicely illustrated by the comparison between Be^{2+} and the larger Mg^{2+}. Mg^{2+}, although hydrated by an extra two water molecules, has a lower ΔH_{hyd}^{θ} than Be^{2+} because the charge density of Mg^{2+} is lower and the six primary shell water molecules are further away.

Simple aqua ion species are not so commonly observed for the heavier metallic elements, especially those in the second and third transition metal series. $[Ru(OH_2)_6]^{2+}$ is the only divalent hexaaqua ion, whereas trivalent hexaaqua ions are known for Mo^{3+}, Ru^{3+}, Rh^{3+}, Ir^{3+} and In^{3+}. Both Ru^{2+} and Ru^{3+} are low spin ($\Delta_o > P$, see below), in contrast to the corresponding hexaaqua iron species. The absence of simple aqua cations for the early heavy transition elements Zr, Hf, Nb, Ta, W, Tc and Re reflects a stability in the higher oxidation states leading to the formation rather of oxo-bridged oligomers or oxo(hydroxo) anions. Thus the simplest Zr^{4+}(aq) species is a tetramer, $[Zr_4(\mu\text{-}OH)_8(OH_2)_{16}]^{8+}$ (see Chapter 4), while Nb^V(aq) and Ta^V(aq) hydrolyse completely, even in strong acid, all the way to the hydrated pentoxides respectively (Chapter 5). There are no purely aqua-derived cations for Tc or Re although thermochemical and $E^{\ominus}$ data for Tc^{2+}(aq) (probably metal–metal bonded) and Tc^{3+}(aq) shows that they ought to exist. The existence of putative $[Os(OH_2)_6]^{3+}$ has been claimed (Chapter 8). It is much less stable than the corresponding hexaammine complex $[Os(NH_3)_6]^{3+}$, which also forms for Os^{2+}, raising pertinent questions as to the somewhat different properties of water as a complexing ligand versus ammonia (see the discussions in Section 1.2.1.2 and Section 9.1.2). The lower oxidation states are more reducing and there is a greater tendency towards metal–metal bond formation promoted largely by more effective d–d overlap of the spatially extended 4d and 5d orbitals. Thus while Cr^{2+} and Co^{2+} exist as monomeric hexaaqua species, the corresponding aqua ions for Mo^{2+} and Rh^{2+} exist as dimeric M—M bonded $[Mo_2(OH_2)_8]^{4+}$ and $[Rh_2(OH_2)_{10}]^{4+}$ cations respectively. The Hg—Hg bonded dimer $[Hg_2(OH_2)_2]^{2+}$, largely responsible for the stability of Hg^I in aqueous solution, provides a further example. For the heavier metals the M—M bonding can extend to oxidation states as high as $+5$. Thus the simplest aqua ions for Mo^{4+} (W^{4+}) and Mo^{5+} (W^{5+}) are not monomeric species but oxo-bridged M—M bonded trinuclear ($M_3O_4^{4+}$(aq)) and dinuclear ($M_2O_4^{2+}$(aq)) species respectively. Trinuclear M—M bonded clusters also characterise the lower oxidation states III and IV for Nb and are largely responsible for these oxidation states being stable in aqueous solution. In contrast, Cr^{4+}(aq), as the monomeric chromyl ion, $[(H_2O)_5Cr=O]^{2+}$, has only a fleeting aqueous existence ($t_{1/2}$, $25\,°C \sim 0.5\,min$). In group 10 the diamagnetic aqua ions for Pd^{2+} and Pt^{2+} reflect the formation of the expected square-planar tetraaqua geometry, contrasting with paramagnetic octahedral $[Ni(OH_2)_6]^{2+}$ (see below). Figure 1.4 illustrates the diversity of structure types known for aqua ions and their derivatives.

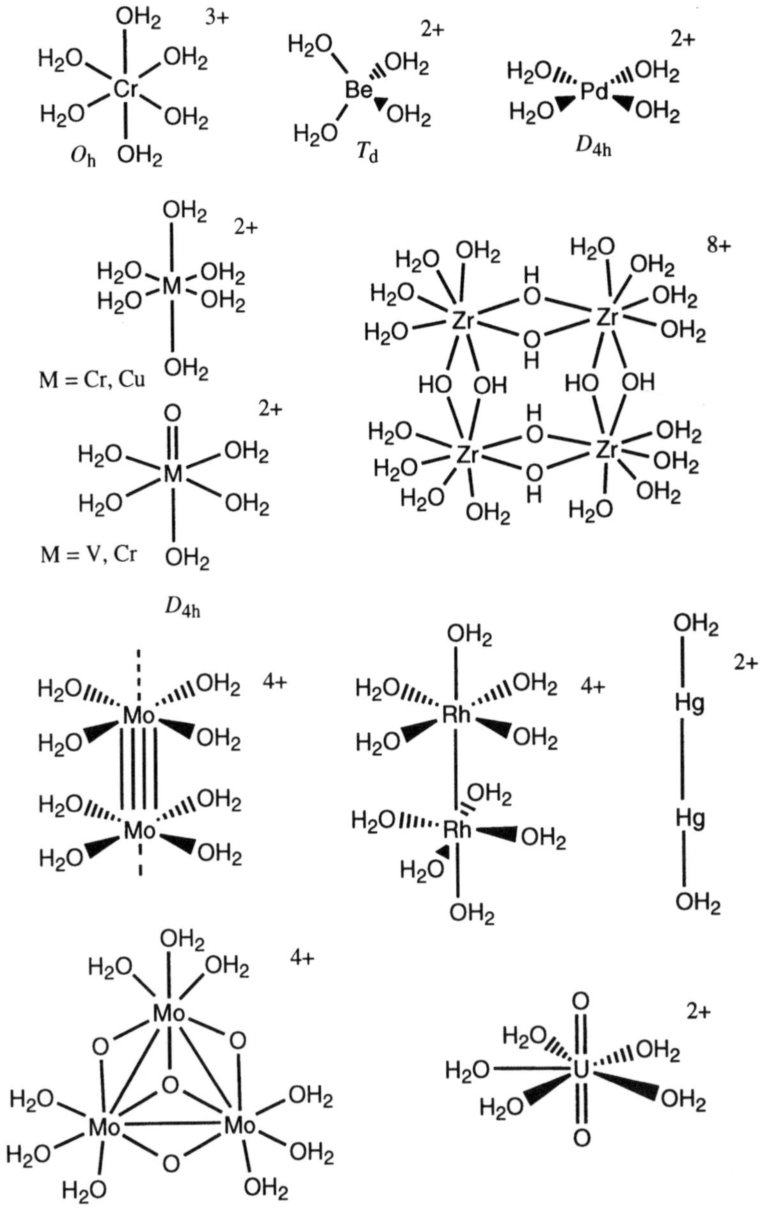

Figure 1.4. Representative structures for various aqua metal cations and their derivatives

1.1.3 Transition Metal Aqua Ions

1.1.3.1 LIGAND FIELD EFFECTS ON STRUCTURE AND REACTIVITY

For the d-transition metal cations the preferred coordination geometry of the primary shell and in turn the $M\!-\!OH_2$ distance(s) and the lability are dictated by ligand field effects.* This arises from the stability of certain electronic configurations that result from the splitting of the valence d orbitals in the presence of particular geometrical arrangements of ligands around the central metal cation. In the crystal field model (Fig. 1.5) the electrons in the $d_{x^2-y^2}$ and d_{z^2} orbitals point directly at the electron density residing on the ligands and are viewed as feeling a greater electrostatic repulsion than those residing in the d_{xy}, d_{yz}, d_{xz} orbitals that point in the directions between the ligands. Thus in an octahedral $[M(OH_2)_6]^{n+}$ ion[†] three degenerate orbitals (the t_{2g} set) are at a lower energy than the other two, termed the e_g set. The energy gap between them, which in reality is the lowest energy barrier for an electron to jump between a t_{2g} and an e_g level, is termed the ligand field splitting energy or Δ_o (o for octahedral). It can be shown that the diagram is 'inverted' for a corresponding tetrahedral field of the same ligands and that $\Delta_t \sim \frac{4}{9}\Delta_o$. The lower Δ_t arises because none of the orbitals point directly at the ligands in a tetrahedral arrangement. Figure 1.5 also shows how the extra stabilisation energy (termed the LFSE) resulting from a particular electron configuration can be calculated. The splitting energy also explains the origin of high- and low-spin forms. Very simply, energy is required to pair up electrons within the same orbital (termed the pairing energy, P), in effect the energy required to overcome the extra repulsion from electrons residing in the same orbital. If the value of Δ_o exceeds this pairing energy then a low-spin complex will be the result. If, however, $P > \Delta_o$ the high-spin form driven by Hund's rule will be formed. All first row transition metal $[M(OH_2)_6]^{2+}$ ions (M = V to Zn) are high spin. However, all tripositive hexaaqua ions, with the exception of Mn^{3+} and Fe^{3+}, are low spin. Ligand field theory uses a molecular orbital (MO) approach to treat the consequences of the M—L interactions on the d orbitals but arrives at the same d-orbital splitting picture for an octahedral complex with a degenerate level of three orbitals found below that containing two. The repulsive doubly degenerate e_g level is now interpreted as in effect 'antibonding'. The trend in Δ for different ligands (the spectrochemical series) is rationalised by the full MO approach (LFT) which introduces a degree of π donor or acceptor bonding (see Appendix 1). This in turn implies a significant degree of covalency in the bonds to certain ligands.

---

* A comprehensive coverage of both crystal and ligand field theory is outside the realm of this book and only a brief introductory overview will be presented here for the purpose of the beginner. Excellent accounts of this topic can be found in most undergraduate inorganic chemistry texts.

[†]Although $[M(OH_2)_6]^{n+}$ is normally referred to as octahedral, it is noted that the symmetry is somewhat lower when the positions of the protons are taken into account.

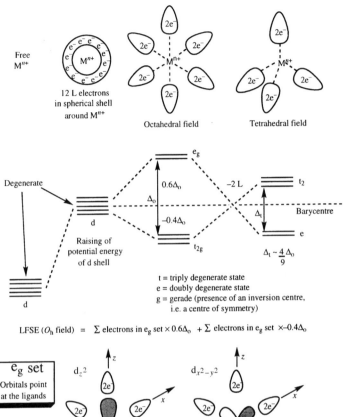

LFSE (O_h field) = Σ electrons in e_g set $\times 0.6\Delta_o$ + Σ electrons in e_g set $\times -0.4\Delta_o$

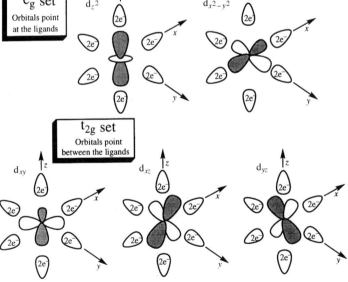

Figure 1.5. The splitting of the five d orbitals of a transition metal cation by the presence of a ligand field

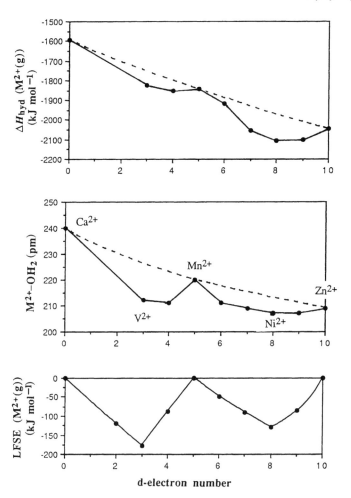

Figure 1.6. Plot of ΔH^{θ}_{hyd} and hydration radius (M^{2+}—OH_2) (pm) vs weak field LFSE (KJ mol^{-1}) for $[M(OH_2)_6]^{2+}$ cations along the first transition series as a function of d-electron number

The ligand field effects show up particularly well whithin the classical 'double-well' feature in plots of ΔH^{θ}_{hyd} versus atomic number for the di- and trivalent transition metal ions (Fig. 1.6). Similar features result from plots of the M—OH$_2$ distance, as determined by X-ray diffraction, versus atomic number, Figs. 1.6 and 1.7 emphasising the correlation of ΔH^{θ}_{hyd} with effective radius of the first hydration shell. One can clearly see the effects of d-orbital splitting in plots such as those in Fig. 1.7, the filling of the more shielded t$_{2g}$ orbitals causing greater

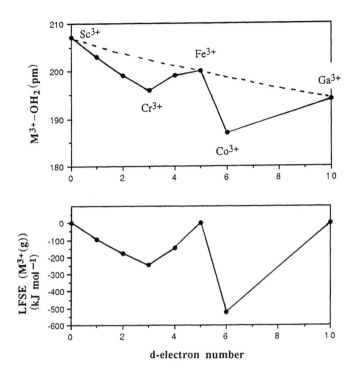

Figure 1.7. Plot of hydration radius (M^{3+}—OH_2) (pm) for $[M(OH_2)_6]^{3+}$ cations vs LFSE (kJ mol^{-1}) along the first transition series as a function of d-electron number (a low-spin d^6 configuration has been assumed)

radius contraction than that predicted by the gradual increase in Z_{eff} (due to the inefficient shielding of the nucleus by the addition of successive d electrons), whereas filling of the antibonding e_g orbitals leads to radius expansion. The low-spin nature of $[Co(OH_2)_6]^{3+}$ is clearly apparent from Fig. 1.7. Figures 1.6 and 1.7 provide powerful experimental evidence that ligand field splitting of the d orbitals is a real phenomenon. However, it can also be seen that at its maximum the LFSE only accounts for ~5–10% of a typical divalent metal ion's total hydration energy despite its significant influence on the hydration number, M—OH_2 distance and, as will become apparent in Section 1.4, the kinetic lability of the primary shell. Above all, the lack of a correlation for the transition metal ions between hydration enthalpy (hydration radius) and the lability of the primary shell should be emphasised. The behaviour of V^{2+} is a good case in point. It has the largest size and thus lowest hydration enthalpy of all the first row divalent transition metal ions and yet its hexaaqua ion is the most inert ($k_{exH_2O}(25\,°C) = 87\,s^{-1}$). This is known to arise from a particular kinetic stability of the singly filled t_{2g}^3 configuration. This is even more apparent in the trivalent ions by the remarkable inertness of $[Cr(OH_2)_6]^{3+}$ ($k_{ex} = 2.4 \times 10^{-6} s^{-1}$, also t_{2g}^3). Here one sees the effect of an extra unit of positive charge. $[Ru(OH_2)_6]^{2+}$, despite a similar size to $[V(OH_2)_6]^{2+}$, is more inert because of the even greater LFSE gain from its low-spin t_{2g}^6 configuration and the 50% increase in Δ_o as it is a second row metal . However, its lower charge makes it still somewhat more labile than Cr^{3+}. The remarkable inertness shown by $[Cr(OH_2)_6]^{3+}$ has indeed led to some ascertions that Cr^{3+} itself is somewhat anomalous (see page 83). For example, one has the surprising finding of the second row analogue $[Mo(OH_2)_6]^{3+}$ being by far the more labile by a factor of $\sim 10^5$ (normally the reverse is true down a group due to the increase in Δ_o and thus LFSE, which more than compensates for the size increase).

The tetragonally distorted octahedral structures for Cr^{2+}(aq) $(t_{2g}^3 e_g^1)$ and $Cu^{2+}(t_{2g}^6 e_g^3)$ and high solvation shell lability (residence time $\sim 10^{-9}$ s) are well documented as resulting from the Jahn–Teller effect, the same being true for Mn^{3+}(aq) $(t_{2g}^3 e_g^1)$ and Ag^{2+}(aq) $(t_{2g}^6 e_g^3)$. Divalent cobalt is known to have a tendency to form tetrahedral complexes, particularly when promoted by the steric demands of the ligand, e.g. the formation of $[CoX_4]^{2-}$ (X = Cl and Br). The reason for this lies in the smallest difference in LFSE between high-spin d^7 octahedral and tetrahedral geometries with the same ligands. Thus, at very high temperatures ($\sim 670\,°C$) and moderate pressures, significant amounts of tetrahedral $[Co(OH_2)_4]^{2+}$ ($\sim 7\%$) can be observed in equilibrium with $[Co(OH_2)_6]^{2+}$ (Chapter 9). The square-planar four coordination for $[Pd(OH_2)_4]^{2+}$ and $[Pt(OH_2)_4]^{2+}$ arises from the favourability of spin pairing which induces, in effect, an extreme Jahn–Teller distortion of the d^8 configuration driven by the large intrinsic ligand field splitting. Spin pairing is more favoured in the second and third row metals by the larger and more 'spatially extended' nature of the 4d and 5d orbitals, and this leads exclusively to 'low-spin' complexes. In contrast, Ni^{2+}(aq) is present as paramagnetic $[Ni(OH_2)_6]^{2+}$ ($\mu_{eff} = 3.3$ BM) as a result of the ligand field strength now insufficient to overcome the larger energy required to pair up the two 3d electrons residing in the doubly degenerate e_g set and lead to spontaneous distortion. These features are illustrated in Fig. 1.8. Selected properties for a range of monomeric aqua metal ions are collected together in Table 1.1.

1.1.3.2 REDOX BEHAVIOUR AND SELF-EXCHANGE RATES

Ligand field effects also play an intimate role in oxidation–reduction reactions of transition metal aqua ions. The intrinsic ability to donate/receive an electron from a redox co-reagent in a truly outer-sphere manner is represented by the rate constant for self-exchange, k_{11} (Table 1.2) [10]. In many cases k_{11} values have been deduced from the observed second-order k values, k_{12}, for outer-sphere cross-redox reactions with reagents having a reliably known self-exchange rate by applying the following Marcus equations [11] where k_{11} and k_{22} are the respective self-exchange rate constants, k_{12}, is the overall reaction equilibrium

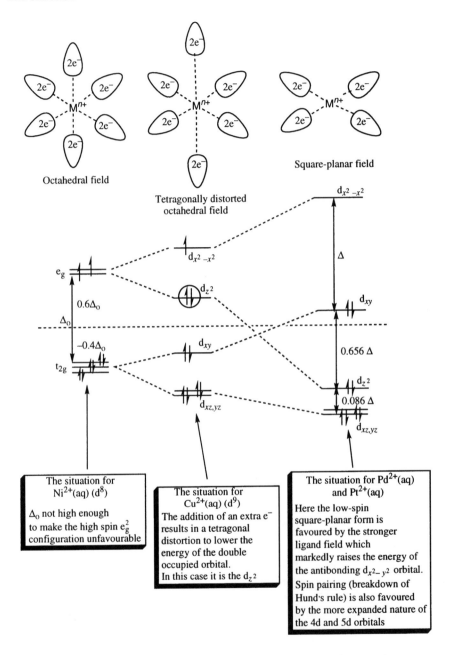

Figure 1.8. Influence of ligand field strength and d-electron number on the geometry adopted by certain d-transition metal aqua ions

Table 1.2. Redox potentials and calculated self-exchange rate constants (25°C) for aqua ion redox couples (compiled from ref. [10]):

$$[M(OH_2)_n]^{x+} + [*M(OH_2)_n]^{(x-1)+} \overset{k_{11}}{\rightleftharpoons} [M(OH_2)_n]^{(x-1)+} + [*M(OH_2)_n]^{x+}$$

Couple	Orbitals	E^θ(V)vs NHE	$k_{11}(M^{-1}s^{-1})$
$[Ti(OH_2)_6]^{4+/3+}$	t_{2g}^0/t_{2g}^1	> 0.11	$\geqslant 10^{-2}$
$[TiOH(OH_2)_5]^{3+/2+}$	t_{2g}^0/t_{2g}^1	> 0.08	$\geqslant 3 \times 10^{-4}$
$VO(OH)^{2+/+}$(aq)	t_{2g}^0/t_{2g}^1	$+1.13$	$\sim 10^{-2}$
$[V(OH_2)_6]^{3+/2+}$	t_{2g}^2/t_{2g}^3	-0.255	1×10^{-2}
$[Cr(OH_2)_6]^{3+/2+}$	$t_{2g}^3/t_{2g}^3 e_g^1$	-0.4	1×10^{-5}
$[Mn(OH_2)_6]^{3+/2+}$	$t_{2g}^3 e_g^1/t_{2g}^3 e_g^2$	$+1.56$	3×10^{-4}
$[Fe(OH_2)_6]^{3+/2+}$	$t_{2g}^3 e_g^2/t_{2g}^4 e_g^2$	$+0.74$	4.0
$[Co(OH_2)_6]^{3+/2+}$	$t_{2g}^6/t_{2g}^5 e_g^2$	$+1.92$	~ 5
$[Ru(OH_2)_6]^{3+/2+}$	t_{2g}^5/t_{2g}^6	$+0.22$	20
$Cu^{2+/+}$(aq)	$t_{2g}^6 e_g^3/t_{2g}^6 e_g^4$	$+0.15$	1×10^{-5}
$Ce^{4+/3+}$(aq)	$4f^0/4f^1$	$+1.72$	$\leqslant 6 \times 10^{-2}$
$U^{4+/3+}$(aq)	$5f^2/5f^3$	-0.63	7×10^{-5}
$NpO_2^{2+/+}$(aq)	$5f^1/5f^2$	$+1.14$	19
$Np^{4+/3+}$(aq)	$5f^3/5f^4$	$+0.17$	2×10^{-2}
$Pu^{4+/3+}$(aq)	$5f^4/5f^5$	$+0.98$	20
$Eu^{3+/2+}$(aq)	$4f^6/4f^7$	-0.43	$\leqslant 2 \times 10^{-4}$
$Yb^{3+/2+}$(aq)	$4f^{13}/4f^{14}$	-1.18	10^{-1}
$Tl^{3+/+}$(aq)	$6s^0/6s^2$	$+1.26$	7×10^{-5}

constant and z is the collision frequency ($\sim 10^{11}$ M^{-1} s^{-1}):

$$k_{12} = (k_{11} k_{22} K_{12} f)^{1/2} \quad \text{where} \quad \log f = \frac{(\log K_{12})^2}{4 \log(k_{11} k_{22}/z^2)} \tag{1.3}$$

If $f = 1$ the following simplified expression is obtained:

$$k_{12} = (k_{11} k_{22} K_{12})^{1/2}$$

for which the logarithmic form is

$$\log k_{12} = 0.5(\log k_{11} + \log k_{22}) + (0.5 \times 16.9)\Delta E^\theta$$

Unless K_{12} is large the work term parameter f can be assumed to be close to unity, in which case the simplified Marcus expression can be used where ΔE^θ is the difference between the E^θ values for the two redox partners. In this form it can be seen that the Marcus equation contains terms relating to the intrinsic kinetic barrier to electron transfer (k_{11} and k_{22}) as well as a contribution from the overall thermodynamic driving force (K_{12} and ΔE^θ). The increase of k_{12} with a favourable driving force is interpreted as reflecting electron transfer occurring earlier along the reaction coordinate (at lower activation energy), preempting the requirement for the optimum preorganisational configuration of the two reactants (both structural and electronic) [12].

On the whole, self-exchange reaction rate constants, such as many of those in Table 1.2, have proved difficult to pin down precisely partly because in reality the precise nature of the individual steps for a given reactant within a given outer-sphere cross-redox reaction may not be the same from one reaction to the next. Also one tends to assume a purely outer-sphere process with optimum preorganisation. Furthermore, it is possible for certain ground state reactants to react via a favourable excited state which might involve, for example, spin equilibria. As a result one is forced to tolerate an error of up to ± 2 orders of magnitude in the reliability of k_{11} values [10]. In a few cases the self-exchange rates are slow enough to be measured independently using isotope labelling methods including the use of NMR (e.g. $[Ru(OH_2)_6]^{3+/2+}$, Chapter 8). The problems are evident by the frequency with which calculated values and the Marcus estimates often differ by many orders of magnitude. For example, values estimated for k_{11} for the $[Co(OH_2)_6]^{3+/2+}$ couple cover 12 orders of magnitude. Clearly apparent, though, from Table 1.2 is that k_{11} values $(M^{-1}s^{-1})$ are usually less than unity for the first row metal ions and the lanthanides and actinides and frequently less than 10^{-2}. On the other hand, the k_{11} value for the one second row transition metal hexaaqua couple $[Ru(OH_2)_6]^{3+/2+}$ is significantly larger $(k_{11}(25°C) = 20\ M^{-1}s^{-1})$ and probably reflective of the more expanded nature of the $4d(t_{2g})$ orbitals (see Section 8.2.2).

The involvement of hydroxo aqua ions is often an indication of the presence of a favourable *inner-sphere* pathway involving a bridged OH^- ligand intermediate [13]. For example, in contrast to their non-importance in water exchange, $TiOH^{2+}(aq)$ and $VOH^{2+}(aq)$ are well established as involved in inner-sphere reactions involving oxidation of $[Ti(OH_2)_6]^{3+}$ and $[V(OH_2)_6]^{3+}$. It is also likely that involvement of the hydroxo form, such as in the case of $MoOH^{2+}(aq)$, can reflect a favourable outer-sphere path to the more hydrolysed higher oxidation state (in this case to dimeric $Mo_2^VO_4^{2+}(aq)$; see Chapter 6). On the other hand, the dominant involvement of $FeOH^{2+}(aq)$ in the reduction of $[Fe(OH_2)_6]^{3+}$ by, for example $Cu^+(aq)$, suggests an inner-sphere pathway. Aspects of the redox chemistry of the various aqua ions is highlighted in the subsequent group chapters.

1.2 STRUCTURAL INVESTIGATIONS OF AQUA IONS AND OF METAL ION HYDRATION

For the d-block transition elements the identification of the primary hydration geometry, usually octahedral six-coordinate, has proved relatively straightforward on the basis of a comparison of the d–d spectrum in solution with that of structurally characterised crystalline hydrates such as in the Tutton salts and alums.* For the rest, including anions and polynuclear species, the lack of such a definitive electronic spectroscopic handle has meant the need to develop techniques for aqueous solutions that are more widely applicable. In this regard much progress has been made in the use of methods such as X-ray diffraction (XRD), neutron diffraction and neutron scattering methods. To this can be added extended X-ray absorption fine structure (EXAFS), IR, Raman and NMR spectroscopies. In addition the current generation of powerful and accessible computers is allowing a number of significant theoretical approaches to emerge.

As Ohtaki and Radnai point out in their review [5] the structure of hydrated ions somewhat depends upon the method of observation. In this regard three categories can be defined depending upon the methods of investigation:

(a) A static structure in which the structure is discussed on the time and space averaged ion–water interactions. The results obtained by X-ray and neutron diffraction and neutron scattering fall into this category.

(b) A structure based on the dynamic properties of coordinated water molecules such as by NMR measurements, e.g. the use of the ^{17}O nucleus (see later). In this regard a primary hydration shell will only be observable provided that the rate of exchange of the primary shell with the rest of the water assemblage is slower than the frequences employed in the NMR measurement. However, the NMR method cannot be used for the investigation of the primary hydration shell in certain paramagnetic systems.

(c) Investigation based on energetics which can discriminate between, for example, the strongly bound water molecules in the primary shell with the more loosely bound waters in the secondary shell. The use of vibrational spectroscopy such as IR and Raman as well as thermodynamic measurements fall into this category.

Clearly ion hydration can manifest itself differently within the various categories and this should always be borne in mind.

1.2.1 Diffraction and Scattering Methods

There are two basic methods. The first is to measure the intensities of elastically scattered X-rays and neutrons by atoms during diffraction measurements whereas the second is the inelastically scattered method. The former method gives directly the number, distance and temperature factors involved with atom–atom

---

* The 'Tutton' salts are formed by divalent hexaaqua metals ions and have the general formula $M^I_2[M^{II}(OH_2)_6](SO_4)_2$, where M^I is a unipositive cation, usually NH_4^+, Rb^+ or Cs^+. The 'alums' are formed by trivalent hexaaqua metal ions and have the general formula $M^I[M^{III}(OH_2)_6](SO_4)_2 \cdot 6H_2O$ where M^I is again usually Rb^+, Cs^+ or NH_4^+. The name 'alum' was given to this family of compounds since the first one to be characterised was that of the trivalent ion Al^{3+}.

interactions. Neutrons are scattered by atomic nuclei giving information on the position of the nuclei. On the other hand, X-rays are scattered by electrons and thus only give the centre of electron density. These quantities are more or less the same for heavy atoms but not the same in the case of hydrogen. Thus for the location of protons neutron scattering is the preferred and often the only reliable method. The replacement of hydrogen atoms by deuteriums is often employed to improve the X-ray scattering power. The quasi-elastic and inelastic neutron scattering methods provide dynamic information predominantly of the translational motion of protons. Within the elastic and quasi- elastic scattering methods it is often important to correct for inelastic collisions between atoms and neutrons since these can involve transfer of specific vibrational energy quanta from the particle to the molecule.

1.2.1.1 STUDIES ON TRANSITION METAL IONS

An IQENS (incoherent quasi-elastic neutron scattering) study of aqueous 1.8 M $Cr(ClO_4)_3$ solutions revealed slow exchanging protons consistent with six water molecules in the primary shell but almost diffusion controlled exchange $(t_H \leqslant 10^{-10}\,s)$ within the secondary shell. Similar behaviour has been found in aqueous solutions of $Rh(ClO_4)_3$. The much slower proton exchange within the primary shell has been attributed to the presence of stronger hydrogen bonding with the secondary shell as a result of cation polarisation.

Enderby and coworkers [6, 7] have pioneered the use of first-order difference neutron scattering methods for the analysis of the structures of both crystalline hydrates and of concentrated aqueous solutions of anions and cations. Central to the method of difference is a comparison of partial structure factors obtained from scattering data using two different isotopes of the cation or anion. The structure factors can be weighted so as to only consider those relating to the ion–water interactions. The results are usually illustrated in the form of a radial distribution plotting the difference function ($\Delta G(r)$), obtained by a Fourier transformation of the various structure factors, as a function of distance (r). Figure 1.9 shows typical data for a 1.4 m solution of $NiCl_2$ in D_2O.

The sharp 'twin peak' structure is characteristic of strongly hydrated ions such as Ni^{2+}(aq). More diffuse scattering from the secondary shell of D_2O can be seen above 400 pm. Since information regarding both Ni—O and Ni—H(D) distances can be obtained, it is possible to investigate the orientation of the water molecule with respect to the M—O bond vector. For concentrated solutions of Ni^{2+} the angle θ (Fig. 1.10) is 42°, indicating a significant tilting away of the water from the M—O plane. However, as the concentration of $NiCl_2$ is reduced, θ is found to decrease effectively to zero at $[NiCl_2] < 0.1$ m. The change in tilt angle θ, if relevant, may be reflective of changing hydrogen-bonding effects in the different solutions and perhaps a change in the interaction with the hydration sphere of the Cl^- anion as the salt concentration is varied. The effect of hydrogen bonding is

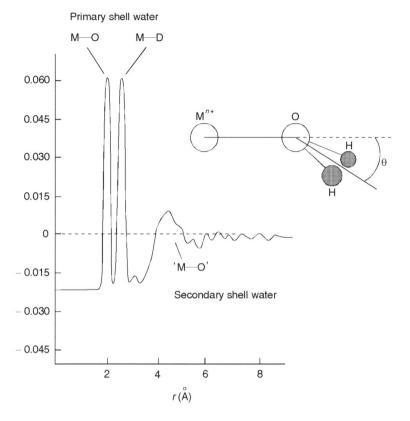

Figure 1.9. Plot of the radial distribution difference function $\Delta G(r)$ (in barns) vs distance r (pm) for a 1.46 m solutions of $NiCl_2$ in D_2O (1 barn = $10^{-28}\,m^2$)

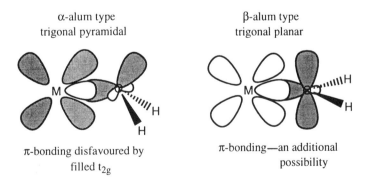

Figure 1.10. Stereochemistry of the $M-OH_2$ interaction in the two alum types

clearly apparent in the X-ray structures of the hydrated sulfates $CrSO_4 \cdot 5H_2O$ and $CuSO_4 \cdot 5H_2O$ [14, 15]. In each case two unique tetragonally distorted MO_6 geometries I and II are present, one having trigonal planar coordinated waters ($\theta = 0$) (I) and the other pyramidal coordinated waters ($\theta > 0$) (II). In the case of I the waters are hydrogen bonded only to SO_4^{2-} whereas in II they are hydrogen bonded to space filling free waters.

1.2.1.2 TRIVALENT ALUMS AND DIVALENT TUTTON SALTS

Good low-temperature neutron diffraction data are now available on the entire series of crystalline trivalent caesium alums, $Cs[M(OH_2)_6](SO_4)_2 \cdot 6H_2O$ [16–18]. Of interest has been the need to establish whether there is a correlation between the preferred orientation of the water molecules and the electronic structure of the cation as opposed to an influence by factors such as hydrogen bonding. There are two important alum modifications, the α type and the β type, each differing in the tilt angle θ of the water molecule with respect to the M—O bond vector (Fig. 1.9), this being close to zero (trigonal planar stereochemistry) for the β alums but $\sim 35 \pm 10°$ (i.e. towards trigonal pyramidal) for the α form. Interestingly there does appear to be a preference for the α modification shown by the low spin t_{2g}^6 ions of group 6, Co, Rh and Ir. All the remaining known caesium alum salts, i.e. those with M = Al, Ga, In, Ti, V, Cr, Mo, Mn, Fe and Ru, conform to the β modification, which appears to be the favoured one for promoting strong hydrogen bonding to the secondary shell. Theoretical studies (see below) indicate a trend towards increasing covalency of the M—OH_2 bonds towards the right of the transition metal series and for the second and third row metals. Thus one possible explanation is that for the low-spin t_{2g}^6 ions, since there can be no possibility of a π-bonding interaction from the O atom of water, the alum structure reflects only a σ-bonding mode and hence the trigonal pyramidal arrangement (Figure 1.10).

All the remaining transition metal trivalent alums have partially filled t_{2g} orbitals and so the trigonal planar arrangement in the β modification could reflect the involvement of M—O π-overlap. However, if this were the case perhaps somewhat shorter M—OH_2 distances might have been expected for the earlier first row trivalent metals with low t_{2g} population, and it certainly does not explain the case of the group 13 alums M = Al, Ga and In, each of which adopt the trigonal planar β modification despite indications of enhanced covalency and the presence for Ga and In of a $t_{2g}^6 e_g^4$ configuration. It is well established that the strongest hydrogen bonds within the water assemblage are those involving the coordinated waters and so one must continue to conclude that the hydrogen bonding, in addition to the largely electrostatic nature of the M—OH_2 bonds, is the major influencing factor on the observed stereochemistry for the majority of first row transition metals and the main group hexa-aqua alums. A similar picture has emerged from low-temperature neutron diffraction data and X-ray

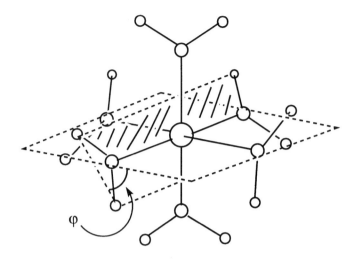

Figure 1.11. Definition of the twist angle angle φ for a metal hexaaqua cation in the β-alum structure viewed down the threefold axis

diffraction data on the corresponding ammonium Tutton salts and hexa-fluorosilicates [19–21].

Studies, however, on two second row β alums (M = Mo and Ru) appear to indicate that the M—OH$_2$ distance is somewhat sensitive to the stereochemistry of the coordinate water molecule and in particular to the twist angle ϕ from the MO$_4$ equatorial plane (Fig. 1.11), possibly reflecting greater covalency and an involvement from the more spatially extended 4d(t_{2g}) orbitals [17]. No such correlation with ϕ is apparent for the first row transition metal β alums, this having been tested for by the successful imposition of different values of the twist angle ϕ by replacing SO$_4{}^{2-}$ in the alum lattice structure with the larger SeO$_4{}^{2-}$ [18].

It is certainly attractive to think of the extremely high E^θ value (+1.92 V) for the [Co(OH$_2$)$_6$]$^{3+/2+}$ couple, contrasting with the much lower value (+0.1 V) for [Co(NH$_3$)$_6$]$^{3+/2+}$, as arising from the respective σ- and π-bonding properties of water versus ammonia as ligand (see section 9.1.2). Ammonia can only be sp^3 hybridised and this, coupled with the slightly lower electronegative of nitrogen versus oxygen, makes ammonia a good σ-donor, thus stabilising Co^{3+} ($t_{2g}^6 e_g^0$) versus Co^{2+} ($t_{2g}^5 e_g^2$). Water, on the other hand, has the possibility of both σ- and π-donation which favours the high-spin Co^{2+} form as creating some available t_{2g} density; thus Co^{2+} is stabilised relative to Co^{3+}. Indeed, in the structure of [Co(OH$_2$)$_6$]SiF$_6$ a trigonal planar coordination ($\theta \sim 0$) of water is observed [21].

Difference neutron scattering methods and X-ray scattering techniques have been used to probe the primary hydration shell of the metal ions of group 1, 2, 13 and 14 (see chapter 2) and also the lanthanides (Chapter 3). Notably in the case of the highly labile group 1 metal ions, hydration numbers are found to be largely dependent upon the H_2O/salt ratio as well as the nature of the counter anion present. Neutron diffraction studies have also played a major part in providing confirmation that there is a change in the hydration number (around Sm to Gd) along the series of lanthanide trivalent ions (Ln^{3+}) from nine (La to Nd) (probably tricapped trigonal prismatic) to eight (Tb and beyond) (possibly square antiprismatic). Indications of such a change were apparent from the work of Spedding and coworkers (see Chapter 3, refs [33–35]) who some years earlier had detected an abrupt change in the trend in partial molar volume for the Ln^{3+} ions along the series around the same region Sm to Gd, the so-called 'gadolinium break'. The difference in partial molar volume correlates exactly with the molar volume of an excluded electrostricted water molecule within the primary/secondary shell interface.

1.2.1.3 SECONDARY HYDRATION

Sandstrom and coworkers have used large-angle X-ray scattering (LAXS) methods to investigate further the nature of secondary hydration. For the trivalent ions Cr^{3+} and Rh^{3+} a well-defined secondary hydration shell of 13 ± 1 waters was found at a distance of 402 ± 20 pm [22]. Thus the picture of the hydration of such ions is probably not far removed from that illustrated in Fig. 1.1. The extent of polarization of water molecules by metal ions depends largely upon the cationic charge (more correctly the charge density) and as such trivalent ions would be expected to have a greater influence on the ordering of secondary shell water molecules than divalent and in turn monovalent ions. Thus, unlike the success of the neutron scattering methods, a corresponding LAXS study on Ni^{2+} (aq) [23] had no difficulty verifying the presence of the primary hydration shell of six waters but failed to reveal evidence of the secondary hydration shell.

A more in-depth discussion of XRD, X-ray scattering and neutron diffraction and scattering methods is outside the scope of this book, but can be found in the excellent review book by Magini *et al.* [2b] and in the various articles by Enderby and coworkers [6, 7, 9, 24] and Sandstrom and coworkers [22, 25].

1.2.2 Spectroscopic Methods

Various spectrosopic methods have been employed in studies of ion hydration. A major development since the late 1970 s has been in the use of X-ray absorption spectroscopy. The extended X-ray absorption fine structure (EXAFS) method gives information on short-range intermolecular interactions which can in some cases involve the secondary hydration shell. Although the EXAFS method gives information only of the number of scattering atoms and their distances it can be used in conjunction with the X-ray absorption near edge structure (XANES) method (which gives information on the electronic structure of the X-ray absorbing atom) to give some degree of angular information.

1.2.2.1   EXTENDED X-RAY ABSORPTION FINE STRUCTURE (EXAFS)

There are many excellent reviews and articles describing the theory of the EXAFS method and its applications [26] so only a very simplified overview will be given here. The essence of the technique is the absorption, by the central ion, of high-energy X-ray photons of sufficient energy near the X-ray edge so as to not only cause quantum jumps of the core electrons but the emission of photo-electrons of sufficient kinetic energy to escape from the atom and radiate outwards. The photoelectrons then experience back-scattering from neighbouring atoms which cause characteristic interference patterns depending upon whether the back-scattered photoelectrons are in or out of phase with the outward radiating photoelectrons. The scattering power is a function of the number of electrons in the neighbouring atom. The resulting characteristic oscillating spectral pattern, plotted as X-ray absorption intensity against the energy of the incident X-ray photon (Fig 1.12), is a superposition of EXAFS from the individual scatterers. Fourier transformation then gives a spherically aver-aged radial distribution function of electron density as a function of distance from the central absorbing atom. From a simulation of the spectral pattern using parameters from compounds of known local structure it is possible to obtain quite reliable data for metal aqua ions regarding the numbering of scattering water molecules and their radial distance. For most purposes it is desirable to use high-energy synchrotron radiation as the incident X-ray source, especially if longer range information, e.g. on the secondary shell or more remote metal atoms at distances $\geqslant$ pm, is desired. The EXAFS method has a large advantage over the normal X-ray diffraction approach because of its high selectivity for the central atom and a high sensitivity, allowing investigations to be carried out in dilution solutions. Also, whereas the normal X-ray diffraction method often suffers by being dominated by scattering from the solvent, which can distort or hide information on the solute, this is not a problem in EXAFS since long-range interactions are eliminated. However, this in turn restricts the information that can be obtained. A further disadvantage is that the EXAFS method itself gives little or no angular information and moreover the interpretation is often more difficult if more than one kind of scattering atom is present.

Although EXAFS is potentially a powerful tool for structural elucidation of species, in aqueous solution and on non-crystalline samples there can be limita-tions in the precision of the structural information one gets, especially if it is largely based upon preconceived models. This has been particularly apparent in

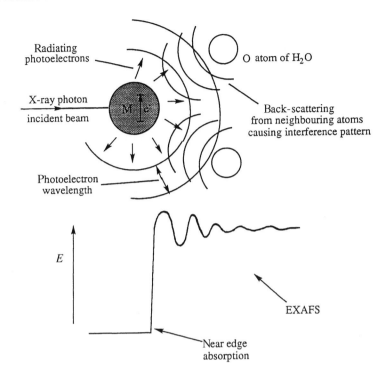

Figure 1.12. The basic principle behind EXAFS

investigations on polynuclear aqua ion species, a good example being studies on the structure of red 'Mo^{4+}(aq)', now known to be the triangular oxo-bridged cluster $[Mo_3O_4(OH_2)_9]^{4+}$ shown in Fig. 1.4. Prior to 1980 the ion was thought to be the di-μ-oxo dimer $[Mo_2(\mu\text{-}O)_2(OH_2)_8]^{4+}$, based largely upon ion-exchange behaviour typical of a 4+ cation and charge/metal determinations suggesting a value of 2.0. An EXAFS study reported in 1980 indeed gave a good fit for the di-μ-oxo dimer structure and so it was concluded, this despite the existence in the literature by that time of a range of derivative complexes known to contain the trinuclear Mo$_3$O$_4$$^{4+}$ core. It was not until later that year that Murmann, in an elegant ^{18}O-labelling study (see Section 6.2.3.1), confirmed that 'Mo^{4+}(aq)' was the trinuclear cluster $[Mo_3O_4(OH_2)_9]^{4+}$, a fact has now been confirmed by subsequent single-crystal X-ray diffraction and ^{17}O NMR measurements (see later and Chapter 6). When the EXAFS data were subsequently reanalysed it was found that, despite an excellent fit for the di-μ-oxo dimer structure, a slightly better fit is obtained if the trinuclear unit, involving back-scattering from two Mo centres ~ 250 pm away, is used for the parameterisation. Thus while EXAFS can give reliable information at short-range distances (~ 150–250 pm) the reliability of fitting falls off at somewhat longer distances $\geqslant 250\,\text{pm}$. This is also the case when the nearest neighbour atoms effectively back-scatter all of the outgoing photo-electrons leaving the more remote scatterers undetected. An example of this is provided by a comparison of the LAXS data on solutions of $[\text{Rh}(\text{OH}_2)_6](\text{ClO}_4)_3$ with that from EXAFS [22]. Thus while the LAXS experiments detected evidence not only of the primary hydration shell of six nearest water molecules (Rh—O) = 202 pm) but also that of a well-defined secondary hydration shell at Rh—O distances of $\sim 402\,\text{pm}$, the EXAFS experiment detected only back-scattering from the primary hydration shell. Owing to the ever-increasing library of parameter sets, the reliability of EXAFS to both simulate and predict local structural information is improving, but it still remains largely a function of the model one uses. EXAFS has nonetheless played a major role in the investigation of the local structure around metal centres in a range of metalloproteins and enzymes, its high sensitivity, atom selectivity and non-requirement for crystalline materials being great virtues. As such it can be said to have played a 'major part in the development of bioinorganic chemistry over the past 25 years'. Another growth area seems to be its use in the study of the surface structure of industrial heterogeneous catalysts.

1.2.2.2   INFRA-RED AND RAMAN SPECTROSCOPY

Both of these techniques played a pioneering role in structural and dynamic investigations of solutions and were dominant during the 1960s and 1970s. A simple schematic representation of the two techniques is shown in Fig. 1.13 [27]. Raman spectroscopy uses a visible laser incident on the sample which leads

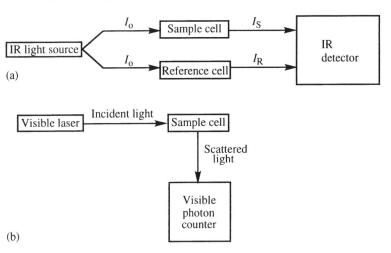

Figure 1.13. Schematic representation of (a) the basic IR experiment and (b) a Raman experiment

to scattering mainly of the same frequency as that impinging on it. However, a small fraction of the light is scattered back at frequencies shifted by amounts corresponding to the energies of different vibrational states. Raman spectroscopy has the advantage that it can be used in aqueous solutions without interference from the bulk water. This is because Raman scattering activity, unlike normal infra-red (IR), requires a change in the polarizability tensor and in this regard water is a poor Raman scatterer. Resonance Raman (RR) makes use of an enhancement in the intensity of a vibrational line when irradiation is absorbed by a visible transition. The enhanced bands can often be related to the chromophore responsible for the light absorption. Another virtue is that the spectrum often becomes simplified since the absorption of only selected vibrations is enhanced and this can in turn allow very dilute samples (mM concentrations) to be measured. Localised heating from the laser in RR experiments can be a problem for sensitive samples, but this has largely been overcome by rapidly spinning the sample. In the normal IR experiment observation of solute vibrations in aqueous solutions requires the subtraction of the large background IR absorption of the bulk water. This can largely impair the information obtained in the regions where water absorbs strongly, e.g. around 2800–4000, 1500–1600 and 800–900 cm^{-1}. One method that has been developed to overcome this is known as the double difference method developed by Lindgren [28]. The aim was to get specific information on ion-affected water molecules. The method involves monitoring OD and OH stretching frequences of isotopically diluted HDO molecules (8 atom %) in the hydration sphere of ions. Four spectra A–D are taken; A (the solute in 8% HDO), B (the solute in D_2O), C (8% HDO in D_2O) and D (pure D_2O). The difference, $(A_{obs}-k_1B_{obs})-k_3(C_{obs}-k_2D_{obs})$, is then calculated where $A-D_{obs}$ are the absorbances of the various solutions at a particular wavelength and k_1-k_3 are adjustable parameters. This method subtracts the contributions from D_2O and HDO molecules in bulk water and leaves behind the ion-perturbed HDO molecules. Allowances are made for HDO produced by H_2O/D_2O exchange in the solid hydrates. In the presence of trivalent ions like Al^{3+}, Cr^{3+} and Rh^{3+} frequency shifts are seen due to the entire primary and secondary shell assemblage. However, the total number of ion-perturbed water molecules detected by this method tends largely to be somewhat underestimated compared with the results obtained from neutron scattering. Nonetheless this method has clearly established the existence of the secondary shell.

Perhaps the most detailed vibrational spectral studies have been those reported on the caesium alums in the series of papers by Best and coworkers [29] in conjunction with data from neutron diffraction [16–18]. The recording of low-temperature single-crystal polarised Raman spectra has allowed an assignment of all the Raman active internal modes v_1, v_2 and v_5 for a range of trivalent hexaaqua cations [30]. A relationship between v_1 (symmetric MO_6) (essentially a measure of the M—O force constant) and the M—O bond length was expected but this is found to be noticably different for the second and third row transition

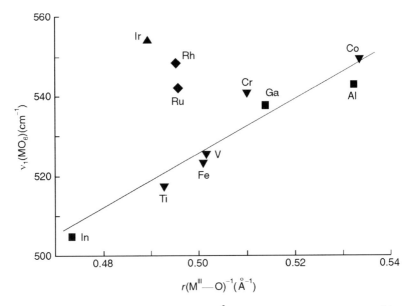

Figure 1.14. Plot of $v_1(MO_6)$ vs $1/(r(M^{III}-O))$ Å^{-1} in the series of trivalent CsM^{3+} alums

metal alums than for the first row (Fig. 1.14). The precise reason behind this remains a mystery. A relationship to the tilt angle θ (different in the case of Rh and Ir due to adoption of the α modification) was viewed unlikely from studies of the CsRu^{3+} alum which has trigonal planar coordination of water despite lying off the line in Fig. 1.14. The external modes of the primary sphere water molecules are found to be highly sensitive to the alum type, the M—O bond length and the stereochemistry of water coordination.

One interesting result from the comparison of solution X-ray diffraction data with Raman measurements is the indication of 'heptahydration' around aqueous Sc^{3+} (see Section 3.1) despite numerous references to it as being [Sc(OH$_2$)$_6$]$^{3+}$. Indeed, seven coordinated pentagonal bipyramidal Sc^{III} is now established in the X-ray structure of the aqua di-μ-hydroxo dimer Sc$_2$(OH)$_2$$^{4+}$(aq) (see also Section 3.1), as yet the only authentic Sc^{III}(aq) species to be crystallised.

Resonance Raman measurements also provide support for formulation of the aqua dimer of V^{III} as [(H$_2$O)$_5$V—O—V(OH$_2$)$_5$]$^{4+}$ rather than as the di-μ-hydroxo form [H$_2$O)$_4$V(μ-OH)$_2$V(OH$_2$)$_4$]$^{4+}$ (see Section 5.1.2). This is based largely on the Raman bands of dimeric V^{III}(aq) being similar to those of authentic {V—O—V}$^{4+}$-containing compounds despite the dihydroxo ('diol') form being the one most commonly encountered in crystal structures.

1.2.2.3 ^{18}O ISOTOPE LABELLING—MASS SPECTROMETRY

Gamsjager and Murmann reviewed the use of the ^{18}O isotopic labelling method in 1983 for the measurement of hydration numbers [31]. The method relies upon conversion of the oxygen, either bound to the metal or from the solvent water itself, into a gaseous form, usually as CO_2 followed by measurement of the $^{18}O/^{16}O$ mass ratio in a mass spectrometer. The advantages of this method are the high accuracy and precision of the mass spectral isotope ratio measurement and the requirement for only low levels of ^{18}O isotope, allowing highly dilute samples (mM) to be handled. Its disadvantage is that it can only be applied to relatively inert systems, i.e. the rate of oxygen exchange must ideally be several orders of magnitude slower than the timescale for isolation and conversion to the gaseous oxygen form. Historically the first formula to be established by the ^{18}O method was that of the dioxo U^{VI} salt $UO_2Cl_2 \cdot 3H_2O$ [32]. The experiment involved dissolving a known amount of the salt in ^{18}O-labelled water and then separation of the water by distillation from the solution and determination of the $^{18}O/^{16}O$ ratio in a sample of CO_2 following its equilibration with the water. During the timescale involved in removal of the water the oxygens of the UO_2^{2+} group do not exchange, allowing their presence to be deduced from the measured $^{18}O/^{16}O$ ratio as different from that in the original sample added to the salt. The number of slow exchanging oxygens is then determined from the uranium content. A range of aqueous species have now been reliably determined using the ^{18}O methods, as the selection in Table 1.3 shows.

Table 1.3. Coordination numbers and formulae for aqueous species established by ^{18}O isotope labelling [31] (ppt-precipitate)

Formula	Number of slowly exchanging oxygens per central atom	Sampling method
$UO_2Cl_2 \cdot 3H_2O$	2.0(terminal)	H_2O
$NpO_2^+(aq)$	2.1 ± 0.2(terminal)	$NpO_2(OH)(OH_2) \cdot H_2O$(ppt)
$VO^{2+}(aq)$	1.0 ± 0.1(terminal)	H_2O
$[VO_4]^{3-}$	4.0 ± 0.1	$[Co(NH_3)_6]VO_4 \cdot H_2O$(ppt)
$[FeO_4]^{2-}$	4.0 ± 0.1	$Ba[FeO_4]$(ppt)
$[Al(OH_2)_6]^{3+}$	6.0 ± 0.5	H_2O
$[Cr(OH_2)_6]^{3+}$	6.0 ± 0.2	H_2O
$[Cr_2(\mu\text{-}OH)_2(OH_2)_8]^{4+}$	5.0 ± 0.07	H_2O
$[Rh(OH_2)_6]^{3+}$	5.9 ± 0.4	H_2O
$[Ir(OH_2)_6]^{3+}$	6.0 ± 0.2	H_2O
$Mo_2O_2(\mu\text{-}O)_2^{2+}(aq)$	0.99 ± 0.02(bridging)	$[Pt(en)_2][Mo_2O_4(EDTA)]$(ppt)
	1.02 ± 0.03(terminal)	$[Pt(en)_2][Mo_2O_4(EDTA)]$(ppt)
$Mo_3O_4^{4+}(aq)$	1.33	$(Et_4N)_5[Mo_3O_4(NCS)_9]$(ppt)

The method has also been adapted for the monitoring of oxygen exchange processes on species with half-lives of $\geqslant$ a few seconds using stop quench methods, which are described in Section 1.4.

1.2.3 NMR Methods

NMR is a powerful method for studying both static and dynamic aspects of metal ion hydration [33]. However, in applying the NMR method one always has to bear in mind the timescale of the experiment itself and this largely depends upon the operating frequency.

1.2.3.1 ^{1}H NMR

Proton exchange is very rapid and as such information on metal ion hydration extracted from ^{1}H NMR spectra is often limited. In proton NMR the ^{1}H nucleus has to remain in a given environment for at least 10^{-4} s to be recognised as a discrete entity. Lowering the solution temperature to slow down the dynamic processes is only a limited option owing to the restrictions imposed by the high freezing point of aqueous solutions. The addition of an organic cosolvent, such as acetone, is sometimes employed to expand the available temperature range, but this has the danger of introducing a change in hydration number and the possibility of separating out dynamic processes in the bulk solvent as different from that of the primary coordination sphere. One can never be sure that such mixed solutions are really representative of pure aqueous solutions and this can introduce errors. Coordination of anions, especially in the mixed solutions, is also frequently the source of underestimated hydration numbers. A further major disadvantage is that high concentrations of dissolved electrolyte (molar) are required to get accurate integrations in addition to giving well-resolved peaks. With diamagnetic metal ions ^{1}H NMR chemical shift differences between the free and bound water can be quite small, although these increase as the temperature is lowered. Paramagnetic metal ions such as Ni^{2+} and Co^{2+} can cause large shift differences between free and bound water (up to ~ 10000 Hz); others, e.g. Mn^{2+}, cause noticeable broadening. This can, however, be put to good use. Co^{2+} and also certain lanthanide ions such as Dy^{3+} have been employed as shift reagents for measurements of hydration on diamagnetic ions such as Al^{3+}, Ga^{3+}, Mg^{2+} and In^{3+} since here the chemical shift of the bound water is not much different from that of the free water. The Co^{2+} interacts more strongly with the free water, thus shifting its position relative to the bound. Hydration numbers on labile metal ions have been estimated based on the effect on the linewidth of free water observed in the presence of Mn^{2+}. The degree of broadening by Mn^{2+} reflects the amount of free water present, which in turn will be a function of the concentration of the added metal ion in the salt and hence its hydration number. However, despite the inherent uncertainties a degree of success has been achieved in assigning hydration numbers on the basis of both ^1H NMR peak areas and chemical shift changes (Fig. 1.15 and Table 1.4).

1.2.3.2 ^{17}O NMR

A more attractive alternative is the ^{17}O nucleus of the oxygen atom of water, which can provide information on the primary hydration number as well as providing direct information on the water exchange dynamics (see Section 1.4).

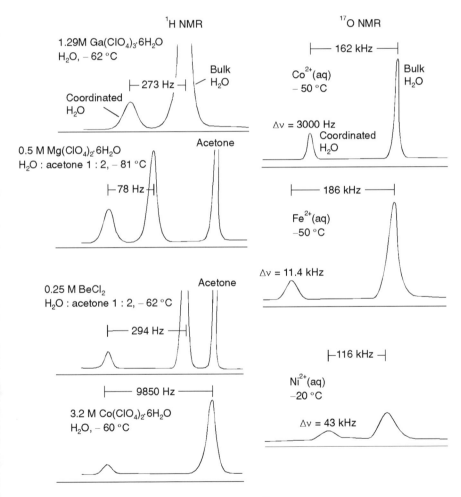

Figure 1.15. Selected ^1H and natural abundance ^{17}O NMR spectra having primary shell water on aqueous metal ions

Table 1.4. Selected metal ion dynamic hydration numbers as determined using NMR measurements (taken from ref. [5])

Salt	Ion	NMR nucleus	Method[a]	Hydration number	Temperature range (°C)
HNO_3	H^+	^1H	cs	2.5	
LiCl	Li^+	^1H	cs	3.0	
LiBr	Li^+	^1H	cs	3.0	
NaOH	Na^+	^1H	cs	3.5	
KF	K^+	^1H	cs	3.0	
RbOH	Rb^+	^1H	cs	3.5	
$BeCl_2$	Be^{2+}	^{17}O	pa	3.95–4.30	18.0–32.7
$Be(ClO_4)_2$	Be^{2+}	^{17}O	cs	3.7–3.9	20
$Be(NO_3)_2$	Be^{2+}	^1H	pa	4	−20–100
$Mg(ClO_4)_2$	Mg^{2+}	^1H	pa	5.6–6.1	−72–82
$Mg(ClO_4)_2$	Mg^{2+}	^1H	cs	6.0	
$Ca(ClO_4)_2$	Ca^{2+}	^1H	cs	6.0	
$Sr(NO_3)_2$	Sr^{2+}	^1H	pa	5.0	
$Sr(NO_3)_2$	Ba^{2+}	^1H	pa	5.7	−20–100
$Fe(ClO_4)_2$	Fe^{2+}	^{17}O		5.85	—
$Fe(ClO_4)_2$	Fe^{2+}	^1H	cs	5.6–5.8	−40–80
$Co(ClO_4)_2$	Co^{2+}	^1H	pa	5.9	−38–63.7
$Ni(ClO_4)_2$	Ni^{2+}	^1H	pa	5.87	−30
$Ni(ClO_4)_2$	Ni^{2+}	^{17}O	cs	6.0	< 127
$Ni(ClO_4)_2$	Ni^{2+}	^{17}O		6.0	
$Zn(ClO_4)_2$	Zn^{2+}	^1H	cs	5.9–6.4	−120
	Zn^{2+}	^1H	pa	6.0	
$Cd(NO_3)_2$	Cd^{2+}	^1H	pa	4.6	−20–100
$Hg(NO_3)_2$	Hg^{2+}	^1H	pa	4.9	−20–100
$Pb(NO_3)_2$	Pb^{2+}	^1H	pa	5.7	−20–100
$Cr(ClO_4)_3$	Cr^{3+}	^{17}O	pa	6.0	20
$Cr(ClO_4)_3$	Cr^{3+}	^{17}O	cs	6.7–6.9	20
$Rh(ClO_4)_3$	Rh^{3+}	^{17}O	pa	5.5	25
$AlCl_3$	Al^{3+}	^1H	pa	6.1	−29.5
$AlCl_3$	Al^{3+}	^{17}O	pa	5.82–6.07	35
$Al(ClO_4)_3$	Al^{3+}	^1H	pa	6.0	−35–52
$Ga(ClO_4)_3$	Ga^{3+}	^{17}O	cs	6.28	35
$Ga(ClO_4)_3$	Ga^{3+}	^{17}O	pa	5.89, 5.9	35
$Ga(ClO_4)_3$	Ga^{3+}	^1H	pa	5.8–6.1	−89–99
$In(ClO_4)_3$	In^{3+}	^1H	pa	5.8–6.0	−35–75
$La(ClO_4)_3$	La^{3+}	^1H	cs	6.0–6.4	−105–120
	Gd^{3+}	^1H		8 or 9	
$Er(ClO_4)_3$	Er^{3+}	^1H	cs	1.0	−75
	Sn^{4+}	^1H	pa	6.0	
$Th(ClO_4)_4$	Th^{4+}	^1H	pa	9.1	−100
	Th^{4+}	^{17}O	pa	10.0	

[a] cs = chemical shift change, pa = peak area integration.

The important ^{17}O NMR parameters are shown below [44].

^{17}O	Spin	5/2
	Natural abundance (%)	0.037
	Receptivity/^{13}C	0.061
	Gyromagnetic ratio	−3.6264
	Quadrupole moment	$−2.6 \times 10^{-2}$

Resonant frequency (MHz) referred to ^1H TMS resonance at 100 MHz

13.5564

^{17}O NMR parameters

1.2.3.3 DYNAMIC HYDRATION NUMBERS

In many cases it is possible to observe ^{17}O NMR peaks due to both bulk water and primary shell water from aqueous electrolyte solutions at natural abundance. High concentrations ($\sim$ molar) are required with Co^{2+}(aq) often employed as a shift reagent. As with ^1H NMR the Co^{2+} interacts more strongly with the free water, shifting it relative to that of the bound water. Low temperatures are again necessary in order to slow down the water exchange dynamics, particularly with regard to more labile metal ions. The 'dynamic' hydration number, as it is termed, is then measured directly from the integration of the peaks which will be proportional to the concentration of free and bound water. Selected ^1H and ^{17}O NMR spectra for a number of metal ions in aqueous media are illustrated in Fig. 1.15.

Table 1.4 lists primary hydration numbers of selected metal cations as determined using NMR peak area and chemical shift measurements [5]. Somewhat lower values are apparent in Table 1.4 for the group 1 and heavier group 2 metals when compared to those from XRD and neutron scattering data. ^{17}O NMR measurements on more dilute solutions requires isotopic enrichment. As a rough guide enrichment of around 5–10 atom % ^{17}O allows adequate signal to noise at mM concentrations of oxygen on a modern high-field 400 MHz NMR spectrometer (resonant frequency of ^{17}O = 54.24 MHz). In the study of enriched dilute solutions a further development has been the use of Mn^{2+} as an effective relaxation agent for the free water signal, thus allowing observation of the weaker bound signals (only slightly broadened if on the slow exchange timescale > 1 ms) which would otherwise be unobservable [35]. This has proved useful for probing solution structures of oligonuclear aqua species by ^{17}O NMR and for the purpose of observing bound water resonances for dynamic measurements of water exchange (see Section 1.4). The efficient broadening of the free water line by

Mn^{2+} arises from a combination of a strong interaction with the electron spin of the Mn^{2+} ion coupled with very fast chemical exchange. The NMR method is viewed, however, as generally less reliable than diffraction methods for determining hydration numbers on the most labile metal ions but, as apparent in Table 1.4, it can give highly reliable values for hydration numbers on the more inert aqua metal ions.

1.2.3.4 ^{17}O NMR—STRUCTURAL ELUCIDATION OF OLIGONUCLEAR SPECIES

In a selected number of cases, e.g. the trinuclear $[M_3O_4(OH_2)_9]^{4+}$ ions for molybdenum and tungsten and the aqua Mo^{III} dimer ion, ^{17}O NMR has proved a powerful solution structural probe [36]. In other cases it is less definitive.

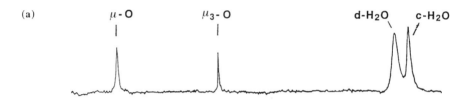

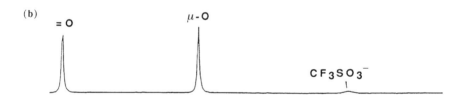

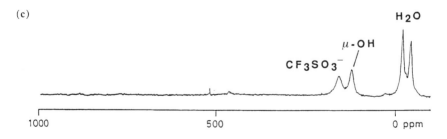

Figure 1.16. Oxygen-17 NMR spectra for 5 atom % enriched solution of three oligomeric aqua molybdenum ions: (a) $[Mo_3^{IV}O_4(OH_2)_9]^{4+}$, (b) $[Mo_2^{V}O_4(OH_2)_6]^{2+}$ and (c) $[Mo_2^{III}(\mu\text{-}OH)_2(OH_2)_8]^{4+}$ (each solution contains 0.1 M Mn^{2+} as a bulk water relaxant)

Figure 1.16(a) shows the ^{17}O NMR spectrum obtained from an enriched (5 atom % ^{17}O) sample of $[Mo_3O_4(OH_2)_9]^{4+}$ (0.2 M) in 2.0 M Hpts. Here integration of the respective ^{17}O resonances assigns the structure directly. All the oxygens are slow exchanging on the NMR timescale, enabling peaks for all four oxygen atom environments to be assigned (Section 6.2.3). It is seen that the alternative di-μ-oxo structure, proposed at the time of the EXAFS measurements, is entirely inconsistent with Fig. 1.16(a). Compare this with the spectrum of the aqua dimer of Mo^V (Fig. 1.16b) (Section 6.2.4). Here only the peaks due to the terminal and bridging oxo groups are observable because the waters are now on the fast exchange timescale with a half-life of < 0.1 ms, resulting in the linewidth of their resonances being indistinguishable from that of the baseline. The spectrum in Fig. 1.16(c) is that of the aqua Mo^{III} dimer ion, which is found to be consistent with a di-μ-hydroxo(diol) arrangement, $Mo_2(OH)_2^{4+}$(aq), and not the single μ-oxo alternative. Here much slower water exchange reflects the presence of the substitution inert d^3 configuration. A further situation is illustrated in Fig. 1.17 for the cases of tetrameric Ru^{IV}(aq) (Section 8.2.3) and trimeric $Nb_3^{III,IV,IV}$(aq) (Section 5.2). For Ru^{IV}(aq) (Fig. 1.17a), two resonances are observable at 25°C, one due to coordinated water and one to bridging oxygen. This is not sufficient information upon which one can assign a structure. For the trimeric Nb(aq) ion (Fig. 1.17b), the presence of two peaks in a 2:1 ratio is reminiscent of those assigned to coordinated water in $[Mo_3O_4(OH_2)_9]^{4+}$ (Fig. 1.16a). As a result, the presence of a similar incomplete cuboidal trinuclear structure (with the presence of a μ_3-Cl cap to account for the lack of the μ_3-oxo resonance) has been tentatively proposed.

1.2.4 Non-spectroscopic Methods

A number of these have been employed based upon transport properties of ions through the electrolyte solution. Several of these methods have already been discussed in the texts by Burgess [1] and so only a brief mention is given here. These largely surround measurements of conductivity and ion mobility. Ion mobility relates to the ability of the ions to move (diffuse) through the electrolyte solution and thus to conduct electricity through the solution. It should be a simple matter to relate ion mobility to the extent of hydration and in turn the hydration number. The single ion conductance relates to the total observed conductance by what is termed its transference number, t_+ [37]:

$$\Lambda_{obs} = \Lambda_o - kc^{1/2}, \qquad \Lambda_o = \Lambda_+ + \Lambda_- \tag{1.4}$$

$$\Lambda_+ = t_+\Lambda_o, \qquad \Lambda_- = t_-\Lambda_o \tag{1.5}$$

The simplest way to measure transference numbers is to add an inert reference substance to the electrolyte solution placed in a cell containing a cathode and an anode. The movement of the solvated ion is then monitored by the changing concentration of the reference compound, often a sugar, in the vicinity of each

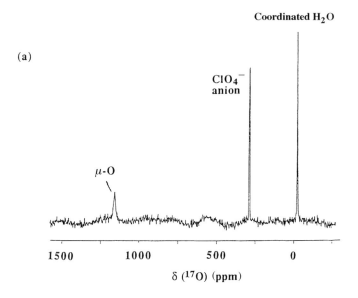

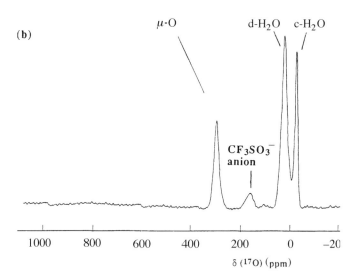

Figure 1.17. Oxygen-17 NMR spectrum for (a) 5 atom % enriched Ru_4^{IV}(aq) (in 2.0 M $HClO_4$) and (b) $Nb_3^{III, IV, IV}$(aq) (in 2.0 M CF_3SO_3H)

electrode. The transference numbers give the individual ionic conductance (1.5) and ion mobility μ_+, μ_-:

$$\mu_\pm = F/\Lambda_\pm \tag{1.6}$$

The simplest approach is then to use Stokes' equation:

$$r = \frac{1}{6\,\pi\eta\mu} \tag{1.7}$$

to calculate an effective radius of the hydrated ion related to its mobility μ and the solvent viscosity η. A hydration number can then be calculated from the estimated volume of the ion itself and that of a solvating water molecule. Not surprisingly ion mobility will reflect both primary and secondary shell hydration and the values obtained reflect this. In an alternative approach Gusev observed a marked discontinuity in the conductance of aqueous $HClO_4$ solutions as the concentration of H^+ and added ionic electrolyte built up [38]. He interpreted the point of discontinuity as when all the available free water had been used up by hydration either of H^+ or added M^{n+} and therefore no water was available for transporting protons. This is in effect another manifestation of the much slower proton exchange that occurs within the more ordered primary and secondary hydration sphere of metal ions as a result of the stronger hydrogen bonding. The water/cation ratio at the discontinuity should in turn be a measure of the cation hydration number. Not surprisingly for highly labile cations, a concentration dependence on the discontinuity is seen, resulting in a range of values.

Estimates of hydration numbers have also been based upon measurements of entropies of crystal hydration, of solvent compressibility, of redox potentials, of the free energy of hydration and of diffusion. The entropy approach is based upon the loss of entropy that occurs on transferring water from the bulk to the solvation shell, the finding being that $T\Delta S^{\theta}$ (kJ mol^{-1}, 25°C) for process (1.8) for a range of metal salt hydrates, e.g. $Na_2SO_4 \cdot 10H_2O$ (7.52), $ZnSO_4 \cdot 6H_2O$ (9.2) and $MgCl_2 \cdot 6H_2O$ (7.11) is very similar to that for process (1.9) (7.52):

$$MX \cdot nH_2O(c) \longrightarrow MX(c) + nH_2O \tag{1.8}$$

$$nH_2O(c) \longrightarrow nH_2O(l) \tag{1.9}$$

Compressibility estimates are based upon the assumption that the compressibility of water molecules within the primary hydration shell is essentially zero. The redox potential approach uses the effect of an added electrolyte on the $E^{\ominus}$ of a given cell couple. The free energy approach simply relates the free energy of hydration of a compound of known hydration number to that of the unknown. Values determined using some of these approaches are listed in Table 1.5.

1.2.5 Computational Methods [39]

The extension of electronic structure methods from treating isolated gas phase systems to modelling what we might term the situation in the 'condensed phase', i.e. that relevant in aqueous solution, remains a great challenge to the computational chemist. With more and more powerful computational capability becom-

Table 1.5. Cation hydration numbers as determined by non-spectroscopic methods [1]

Method	Li^+	Na^+	Cs^+	Mg^{2+}	Ca^{2+}	Ba^{2+}	Fe^{2+}	Zn^{2+}	Cr^{3+}	Al^{3+}
Transport numbers	13–22	7–13	4	12–14	8–12	3–5		10–13		
Mobilities	3–21	2–10		10–13	7–11	5–9	10–13	10–13		
Gusev conductivities	2–3	2–4	6	8	8	8				
Diffusion	5	3	1	9	9	8	12	11	17	13
Entropies	5	4	3	13	10	8	12	12	21	
Compressibilities	3	4							31	
Activity coefficients				5	4	3	12			12
cf. NMR peak areas				6			6	6	6	6

ing routinely within the reach of a growing number of physicists and chemists in recent times there has been an explosion in the number of reported computational studies concerned with the structure and dynamics of hydrated metal ions. The challenge is to model realistically the situation for an ion when surrounded by its solvating sheath of hydrogen-bonded water molecules, i.e. in what we would term the condensed phase. Two principal approaches have been adopted:

(a) Quantum mechanical treatments, e.g. the *ab initio* self-consistent field (SCF) approach usually up to the Hartree–Fock level employing large basis sets wherein corrections for the effects of solvation, such as measured by solute polarisation effects, are later added [40–42] or the use of density functional theory (DFT) methods [43, 44] based upon functions derived from quantum mechanical calculations of the local electron density. A further approach has been to use SCF methods that combine continuum models of solvation with a DFT treatment for the solute [45, 46]. The *ab initio* approach based upon density functionals is proving attractive, not least because it is less demanding on computing power and since parameters that relate directly to solvation effects such as electron correlations, possible if gradient correction are included, can be largely built in.

(b) Monte Carlo (MC) and molecular dynamics (MD) simulation methods [41, 47, 48] employing classical force fields based upon calculated *ab initio* atom–atom pair potentials and the explicit treatment of solvent molecules from the outset.

Information obtained from the two types of approach has been of interest to compare and has often been used in tandem. Such combined approaches, e.g. the use of *ab initio* methods to calculate reliable pair potentials which are then used in MC/MD simulations, have been applied with some success for probing the nature of the hydration sphere around the most labile and mobile of cations including H^+ (H_3O^+) and the group 1 and 2 metal ions (see discussions in Chapter 2) and the trivalent lanthanides (Chapter 3). There is growing interest and optimism in the power of the simulation method (see below). Calculations based upon semi-empirical methods have been employed in some cases. The reader can be referred to articles by Sandstrom and coworkers [49] (*ab initio* SCF), Deeth [43] (DFT methods), Hillier and coworkers [46] (combined *at initio* SCF and DFT approach), Rode and coworkers [41], Foglia, Helm and Merbach and coworkers [50] and Heinzinger and coworkers (MC and MD simulations) [47] and references therein.

1.2.5.1   QUANTUM MECHANICAL *AB INITIO* APPROACHES

As stated above, *ab initio* methods have been widely employed to investigate the nature of the hydration sphere around metal ions. Although much interest has focused on the labile s- and p-block metals there have been a number of studies concerning the metals of the d block.

Sandstrom and coworkers have applied large basis set *ab initio* SCF methods in an attempt to provide a theoretical basis for ligand field effects [49] and the dynamics of the water exchange process (see Section 1.4) on the hexahydrated divalent and trivalent metal ions of the first and second transition metal series. Despite the model being necessarily restricted to a single metal-aqua system effectively in the gas phase the agreement of theory with experiment, particularly in correlating trends across the period, is highly encouraging. Calculated $M-OH_2$ distances show a good correlation with the trend in experimental values based largely on those from the respective ammonium Tutton salts [19–21] and caesium alums [16–18]. The calculated bond lengths were, however, consistently larger than the experimental values by around 0.05 Å, a feature attributed to possible shortening of the experimental bond lengths due to greater polarization of the primary shell water molecules as a result of strong hydrogen bonding with the secondary shell. Additional corrections for, for example, real temperature, electron correlation and relativistic effects (see later) were also recognised as possibly responsible for the shorter experimental values.

The calculations extended to an assessment of the possible preferred orientation of six primary shell water molecules, modelled assuming trigonal planar coordination in T_h symmetry, 'all-vertical' D_{3d} symmetry and 'all-horizontal' D_{3d} symmetry (Fig. 1.18). The energy difference between the three orientations, particularly the two of D_{3d} symmetry, was found to be relatively small ($< 10 kJ mol^{-1}$) and no clear pattern emerged. $[Ti(OH_2)_6]^{3+}$ and $[Fe(OH_2)_6]^{2+}$ were found to be somewhat more stable in the all-vertical D_{3d} orientation whereas $[V(OH_2)_6]^{3+}$ was very marginally more stable in all-horizontal D_{3d} symmetry. In the X-ray structure of $(H_5O_2)[V(OH_2)_6]$ $(CF_3SO_3)_4$ the $V-OH_2$ orientation is indeed all-horizontal D_{3d} (Section 5.1.2), although in view of significant hydrogen-bonding effects some caution is needed in making general extrapolations. It was shown for $[V(OH_2)_6]^{3+}$, however, that

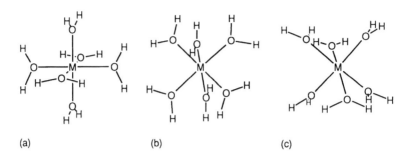

(a)   (b)   (c)

Figure 1.18. Hexaaqua ions with planarly coordinated water ligands in (a) T_h symmetry, (b) 'all-vertical' D_{3d} symmetry and (c) 'all horizontal' D_{3d} symmetry

tilting of the water molecule towards a pyramidal lone pair arrangement caused a significant increase in the energy difference between the T_h and D_{3d} orientations. Finally, the linear correlation of binding energy with inverse $M-OH_2$ distance shows that the bonding in the first row divalent and trivalent hexaaqua ions is largely dominated by electrostatics, although Mulliken population analyses (variations in metal atomic charges) seem consistent with greater covalency in the bonds to the trivalent ions, especially towards the right of the series and for those of the second row. For the 4d metal ions the lower binding energies were found to be largely compensated by greater ligand field splitting from the more spatially extended 4d orbitals.

1.2.5.2 MONTE CARLO (MC) AND MOLECULAR DYNAMICS (MD) SIMULATION METHODS

A common approach has been to evaluate atom–atom pair potentials using extended basis set SCF LCAO MO calculations followed by their use in MC/MD simulations of the ion–water interaction via the treatment of electron polarisation effects. The simulation method largely concentrates on providing a geometry optimisation for locally observed structural minima. The quality of the geometry optimisation is then often checked by performing additional SCF or DFT calculations. The validity of the approach largely depends upon the availability of good pair potential functions for each atom pair. There are few reliable methods of checking this other than ability to reproduce measurable properties such as atom–atom distance information, if available, from neutron scattering or X-ray diffraction data. A model extrapolating to the bulk condition is arrived at by a special treatment, which includes use of the periodic boundary condition in which a small system containing a limited number of ions and water molecules is repeatedly set up to build an infinite system.

The potential function describing water–water intermolecular interactions has been the subject of much investigation using simulation methods. Since the attempts of Bjerrum in the early 1950s [61] many rigid models have been proposed to describe both the structure and dynamic properties of water and aqueous solutions. Of these one can highlight the Ben–Naim–Stillinger (BNS) [62], Stillinger–Rahman (ST2 model) [63] and Rowlinson [64] models, as based on a four-point charge treatment, and the three-point charge system of Matsuoka, Clementi and Yoshimine [65]. Flexible models have also been introduced in which the bond length and angles between the hydrogen and oxygen atoms are chargeable due to vibrations. Information obtained from MD simulations can be summarised below:

(a) The derivation of radial distribution functions, $G(r)$, as a function of r, via calculated atom–atom distances, which can be compared with those obtained from X-ray and neutron diffraction (scattering) measurements.
(b) The angular distributions of water molecules around ions in the system.
(c) The numbers of water molecules in the first and second hydration shells provided that a suitable definition of the hydration is set. (note that hydration numbers are not necessary identical to those determined by diffraction methods since the concept of the coordination shell is not always fully expressed in MD/MC simulations such that peaks in the computed radial distribution functions may not be of a symmetrical Gaussian type.
(d) Calculations of the time evolution of interatomic distances and bond angles, from which the rigidity of the hydration shell and its symmetry can be discussed.
(e) Calculations of the time evolution of distances and angles in a specific atomic group. In this regard it is conceivable that a fast enough computer might ultimately one day be capable of handling the vast amount of data needed for a full evaluation of the mechanisms governing the relatively slow water substitution processes on metal ions when compared to the rapid timescale of the water dynamic processes.
(f) The derivation of self-diffusion coefficients.
(g) Important conclusions about the collective motion of ions and their hydrates from an evaluation of spectral densities of hindered translational motions of ions and molecules.
(h) Calculations of intramolecular vibrations from flexible models allowing comparisons with infra-red and Raman data.

The reader can be referred to the articles by Heinzinger and coworkers [57] for an overview of recent progress with MD simulations.

Monte Carlo (MC) simulations differ from MD simulations in giving only static structural and energetic information since atomic configurations are created randomly, thus losing any time-dependent information. The problem is that physically irrelevant configurations can occur but the procedure is able to distinguish these from acceptable configurations, allowing them to be discarded in the course of the calculation. The classical formulation of the method has been described by Metropolis *et al.* [56]. The adoption of good interatomic potentials is, however, very important in order to obtain reliable results since the energy and the acceptance criterion are directly linked [57]. Often the assumption of pairwise additivity of the interaction energies has proved insufficient and Heinzinger in particular has defended the use of three-body potentials despite the extra demands on computational capacity. The hydration around Be^{2+} provides a good example of the success of the three-body approach (see Section 2.3.1.1). The reader may be referred to reviews on the MC method by Wood and Erpenbeck [58]. Beveridge *et al.* [59] have collected results from MC simulations on aqueous solutions.

MC and MD methods have both been successfully employed to simulate the experimental radial distribution functions for transition metal M^{2+} ions [60]. Merbach and coworkers have applied MD simulations to the problem of the hydration sphere around the Ln^{3+} ions [50]. By taking account of water polarisation the results have provided further support for a changeover in the hydration number along the series Nd^{3+}, Sm^{3+}, Dy^{3+} and Tb^{3+} from nine, around Sm^{3+}, to eight (see Chapter 3).

A further recent development has been the use of the so-called reverse Monte Carlo method (RMC) invented by McGreevy and Pusztai [61]. In the RMC simulations pair potentials and energy criteria are completely omitted and a new criterion is set for the fit between the radial function $G(r)$ calculated from the randomly generated configurations and that for a diffraction experiment. Despite the requirement for more extensive calculations the lack of dependency on dubious pair potentials is a distinct advantage and the set of accepted configurations are found to reproduce well the experimental pair correlation functions. Reviews of the method for the treatment of ion hydration have been presented by Howe and Radnai [62].

1.3 HYDROLYSIS OF METAL IONS

Baes and Mesmer [3] have provided a comprehensive review of methods, difficulties, uncertainities, ionic strength effects and results obtained in their essential reference book *The Hydrolysis of Cations*. We will pay particular attention in this book not to the methods of obtaining hydrolysis constants, which have been covered at length elsewhere, but to the significance of values obtained in the context of structure and reactivity. At this point, though, it will be important to define fully the various terms and nomenclature used for the purpose of discussion throughout the book.

Hydrolysis literally means 'breakdown by water'. However, in the context of aqueous metal ions, hydrolysis occurs when the 'acidity' of the protons on the water molecule reaches a level when a free water molecule itself becomes a sufficient Bronsted base to effectively remove a proton forming a hydroxy metal species and H_3O^+. The following equations are normally used to express the various equilibria involved:

$$M^{n+}(aq) + H_2O \rightleftharpoons M(OH)(aq)^{(n-1)+} + H^+ \qquad (K_{11}) \qquad (1.10)$$

$$M^{n+}(aq) + OH^- \rightleftharpoons M(OH)(aq)^{(n-1)+} \qquad (K_{OH}) \qquad (1.11)$$

$$H_2O \rightleftharpoons OH^- + H^+ \qquad (K_w) \qquad (1.12)$$

metal aqua   metal hydroxo   (1.13)

$$M^{n+}(OH_2)(aq) + H_2O \rightleftharpoons M(OH)(aq)^{(n-1)+} + H_3O^+ \qquad (K_{11*}) \qquad (1.14)$$

Since equation (1.13) is usually representative of the hydrolysis reaction occurring within the hydrogen-bonded aqueous matrix the reaction (1.14) is probably more representative. In the extreme case of very high cationic charge the proton on the hydroxide ligand can remain acidic enough for a second proton abstraction to occur forming a terminal oxo species:

$$(aq)M^{n+}=O + H_3O^+$$

(1.15)

This is particularly promoted if a strong M=O multiple bond is the result. Examples are countless 'oxo' species such as RuO_4, oxo anions such as ReO_4^-, MoO_4^{2-}, $Cr_2O_7^{2-}$ and also cationic M=O species such as VO^{2+}, *cis*-VO_2^+, $Mo_2O_4^{2+}$ and *trans*-UO_2^{2+}. In an attempt to illustrate the conditions under which H_2O, OH^- and O^{2-} would be common ligands at a metal centre in aqueous solution Jorgensen came up with what he termed a 'charge–pH' diagram, Fig. 1.19 [63]. The diagram accounts for the formation of oxo anions CrO_4^{2-}

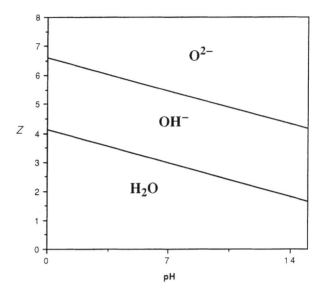

Figure 1.19. Qualitative regions of pH and cationic charge wherein H_2O, OH^- or O^{2-} would be common ligands on metal centres in aqueous solution

and MnO_4^- for $Cr^{VI}(aq)$ and $Mn^{VII}(aq)$, whereas $Cr^{III}(aq)$ and $Mn^{II}(aq)$ exist as $[Cr(OH_2)_6]^{3+}$ and $[Mn(OH_2)_6]^{2+}$ respectively, and shows why the Bronsted acid strength decreases along the series $Cl^{VII}O_3(OH)$, $S^{VI}O_2(OH)_2$, $P^VO(OH)_3$ and $Si^{IV}(OH)_4$. This simple rule of thumb does not, however, take into account the effect of size, which will somewhat influence the strength of M=O bonds.

One can also view a hydrolysis reaction as involving attack by OH^- on the metal ion to form the hydroxo species (1.11). This is often a useful representation in terms of the resulting number of OH^- ligands per M. K_{OH} and K_{11} are related by the ionic product of water, $K_w (= K_{11}/K_{OH})$, which is $\approx 10^{-14}$. The hydrolysis constant K_{11} (1.10) is the one normally quoted, as in Baes and Mesmer's book [3], although in many cases it is frequently quoted to be constant of reaction (1.14). Sometimes the symbol K_a (acidity constant) is used for K_{11} since in effect it represents an 'acid dissociation' process. Equation (1.11) can be expressed in a more general form, relating to an overall stability constant K_{xy} (or β_{xy}), to reflect more than one deprotonation and/or the formation of oligonuclear species:

$$xM^{n+}(aq) + yH_2O \rightleftharpoons M_x(OH)_y(aq)^{(xn-y)+} + yH^+ \qquad (K_{xy}) \qquad (1.16)$$

The true thermodynamic K_{xy} value should really reflect the ratio of activities of the species involved and not the concentrations of the species which are the quantities actually measured. Thus, strictly speaking K_{xy} should be expressed as

$$K_{xy} = \frac{[M_x(OH)_y^{(xn-y)+}][H^+]^y}{[M^{n+}]^x} \frac{\gamma_{M_x(OH)_y}\gamma_{H^+}^y}{\gamma_{M^{n+}}^x} \qquad (1.17)$$

where

$$Q_{xy} = \frac{[M_x(OH)_y^{(xn-y)+}][H^+]^y}{[M^{n+}]^x}$$

with inclusion of activity coefficients γ for each species. Only at infinite dilution can one ignore the γ values as being unity. Since much of the experimental data relates to real concentrations the quoted hydrolysis constants are in effect formation quotient Q_{xy} values pertaining to the particular conditions of the experiment.

1.3.1 The Determination of Hydrolysis Constants

1.3.1.1 THE POTENTIOMETRIC TITRATION METHOD

The most common method of determination of hydrolysis (protonation) con-stants is by titration of a solution of the metal aqua ion with OH^- ions and monitoring the pH with a glass electrode as a function of OH^- added. Titrations are normally carried out in a stirred cell set up as in Fig. 1.20 under an inert atmosphere in CO_2 free water. From knowledge of the K_w value under the conditions of study (usually determined in separate studies by titrating a strong acid) one can evaluate K_{xy} values for all the stepwise hydrolysis processes, e.g. for reaction (1.10), from (1.11) and (1.12). The normal procedure is to perform titrations in a medium of constant ionic strength (I),* several orders of magnitude higher than the concentrations of the titrants, such that the activity coefficients can be assumed at least to be constant and therefore Q_{xy} values should be

Glass electrode	Metal aqua ion solution + H⁺	Electrolyte in excess	Cl⁻ + electrolyte in excess	Reference electrode Ag/AgCl or Hg/Hg₂Cl₂

Figure 1.20. Typical cell used in potentiometric determinations of hydrolysis constants

---

*Ionic strength is normally defined by the expression $I = \sum c_i z_i^2$ where c_i is the concentration of each metal ion in question and z_i is the ionic charge. For a 1:1 electrolyte I equals the molarity. However, for a 2:1 electrolyte the two are different, viz. for a 2:1, $I = 3$ M; for a 3:1, $I = 5$ M and for a 4:1 electrolyte, $I = 10$ M, etc.

constant also. This is certainly the preferred procedure when dealing with more than one hydrolysis product. In an attempt to estimate the thermodynamic K_{xy} values Baes and Mesmer [3] have applied a modified form of the Debye–Huckel theory to take into account changes in the ionic activity coefficients as a result of changes in the composition of the solution in the presence of an excess of supporting electrolyte [64]. For a hydrolysing cation at a low concentration in the presence of a 1:1 supporting electrolyte they derived the expression

$$\log Q_{xy} = \log K_{xy} + \frac{\Delta z^2_{xy} S I^{1/2}}{1 + I^{1/2}} - (\Delta B_{xy} + 0.0157 y \phi) I \qquad (1.18)$$

where the quantity $\Delta z^2{}_{xy}$ is the square of the charge on each species summed over the entire formation reactions to give $M_x(OH)_y^{z_{xy}}$:

$$\Delta z^2_{xy} = z^2_{xy} + y - x z^2_M$$

and S is the Debye–Huckel limiting slope ($0.511 \, kg^{1/2} \, mol^{-1/2}$ at $25°\,C$) and I is the ionic strength. The quantity ΔB_{xy} is a similar sum of interaction coefficients B_{MX}:

$$\Delta B_{xy} = B_{(xy)X} + y B_{HX} - x B_{MX}$$

The last term in equation (1.18) contains the osmotic coefficient ϕ and reflects the effect of the water activity on Q_{xy} according to Robinson and Stokes [65].[†] Some of the values quoted for species in subsequent chapters will be those estimated by this approach, whereas others will be uncorrected concentration-related Q_{xy} values. For the purpose of simplification, unless otherwise indicated, the symbol K_{xy}, or more specifically $pK_{xy}(-\log_{10} K_{xy})$, will be used in reporting hydrolysis constants for the various metal ion hydrolysis products throughout the book.

It is often useful to express the extent of the hydrolysis process with pH by what is termed the ligand number n [3, 66]. This relates to the average number of OH^- ligands attached to M^{n+}. For an analytical excess of acid or base present (m_H or m_{OH}) the ligand number is given by

$$n = ([H^+] - [OH^-] - m_H)/m_M \qquad \text{(acid)} \qquad (1.19)$$

$$n = ([H^+] - [OH^-] - m_{OH})/m_M \qquad \text{(basic)}$$

Where m_M is the total metal ion content, usually expressed in terms of molality ($mol \, kg^{-1}$). Values for m_H and m_{OH} for the acid and alkaline parts of the titration curve can be calculated by the Gran equation [77]

$$m^0(M_1 + M_2) 10^{(E - E_j)/k} = 10^{E_0/k}(m_H^0 M_1 - m_B^0 M_2) \qquad (1.20)$$

[†] $\log a_w = -2\,(0.018/2.3)\phi I.$

where M_1 and M_2 are the masses of water in the metal ion solution and in the solution of base, $k = RT \ln(10)/F$, m_H^0 is the initial molality of H^+ in the metal ion solution, m_B^0 is the molality of OH^- in the base solution and m^0 is the standard value of the molality equal to 1 mol kg^{-1} of water. From equation (1.20) the cell potential E_0 and m_H^0 (m_H^0 for the alkaline part) are evaluated by iteration. It is normal to determine preliminary values of E_0 firstly from the acid part of the titration curve and then use to estimate preliminary values for the protonation constants Q_{xy} from the alkaline part. Both are then later refined to constant values over the entire curve. Corrections for cell junction potentials should also be made. It is also important to perform titrations at different m_M in order to check for the presence of polynuclear species. Values for $Q_{xy}(K_{xy})$ are then evaluated from the n vs pH plots by an iterative procedure in terms of various simultaneous hydrolytic equilibria [see the following examples for the case of two successive deprotonations on a trivalent metal ion:]

$$[M(OH_2)_6]^{3+} \overset{K_1}{\rightleftharpoons} [M(OH_2)_5(OH)]^{2+} + H^+ \tag{1.21}$$

$$[M(OH_2)_5(OH)]^{2+} \overset{K_2}{\rightleftharpoons} [M(OH_2)_4(OH)_2]^+ + H^+ \tag{1.22}$$

$$n = \frac{[M(OH)] + 2[M(OH)_2]}{[M] + [M(OH)] + [M(OH)_2]} \tag{1.23}$$

$$n = \frac{K_1[H^+]^{-1} + 2K_1K_2[H^+]^{-2}}{1 + K_1[H^+]^{-1} + K_1K_2[H^+]^{-2}} \tag{1.24}$$

As defined in (1.16) and (1.17) the stability constants for (1.21) and (1.22) will be $K_{11} = K_1$ and $K_{12} = K_1K_2$.

1.3.1.2 THE SPECTROPHOTOMETRIC METHOD

In certain cases extremely low pK_{xy} values (< 1) are relevant, particularly with cationic charges $> 3+$, and here it is not possible to determine hydrolysis constants by potentiometry using a glass electrode owing to the high background $[H^+]$ values that need to be present. If the hydrolysis products have different optical absorption spectral characteristics, e.g. the appearance of OH^- to M^{n+} charge transfer bands, it is sometimes possible to measure equilibrium concentrations of M^{n+} and $M_x(OH)_y^{(xn-y)+}$ and hence Q_{xy} by spectrophotometry. Realistically it is very difficult to determine more than one protonation constant by this method unless each species absorbs significantly differently from each other, which is not usually the case. Furthermore, one is often forced to work at much higher concentrations of metal ion which has the danger of introducing complications from hydrolytic polymerisation (see below). However, if only one hydrolytic species, e.g. $M(OH)^{(n-1)+}$, is relevant under the appropriate choice of conditions it is possible to determine the relevant K_{11} value from

$$\varepsilon^{\lambda}_{obs} = \frac{\varepsilon^{\lambda}(M^{n+})[H^+] + \varepsilon^{\lambda}(MOH^{(n-1)+})K_{11}}{[H^+] + K_{11}} \qquad (1.25)$$

where $\varepsilon^{\lambda}_{obs}$ is the observed molar extinction coefficient at wavelength λ based on the total metal content, $\varepsilon^{\lambda}(M^{n+})$ is that of the fully protonated aqua species and $\varepsilon^{\lambda}(MOH^{(n-1)+})$ is that of the hydrolysis product. An example of the use of this method is illustrated in Section 6.2.3 (Fig. 6.27) for the determination of the hydrolysis constant K_{11} for the trinuclear oxo-bridged cluster ion $[Mo_3O_4(OH_2)_9]^{4+}$. Despite the already extensively 'hydrolysed' nature of this species the water molecules remain very acidic, $K_{11} = 0.43$ M. In view of the high concentrations needed to observe the absorbance change between $Mo_3O_4^{4+}$(aq) and $Mo_3O_4(OH)^{3+}$(aq), direct titration with OH^- is viewed as likely to cause extensive hydrolytic polymerisation. Therefore the variation of $[H^+]$ is carried out by accurate dilution of an acidified solution of the aqua cluster ion into the solution of the sodium or lithium salt of the same acid. Here water itself is effectively behaving as the base to alter the 'pH' ($-\log_{10}[H^+]$) in the range -0.5–1.5. This method also has the advantage that a constant excess of H^+ is always maintained over the metal ion which suppresses hydrolytic polymerisation as the bulk 'pH' is raised.

1.3.2   Factors Affecting the Magnitude of the Hydrolysis Constant

For divalent metal ions a simple correlation exists between the pK_{11} value for loss of the first coordinated water proton and $Z^2/(r_{M-OH_2}) \times n$ (Fig. 1.21), where Z is the cationic charge and n is the assumed primary hydration number. Z^2/r measures the Coulombic attractive force for the water molecule and n will determine how much of the attractive force is shared out by each water molecule. The smaller the value of n the larger is the attractive force felt by each water molecule and this should correlate with an increase in the acidity. Indeed, the pK_{11} value for Be^{2+} correlates with the known hydration number of four (Section 2.3.1). Noteworthy in Fig. 1.21 is that Cu^{2+} only falls on the line if a hydration number of four is assumed consistent with the four strongly interacting equatorial waters of the tetragonally distorted structure. Similarly Pb^{2+} and Sn^{2+}, Hg^{2+} lie close to the line only if hydration numbers of four and two respectively are assumed, but deviate markedly for a regular hydration number of six. For Pb^{2+} and Sn^{2+} the high acidity of the coordinated waters is entirely consistent with the low hydration numbers relevant from the solution XRD measurements. For Hg^{2+} there is good evidence from neutron scattering of

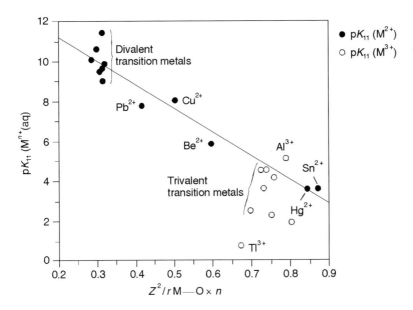

Figure 1.21. Dependence of pK_{11} for aqua metal ions with cationic charge Z, M—OH$_2$ distance and the primary hydration number n

an effective coordination number of two for the hydroxo species Hg(OH)$^+$(aq) and Hg(OH)$_2$(aq) (Section 12.2).

For the trivalent ions there seems no simple correlation of pK_{11} with $Z^2/(r_{M-OH_2}) \times n$ and here other factors, such as a degree of covalency within the M—OH$_2$ bonds, may be involved. The ions Al^{3+}, Cr^{3+}, Rh^{3+} and Ir^{3+} lie close to the correlation line for the divalent ions, possibly fortuitously. However, most first row transition metal trivalent ions have pK_a's around ~ 2 and show no correlation at all.

Baes and Mesmer (ref. [3], page 407) plotted pK_{11} versus the charge distance ratio and found a roughly linear correlation for cations subdivided into four groups A–D:

Group A: those most resistant to hydrolysis for their size and charge, 'hard' pretransition element ions and lanthanides: Mg^{2+}, Ca^{2+}, Sr^{2+}, Ba^{2+}, Al^{3+}, Y^{3+}, Ln^{3+}

Group B: those somewhat less resistant for their size and charge: Li$^+$, Na$^+$, K$^+$, Be^{2+}, Mn^{2+}, Fe^{2+}, Co^{2+}, Ni^{2+}, Cu^{2+}, Zn^{2+}, Cd^{2+}, Sc^{3+}, Ti^{3+}, V^{3+}, Cr^{3+}, Fe^{3+}, Rh^{3+}, Ga^{3+}, In^{3+}, Ce^{4+}, Th^{4+}, Pa^{4+}, U^{4+}, Np^{4+}, Pu^{4+}.

Group C: low resistance to hydrolysis, 'soft' post-transition elements with filled d shells: Ag$^+$, Tl$^+$, Pb^{2+}, Tl^{3+}, Bi^{3+}

Group D: an anomalously low resistance to hydrolysis: Sn^{2+}, Hg^{2+} and Pd^{2+}

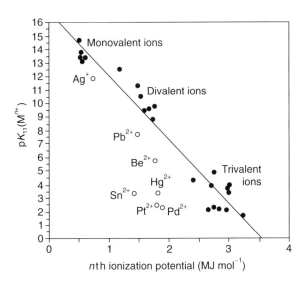

Figure 1.22. Correlation of pK_{11} for M^{n+}(aq) ions with the nth ionization potential for gaseous M^{n+}

The high acidity of the aqua ions in groups C and D probably results from a high degree of covalency in the M—OH_2 bonds coupled with a low hydration number.

A good correlation was also found by Kepert [68] in plots of pK_{11} versus the ultimate ionisation potential of the gaseous metal ion (Fig. 1.22); this was defined as the first ionisation potential for an alkali metal, the third for a trivalent metal, etc., and considered a measure of the intrinsic 'acid strength' of the ion. It is noted that the divalent ions Be^{2+}, Sn^{2+}, Hg^{2+}, Pb^{2+}, Pd^{2+} and Pt^{2+} along with Ag^+ all lie well off the line (open circles in Fig. 1.22), implying an unusually high acidity resulting from the low coordination number (between two and four) for their respective hydroxo aqua complexes.

Barnum [79] has examined the formation constants and standard free energies of formation for both mononuclear and polynuclear hydroxy complexes of metal ions throughout the periodic table and has detected some empirical correlations. For mononuclear complexes the following relationship can be derived:

$$\Delta G_f^\circ\{M(OH)_y\} = \Delta G_f^\circ\{M\} + By + Cy^2 + D/y \tag{1.26}$$

where B, C and D are empirical parameters and y is the number of coordinated hydroxides. Using the empirical parameters determined for known systems it was possible to estimate free energies and hence formation constants including those on previously unknown systems. Predicted values for pK_{xy} were claimed to have an uncertainty of only $\sim \pm 0.5$. A correlation of B (effectively proportional to pK_{11}) with Pauling electronegativity was also found. Here Be^{2+}, Sn^{2+}, Hg^{2+} and

Pd^{2+} lie below the correlation line, indicating much higher acidity than expected due to the low hydration number in the resulting hydroxy species. However, Sc^{3+} also lies below the line for the trivalent ions and here a lower hydration number does not appear to be the explanation (Section 3.1).

It has become further apparent from the data compiled by Baes and Mesmer that to a large extent the formation constants for complexes with two, three or four hydroxide ions per metal ion, as well as those for polynuclear hydroxy complexes (Section 1.3.3 below), are a reflection of the first ionisation constant of the aqua metal ion, i.e. all subsequent hydrolysis constants are related to K_{11} (ref. [3], page 427). Thus log K_q for the following equation was found to be approximately constant for a wide range of metal ions:

$$MOH \stackrel{K_q}{\rightleftharpoons} \left(\frac{1}{y}\right) M_x(OH)_y + \left(\frac{y-x}{y}\right) M$$

$$\log K_{xy} = y(\log K_q + \log K_{11})$$

(1.27)

Barnum found that K_2 values do vary over 2–3 orders of magnitude but no clear correlation occurred with parameters B, C, x, y, $\Delta G_f^\circ\{M\}$ or electronegativity. So K_q can be considered constant only in the context of comparisons with K_{xy} values which vary over some 30 orders of magnitude. Nonetheless, based on the experimental values for log K_{xy} on a wide range of systems it was possible to evaluate mean log K_q values for use in evaluating log K_{xy} values for hitherto unknown $M_x(OH)_y$ species. This approach, however, failed for $M_2(OH)^{n+}$ species owing to the variation in log K_q over too wide a range. The plots of log K_{xy} versus log K_{11} for different y pass through an isosbestic point at log $K_{11} = -1.15$. Thus metal ions having pK_{11} around 1.15 should have the same formation constant (≈ 0) for all hydrolytic polymers. An interesting metal ion in this regard is Bi^{3+}, for which pK_{11} is quoted as 1.09. Indeed, the value of '$pK_{6,12}$' for formation of $Bi_6O_4(OH)_4^{6+}$(aq) (Section 2.6.3) is quoted as 0.53 (ref. [3], page 381).

As was highlighted briefly in Section 1.1, hydroxo species can markedly influence the reaction pathways adopted in redox reactions involving aqua metal ions. It will also become apparent in the discussion in Section 1.4 and throughout the book that hydroxo and oxo complexes also play a major part in influencing the rates and mechanisms of water ligand exchange, substitution and redox reactions.

1.3.3 Hydrolytic Polymerisation

1.3.3.1 BASIC MECHANISMS

Frequently hydrolysis leads on to the formation of polynuclear hydroxo- and even oxo-bridged species. This becomes particularly important for $m_M > 10^{-5}$ m, as judged by the data in Baes and Mesmer [3]. Hydrolytic polymerisation becomes so acute for cationic charges $> 3+$ that simple monomeric $M^{n+}(aq)$ species are not observed even in strongly acidic solution and the 'simplest' aqua species can be polymeric, cf. $[M_4(OH)_8(OH_2)_{16}]^{8+}$ $(M = Zr^{IV})$ and $[M_3O_4(OH_2)_9]^{4+}$ $(M = Mo^{IV}, W^{IV})$. The fate of metal–hydroxo compounds is twofold. Either the acidity of the metal centre remains high enough to cause further deprotonation of the OH group by a water molecule, leading to formation of a terminal oxo moiety as already discussed (1.16), or the OH group can remain nucleophilic enough to attack an adjacent metal centre forming the OH-bridged dimer compound:

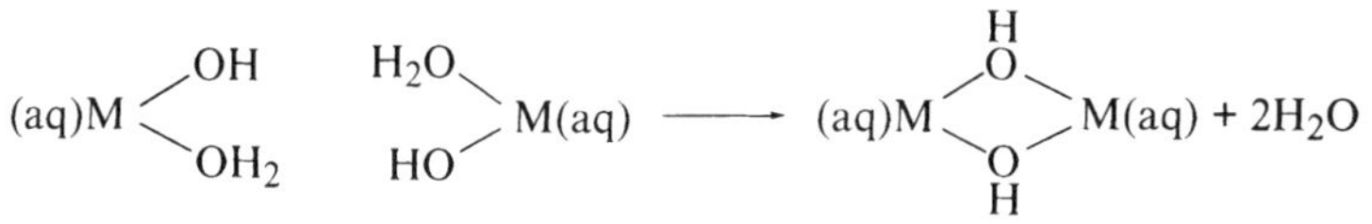

$$\text{(aq)M}\underset{\text{OH}_2}{\overset{\text{OH}}{<}} \quad \underset{\text{HO}}{\overset{\text{H}_2\text{O}}{>}}\text{M(aq)} \longrightarrow \text{(aq)M}\underset{\overset{|}{\text{O}}\atop\text{H}}{\overset{\overset{\text{H}}{|}\atop\text{O}}{<}}\text{M(aq)} + 2\text{H}_2\text{O} \qquad (1.28)$$

An important finding regarding the mechanism of olation (1.28) has been the characterisation of a number of $H_3O_2^-$ bridged ligand species by Ardon, Bino and Springborg [70] in the reactions between certain terminal hydroxo and aqua metal complexes:

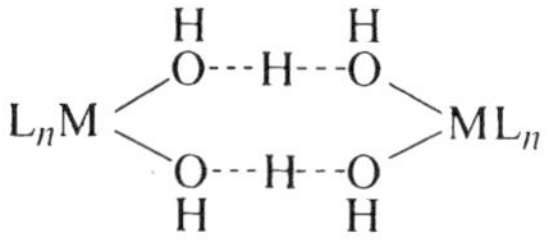

$$L_nM\underset{\overset{\text{O}\cdots\text{H}\cdots\text{O}}{\overset{|}{\text{H}}\quad\overset{|}{\text{H}}}}{\overset{\overset{\text{H}}{|}\quad\overset{\text{H}}{|}\atop\text{O}\cdots\text{H}\cdots\text{O}}{<}}\text{ML}_n \qquad (1.28)$$

Such species have been discussed in the context of being an intermediate in the olation reaction from mononuclear cis-aqua-hydroxo complexes to the diol (di-μ-hydroxo bridged) dimer [70, 71]. The formation of $H_3O_2^-$ bridged dimers has indeed been proposed to account for anomalously high K_{11} values observed in several mono-ol bridged aqua dimer complexes such as those of Cr^{III} (see Section 6.1.2) and Ir^{III} (Section 9.3). This is believed to be the result of stabilisation of the hydroxo ligand via $H_3O_2^-$ ligand bridge formation (see Fig. 1.23). A possible mechanism for conversion of the $H_3O_2^-$-bridged species to the di-μ-OH-bridged form is shown below (1.30) involving a hydrogen bond assisted dissociation of water followed by attack by the OH group:

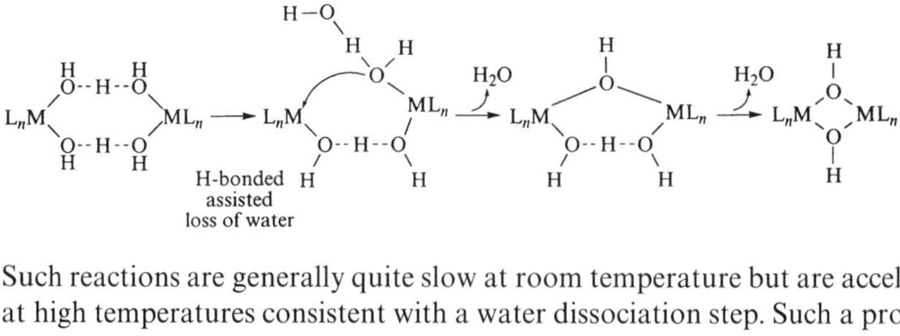

$$L_nM\underset{\overset{\text{O}\cdots\text{H}\cdots\text{O}}{\text{H}\quad\text{H}}}{\overset{\overset{\text{H}}{\text{O}\cdots\text{H}\cdots\text{O}}{\text{H}}}{<}}\text{ML}_n \longrightarrow \cdots \quad (1.30)$$

$$\text{H-bonded} \atop \text{assisted} \atop \text{loss of water}$$

Such reactions are generally quite slow at room temperature but are accelerated at high temperatures consistent with a water dissociation step. Such a process is believed to be involved in the high-temperature Cr^{III} promoted 'tanning' process to produce hardened leather from skin hide. Here hydroxoaqua Cr^{III} species, probably bound to proteins in the skin hide, undergo olation (di-μ-OH bridge formation) and then oxolation (single μ-O bridge) via successive dehydration reactions [72]. Figure 1.23 shows how various single and double hydroxo- and oxo-bridged digomeric structures may form. Rate and equilibrium data for the mono-ol/diol interconversion (contained within the box in Fig. 1.23) have been determined for several systems for $M = M^{3+} = Cr^{3+}$, Rh^{3+} and Ir^{3+} (Section 6.1.2, 9.2.2 and 9.3).

A wide range of polynuclear oxo(hydroxo)-bridged structures can be formed (Fig. 1.24). For the heavier metals the formation of certain structures may well be driven by metal–metal bond formation. Cyclic structures may also be more thermodynamically stable than their linear counterparts. It remains somewhat of a challenge to chemists to account qualitatively for the hydrolysis behaviour of metal cations and to rationalise why certain structures form. In particular, in the rapidly expanding area of sol-gel processing of certain oxide ceramics, it has become important to rationalise the circumstances under which oxo-bridge formation occurs as opposed to hydroxo-bridge formation in deciding whether a hydrolysis reaction will give rise to a hydroxide, an oxyhydroxide or an oxide (all distinctly different in their molecular structure and properties).

1.3.3.2 THE PARTIAL CHARGE MODEL

In an attempt to provide a rationale and a quantitative explanation for Jorgensen's charge–pH diagram (Fig. 1.19), Livage and coworkers have developed the partial charge model (PCM) [71, 73]. This is largely based around the original principal of Sanderson [74] and states that when two atoms combine partial electron transfer occurs so that each atom acquires a positive or negative partial charge δ_i. It is usually assumed that the electronegativity χ_i of an atom changes linearly with its charge:

$$\chi_i = \chi_i^0 + \eta_i \delta_i \tag{1.31}$$

where χ_i^0 is defined as the electronegativity of the neutral atom and η_i is the hardness defined as

$$\eta_i = k\sqrt{\chi_i^0} \tag{1.32}$$

where k is a constant which depends upon the electronegativity scale (1.36 for the Pauling scale). According to the electronegativity equalisation principle as stated by Sanderson, the charge transfer should stop when the electronegativities of all the constituent atoms becomes equal to the mean electronegativity, χ_{mol}, given by

$$\chi_{mol} = \frac{\sum p_i \sqrt{\chi_i^0} + 1.36z}{\sum (p_i/\sqrt{\chi_i^0})} \tag{1.33}$$

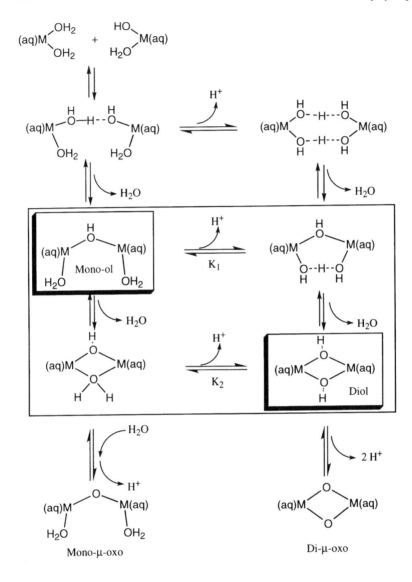

Figure 1.23. Scheme showing the mechanisms involved in the formation of hydroxo- and oxo-bridged hydrolytic dimers from aqua hydroxy monomers

where p_i is the stoichiometry of the ith atom within the molecule and z is the total charge of the ionic species. Electronegativity equivalence in solution is nothing more than the principle of chemical potential equalisation in the equilibrium state. The partial charge δ_i can be deduced from equations (1.31), (1.32) and (1.33),

Structure	Examples
	X = OH $[Cr_2(OH)_2(OH_2)_8]^{4+}$
	$[Rh_2(OH)_2(OH_2)_8]^{4+}$
	X = OH $[Cr_2(OH)(OH_2)_{10}]^{5+}$
	$[Ir_2(OH)(OH_2)_{10}]^{5+}$
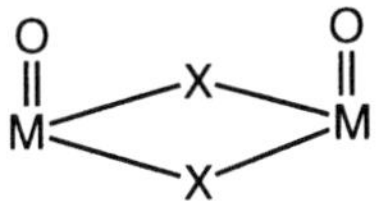	X = O $[M_2O_4(OH_2)_6]^{2+}$ M = Mo^V, W^V
	X = O $[M_3O_4(OH_2)_9]^{4+}$ M = Mo^{IV}, W^{IV}
	X = S,Se $[M_3X_4(OH_2)_9]^{4+}$ M = Mo^{IV}, W^{IV}
	X = OH $[M_3(OH)_4(OH_2)_9]^{5+}$ M = Cr^{III}, Mo^{III}
	X = OH $[M_3(OH)_4(OH_2)_n]^{2+}$ M = Sn^{II}
	X = OH $[M_3(OH)_4(OH_2)_{10}]^{5+}$ M = Mo^{III}
	possibly Rh^{III}
	$[M_3(OH)_4(OH_2)_{10}]^{6+}$ M$_3$ = Mo^{III,III,IV}
	M$_3$ = W^{III,III,IV}
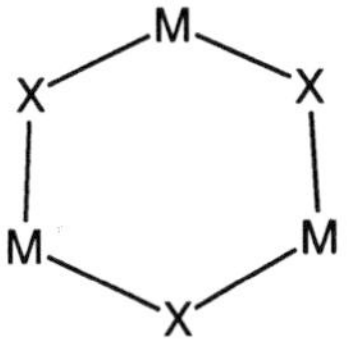	X = OH $[M_3(OH)_3(OH_2)_6]^{3+}$ M = Pt^{II}, Be^{II}

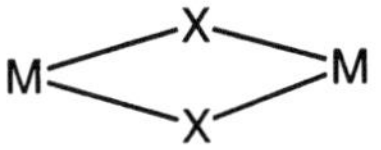

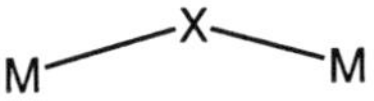

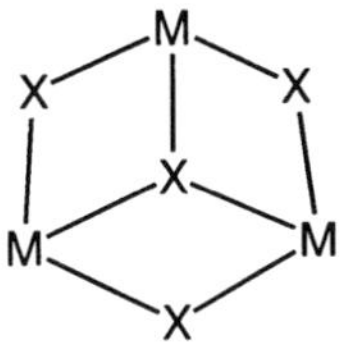

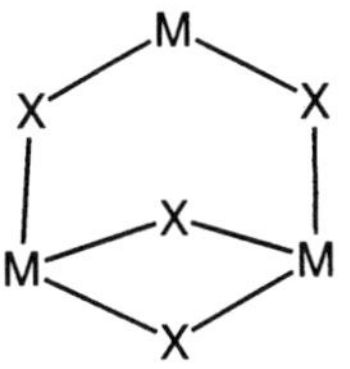

Figure 1.24(a). A selection of the structures adopted in polynuclear oxo(hydroxo)aqua complexes

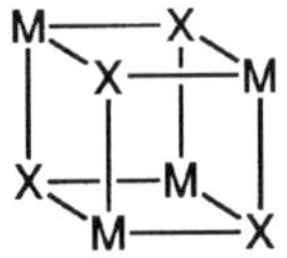

X = OH $[M_4(OH)_4(OH_2)_4]^{4+}$ M = Ni^{II}, Co^{II}, Pb^{II}

X = S,Se $[Mo_4X_4(OH_2)_{12}]^{4+/5+/6+}$

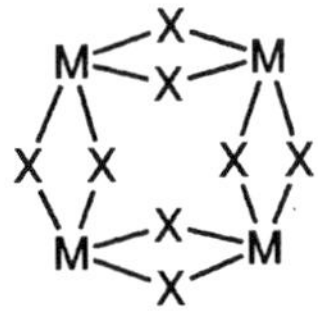

X = OH $[Zr_4(OH)_8(OH_2)_{16}]^{8+}$

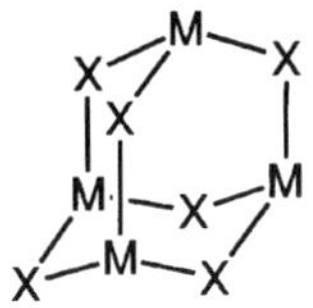

X = O,OH $[Ru_4O_{6-n}(OH)_n(OH_2)_{12}]^{(4+n)+}$

(Ru^{IV}(aq))

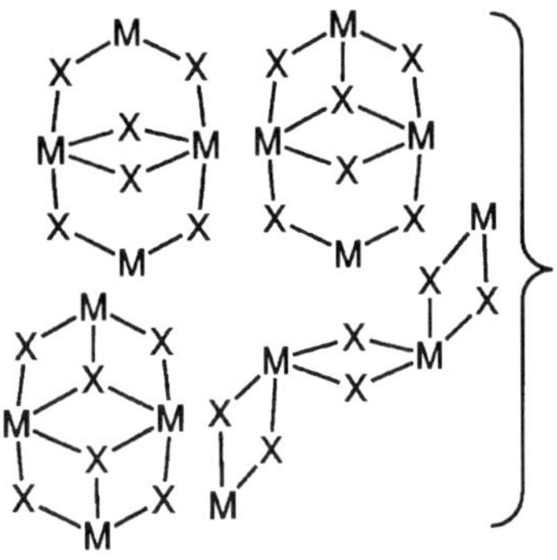

X = OH

structures possibly relevant to $[Cr_4(OH)_6(OH_2)_{12}]^{6+}$

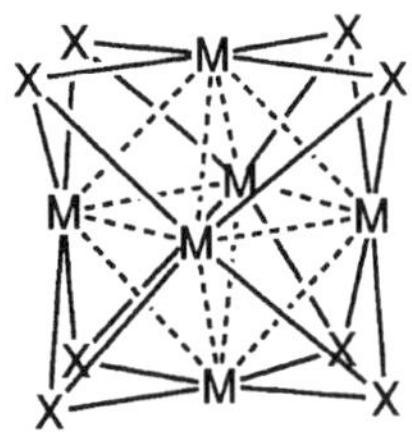

X = O,OH $[Bi_6O_4(OH)_4(OH_2)_n]^{6+}$

Figure 1.24(b). A selection of the structures adopted in polynuclear oxo(hydroxo)aqua complexes

leading to

$$\delta_i = \frac{\chi_{\text{mol}} - \chi_i^0}{k\sqrt{\chi_i^0}} \tag{1.34}$$

The partial charge is thus simply calculated knowing the electronegativity χ_i^0 of all neutral atoms, the stoichiometric composition for the ionic species and its charge z.

A general hydrolysis reaction can be expressed by

$$[M(OH_2)_n]^{z+} + H_2O \rightleftharpoons [M(OH)_h(OH_2)_{n-h}]^{(z-h)+} + hH^+ \tag{1.35}$$

According to Livage and coworkers this equilibrium proceeds until equalisation of the electronegativity of the metal complex and that of the aqueous solution occurs, the electronegativity of water itself, χ_w, varying with pH according to

$$\chi_w = 2.732 - 0.035\,\text{pH} \tag{1.36}$$

Charge conservation in the hydroxo complex species leads to

$$\delta(M) + n\delta(H_2O) - h\delta(H) = z - h$$

or $\tag{1.37}$

$$h = \frac{z - \delta\{M(OH_2)_n\}}{1 - \delta(H)}$$

where the partial charges may be computed using equation (1.32). The hydrolysis ratio, h, can thus be determined from a knowledge of z and the pH and (1.37) allows the speciation of the solution with pH to be determined.

Livage has attempted to apply the PCM in order to examine the role of the electronegativity of the metal ion in determining the subsequent reactivity of hydrolytic precursors, such as $[M(OH)_h(OH_2)_{n-h}]^{(z-h)+}$ in (1.35), towards complete hydrolytic precipitation. It was found that a relationship existed between calculated partial charges on the water molecules, $\delta(H_2O)$, and the tendency of the metal to form either an oxide precipitate or a mixed hydroxide/oxyhydroxide precipitate. The $\delta(H_2O)$ values were calculated from (1.34). Central to this argument was the rationale that oxide formation requires the conversion of bridging hydroxo groups to bridging oxo via the proton transfer process

$$
\begin{array}{c}
\overset{\delta+}{H}\;\;\overset{\delta-}{OH} \\
|\quad\;\; | \\
-M-O-M- \\
\delta-\;\;\delta+
\end{array}
\rightleftharpoons
\begin{array}{c}
\overset{\delta+}{H}\;\;H \\
H-\overset{|}{O}\;\delta- \\
| \\
-M-O-M- \\
\delta-\;\;\delta+
\end{array}
\longrightarrow
-M-O-M-+H_2O
\tag{1.38}
$$

If $\delta(H_2O)$ is negative then the H_2O molecule formed in (1.38) is attracted to the metal centre, preventing water elimination and allowing the proton transfer process to be reversible. The end product is thus a hydroxide. However, if $\delta(H_2O)$

is positive then according to Livage and coworkers the water is repelled by the metal centre, allowing the oxo bridge formation to be irreversible and compete with olation. The result is an oxide or an oxyhydroxide. Thus aged hydroxide gels of electronegative metals such as Zr^{IV} give oxide gels, e.g. ZrO_2, whereas electropositive metals such as Mn^{II} give hydroxide gels. These are fairly clear-cut examples. The problem is that the majority of systems fall in the grey area in between. A further obvious drawback with the PCM, like others that are reliant on electronegativity scales, is which scale to adopt for calculating the partial charges. In the case of small δ_i values adopting a different electronegativity scale can actually change the resulting sign of the calculated partial charge. With this in mind it is probably best to view the PCM approach as providing a guide to reactivity rather than a precise indicator of the hydrolysis mechanism.

Livage has argued, however, that the extent of hydrolytic polymerisation can be rationalised to an extent by the use of partial charge calculations of the type described. For example, attack by OH^- on another metal centre will only occur if $\delta(OH)$ is negative. As soon as the partial charge becomes positive hydrolytic polymerisation stops and the hydroxy species is stable in solution. Thus $Cr^{3+}(aq)$ forms the stable hydroxo-bridged dimer $Cr_2(OH)_2{}^{4+}(aq)$ because the partial charge on the bridging OH groups, $\delta(OH_b)$, now achieves a positive value ($+0.01$). Further oligomerisation is of course well known, but this occurs largely through condensation of the diol dimer units via the formation of OH groups at the terminal positions (OH_t) (Section 6.1.2). Here, at the higher pH's involved, $\delta(OH_t)$ will be negative. On the other hand, the corresponding formation for $Ni^{2+}(aq)$ of $Ni_2(OH)_2{}^{2+}(aq)$ has $\delta(OH_b)$ still negative, and this explains why further polymerisation to $Ni_4(OH)_4{}^{4+}(aq)$ is the result. This appears also the case for Co^{2+} (Section 9.1.1) and Pb^{2+} (Chapter 2) (Fig. 1.24). In the tetramer $\delta(OH_b)$ is now positive and this explains why hydrolytic polymerisation stops.

A similar argument can be used to explain why $[Zr(OH)_2(OH_2)_6]^{2+}$ precursors spontaneously oligomerise until tetrameric $Zr_4(OH)_8{}^{8+}(aq)$ is formed (Section 4.2). Only at this point do calculations reveal that $\delta(OH_b)$ has a positive value. This also implies that hypothetical monomeric 'zirconyl' $Zr=O^{2+}(aq)$ is never formed; rather it exists as the dihydroxo species which rapidly oligomerises. The strength of the $M=O$ π interaction (1.15) is probably also of crucial importance in defining which of the two tautomeric forms, $\{OH-M-OH\}^{n+}$ or $\{O=M-OH_2\}^{n+}$, is preferred. For $Ti^{4+}(aq)$ both forms appear to be present in a dynamic equilibrium (Section 4.1), whereas for $V^{4+}(aq)$, $Mo(W)^{6+/5+}(aq)$ and the actinide $M^{6+/5+}(aq)$ ions, the oxo-aqua species is formed exclusively.

1.4 WATER LIGAND SUBSTITUTION REACTIONS

1.4.1 Stability Constants for Complex Formation

Stability constants have long been employed as an effective measure of the affinity of a ligand towards replacing coordinated water on a metal ion in aqueous solution from the earliest systematic study of transition metal ammonia complexes by Bjerrum in his classic dissertation of 1941 leading on to the work on metal chelates in 1945 by Calvin and Wilson.

Since discussions on the practical techniques for the measurement and for the correct formulation of stability constants have appeared in a number of excellent texts [75] and compilations [76] there is little need for a further detailed discussion here. However, we should seek to briefly define the term 'stability constant' for the purpose of clarifying the precise origin of the values to be referred to in discussions throughout the book.

Very simply the stability constant, as opposed to merely an equilibrium constant for the metal complexation reaction, becomes important when one has more than one metal–ligand complex or different protonated forms of the same metal–ligand complex as being relevant in equilibrium. For the case of the formation of a 1:1 M–L complex the equilibrium constant K_1 equals the stability constant β_1:

$$M + L \xrightleftharpoons{K_1} ML, \quad K_1(=\beta_1) = \frac{a_{ML}}{a_M a_L} = \frac{[ML]}{[M][L]} \quad (1.39)$$

$$\text{(at constant I)}$$

However, for a 2:1 complex K_2 and β_2 are now not the same. β_2 is defined as $K_1 K_2$ according to

$$ML + L \xrightleftharpoons{K_2} ML_2, \quad K_2 = \frac{[ML_2]}{[ML][L]}$$

but

$$\beta_2 = \frac{[ML_2]}{[M][L]^2} = \frac{K_2[ML][L]}{[M][L]^2} = K_1 K_2 \quad (1.40)$$

(note that β_n relates each complex back to the equilibrium concentrations of M and L) and

$$\beta_n = K_1 K_2 K_3 \cdots K_n \quad (1.41)$$

Similar expressions were used to define stepwise hydrolysis constants, e.g. K_{11} and K_{22} (see Section 1.3).

Strictly speaking, equilibrium constants (K_n) and therefore stability constants (β_n) should reflect the equilibrium activities of the components. Determining activation coefficients for every reactant under every conceivable condition is unfeasable and indeed unnecessary. By working at a constant ionic strength of an electrolyte, usually in at least a hundred fold excess over the equilibrating complexes, the activities can be approximated to concentrations. Also by being able to use concentration quantities the values obtained from different methods, e.g. potentiometry versus spectrophotometry, can be directly compared.

The most common and probably most versatile measurement of stability constants, in conjunction with protonation constants, is by potentiometry. Here one titrates a mixture of different mole ratios of metal and added ligand, usually in the presence of an excess of added H^+, with OH^- ions. The pH is monitored as a function of added $[OH^-]$ with a glass electrode in a calibrated cell. The ionic strength (I) is usually kept constant at around 0.1 M. By determining the concentration at equilibrium of just one of the component species, namely $[H^+]$, from the pH measurement, the equilibrium concentrations of all the other component species and therefore the equilibrium constant can be determined if the precise analytical composition of the solution is known. In effect protonation and stability constants are measured in terms of a series of simultaneous multiple equilibria wherein protons and metal ions compete for donor atoms on the ligand. Eventually at a high enough pH OH^- ions themselves can eventually become involved in the equilibria by competing with the ligand for the metal ion. Therefore one has at the low pH extreme $[M(OH_2)_n]^{z+}$ ions plus protonated L and at the high pH extreme $M(OH)_n(aq)^{(z-n)+}$ and L. In between could be whole range of species such as ML, ML_2, MLH, ML(LH), $M(LH)_2$, etc. The following equations comprise a series of determined equilibria in aqueous solution for, for example, the formation of an $EDTA^{4-}$ complex on an alkaline earth metal like Ca^{2+}:

$$M^{2+} + L^{4-} \rightleftharpoons ML^{2-}$$

$$K^M_{ML} = \frac{[ML^{2-}]}{[M^{2+}][L^{4-}]}, \qquad \beta_{ML} = \frac{[ML^{2-}]}{[M^{2+}][L^{4-}]}$$

$$H^+ + ML^{2-} \rightleftharpoons MLH^- \tag{1.42}$$

$$K^H_{MLH} = \frac{[MLH^-]}{[H^+][ML^{2-}]}, \qquad \beta_{MLH} = \frac{[MLH^-]}{[H^+][M^{2+}][L^{4-}]}$$

$$K^M_{MLH} = \frac{[MLH^-]}{[M^{2+}][HL^{3-}]} \quad \text{(involving protonated L)}$$

Any method that can allow the accurate monitoring of the equilibrium concentration of one of the components in a solution of known analytical composition can be used to measure equilibrium constants. In addition to potentiometry one can employ spectrophotometry (absorbance measurements of the component at a particular wavelength), as well as others, e.g. NMR, polarography (reduction potentials), ion exchange, specific ion EMF (ion-selective electrode measurement), colorimetry and ionic conductivity (see ref. [75] for references and other examples). In cases where very high stability constants are relevant for a particular ligand a number of methods based upon competition between different ligands have also been devised [77].

Stability constants reflect the affinity of a ligand for a particular metal ion and indeed the type of bonding present. For the alkali, alkaline earth and lanthanide elements, where the bonding is largely electrostatic (ionic), the strongest complexes formed are those with hard base donors such as F^-, RCO_2^-, RO^-, R_2N^-,

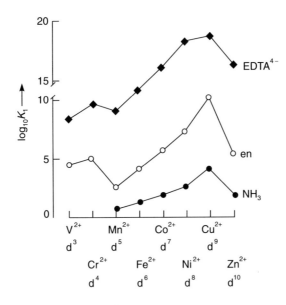

Figure 1.25. Stability constants for M^{2+}(aq) ligand complex formation along the first row transition metals: the Irvine–Williams effect

etc. Also since ligand field effects are absent or minimal, chelating ligands (e.g. EDTA) or macrocyclic ligands (e.g. cryptands and crown ethers) are required for the highest constants. On the other hand, the constants for many of the transition metal ions, particularly those of the second and third rows, reflect considerable covalency in the bonds with some of the strongest complexes being formed with monodentate ligands, e.g. CN^-. However, for many first row transition metal ions the highest constants are still invariably seen for chelate and macrocyclic complexes maximising at Cu^{2+} ($\leqslant$ four coordinated donors) within a gradual increase along the period left to right from Mn^{2+} to Zn^{2+} paralleling the increasing Z_{eff}, the so-called Irving–Williams effect (Fig. 1.25) [78]. The maximisation at Cu^{2+} arises from the Jahn–Teller distortion which leads to a strengthening of the ligand field for the four donors in the equatorial plane at the expense of a weaker field for the two axial donors (see Section 11.1.2).

1.4.2 Reaction Kinetics and Dynamics: A Survey of Water Exchange Rates on Aqua Ions

Some 20 orders of magnitude cover the present range of water exchange rates on aqua metal ions from the most labile, Cs^+ (residence time, 25 °C $\sim 10^{-10}$ s), to the most inert, Ir^{3+} (residence time $\sim 10^{10}$ s) (Fig. 1.26). It is recognised that these factors encompass at the one extreme the size of the metal ion and the magnitude

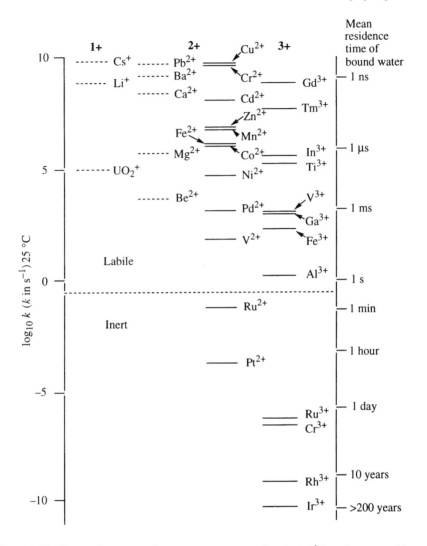

Figure 1.26. Range of water exchange rate constants (log $k_{ex}(s^{-1})$) and mean residence times (s) for primary shell water molecules on aqua metal ions at 25 °C (the dotted line represents Taube's inert/labile boundary [79])

of the cationic charge (electrostatic), thus somewhat correlating with the extent of hydration, and at the other extreme the presence of ligand field effects (transition metal cations). Cs^+ is the largest singly positive metal cation and as such possesses the lowest charge density. On the other hand, Ir^{3+} is a third row transition metal ion with the maximum LFSE for its octahedral field of coordinated waters ($-2.4\Delta_o$ with Δ_o appreciably large for a third row metal) as a result of its low-spin t_{2g}^6 configuration (see below). Hence water exchange has to overcome an extremely large activation barrier. Indeed, for the first row transition metal divalent ions water exchange rate constants, plotted as $\log k_{ex}$, correlate with crystal field activation energy as proportional to the CFSE or LFSE associated with the particular electronic configuration (Fig. 1.27). It may be noted that $\sim 100\,\mathrm{kJ\,mol^{-1}}$ of LFSE amounts to a kinetic effect between $Ca^{2+}(d^0)$ and $V^{2+}(t_{2g}^3)$ in excess of *six* orders of magnitude. The resulting inertness of V^{2+}, contrasting with the general trend with increasing size throughout Fig. 1.26, has already been eluded to, as has the high lability of $Cr^{2+}(aq)$ and $Cu^{2+}(aq)$ due to the two weakly bonded axial waters of the tetragonally distorted structure in rapid fluctionality with the equatorial waters, the dynamic Jahn–Teller effect (Section 11.1.2). In the case of relatively inert systems, like many of the transition metal ions, one can easily talk of a discrete primary hydration shell

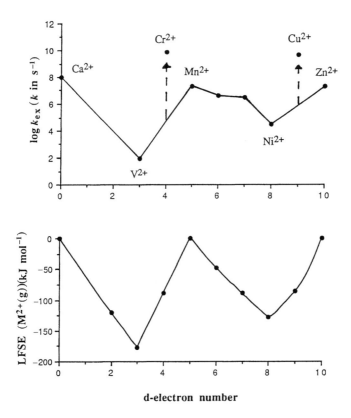

Figure 1.27. Correlation of $\log k_{ex}$ (s^{-1}) with LFSE (kJ mol^{-1}) for the first row transition metal $[M(OH_2)_6]^{2+}$ cations

undergoing slow exchange. In the case of large univalent ions like Cs^+, however, a discrete primary shell may have less significance given a lability comparable to that of the secondary shell. In another striking comparison, the oxo groups of ClO_4^- reflect the high M—O bond strength exemplified by an extremely long half-life (> 100 years) for exchange with bulk water at $25\,^\circ C$. This may be compared to the diffusion controlled rate ($t_{1/2} < 10^{-10}$ s) for water exchange on the hydrated Cl^- ion. Both acid and base catalysed pathways for water exchange on oxo anions have been detected.

1.4.3   Measurement of Water Exchange Rates on Aqua Ions

The standard methods employ the use of the two isotopes of oxygen: ^{17}O and ^{18}O.

1.4.3.1   THE ^{18}O LABELLING METHOD

This method has been pioneered by workers such as Hunt, Plane and Taube [80] and Gamsjager and Murmann [31]. The technique, as describe in Section 1.2, involves measurement of the oxygen/water exchange by converting the oxygen in the species into a gaseous form, usually CO_2 and O_2 itself, and measuring the $^{18}O/^{16}O$ ratio in a mass spectometer. The earliest method described by Cohn and Urey in 1938 [81] involved distillation or direct sampling of the aqueous matrix followed by isotopic equilibration with normal CO_2. The equilibration process, which normally takes three days at room temperature, can be speeded up in the gas phase over a heated Pt wire [82] or by using an electrical discharge [83]. The value of K for the equilibrium reaction

$$K = \frac{[C^{18}O_2]^{1/2}[H_2{}^{16}O]}{[C^{16}O_2]^{1/2}[H_2{}^{18}O]} \tag{1.43}$$

is known as a function of temperature, $\log K = 16.6\,T^{-1} - 0.015\,69$. For the purpose of bulk H_2O sampling the exchange reaction requires to be quenched after set time intervals. This can be achieved in a number of ways, e.g. by a pH jump to a region where exchange is much slower or by precipitation of the oxygen-containing compound in the form of a solid derivative. Here it is assumed that heterogeneous exchange is always several orders of magnitude slower (often the case), although induced exchange by precipitative or separative methods is normally checked for. The solid compound is then washed and dried for conversion into a gaseous oxygen form. Conversion to CO_2 can be achieved by heating with anhydrous species such as guanidinium salts [84], *para*-cyanogen, AgCN or $Hg(CN)_2$ [95], usually within a sealed tube at around $\sim 400\,^\circ C$ and under reduced pressure ($< 10^{-4}$ torr). Similarly, one can analyse samples of water. Some of these methods have proved successful on the mg scale. Another method has involved the oxidation of the water to O_2 using alkaline peroxydisulfate as well as high-temperature fusion with ultra-pure graphite at $1400\,^\circ C$, giving mainly CO [31]. Exchange reactions with half-lives as low as 10 s can be followed by use of a multimixing stopped-flow apparatus. Here the first two syringes mix the species and the $H_2^{18}O$, each adjusted to the required pH, ionic strength and temperature with the third syringe used to add the precipitating agent after a present time interval. The method would also seem adaptable to a pH jump quench method. Infra-red spectroscopy can often be used to give an indication of the end of the exchange process, although it is probably not of sufficient sensitivity for accurate monitoring of the exchange process itself. A further disadvantage would be the requirement for higher enrichment levels. The accuracy of the mass spectral determination of the isotope ratio can be around 1 in 10 000, which is by far superior to any other method. Compared to the use of ^{17}O (for NMR), ^{18}O is much cheaper to buy (around a factor of 10–100 per atom % enrichment) and to use since mg amounts and enrichments of only $\sim 10 \times$ natural abundance can suffice owing to the accuracy of the isotope ratio determination. The main source of experimental error is in sampling, sampling-induced exchange and ensuring that the gaseous oxygen form analysed is pure and representative. The ^{18}O method has proved highly successful for the determination of oxygen exchange rates on many oxoanions [31] as well as on Cr^{III} aqua complexes (paramagnetic) including the hexaaqua ion $[Cr(OH_2)_6]^{3+}$ [80, 84, 86]. A wide variation of rates between different species is seen in Table 1.6. It is likely that the origin of these variations lies in the existence of a range of different exchange mechanisms depending upon the species involved.

1.4.3.2   NMR METHODS [86, 88]

The ^{17}O NMR method, however, is the only one that can be applied to labile aqua metal ions with half-lives of < 10 s and particularly where there are

Table 1.6. Half-lives for water exchange on terminal oxo species determined by ^{18}O labelling (taken from ref. [31])

Species	$t_{1/2}(s)$ (medium)	Temperature(°C)	Comments
$U^{VI}O_2^{2+}$ (aq)	$10^{10}(0.8 \text{ M HClO}_4)$	25	Base catalysed
$U^{V}O_2^{+}$ (aq)	$> 10^{-2}(1.0 \text{ M HClO}_4)$	25	Catalyses exchange on UO_2^{2+} in strong acid solution
$Pu^{VI}O_2^{2+}$ (aq)	$\geqslant 10^7$	83	Catalysed by $Pu^{V}O_2^{+}$ (aq)
$V^{IV}O^{2+}$ (aq)	$10^5(1.0 \text{ M HClO}_4)$	25	Base catalysed
$V^{V}O_2^{+}$ (aq)	$0.15(1.0 \text{ M HClO}_4)$	0	[H$^+$] dependence not confirmed
$Mo_2^{V}O_4^{2+}$ (aq)	4s	0	No [H$^+$] dependence found

a number of different exchanging oxygen sites (e.g. on polynuclear species). For slow exchanging oxygen sites on diamagnetic metal ions one can monitor the exchange very simply by conventional means by following the growth or decay to equilibrium of the ^{17}O resonance of the particular oxygen, hydroxide or water site. It is often good practice to measure the rate both ways as a check. Given a choice following the decay of the peak height (area) of an enriched sample to equilibrium in natural abundance water is to be preferred owing to the presence of more reliable data over the crucial first half-life. The measuring accuracy is, however, still somewhat inferior to the mass spectral method. Half-lives of around a few seconds can be followed by rapid injection methods of the type pioneered by Merbach and coworkers [89]. Water exchange rates measured by the ^{17}O NMR isotope dilution method include that on $[Ru(OH_2)_6]^{2+}$ (by fast injection), $[Ru(OH_2)_6]^{3+}$, $[Rh(OH_2)_6]^{3+}$, $[Ir(OH_2)_6]^{3+}$ and $[Pt(OH_2)_4]^{2+}$.

For faster exchanging metal ions with half-lives (25 °C) of < 1 s one can employ a number of line-broadening methods. [98]. For diamagnetic metal aqua ions with half-lives of between 1 s and 1 ms the linewidth of the bound water ^{17}O NMR resonance can be measured as a function of various parameters such as temperature and $[H^+]$ at a constant ionic strength. Mn^{2+} can act as a convenient relaxation agent for the intense line of free water [35]. The bound water linewidth can be related to a measure of the transverse relaxation time T_2 (for bound water both T_1 and T_2 are comparable) which is a function of two components, the quadrupolar relaxation time T_{2Q} and the bulk water exchange rate constant k_{ex} according to

$$\frac{1}{T_{2\,obs}} = \frac{1}{T_{2Q}} + k_{ex} \tag{1.44}$$

The temperature dependence of k_{ex} is fitted to the normal Eyring expression

$$k_{ex} = \frac{k_B T}{h} \exp\left(\frac{-\Delta H^{\ddagger}_{ex}}{RT} + \frac{\Delta S^{\ddagger}_{ex}}{R}\right) \tag{1.45}$$

for the purpose of estimating $\Delta H^{\ddagger}_{ex}$ and $\Delta S^{\ddagger}_{ex}$, whereas an Arrhenius expression

$$\frac{1}{T_{2Q}} = \frac{1}{T_{2Q}^{298.2}} \exp\left[\frac{E_Q}{R}\left(\frac{1}{T} - \frac{1}{298.2}\right)\right] \tag{1.46}$$

is normally used for T_{2Q}. $1/T_{2Q}$ is proportional to the correlation time relating to the tumbling frequency of the water molecules. This depends upon the viscosity of the medium and as such the use of media of low viscosity is preferred for the narrowest lines, reproducibility and the avoidance of medium effects. In terms of non-complexing anionic media the preferred order appears to be $ClO_4^- > BF_4^- > PF_6^- > tfms^- > pts^-$.* A particularly nice example where two different coordinated water ligands are exchanging on different time domains is that of the incomplete cuboidal Mo^{IV} cluster ion, $[Mo_3S_4(OH_2)_9]^{4+}$ (Section 6.2.3) [90]. The plot of ln $(1/T_{2obs})$ as a function of reciprocal absolute temperature is shown in Fig. 1.28.

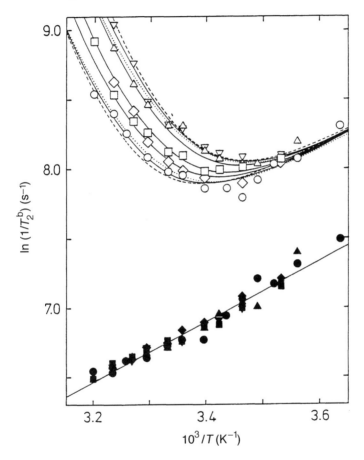

Figure 1.28. A Plot of ln $(1/T_{2obs})$ (^{17}O) versus $10^3/T$ (K) for the coordinated waters of $[Mo_3S_4(OH_2)_9]^{4+}$ at $I = 2.0\,M\ ClO_4^-$ as a function of $[H^+]$. ($\circ, \bullet$) 1.56 M ($\square, \blacksquare$) 1.18 M, ($\blacklozenge, \lozenge$) 0.945 M, ($\triangle, \blacktriangle$) 0.664 M and ($\triangledown, \blacktriangledown$) 0.545 M [100]. Filled symbols (c—H_2O), open symbols (d—H_2O)

---

*This order shows up also in the amount of broadening seen on the addition of 0.1 M Mn^{2+} and reflects the strength of ion pairing and perhaps even complexation within the inner coordination sphere of the relaxant. Thus the resonance of ClO_4^- is broadened, but only sufficiently to remove resolution of the quartet from coupling to the $^{35,37}Cl$ nuclei, $CF_3SO_3^-$ is significantly broadened (linewidth at $RT \sim$ 10–50 times that in the absence of added Mn^{2+}; Fig. 1.16b and c and Fig. 1.17b) whereas pts$^-$ is completely broadened into the baseline (see Fig. 1.16a).

The c waters, approximately *trans* to the capping sulfide group, exchange much more slowly than the d-waters, which are approximately *trans* to the bridging sulfides, the latter also showing an exchange rate acceleration with decreasing $[H^+]$. The c waters only show the quadrupolar contribution which works to sharpen the lines as the temperature is increased. In contrast, at temperatures depending upon the $[H^+]$, the linewidth of the d-water resonance goes through a minimum and then increases with temperature due to the chemical exchange. The fit is basically to two intersecting straight lines going in opposite directions with temperatures. The $[H^+]$ dependence of the d-water exchange is of the form

$$k_{ex} = \frac{k_{OH} K_{11}}{[H^+] + K_{11}} \tag{1.47}$$

indicating dominant involvement of $Mo_3S_4(OH)^{3+}(aq)$, with hydrolysis constant 0.18 M, the significance of which suggests that the relevant deprotonation occurs at an adjacent water ligand on the same Mo centre. This is believed to be a rare example of the *cis* conjugate base effect (Sections 1.4.6 and 6.2.3). At sufficiently high values of k_{ex} the linewidth becomes indistinguishable from that of the baseline (around a half-life at 25 °C of ~ 0.1 ms). This is the situation arising in the case of the more labile aqua Mo^V dimer ion $[Mo_2O_4(OH_2)_6]^{2+}$. The much slower c-water exchange on $Mo_3S_4^{4+}(aq)$ can be studied at 25 °C by conventional fast injection isotopic enrichment, and interestingly it shows no corresponding $[H^+]$ dependence. Other labile diamagnetic metal ions that have had their water exchange rates measured by the line-broadening technique include $[Be(OH_2)_4]^{2+}$, $[Al(OH_2)_6]^{3+}$, $[Ga(OH_2)_6]^{3+}$ and $[Pd(OH_2)_4]^{2+}$.

For faster exchanging metal ions that are paramagnetic Swift and Connick [88] developed a method based upon an analysis of both the chemical shift and relaxation times for the ^{17}O NMR nuclei of both free and bound water and showed that the former could be used for the measurement of water exchange rates for labile paramagnetic metal ions such as the those of the first row divalent and trivalent transition metals. The relaxation of the ^{17}O nucleus of waters bound to a paramagnetic metal ion has been shown to be due to scalar coupling between the magnetic moment of the ^{17}O nucleus and the unpaired electrons. Swift and Connick demonstrated that it was possible to monitor the linewidth of the free water ^{17}O resonances as a function of temperature over a very wide temperature range owing to their intrinsically longer relaxation times. They also demonstrated that at sufficiently high temperatures the linewidths eventually decrease again (Fig. 1.29), in accordance with the following full expression:

$$\frac{1}{T_{2r}} = \frac{1}{\tau_m} \left[\frac{T_{2m}^{-2} + (T_{2m}\tau_m)^{-1} + \Delta\omega_m^2}{(T_{2m}^{-1} + \tau_m^{-1})^2 + \Delta\omega_m^2} \right] + \frac{1}{T_{2os}} \tag{1.48}$$

where T_{2r} is the reduced bulk water relaxation time, related to the observed

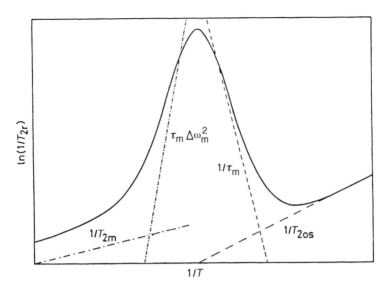

Figure 1.29. The full Swift–Connick temperature dependence of $\ln (1/T_{2\text{obs}})$ for the ^{17}O nucleus of a water molecule in the presence of a fast exchanging metal ion (the dashed lines represent the four regions where predominant relaxation contributions lead to a simplification of equation (1.48))

relaxation time $T_{2\text{obs}}$ by

$$\frac{1}{T_{2r}} = \frac{1}{p_m}\left(\frac{1}{T_{2\text{obs}}} - \frac{1}{T_{2s}}\right) \tag{1.49}$$

T_{2m} is the relaxation time for bound waters in the absence of exchange, $\Delta\omega_m$ is the chemical shift between the bulk and bound waters in the absence of exchange, τ_m is the residence time of bound water and $T_{2\text{os}}$ is the contribution due to the simple outer-sphere interaction. In equation (1.49) T_{2s} is the contribution to the bulk water relaxation from the pure solvent and p_m is the ratio of the number of moles of water bound to the metal to the total moles of water. Finally $\Delta\omega_r$, the reduced chemical shift of the bulk water, is given by

$$\Delta\omega_r = \frac{\Delta\omega_s}{p_m} = \frac{\Delta\omega_m}{(\tau_m/T_{2m} + 1)^2 + \tau_m^2 \Delta\omega_m^2} + \Delta\omega_{os} \tag{1.50}$$

where $\Delta\omega_s$ is the chemical shift of the pure bulk solvent and $\Delta\omega_{os}$ is the chemical shift difference between pure solvent and bulk solvent in the presence of the paramagnetic metal ion, the correction being due to the simple outer-sphere interaction. The region wherein the above diamagnetic metal ions undergo exchange is given simply by $1/\tau_m$, termed the 'slow' exchange domain. The water exchange rate constant k_{ex} is simply given by $1/\tau_m$. Water exchange on the labile paramagnetic metal hexaaqua ions $Mn^{2+}, Fe^{2+}, Co^{2+}$ and Ni^{2+} has been investigated using the full Swift–Connick equation (1.48) in the region of data around the 'maximum' and into the region dominated by the term $T_{2m}\tau_m$ in Fig. 1.29. The Swift–Connick treatment has also been applied successfully for the determination of the water exchange rate constants on the highly labile paramagnetic trivalent lanthanide ions (Section 3.2).

Despite the wide range of timescales available for monitoring rates using NMR it is not possible to cover the entire range. Occasionally the exchange half-life falls in the range that makes it unamenable to study. This is when the rate is too slow and the activation parameters such that it cannot be brought on to the line-broadening timescale at elevated temperatures while at the same time being too fast for a rapid mixing conventional isotopic dilution approach. This occurs for half-lives (at 25 °C) in the range $\sim 10\,s > t_{1/2} > 10^{-2}\,s$. For paramagnetic metal ions one can simply put an upper limit on the half-life of 10 s. Unfortunately, paramagnetic $[Mo(OH_2)_6]^{3+}$ (Section 6.2.2) is one such metal ion whose rate of water exchange falls within this so-called 'blind' region.

1.4.3.3   ESTIMATES FROM COMPLEX FORMATION USING RELAXATION METHODS

Although the ever-increasing routine availability of high-field NMR instruments is lowering the upper rate limits for monitoring of solvent exchange processes, most of the quoted rate contants for water exchange on the most labile of the diamagnetic metal ions, such as those of groups 2 and 12 and the heavier members of groups 13, 14 and 15, are still based largely on extrapolation from rate constants for complex formation reactions as determined by rapid mixing techniques such as stopped-flow spectrophotometry, or if an equilibration reaction is relevant the use of relaxation methods such as temperature jump (if spectroscopic monitoring is possible) or, for example, ultrasound absorption (if not). This is because of the general observation that the more *labile* a metal ion is the more *insensitive* are the observed reaction rates to the nature of the incoming ligand nucleophile. The result is such that they become largely comparable to, or certainly within an order of magnitude of, the rate constant for water exchange. Many of the estimates from ultrasound absorption have been based on the values of rate constants determined for $SO_4{}^{2-}$ complexation reactions, such as for the following: $Be^{2+} (\sim 10^2\,s^{-1})$, $Mg^{2+} (\sim 10^5\,s^{-1})$, $Ca^{2+} (\sim 10^8\,s^{-1})$, $Zn^{2+} (\sim 10^7\,s^{-1})$, $Cd^{2+} (\sim 10^8\,s^{-1})$, $Hg^{2+} (\sim 10^8 - 10^9\,s^{-1})$ and the Ln^{3+} ions $(\sim 10^8\,s^{-1})$ [91]. The ultrasound absorption method has not been without its critics over the years. However, where water exchange rate constants have been determined by other means (e.g. by NMR or T-jump) the values have been largely comparable to those from the ultrasound technique. The reader is here referred to the excellent 1978 article by Margerum *et al.* [92] for a critical compilation of the available kinetic data (at that time) for water exchange and ligand substitution reactions on both labile and inert aqua metal ions. Much of the data and most of the mechanistic conclusions reached have stood the test of time over the past 18 or so years, although most water exchange kinetic data obtained by NMR prior to 1970 have now been remeasured using more powerful modern instruments and should therefore be viewed with caution. In the 1978 review it was stated, for example, that 'many metal ion substitution reactions seem to be insensitive to the nature of the ligands in the second coordination sphere', i.e. controlled by the rate of water ligand dissociation. With the exception of a few clear-cut examples (see Section 1.4.4) below, one is coming around to the view that at least for reactions at many octahedral aqua complexes this is a fairly accurate picture.

1.4.4 Kinetic Studies of Water Ligand Replacement Reactions

1.4.4.1 THE PROBLEM OF MECHANISTIC CLASSIFICATION

The rates of aqua metal ion substitution reactions themselves give little of an insight into mechanism. Much more informative is the monitoring of rates as a function of the type of entering ligand, leaving ligand, pH, temperature and pressure. The challenge remains one of careful and meaningful interpretation, a subject of continuing debate and controversy. It would be sensible to assume that the focus of this debate will change further in the future as more experimental and indeed a growing body of theoretical data are gathered.

It is intriguing to think that only as recently as the 1950s and 1960s was any attention at all given to the possible systemisation and categorisation of reactions occurring at metal centres. In 1966 Langford and Gray proposed a classification of ligand subsitution reactions on metal ions and complexes as being of three basic types: associative (A), dissociative (D) and interchange (I) [93], as distinct from the already well-established S_N1 or S_N2 dichotomy used for reactions on carbon. These labels were based largely on the obvious realisation that, unlike carbon, metal ions are centres of high cationic charge which will lead invariably to a degree of preassociation with anionic or indeed any dipolar nucleophilic incoming ligand molecules. In the extreme D and A mechanisms a discernable intermediate,* of reduced or expanded coordination number, is formed. In the I mechanism there is no discernable intermediate. There is a growing acceptance now that the truly extreme A or D situation probably may never be strictly relevant for a substitution/exchange reaction on a metal ion in aqueous solution, owing to the fact that the leaving or entering ligand always remains associated to some degree with the entire assemblage of hydrogen-bonded water molecules in

---

*An intermediate may be defined as a species able to undergo several molecular collisions and therefore an ability to distinguish between different incoming nucleophiles. Within the D mechanism the intermediate can be considered as having lost all memory of the departed ligand and thus an ability to treat all ligands held within the secondary shell assemblage as independently reacting species.

the primary and secondary hydration shell throughout the reaction coordinate to the transition state. Thus one might argue that the interchange I label is the only one that needs to be considered for aqueous reactions, as has been suggested by Swaddle [94] and Lay [95]. Indeed, Langford and Gray proposed a further subdivision of the I classification to include I_A and I_D labels as defining whether bond making with the entering ligand or bond breaking with the leaving ligand were important processes along the reaction coordinate to the transition state. Swaddle has argued that adopting these labels is highly restrictive as they introduce an interpretation based largely upon the chosen operational criteria, i.e. a classification based upon only one measurable property, i.e. a sensitivity or insensitivity, as it were, to the nature of the entering ligand [94, 96]. Swaddle [97] and Lay [95] have argued that we are probably dealing with a continuous range of phenomena that is a continuum in the strengths of interactions between central metal and active ligands as one goes along a series of reactions, without necessarily any clear break in behaviour. This view will become apparent when we discuss individual cases throughout the book and some general trends in behaviour here below.

A parameter that has been much argued for as a powerful indicator of the mechanisms of ligand exchange and substitution reactions on metal ions is the activation volume, $\Delta V^{\ddagger}$. $\Delta V^{\ddagger}$ is measured from the pressure dependence of the logarithm of the rate constant for the substitution reaction according to

$$\ln k = \ln k_0 + \frac{\Delta V^{\ddagger}}{RT}P + \frac{\Delta \beta^{\ddagger}}{2RT}P^2 \tag{1.51}$$

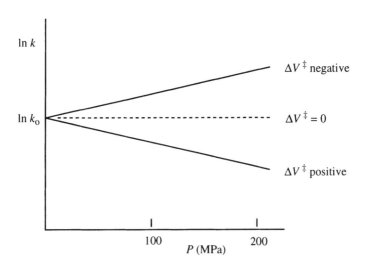

Figure 1.30. Pressure dependence of reaction rate constants and sign of the activation volume

where $\ln k_0$ is the value at zero pressure. It is normally expressed in units of cm^3 mol^{-1}. $\Delta\beta^{\ddagger}$ measures the pressure dependence of $\Delta V^{\ddagger}$ itself and for reactions in aqueous solution it is more often than not very small. Thus $\Delta V^{\ddagger}$ can often be evaluated directly from the slope. An increase in rate with pressure corresponds to a negative $\Delta V^{\ddagger}$ and vice versa (Fig. 1.30). $\Delta V^{\ddagger}$ is a small coefficient, meaning that quite high pressures of several kbar (100 MPa) are required for an appreciable effect on the rate to be seen. However, equipment for spectrophotometric monitoring (even under rapid mixing conditions) and even probes for NMR measurement under such pressures are now in routine use in a growing number of laboratories around the world and much kinetic data have now been accumulated [89, 98]. The problem now seems to be one of careful and meaningful evaluation. The simplest mechanistic indicator of a substitution reaction on an aqua metal ion should be the water exchange reaction itself, since solvational changes and electrostriction factors arising from charge neutralisation are largely absent. Activation volumes have now been measured for water exchange reactions on a wide range of simple mononuclear M^{n+}(aq) ions and some interesting trends have emerged (Table 1.7). The vast majority of these values have been compiled using ^{17}O NMR by Merbach and coworkers in their laboratories in Lausanne, Switzerland [97b, 99]. For the first row divalent and trivalent metal ions a trend is seen from negative values of $\Delta V^{\ddagger}$ for the early elements (Ti^{3+}, V$^{2+/3+}$, Cr^{3+}, Mn^{2+}, Fe^{3+}) to positive values for the later ones (Fe^{2+}, Co^{2+}, Ni^{2+} and Ga^{3+}). Thus for ions of the same charge an increase in electron density and a decrease in size correlates with a trend towards more positive values of $\Delta V^{\ddagger}$. The challenge is one of interpretation. Merbach has argued [87b, 89, 99] that the trend reflects increasing dissociative character in the water exchange reactions, indeed perhaps evidence of a changeover from associative (probably I_A) (negative $\Delta V^{\ddagger}$ values) to dissociative (probably I_D) (positive $\Delta V^{\ddagger}$ values). A necessary presumption within this argument is that the volume change within the coordination sphere of the complex itself can be largely neglected, allowing the sign of the activation volume to be directly diagnostic. This follows the assumption originally made by Stranks [100]; i.e. an associatively activated process will have a negative activation volume corresponding to the loss, from the initial state, of what will be the effective partial molar volume of the entering water ligand on forming the transition state. Conversely, a dissociatively activated process will involve the liberation, at the transition state, of the equivalent effective partial molar volume of a leaving water molecule (Fig. 1.31). For limiting A or D mechanisms $\Delta V^{\ddagger}$ should relate directly to the full partial molar volume of a water molecule electrostricted within the secondary hydration shell. Interchange I_D or I_A processes will have positive or negative $\Delta V^{\ddagger}$ values as required, but of a magnitude somewhat less than that corresponding to the full partial molar volume of the electrostricted water.

This hypothesis is, however, largely at odds with the known change in M—L distance (hence volume of ML_n) on changing the coordination number at M (i.e.

Table 1.7. Experimental volumes of activation ($cm^3 mol^{-1}$) for water exchange on mononuclear aqua metal ions

Be^{2+} −13.6											
Mg^{2+} —										Al^{3+} +5.7	
										Ga^{3+} +5.0	
Ca^{2+} —	Sc^{3+} —	Ti^{3+} −12.1	V^{3+} −8.9	Cr^{3+} −9.6	Mn^{3+} —	Fe^{3+} −5.4	Co^{3+} —				
			V^{2+} −4.1	Cr^{2+} —	Mn^{2+} −5.4	Fe^{2+} +3.8	Co^{2+} +6.1	Ni^{2+} +7.2	Cu^{2+} +ve[b]	Zn^{2+} +ve	
Sr^{2+} —	Y^{3+} —			Mo^{3+} −ve[c]		Ru^{3+} −8.3	Rh^{3+} −4.1				In^{3+} —
						Ru^{2+} −0.1		Pd^{2+} −2.2	Ag^{2+} —	Cd^{2+} −ve	
Ba^{2+} —	Ln^{3+a} −6						Ir^{3+} −5.7	Pt^{2+} −4.6		Hg^{2+} —	Tl^{3+} —

[a] For the elements Tb^{3+} to Tm^{3+}.
[b] Based on the activation volume of +8.3 $cm^3 mol^{-1}$ for MeOH exchange on $[Cu(MeOH)_6]^{2+}$.
[c] Based on the activation volume of −11.4 $cm^3 mol^{-1}$ for the 1:1 complex formation reaction of $[Mo(OH_2)_6]^{3+}$ with NCS^-.

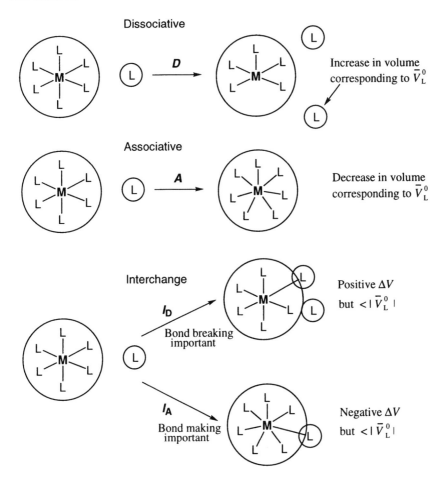

Figure 1.31. Volume changes accompanying ligand substitution reactions assuming a negligible change in volume of the first coordination sphere

ionic radii increase with increasing coordination number, as the compilation by Shannon [101], Appendix II, clearly shows). For ions of a given charge, Swaddle [102] has described an empirical method for calculating partial molar volumes of $[M(OH_2)_n]^{z+}$ ions using

$$\bar{V}^0_{abs} = 2.523 \times 10^{-6}(r_M + \Delta r)^3 - 18.07\,n - \frac{417.5\,z^2}{r_M + \Delta r} \qquad (1.52)$$

where $\bar{V}^0_{abs}$ is the absolute volume relative to $\bar{V}^0_{abs}(H^+) = -5.4\,\mathrm{cm^3\,mol^{-1}}$, r_M is the ionic radius of the metal ion (pm), Δr is the apparent radius of a coordinated water molecule, determined empirically, and 18.07 is the partial molar volume of water. The inclusion of the final term takes into account solvent electrostriction around the charged ion. The approach satisfactorily reproduced experimental partial molar volumes for a range of hydrated metal ions encompassing the first transition series, several lanthanides and those from the s and p blocks. Using appropriately adjusted values of r_M and n to account for the change in M—L distance with coordination number, Swaddle used equation (1.52) to estimate the limiting volume changes for the extreme A or D water exchange mechanisms. The limiting $\Delta V^{\ddagger}$ values, which are a function of both r_M and z^{+}, were typically $\sim 13.5\,\text{cm}^3\,\text{mol}^{-1}$ for a typical first row M^{3+} ion and $13.1\,\text{cm}^3\,\text{mol}^{-1}$ for a typical M^{2+} ion. In addition one might argue that as the radius of the metal ion decreases, i.e. towards the right along the first transition metal period, the packing of water around a metal ion could eventually approach a 'close-packed' arrangement and thus a much lower dissociative limit of $\sim +9\,\text{cm}^3\,\text{mol}^{-1}$ might be relevant.* For the more expanded situation of an associatively activated limiting process a similar restriction may not apply and the value of $\sim -13\,\text{cm}^3\,\text{mol}^{-1}$ may still be valid for a first row metal. From Table 1.7, one sees that $\Delta V^{\ddagger}$ for water exchange on Ti^{3+} ($-12.1\,\text{cm}^3\,\text{mol}^{-1}$) is indeed underestimated by the close-packed model approach and closer to the empirical limit of $\sim -13\,\text{cm}^3\,\text{mol}^{-1}$. On the other hand, $\Delta V^{\ddagger}$ for water exchange on Ni^{2+} is $+7.2\,\text{cm}^3\,\text{mol}^{-1}$, significantly below both the $+9$ and $+13$ limiting values. One might conclude therefore from the $\Delta V^{\ddagger}$ data that water exchange on Ti^{3+} is associatively activated, possibly I_A approaching A, while that on Ni^{2+} is dissociatively activated, possibly I_D approaching D, but as stated these labels are highly restrictive and one should look for better criteria such as the degree of selectivity by a metal ion for a group of ligands as reflecting the degree of incoming ligand involvement (see Section 1.4.6).

Swaddle and coworkers have also shown that a correlation exists between the absolute partial molar volumes, $\bar{V}^0_{abs}$, for a number of aqua pentammine complexes and the observed $\Delta V^{\ddagger}$ value for water exchange [104]. The larger the $\bar{V}^0_{abs}$ the more negative is $\Delta V^{\ddagger}$. In terms of the following equation:

$$V^{\ddagger}(\text{essentially constant}) = \bar{V}^0_{abs} + \Delta V^{\ddagger} \qquad (1.53)$$

this amounts to an essentially invariant value of $V^{\ddagger}$, the volume of the transition state, in each case (Table 1.8). Thus the observed $\Delta V^{\ddagger}$ values appear to reflect predominantly initial state properties. This was interpreted as the transition

---

* It had been earlier noted that in the close-packed structure of ice the octahedral interstitial sites were of a cavity size similar to the ionic radius of most first row M^{2+} ions (57 pm). Within such a close-packed structure the effective volume of a water molecule is $\sim 9\,\text{cm}^3\,\text{mol}^{-1}$. Thus the volume change to the system on the expulsion of a water molecule from a close-packing arrangement to that in the free state might be estimated as $\sim 18 - 9 = \sim 9\,\text{cm}^3\,\text{mol}^{-1}$ [102, 103].

Table 1.8. Values for $\bar{V}^0_{abs}$ and $\Delta V^\ddagger$ for water exchange on selected metal aqua ion perchlorates

	$\bar{V}^0_{abs}$ (cm^3 mol^{-1})	$\Delta V^\ddagger$ (cm^3 mol^{-1})	$V^\ddagger$ (cm^3 mol^{-1})
$[M(OH_2)_6]^{2+}$			
Mn	− 17.4	− 5.4	− 22.8
Fe	− 25.3	+ 3.8	− 21.5
Co	− 25.4	+ 6.1	− 19.3
Ni	− 28.4	+ 7.2	− 21.2
$[M(NH_3)_5(OH_2)]^{3+}$			
Co	54.9	+ 1.2	56.1
Rh	61.2	− 4.1	57.1
Ir	61.2	− 3.2	58.0
Ru	63.2	− 4.0	59.2
Cr	65.3	− 5.8	59.5

states involving a cooperative movement of the entire assemblage of primary and secondary hydration shells so as to become essentially insensitive to the nature of the metal centre, whereas the molar volume of the ground state static system directly reflects the properties of that system. Thus, not unsurprisingly, those metal aqua ions and complexes with the larger $\bar{V}^0_{abs}$ values are more expanded and thus likely to allow a greater penetration of the entering ligand nucleophile along the reaction coordinate to the transition state.

Until quite recently there have been few meaningful computational studies that have attempted to model the simple water exchange process. In 1983 Connick and Alder [105] reported the results of calculations carried out on a two-dimensional model consisting of 90 particles which included one metal ion around which four solvent molecules were in a close-packed arrangement. The conclusions reinforced Swaddle's evidence of a cooperative movement of all solvent molecules in the system, which controlled the progress to the transition state rather than the individual properties of the metal centre.

In 1994 Sandstrom and coworkers reported attempts to model the various mechanistic pathways using large basis set SCF computations on isolated penta-(in trigonal bipyramidal or square pyramidal geometry), hexa- and heptahydrated metal ion complexes [106]. Figure 1.32 shows the models used for defining the refined parameters and the coordinate system. A dissociative process was modelled using the dissociation energy of the sixth water ligand (ΔE_6):

$$\Delta E_6 = E\{[M(OH_2)_5]^{n+}\} + E\{H_2O\} - E\{[M(OH_2)_6]^{n+}\} \qquad (1.54)$$

in a gas phase process as the activation energy in an Arrhenius plot vs the

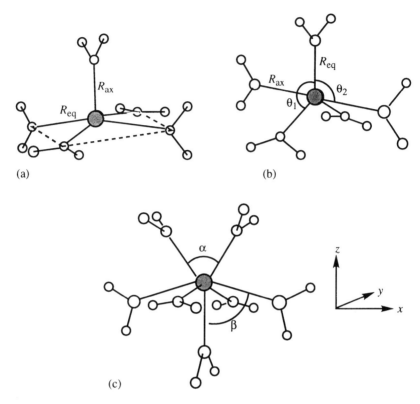

(a)   (b)

(c)

Figure 1.32. The various hydrated ion models defining the refined parameters and the coordinate system used for the large basis set SCF calculations: the pentahydrated ion in (a) square-based pyramidal geometry, (b) trigonal bipyramidal geometry and (c) the heptahydrated ion

experimental water exchange rate constant k_{ex}:

$$k_{ex} = A \exp\left(\frac{-\Delta E_6}{RT}\right) \tag{1.55}$$

Energy contributions from solvation effects were assumed constant for a series of ions of similar charge. The alternative associative mechanism was investigated by using the binding energy of the seventh water ligand (ΔE_7) in a similar Arrhenius plot. A linear correlation with ΔE_6 is found for the experimental $\log k$ for water exchange on all the first row divalent ions including V^{2+} and Mn^{2+} for which negative volumes of activation have been found. In particular, no corresponding correlation with ΔE_7 was found for the alternative associative process. Sandstrom himself points out [106] that the modelling of the energetics may lead to a different profile from that of the volume changes occurring during the water exchange process; i.e. the compaction necessary to achieve the transition state may not necessarily correspond to the energetics of the bond-making/breaking process. Thus one could argue that the correlation shown by the divalent ions is one based solely on the gas phase energetics of a final pentacoordinated complex. The apparent changeover in sign of $\Delta V^{\ddagger}$ (negative to positive) along the 3d divalent metal ions was nonetheless viewed as reflecting a deeper penetration by the incoming water ligand along the trifold axis of the MO_6 octahedron (Fig. 1.33) in the case of the more expanded V^{2+} (low t_{2g} population) and Mn^{2+} (high spin d^5, zero LFSE) ions before bond breaking occurs, this becoming less as both the size and number of available t_{2g} orbitals decrease along the series. This is largely in keeping with the idea of a continuium of reaction coordinates wherein the degree of penetration by the entering ligand is largely a function of the ground state partial molar volume of the metal ion, the electron density and the properties of the entering ligand in question. Thus the $\Delta V_{obs}^{\ddagger}$ values for V^{2+}(aq) and Mn^{2+}(aq) reflect a low d-electron population and a larger ground state partial molar volume [104].

In an independent study, Rotzinger has used *ab initio* methods at the Hartree–Fock level to calculate energies and structures for both transition state and

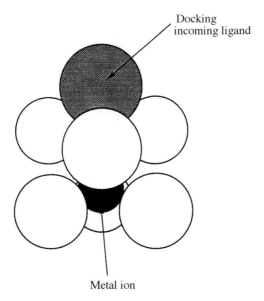

Figure 1.33. Illustration of the docking of an entering water ligand from the secondary shell on to octahedral face of an $[M(OH_2)_6]^{n+}$ ion during the initial stages of water exchange [106b]

intermediate configurations arising from the various water exchange mechanisms [107]. A relevant transition state was defined as that characterised by not more than one imaginary vibrational frequency as opposed to an intermediate which possesses no imaginary frequency. Species with more than one imaginary frequency were viewed as not representing any chemically relevant form on the potential energy surface. The validity of the approximation, in which dynamic electron correlations were neglected, was verified using density functional theory. The examples chosen to model were $[Ti(OH_2)_6]^{3+}$ (an example of an A process), $[V(OH_2)_6]^{2+}$ (a concerted process) and $[Ni(OH_2)_6]^{2+}$ (a D process). As an approximation all second sphere water molecules (except one), bulk water and the anions were neglected. Activation energies for the associative (A) and the concerted (I_a, I, I_d) pathways were computed based on (1.56). For the dissociative pathway both of the following processes were viewed as valid:

$$[M(OH_2)_6 \cdot OH_2]^{n+} \longrightarrow \{[M(OH_2)_7]^{n+}\}^{\ddagger} \qquad (1.56)$$

$$[M(OH_2)_6]^{n+} \longrightarrow \{[M(OH_2)_6]^{n+}\}^{\ddagger} \qquad (1.57)$$

Bond length changes occurring during the activation process, from computed transition state structures, were compared with the activation volumes and viewed as providing a unique assignment of mechanism. Much of the discussion centred on the controversy surrounding $[V(OH_2)_6]^{2+}$, its negative activation volume and the proposal above [106] of a common (dissociative) mechanism for the entire divalent transition metal series. Rotzinger's computed activation energies for all three ions studied were in close agreement with the experimental values, but interestingly completely independent of the chosen mechanistic pathway. It was argued therefore that a mechanistic diagnosis based solely on computed activation energies was probably invalid. The success of Sandstrom's energy correlations, which included $[V(OH_2)_6]^{2+}$, was found to be largely due to similar computed activation energies for both the D and I_d mechanisms. Furthermore, Rotzinger argued [107] that it was important to model the structures of the transition states and not merely those likely to represent intermediates. Attempts to model a dissociative process,

$$[V(OH_2)_6]^{n+} \longrightarrow \{[V(OH_2)_5 \cdots OH_2]^{n+}\}^{\ddagger} \qquad (1.58)$$

$$[V(OH_2)_6 \cdot OH_2]^{n+} \longrightarrow \{[V(OH_2)_5 \cdots (OH_2)_2]^{n+}\}^{\ddagger} \qquad (1.59)$$

resulted in an overall bond elongation (positive $\Delta\Sigma$). Only the concerned process resulted in a negative $\Delta\Sigma$ correlating with the observed negative $\Delta V^{\ddagger}$. A similar treatment for $[Ni(OH_2)_6]^{2+}$ gave a positive $\Delta\Sigma$ value. It was thus concluded that a greater penetration of the entering ligand is relevant to the transition state structure in the case of $[V(OH_2)_6]^{2+}$ as opposed to $[Ni(OH_2)_6]^{2+}$.

Despite these indications the fact remains that despite the observed negative $\Delta V^{\ddagger}$ the rates of substitution and substitution-controlled redox reactions on $V^{2+}(aq)$ are largely insensitive to the nature of the entering ligand, (see Section

5.1.2) moreover similar to the rate of water exchange, implying a mechanism controlled by water ligand dissociation. More kinetic data are, however, needed. A continuing and lively debate is thus ensured for some time yet.

Sandstrom's theoretical treatment on the other hand, shows a good correlation of $\log k_{ex}$ with ΔE_7 for the trivalent ions Ti^{3+}, V^{3+}, Fe^{3+} and Cr^{3+}, but no corresponding correlation with ΔE_6 implying here a more associatively activated process consistent, in this case, with their negative activation volumes for water exchange. Indeed, the $-12.1\,cm^3\,mol^{-1}$ value for Ti^{3+} correlates with a high relative stability of its heptahydrated form [106]. By obtaining calculated energies for both transition states and intermediates Rotzinger showed that the seven-coordinated intermediate in the case of Ti^{3+} was quite long lived (2 ns). The proposal of an A mechanism was also backed up by calculations showing an invariance of spectator Ti—$O(OH_2)$ bond lengths to the transition state pointing to the observed $\Delta V^{\ddagger}$ as arising largely therefore from the take up of the seventh water [107].

The situation for Cr^{3+} is interestingly. The calculations by Sandstrom and coworkers reveal that the size of the heptahydrates for V^{3+}, Fe^{3+} and Cr^{3+} are quite similar, but Cr^{3+} has the smallest size hexahydrate and thus the highest solvation energy [106]. Thus the extremely small rate constant and anomalously high enthalpy of activation ($108\,kJ\,mol^{-1}$) may not only be reflective of the difficulty of perturbing a small octahedral coordination sphere with high LFSE (-1.2Δ) but also reflective of the high solvation energy and the need to break a stronger hydrogen-bonded network. The result is that the favourable associative activation pathway on Cr^{3+}, reflected by the negative $\Delta V^{\ddagger}$ value, is somewhat suppressed. High-spin $Fe^{3+}(aq)$, because it is more expanded (zero LFSE), is better suited to an associative activation process. This is reflected in the trend in $\Delta H^{\ddagger}$ for water exchange of $Cr^{3+} > Fe^{3+} > V^{3+} > Ti^{3+}$ and in the $\log k$ correlation of $Cr^{3+} < Fe^{3+} < V^{3+} < Ti^{3+}$. Similarly, the larger Mo^{3+} is better suited to associative activation and this is probably largely responsible for the surprising trend in lability $Mo^{3+} \gg Cr^{3+}$ (factor of $\sim 10^5$) as working against the normal consequence of a 50% increase in Δ down a group. Ga^{3+} ($3d^{10}$), on the other hand, lies off the correlation line for the trivalent ions consistent with its positive $\Delta V^{\ddagger}$ and dissociative tendency.

Prior to the measurement of activation volumes the prevailing view was that water exchange and ligand substitution reactions were dissociative for octahedral divalent ions and mainly associative for most trivalent ions due to their stronger electrostatic attraction for the entering ligand nucleophile. These energy correlation results seem largely to be in keeping with this view at least for the above metal ions in question. The situation for a number of other trivalent aqua ions such as Co^{3+}, Ru^{3+}, Rh^{3+} and Ir^{3+}, however, seems not so clear cut and requires separate discussion (see Chapters 8 and 9). An extension of the above theoretical correlations to include the 4d trivalent metal aqua ions Mo^{3+} and Ru^{3+} and the low-spin group 9 ions Co^{3+}, Rh^{3+} and Ir^{3+} would be of interest.

Reactions on a wide range of square planar Pd^{II} and Pt^{II} complexes are established as proceeding by an associative pathway and discrete penta-coordinated intermediates have been established in a number of cases. There is no reason to think that the square planar aqua ions $[Pd(OH_2)_4]^{2+}$ and $[Pt(OH_2)_4]^{2+}$ behave any differently. However, $\Delta V^{\ddagger}$ for water exchange has been measured on each as respectively $-2.2\,cm^3\,mol^{-1}$ (Pd^{2+}) and $-4.6\,cm^3\,mol^{-1}$ (Pt^{2+}), (Table 1.7), well below the limiting associative value. It is likely that this is a reflection of the significant volume expansion that results on forming the penta-coordinated square pyramidal and eventually trigonal bipyramidal inter-mediates. Furthermore, the penta-coordinated intermediate with five water molecules coordinated may be less well solvated and also less strongly hydrogen bonded with the solvent water assemblage, both being factors contributing to a volume increase in the system along the pathway to the transition state.

The above discussion points to a clear caution in the use of $\Delta V^{\ddagger}_{obs}$ values in isola-tion as indicators of the mechanistic pathway. It is better to use a variety of independent indicators as will be highlighted below and in the subsequent chapters.

1.4.5   The Eigen–Wilkins Preassociation Mechanism

Manfred Eigen pioneered the development of rapid mixing and relaxation methods for studying reactions on labile metal ions [91]. Reactions on Ni^{2+} were found to fall into the range suitable for stopped-flow monitoring (see Fig. 1.43 in Section 1.5.3.3). Wilkins has reviewed the substitution data on $[Ni(OH_2)_6]^{2+}$ (Section 10.1). Second-order rate constants were found to show a large entering group dependence. However, no correlation with nucleophilicity (see Section 1.4.6) could be ascertained, as judged by the lack of a correlation with respective pK_a values for the protonated ligands. Eventually it was shown by a wider study that the observed variation in k depended largely on the charge on the entering ligand and this led on to the now classical Eigen–Wilkins mechanism wherein the interchange of ligands takes place within a preformed outer-sphere encounter complex [108]:

$$[M(OH_2)_n]^{z+} + L^{z-} \underset{}{\overset{K_{os}}{\rightleftharpoons}} [M(OH_2)_n]^{z+}, L^{z-}$$

outer-sphere preassociation

$$\downarrow k_1 \qquad\qquad\qquad (1.60)$$

$$[M(OH_2)_{n-1}L]^{(z-z-)+} + H_2O \longleftarrow [M(OH_2)_{n-1}]^{z+}, L^{z-}$$
$$\vdots$$
$$OH_2$$

(illustrated for the case when bond breaking is important)

$$Rate = \frac{k_1 K_{os}[L^{z-}][M(OH_2)_n^{z+}]}{1 + K_{os}[L^{z-}]} = k_{obs}[M(OH_2)_n^{z+}] \qquad (1.61)$$

For the condition $K_{os}[L^{z-}] \ll 1$ (as often the case with $z- = 1-$),

$$k_{obs} \text{ reduces to } k_I K_{os}[L^{z-}] \qquad (1.62)$$

Within this encounter complex the entering ligand L^{z-} is perceived as residing within the secondary solvation shell of M^{z+} largely accompanied by solvational changes as a result of charge neutralisation and the general disruption of the secondary shell assemblage. It was thus found for reactions on Ni^{2+} that the variation of rate constants, k_{obs}, reflected the variation in the encounter complex formation constant K_{os} rather than in k_I. Values of k_I are found to be largely constant and moreover close to the first-order rate constant for water exchange, $3.6 \times 10^{-4} s^{-1} (25 °C)$, implying a process largely controlled by dissociation of a water ligand (1.60). Here the activation volume for water exchange is, as it happens, duly diagnostic and supportive of the dissociative process. When one analyses complex formation data on Ti^{3+}, (Section 4.1), on the other hand, a similar variation in k_{obs} is seen, but in this case a correlation is seen with Bronsted basically of the ligand (pK_a value for HL) and not simply charge. Thus the kinetic data support an associative mechanism for reactions on Ti^{3+}, as suggested by the activation volume for water exchange.

Eigen [109] and Fuoss [110] have developed equations for the estimation of K_{os} based on extended Debye–Huckel theory and a hard sphere model for the reacting species:

$$K_{os} = \frac{4\pi N a^3}{3000} \exp\left(\frac{-U}{kT}\right) \qquad (1.63)$$

where

$$U = \frac{z_1 z_2}{\varepsilon}\left[\frac{1}{a(1 + \kappa a)}\right] \quad \text{and} \quad \kappa = \sqrt{\frac{8\pi N e^2 I}{1000 \varepsilon k_B T}} . \qquad (1.64)$$

and N is Avogadro's number (6.022×10^{23}), a is the distance of closest approach (cm), e is the electron charge, k_B is the Boltzmann constant, ε is the bulk solvent dielectric constant, I is the ionic strength and z_1 and z_2 are the charges on the species. Values of $\sim 1 M^{-1}$ for a $2+/1-$ reactant pair and $\sim 2 M^{-1}$ for a $3+/1-$ pair at ionic strength 1.0 M seem to be largely in keeping with the calculated values. Apparent from equations (1.63) and (1.64) is that K_{os} values increase with the product z^+, z^-, decreasing ionic strength and decreasing dielectric constant. For highly charged reacting pairs, e.g. $2+/2-$, the value of K_{os} can be large enough to give rise to saturation kinetics ($K_{os}[L^{z-}] \gg 1$) as $[L^{z-}]$ is increased. This has led to a difficulty in distinguishing the Eigen–Wilkins interchange mechanism from that of an extreme dissociative (D) process which can be reflected by a similar rate expression involving a dependence on $[L^{z-}]$ appearing in both the numerator and denominator leading to saturation kinetics at sufficient high values of $[L^{z-}]$. The D mechanism might be indicated if an Eigen–Wilkins analysis fails to give reasonable values for K_{os} for the reactant pair in question. However, this is not sufficient proof on its own and other indicators would be required. For example, an extreme D process would be expected to have a rate constant for the water dissociation step matching that of water exchange. Any discrepancy, i.e. a smaller value, even within a factor of 2 or 3 of that for water exchange, would imply that the putative five-coordinated intermediate cannot be long lived on the timescale of the relaxation of the coordination sphere. This is exactly the situation for $[Co(CN)_5(OH_2)]^{2-}$, a species long believed to be the archetypal example of the D process [111], but now believed to be in the interchange category [112], admittedly with a strong leaving group (H_2O) dependence. The activation volume for water exchange incidently is $+ 7 \, cm^3 \, mol^{-1}$, identical to that on $[Ni(OH_2)_6]^{2+}$. Water ligand replacement on $[RhCl_5(OH_2)]^{2-}$ (Section 9.2.2) remains, however, a strong candidate for a D process [123]. Indeed, Rh^{3+} is an intriguing case. The activation volume for water exchange on $[Rh(OH_2)_6]^{3+}$ is $-4.2 \, cm^3 \, mol^{-1}$ yet substitution reaction rate constants, where $Rh\!-\!OH_2$ bond cleavage is involved, are found to be largely independent of the nature of the entering ligand and moreover close to the water exchange value (Table 9.7), implying a dissociative process. A similar situation is apparent for complexation reactions on $[Ru(OH_2)_6]^{2+}$ (rates largely entering ligand independent) for which an activation volume close to zero is relevant. For water exchange on $[Rh(OH_2)_6]^{3+}$ the negative activation volume could be reflective of a significant volume contraction on forming the five-coordinated intermediate, in this case sufficient to change the sign of $\Delta V_{obs}^{\ddagger}$. The water exchange reaction is also known to be strongly promoted through the conjugate base $[Rh(OH_2)_5OH]^{2+}$ (see Section 1.4.6). Complexation reactions on Rh^{3+}(aq) are also similarly promoted through the hydroxy species. The obvious conclusion is that ligand substitution reactions on Rh^{3+}(aq) complexes, involving $Rh\!-\!OH_2$ cleavage, are dissociative and that the conjugate base labilising effect stems from a further stablisation of the five-coordinated intermediate via σ- and π-donation from OH (Fig. 1.34).

Could it be concluded therefore that all low-spin t_{2g}^6 metal ions, irrespective of whether divalent or trivalent, react dissociatively? Such a premise is perfectly reasonable given that the filled t_{2g} orbitals are those lying along the line of approach of the entering ligand. Thus the relative insensitivity to the nature of the entering group, as shown by reactions of low-spin Co^{III} aqua complexes, Rh^{3+}(aq) and Ru^{2+}(aq), is largely to be expected.

1.4.6   Reactions of Hydroxy Aqua ions

1.4.6.1   THE 'CONJUGATE BASE' LABILISING EFFECT

It is now well established [114] that the rates of water ligand exchange and substitution reactions on monodeprotonated aqua metal ions

$$[M(OH_2)_n]^{x+} \rightleftharpoons [M(OH_2)_{n-1}OH]^{(x-1)+} + H^+ \qquad (1.65)$$

are more often than not faster than equivalent substitution rates on the fully protonated species. Reference to the data below in Table 1.9 shows that the rate acceleration can amount to > 4 orders of magnitude. It has been argued that this largely amounts to a charge neutralisation effect and the lower charge on the hydroxo complex explains the lability. Volumes of activation for water exchange moreover tend to be much more positive for the hydroxo aqua ions (Table 1.10), suggesting that reactions on the monodeprotonated aqua ions are more dissociative in character. Although this behaviour would be largely in keeping with reactions of essentially lower charge dipositive ions (see below) one cannot rule out the alternative explanation of a specific electronic effect from the OH ligand termed the 'conjugate base' labilising effect. This premise is based largely on the evidence accumulated from base hydrolysis reactions involving ammine and amine complexes of Co^{III} in favour of the S_N1_{CB} or D_{CB} mechanism [115]. In these reactions direct attack of the OH^- nucleophile is not relevant. Instead, one of the

Table 1.9. Water exchange rate constants on some aqua ions and their monohydroxo-aqua forms at 25 °C

$M^{n+}(aq)$	$k(M^{n+})$	$k(MOH^{(n-1)+})$	Ratio $\dfrac{k(MOH^{(n-1)+})}{k(M^{n+})}$	pK_{11}
$[Mo_3S_4(OH_2)_9]^{4+}$	<1	7.5×10^3	> 7500	0.66
$[Rh(OH_2)_6]^{3+}$	2.2×10^{-9}	4.2×10^{-5}	19 100	3.5
$[Ir(OH_2)_6]^{3+}$	1.1×10^{-10}	5.6×10^{-7}	5 100	4.45
$[Fe(OH_2)_6]^{3+}$	1.6×10^2	1.2×10^5	750	2.19
$[Ga(OH_2)_6]^{3+}$	4.0×10^2	1.1×10^5	275	3.9
$[Ru(OH_2)_6]^{3+}$	3.5×10^{-6}	5.9×10^{-4}	170	2.7
$[Cr(OH_2)_6]^{3+}$	2.4×10^{-6}	1.8×10^{-4}	75	4.0

Table 1.10. Activation enthalpies and volumes for water exchange on $M^{3+}(aq)$ and $MOH^{2+}(aq)$

M	Reaction on $[M(OH_2)_6]^{3+}$		Reaction on $[M(OH_2)_5OH]^{2+}$	
	$\Delta H^{\ddagger}$ (kJ mol^{-1})	$\Delta V^{\ddagger}$ (cm^3 mol^{-1})	$\Delta H^{\ddagger}$ (kJ mol^{-1})	$\Delta V^{\ddagger}$ (cm^3 mol^{-1})
Cr	108.6	-9.6	110	$+2.7$
Ru	89.8	-8.3	95.8	$+0.9$
Fe	63.9	-5.4	42.5	$+7.0$
Rh	131	-4.2	103	$+1.5$
Ir	131	-5.7	n.a.	$+1.3$
Ga	67.1	$+5.0$	58.9	$+6.2$

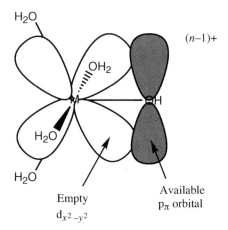

Figure 1.34. Stabilisation of a trigonal bipyramidal intermediate by π donation from OH in the hydroxo-aqua complex $[M(OH_2)_4(OH)]^{(n-1)+}$

ligand N—H protons is removed by OH^- forming an 'amido' complex which is believed to promote the dissociative D_{CB} mechanism by stabilising the resulting trigonal bipyramidal intermediate via π-donation into the resulting empty $d_{x^2-y^2}$ orbital of the Co^{III}. For reactions on hydroxo-aqua metal complexes the OH-ligand itself could also act as the effective π-donating ligand stabilising trigonal bipyramidal '$[M(OH_2)_4(OH)]^{n+}$' (Fig. 1.34). Such a species, however, probably has only a fleeting existence in aqueous solution and moreover its formation is not easy to substantiate since it will remain strongly associated with the assemblage of hydrogen-bonded waters surrounding it. The water ligand dissociating could conceivably be *cis* or *trans* to the OH group, a fact not possible to establish because of the rapid proton exchange occurring within the primary hydration shell. In the hydroxo ligand-promoted water exchange at the d-H_2O ligands on the cluster $[Mo^{IV}_3X_4(OH_2)_9]^{4+}$ (X = O or S), however, only a *cis*-conjugate base effect can be relevant (Fig. 1.35a) [90]. Support for the conjugate base effect, rather than merely a charge reduction effect, comes from three other key observations. Firstly, the rate constant (25 °C) for the single-coordinated water on the 2 + charged triangular Mo^{IV} acetato-bridged cluster complex (Fig. 1.35b) is $1.8 \times 10^{-6} s^{-1}$ [116], some 10^4 times slower than that for the slow exchanging c-H_2O on $[Mo^{IV}_3O_4(OH_2)_9]^{4+}$ and nearly 10^8 times slower than exchange of the d-H_2O on the same species which is activated through the hydroxo form (see Section 6.2.3). Secondly, bridging OH groups can also exert a labilising effect such that two significantly different water exchange rates have been detected for the aqua dimer of Cr^{III}, $[Cr_2(OH)_2(OH_2)_8]^{4+}$, arising from

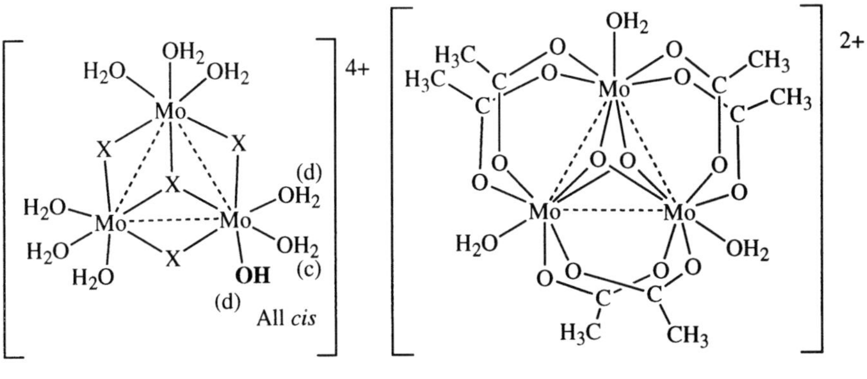

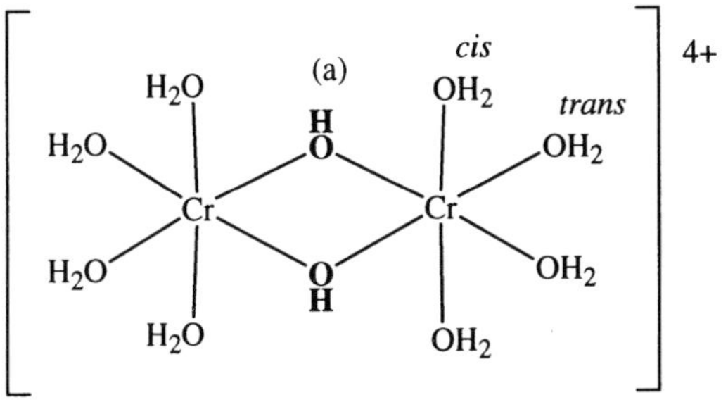

Figure 1.35. Structure/reactivity relationships in oligomeric aqua metal complexes

waters *trans* to the bridging OH as opposed to waters *trans* to another water (Fig. 1.35c) [117]. Here the ratio of rate constants (25 °C) is only a factor of ~ 6. Thirdly, within the series of Cr^{III} complexes $[Cr(OH_2)_5X]^{2+}$ (X = various mono-anionic ligands) Bracken and Baldwin have found that one of the water molecules is more labile than the other four, a result that cannot be explained by mere charge reduction [118]. Within the aqua Cr^{III} dimer the more labile waters *trans* to the bridging OH groups are not manifest in longer $M-OH_2$ bonds (Section 6.1.2). Similarly, no difference in bond lengths to the two types of water (c or d) is seen in the crystal structures of $[Mo_3X_4(OH_2)_9]^{4+}$ (X = O or S) (Fig. 1.35). This might imply that competing and overriding *cis* effects may be just as important to consider in structure/reactivity relationships in reactions at hydroxo-aqua metal complexes.

1.4.7 Selectivity Indicators

1.4.7.1 METAL ION SUBSTRATES

As an alternative to the rigid Langford and Gray classifications there have been a number of attempts to come up with a measure of the selectivity of an aqua metal ion (the substrate) for a range of different entering group nucleophiles. These would seek to measure the relative importance of bond making with the entering group along the reaction coordinate to the transition state. In 1975 Sasaki and Sykes proposed a scale based upon the ratio of rate constants (R) for complexation with the ligands NCS^- and Cl^-, these two chosen largely because of available rate data for a wide range of metal ions and because they are viewed as being sufficiently different in their nucleophilicity so as to give a meaningful measure of an entering group selectivity [119]. Associatively activated reactions would be indicated if $R \geqslant 10$, whereas dissociatively activated reactions would be indicated by values around unity. Some of the data are compiled in Table 1.11. Values well in excess of 10 are seen for Mo^{3+}, V^{3+}, Cr^{3+} and Fe^{3+}, in keeping with the associatively activated reactions on these metal ions (as discussed above), whereas hydroxo-aqua complexes and the aquapentammine complexes of Rh^{III} and Co^{III} fall into the dissociative category. $[Cr(NH_3)_5OH_2]^{3+}$ appears to be somewhat borderline. The behaviour of $[Co(OH_2)_6]^{3+}$ is, however, some-

Table 1.11. Ratio of rate constants for NCS^- complexation relative to Cl^- and activation volumes for water exchange on $[ML_5(OH_2)]^{z+}$ ions [119]

$[ML_5(OH_2)]^{z+}$	R^a	$\Delta V^{\ddagger}H_2O$ (cm^3 mol^{-1})
$[Mo(OH_2)_6]^{3+}$	62	Not available
$[V(OH_2)_6]^{3+}$	$\geqslant 36$	-8.9
$[Cr(OH_2)_6]^{3+}$	55	-9.6
$[Co(OH_2)_6]^{3+}$	$\geqslant 43$	Not available
$[Fe(OH_2)_6]^{3+}$	14	-5.4
$[Cr(NH_3)_5OH_2]^{3+}$	6	-5.8
$[Co(OH_2)_5OH]^{2+}$	2	Not available
$[Cr(OH_2)_5OH]^{2+}$	1	$+2.7$
$[Rh(NH_3)_5OH_2]^{3+}$	0.6	-4.1
$[Fe(OH_2)_5OH]^{2+}$	0.6	$+7.0$
$[Co(NH_3)_5OH_2]^{3+}$	0.5	$+1.2$
Others		
$[Mo_3S_4(OH_2)_9]^{4+}$	1.2	Not available
$[Mo_3S_4(OH_2)_8OH]^{3+}$	1.4	Not available

^{a} Ratio of second-order rate constants, $M^{-1}s^{-1}$, at 25 °C.

Table 1.12. Selectivity parameter S, relative to $[Cr(OH_2)_6]^{3+}$, for the reaction of some cationic substrates with mononegative anions in aqueous solution [84, 97, 119, 120]

Substrate	S^a
$[Mo(OH_2)_6]^{3+}$	1.1^b
$[Cr(OH_2)_6]^{3+}$	1.0
$[Fe(OH_2)_6]^{3+}$	0.7
$[Cr(OH_2)_5OH]^{2+}$	0.4
$[Cr(NH_3)_5OH_2]^{3+}$	0.3
$[Fe(OH_2)_5OH]^{2+}$	0.1
$[Co(NH_3)_5OH_2]^{3+}$	-0.1

$^a \pm 0.1$, rate constants have dimensions $M^{-1} s^{-1}$.
^{b} Only two data points.

what exceptional and not in keeping with the general behaviour of low-spin t_{2g}^6 metal ions. $[Co(OH_2)_5OH]^{2+}$, on the other hand, duly falls into the dissociative category. The absence (thus far) of measurable activation volumes for water exchange on $[Mo(OH_2)_6]^{3+}$ and on both $[Co(OH_2)_6]^{3+}$ and $[Co(OH_2)_5OH]^{2+}$ is a pity.

Swaddle has proposed a selectivity scale based upon parameter S

$$S = \frac{\delta \ln k^A}{\delta \ln k^B} \tag{1.66}$$

which relates the behaviour of different metal ions towards a more wider range of ligands [94, 97]. S measures the selectivity of a substrate A (a metal ion complex) relative to another, B; $\delta \ln k'$ represents the change in ln (rate constant) as the incoming ligand (of similar charge) is varied. Since S is a relative quantity it was decided that a suitable reference compound B might be $[Cr(OH_2)_6]^{3+}$ (Table 1.12). The reason why not more metal ions appear in this table is due to a lack of kinetic data on a wide enough range of mononegative ligands. The choice of mononegative nucleophiles is moreover limited to those with negligible Bronsted basicity because of the 'proton ambiguity' problem [121]. This arises because reaction of $[M(OH_2)_6]^{3+}$ with L^- cannot be separated, by kinetic evaluation, from reaction of $[M(OH_2)_5OH]^{2+}$ with HL because of rapid proton transfer. The result is the product of constants for the two reactant pair reactions that cannot be separated unless some assumptions are made. Furthermore, for meaningful comparisons rate constants in equation (1.66) should be first-order k_1's to allow for any ion pair preassociation. Here lies an additional problem. Accurate and reliable resolution of k_{obs} into k_1 and K_{os} is only rarely possible for aqueous reactions and calculated K_{os} values, as mentioned above, are prone to uncertainty since they are based on 'hard sphere' models. However, provided one confines the comparison to ions of similar charge type second-order k's can be used as in the comparison in Table 1.11.

What Table 1.11 tries to show is that it is unhelpful often to draw a line rigidly defining the substrates as having associative or dissociative tendency but to picture them along a continuous scale of relative affinities. Only at the extreme ends can one talk confidently of an associative or dissociative mechanism as being important based solely on the properties of the metal centre. One goal of future research should be to expand the list of substrates in tables 1.11 or 1.12 to include as many examples as possible. Here perhaps lies the definitive scale of comparison for different metal aqua complexes.

1.4.7.2   NUCLEOPHILICITY SCALES [121a]

One desires an independent measured scale for nucleophilicity of an incoming ligand so that one can have a degree of confidence in the measure of the selectivity as shown by the aqua metal ion. A simple measure of the nucleophilicity of L is the measure of its relative affinity for the simplest of all cations, namely H^+ as measured by the Bronsted basicity or simply the pK_a of HL. Thus it is seen that a correlation of rate exists with pK_a of HL for substitution on $[Ti(OH_2)_6]^{3+}$ (associative) but not on $[Ni(OH_2)_6]^{2+}$ (dissociative). Other factors affecting nucleophilicity are oxidisability (a more easily oxidisable ligand being more willing to give up its electrons as measured by the standard reduction potential), polarizability and solvation energy (a more highly solvated ligand will be a poorer nucleophile and will require solvation changes to accompany metal complex formation). Several other scales have been proposed but few have received universal acceptance by inorganic chemists. It seems that each metal ion must be treated individually.

1.4.7.3   HARD–SOFT ACID–BASE THEORY

This idea was first put forward by Pearson in 1968 [122] in response to the earlier suggestion of separating metal ions into (a) and (b) categories by Ahrland, Chatt and Davies [123]. Pearson expanded the idea to talk of hard (a) and soft (b) acids and bases. The basic principal was that the strongest interaction, which could be measurable by a rate of complex formation or the magnitude of a stability constant (Section 1.4.1), would be found for a hard acid (M^{n+} ion) with a hard base (L^-) and likewise for a soft–soft matching pair. This theory remains attractive as one of the few to treat metal centres and ligands largely as independent reacting entities without the necessity to define the type of interaction. Indeed, for a soft–soft interaction such as I^- reacting at, for example, square planar Pd^{2+}(aq), it might be premature to assume that it necessarily involves nucleophilic attack by I^- at the metal. The Pd^{2+} centre with its filled d_z^2 orbital could equally be envisaged as the effective nucleophile here attacking, or perhaps more correctly leading to polarization of, the diffuse I^- ion (Fig. 1.36). It is for this reason that Pearson's theory has stood up well to the test of time.

One of the problems with the hard–soft acid–base theory is that it is largely qualitative. With this in mind Pearson has more recently attempted to establish a quantitative scale of 'hardness' and 'softness' based on ionization potentials (I) and electron affinities (A) [124]. The absolute hardness is defined as $\eta = (I - A)/2$ and softness as $\sigma = 1/\eta$. Table 1.13 lists Pearson's estimates for absolute hardness

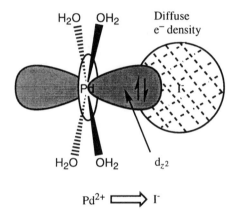

Figure 1.36. A 'soft–soft' Pearson interaction; I^- complexation at Pd^{2+}(aq)

Table 1.13. Pearson's attempted quantification of absolute hardness [124]

Acid	η	χ	Base	η	χ
Ca^{2+}	19.52	31.39	H_2O	9.5	3.1
Fe^{3+}	12.08	42.73	NH_3	8.2	2.6
Zn^{2+}	10.88	28.84	Me_2O	8.0	2.0
Ru^{3+}	10.7	39.2	CH_3CN	7.5	4.7
Cr^{3+}	9.1	40.0	F^-	7.01	10.41
Mn^{2+}	9.02	24.66	Me_3N	6.3	1.5
Co^{3+}	8.9	42.4	Me_3P	5.9	2.8
Co^{2+}	8.22	25.28	Me_2NCHO	5.8	3.4
Pt^{2+}	8.0	27.2	OH^-	5.67	7.5
Os^{3+}	7.5	35.2	Pyridine	5.0	4.4
Fe^{2+}	7.24	23.42	Cl^-	4.7	8.31
Cr^{2+}	7.23	23.73	Br^-	4.24	7.6
Pd^{2+}	6.75	26.18	I^-	3.7	6.76

General categorisation as previously quoted in textbooks

Hard acids:	H^+, Li^+, Mg^{2+}, Cr^{3+}, Co^{3+}, Fe^{3+}
Soft acids	Cu^+, Ag^+, Pd^{2+}, Pt^{2+}, Hg^{2+}, Tl^{3+}
Borderline:	Mn^{2+}, Fe^{2+}, Zn^{2+}, Pb^{2+}

Hard bases:	F^-, Cl^-, H_2O, NH_3, OH^-, $CH_3CO_2^-$
Soft bases:	I^-, CO, PPh_3, C_2H_4, RSH
Borderline:	N_3^-, pyridine, NO_2^-, Br^-, SO_3^{2-}

along with the absolute electronegativity using Mulliken's scale ($\chi = (I + A)/2$). The strength of the Lewis acid–Lewis base interaction was assumed to be related to the fractional electron transfer given by $\Delta N = (\chi_1 - \chi_2)/2 \, (\eta_1 + \eta_2)$.

1.5 THE PREPARATION OF AQUA IONS AND DERIVATIVES: SYNTHETIC METHODS AND HANDLING TECHNIQUES

1.5.1 Synthetic Routes and Methods

The preparation of aqueous solutions on many simple monomeric aqua ions, $[M(OH_2)_n]^{z+}$, can be achieved by mere dissolution of a simple hydrated salt of the metal. Examples in this category are the alkali and alkaline earth metals, lanthanides, actinides and most divalent and trivalent first row transition metals:

Divalent ions:

$$MX_2 \cdot nH_2O \ (X = Cl^-, \ Br^-, \ NO_3^-, \ ClO_4^-, \ \text{etc.})$$
$$M_2^I[M(OH_2)_6](SO_4)_2 \ \text{(Tutton salt)} \xrightarrow{\ H_2O\ } [M(OH_2)_6]^{2+}$$
$$(M = Fe, Co, Ni) \qquad\qquad (M = \text{alkaline earths,}$$
$$Mn, Fe, Co, Ni, Cu)$$

$$(1.67)$$

To this one can add zinc, cadmium, mercury and the p-block elements aluminium, gallium, indium and thallium. In the presence of potential complexing ions, e.g. chloride, bromide, sulfate and nitrate, it is often good practice to purify the species by cation-exchange chromatography, usually on a polystyrene-based strongly acid resin such as Dowex or Amberlite. This also has the advantage of introducing the non-complexing counter-anion of choice (X^-) usually ClO_4^-, $CF_3SO_3^-$ or pts$^-$:

$$M^{2+}(aq) \xrightarrow[\substack{\text{Dowex chromatography} \\ \text{HX of choice}}]{} \begin{array}{c} \text{pure solution of } [M(OH_2)_6]^{2+} \\ \text{in aqueous HX} \end{array} \quad (1.68)$$

In the case of trivalent and higher valent ions the use of acidified solutions and cation-exchange chromatography are essential in order to suppress and then remove the products of hydrolytic polymerisation:

Trivalent ions:

$$MX_3 \cdot nH_2O \; (X = Cl^-, Br^-, NO_3^-, ClO_4^-, \text{etc.}) \xrightarrow[H^+]{H_2O} [M(OH_2)_6]^{3+}$$

$$M^I[M(OH_2)_6](SO_4)_2 \quad \text{(alum)}$$

$$(M = Ti, V, Cr, Fe, Al, \quad (1.69)$$
$$Ga, In, Tl)$$

$$M^{3+}(aq) \xrightleftharpoons[\substack{\text{Dowex chromatography}\\ \text{HX of choice}}]{} \text{pure solution of } [M(OH_2)_6]^{3+} \text{ in aqueous HX} \quad (1.70)$$

This is illustrated for $Fe^{3+}(aq)$ wherein, despite a pK_{11} value of only 2.16, acidic solutions containing less than $[H^+] = 2.0\,M$ become invariably contaminated with intensely yellow hydrolysis products. This arises due to rapid favourable equilibria to the oligomeric forms due to the slow rate of iron hydroxo (oxo)-bridged cleavage (see Chapter 8). Separation of the hydrolytic polymers by cation-exchange chromatography is straightforward for inert species such as $Cr^{3+}(aq)$ because of the extremely slow equilibria in both directions and because all of the oligomers have cationic charges in excess of 3 +. In the case of iron and other labile metal ions the lability and rapid equilibration processes make such a separation impossible.

Preparation of certain anhydrous triflates, $M(CF_3SO_3)_3$:

$$MCl_3 \; \text{anhydrous} \xrightleftharpoons[\substack{CF_3SO_3H\\ \text{pure}}]{HCl \,\text{(vacuum line)}} M(CF_3SO_3)_3 \xrightarrow[CF_3SO_3H]{H_2O} [M(OH_2)_6]^{3+} \quad (1.71)$$
$$(M = Ti, V)$$

is attractive as rapid aquation in aqueous triflic acid readily produces aqueous solutions of the aqua ion ready for use in a suitable non-complexing anionic medium. For Ti^{3+} and V^{3+}, preparation of the triflate merely involves treatment of the anhydrous trichloride with pure triflic acid followed by removal of the 'weaker acid' HCl and the excess triflic acid at $\sim 50\,°C$ on a vacuum line at 10^{-3} torr. The solid residue can be used as obtained [125].

The preparation of certain aqua ions makes use of redox reactions involving an aqua ion or precursor in either a lower or a higher oxidation state of the metal. These can involve chemical or electrochemical reduction or oxidation. Some examples are illustrated on the next page:

(a) Chemical reductions:

e..g. the preparation of $[VO(OH_2)_5]^{2+}$: $VOSO_4 \cdot 5H_2O$ (solid)

$$VO_3^-$$
metavanadate $\xrightarrow[H_2O]{H^+}$ V_2O_5(aq) $\xrightarrow[H^+/H_2O]{\text{bubble } SO_2}$ $[VO(OH_2)_5](SO_4)$ $\binom{\text{blue}}{\text{soln.}}$ (1.72)
or $V_{10}O_{28}^{6-}$ red
decavanadate   solid

with vertical arrow: $\Big\| H^+/H_2O$ (from solid to product)

with vertical arrow: $\Big\|$ Dowex aqueous HX

pure
$[VO(OH_2)_5]^{2+}$ blue
Preparation of $[Cr(OH_2)_6]^{3+}$:   in aqueous HX

$$[Cr(OH_2)_6]^{3+} \xrightleftharpoons[\substack{H/H_2O \\ \text{(Jones reductor)}}]{Zn/Hg} [Cr(OH_2)_6]^{2+}$$ (1.73)
blue-violet   sky blue

(b) Reduction using a stirred mercury pool electrolysis cell:

Preparation of $[V(OH_2)_6]^{3+}$ and $[V(OH_2)_6]^{3+}$ from $[VO(OH_2)_5]^{2+}$

$$[VO(OH_2)_5]^{2+} \xrightleftharpoons[H^+/H_2O]{e^-} [V(OH_2)_6]^{3+} \xrightleftharpoons{e^-} [V(OH_2)_6]^{2+}$$ (1.74)
blue

Dowex
ion-exchange
chromatography
in aqueous HX

pure   pure
$[V(OH_2)_6]^{3+}$ $[V(OH_2)_6]^{2+}$
in aqueous HX   in aqueous HX
gray-blue   lilac

In (1.72) SO_2 is conveniently used to reduce V^V(aq) down only to V^{4+}(aq) (as the blue vanadyl ion $[O{=}V(OH_2)_5]^{2+}$). The oxidation product of SO_2 is conveniently SO_4^{2-}. Amalgamated zinc in the form of a column (known as the Jones reductor) is a very useful reducing agent in acidic solution for the preparation of certain low valent ions such as $[Cr(OH_2)_6]^{2+}$ (1.73), $(V(OH_2)_6]^{2+}$ and Eu^{2+}(aq). An obvious drawback is the presence of an unknown quantity of Zn^{2+} ions which can be removed by cation-exchange chromatography at the risk of reducing the yield via oxidation. For this reason the mercury pool electrolysis method (1.74), (Fig. 1.37), which introduces little or no contaminants, is often preferred. Solutions so prepared are ready for standardisation and immediate use. In the illustrated case of reduction of VO^{2+}(aq) the electrolysis can be stopped at either the V^{3+} or V^{2+} stage depending upon the required ion. The use of SO_4^{2-} can slow down reduction to V^{2+} and provides a suitable route for isolation of the V^{3+}

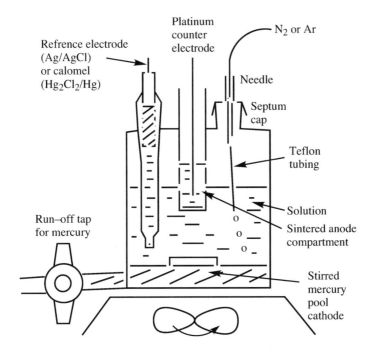

Figure 1.37. Electrochemical cell for a stirred mercury pool electrochemical reduction

alum (following the addition of $M^I_2SO_4$) (see Chapter 5). Further exhaustive reduction, however, eventually yields a solution of $V^{2+}(aq)$ from which the Tutton salt $M^I_2V(SO_4)_2 \cdot 6H_2O$ can be crystallised following the similar addition of $M^I_2SO_4$.

The anodic equivalent to a mercury pool is a high surface area platinum gauze. One type of cell design is illustrated in Chapter 9, Fig. 9.19, for the preparation of solutions of oligomeric $Ir^V(aq)$. In Fig. 1.38 the jacketted cell illustrated here is of the type suitable for the oxidation of $Co^{2+}(aq)$ to prepare $Co^{3+}(aq)$:

(c) Electrochemical oxidation (Pt gauze electrode): Preparation of $[Co(OH_2)_6]^{3+}$:

$$[Co(OH_2)_6]^{2+} \xrightarrow[H^+/H_2O]{-e^-} [Co(OH_2)_6]^{3+} \qquad (1.75)$$
$$\text{red} \qquad\qquad\qquad \text{blue}$$

When no further oxidation at the metal is a problem at high oxidising potentials (also the case for extreme reducing potentials) the electrolysis is best conducted at constant current (galvanostatically). This allows a high current density to be used, particularly, if it is desirable to conduct the electrolysis quickly for fear of product solution ageing. Use of an ice-cooled water jacket can dissipate solution heating as well as slowing down product decomposition. In the case of $Co^{3+}(aq)$ it is the spontaneous oxidation of water ($E^\theta = +1.92$ V). The final example illustrates the preparation of the divalent silver aqua ion via chemical oxidation of $Ag^+(aq)$ with strong oxidants such as peroxydisulfate, $S_2O_8^{2-}$ ($E^\theta = +2.00$ V):

(d) Chemical oxidation: Preparation of $Ag^{2+}(aq)$:

$$[Ag(OH_2)_4]^+ \xrightarrow[\text{H}^+/\text{H}_2\text{O}]{S_2O_8^{2-}} Ag^{2+}(aq) \tag{1.76}$$

$Ag^{2+}(aq)$ can also be prepared by anodic electrolysis using a cell similar to that in Fig. 1.38.

For second and third row transition metals the situation is more complicated. The precursor species are often kinetically inert, air sensitive and frequently susceptible to metal–metal bond formation reactions and/or hydrolytic polymerisation. The consequencies of this are twofold: (a) simple $[M(OH_2)_n]^{z+}$ ions are not always formed and (b) the preparative route is not easy to predict and is often unique for each species requiring the appropriate precursor. The classic example is that of the aqua ion species of molybdenum (see Section 6.2):

$Mo^{2+}(aq)$: not $[Mo(OH_2)_6]^{2+}$ (cf. Cr)
      but **the quadruply M—M bonded dimer:** $[Mo_2(OH_2)_8]^{4+}$

The preparative route requires a quadruply M—M bounded dimer precursor:

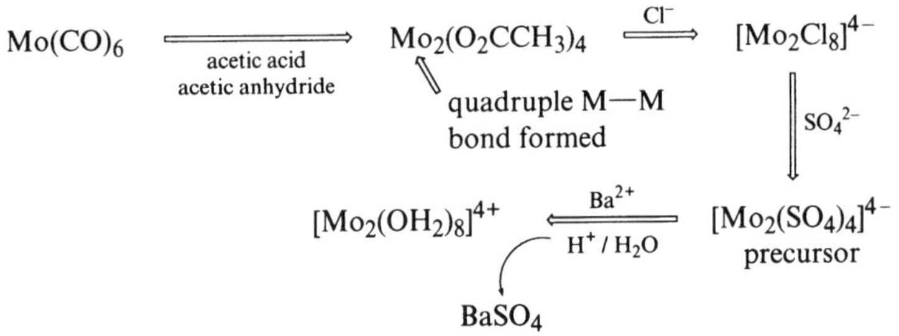

$$Mo(CO)_6 \xrightarrow[\substack{\text{acetic acid}\\\text{acetic anhydride}}]{} Mo_2(O_2CCH_3)_4 \xrightarrow{Cl^-} [Mo_2Cl_8]^{4-} \tag{1.77}$$

quadruple M—M
bond formed

$$[Mo_2Cl_8]^{4-} \Big\updownarrow SO_4^{2-}$$

$$[Mo_2(OH_2)_8]^{4+} \xleftarrow[\text{H}^+/\text{H}_2\text{O}]{Ba^{2+}} [Mo_2(SO_4)_4]^{4-}$$
                                                                    precursor

BaSO$_4$

Mo^{3+} (aq): Three forms can be made depending upon the preparative route:

(a) Monomeric $[Mo(OH_2)_6]^{3+}$

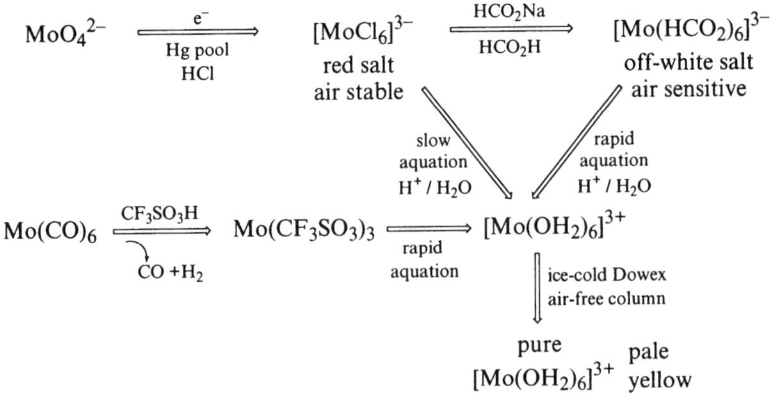

$$MoO_4^{2-} \xrightarrow[\substack{\text{Hg pool}\\\text{HCl}}]{e^-} [MoCl_6]^{3-} \xrightarrow[\text{HCO}_2\text{H}]{\text{HCO}_2\text{Na}} [Mo(HCO_2)_6]^{3-}$$

red salt   off-white salt
air stable   air sensitive

slow   rapid
aquation   aquation   (1.78)
H^+/H_2O H^+/H_2O

$$Mo(CO)_6 \xrightarrow[\substack{\\\text{CO}+\text{H}_2}]{\text{CF}_3\text{SO}_3\text{H}} Mo(CF_3SO_3)_3 \xrightarrow[\substack{\text{rapid}\\\text{aquation}}]{} [Mo(OH_2)_6]^{3+}$$

ice-cold Dowex
air-free column

pure   pale
$[Mo(OH_2)_6]^{3+}$ yellow

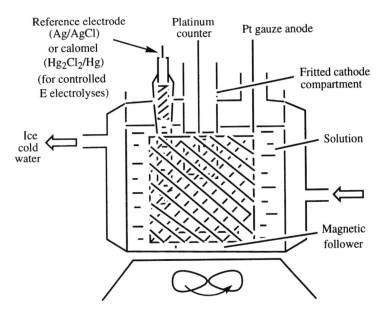

Figure 1.38. Platinum gauze electrochemical cell for oxidations

(b) Dimeric: $[(H_2O)_4Mo\overset{\overset{\displaystyle H}{\overset{\displaystyle |}{O}}}{\underset{\underset{\displaystyle H}{\underset{\displaystyle |}{O}}}{}}Mo(OH_2)_4]^{4+}$

Dimeric precursor is the Mo^V dimeric aqua ion:

(1.79)

$$Mo^{VI}O_4^{2-} \xrightleftharpoons[\text{H}^+ / \text{H}_2\text{O}]{\text{N}_2\text{H}_4} [(H_2O)_3Mo\overset{\overset{\displaystyle O}{\|}}{\underset{\underset{\displaystyle O}{}}{}}\overset{O}{\underset{}{}}\overset{\overset{\displaystyle O}{\|}}{Mo(OH_2)_3}]^{2+}$$

$$\text{'Mo}^{5+}\text{(aq)'} \qquad \text{yellow}$$

Zn / Hg ⟍ H$^+$ / H$_2$O   Zn / Hg
                             H$^+$ / H$_2$O

$$[(H_2O)_4Mo\overset{\overset{\displaystyle H}{\overset{\displaystyle |}{O}}}{\underset{\underset{\displaystyle H}{\underset{\displaystyle |}{O}}}{}}Mo(OH_2)_4]^{4+}$$

blue-green

(c) Trimeric: $[Mo_3(OH)_4(OH_2)_{9 \text{ or } 10}]^{5+}$

Route goes via the **trimeric Mo^{4+} aq ion:**

$$
\begin{array}{ccc}
\text{Mo}^{III}\text{(aq)} & & (OH_2)_3 \quad 4+ \\
\underset{\substack{\| \quad O \quad \| \\ [(H_2O)_3Mo \diagdown \diagup Mo(OH_2)_3]^{2+} \\ O}}{O \diagdown \diagup O} & \xrightarrow[H^+/H_2O]{\Delta} & \underset{\substack{| \quad O \quad | \\ [(H_2O)_3Mo \diagdown \diagup Mo(OH_2)_3 \\ O}}{\overset{\text{Mo}}{O \diagup | \diagdown O}} \\
\text{'Mo}^{5+}\text{(aq)'} & & \text{'Mo}^{4+}\text{(aq)'} \quad \text{red}
\end{array} \qquad (1.80)
$$

(note that Mo^{III}(aq) can be of any form) $H^+ / H_2O \parallel {e^- \text{ or} \atop Zn/Hg}$

$$
\begin{array}{ccc}
(OH_2)_4 \quad 5+ & & (OH_2)_3 \quad 5+ \\
\text{Mo} & & \text{Mo} \\
HO \diagup H \diagdown OH & \xrightleftharpoons[(H^+)]{H_2O} & HO \diagup | \diagdown OH \\
| \quad O \quad | & & | \quad OH \quad | \\
[(H_2O)_3Mo \diagdown \diagup Mo(OH_2)_3 & & [(H_2O)_3Mo \diagdown \diagup Mo(OH_2)_3 \\
O & & O \\
H & & H \\
 & \text{green} &
\end{array}
$$

In (1.78) use of the more labile $[Mo(HCO_2)_6]^{3-}$ is preferred for preparing monomeric $[Mo(OH_2)_6]^{3+}$ since it aquates rapidly, restricting hydrolytic polymerisation, and because the weak acid HCO_2H is very easily removed on the cation-exchange column. Note also that the use of $Mo(CF_3SO_3)_3$ is an alternative precursor but unlike in the case of Ti^{3+} and V^{3+} [125] the triflate of Mo^{III} has to be prepared via refluxing $Mo(CO)_6$ under air-free conditions in triflic acid [126]. This is because '$MoCl_3$' unlike $TiCl_3$ and VCl_3, exists as a covalent M—M bonded cluster species. The structural diversity apparent for the various aqua ions of Mo^{III}, Mo^{IV} and Mo^{V} leads to rather slow redox interconversions. For example, air oxidation of trimeric aqua Mo^{IV} is slow because of the necessitated dissociation of one Mo atom in the reassembly of dimeric Mo^{V}. Similarly, solutions of dimeric aqua Mo^{III}, $Mo_2(OH)_2^{4+}$(aq), rapidly air-oxidise to yellow dimeric Mo^{V} but under restricted access to oxygen turn red-brown owing to the formation of trimeric aqua Mo^{IV}. This is because the dimeric aqua Mo^{V} is then formed in the presence of Mo^{III}, allowing time for it to react in a comproportionation reaction to give trimeric aqua Mo^{IV} as shown in (1.81).

The 'incomplete cuboidal' terminology often used to describe trimeric Mo^{IV}(aq) species is clearly apparent in (1.81).

A further method, developed for the preparation of concentrated solutions of the redox and Cl$^-$ stable cluster species $Mo_3S_4^{4+}$(aq) and $Mo_3(S)O_3^{4+}$(aq) in ClO_4^- media for ^{17}O NMR kinetic measurements of water exchange (Section 6.2.3), involves elution from a cation-exchange column with HCl, forming a mixture of aqua-chloro complexes, which is then evaporated to dryness to

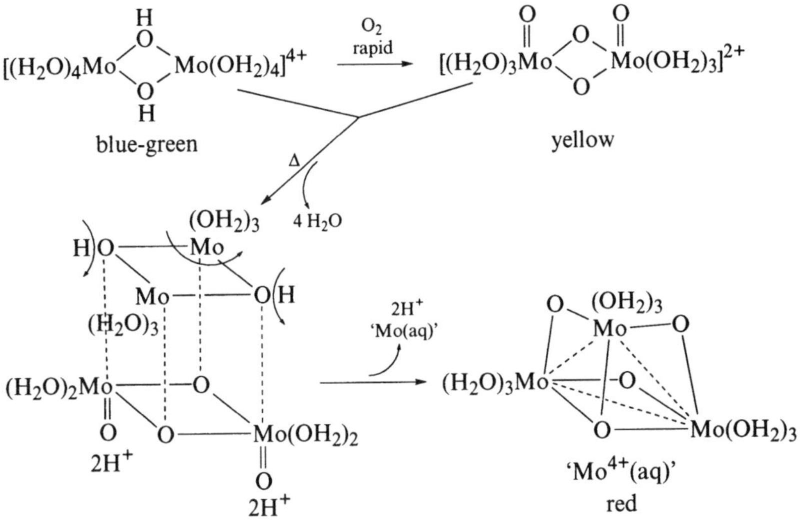

$$[(H_2O)_4Mo\underset{\underset{H}{O}}{\overset{\overset{H}{O}}{<}}Mo(OH_2)_4]^{4+} \xrightarrow[\text{rapid}]{O_2} [(H_2O)_3Mo\underset{O}{\overset{\overset{O}{\|}\quad\overset{O}{\|}}{<}}Mo(OH_2)_3]^{2+}$$

blue-green   yellow

(1.81)

[cluster structure diagram with $(OH_2)_3$, Mo, $(H_2O)_3$, $(H_2O)_2Mo$, $Mo(OH_2)_2$, $2H^+$ labels]

$\xrightarrow[\text{'Mo(aq)'}]{2H^+}$ $(H_2O)_3Mo$ [cluster] $Mo(OH_2)_3$

'Mo^{4+}(aq)'
red

leave a residue which can be dissolved in the required deoxygenated medium of choice, $HClO_4/NaClO_4$ (2.0 M) [90]. Here direct elution from the cation column with $HClO_4$ cannot produce high enough stock concentrations for the NMR studies (see below). A major drawback, however, is that amounts of Cl^- are always present. If the species is not redox active with Ag^+ the Cl^- can simply be removed as AgCl by adding AgX ($X = CF_3SO_3^-$ or pts$^-$, etc.). A further drawback is that the aqua cluster must be stable in the presence of a high concentration of Cl^-. This is not always the case, as exemplified by the behaviour of the labile tetrameric Ru^{4+}(aq) ion (see below and Section 8.2.3). On the other hand, the chloride residue method has allowed concentrated aqueous solutions of a niobium 'aqua ion' cluster species to be prepared, leading to ^{17}O NMR characterisation (see Fig. 1.17 and Section 5.2).

For tungsten only the W^{4+} and W^{5+} states form aqua ions and they have identical trimeric and dimeric oxo-bridged structures respectively to those formed by molybdenum. In this case the aquation of a simple salt in the required oxidation state is the method of choice:

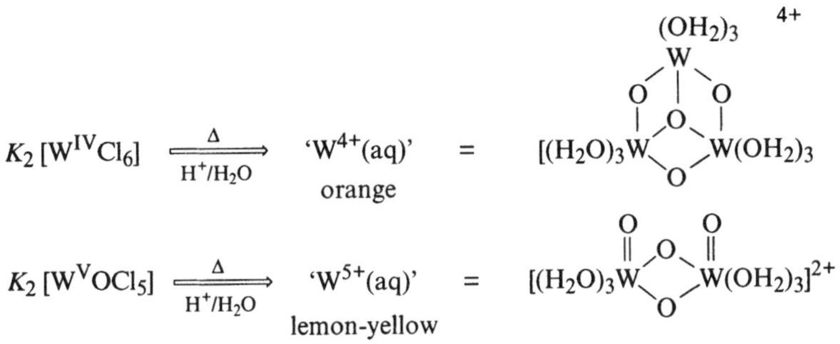

$$K_2[W^{IV}Cl_6] \xrightleftharpoons[H^+/H_2O]{\Delta} \text{'}W^{4+}\text{(aq)'} = [(H_2O)_3W\underset{O}{\overset{\overset{O}{|}}{<}}W(OH_2)_3]^{4+}$$

orange

(1.82)

$$K_2[W^VOCl_5] \xrightleftharpoons[H^+/H_2O]{\Delta} \text{'}W^{5+}\text{(aq)'} = [(H_2O)_3W\underset{O}{\overset{\overset{O}{\|}\quad\overset{O}{\|}}{<}}W(OH_2)_3]^{2+}$$

lemon-yellow

since similarly favourable comproportionation equilibria are not relevant (Section 6.3). The routes described in (1.82) also work for the corresponding molybdenum species.

Ruthenium, like molybdenum, is an element with a rich redox chemistry and this is reflected in a rich aqua ion chemistry also. Like molybdenum and tungsten there are no simple routes via aquation of a simple commercially available salt. Commercial 'ruthenium trichloride' is not $RuCl_3 \cdot nH_2O$, as the label often implies, but a mixture of Ru^{III} and Ru^{IV} oxo-halo and even nitrido species. The simplest approach is to oxidise '$RuCl_3 \cdot nH_2O$' to the tetraoxide, the reduction of which under the appropriate conditions, and with the appropriate reagents, can generate oxo or aqua ion species representing all of the lower valent states, Ru^{VII} down to Ru^{II}:

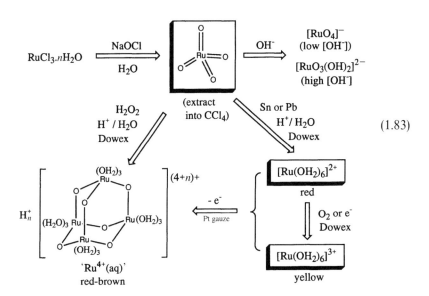

$$(1.83)$$

For iron the formation of $[Fe(OH_2)_6]^{3+/2+}$ in acid solution and oxo Fe^{VI} (as tetrahedral $[FeO_4]^{2-}$) in strong alkali has some similarities with ruthenium, but in contrast there is no Fe^{IV} aqueous chemistry in either acidic or alkaline media and no formation of FeO_4. Ruthenium also contrasts with osmium which, despite forming a similar tetraoxide OsO_4, has surprisingly no corresponding cationic aqueous chemistry of water-derived species of the lower valent states (see Section 8.3).

For rhodium only Rh^{2+}(aq) and Rh^{3+}(aq) are relevant, but these strongly contrast with cobalt in terms of redox behaviour (Chapter 9):

Rh^{2+}(aq): not $[Rh(OH_2)_6]^{2+}$, cf. Co
but the **M—M bonded dimer** $[Rh_2(OH_2)_{10}]^{4+}$

Note that there are no simple Rh$_2^{4+}$ salts for aquation and Rh^{II}(aq) is readily oxidised to Rh^{III}(aq), even with H$_2$O as ligand.

Preparation of Rh$_2^{4+}$(aq) requires reduction of Rh^{III}(aq) complexes with Cr^{2+}(aq):

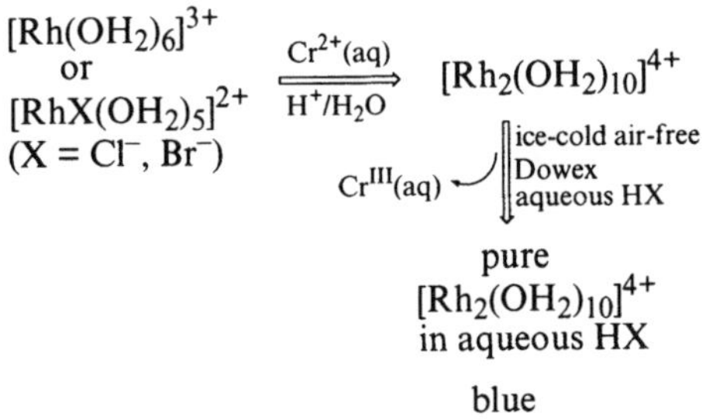

$$[Rh(OH_2)_6]^{3+} \atop \text{or} \atop [RhX(OH_2)_5]^{2+} \atop (X = Cl^-, Br^-) \quad \xrightarrow[{H^+/H_2O}]{Cr^{2+}(aq)} \quad [Rh_2(OH_2)_{10}]^{4+}$$

$$Cr^{III}(aq) \curvearrowleft \Big| \begin{array}{l} \text{ice-cold air-free} \\ \text{Dowex} \\ \text{aqueous HX} \end{array}$$

$$\begin{array}{c} \text{pure} \\ [Rh_2(OH_2)_{10}]^{4+} \\ \text{in aqueous HX} \end{array} \qquad (1.84)$$

blue

Rh^{3+}(aq): $[Rh(OH_2)_6]^{3+}$

Preparation involves heating RhCl$_3$·nH$_2$O in 'fuming' HClO$_4$ followed by dilution and Dowex chromatography:

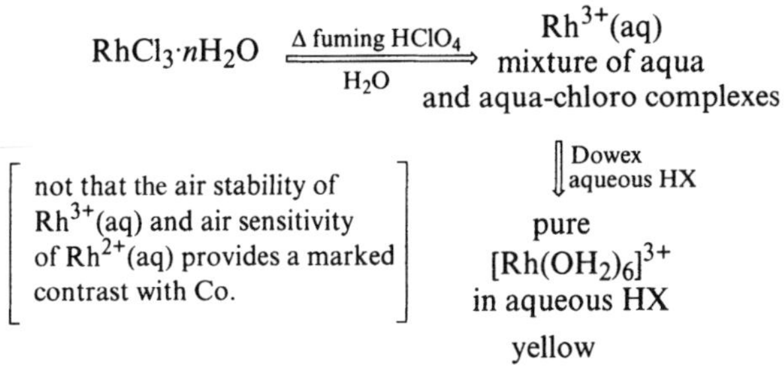

$$RhCl_3 \cdot nH_2O \xrightarrow[{H_2O}]{\Delta \text{ fuming HClO}_4} \begin{array}{c} Rh^{3+}(aq) \\ \text{mixture of aqua} \\ \text{and aqua-chloro complexes} \end{array}$$

$$\left[\begin{array}{l} \text{not that the air stability of} \\ Rh^{3+}(aq) \text{ and air sensitivity} \\ \text{of } Rh^{2+}(aq) \text{ provides a marked} \\ \text{contrast with Co.} \end{array}\right] \qquad \begin{array}{c} \Big| \begin{array}{l} \text{Dowex} \\ \text{aqueous HX} \end{array} \\ \text{pure} \\ [Rh(OH_2)_6]^{3+} \\ \text{in aqueous HX} \end{array} \qquad (1.85)$$

yellow

Examples are now given of the use of **'active' hydroxides** (inert metal ions) in the isolation of the products of hydrolytic polymerisation:

Example 1: Cr^{3+}(aq)

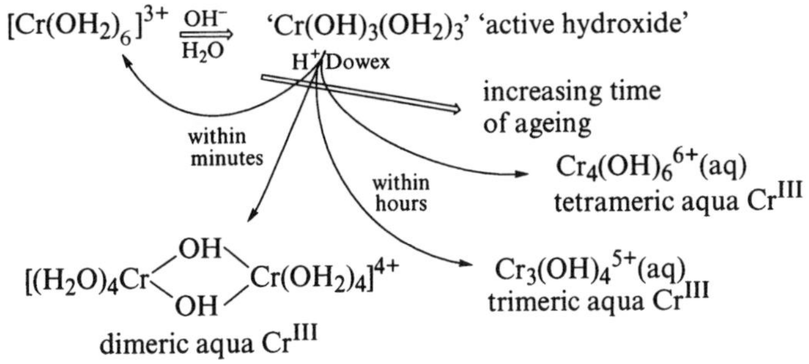

$$[Cr(OH_2)_6]^{3+} \xrightleftharpoons[{H_2O}]{OH^-} \text{'Cr(OH)}_3(OH_2)_3\text{'} \text{ 'active hydroxide'}$$

H$^+$/Dowex

increasing time of ageing

within minutes

within hours

$$Cr_4(OH)_6^{6+}(aq)$$
tetrameric aqua Cr^{III} $\qquad$ (1.86)

$$[(H_2O)_4Cr{\overset{OH}{\underset{OH}{<}}}Cr(OH_2)_4]^{4+}$$
dimeric aqua Cr^{III}

$$Cr_3(OH)_4^{5+}(aq)$$
trimeric aqua Cr^{III}

(note that similar behaviour is apparent for Rh^{3+}(aq) but with longer ageing times.)

Example 2: Ir^{3+}(aq)
The preparation of $[Ir(OH_2)_6]^{3+}$ requires the formation of the 'active' hydroxide. Here use is made of the greater lability of Ir—Cl bonds in alkaline solution:

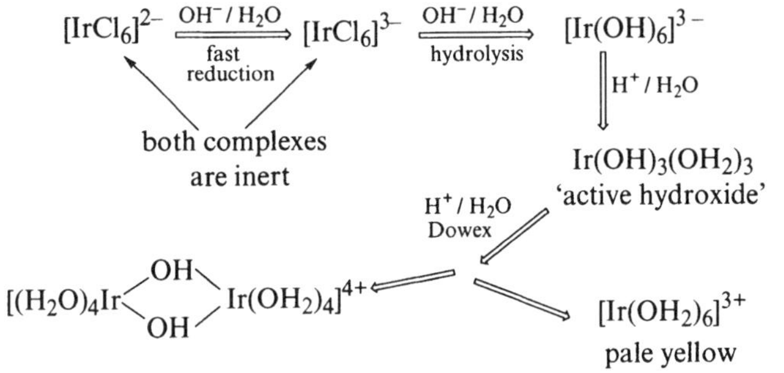

$$[IrCl_6]^{2-} \underset{\substack{\text{fast} \\ \text{reduction}}}{\overset{OH^-/H_2O}{\rightleftharpoons}} [IrCl_6]^{3-} \xrightarrow[\text{hydrolysis}]{OH^-/H_2O} [Ir(OH)_6]^{3-}$$

both complexes are inert

$\Big\Updownarrow H^+/H_2O$

$Ir(OH)_3(OH_2)_3$ 'active hydroxide' $\qquad$ (1.87)

H^+/H_2O Dowex

$[(H_2O)_4Ir\overset{OH}{\underset{OH}{\diagdown\diagup}}Ir(OH_2)_4]^{4+} \rightleftharpoons \qquad \Longrightarrow [Ir(OH_2)_6]^{3+}$
pale yellow

The use of different preparative routes for palladium and platinum aqua ions (see Chapter 10) reflects some differences between the two metals. For palladium the active hydroxide route or direct H^+ catalysed aquation can be used:

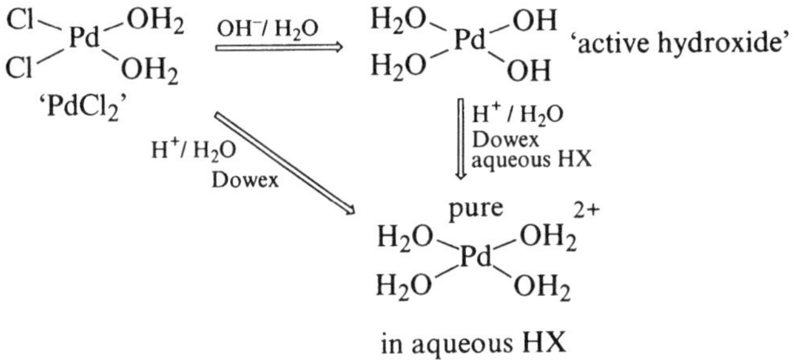

$$\underset{Cl}{\overset{Cl}{\diagdown}}Pd\underset{OH_2}{\overset{OH_2}{\diagup}} \xrightarrow{OH^-/H_2O} \underset{H_2O}{\overset{H_2O}{\diagdown}}Pd\underset{OH}{\overset{OH}{\diagup}} \text{ 'active hydroxide'}$$

'PdCl$_2$'

H^+/H_2O Dowex

$\Big\Updownarrow \begin{array}{l} H^+/H_2O \\ \text{Dowex} \\ \text{aqueous HX} \end{array}$

$\qquad\qquad (1.88)$

pure $\qquad$ 2+
$\underset{H_2O}{\overset{H_2O}{\diagdown}}Pd\underset{OH_2}{\overset{OH_2}{\diagup}}$

in aqueous HX

For platinum the complexes $[PtCl(OH_2)_3]^+$ and $[PtCl(OH)(OH_2)_2]$ are very inert so the preferred method is Ag^+(aq) catalysed aquation in acid solution to remove final Cl^-:

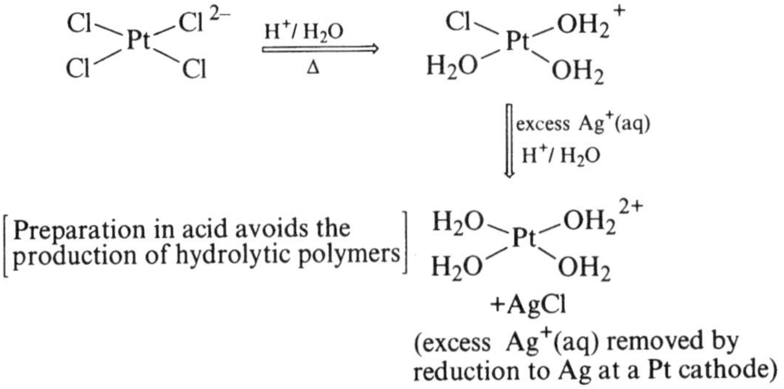

$$\underset{Cl}{\overset{Cl}{\diagdown}}Pt\underset{Cl}{\overset{Cl}{\diagup}}{}^{2-} \xrightarrow[\Delta]{H^+/H_2O} \underset{H_2O}{\overset{Cl}{\diagdown}}Pt\underset{OH_2}{\overset{OH_2}{\diagup}}{}^{+}$$

$\Big\Updownarrow \begin{array}{l} \text{excess } Ag^+(aq) \\ H^+/H_2O \end{array}$

$\qquad\qquad (1.89)$

$\left[\begin{array}{l}\text{Preparation in acid avoids the} \\ \text{production of hydrolytic polymers}\end{array}\right]$ $\underset{H_2O}{\overset{H_2O}{\diagdown}}Pt\underset{OH_2}{\overset{OH_2}{\diagup}}{}^{2+}$

$+AgCl$
(excess Ag^+(aq) removed by reduction to Ag at a Pt cathode)

Aqua ions for the lanthanides can be made by simple dissolution of hydrated salts since exchanges rates are fast and the complexes are highly labile:

Trivalent lanthanides:

$$LnX_3 \cdot nH_2O \ (X = Cl^-, Br^-, NO_3^-, ClO_4^-) \xrightarrow{H_2O} [LnOH_2)_n]^{3+}$$

$$n = 9 \ (La, Ce, Pr, Nd, Pm)$$
$$n = 8 \ (Tb, Dy, Ho, Er, Tm, Yb, Lu)$$

Divalent lanthanides: (1.90)

$$[Ln(OH_2)_n]^{3+} \xrightarrow[H^+/H_2O]{e^-} [Ln(OH_2)_n]^{2+}$$

$$Ln = Eu \ (Hg \ pool \ or \ Zn \ / \ Hg)$$ (1.91)
$$Ln = Yb \ (Hg \ pool \ or \ Na \ / \ Hg)$$

Tetravalent lanthanides:

$$\begin{pmatrix} (NH_4)_2[Ce(NO_3)_6] \\ (NH_4)_2[Ce(SO_4)_3] \end{pmatrix} \xrightarrow[H^+/H_2O]{} [Ce(OH_2)_9]^{4+} \qquad (1.92)$$

$$[Ce(OH_2)_9]^{3+} \xrightarrow[\substack{Pt \ gauze \\ H^+/H_2O}]{e^-} [Ce(OH_2)_9]^{4+}$$

(1.93)

1.5.2 Tools of the Trade: Purification and Manipulation Techniques

1.5.2.1 CATION-EXCHANGE CHROMATOGRAPHY

Cation-exchange chromatography has been highlighted as important for the separation and purification of many of the aqua ion species described above. Two types of cation-exchange resin are in common use, the sulfonated cross-linked polystyrene-based resins such as those marketed under the names Dowex® 50W (Dow Chemical Co.), Amberlite® IR (Rohm and Haas Co.) and BioRad AG and the polysaccharide-based resins under the name Sephadex® SP C. For monovalent and divalent ions and a few trivalent ions, where hydrolysis is not extensive and neutral and low acidity solutions can be tolerated, use of the Sephadex resins is advantageous since it combines cation exchange with gel filtration, helping the separation of species on the basis of size as well as charge. However, prolonged exposure to 1.0 M [H$^+$] or higher can cause extensive hydrolytic break-up of the sugar polymer and for separations in strongly acid and strongly alkaline solutions use of the hydrolytically stable polystyrene based resins is essential. The most commonly used cation resin in this regard is Dowex 50W. A diagrammatical representation of the chemical structure within Dowex resin is shown in Fig. 1.39. Polysulfonated polystyrene itself is water soluble, forming linear chains. The

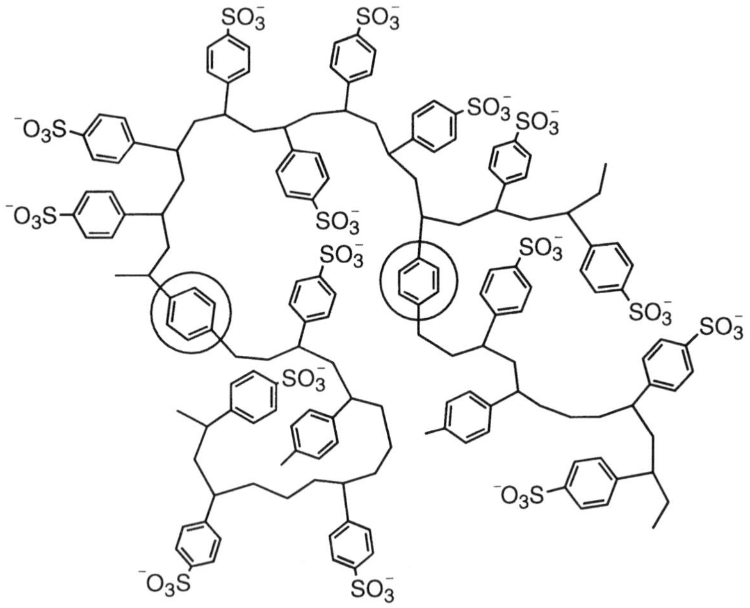

Figure 1.39. A section of the polymeric structure within Dowex 50W sulfonated polystyrene resin

three-dimensional insoluble material is made by copolymerisation of styrene and divinyl benzene. The % cross-linking reflects directly the amount of divinyl benzene used in the copolymerisation process. Upon contact with water the resin material swells as water molecules are taken up. This is particularly noticable in the highly hydrated Sephadex resins which swell to form a gel-like material. In the polystyrene-based resins the swelling factor is related to the degree of cross-linking. The lower the % cross-linking the greater is the swelling factor. High swelling factor resins such as Dowex 50W X2 shrink noticeably upon cation loading and upon contact with aqueous solutions of high ionic strength. This is due to the extrusion of water molecules from the resin as a result of their greater affinity (through hydration) for the dissolved ions in the aqueous phase.

Polystyrene-based resins such as Dowex are normally used in the form of a column but a stirred slurry of resin in contact with the species of interest can also be employed. The resin is usually prepared for use in its fully protonated (H^+) form by passing a solution of $\geqslant 2.0\,M\,H^+$ down a column of the resin, which is then washed with distilled water until the washings are back to neutral pH.

Various methods for recovering and regenerating Dowex resins have been documented. The polystyrene polymer will survive the extreme conditions of pH from $10.0 M$ H^+ to $> 4.0 M$ OH^-; however, strongly oxidising concentrated acids should be avoided. A common method of purification is to stir the resin in an alkaline 10% H_2O_2 solution for a few hours, gently warming if necessary, followed by repeated decantation until bubbles of O_2 are no longer seen. The resin is then poured into a column where it is further washed by passing down successively $200 \, cm^3$ each of $2.0 M$ NaOH, distilled water, $5.0 M$ HCl and then finally distilled water again until the washings are back to neutral pH.

The principle of cation-exchange chromatography is very simple. The cations take part in a dynamic equilibrium between the sulfonate groups on the resin and the hydrating water molecules in the aqueous phase. The relative affinity of dissolved cations for the resin depends on the magnitude of the cationic charge or more correctly the charge density, since the interactions are viewed as largely electrostatic in nature. Thus H^+ ions have a relatively low affinity for the resin, but this may be higher than the alkali metal ions for example. However, the equilibria can be driven over to either the H^+ or M^+ form of the resin by using concentrated solutions of the metal ion or of H^+. The metal ion solutions are usually loaded at low $[H^+]$, allowing them to displace the $[H^+]$ on the column. The progress down the column depends on the cationic charge so the higher charged ions are retained more strongly and appear further up the column (Fig. 1.40).

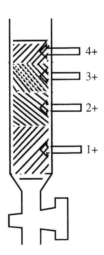

Figure 1.40. Separation of cations on the basis of charge on a typical cation-exchange resin column

Elution of the successive bands can be achieved with increasing concentrations of $[H^+]$. There is no rule of thumb. Trial and error is the only way, but a rough guide is that $1+$ ions elute from a Dowex 50W X2 column with ~ 0.01–0.05 M H^+, $2+$ ions with ~ 0.2–0.5 M H^+, $3+$ ions with ~ 1.0 M H^+ and $4+$ ions with $\geqslant 2.0$ H^+, etc. Use of higher $[H^+]$ concentrations on saturated columns can be useful for the purpose of concentrating already pure stock solutions. However, in some cases the $[H^+]$ elution method does not work. A good example is tetrameric $Ru^{4+}(aq)$. Here protonation of oxo bridges, resulting in an increase in effective cationic charge of the species, accompanies an increase in $[H^+]$ such that elution with $[H^+]$ alone cannot be used. In this case elution is carried out at low $[H^+]$ (~ 0.1 M), with a solution of a suitable highly charged displacing cation such as La^{3+} (Section 8.2.3).

It is good practice to determine the number of H^+ sites per gram in a sample of dry resin, termed the capacity. This can be achieved by treating the H^+ form of a weighted quantity of dried resin with an excess of OH^- ions and back titration. Sometimes this type of information is supplied. As well as its use in purification and isolation, cation-exchange chromatography can also be used for the standardisation of stock salt solutions and for the determination of the cationic charge per metal.

Metal salts are frequently obtained with varying degrees of hydration (wet!). Determination of the number of hydrating waters is required if a stock solution of the salts is to be accurately made up. Very simply a weighed amount of the hydrated salt is made up to an approximately known stock concentration. Dilution, if necessary, to around 0.1 M is performed and then a measured amount, say about 25 cm^3, is loaded on to a thoroughly washed 25 cm $\times$ 1.5 cm column of cation resin in the H^+ form. Here a coarse resin (50–100 mesh size) with a high degree of cross-linking, such as Dowex 50W X8 50–100 (8% cross-linked), Amberlite IR120 or BioRad AG X8, is used since a reasonably rapid flow rate is useful here. It is important to ensure that saturation is not achieved and that the flow rate is sufficiently slow so as to retain all of the M^{n+} ion on the column. The displaced H^+ (one mole for every mole of M^+, two moles for every mole of M^{2+}, etc.) is then titrated with standard 0.1 M NaOH. The same method can be used to standardise solutions of aqua ions themselves that elute with acid from a cation-exchange column since the quantity of background H^+ can be determined if the amount of the $M^{n+}(aq)$ ion (determined by other means, e.g. spectroscopic) is known. Sometimes the background $[H^+]$ can be estimated by accurate dilution of the acidic eluent into the normal glass electrode pH measuring range (pH $\geqslant 1$).

Determination of the charge/metal has become an important analytical application for cation-exchange chromatography in the elucidation of the degree of polymerisation in certain inert hydrolytic polymers. The classical example is that of $Cr^{III}(aq)$, although the method was first developed by Cady and Connick [137] in the 1950 s for the investigation of the charge on the ruthenium ions $Ru^{3+}(aq)$,

$RuCl^{2+}$(aq) and $RuCl_2^+$(aq). Its use for investigating the hydrolytic polymers of Cr^{III} is discussed in Section 6.1.2. It is important and sometimes essential in charge/metal studies to keep the background H^+ as low as can be tolerated so that it does not swamp the amount of H^+ generated by the cation charge displacement, leading to errors. For this reason the Cr^{III} hydrolytic polymers were separated using an $HClO_4/NaClO_4$ displacing eluent on Sephadex SP C25 cation-exchange resin, thereby keeping the background $[H^+]$ as low as possible.

The preferred type of Dowex resin depends very much on the use. For separating charged species $\geqslant 4+$ from lower charged species, for large-scale separations or for concentrating dilute samples, the best polystyrene-based resin is one with a small mesh size and a high swelling factor, e.g. Dowex 50W X2 200–400. It will be the one most commonly referred to in the preparative procedures. Since energy is required to swell the resin, hydrated ions occupying the smallest volume will tend to be selected as the swelling increases. Hence the selectivity for lower charged ions should increase with the degree of cross-linking. As a result a resin such as Dowex 50W X8 (8%) may be more desirable for separating lower charged species, e.g. $3+$ ions from $2+$ ions and $2+$ from $1+$, etc., at least in the first instance. If a fast flow rate is desirable, e.g. in a charge/metal determination, a salt solution standardisation or where the cation separation is clear cut, a coarse resin size (e.g. 50–120 mesh) can be employed. The choice of mesh size depends upon the balance between flow rate vs a lowering of surface area and column capacity.

Basically, as in any form of column chromatography, one desires, for good separation, (a) a high ion-exchange rate (fast attainment of equilibration), (b) a low migration rate, (c) a small particle size, (d) a low flow rate and (e) an adequate column length. Choosing the dimensions of the column is also important, i.e. short and fat vs long and thin. There is no real rule of thumb. The proof is in the result and it is worth while experimenting with different resins and column dimensions. A general rule, however, is that one should avoid overloading a column and only saturate for the purpose of concentrating or changing the acid counter-anion unless the separation is clear cut (i.e. where an immovable polymer is the only contaminant). For good separations a general guide is not to use more than half of the resin bed volume for loading.

1.5.2.2   Air-free handling techniques [128]

Using rubber septum caps, stainless steel needles, narrow gauge teflon tubing and gas-tight syringes it is possible to manipulate and transfer air-sensitive aqueous solutions from one vessels to another or for the purpose of characterisation/ standardisation or for running a reaction. Some of the basic techniques and tools are illustrated in Fig. 1.41. For the purpose of simple transference of a solution the syphoning method, using a positive pressure of inert gas (purified N_2 or Ar), is the

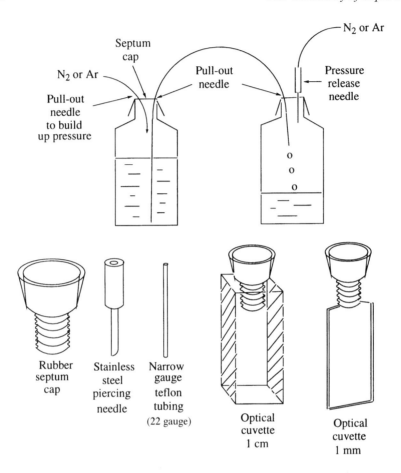

Figure 1.41. Basic syphoning method and some of the tools used

most convenient for avoiding air contamination. The stainless steel needles have an inner diameter to allow the narrow-gauge teflon tubing to pass comfortably through it, allowing release of pressure around the side. Withdrawing the needle to leave the tubing in the rubber allows a positive pressure of inert gas to build up, which can be used in the syphoning to another vessel or into an optical cuvette, NMR tube, etc. To seal up a sample for long-term storage the needle is withdrawn to build up a little pressure and then the tubing is removed. The hole in the cap can then be sealed with rubber adhesive.

A convenient way to deoxygenate aqueous solutions is to purge with the inert gas using teflon tubing. This is more convenient than distillation which is time consuming unless a special continuous flushing still is set up and then one only produces the pure water solvent. The distillation method is also expensive and wasteful on inert gas. It is better to prepare doubly distilled water first, use it for making up the required solution and then deoxygenate by purging and seal. The purging time depends on the volume of solution and the rate of gas flow. It is always better to be generous, allowing four hours for N_2 purging of a litre of solution, three hours for $500\,cm^3$, two hours for $250\,cm^3$ and one hours for $100cm^3$ as a rough guide for a gas flow of 10 bubbles per second. Use of argon at the same flow roughly halves the purging time.

For cation-exchange chromatography of air-sensitive species deoxygenating a slurry of the resin within an ice-cold water jacketted column by purging with the inert gas works as a convenient method (Fig. 1.42). The jacketted column itself can be made simply by modifying a normal water condenser. The teflon tubing is passed through a stainless steel needle in a septum cap in the reservoir funnel through the tap and down through the slurry of resin to touch the sinter at the bottom. It is then connected to the gas cylinder. At the end of the purging time the tubing is withdrawn back into the reservoir, allowing the resin to slowly settle into a tightly packed bed ready for the loading of the air-sensitive aqueous solution. Transfer of the air-sensitive solution to the reservoir is performed by syphoning. The jacketted column in Fig. 1.42 could contain either cation-exchange resin or a slurry of amalgamated zinc (Jones reductor).

1.5.2.3 AIR-FREE FAST REACTION TECHNIQUES

Figure 1.43 shows a diagrammatical representation of the drive syringe set up at the heart of a typical stopped-flow rapid-mixing spectrophotometer. The stopped-flow spectrophotometer first developed by Eigen and then by Wilkins has proved a versatile piece of equipment for the monitoring of the many ligand substitution and redox reactions that take place in the second to 0.1 ms timescale. The principal of stopped-flow monitoring is simple. The two reactant stock solutions are loaded into the separate drive syringes which are then driven simultaneously by a pneumatic ram. Following mixing, the two solutions then pass into a monitoring cell usually containing two optically transparent windows for absorbance measurements, sometimes with a third window at 90° if the monitoring of fluorescence is required. The moving solution then flows into a stop syringe, pushing its barrel up against a plate which sets off an electronic trigger activating a data capture device. The entire process from syringe drive to acquisition time usually takes no more than a few ms. The absorbance from a disappearing reactant, an intermediate or an appearing product species have all been followed for the purpose of monitoring the kinetics of the reaction process.

To be versatile the stopped-flow instrument has to be capable of handling air-sensitive reactants. Figure 1.43 shows the loading of the two drive syringes by

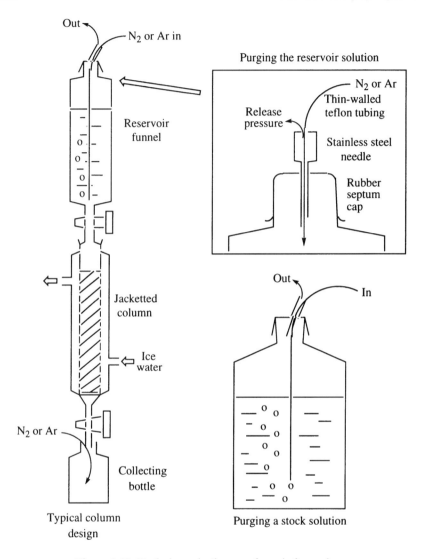

Figure 1.42. Techniques in the use of an air-free column

simple syphoning from purged stock solutions of the two reagents. To ensure no leakage in of air around the syringe barrels these themselves are often purged with inert gas through special entry ports as shown. Flushing out the system with a little of the reducing ion solution serves further to deoxygenate the system prior

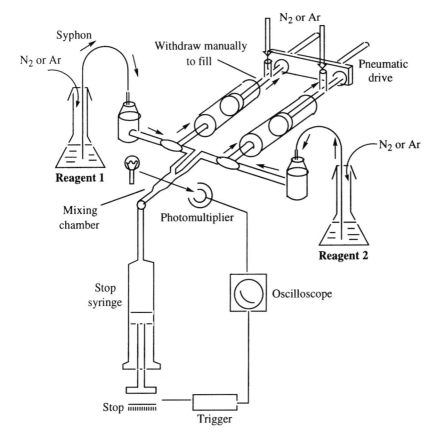

Figure 1.43. A stopped-flow spectrophotometer set up for handling air-sensitive solutions

to the run. Fast reactions often have the added advantage of taking place on a timescale faster than the reaction with O_2. A nice example of the use of the air-sensitive stopped-flow set-up shown in Fig. 1.43 is in the monitoring of the rapid outer-sphere oxidation of $[Mo(OH_2)_6]^{3+}$ by $[IrCl_6]^{2-}$ [129] (see also Section 6.2.3). Here air-sensitive stock solutions of $Mo^{3+}(aq)$ are stored in ice-cold flask and then $2\,cm^3$ amounts are syphoned into the drive syringes via narrow-gauge teflon tubing. A range of redox reactions involving reduction of co-reagents by the air-sensitive ions $[Cr(OH_2)_6]^{2+}$, $[V(OH_2)_6]^{2+}$ and $[Eu(OH_2)_9]^{2+}$ have been followed using air-free rapid mixing stopped-flow spectrophotometry.

REFERENCES

[1] (a) Burgess, J., *Metal Ions in Solution*, Ellis Horwood, Chichester, 1978.
     (b) Burgess, J., *Ions in Solution*, Ellis Horwood, Chichester, 1986.
[2] (a) Marcus, Y., *Ion Solvation*, Ellis Horwood, Chichester, 1986.
     (b) Magini. M., Licheri, G., Paschina, G., Piccaluga, G., and Pinna, G., *X-ray Diffraction of Ions in Aqueous Solution: Hydration and Complex Formation*, CRC Press, Boca Raton, Florida, 1988.
[3] Baes, C. F., and Mesmer, R. E., *The Hydrolysis of Cations*, Krieger, 1986, pp. 129–38.
[4] Marcus, Y., *Chem. Revs.*, **88**, 1475 (1988).
[5] Ohtaki, H., and Radnai, T., 'Structure and dynamics of hydrated ions', *Chem. Revs.*, **93**, 1157 (1993).
[6] Johansson, G., 'Structure of metal complexes in solution', *Adv. Inorg. Chem.*, **39**, 178 (1992).
[7] (a) Enderby, J. E., and Neilson, G. W., *Adv. Inorg. Chem.*, **34**, 195 (1989).
     (b) Enderby, J. E., 'The Solvation of Ions', *J. Phys. Condens. Matter*, F87–90 (1991).
     (c) Enderby, J. E., Cummings, S., Herkman, G. J., Neilson, G. W., Salmon, P. S., and Skipper, N., 'Diffraction and the study of aqua ions', *J. Phys. Chem.*, **91**, 5851 (1987).
[8] Philips, C. S. G., and Williams, R. J. P., *Inorganic Chemistry*, Claredon Press, Oxford, 1965.
[9] (a) Powell, D. H., Neilson, G. W., and Enderby, J. E., *J. Phys. Condens Matter*, **5**, 5723 (1993) and *New. J. Chem.*, **19**, 27 (1995).
     (b) Powell, D. H., Barnes, A. C., Enderby, J. E., Neilson, G. E., and Salmon, P. S., *Farad. Disc. Chem. Soc.*, **85**, 137 (1988).
     (c) Cummings, S., Enderby, J. E., Neilson, G. W., Newsome, J. R., Howe, R. A., Howells, W. S., and Soper, A. K., *Nature, Lond.*, **287**, 714 (1980).
[10] Creutz, C., and Sutin, N., in *Inorganic Reactions and Methods* (ed. J. J. Juckerman), Vol. 15, VCH, Deerfield Beach, Florida, 1986, pp. 13–51 (and refs. therein).
[11] (a) Marcus, R. A., *Ann. Rev. Phys. Chem.*, **15**, 555 (1964) and *J. Chem. Phys.* **43**, 670 (1965).
     (b) Hush, N. S., *Trans. Farad. Soc.*, **57**, 557 (1961), *Electrochim. Acta*, **13**, 1005 (1968) and *Prog. Inorg. Chem.*, **8**, 391 (1967).
     (c) Levich, V. G., *Adv. Electrochem. Eng.*, **4**, 249 (1966).
[12] (a) Sutin, N., *Acc. Chem. Res.*, **1**, 225 (1968), *Prog. Inorg. Chem.*, **30**, 441 (1983) and *J. Phys. Chem.*, **90**, 3465 (1986).
     (b) Ratner, M. A., and Levine, R. D., *J. Am. Chem. Soc.*, **102**, 4898 (1980).
     (c) Marcus, R. A., and Sutin, N., *Biochim. Biophys. Acta*, **811**, 265 (1985).
[13] (a) Cannon, R. D., *Electron Transfer Reactions*, Butterworths, London, 1980.
     (b) Sykes, A. G., *Kinetic of Inorganic Reactions*, Pergamon, Oxford, 1966.
[14] Vaalsta, T. P., and Meslen, E. N., *Acta. Cryst.*, **B43**, 448 (1987).
[15] Varghese, J. N., and Maslen, E. N., *Acta. Cryst.*, **B41**, 184 (1985).
[16] (a) Beattie, J. K., Best, S. P., Skelton, B. W., and White, A. H., *J. Chem. Soc. Dalton Trans.*, 2105 (1981).
     (b) Cromer, D. T., Kay, M. I., and Larson, A. C., *Acta. Cryst.*, **21**, 383 (1966).
     (c) Sygusch, J., *Acta. Cryst.*, **30**, 662 (1974).
     (d) Armstrong, R. S., Beattie, J. K., Best, S. P., Skelton, B. W., and White, A. H., *J. Chem. Soc. Dalton Trans.*, 1973 (1983).

(e) Best, S. P., and Forsyth, J. B., *J. Chem. Soc. Dalton Trans.*, 1721 (1991) and 3507 (1990).

[17] Best, S. P., Forsyth, J. B., and Tregannapiggott, P. L., *J. Chem. Soc. Dalton Trans.*, 2711 (1993).

[18] (a) Armstrong, R. S., Beattie, J. K., Best, S. P., Braithwaite, G. P., Delfavero, P., Skelton, B. W., and White, A. H., *Aust. J. Chem.*, **43**, 393 (1990).

(b) Best, S. P. and Forsyth, J. B., *J. Chem. Soc. Dalton Trans.*, 395 (1990).

[19] (a) Iversen, B. B., Larsen, F. K., Reynolds, P. A., and Figgis, B. N., *Acta. Chem. Scand.*, **48**, 800 (1994).

(b) Figgis, B. N., Reynolds, P. A., and Cable, J. W., *Mol. Phys.*, **80**, 1377 (1993).

(c) Figgis, B. B., Kucharski, E. S., Reynolds, P. A., and Tasset, F., *Acta. Cryst.*, **C45**, 942 (1989).

(d) Figgis, B. N., Kucharski, E. S., and Forsyth, J. B., *Acta. Cryst.*, **C47**, 419 (1991).

(e) Deeth, R. J., Figgis, B. N., Forsyth, J. B., Kucharski, E. S., and Reynolds, P. A., *Aust. J. Chem.*, **41**, 1289 (1988).

(f) Fender, B. E. F., Figgis, B. N., and Forsyth, J. B., *Aust. J. Chem.*, **39**, 1023 (1986).

[20] (a) Montgomery, H., Chastia, R. V., Natt, J. J., Witkowska, A. M., and Lingafelter, E. C., *Acta Cryst.*, **22**, 775 (1967).

(b) Montgomery, H., Chastain, R. V., and Lingafelter, E. C., *Acta Cryst.*, **20**, 731 (1966).

(c) Montgomery, H., and Lingafelter, E. C., *Acta Cryst.*, **17**, 1295, 1478 (1964) and **20**, 659 (1966).

[21] (a) Cotton, F. A., Daniels, L. M., Murillo, C. A., and Quesada, J. F., *Inorg. Chem.*, **32**, 4861 (1993).

(b) Cotton, F. A., Falvello, L. R., Murillo, C. A., and Quesada, J. F., *J. Solid State Chem.*, **96**, 192 (1992).

[22] Read, M. C., and Sandstrom, M., *Acta. Chem. Scand.*, **46**, 1177 (1993).

[23] Kristiansson, O., Lindgren, J., and de Villepin, J., *J. Phys. Chem.*, **92**, 2680 (1988).

[24] Powell, D. H., Neilson, G. W., and Enderby, J. E., *J. Phys. Condens. Matter*, **1**, 8721 (1989).

[25] (a) Salmon, P. S., Herdman, G. J., Lindgren, J., Read, M. C., and Sandstrom, M., *J. Phys. Condens. Matter*, **1**, 3459 (1989).

(b) Newsome, J. R., Neilson, G. W., Enderby, J. E., and Sandstrom, M., *Chem. Phys. Lett.*, **82**, 399 (1981).

[26] (a) Lee, P. A., Citrin, P. H., Eisenberger, P., Kincaid, B. M., *Rev. Mod. Phys.*, **53**, 769 (1981).

(b) Sham, T. K., *Acc. Chem. Res.*, **19**, 99 (1986).

(c) Teo, B. K., and Joy, D. C. (eds.), *EXAFS Spectroscopy: Techniques and Applications*, Plenum, New York (1981).

(d) Yamaguchi, T., Lindquist, O., Boyce, J. B., and Claeson, T., in *EXAFS and Near Edge Structure III*, (eds. K. O. Hodgson, B. Hedman, J. E. Pennerhahn), Springer-Verlag, Berlin, 1985.

[27] Ebsworth, E. A. V., Rankin, D. W., and Craddock, S., *Structural Methods in Inorganic Chemistry*, 2nd edn, Blackwell, Palo Alto, California, 1987.

[28] Bergstrom, P.-A., Lindgren, J., Read, M. C., and Sandstrom, M., *J. Phys. Chem.*, **95**, 7650 (1991).

[29] (a) Best, S. P., Beattie, J. K., and Armstrong, R. S., *J. Chem. Soc. Dalton Trans.*, 2611 (1984) and 1655 (1982).

(b) Best, S. P., and Clark, R. J. H., *Chem. Phys. Lett.*, **122**, 401 (1985).

(c) Best, S. P., Beattie, J. K., Armstrong, R. S., and Braithwaite, G. P., *J. Chem. Soc. Dalton Trans.*, 1771 (1989).

[30] Best, S. P., Armstrong, R. S., and Beattie, J. K., *J. Chem. Soc. Dalton Trans.*, 299 (1992).

[31] Gamsjager, H., and Murmann, R. K., *Adv. Inorg. Bioinorg. Mech.*, **2**, 317 (1983) (and refs therein).

[32] Crandall, H. W., *J. Chem. Phys.*, **17**, 602 (1949).

[33] (a) Banci, L., Bertini, I., and Luchinat, C., *Mag. Reson. Rev.*, **11**, 1, (1986).
(b) Fratiello, A., *Prog. Inorg. Chem.*, **17**, 57 (1972).

[34] Harris, R. K., and Mann, B. E. (eds.), *NMR and the Periodic Table*, Academic Press, New York, 1978.

[35] (a) Helm, L., Elding, L. I., and Merbach, A. E., *Helv. Chim. Acta*, **67**, 1453 (1984).
(b) Hugi-Cleary, D., Helm, L., and Merbach, A. E., *J. Am. Chem. Soc.*, **109**, 4444 (1987).

[36] Richens, D. T., Helm, L., Pittet, P. -A., and Merbach, A. E., *Inorg. Chim. Acta*, **132**, 85 (1987).

[37] (a) Feakins, D., O'Neill, R., and Waghorne, E., *Pure. Appl. Chem.*, **54**, 2317 (1982).
(b) Rutgers, A. T., and Hendriks, Y., *Trans. Farad. Soc.*, **58**, 2084 (1962).

[38] Gusev, N. I., *Russ. J. Inorg. Chem.*, **45**, 1268, 1455, 1575 (1971).

[39] Radnai, T., 'Structure and dynamics of solvated ions—new tendencies in research', *J. Mol. Liquids*, **65**, 229 (1995) (and refs. therein).

[40] (a) Wong, M. W., Frisch, M. J., and Wiberg, K. B., *J. Am. Chem. Soc.*, **113**, 4776 (1991) (and refs. therein).
(b) Wong, M. W., Wiberg, K. B., and Frisch, M. J., *J. Am. Chem. Soc.*, **114**, 799 (1992).

[41] (a) Rode, B. M., Islam, S. M., and Yongyai, Y., 'Computational methods in solution chemistry', *Pure. Appl. Chem.*, **63**, 1725 (1991) (and refs. therein).
(b) Rode, B. M., and Ialam, S. M., *J. Chem. Soc. Faraday Trans.*, 417 (1992) (and refs. therein).

[42] Yamaguchi, T., Ohtaki, H., Spohr, E., Palinkas, G., Heinzinger, K., and Probst, M. M., *Z. Naturforsch*, **A41**, 1175 (1986).

[43] Deeth, R. J., *Structure and Bonding*, **82**, 1 (1995).

[44] (a) Becke, A. D., *J. Chem. Phys.*, **96**, 2155 (1991), **97**, 9173 (1992) and **98**, 1372 (1993).
(b) Johnson, B. G., Gill, P. M. W., and Pople, J. A., *J. Chem. Phys.*, **98**, 5612 (1993).

[45] Tawa, G. J., Martin, R. L., Pratt, L. R., and Russo, T. V., *J. Phys. Chem.*, **100**, 1515 (1996).

[46] Hall, R. J., Davidson, M. M., Burton, N. A., and Hillier, I. H., *J. Phys. Chem.*, **99**, 921 (1995) (and refs. therein).

[47] (a) Heinzinger, K., *Computer Modelling of Fluids, Polymers and Solids*, Kluwer Academic Publishing, Dordrecht, 1990, p. 357.
(b) Heinzinger, K., and Palankas, G., *Interactions of Water in Ionic and Nonionic Hydrates*, Springer, Berlin, 1987, p. 1.
(c) Heinzinger, K., Bopp, P., and Jancso, G., *Acta Chim. Hung.*, **121**, 27 (1986).
(d) Heinzinger, K., *Pure Appl. Chem.*, **57**, 1031 (1985).

[48] Friedman, H. L., Raineri, F. O., and Hua, X., *Pure Appl. Chem.*, **63**, 1347 (1991).

[49] (a) Akesson, R., Pettersson, L. G. M., Sandstrom, M., Siegbahn, P., and Wahlgren, U., *J. Phys. Chem.*, **96**, 10773 (1992).
(b) Akesson, R., Pettersson, L. G. M., Sandstrom, M., and Wahlgren, U., *J. Am. Chem. Soc.*, **116**, 8691 (1994).

[50] (a) Kowall, T., Foglia, F., Helm, L., and Merbach, A. E., *J. Phys. Chem.*, **99**, 13078 (1995) and *J. Am. Chem. Soc.*, **117**, 3790 (1995).
(b) Helm, L., Foglia, F., Kowall, T., and Merbach, A. E., *J. Phys. Condens. Matter*, **6**, 137 (1994).
(c) Galera, S., Lluch, J. M., Oliva, A., Bertran, J., Foglia, F., Helm, L., and Merbach, A. E., *New. J. Chem.*, **17**, 773 (1993).

[51] Bjerrum, N., *Kungl. Dansk. Videnskab. Selskab. Mat.-Fys. Medd.*, **27**, 1 (1951).

[52] Ben–Naim, A., and Stillinger, F. H., in *Water and Aqueous Solutions*, (ed. R. A. Horne) Wiley Interscience, New York, 1972, p. 295.

[53] Stillinger, F. A., and Rahman, A., *J. Chem. Phys.*, **60**, 1545 (1974).

[54] Rowlinson, J. S., *Trans. Faraday. Soc.*, **47**, 120 (1951).

[55] Matsuoka, O., Clementi, E., and Yoshimine, M., *J. Chem. Phys.*, **64**, 1351 (1976).

[56] Metropolis, N., Rosenbluth, A. W., Teller, A. H., and Teller, E., *J. Chem. Phys.*, **21**, 1087 (1953).

[57] Floris, F., Persico, M., Tani, A., and Tomasi, J., *Chem. Phys.*, **195**, 207 (1995).

[58] Wood, W. W., and Erpenbeck, J., *J. Ann. Rev. Phys. Chem.*, **27**, 319 (1976).

[59] Beveridge, D. L., Mezei, M., Mehrotra, P. K., Marchese, F. T., Rahvi-Shanker, G., Vasu, T., and Swamanathan, S., in *Molecular-Based Studies of Fluids*, (eds. J. M. Haile and G. A. Mansoori), ACS Publications, Washington, DC, 1983.

[60] (a) Yongyos, Y. P., Kokpol, S., and Rode, B. M., *Chem. Phys.*, **156**, 403 (1991).
(b) Bounds, D., *Mol. Phys.*, **54**, 1335 (1985).
(c) Kneifel, C. L., Friedman, H. L., and Newton, M. D., *Z. Naturforsch, A***44**, 385 (1989).
(d) Rustad, J. R., Hay, B. P., and Halley, J. P., *J. Chem. Phys.*, **102**, 1 (1995).

[61] McCreevy, R. L., and Pusztai, L., *Mol. Simul.*, **1**, 359 (1988).

[62] (a) Howe, M. A., *J. Phys. Condens. Matter*, **2**, 741 (1990).
(b) Radnai, T., *J. Mol. Liquids*, **65**, 229 (1995).

[63] Jorgensen, C. K., *Inorganic Complexes*, Academic Press, London, 1963.

[64] Pitzer, K. S., and Brewer, L. (revised edn.), Lewis, G. N., and Randall, M., *Thermodynamics*, McGraw-Hill, New York, 1961.

[65] Robinson, R. A., and Stokes, R. H., *Electrolyte Solutions*, 2nd edn (revised), Butterworths, London, 1968, p. 29.

[66] Baes, C. F., *Inorg. Chem.*, **4**, 588 (1965).

[67] Gran, G., *Analyst*, **77**, 661 (1952).

[68] Kepert, D. L., 'Metal ions in water', *Proceedings of the Royal Australian Chemical Institute, June 1970*, 135 (1970).

[69] Barnum, D. W., 'Hydrolysis of Cations. Formation constants and standard free energies of formation of hydroxy complexes', *Inorg. Chem.*, **22**, 2297 (1983).

[70] (a) Ardon, M., and Bino, A., *Structure and Bonding*, **65**, 1 (1987).
(b) Springborg, J., *Adv. Inorg. Chem.*, **32**, 55 (1988)

[71] Henry, M., Jolivet, J.-P., and Livage, J., *Structure and Bonding*, **77**, 153–206 (1992).

[72] Gustavson, K. H., *The Chemistry of the Tanning Process*, Academic, New York, 1956.

[73] (a) Livage, J., Henry, M., Jolivet, J.-P., and Sanchez, C., *MRS Bull. Jan. 1990*, 18 (1990).
(b) Livage, J., Henry, M., and Sanchez, C., *Prog. Solid State Chem.*, **18**, 259 (1988).

[74] Sanderson, R. T., *Chemical Bonds and Bond Energy*, 2nd edn, Academic Press, New York, 1976.

[75] (a) Martell, A. E., and Motekaitis, R. J., *Determination and Use of Stability Constants*, VCH, Weinheim, 1988.

(b) Rossotti, F. J. C., and Rossotti, H., *The Determination of Stability Constants*, McGraw-Hill, New York, 1961.

[76] (a) Perrin, D., *Stability Constants of Metal Ion Complexes, Part B, Organic Ligands*, IUPAC Chemical Data Series, Pergamon, Oxford, 1982.
      (b) Martell, A. E., and Smith, R. M., *Critical Stability Constants*, Vols. 1–6, Plenum, New York, 1974, 1975, 1976, 1977, 1982 and 1985.
      (c) Rossotti, H., *The Study of Ionic Equilibria*, Longman, New York, 1978.
      (d) Cram, D. J., *Angew. Chem. Intl. Ed. Eng.*, **25**, 1039 (1986).

[77] (a) Anderegg, G., *Pure Appl. Chem.*, **54**, 2693 (1982).
      (b) Harris. W. R., and Martell, A. E., *Inorg. Chem.*, **15**, 713 (1976).
      (c) Harris, W. R., Motekaitis, R. J., and Martell, A. E., *Inorg. Chem.*, **14**, 974 (1975).

[78] Irving, H., and Williams, R. J. P., *J. Chem. Soc.*, 3192 (1953).

[79] Taube, H., *Chem. Revs.*, **50**, 69 (1952).

[80] (a) Hunt, J. P., and Taube, H., *J. Chem. Phys.*, **18**, 757 (1950), **19**, 602 (1951).
      (b) Plane, R. A., and Taube, H., *J. Chem. Phys.*, **56**, 33 (1952).
      (c) Hunt, J. P., and Plane, R. A., *J. Am. Chem. Soc.*, **76**, 5960 (1954).
      (d) Plane, R. A., and Hunt, J. P., *J. Am. Chem. Soc.*, **79**, 3343 (1957).

[81] Cohn, M., and Urey, H. C., *J. Am. Chem. Soc.*, **60**, 679 (1938).

[82] Dostrovsky, I., and Klein, F. S., *Anal. Chem.*, **24**, 414 (1952).

[83] Falcone, A. B., *Anal. Biochem.*, **2**, 147 (1961).

[84] Stranks, D. R., and Swaddle, T. W., *J. Am. Chem. Soc.*, **93**, 2783 (1971).

[85] (a) Anbar, M., and Guttmann, S., *Int. J. Appl. Radiation Isotopes*, **4**, 233 (1959).
      (b) Shahashiri, B. Z., and Gordon, G., *Talanta*, **13**, 142 (1966).
      (c) Rodgers, K. R., Murmann, R. K., Schlemper, E. O., and Shelton, M. E., *Inorg. Chem.*, **24**, 1313 (1985).
      (d) Webster, L. A., Wahl, M. H., and Urey, H. C., *J. Chem. Phys.*, **3**, 129 (1935).

[86] Xu, F.-C., Krouse, H. R., and Swaddle, T. W., *Inorg. Chem.*, **24**, 267 (1985).

[87] (a) Lincoln, S. F., *Prog. React. Kinet.*, **9**, 1 (1977).
      (b) Merbach, A. E., and Akitt, J. W., 'High resolution variable pressure NMR for chemical kinetics, in *NMR Basic Principles and Progress*, Vol. 24, Springer-Verlag, Berlin (1990), p. 189 (and refs. therein).

[88] Swift, T. J., and Connick, R. E., *J. Chem. Phys.*, **37**, 307 (1962) and **41**, 2553 (1964).

[89] (a) Ducommun, Y., and Merbach, A. E. 'Solvent exchange reactions', in *Inorganic High Pressure Chemistry, Kinetics and Mechanisms* (ed. R. van-Eldik), Elsevier, Amsterdam, 1986.
      (b) Merbach, A. E., *Pure. Appl. Chem.*, **21**, 1479 (1982) and **59**, 161 (1987).
      (c) van-Eldik, R., and Merbach, A. E., *Comm. Inorg. Chem.*, **12**, 341 (1992).
      (d) Bernhard, P., Helm, L., Ludi, A., and Merbach, A. E., *J. Am. Chem. Soc.*, **107**, 312 (1985).

[90] Richens, D. T., Pittet, P.-A., Merbach, A. E., Humanes, M., Lamprecht, G. J., Ooi, B.-L., and Sykes, A. G., *J. Chem. Soc. Dalton Trans.*, 2305 (1993).

[91] Eigen, M., *Pure Appl. Chem.*, **6**, 105 (1963).

[92] Margerum, D. W., Cayley, G. R., Weatherburn, D. C., and Pagenkopf, G. K., in *Coordination Chemistry* (ed. A. E. Martell), ACS Monograph 174, Washington, DC, 1978, Ch. 1.

[93] Langford, C. H., and Gray, H. B., *Ligand Substitution Processes*, Benjamin Inc., New York, 1966.

[94] Swaddle, T. W., *Comm. Inorg. Chem.*, **12**, 237 (1991).

[95] Lay, P. A., *Comm. Inorg. Chem.*, **9**, 235 (1991).

[96] (a) Swaddle, T. W., *Rev. Phys. Chem. Japan*, **50** 230 (1980).

      (b) Swaddle, T. W., *Coord. Chem. Revs.*, **14**, 217 (1974).

[97] Swaddle, T. W., *Adv. Inorg. Bioinorg. Mech.*, **2**, 96 (1983).

[98] van Eldik, R., Asano, T., and LeNoble, W. J., *Chem. Revs.*, **89**, 549 (1989).

[99] Ducommun, Y., Newman, K. E., and Merbach, A. E., *Inorg. Chem.*, **19**, 3696 (1980).

[100] Stranks, D. R., *Pure Appl. Chem.*, **38**, 303 (1974).

[101] Shannon, R. D., *Acta Cryst.*, **A32**, 751 (1976).

[102] (a) Swaddle, T. W., *Inorg. Chem.*, **22**, 2663 (1983).

      (b) Swaddle, T. W., and Mak, M. K. S., *Can. J. Chem.*, **61**, 473 (1983).

[103] (a) Swaddle, T. W., in *Mechanistic Aspects of Inorganic Reactions* (eds. D. B. Rorabacher and J. F. Endicott), *ACS Sym. Ser.* **198**, 39 (1982).

      (b) Swaddle, T. W., *Inorg. Chem.*, **19**, 3203 (1980).

[104] (a) Doine, H., Ishihara, K., Krouse, H. R., and Swaddle, T. W., *Inorg. Chem.*, **26**, 3240 (1987).

      (b) Swaddle, T. W., *J. Chem. Soc. Chem. Commun.*, 832 (1982).

[105] Connick, R. E. and Alder, B. J., *J. Chem. Phys.*, **87**, 2764 (1983).

[106] (a) Akesson, R., Pettersson, L. G. M., Sandstrom, M., and Wahlgren, U., *J. Am. Chem. Soc.*, **116**, 8705 (1994).

      (b) Akesson, R., Pettersson, L. G. M., Sandstrom, M., Siegbahn, P. E. G., and Wahlgren, U., *J. Phys. Chem.*, **96**, 10773 (1992) and **97**, 3765 (1993).

[107] Rotzinger, F. P., *J. Am. Chem. Soc.*, **118**, 6760 (1996).

[108] Wilkins, R. G., and Eigen, M., *Adv. Chem. Ser.*, **49**, 55 (1965).

[109] Eigen, M., *Z. Electrochem.*, **64**, 115 (1960).

[110] Fuoss, R. M., *J. Am. Chem. Soc.*, **80**, 5059 (1958).

[111] (a) Haim, A., and Wilmarth, W. K., *Inorg. Chem.*, **1**, 573, 583 (1962).

      (b) Haim, A., Grassi, R. J., and Wilmarth, W. K., *Adv. Chem. Ser.*, **49**, 61 (1965).

[112] Bradley, S. M., Doine, H., Krouse, H., Sisley, M. J., and Swaddle, T. W., *Aust. J. Chem.*, **41**, 1323 (1988).

[113] Robb, D., Steyn, M. M. De V., and Kruger, H., *Inorg. Chim. Acta*, **3**, 383 (1969).

[114] (a) Wilkins, R. G., *The Kinetics and Mechanism of Reactions of Transition Metal Complexes*, VCH, Weinheim, 1991.

      (b) Basolo, F., and Pearson, R. G., *Mechanisms of Inorganic Reactions*, 2nd edn, Wiley, New York, 1967.

[115] (a) Jackson, W. G., McGregor, B. C., and Jurisson, S. S., *Inorg. Chem.*, **26**, 1286 (1987).

      (b) Brasch, N. E., Buckingham, D. A., Clark, C. R., and Finnie, K. S., *Inorg. Chem.*, **28**, 4567 (1989).

      (c) Tobe, M. L., *Adv. Inorg. Bioinorg. Mech.*, **2**, 1 (1983).

[116] Powell, G., and Richens, D. T., *Inorg. Chem.*, **32**, 4021 (1993).

[117] Crimp, S., Spiccia, L., Krouse, H. R., and Swaddle, T. W., *Inorg. Chem.*, **33**, 465 (1994).

[118] Bracken, D. E., and Baldwin, H. W., *Inorg. Chem.*, **13**, 1325 (1974).

[119] Sasaki, Y., and Sykes, A. G., *J. Chem. Soc. Dalton Trans.*, 1048 (1975).

[120] Swaddle, T. W., and Guastalla, G., *Inorg. Chem.*, **8**, 1604 (1969).

[121] (a) Jordan, R. B., *Reaction Mechanisms of Inorganic and Organometallic Systems*, Oxford University Press, New York, pp. 77–80 (1991).

      (b) See, for example, Grant, M. W., and Jordan, R. B., *Inorg. Chem.*, **20**, 55 (1981) (for a good compilation of examples).

[122] Pearson, R. G., *J. Chem. Educ.*, **45**, 643 (1968).

[123] Ahrland, S., Chatt, J., and Davies, N. R., *Quart. Rev. Chem. Soc.*, **12**, 265 (1958).

[124]   Pearson, R. G., *Inorg. Chem.*, **27**, 734 (1988).
[125]   Hugi, A. D., Helm, L., and Merbach, A. E., *Inorg. Chem.*, **26**, 1763 (1987) and *Helv. Chim. Acta*, **68**, 508 (1985).
[126]   Mayer, J. M., and Abbott, E. H., *Inorg. Chem.*, **22**, 2774 (1983).
[127]   Cady, H. H. and Connick, R. E., *J. Am. Chem. Soc.*, **80**, 2647 (1958).
[128]   (a) Richens, D. T., and Sykes, A. G., *Comm. Inorg. Chem.*, **1**, 141 (1981).
        (b) Richens, D. T., and Sykes, A. G., *Inorg. Syn.*, **23**, 130–139 (1985).
[129]   Richens, D. T., Harmer, M. A. and Sykes, A. G., *J. Chem. Soc. Dalton Trans.*, 2099 (1984).

Chapter 2

Main Group Elements: Groups 1, 2, 13, 14, 15, 16, 17 and 18

This chapter deals with those chemical elements that have only valence s and p electrons. This gives rise to properties largely distinct from those of the d-electron transition elements (Chapters 4 to 12). Scandium, yttrium and the lanthanides and actinides are also a special case and are discussed in Chapter 3. Here we start by considering what can be regarded as the simplest cation of all, namely the proton itself, H^+.

2.1 THE HYDRATED PROTON

The reaction between H^+ and H_2O, to give the hydronium ion H_3O^+, can be considered the most fundamental of Lewis acid and base reactions:

$$H^+ + H_2O \rightarrow H_3O^+, \qquad \Delta H^\theta \sim -720 \text{ kJ mol}^{-1} \qquad (2.1)$$

Protons are strongly hydrated in water

$$H^+(g) \rightarrow H_3O^+(aq), \qquad \Delta H^\theta \sim -1090 \text{ kJ mol}^{-1} \qquad (2.2)$$

reflecting the high charge density and direct participation in the extensive hydrogen-bonded network of water and one freely discusses aqueous solutions of acids as containing ions such as H_3O^+, $H_5O_2^+$, $H_7O_3^+$, etc., which are in essence $[H(OH_2)_n]^+$ ($n = 1, 2, 3,$ etc.). Indeed, it follows from (2.1) and (2.2) that the enthalpy of hydration of H_3O^+ is still highly exothermic, $\sim -370 \text{ kJ mol}^{-1}$, and roughly intermediate between the values calculated for Na^+ (-405 kJ mol^{-1}) and K^+ (-325 kJ mol^{-1}). Occasionally specific values of n, e.g. $H_5O_2^+$, can be found in the isolated structures of some acid salts, e.g. $[H_5O_2][V(OH_2)_6]$ $(CF_3SO_3)_4$ (see Chapter 5). However, values > 2 are relevant to the hydration of H^+ in bulk water and indeed one normally talks of the degree of hydration around the hydronium ion H_3O^+ itself.

Hydrogen can be considered to be the classic example of amphotericity in the sense that it forms a stable hydrated cation, $[H(OH_2)_n]^+$, etc., and an oxo anion, OH^-. Plots of effective ionic radius for six-coordination against oxidation number indeed place the effective radius for H^+ within the range estimated for an amphoteric metal ion with oxidation number $+1$ (~ 16–20 pm). The value for the next smallest singly positive ion Li^+ (76 pm) highlights just how small would be the effective ionic radius of the isolated proton if it existed in aqueous solution.

A number of experimental and theoretical studies have sought to characterise the interaction between water and the hydronium ion. Neutron and X-ray diffraction data appear to show evidence of hydrogen bonding of H_3O^+ to four water molecules in the immediate vicinity with a $H_3O^+...OH_2$ distance of 252 pm [1]. Extensive *ab initio* MO calculations on the gas phase hydration of H_3O^+ led to an O...O distance of 253 pm, but formulation as $H_3O^+ \cdot 3H_2O$ [2]. Using a Monte Carlo type simulation method based on *ab initio* atom–atom pair

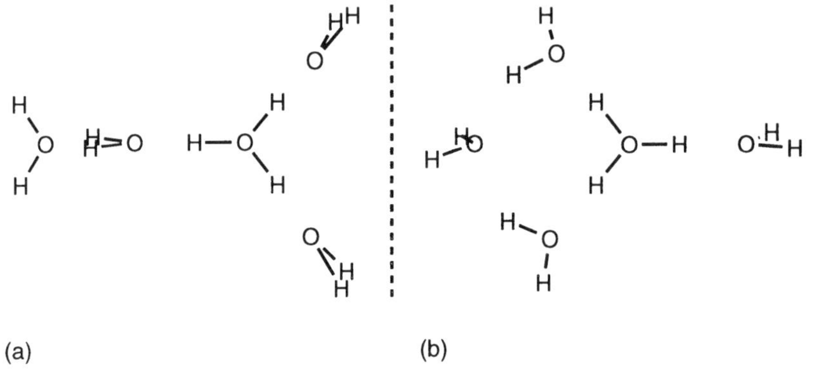

(a)   (b)

Figure 2.1. Calculated hydration structures for $[H_3O(OH_2)_4]^+$

potentials, evaluated by SCF LCAO MO calculations using extended basis sets, the water–hydronium ion interaction has been modelled at 300 K [3]. The assumption was made, given the short lifetime of hydronium, 10^{-12}–10^{-11} s [4], that the water molecules in the ion field can relax towards equilibrium configurations before the solvated proton migrates to neighbouring water molecules. The simulation showed that the effective size of the hydronium, as measured by the O...O separation of 248 pm, is intermediate between that of Na^+ and K^+, thus paralleling the respective values of the heats of hydration. Four water molecules complete the hydration sphere around H_3O^+ in agreement with the X-ray and neutron scattering data, three of them hydrogen bonded to the ion at any one time and a fourth in close proximity (Fig. 2.1). The competition for hydrogen bonding of the fourth water leads to some distortion, which induces breaking and reforming of bonds and in turn mobility. Further studies using a combination of Monte Carlo simulations and DFT quantum mechanical computations suggest two equal energy situations (Fig. 2.1), which may be interchangeable [5].

Similar studies of the hydration around the OH^- ion indicates that there may be as many as five water molecules in the first hydration sphere, with four of them hydrogen bonded to the ion at any one time [6]. The competition for hydrogen bonding of the fifth water leads again to distortion and a probable role in mobilising the ion. At an OH^-...OH_2 distance of 264 pm an interaction energy minimum of -104.3 kJ mol^{-1} was found. The fact that this energy compares highly favourably with the experimental hydration energy of OH^- (-100.3 kJ mol^{-1}) [7] as well as with early *ab initio* results [8] provides further encouragement for the validity of the Monte Carlo simulation approach.

2.2 GROUP 1 METAL IONS: LITHIUM, SODIUM, POTASSIUM, RUBIDIUM AND CAESIUM

The group 1 or alkali metals are relatively poorly solvated and thus their water ligands are highly labile with residence times ranging from $\sim$ 1 ns (Li^+) to $\sim$ 0.1 ns (Cs^+). From neutron scattering experiments a well-defined hydration shell is seen around Li^+ whereas the larger K^+, Rb^+ and Cs^+ ions are so weakly hydrated that no specific dynamical significance can be attributed to water molecules located in the primary shell (see Fig. 2.2). Na^+ has intermediate behaviour between Li^+ and the rest. Hydration enthalpies decrease with the increase in effective ionic radius from $-498\,kJ\,mol^{-1}$ (Li^+, radius for six-coordination 76 pm) to $-251\,kJ\,mol^{-1}$ (Cs^+, radius 167 pm). It should be borne in mind that ionic radii are themselves highly sensitive to the coordination (hydration) number, which in turn changes with effective size of the metal ion. Hydration numbers for the group 1 metals are also highly sensitive to the metal ion concentration. As a result the use of computational, molecular dynamics and molecular simulation approaches in addition to experimental neutron scattering and X-ray diffraction studies are playing a major role in deducing information about the hydration sphere under the various conditions. For the larger metal ions the hydration sphere is highly diffuse and at best poorly defined.

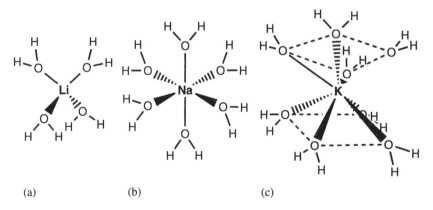

(a) (b) (c)

Figure 2.2. Possible hydration structures for the group 1 metal cations in concentrated aqueous solution: (b) present also in $Na_2[MO_4]\cdot10H_2O$ (M = S, Se, Cr, Mo and W) [9] and (c) present in $CaK[AsO_4]\cdot8H_2O$ [10] (note that hydration structures for Rb^+ and Cs^+ may not be dissimilar to that for K^+)

2.2.1   Lithium

2.2.1.1   CHEMISTRY OF $Li^+(aq)$

Hydrated Li^+ salts have not been found with more than four water molecules and there have been suggestions that the preferred hydration number in solution is four (presumably tetrahedral as in the case of $[Li(NH_3)_4]^+$ in $LiI \cdot 4NH_3$). Generally, however, the hydration number of Li^+ in solution appears variable, ranging between 3.3 and 6.0 (Li—O = 195–228 pm) depending upon the H_2O/salt ratio [11]. Difference neutron scattering measurements show that the hydration number approaches six towards the dilute limit but drops below four at higher concentrations. Anion effects may also be responsible for the varying hydration numbers observed. A hydration number of four has been deduced from X-ray and neutron diffraction studies on aqueous solutions of LiCl and LiBr [12–14]. Kamada and Uemura [14] reported the Li—$O(OH_2)$ distance as 196 ± 2 pm from neutron diffraction measurements, which was moreover reproduced with a non-linear dependence on increasing hydration number, although little attention was paid to the effects of Li^+ concentration and neighbouring anions. More recent difference neutron scattering measurements on aqueous solutions of $LiClO_4 < 3.0$ m have been interpreted as indicating discrete $[Li(OH_2)_6]^+$ ions [15]. Molecular dynamics simulations of a 2.2 m solution of LiI have also indicated octahedral coordination by six waters in the first hydration shell [16]. It has been suggested that the degree of anion hydration may play a significant role in defining how much free water is 'available' to bind to a given cation in aqueous solution. Thus the larger and thus less well solvated I^- and ClO_4^- ions lead to the existence of a higher effective hydration number for Li^+ in solutions of LiI and $LiClO_4$ than in equivalent solutions of LiCl or LiBr, even for concentrated solutions of the salts. Hydration numbers for the group

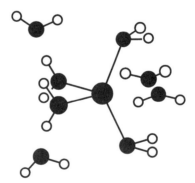

Figure 2.3. Hydration structure for Li^+ in concentrated aqueous solutions from *ab initio* calculations at the Hartree–Fock level

1 metals are also found to largely depend upon the measuring technique. A number of these were highlighted in Chapter 1. For example, a hydration number of three for Li^+ has been deduced from ^1H NMR chemical shifts [17]. A number of other studies have reported on the modelling of the hydration around Li^+. *Ab initio* calculations at the Hartree–Fock level aimed at modelling the symmetric $Li-O(OH_2)$ vibration observed at $255 \, cm^{-1}$ in the Raman spectrum of concentrated $Li^+(aq)$ solutions indicated a preferred minimum energy structure having a hydration number of four along with a loosely defined secondary shell in which a further four water molecules were closely associated with $[Li(OH_2)_4]^+$ [18] (Fig. 2.3).

2.2.2 Sodium

2.2.2.1 CHEMISTRY OF $NA^+(AQ)$

The hydration number of Na^+ is generally quoted over the range 4–8 ($Na-O = 240$–$250 \, pm$) on the basis of diffraction data [19]. $[Na(OH_2)_6]^+$ is relevant to the X-ray structure of $Na_2SO_4 \cdot 10H_2O$ [9] (av. $Na-O(OH_2) = 243 \, pm$), known as Glauber's salt or mirabilite, as well as to the isomorphous series $Na_2MO_4 \cdot 10H_2O$ (M = Se, Cr, Mo and W). The hydration geometry is slightly distorted from regular octahedral. $[Na(OH_2)_6]^+$ is likewise present in the structure of $Na_2B_4O_7 \cdot 10H_2O$ (borax) (av. $Na-O(OH_2) = 242 \, pm$) [20] and given the preference for six-coordination by Na^+; e.g. in the common rock salt structure of NaCl, the retention of a hydration number of six for Na^+ in aqueous solution is highly attractive. Indeed, average hydration numbers from analysis of the peak area under the radial distribution function obtained from computer simulations are around six [21] and there have even been attempts at fully optimising the geometry of the putative $[Na(OH_2)_6]^+$ ion and its respective cation–anion pairs in solution using DFT methods [22]. A 2.2 m aqueous solution of $NaClO_4$ has been simulated by Heinzinger *et al.* in a molecular dynamics study using 200 water molecules, 8 sodium and 8 perchlorate ions [23]. Assuming tetrahedral hydration around ClO_4^- they computed a hydration number around Na^+ of 6.5 ($Na-O(OH_2) = 236 \, pm$) with evidence of a distinct secondary hydration shell. $Na-O(OH_2)$ bond lengths arising from computer simulations generally tend to be somewhat shorter (range 230–240 pm) when compared to those obtained by diffraction methods.

2.2.3 Potassium

2.2.3.1 CHEMISTRY OF $K^+(AQ)$

Although one might predict a higher hydration number than six for the larger K^+ the situation is offset by weaker hydration and a somewhat less well-defined hydration shell. Square antiprismatic $[K(OH_2)_8]^+$ ($K-O(OH_2) = \, pm$) has nonetheless been found in the salt $CaK[AsO_4] \cdot 8H_2O$ [10]. In practice, however, the hydration number of K^+ in aqueous solution is difficult to ascertain for the simple reason that the K—O distance is comparable to the H_2O—H_2O distance in bulk water and thus an evaluation requires an assumed knowledge of the water structure. Values have been quoted between 5.3 and 8.0 (K—$O(OH_2) = 260$–295 pm) [19, 24] based on X-ray and neutron scattering measurements. Computer simulations give a range of hydration numbers between 6.3 and 7.8 (K—$O(OH_2) = 265$–286 pm) [25]. These have the advantage that no assumption is required about the bulk water structure. Values obtained from NMR measurements, on the other hand, tend to be somewhat smaller than those for Na^+ since the K^+–H_2O interactions are electrostatically much weaker [17].

Potassium, along with sodium, magnesium and calcium, are bulk metals constituting around 1% of the human body weight. The level of K^+(aq) inside a mammalian blood cell, for example, is around 10 times higher (105 mM) than the level of Na^+(aq) (10 mM), whereas roughly the reverse is true outside the cell. Maintenance of these levels is crucial in order to drive biochemical processes across the cell membrane through the so-called 'Na^+/K^+ ion pump' mechanism linked to Mg^{2+} promoted ATP hydrolysis. In order to maintain the required concentration gradient nature has designed sophisticated complexing molecules (the ionophores) which pick up and transport the two ions in opposite directions through the hydrophobic lipid bilayer which constitutes the cell membrane. The selection is made via a match between the cavity size of the complexant and the effective ionic size (radius) of M^+. This has since led to the development of simpler synthetic analogues of these molecules such as the crown ethers, which can be used to solubilise the different group 1 M^+(aq) ions in non-polar media.

2.2.4 Rubidium and Caesium

2.2.4.1 CHEMISTRY OF RB^+(AQ) AND CS^+(AQ)

Rb^+ unfortunately emits strong fluorescent X-rays upon irradiation of samples from the common Mo source and as a result there is no diffraction data available on this ion. X-ray diffraction data is, however, available for Cs^+. Values for the hydration number are quoted between 6 and 8 [13, 26] (Cs—$O = 295$–321 pm) depending upon the system. The existence of eight-coordination of Cl^- around Cs^+ in CsCl lends some support for a hydration number of eight in aqueous solution. Computer simulation shows a spread of values between 5.3 and 8.2 (Cs—$O(OH_2) = 303$–320 pm) [27], paralleling the diffraction data.

Since the defined hydration sphere on these highly labile metal ions is as much a function of the method of measurement as it is of the metal ion concentration the assignment of a discrete hydration number to each of the group 1 M^+ ions is largely meaningless. At best, quoting a range of values in solution relevant to the conditions of the measurement is probably all that can ever realistically be done.

2.3 GROUP 2 METAL IONS: BERYLLIUM, MAGNESIUM, CALCIUM, STRONTIUM, BARIUM AND RADIUM

The higher charge for the group 2 metal ions leads to a well-defined hydration shell detectable around M^{2+}, certainly in the case of beryllium, magnesium, calcium and strontium. For beryllium and magnesium the exchange rate of the primary shell with bulk water is slow enough to be monitorable by NMR techniques whereas for the more labile and less well hydrated heavier members direct measurement of the much faster exchange is more difficult and requires data from the monitoring of complex formation reactions using ultrasound absorption (Section 1.4). Because of its very small size and correspondingly high charge density Be^{2+} compounds show a high degree of covalency and indeed some of the properties of $Be^{2+}(aq)$ might be attributable to this tendency.

2.3.1 Beryllium

2.3.1.1 THE TETRAHEDRAL $[Be(OH_2)_4]^{2+}$ ION: PREPARATION AND STRUCTURE

The small Be^{2+} ion is strongly hydrated in aqueous solution and has received much detailed study. It has by far the most exothermic hydration enthalpy $(-2487\,kJ\,mol^{-1})$ of any of the sp block metal ions. Givens its small size, and the fact that the coordination number in complexes of Be^{2+} is invariably four [28], it comes as no surprise to find strong evidence, e.g. from X-ray diffraction [29], ^1H NMR [30], ^{17}O NMR [31–33] and molecular dynamic simulation studies [29, 34], for a descrete hydration shell containing only four water molecules around the cation in a regular tetrahedral arrangement (Fig. 2.4). Discrete $[Be(OH_2)_4]^{2+}$ is characterised in a range of solid hydrated salts such as $BeSO_4 \cdot 4H_2O$ [35] and $BeX_2 \cdot 4H_2O$ $(X = NO_3^-)$ [36], IO_3^- [37], IO_4^- [38], Cl^- and Br^- [39]. Simple dissolution of pure beryllium metal [40] or the oxide BeO [33, 41] followed by crystallisation from a concentrated aqueous solution can suffice for the preparation of Be^{2+} salts.

Metallurgical grade beryllium (99.999% purity) has been used to prepare solutions for quantitative studies (see below). The solid state structure of $BeSO_4 \cdot 4H_2O$ has also been studied by neutron diffraction (ND) [42]. As has been mentioned in Section 1.2, the $Be—O(OH_2)$ distances obtained from ND tend to be longer (161.8(4) pm) than in the case of those from XRD (161.0(4) pm). The water molecules are tilted to the $Be—O$ bond vector. Some distortion from regular tetrahedral geometry is seen with $O—Be—O$ angles respectively $117.5(3)°$ and $105.6(3)°$ (X-ray) and $117.0(2)°$ and $105.8(2)°$ (neutron), probably arising from hydrogen-bonding interactions to the SO_4^{2-} ions. A wide variation of $Be—O$ distances (161–175 pm) has been found from the various measurements/simulations. The value determined from solution XRD measurements is 169 pm [29]. A number of *ab initio* MO studies at the RHF level have been

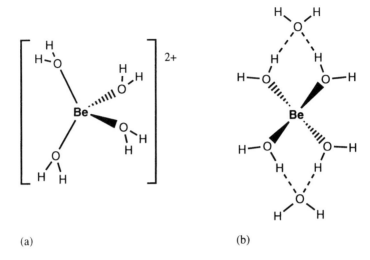

(a)   (b)

Figure 2.4. (a) Tetrahedral hydration around Be^{2+} in $BeSO_4 \cdot 4H_2O$ and (b) the minimum energy structure for '$Be^{2+} (OH_2)_4 \cdot 2H_2O$'

reported. Gradient-optimized *ab initio* calculations by Bock and Glusker have examined the local minima within the potential energy surfaces of various tetra-, penta- and hexahydrated Be^{2+} arrangements [43]. In each case the lowest energy configuration has only four water molecules in the primary hydration shell with any remaining waters relegated to the secondary shell. Furthermore, the lowest energy configuration for '$Be^{2+}(OH_2)_4 \cdot 2H_2O$' was with the two secondary shell waters hydrogen-bonded to primary shell waters on opposite sides of the tetrahedron with $Be-O(OH_2) = 164.5$ pm (Fig. 2.4). A molecular dynamics simulation of a 1.1 m Be^{2+} solution has been performed by Heinzinger and coworkers [34] using a three-body potential for the $Be-OH_2$ interaction. It was believed that the discrepancy apparent from coordination numbers obtained from earlier MD simulations (six) and X-ray scattering (four) [29] was due to the inadequacies of the assumption of pairwise additivity of the interaction energies between Be^{2+} and H_2O. Heinzinger's computational approach employing a three-body potential representing the interacting complex '$H_2O-Be-OH_2$' gave rise to a preferred primary hydration number of four, $Be-O(OH_2) = 175$ pm (from peak analysis of the simulated radial distribution function), along with clear evidence for the existence of a well-defined secondary shell ($Be-O \sim 400$ pm), in good agreement with the XRD data. An indication that the assumption of pairwise additivity of the $Be^{2+}-H_2O$ interaction energies was not as justified as for other cations became apparent from quantum chemical calculations. Given this unique behaviour for Be^{2+} one wonders whether this reflects in part a considerable covalency in the $Be-OH_2$ bonds.

2.3.1.2 HYDROLYSIS AND WATER EXCHANGE ON $[Be(OH_2)_4]^{2+}$

A number of studies have reported on the hydrolysis behaviour of $Be^{2+}(aq)$ [40, 44–51]. The small size and low hydration number of four leads to anomalously high acidity for Be^{2+} and hydrolytic tendency amongst the M^{2+} ions of group 2. The first extensive study was carried out in $I = 3.0$ M, $NaClO_4$ media at 25 °C by Kalihana and Sillen in 1956 [44]. They established that the predominant species was the trimer $Be_3(OH)_3^{3+}(aq)$. The minor species were $Be(OH)_2(aq)$ and $Be_2(OH)^{3+}(aq)$. No evidence for the monohydroxo ion $[Be(OH_2)_3(OH)]^+$ could be found. Carell and Olin [45] later validated Kalihana and Sillen's model. Schwarzenbach and Wenger [46], however, performed a rapid mixing study to catch the putative $BeOH^+(aq)$ (5 ms after mixing) before the onset of hydrolytic polymerisation. On the basis of the findings Baes and Mesmer [47] estimated a pK_{11} value of 5.4 ± 0.1. In 1983 Sylva and coworkers reported potentiometric investigations on $Be^{2+}(aq)$ in 0.1 M KNO_3 media at 25°C [48]. A similar distribution of products was found with reported constants, $Be(OH)_2(aq)$ ($pK_{12} = 11.320 \pm 0.008$), $Be_2(OH)^{3+}(aq)$ ($pK_{21} = 2.955 \pm 0.007$) and $Be_3(OH)_3^{3+}$ (aq) ($pK_{33} = 8.804 \pm 0.002$). The presence of $Be(OH)_2$ was required to explain the data at low metal concentrations (0.358–2.684 mM). Bruno [40] has provided a more recent survey of the available data and from a study conducted in $I = 3.0$ M, $NaClO_4$ at 25°C proposed the existence of higher oligomers, $Be_5(OH)_6^{4+}(aq)$ ($pK_{56} = 18.81 \pm 0.03$) and $Be_6(OH)_8^{4+}(aq)$ ($pK_{68} = 26.70 \pm 0.05$) in addition to $Be(OH)_2(aq)$ ($pK_{12} = 11.09 \pm 0.04$), $Be_2(OH)^{3+}(aq)$ ($pK_{21} = 3.23 \pm 0.05$) and $Be_3(OH)_3^{3+}(aq)$ ($pK_{33} = 8.656 \pm 0.002$). The available data from a range of studies are summarised in Table 2.1. The data at infinite

Table 2.1. Hydrolysis data for $[Be(OH_2)_4]^{2+}(aq)$ in various media (taken from ref. [40])

Medium	$(x, y) = (1,2)^b$	(2,1)	(3,3)	(5,6)	(5,7)	(6,8)	(6,9)	Ref.
				$pK_{xy}{}^a$				
0.5 M NaClO$_4$		3.20	8.81					[49]
3.0 M NaClO$_4$	10.9	3.24	8.66					[44]
	11.16	3.16	8.66					[50]
	11.09	3.23	8.66	8.84		26.69		[40]
1.0 m NaCl		3.42	8.91		25.33	27.46		[47]
3.0 m KCl		3.18	8.91					[51]
0.1 M KNO$_3$	11.32	2.96	8.81					[48]
2.0 M KNO$_3$		3.28	8.90			27.50	34.50	[51]
0 (infinite dilution)		3.47	8.86	19.5		26.3		[40]
($\pm$ s.d.)		± 0.05	± 0.05	± 0.1		± 0.1		

$^a pK_{xy}$ refers to $x \, Be^{2+} + y H_2O \rightleftharpoons Be_x(OH)_y{}^{(2x-y)+} + y H^+$.
$^b (1,2)$ refers to $Be(OH)_2$, (3,3) to $Be_3(OH)_3^{3+}(aq)$, etc.

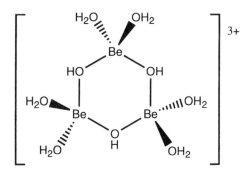

Figure 2.5. Structure of the cationic unit of $Be_3(OH)_3^{3+}$

dilution were derived via a correlation of the data in different ionic media using the Bronsted–Guggenheim–Scatchard specific ion interaction theory [40].

The trimer $Be_3(OH)_3^{3+}$(aq) is believed to be a cyclic species (Fig. 2.5), on the basis of infra-red and Raman [52] and ^1H NMR [53] studies of the aqueous solutions deriving from the hydrolysis of $BeCl_2$. Cyclic $Be_3(OH)_3^{3+}$ units have also been established in the solid state [54].

Early estimates of the rate of water ligand replacement on $[Be(OH_2)_4]^{2+}$ were obtained from monitoring the complexation of SO_4^{2-} by ultrasound absorption [55]. The residence time was found to be around 2–10 ms, thus well within the range of monitoring by normal line-broadening NMR methods. Subsequently variable-temperature ^{17}O NMR studies of the exchange process between the primary shell of four water molecules and bulk water were carried out by Connick and Fiat [31], Frahm and Fuldner [32], Neely [56] and more recently by Merbach and coworkers [33], who also performed a variable pressure study. At the time of the early measurements the Be^{2+}(aq) exchange process attracted much interest as the first example of a measured proton exchange ($k = 8.0 \times 10^4 \, s^{-1}$), as determined using ^1H NMR [32], which was faster than that for water exchange determined using ^{17}O NMR [31,33]. With the help of bulk water Shift reagents such as Co^{2+}[31] or Eu^{3+} [32] or relaxation agents such as Mn^{2+} [33], the ^{17}O NMR resonance of the coordinated waters is easily seen at $+3$ ppm from bulk water (Fig. 2.6), providing further conclusive evidence for the well-defined primary hydration shell. Peak integration has verified that the hydration shell consists of four water molecules. Table 2.2 lists data obtained from the various water exchange kinetic studies.

The studies by Merbach and coworkers are probably the most reliable and question the large negative activation entropies obtained in the earlier studies. However, of particular interest is the $\Delta V_{ex}^{\ddagger}$ value of $-13.6 \, cm^3 \, mol^{-1}$ [33], the most negative value so far obtained for water exchange on an aqua metal ion and

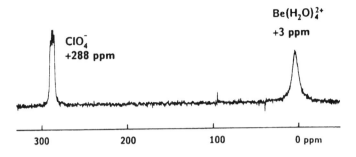

Figure 2.6. ^{17}O NMR spectrum of a solution of 0.1 m $Be(ClO_4)_2$ in 0.4 atom % enriched $H_2^{17}O$ containing 0.1 m $Mn(ClO_4)_2$ as bulk water relaxation agent

Table 2.2 Rate constants and activation parameters for the water exchange process on $[Be(OH_2)_4]^{2+}$ at 25 °C determined by ^{17}O NMR

Medium	$k_{ex}(25\,°C)$ (s^{-1})	$\Delta H^{\ddagger}_{ex}$ $(kJ\,mol^{-1})$	$\Delta S^{\ddagger}_{ex}$ $(J\,K^{-1}\,mol^{-1})$	$\Delta V^{\ddagger}_{ex}$ $(cm^3\,mol^{-1})$	Ref.
1.0 m ClO_4^-	7.33×10^2	59.2 ± 1.5	$+8.4 \pm 4.5$	-13.6 ± 0.5	[33]
	1.3×10^3	41	-43	—	[32]
	2.1×10^3	35	-63	—	[31, 56]

close to the value estimated by Swaddle's empirical calculations (see Section 1.4) for a limiting associative process. Thus water exchange on $[Be(OH_2)_4]^{2+}$, because of the hydration number of four and small size of water, may represent a rare example of the limiting A process. One can calculate the expected $\Delta V^{\ddagger}_{lim}(A)$ value for Be^{2+} by assuming it is the difference between the partial molar volumes for the aqua ion of increased coordination number and that of the aqua ion in its initial state. Using Shannon's values (Appendix 2) for the radii of Be^{2+} (45 pm for five-coordination) and H_2O (238.7 pm) and Swaddle's equation (1.54), $\Delta V^{\ddagger}_{lim}(A)$ for Be^{2+} is calculated to be $-12.9\,cm^3\,mol^{-1}$ [57]. The agreement between experiment and calculation is gratifying. Very little complex formation kinetic data is available on Be^{2+} for detailed comparisons and further studies should be encouraged given the scarcity of data generally on tetrahedral species. Additional support for the associative mechanism comes from the negative $\Delta V^{\ddagger}$ and $\Delta S^{\ddagger}$ values and low $\Delta H^{\ddagger}$ value from kinetic data for the 1:1 complex forming reaction with 4-isopropyltropolone (Hipt) (Table 2.3). Likewise, kinetic data for the reaction of $[Be(OH_2)_4]^{2+}$ with HF also supports an associative mode of activa-

Table 2.3. Complex formation data for $[Be(OH_2)_4]^{2+}$ at 25 °C

Incoming ligand	k^a $(M^{-1}s^{-1})$	$\Delta H_f^{\ddagger}$ $(kJ\,mol^{-1})$	$\Delta S_f^{\ddagger}$ $(K^{-1}\,mol^{-1})$	$\Delta V_f^{\ddagger}$ $(cm^3\,mol^{-1})$	Ref.
H_2O	52.77^b	59.2 ± 1.5	$+8.4 \pm 4.5$	-13.6 ± 0.5	[33]
HF	73.0				[58]
F^-	720				[58]
Hipt	58.1 ± 0.5	$38.1 \pm b0.4$	-83.5 ± 1.3	-7.1 ± 0.2	[41]

	k_b^a $(M^{-1}s^{-1})$	$\Delta H_b^{\ddagger}$ $(kJ\,mol^{-1})$	$\Delta S_b^{\ddagger}$ $(JK\,mol^{-1})$	$\Delta V_b^{\ddagger}$ $(cm^3\,mol^{-1})$	Ref.
HF					[58]
Hipt	49.8 ± 0.5	35.7 ± 0.4	-92.7 ± 1.2	-12.4 ± 0.2	[41]

$^a k_f$ and k_b are the forward and reverse second-order rate constants respectively.
^{b} Second-order rate constant for water exchange evaluated as $4k_{ex}/55.56$.

tion (with a second order rate constant faster than the equivalent value for water exchange) [58].

The proposal of a rare limiting associative mechanism for water exchange (replacement) on Be^{2+}(aq) has attracted interest from a number of theoretical studies. Winter and coworkers have used *ab initio* calculations at the Hartree–Fock level to compute optimum geometries for $[Be(OH_2)_4]^{2+}$ and $[Be(OH_2)_5]^{2+}$ in an attempt to model the activation parameters $\Delta H^{\ddagger}$ and $\Delta V^{\ddagger}$ for the associative water exchange process [59]. The intermediate structures (Fig. 2.7) were determined by optimizing all degrees of freedom for the pentahydrated complex except the Be—O distance for the incoming water molecule. The final geometry was constrained to trigonal bipyramidal with both axial bonds equal (Be—O = 200 pm). while optimizing all other degrees of freedom. With only one water molecule within the secondary shell they reproduced the experimental $\Delta H^{\ddagger}$

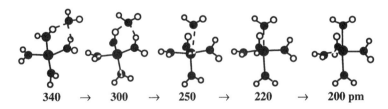

| 340 | $\rightarrow$ | 300 | $\rightarrow$ | 250 | $\rightarrow$ | 220 | $\rightarrow$ | 200 pm |

Figure 2.7. Formation of the pentahydrated transition state from a second shell water as a function of Be—O distance (pm) for the incoming water molecule

value very closely (calculated value of $62.6\,kJ\,mol^{-1}$). However, the calculated $\Delta V^{\ddagger}$ was $\sim 0\,cm^3\,mol^{-1}$. This discrepancy was assigned to the effects of the restricted secondary shell which was viewed as unable to confine the leaving water. Including a continuum model for the solvent was not found to improve the agreement.

Bock and Glusker [43] used *ab initio* methods to look at a range of $Be^{2+}-H_2O$ clusters and showed that the next most favourable energy situation for Be^{2+}(aq), above that of the $4(+2)$ cluster (Fig. 2.4), was not a water cluster with primary hydration number > 4 but the $3(+3)$ cluster, '$Be^{2+}(OH_2)_3 \cdot 3H_2O$', which interestingly models the dissociative situation. With this in mind, and since size and electrostriction are key factors for Be^{2+}, it was no surprise to find for more sterically demanding $[BeL_4]^{2+}$ complexes, e.g. with L = trimethylphosphate, an exchange process independent of the solvent concentration [60] and for L = tetramethylurea $\Delta V^{\ddagger}$ values $\sim +10.0\,cm^3\,mol^{-1}$ [33], implying an activation process now dominated by bond breaking. As one increases the steric bulk of the entering ligand it is not inconceivable therefore that a point will be reached when the only activation process open to $[Be(OH_2)_4]^{2+}$ is one largely dissociative in character. Thus, while the water exchange process itself appears close to what we might term the 'associative limit', it is clear that the properties of both *entering* and *leaving* ligands can play a crucial role in defining the mechanisms governing ligand replacement reactions at small, highly electrostricted, metal ions such as Be^{2+}.

2.3.2 Magnesium

2.3.2.1 CHEMISTRY OF THE $[Mg(OH_2)_6]^{2+}$ ION

A survey of the structural properties of solid hydrates of magnesium(II) have revealed a propensity for the larger Mg^{2+} to form the regular octahedral hexahydrate $[Mg(OH_2)_6]^{2+}$. The X-ray crystal structures of $MgSO_4 \cdot 7H_2O$ (Epsom salt) [61], $Mg(NO_3)_2 \cdot 6H_2O$ [62], $MgSO_3 \cdot 6H_2O$ [63], $Mg_2CdCl_6 \cdot 12H_2O$ [64], $MgCO_3 \cdot 3H_2O$ [65] and $MgCl_2 \cdot 6H_2O \cdot 12$-crown-4 (12-crown-4 not coordinated) [66] all provide examples of the $[Mg(OH_2)_6]^{2+}$ ion. In all some 118 structures containing octahedral Mg^{2+} were known to the Cambridge Structure Database in 1994 of which 31 contained $[Mg(OH_2)_6]^{2+}$. Indeed, the propensity for Mg^{2+} to stick rigidly to octahedral coordination points to a largely structural role for Mg^{2+} in biology as distinct from Zn^{2+} which adopts a more active role in catalysis through a more flexible geometry. *Ab initio* calculations also point to favourable six-coordinate hydration for Mg^{2+}. Indeed, attempts to find a stable heptahydrated structure for Mg^{2+} results in optimized structures consisting of six water molecules in the primary shell with the seventh water in the secondary shell and hydrogen-bonded to two of the primary shell waters [67]. This behaviour is exemplified experimentally by ready loss of the 'seventh' water from the heptahydrate $MgSO_4 \cdot 7H_2O$ at $48.3\,°C$ [61].

The formula $[Mg(OH_2)_6]^{2+}$ has been established in aqueous solutions of the perchlorate salt by ^1H NMR at temperatures below $-70\,°C$ [68]. Using the difference neutron scattering method, based largely around comparisons with Ni^{2+}, retention of the octahedral hydration around Mg^{2+} in solution is also strongly indicated [13, 69]. XRD measurements are also in support of the retention of $[Mg(OH_2)_6]^{2+}$ in aqueous solution (Mg—O(OH$_2$) = 200–215 pm) [70]. Heinzinger and coworkers have successfully simulated the difference neutron data on a 1.1 m solution of $MgCl_2$ using MD methods [71]. These results give validity to the calculated pair potentials from *ab initio* calculations. A well-defined first and second hydration shell is seen with Mg—O(OH$_2$) ~ 200 pm. The Mg—OH$_2$ bonds of $[Mg(OH_2)_6]^{2+}$ appear to be less covalent than those of $[Be(OH_2)_4]^{2+}$ on the basis of the transfer of charge from metal ion to oxygen, a feature exemplified in the weaker hydrogen-bonding between the first and second hydration shells. As a result $[Mg(OH_2)_6]^{2+}$ has a much lower hydrolytic tendency ($pK_{11} = 11.44$ [72]). In addition to monomeric $Mg(OH)^+(aq)$ the formation of tetrameric $Mg_4(OH)_4^{4+}(aq)$ ($pK_{44} = 39.7$) is required in order to account for the potentiometric titration data obtained from concentrated solutions of $Mg^{2+}(aq)$ ($\geqslant 1.0$ M) [73]. The quoted water exchange rate constant still stems from the 1970 work of Neely and Connick with $k_{ex} = 5.3 \times 10^5 \, s^{-1}$, $\Delta H^{\ddagger} = 63 \, kJ \, mol^{-1}$, $\Delta S^{\ddagger} \sim 8 \, J \, K^{-1} \, mol^{-1}$ [56, 74].

2.3.3   Calcium

2.3.3.1   CHEMISTRY OF CA^{2+}(AQ)

Hydrated calcium(II) salts are quite widespread and the coordination numbers somewhat variable. Enderby and coworkers reported on difference neutron scattering measurements on $CaCl_2$ in D_2O in a *Nature* paper in 1982 [75]. The hydration number is strongly concentration dependent varying from 10.0 ± 0.6 (1 m Ca^{2+}) to 7.2 ± 0.2 (2.8 m) and 6.4 ± 0.3 (4.5 m) (Ca—O(OH$_2$) = 241–246 pm). The plane of the water molecules are noticeably tilted to the Ca—O bond vector ($\phi = 34$–$38 \pm 9°$). On the basis of X-ray diffraction data on Ca^{2+} hydrates, coordination numbers greater than six are quite common [76] (Ca—O(OH$_2$) = 233–241 pm). In the salt CaK[AsO$_4$] the Ca^{2+} ions are, like the K^+ ions, surrounded by eight nearest-neighbour water molecules in a square antiprismatic arrangement [10]. The coordination polyhedron around Ca^{2+} shares a face of four water molecules and two edges with the coordination polyhedron of water around K^+. Here the tilt of the water molecules to the Ca—O vector is largely controlled by the sharing with the K^+ ions and the network of hydrogen bonds to the AsO$_4^{3-}$ ions. On the other hand, dodecahedral $[Ca(OH_2)_8]^{2+}$ units are present in the crystal structure of the cubic hydrate $2CaCl_2 \cdot 11HgCl_2 \cdot 16H_2O$ (Fig. 2.8) [77]. The $[Ca(OH_2)_8]^{2+}$ units are characterised by four unique Ca—O(OH$_2$) distances at 236(8), 236(5), 253(9) and 247(7) pm

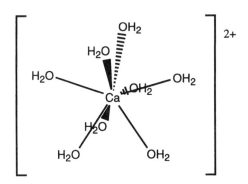

Figure 2.8. The dodecahedral $[Ca(OH_2)_8]^{2+}$ unit present in the crystal structure of the cubic hydrate $2CaCl_2 \cdot 11HgCl_2 \cdot 16H_2O$

and lie in the spaces between layers of icosahedral mercury-rich Hg_6Cl_{13} and Hg_5Cl_{13} clusters. In $CaX_2 \cdot 6H_2O$ (X = Cl, Br) the Ca^{2+} is surrounded by the six water molecules and three halide ions [78]. $CaBr_2 \cdot 10H_2O$ (hexamethylene tetramine), however, contains $[Ca(OH_2)_6]^{2+}$ [79].

Heinzinger and coworkers have performed MD simulations of a 1.1 m $CaCl_2$ solution using ion–water interaction pair potentials derived from *ab initio* calculations [80]. They also performed an X-ray scattering study for comparison. The XRD studies gave hydration number of 6.9, which compares with earlier XRD values of 6 from a number of investigations [69, 81]. The MD simulations, however, gave rise to a larger hydration number of 9.0, a value viewed as possibly better representative of the average Ca^{2+}–H_2O interaction and distribution of particles as a function of space coordinates and time. Ca—O(OH$_2$) values generally fall in the narrow range 239–242 pm. The apparent sensitivity of hydration number to the counter-anion and concentration is nonetheless a clear reflection of the much weaker hydration around Ca^{2+} and a much closer similarity now to the group 1 metals than to its lighter partners Mg^{2+} and Be^{2+}.

The much higher lability of the hydration sphere is reflected in the fact that the water exchange process on Ca^{2+} and indeed that on Sr^{2+} and Ba^{2+} is too fast to be measured by dynamic NMR. An ultrasound absorption study of the acetate complexation process in 1974 puts the water exchange rate constant on Ca^{2+}(aq) at around 6–$9 \times 10^8 \, s^{-1}$ [82]. Eigen's 1963 paper, on the other hand, quotes a value no more accurate than $\sim 10^8 \, s^{-1}$ [83].

The differing behaviour between Ca^{2+} and Mg^{2+} is reflected in their biofunctions. While the smaller Mg^{2+} is mainly located within cells largely in a rigid structural role the larger and more labile Ca^{2+}(aq), like the group 1 metals Na^+ and K^+, is mobile and well established as playing a key role in triggering dynamic processes in higher organisms such as nerve impulses and muscle contraction.

Mg^{2+} is perhaps best known for its association with ATP hydrolysis (Mg–ATPase activity), bacterial nitrogen fixation (linked also to Mg–ATPase activity) and in green plant photosynthesis (Mg–porphyrin complex).

2.3.4 Strontium and Barium

2.3.4.1 CHEMISTRY OF SR^{2+}(AQ) AND BA^{2+}(AQ)

As expected, the extent of hydration continues to fall off as one progresses down the group 2 element series. Hydrates containing $\geq$ six water molecules are rare for both Sr^{2+} and Ba^{2+} in contrast to Mg^{2+} and Ca^{2+}. Data from difference neutron scattering measurements by Nielson and Broadbent from a 3 m Sr^{2+} solution in D_2O showed very weak hydration such that it was now not possible to resolve the individual Sr^{2+}–O and Sr^{2+}–D peaks. Fourier transformation of the difference intensity function, however, led to estimates for the Sr—O and Sr—D distances as 265 and 320–330 pm respectively [84]. Values somewhat in excess of six have nonetheless been claimed from this and several other XRD and neutron diffraction studies [85, 86] (Sr—O = 250–265 pm). Heinzinger's MD simulation study of a 1.1 m $SrCl_2$ solution gives a value for the hydration number as 9.8 [87]. This value, as with that determined for Ca^{2+}, is perfectly reasonable given the larger size of Sr^{2+} and insofar as one can define a hydration shell.

Since barium extensively absorbs X-rays XRD methods are not suitable for structural studies of the hydration around Ba^{2+}. Nonetheless, attempts have been made and in one such study a hydration number of 9.5 was claimed (Ba—$O(OH_2)$ = 290 pm) [85], which is similar to a value of 9.7 deduced from NMR measurements [88]. Such a hydration number for Ba^{2+} is perfectly consistent with values found in coordination compounds.

Perhaps the best-studied hydrated salts of Sr^{2+} and Ba^{2+} are the soluble octahydrated hydroxides isolated from aqueous dimethylsulfoxide or aqueous pyridine. A redetermination of the X-ray structures of both salts by Sandstrom and coworkers led to the following parameters: $[Sr(OH_2)_8](OH)_2$: average Sr—O, (eight primary shell waters) = 262 pm, (twelve secondary shell waters plus OH^-) $\sim$ 476–480 pm); $[Ba(OH_2)_8](OH)_2$: average Ba—O, (eight primary shell waters) = 279 pm, (twelve secondary shell waters plus OH^-) $\sim$ 476–480 pm. The $[M(OH_2)_8]^{2+}$ units in $[Sr(OH_2)_8]$ $(OH)_2$ take up a highly distorted square antiprismatic arrangement approaching dodecahedral whereas in $[Ba(OH_2)_8](OH)_2$ the $[M(OH_2)_8]^{2+}$ units are more towards regular square antiprismatic (Fig. 2.9) [89]. The OH^- ions have an unusual hydrogen bond arrangement, with five bonds accepted and one donated. The coordinated waters are tilted to the plane of the M—O bond vector. These workers also reinvestigated the hydration of Sr^{2+} and Ba^{2+} in solution using both LAXS and EXAFS methods [89]. The LAXS experiments, which can give a weighted contribution

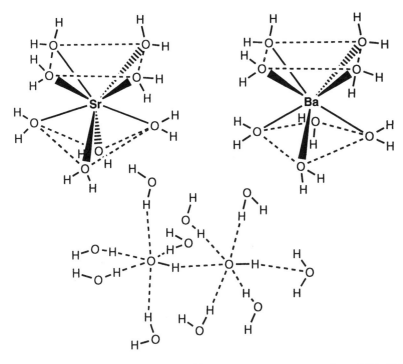

Figure 2.9. Structures of the $[M(OH_2)_8]^{2+}$ units in $[Sr(OH_2)_8]$ $(OH)_2$ and $[Ba(OH_2)_8]$ $(OH)_2$ shown together with the unusual hydrogen-bonding network around OH^-

over all interatomic distances, showed severe overlap between the M—O distance and the hydrogen bonded O—O distance for the aqueous solution and required elimination of the background contribution by subtraction within a difference procedure. The two techniques found independent evidence in support of a primary hydration number of 8.1(3) for both ions with mean bond lengths of Sr—O(OH$_2$) = 263(2) pm and Ba—O(OH$_2$) = 281(3) pm. The second hydration sphere was very diffuse with around 13 water molecules at similar distances as in the X-ray structures of the hydroxides with a further 2–3 at a closer distance.

With the aid of the above M—O distances for Sr^{2+}(aq) and Ba^{2+}(aq) one is in a position to construct a table of hydration data for the group 2 metal ions (Table 2.4). A clear linear correlation results when $-\Delta H^{\circ}_{hyd}$ is plotted against the reciprocal of M—O distances (Fig. 2.10), indicative of the simple Coulombic relationship of Fig. 1.2. The somewhat lower M—O distances of 162 pm for some Be^{2+} compounds are probably a reflection of significant covalency. Including a value of -1266 kJ mol^{-1} for the hydration enthalpy of Ra^{2+} along with M—O

Table 2.4. Hydration data for the group 2 metal ions (from ref. [89])

$M^{2+}(aq)$	$M—O(OH_2)_{av}$ [a] (pm)	$-\Delta H_{hyd}^\circ$ (kJ mol^{-1})	ΔV_i° [b] (cm^3 mol^{-1})	pK_{11} [c]
Ba	167	2494	−12.0	5.4
Mg	209	1921	−21.2	11.4
Ca	242	1577	−17.9	12.85[d]
Sr	263	1443	−18.2	13.29[d]
Ba	281	1305	−12.5	13.47[d]

[a] Mean value determined from XRD, ND, LAXS and EXAFS data.
[b] Partial molar hydration volumes relative to that for $H^+ = 0$ (taken from ref. [90]).
[c] Values from ref. [47].
[d] Soluble dihydroxide for M^{2+}.

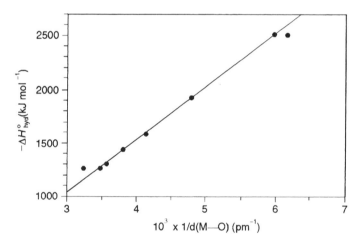

Figure 2.10. Plot of $\Delta H_{hyd}^\circ (M^{2+})$ (kJ mol^{-1}) vs the reciprocal of the M—O(OH$_2$) distance in hydrates (pm) for the group 2 metal ions

distances for eight- and twelve-coordinated Ra^{2+} compounds (mean values of 287 and 309 pm) allows an estimate of the hydration number of Ra^{2+}(aq) from Fig. 2.10 as 8.5 ± 0.5. Estimates of the partial molar hydration volumes, ΔV_i°, have been used to predict average hydration numbers around each cation. Assuming a value of six for Mg^{2+} and eight for Sr^{2+} and Ba^{2+}, the trends predict a value for Ca^{2+} around seven [89, 90]. Although examples of eight-coordinated Ca^{2+} are known in some hydrates, six-coordination is also not uncommon and the value of seven might reflect a tendency for Ca^{2+} to adopt either depending upon other factors such as crystal packing and the water/Ca^{2+} ratio. A similar situation may be relevant for Sc^{3+} (see Section 3.1).

2.4 GROUP 13 ELEMENTS: BORON, ALUMINIUM, GALLIUM, INDIUM AND THALLIUM

2.4.1 Boron

2.4.1.1 CHEMISTRY OF THE TETRAHYDROXOBORATE ION $[B(OH)_4]^-$

An extensive range of hydroxooxoborates, both mononuclear and polynuclear, exists as anions in a variety of metal ion salts. Many can be made to form glasses like the silicates with their properties depending upon the salt composition. The radius of B^{3+} (11 pm) is so small that true aqua ion cannot exist (the coordinated water molecules would be so acidic that complete acid dissociation in water would occur). In many cases extensive deprotonation to borates or hydroxoborates is the result. However, the nearest aqua ion equivalent, the tetrahydroxoborate ion, $[B(OH)_4]^-$, has been found in the crystal structures of a number of borate salts as well as certain crystallised forms of sodalite [91]. In $Ba[B(OH)_4]_2$ [92] the $[B(OH)_4]^-$ ion has the expected regular tetrahedral shape with an average B—OH distance of 147.1 pm (Fig. 2.11). The structure also consists of an irregular geometry of nine oxygens around Ba^{2+}. $[B(OH)_4]^-$ is characterised by the expected vibrations for tetrahedral symmetry at 941 cm^{-1} (v_3, strong in the infra-red), 741 cm^{-1} (v_1, strong in Raman, weak in the infra-red) and 535 cm^{-1} (v_4, weak in both the Raman and the infra-red). In aqueous media the ion is the dominant form at low boron concentrations (< 0.025 M) and also predominates above pH 11 [47]. At higher boron concentrations and/or lower values of pH a variety of polynuclear hydroxooxoborates are formed. The most important of these is the pentahydroxoooxoborate $[B_5O_6(OH)_4]^-$ which has a spiro structure. Monocyclic $[B_3O_3(OH)_4]^-$ and bicyclic $[B_4O_5(OH)_4]^{2-}$ ions are also known and are relevant to hydrolysed solutions of $B(OH)_3$ (see below). Improved

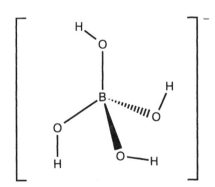

Figure 2.11. The structure of the $[B(OH)_4]^-$ ion in $Ba[B(OH)_4]_2$

synthetic routes to salts of the anion $[B_5O_6(OH)_4]^-$ have been described [93]. Vibrational spectroscopy remains a technique of choice for distinguishing between the various polynuclear species.

The NMR active nucleus ^{11}B $(I = 3/2)$ has been widely used to follow polymerisation and hydrolytic reactions of aqueous boron species under varying conditions [94]. The protonated form of $[B(OH)_4]^-$ is the Lewis acid $B(OH)_3$ (boric acid). The hydrolysis of $B(OH)_3$ above pH 7 and at low boron concentrations, to $[B(OH)_4]^-$, has been extensively studied [95, 96]. Baes and Mesmer [47] calculated the following values for 25 °C: $pK_{11} = 9.235$ ($[B(OH)_4]^-$), $pK_{21} = 9.36$ ($[B_2O(OH)_5]^-$), $pK_{31} = 7.03$ ($[B_3O_3(OH)_4]^-$) and $pK_{42} = 16.3$ ($[B_4O_5(OH)_4]^{2-}$). The anion $[B_5O_6(OH)_4]^-$ was not discussed. Ishihara *et al.* [97] have carried out a variable temperature (20–52 °C) and pressure (0.1–250 MPa) ^{11}B NMR band shape analysis study of the kinetics of the interchange between boric acid, $B(OH)_3$, and $[B(OH)_4]^-$ in aqueous solutions at pH 8–10:

$$B(OH)_3 \underset{k_y}{\overset{k_x}{\rightleftharpoons}} B(OH)_4^- \qquad (2.3)$$

The observed first-order rate constant to the right (k_X) is related to the second-order rate constant (k_{ex}) for the exchange

$$*B(OH)_3 + B(OH)_4^- \overset{k_{ex}}{\rightleftharpoons} *B(OH)_4^- + B(OH)_3 \qquad (2.4)$$

by the rate law

$$k_X = \frac{k_{ex}[B]_T[OH^-]}{K_b + [OH^-]} \quad \text{where } K_b = \frac{[B(OH)_3][OH^-]}{[B(OH)_4^-]} \qquad (2.5)$$

At 25.8 °C, $k_{ex} = 2.58 \times 10^6 \, M^{-1} s^{-1}$, $\Delta H_{ex}^{\ddagger} = 20.1 \pm 1.0 \, kJ \, mol^{-1}$, $\Delta S_{ex}^{\ddagger} = -55.0 \pm 3.1 \, J \, K^{-1} \, mol^{-1}$, $\Delta V_{ex}^{\ddagger} = -9.9 \pm 0.5 \, cm^3 mol^{-1}$ and $K_b = 1.72 \times 10^{-5} \, M$. The negative $\Delta S_{ex}^{\ddagger}$ and $\Delta V_{ex}^{\ddagger}$ values are consistent with associative attack of $[B(OH)_4]^-$ on the planar $B(OH)_3$ to form a dimeric transition state (Fig. 2.12).

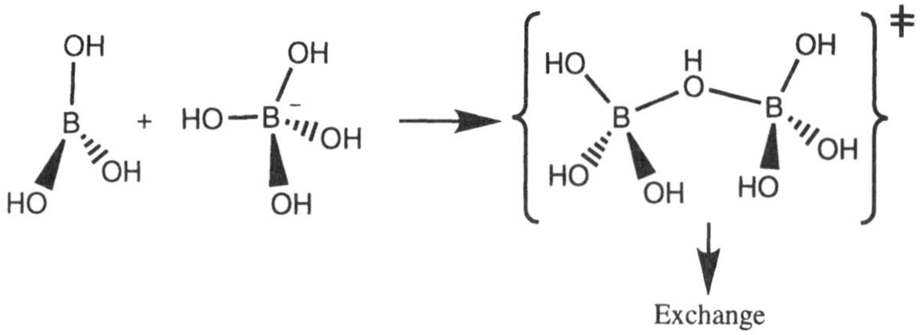

Figure 2.12. Dimeric Transition state in the Reaction of $[B(OH)_4]^-$ with $B(OH)_3$

The existence of such a mechanism is of interest since oxygen isotopic exchange reactions with bulk water at most kinds of oxoanion, $[XO_m]^{n-}$, appear to involve dissociative X—O bond cleavage. Molybdate and tungstate $[MO_4]^{2-}$ anions (Section 6.2.5) may be an exception and *ortho*-phosphate, PO_4^{3-}, is another. It is no surprise therefore to find that these particular oxo anions have a strong tendency to form a range of stable oligomeric species.

2.4.2 Aluminium

2.4.2.1 CHEMISTRY OF THE $[AL(OH_2)_6]^{3+}$ ION

The regular octahedral $[Al(OH_2)_6]^{3+}$ ion has been characterised in a range of hydrated Al^{III} salts [98–100]. The series of isomorphous mixed metal sulfates $M^IM^{III}(SO_4)_2 \cdot 6H_2O$ all have in common an $[M(OH_2)_6]^{3+}$ ion and are termed 'alums' after the Al^{III} derivative which was the first to be characterised. A variety of XRD and neutron diffraction measurements have since been carried out on hydrated Al^{III} salts and in solution. Typical Al—$O(OH_2)$ distances are between 187 and 190 pm. Single-crystal Raman measurements place the v_1 (sym. Al—O str) frequency at 524 cm^{-1}. The Raman spectra from solutions of $[Al(OH_2)_6]^{3+}$ ($v_1 = 542$ cm^{-1}, $v_2 = 473$ and $v_5 = 347$) have been analysed as providing the archetypal example of regular octahedral six-coordination (see Section 3.1)

Water exchange on $[Al(OH_2)_6]^{3+}$ is slow enough to allow distinct signals from the free and coordinated waters to be observable at room temperature by ^{17}O NMR and at low temperatures ($\sim -40\,^\circ C$) by ^1H NMR. The ^{17}O resonance of the coordinated waters is only 6 ppm downfield from free water which, for observational purposes, has necessitated the use either of a free water Shift reagent (Co^{2+}) [101] or a free water relaxant (Mn^{2+}) [74, 102]. Integration of the respective peak areas of free and bound water has been used to confirm the hydration number of six around Al^{3+}. This has been well documented as an archetypal example. The hydration number has also been confirmed as 6 from ^{18}O labelling experiments [102], although the exchange was too rapid to be measured. Merbach and coworkers found that at temperatures around $-20\,^\circ C$ the water exchange was slow enough to be monitored conventionally using the growth of the coordinated water ^{17}O NMR signal following rapid injection of $Al^{3+}(aq)$ into an enriched $H_2{}^{17}O$ sample in the presence of added $Mn^{2+}(aq)$ [103]. At temperatures above 30 °C line broadening of the coordinated water resonance occurs in the fast exchanging region (see Section 1.4.3). Figure 2.13 shows the unique situation for $[Al(OH_2)_6]^{3+}$ of rate data for the water exchange process obtainable using both kinetic methods over a very wide temperature range. The parameters obtained are k_{ex} (25 °C) $= (1.29 \pm 0.04)$ s^{-1}, $\Delta H^\ddagger = (84.7 \pm 0.32)$ kJ mol^{-1}, $\Delta S^\ddagger = (+41.6 \pm 0.8)$ J K^{-1} mol^{-1} and $\Delta V^\ddagger = (+5.7 \pm 0.2)$ cm^3 mol^{-1}. Despite the 3+ charge it is believed that the Al^{3+} coordination sphere is too tightly packed to facilitate associative attack by an

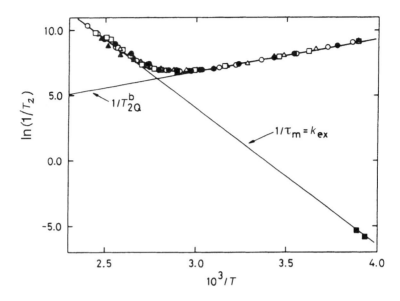

Figure 2.13. Temperature dependence of $1/T_2$ for the coordinated water ligands on $[\text{Al(OH}_2)_6]^{3+}$ ($\blacksquare = k_{\text{ex}}$ value determined by conventional means following rapid injection) [103]. (Reproduced by permission of Verlag Helvetica Chimica Acta AG)

incoming water and so the parameters have been interpreted as indicative of a mechanism in this case largely controlled by water ligand dissociation (see Section 1.4.4).

Substitution reactions on $[\text{Al(OH}_2)_6]^{3+}$ have been analysed assuming an Eigen–Wilkins preassociation (Section 1.4.5). Interchange rate constants (k_1) obtained [104] are fairly constant and similar in magnitude to k_{ex}, largely in keeping with the dissociative mechanism. The larger rate constant for acetate is believed to relate to its ability to promote hydrolytic polymerisation (see below).

2.4.2.2   THE HYDROLYSIS OF $\text{AL}^{3+}(\text{AQ})$

The small ionic size of Al^{3+} leads to extensive hydrolysis of its aqueous solutions. Extensive studies by Akitt *et al.* in the 1970 s [105] and then Akitt and Farthing in the early 1980 s [106] using ^1H, ^{27}Al and ^{17}O NMR have investigated the nature of hydrolysed $\text{Al}^{3+}(\text{aq})$ solutions. The major component of solutions $< 2 \times 10^{-2}\,\text{M}$ in the pH range 5–6 was found to be the hydrolytic tridecamer $[\text{Al}_{13}\text{O}_4(\text{OH})_{24}(\text{OH}_2)_{12}]^{7+}$ (Fig. 2.14) [105–107]. The other significant hydrolytic species appears to be the trimer, $\text{Al}_3(\text{OH})_4{}^{5+}(\text{aq})$, with smaller amounts of the diol dimer, $\text{Al}_2(\text{OH})_2{}^{4+}(\text{aq})$ [107, 108]. Hydrolysis constant, $\text{p}K_{xy}$, values, evaluated for $\text{Al}^{3+}(\text{aq})$ species by Baes and Mesmer [47] are $\text{p}K_{11} = 4.97$

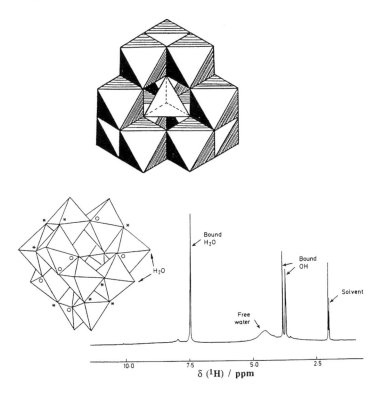

Figure 2.14. Structure of the $[Al_{13}O_4(OH)_{24}(OH_2)_{12}]^{7+}$ ion [110] and the ^1H NMR spectrum in d^6-acetone [111] (Reproduced by permission of The Royal Society of Chemistry)

$(AlOH^{2+}(aq))$, $pK_{12} = 9.3$ $(Al(OH)_2^+(aq))$, $pK_{22} = 7.7$ $(Al_2(OH)_2^{4+}(aq))$, $pK_{34} = 13.94$ $(Al_3(OH)_4^{5+}(aq))$, $pK_{13,32} = 98.73$ (tridecamer) and $pK_{14} = 23.0$ $([Al(OH)_4]^-)$. In acetate solution, the buffering effect leads to an appropriate concentration of monohydroxo species such that the principal product is now the diol dimer [109].

2.4.2.3 THE $[AL_{13}O_4(OH)_{24}(OH_2)_{12}]^{7+}$ ION

The hydrolytic tridecamer is perhaps one of the most illustrated oxo-hydroxo hydrolytic polymers. It is the Al^{III} cation present in 'basic aluminium sulfate' $Na[Al_{13}O_4(OH)_{24}(OH_2)_{12}](SO_4)_4 \cdot xH_2O$ and has existence as a hydrated perchlorate salt. The structure is based on a modified 'Keggin-type' arrangement [110] consisting of a central tetrahedral AlO_4^{5-} unit surrounded at its corners by four sets of three octahedrally coordinated aluminiums, each of which are joined by common edges. Thus the structure can be formulated as $[AlO_4(\{Al(OH)_2(OH_2)\}_3)_4]^{7+}$. This species provides a pertinent example of the use of NMR in the structural elucidation of hydrolytic polymers (see Section 1.2.3). In addition to the 12 apical waters there are two types of OH groups (Fig. 2.14), each giving a separate sharp resonance in the ^1H NMR spectrum in aqueous d^6-acetone at $-30\,^{\circ}C$ [111]. The tetrahedral and octahedral Al sites can be readily distinguished in solution in the ^{27}Al NMR spectrum. The central AlO_4 sites appears as a sharp resonance $+62.5\,ppm$ downfield of the singlet for $[Al(OH_2)_6]^{3+}$. Octahedral and indeed tetrahedral sites of other hydrolytic polymers often appear as very broad poorly defined features. The diol dimer, $Al_2(OH)_2^{4+}(aq)$, for example, shows only a broad ^{27}Al NMR resonance around $+(3-4)\,ppm$ in the acetate hydrolysed solutions [109]. Typical ^{27}Al NMR chemical shifts for tetrahedral and octahedral sites are in the range $+(60-70)\,ppm$ and $+(3-12)\,ppm$ from $[Al(OH_2)_6]^{3+}$ [107, 109].

It is not difficult to rationalise why further polymerisation occurs beyond initial formation of the diol dimer as the pH is raised since the dimer has a higher charge $(4+)$ than either $[Al(OH_2)_6]^{3+}$ and the precursor $[AlOH(OH_2)_5]^{2+}$ and this, coupled with the high effective positive charge on the small Al^{III} centre, readily leads to further deprotonation of coordinated water and to the formation of further OH^- bridges. Furthermore, PCM calculations (Section 1.3.3) show that the bridging OH^- groups on $Al_2(OH)_2^{4+}(aq)$ are themselves nucleophilic enough to attack other hydroxo aqua Al^{III} units (negative $\delta(OH_b)$) [112], leading to formation of $Al_3(OH)_4^{5+}(aq)$ and ultimately the hydrolytic tridecamer. The tridecamer structure is believed to arise from deprotonation of the μ_3-OH group of the trimer above followed by further water hydrolysis and the assembly of four trimer units, via the resulting nucleophilic μ_3-O^{2-} atom, around a central tetrahedral Al^{III}. The tridecamer structure probably owes its existence to the ready adoption of both octahedral and tetrahedral geometries by Al^{III} depending upon the pH. Estimates of the molar volume of the tridecamer have been reported [113].

As exemplified in the tridecamer, Al^{III}, like Si^{IV}, readily adopts a tetrahedral environment when surrounded by O^{2-} ions. Thus Al atoms can be readily incorporated into silicate structures forming aluminosilicates, of which the most well known and most interesting are the clays and zeolites.

2.4.3 Gallium and Indium

2.4.3.1 CHEMISTRY OF $[Ga(OH_2)_6]^{3+}$ AND $[In(OH_2)_6]^{3+}$

Both $[Ga(OH_2)_6]^{3+}$ and $[In(OH_2)_6]^{3+}$ have been structurally characterised in caesium alums [114]; the indium ion also in the selenate alum [115] and XRD measurements in solution has confirmed retention of the hexahydrated structure

[116]. In addition [71]Ga [117] and ^{17}O [101, 118] NMR has been used to confirm the presence of $[Ga(OH_2)_6]^{3+}$ in solution and the latter technique has been used to follow the water exchange process. The studies by Merbach and coworkers give parametres: $k_{ex} = 4.0 \times 10^2 \, s^{-1}$, $\Delta H^{\ddagger} = 67.1 \, kJ \, mol^{-1}$, $\Delta S^{\ddagger} = +30.1 \, J \, K^{-1} \, mol^{-1}$ and $\Delta V^{\ddagger} = +5.0 \, cm^3 \, mol^{-1}$ [118]. $GaOH^{2+}(aq)$ makes a significant contribution to the water exchange process although a detailed evaluation has proved difficult due to the wide range of values, 2.6–4.3, quoted for the hydrolysis constant, K_{11} or $-\log Q_{11}$, on $Ga^{3+}(aq)$ from various studies. The water exchange study only gives the product kK_{11}. Baes and Mesmer estimate pK_{11} to be 2.6 [47]. Since then a more recent investigation of the ionic strength dependence in nitrate media by Brown reported values for zero ionic strength as $pK_{11} = 2.56$ and $pK_{12} = 6.37$ [119]. Values determined experimentally at 0.1 M KNO_3 were reported as $-\log Q_{11} = 3.16$ and $-\log Q_{12} = 7.07$. Assuming pK_{11} to be 2.6 gives a value for the water exchange process on $GaOH^{2+}(aq)$ as $2 \times 10^5 \, s^{-1}$. The 10^3 times faster rate for $GaOH^{2+}(aq)$ is believed to stem from a labilisation of the leaving water ligand by a combination of a lower $2+$ charge (weaker electrostatics) and electron donation from the OH^- ligand (Section 1.4).

A number of kinetic studies of ligand substitution reactions have been reported on $Ga^{3+}(aq)$ with evidence for a significant contribution from the more labile $GaOH^{2+}(aq)$ [104, 120]. The data have been treated assuming an Eigen–Wilkins preassociation of reactants with values of K_{os} calculated using the Fuoss equation (Section 1.4). Interchange rate constants, k_I, are found to cover a fairly narrow range of values (~ 10–$130 \, s^{-1}$), providing further evidence for a water exchange process largely controlled by water ligand dissociation. The faster rate versus $Al^{3+}(aq)$ can be seen to derive largely from a lower $\Delta H^{\ddagger}$ value, reflecting longer and thus weaker largely electrostatic Ga—OH_2 bonds [118]. Earlier proposals of a changeover to an S_N2 process [101] seem unfounded.

Solutions of $[Ga(OH_2)_6]^{3+}$ are usually taken as the reference for [71]Ga NMR ($I = 1/2$) measurements. The technique has been used to study the nature of the products deriving from the hydrolysis of Ga^{3+} solutions [117]. Tetrahedral $[Ga(OH)_4]^-$ forms in strongly alkaline solutions and resonates downfield of $[Ga(OH_2)_6]^{3+}$ at $+192 \, ppm$ [121]. The ready formation of tetrahedral $Ga(OH)_4$ units leads one to suppose that Ga^{3+} should form the same hydrolytic tridecamer as in the case of Al^{3+}. Indeed, the hydrolytic polymer originally reported as '$Ga_{26}(OH)_{65}^{13+}(aq)$' from 0.1 M aqueous solutions of Ga^{III} sulfate has now been correctly reformulated as $[Ga_{13}O_4(OH)_{24}(OH_2)_{12}]^{7+}$ on the basis of [71]Ga NMR studies [122]. Ga^{3+} is moreover present at the central tetrahedral site in the mixed tridecamer $[Al_{12}GaO_4(OH)_{24}(OH_2)_{12}]^{7+}$ [123] and there have been EPR studies of the Mn^{III} sites in a number of substituted Mn_xAl_{13-x} tridecamers [124]. ^{1}H NMR studies have demonstrated that $Ga^{3+}(aq)$ retains its coordination sphere of six water molecules even in the presence of a large excess of acetone [125].

Both ^{1}H and ^{115}In NMR measurements have identified $[In(OH_2)_6]^{3+}$ as the cation present in aqueous perchloric acid solutions of In^{III} [126]. XRD studies of a 1.7 M In^{3+}(aq) solution quote the $In—O(OH_2)$ bond length as 215 pm [116]. This distance is significantly shorter than the sum of the effective sizes of In^{3+} and H_2O and suggests some covalency. A wide range of quoted $-\log Q_{11}$ values (formation of $InOH^{2+}$(aq)) have been quoted, ranging from 3.7 to 4.4 [47, 127]. Sylva and coworkers [127] have more recently reported values in 0.5 M nitrate media as: $pK_{11} = 4.31 \pm 0.003$, pK_{12} (formation of $In(OH)_2{}^+$(aq)) = 9.35 ± 0.01 and $pK_{46} = 7.32 \pm 0.006$. The tetrameric cation is relevant at indium concentrations ≥ 0.1 M. Baes and Mesmer [47] estimate pK_{11} for $[In(OH_2)_6]^{3+}$ to be around 4.0. LAXS and XRD studies carried out on hydrolysed 4.0–4.8 M aqueous solutions of $[In(OH_2)_6]$ $(NO_3)_3$ confirm the presence of a dominant tetrameric species, formulated as $[In_4(OH)_6(OH_2)_{12}]^{6+}$, and described in terms of an adamantanoid structure (Fig. 2.15), with a unique $In—In$ distance of 389 pm [116, 128]. The existence of trimeric $In_3(OH)_4{}^{5+}$(aq) has also been envoked within earlier potentiometric results on hydrolysed In^{3+}(aq) solutions [129]. The strong preference for octahedral coordination is the probable reason why the tridecamer structure, requiring one tetrahedral M^{3+}, has not been seen in the case of In^{3+}.

$In(OH)_3$ is known to dissolve in aqueous alkali but the nature of the hydroxy anionic species that result remains uncertain. The formation of both $[H_2InO_3]^-$ and $[In(OH)_4]^-$ have been proposed for the range up to 1.0 M $[OH^-]$ [130]. The solubility of $In(OH)_3$ then passes through a maximum around $[OH^-]$, 11.3 M above which the anion $[In(OH)_6]^{3-}$ can be precipitated [131].

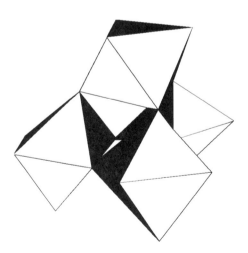

Figure 2.15. Coordination   polyhedral   structure   for   the   tetrameric   ion $[In_4(OH)_6(OH_2)_{12}]^{6+}$

The water exchange rate constant $(25\,°C)$ on $[In(OH_2)_6]^{3+}$ has been reported from an NMR study as $k_{ex} = 4 \times 10^4\,s^{-1}$, $\Delta H^{\ddagger} = 19\,kJ\,mol^{-1}$ and $\Delta S^{\ddagger} = -77\,J\,K^{-1}\,mol^{-1}$ [132]. The rate constant confirms a trend in increasing lability down the group, probably largely as a result of the increase in M^{3+} radius and corresponding decrease in cationic charge density. The low activation enthalpy and highly negative entropy might be considered to provide evidence of increased incoming ligand participation for the larger In^{3+}, contrasting with the mechanism believed to operate on $[Al(OH_2)_6]^{3+}$ and $[Ga(OH_2)_6]^{3+}$. However, further supportive kinetic data are needed.

2.4.4 Thallium

2.4.4.1 THALLIUM (I): CHEMISTRY OF TL^+ (AQ)

Thallium is unlike its group 13 members in having a stable $+1$ state in aqueous solution due to retention of its relativistically contracted 6s electrons. $Tl^+(aq)$ is formed from the aqueous dissolution of a number of hydrated Tl^I salts and can be obtained by simple dissolution of Tl_2O in the appropriate aqueous acid. It is readily obtainable from $[Tl(OH_2)_6]^{3+}$ (see below) by reduction. The E^{θ} value for $Tl^{3+}(aq)$ to $Tl^+(aq)$ $(+1.25\,V)$ shows that it is the stable ion for thallium in aqueous solution, but surprisingly little is known of the hydration sphere. There are no XRD studies from which to compute $Tl—O(OH_2)$ distances. It is often quoted as being hexaaqua $[Tl(OH_2)_6]^+$ [133] despite the possible presence of the stereochemically active lone pair of electrons (see Section 2.5). As a result one might predict a lower hydration number than six coupled with an anomalously high acidity. Solid $TlOH$ can be obtained from the metal and aqueous ethanol in the presence of oxygen. It dissolves to give strongly alkaline solutions. Baes and Mesmer have estimated pK_{11} for $Tl^+(aq)$ to be ~ 13.2 [47], which is reminiscent of the lighter group 1 and 2 metals and indeed supportive of a regular hydration number of ~ 6. However, $Tl^+(aq)$ resembles $Ag^+(aq)$ (Section 12.2) and not the group 1 $M^+(aq)$ ions in forming sparingly soluble $TlCl$, a feature believed to reflect a significant degree of covalency in the $Tl—Cl$ bonds. $TlCl$ itself is however, rather unstable in solution $(\log Q = 4.6 \ (I = 0, \ KCl))$ [134] and generally there have been few detailed reports of complex formation reactions involving $Tl^+(aq)$.

2.4.4.2 THALLIUM (II)

The $Tl^{2+}(aq)$ ion has been detected by pulse radiolysis on aqueous solutions of Tl^I sulfate [135] and in γ-irradiated frozen aqueous solutions [136]. $Tl^{2+}(aq)$ has been invoked as an intermediate in the photochemical reduction of aqueous Tl^{3+} solutions [137] but its role in redox reactions has been disputed [138]. There certainly seems to be no stable equilibrium aqueous phase corresponding to Tl^{2+}.

2.4.4.3   THALLIUM (III): CHEMISTRY OF $[Tl(OH_2)_6]^{3+}$

The octahedral hexaaqua structure for Tl^{3+} has been characterised from solution XRD studies [116, 139] and it is apparently present in a number of
hydrated salts. The perchlorate can be prepared by the reaction of $TlCl_3$
with $AgClO_4$ or by anodic electrolysis of $TlClO_4$ solutions. The hydrate
$Tl(ClO_4)_3 \cdot 6H_2O$ contains regular octahedral $[Tl(OH_2)_6]^{3+}$ ions [140]. The
rather large size of Tl^{3+}(aq), $Tl—O(OH_2) = 223$ pm, coupled with its very high
lability and high acidity (see below), may be reasons why the ion does not form an
alum in contrast to its lighter partners. The $Tl—OH_2$ distance is nonetheless
short when compared to the sum of the individual radii of Tl^{3+} and H_2O and
this, together with the anomalously low pK_{11} value of 0.6 [47, 141], has been
interpreted in terms of significant degree of covalency in the $Tl—O$ bonds.

Much less is known of the solution chemistry of Tl^{3+}(aq) compared to its
lighter group partners. Its water exchange rate constant has not been measured
directly but values of $\sim 10^7$–10^8 s^{-1} have been deduced from complexation
studies employing ultrasound absorption. An early theoretical estimate put the
value around 3×10^9 s^{-1} [142]. Glaser and coworkers have pioneered the use of
^{205}Tl NMR to study the complex halide ion-exchange process occurring within
$[TlX_{6-x}(OH_2)_x]^{(x-3)+}$ species ($X = Cl, Br$) in aqueous 3 M $HClO_4$ media. The
kinetic data on the exchange was obtained by an analysis of the ^{205}Tl NMR
linewidth under solution conditions wherein different $[TlX_{6-x}(OH_2)_x]^{(x-3)+}$
complexes were present (e.g. varying halide/Tl^{III} ratios). Rate constants for
the halide ion-exchange processes corresponding to $[Tl(OH_2)_6]^{3+}$ with
$[TlX(OH_2)_5]^{2+}$ and $[TlX_2(OH_2)_4]^+$ were similar for $X = Cl$ or Br and pointed to
a k_{ex} value in the range $(1$–$6) \times 10^8$ s^{-1} [143]. Structural information on the
various $[TlX_{6-x}(OH_2)_x]^{(x-3)+}$ complexes ($X = Cl, Br$ and CN) has been obtained
from a combined LAXS, EXAFS and Raman study. The findings confirmed
octahedral geometry for $[Tl(OH_2)_6]^{3+}$ ($Tl—O = 221$ pm), $[TlX(OH_2)_5]^{2+}$
($Tl—O = 224$ ppm), $trans$-$[TlX_2(OH_2)_4]^+$ ($Tl—O = 233$ ppm) and $[TlX_5$
$(OH_2)]^{2-}$, but trigonal bipyramidal geometry for neutral $[TlX_3(OH_2)_2]$
($Tl—O = 237$ ppm) [144].

Due to the high acidity constant K_{11}, Tl^{3+}(aq) is only stable in strongly acidic
solution. Mononuclear hydrolysis products $TlOH^{2+}$(aq) and $Tl(OH)_2{}^+$(aq)
appear to be the only forms relevant prior to precipitation of the hydrous Tl_2O_3
around pH 1. The absence of polymeric hydrolysis products has been explained
on the basis of the fact that the ion $[TlX(OH_2)_5]^{2+}$ ($X = Cl, Br$) retains very high
acidity ($K_{11} \sim 0.016$ M) whereas $[TlX_2(OH_2)_4]^+$ is not acidic at all [145]. It has
been proposed that only two water molecules are strongly coordinated to
Tl^{3+}(aq) such that on formation of $Tl(OH)_2{}^+$(aq) the remaining weakly bonded
waters have partial charges (Section 1.3.2) at oxygen that are too negative to
allow further deprotonation by water leading to oligomerisation. One can
therefore picture the $Tl—O$ bonds in the Tl^{3+}(aq) ion and in $Tl(OH)_2{}^+$(aq), as

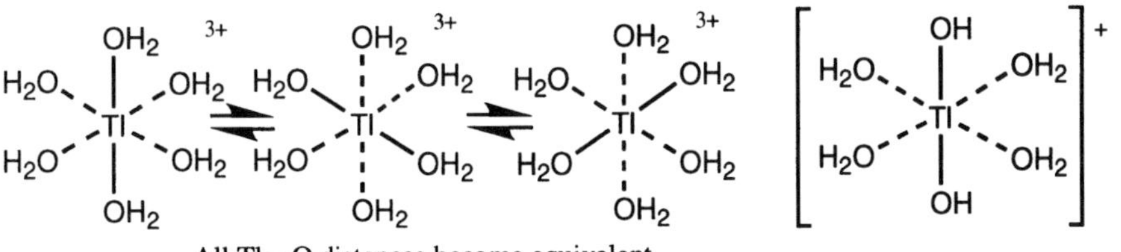

All Tl—O distances become equivalent

Figure 2.16. Dynamic solution structures for $[Tl(OH_2)_6]^{3+}$ and *trans-* $[Tl(OH)_2(OH_2)_4]^+$

depicted in Fig. 2.16. A similar situation is believed to operate for the isoelectronic $Hg^{2+}(aq)$ ion (see Section 12.3.2).

The high acidity of $[Tl(OH_2)_6]^{3+}$ has been used to promote hydrolysis reactions on bound ligand substrates such as in the case of thiolurethanes [146] and organo isothiocyanates [147]. These are rather unique reactions promoted by Tl^{III} because of its 'soft' nature and tendency to bind to sulfur donors.

$[Tl(OH_2)_6]^{3+}$ is a powerful oxidant with interest stemming from the two-electron nature of the reduction process to $Tl^+(aq)$. Kinetic studies describing the oxidation of a wide variety of substrates including carbohydrates [148], SO_2 [133], ascorbic acid [149], various carboxylic acids [150] and thiocyanate [151] have been documented. In most cases the reactions are believed to be inner sphere in nature, involving initial preassociation and then substitution into the coordination sphere of the Tl^{3+} prior to the electron transfer step [133].

2.5 GROUP 14 ELEMENTS: SILICON, GERMANIUM, TIN AND LEAD

2.5.1 Silicon

There are no aqua ions of silicon because of the extremely small effective radius of Si^{4+}. Indeed, Si^{4+} is a strong enough 'acid' to deprotonate bound OH^-, leading quickly to the formation of the hydrous dioxide silica, $SiO_2 \cdot nH_2O$, and with M^{n+}, OH^- to extensive polymeric oxo-bridged structures deriving from SiO_4 tetrahedra, the silicates which make up much of the earth's crust. 'Water glass' is a highly viscous aqueous solution of sodium silicate $(Na_2SiO_3 \cdot nH_2O)$ consisting of mobile Na^+ and OH^- ions within an extended hydrogen-bonded network of oxo-bridged $[SiO_4]^{4-}$ ions [152]. The solutions contain linear, cyclic or cage structures, the proportions depending upon the pH, concentration and temperature. The lifetime of free $[SiO_4]^{4-}$ in solution is in the millisecond range [153]. The compound 'Si(OH)$_4$' has only a fleeting existence in hydrolysed Si^{IV} solutions before quickly transforming into various phases of hydrated silica.

2.5.2 Germanium

2.5.2.1 GERMANIUM (II)

Ge^{2+}(aq) or possibly $Ge(OH)^+$(aq) is said to exist in dilute air-free aqueous suspensions of the yellow hydrous monooxide [47]. However, both are unstable with respect to the ready formation of $GeO_2 \cdot nH_2O$.

2.5.2.2 GERMANIUM (IV)

Although the oxide chemistry of Ge^{IV} is largely reminiscent of Si^{IV}, Ge^{IV} has a greater tendency towards octahedral coordination and as such its aqueous chemistry is somewhat intermediate between that of silicon and tin in that discrete mixed oxo-hydroxo anions are known. The major Ge^{IV} species are anionic and appear to be $[GeO(OH)_3]^-$ and octameric $[Ge_8O_{16}(OH)_3]^{3-}$ with smaller amounts of $[GeO_2(OH)_2]^{2-}$ [47, 154].

2.5.3 Tin

2.5.3.1 TIN(II): CHEMISTRY OF SN^{2+}(AQ)

Tin(II) is the first group 14 oxidation state to form a true hydrated cation. However, X-ray scattering and EXAFS measurements on 3 M aqueous acidic Sn^{2+} perchlorate solutions reveals a highly asymmetric hydration sphere consisting of three or possibly four water ligands at a distance of around 233–234 pm in a distorted trigonal pyramidal arrangement with the active lone pair occupying the remaining special position [155]. A number of further weakly bonded waters may also be relevant at distances of around 280–290 pm. These results have been discussed in terms of a monomeric hydrated Sn^{2+}(aq) ion. However, the ready formation in these solutions of oligonuclear species such as trimeric $Sn_3(OH)_4^{2+}$(aq) (predominant form) along with some $Sn_2(OH)_2^{4+}$(aq) is indicated from the earlier potentiometric work of Tobias [156]. The cyclic trimeric unit $Sn_3O(OH)_2^{2+}$ has been detected in basic salts such as $[Sn_3O(OH)_2](SO_4)$ [156], leading to the ascertion that similar cyclic units might be present in the oligomeric Sn_3^{2+}(aq) solutions. In the structure of $[Sn_3O(OH)_2](SO_4)$ there are two basic types of Sn atom, one trigonally coordinated to all three ring oxygens and the other square pyramidally coordinated to ring oxygens and SO_4^{2-} ions. The SO_4^{2-} ions occupy two different positions within the unit cell arrangement (Fig. 2.17). The $Sn_3(OH)_4^{2+}$ unit could arise by H_2O effectively replacing the

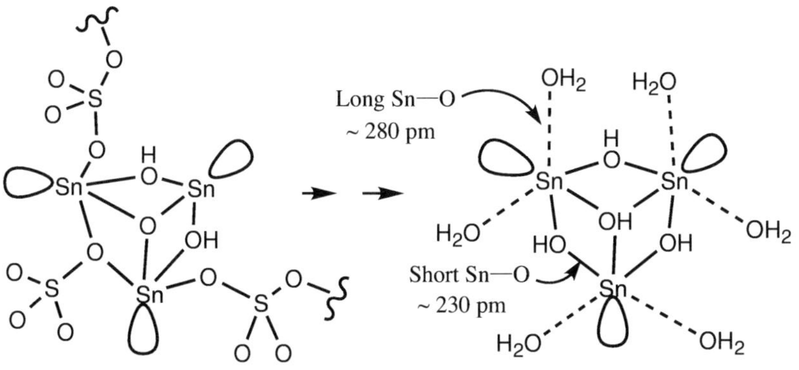

Figure 2.17. (a) a portion of the chain structure in the unit cell of $[Sn_3O(OH)_2]$ (SO_4) showing the lone pairs and (b) a possible structure for the trimeric $Sn_3(OH)_4^{2+}$ (aq) cation

bridging SO_4^{2-} ions. The lone pairs point outwards from the trimer. The structure of $[Sn_3O(OH)_2]$ (SO_4) reveals a range of Sn–O distances for the two types of tin atoms [157]. The trigonal Sn—O distances are in the range 206–218 pm whereas the square pyramidal tin atoms have a wider range of Sn—O distances from 206 to 249 pm. It has proved difficult, however, to match up both Sn—O and Sn—Sn distances in the basic sulfate with those from XRD data on the hydrolysed trimer in solution [158]. In the latter the Sn—O and Sn—Sn distances would be perfectly consistent with an expanded cyclic $Sn_3(OH)_4^{2+}$ structures shown in Fig. 2.17 with the longer Sn—O distance of 280–290 pm assigned to the weakly bonded waters.

2.5.3.2 TIN (IV): CHEMISTRY OF SN^{4+}(AQ)

There appears to be good ^{119}Sn NMR evidence in support of the existence of monomeric octahedral aqua-chloro (bromo) species $[SnX_n(OH_2)_{6-n}]^{(4-n)+}$, in addition to putative $[Sn(OH_2)_6]^{4+}$ itself, in dilute acidic aqueous solutions of Sn^{4+} in $HClO_4$ in the presence of Cl^- and Br^- [159]. In alkaline solution the species $[Sn(OH)_6]^{2-}$ appears to have existence.

2.5.4 Lead

2.5.4.1 LEAD (II): CHEMISTRY OF PB^{2+}(AQ)

In constrast to the reducing properties of divalent tin the stable oxidation state for lead in aqueous media is Pb^{II} for the same reasons as described for Tl^+. However, there is no evidence for a simple monomeric Pb^{2+}(aq) ion. The two

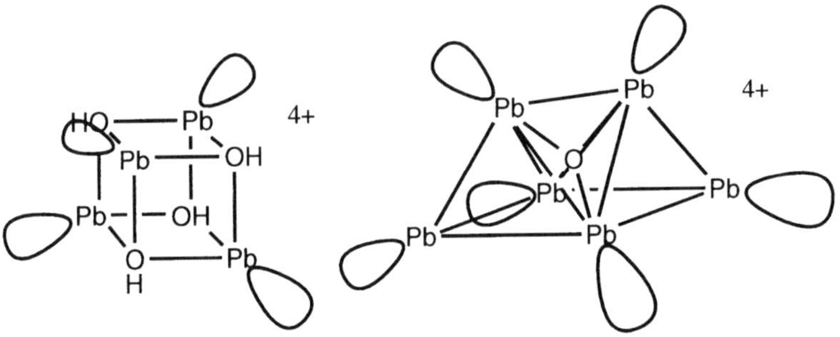

Figure 2.18. Structures of cuboidal $Pb_4(OH)_4^{4+}$ (aq) and hexameric $Pb_6O(OH)_6^{4+}$(aq) (trigonal face capping OH ligands are omitted for clarity)

species that appear to have been characterised are oligonuclear. Cuboidal $Pb_4(OH)_4^{4+}$(aq) has been characterised in solid state structures of $[Pb_4(OH)_4]$ $(ClO_4)_4$ and $[(Pb_4(OH)_4)_3(CO_3)]$ $(ClO_4)_{10} \cdot 6H_2O$ [160] and $Pb_6O(OH)_6^{4+}$(aq) in the α and β forms of $[Pb_6O(OH)_6]$ $(ClO_4)_6 \cdot H_2O$ [161]. X-ray scattering measurements in solution have proved difficult owing to the high X-ray absorption by Pb atoms, but the available data [162] appears to be consistent with the presence of both of these oligomeric structures in aqueous solutions of Pb^{2+} depending upon the OH^-/Pb ratio. In the structure of the $Pb_6O(OH)_6^{4+}$ the oxide atom is tetrahedrally coordinated to four Pb atoms with six OH groups capping the six outer trigonal faces (Fig. 2.18). The lone pairs again point outwards from the cage, largely restricting its hydration. In the two hexameric perchlorate salts a single weakly bound water ligand is present at 274 pm from one corner lead atom [161]. The Pb—Pb distances in the hexamer show a wide variation from 344 to 419 pm whereas in the cuboidal cation they are at 381 pm. The existence of a simple monomeric Pb^{2+}(aq) ion has not been confirmed.

2.5.4.2   LEAD(IV): CHEMISTRY OF PB^{4+}(AQ)

Tetravalent lead is oxidising in aqueous media. The Pb^{IV}/Pb^{II} redox change (in aqueous sulfate media) is of course central to the electrochemical processes operating in the rechargeable lead acid battery. Pb^{4+}(aq) is extensively hydrolysed in aqueous media although the nature of the soluble hydrolysis products remains unknown. There is little reliable data on the solution equilibria or whether soluble cations exist, since the formation of hydrous PbO_2 quickly dominates. In alkaline solutions of Pb^{IV} the hexahydroxo ion $[Pb(OH)_6]^{2-}$ is said to persist.

2.6 GROUP 15 ELEMENTS: ARSENIC, ANTIMONY AND BISMUTH

2.6.1 Arsenic

The oxidation states relevant to arsenic in aqueous media are As^{III} and As^V. As^V resembles P^V in its properties whereas As^{III} resembles Bi^{III} (see below). Like $B(OH)_3$ (Section 2.3.1), pyramidal $As(OH)_3$, despite the presence of the stereochemically active lone pair, shows Lewis acid behaviour since it can readily expand its coordination geometry. Thus it readily forms adducts with a wide variety of Lewis base donors including OH^- to form $As(OH)_4^-$(aq). In dilute aqueous solutions $< 0.1\,m$ the principal species are $As(OH)_3$(aq), $As(OH)_4^-$(aq), $AsO_2(OH)^{2-}$(aq) and AsO_3^{3-}(aq). The following hydrolysis constants for As^{III} and As^V species have been evaluated by Baes and Mesmer [47]:

$$As(OH)_3(aq) + H_2O \rightleftharpoons As(OH)_4^- + H^+, \quad pK = 9.29 \qquad (2.6)$$

$$H_3AsO_4 \rightleftharpoons H_2AsO_4^- + H^+, \quad pK = 2.24 \qquad (2.7)$$

$$H_3AsO_4 \rightleftharpoons HAsO_4^{2-} + 2H^+, \quad pK = 9.20 \qquad (2.8)$$

$$H_3AsO_4 \rightleftharpoons AsO_4^{3-} + 3H^+, \quad pK = 20.7 \qquad (2.9)$$

The kinetics of oxygen exchange from water to AsO_4^{3-} has been monitored by ^{18}O labelling [163, 164]. The process in very complex involving exchange with water and between various arsenate anions themselves and it is also catalysed by trace amounts of $H_2AsO_3^-$, which exchanges much faster, probably via a favourable associatively activated transition state (Fig. 2.19). Table 2.5 summarises some of the available data.

2.6.2 Antimony

The oxidation states of relevance to aqueous solution are Sb^{III} and Sb^V. In water interestingly neither state seems particularly stable in the absence of either reducing or oxidising conditions.

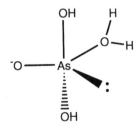

Figure 2.19. Associatively activated transition state for water exchange on $H_2AsO_3^-$

Table 2.5. Kinetic data for oxygen exchange on various arsenic anions at 30 °C from ^{18}O labelling [163]

Reactants	k_{ex}	$\Delta H^{\ddagger}$ (kJ mol^{-1})	$\Delta S^{\ddagger}$ (J K^{-1} mol^{-1})	Ref.
$AsO_4^{3-} + H_2O$	$1.5 \times 10^{-6} s^{-1}$			[164]
$HAsO_4^{2-} + H_2O$	$1.2 \times 10^{-5} s^{-1}$	92	-46	[164]
$H_2AsO_4^- + H_2O$	$1.0 \times 10^{-4} s^{-1}$			[164]
$H_2AsO_3^- + H_2O$	$1.67 \times 10^2 s^{-1}$	25	-120	[165]
$HAsO_4^{2-} + HAsO_4^{2-}$	$8.5 \times 10^{-6} M^{-1} s^{-1}$		-46	[164]
$H_2AsO_4^- + H_3AsO_3$	$6.8 M^{-1} s^{-1}$	52	-65	[166]
$H_2AsO_3^- + H_2AsO_3^-$	$6.9 \times 10^1 M^{-1} s^{-1}$	33	-102	[165]

2.6.2.1 ANTIMONY(III)

Sb^{III} is somewhat less acidic than Sb^V and the formation of a cation $Sb(OH)_2^+$(aq) below pH 2 has been fitted to the solubility profile of Sb_2O_3 [167, 168]. Fully protonated Sb^{3+}(aq) ions may have existence in dilute Sb^{III} solutions in concentrated acidic solutions, but this has not been substantiated and little remains known of the hydration sphere around cationic Sb^{III} ions. Above pH 11 the anion $Sb(OH)_4^-$(aq) (pK_{11} from $Sb(OH)_3$(aq) = 11.82) is formed [47, 168].

2.6.2.2 ANTIMONY(V)

Sb^V, however, differs from P^V and As^V in a preference for octahedral coordination and the anion $[Sb(OH)_6]^-$ is said to have existence. Antimonic acid has been formulated as either $H[Sb(OH)_6]$ (or $H_3O[Sb(OH)_6]$ or $Sb(OH)_5$). In view of the expected high Lewis acidity of $Sb(OH)_5$ a more likely formulation is $Sb(OH)_5 \cdot (OH_2)$. At low antimony concentrations (10^{-5} m) only the mononuclear species $Sb(OH)_5 \cdot (OH_2)$ and $[Sb(OH)_6]^-$ seem to be important, with the latter the dominant form above pH 4 [47]. However, at 0.1 m concentrations the titration data seem compatible with the formation of a range of dodecameric species in addition to $[Sb(OH)_6]^-$ above pH 6 [169]. Many solid antimonate (V) salts such as $Na_2H_2Sb_2O_7 \cdot 5H_2O$ and $Mg(SbO_3)_2 \cdot 12H_2O$ have been reformulated on the basis of X-ray evidence to consist of $Na[Sb(OH)_6]$ and $[Mg(OH_2)_6][Sb(OH)_6]_2$.

2.6.3 Bismuth

2.6.3.1 BISMUTH(III): CHEMISTRY OF Bi^{3+}(AQ)

The simple monomeric Bi^{3+}(aq) ion has recently been characterised from acidic solutions as a triflate salt and shown to have nine primary shell water molecules coordinated in a tricapped trigonal prismatic arrangement (Fig. 2.20) [170].

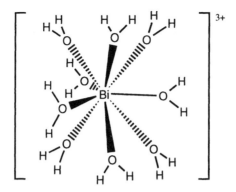

Figure 2.20. The structure of the $[Bi(OH_2)_9]^{3+}$ ion in $Bi(O_3SCF_3)_3 \cdot 9H_2O$

Similarities with the lanthanides become immediately apparent (see Section 3.2). A feature of Fig. 2.20 is the absence of lone pair stereoactivity. However, this is often seen to be the case when high coordination numbers $\geq$ six are relevant. The Bi—O distances are 244.8 pm (waters at the corners of the trigonal prism) and 257.7 pm (face capping waters). The ratio (1.05) is in keeping with other regular tricapped trigonal prismatic geometries. The effective ionic radius of Bi^{3+} can be estimated to be 115 pm, which is very similar to the Shannon radii [171] for nine-coordination at Nd^{3+} (116.3 pm) and Sm^{3+} (113.2 pm) for which hydration by nine water molecules in a similar tricapped trigonal prismatic geometry has been established (Section 3.2). Estimates for the pK_{11} value for $BiOH^{2+}(aq)$ have been given as 1.09 [47]. The ability to isolate the simple monomeric $[Bi(OH_2)_9]^{3+}$ ion seems unique to the triflate solutions since the use of even more dilute Bi^{3+} solutions in $HClO_4$ or HNO_3 lead to crystallisation of hexameric 'bismuthyl' oxo-hydroxo species such as $Bi_6O_4(OH)_4^{6+}$, as illustrated in Chapter 1, Fig. 1.24. These findings clearly show that different anions can exert profound effects with regard to the extent of hydrolytic polymerisation processes on aqua cations. In this regard it may prove of interest to redetermine the pK_{xy} values for Bi^{3+} in triflate media. Likewise it might be of interest to reinvestigate the hydration behaviour of Pb^{II} (radius for six coordination 119 pm) in triflate media for the possible existence of $(Pb(OH_2)_9]^{2+}$.

The presence of the $Bi_6O_4(OH)_4^{6+}(aq)$ ion in acidic aqueous solutions of $Bi(ClO_4)_3$ and $Bi(NO_3)_3$ has been verified by XRD [172] as well as in solid salts such as $Bi_6O_4(OH)_4(ClO_4)_6 \cdot 7H_2O$ [172] and $Bi_6O_4(OH)_4(NO_3)_6 \cdot nH_2O$ ($n = 1$ or 4) [173]. Here lone pair activity is seen. As in the case of the Pb^{II} species the lone pairs are directed away from the cage with each Bi coordinated to four O (or OH) groups completing a distorted trigonal bipyramidal arrangement. Two Bi_9 species, formally $Bi_9(OH)_{20}^{7+}(aq)$ and $Bi_9(OH)_{22}^{5+}(aq)$, also appear relevant in the pH range above 3.0 from potentiometric data in the region of n between 2.0 and 2.5. The compound 'bismuth citrate' is well established in the treatment of stomach ulcers. A number of structural studies have shown that bismuthyl 'oxo' complexes containing between 2 and 12 Bi atoms can be isolated from the aqueous clinical preparations [174].

2.6.3.2 BISMUTH(V)

The strongly oxidising nature of $Bi^V(aq)$ solutions (E^θ of $Bi^V/Bi^{III} = +2.03$ V) illustrates the increasing stability of the $5d^{10}6s^2$ configuration (two below the maximum) along the p-block elements following mercury. This has been assigned to relativistic contraction of the 6s shell enhanced by the continuing increase in Z_{eff} along the seventh period. As a result $Bi^V(aq)$ rivals $[S_2O_8]^{2-}$ ($E^\theta = +2.01$ V) and ozone ($E^\theta = +2.07$ V) as being amongst the most powerful oxidants in aqueous solution. Heating Na_2O_2 and Bi_2O_3 gives a yellow-brown solid formulated as $NaBiO_3$ (sodium bismuthate). This dissolves in 0.5 M $HClO_4$ to give a strongly oxidising solution which is stable in the absence of light for several days [175]. By analogy with antimony and tellurium (Section 2.7.3) the soluble Bi^V species could be $H[Bi(OH)_6]$.

2.6.3.3 CORNER-SHARED HETEROMETALLIC CUBE AQUA IONS, $[Mo_6M'S_8(OH_2)_{18}]^{8+}$

A feature of the p-block elements In, Tl, Sn, Pb, Sb and Bi, like Hg (Section 12.3), is an ability to react with the trinuclear $[Mo_3S_4(OH_2)_9]^{4+}$ ion (Section 6.2.3) to form corner-shared double cube aqua ions (Fig. 2.21), with the heteroatom at the centre [176]. This arises via favourable coordination of the p-block metals with the electron-rich and facially-coordinating bridging sulfido ligands of the $Mo_3S_4^{4+}$(aq) ion. In and Sn are unique in also existing as single cube ions $[Mo_3(SnCl_3)S_4(OH_2)_9]^{3+}$ and $[Mo_3In(pts)_2S_4(OH_2)_9]^{3+}$. The bismuth

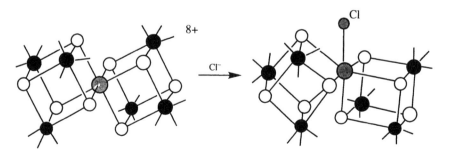

Figure 2.21. $[Mo_6BiS_8(OH_2)_{18}]^{8+}$ and proposed monochlorocomplex

derivative $[Mo_6BiS_8(OH_2)_{18}]^{8+}$ is also unique in strongly complexing one chloride $(K > 40\,M^{-1})$, believed to arise from complexation at bismuth to form a seven-coordinated species [177] (Fig. 2.20). The monochloro species is characterised by an intense near infra-red transition at 795 nm $(e \sim 16\,000\,M^{-1}\,cm^{-1}$ per $Mo_6)$ assignable to a Bi—Cl charge transfer.

2.7 GROUP 16 ELEMENTS: SULFUR, SELENIUM AND TELLURIUM

None of the members of group 16 can be considered to be truly metallic and as such one would not expect to find strong evidence for cationic aqueous species. As with other p-block groups the degree of metallic or semi-metallic behaviour increases down the group. The only true aqua cation may be that formed by Te^{IV}, discussed as $Te(OH)_3^+(aq)$. The semi-metallic behaviour for tellurium is further reflected in the highest $+6$ state by the formation of an octahedral hexahydroxo species but for sulfur and selenium only lower coordination oxo anions. Much less is known of the aqueous chemistry of polonium except in its resemblance to tellurium. Gamsjager and Murmann's 1983 review [163] provides an excellent account of the use of ^{18}O methods in the study of water exchange on oxo anions including those of the group 16 and 17 elements, so that only a few examples need be mentioned here.

2.7.1 Sulfur

A modelling of the structural, thermodynamic and dynamic properties of the tetrahedral SO_4^{2-} ion in water has been described [178]. Water exchange is immeasurably slow in neutral or alkaline solution. This was shown by no exchange detectable from a 90 atom % ^{18}O-enriched sample of sodium sulfate in water and in 1 M NaOH kept at 10 °C for 63 days [179]. A $t_{1/2}$ of > 100 years at 100 °C can be estimated. In acidic solution the reaction rate obeys the rate law: rate $= k[H^+][HSO_4^-]$ [180]. By varying the proportions of H_2SO_4 and water it was found that the rate was proportional to the concentration of sulfuric acid and the following mechanism was suggested:

$$H^+ + HSO_4^- \underset{}{\overset{\text{fast}}{\rightleftharpoons}} H_2SO_4 \qquad (2.10)$$

$$H_2SO_4 \overset{\text{slow}}{\rightleftharpoons} H_2O + SO_3 \qquad (2.11)$$

2.7.2 Selenium

Water exchange on the selenate ion occur much faster than on sulfate and is again acid catalysed. The rate law found is: rate $= k_1[H^+][HSeO_4^-]$

$+ k_2 [\mathrm{H}^+]^2 [\mathrm{HSeO_4}^-]^2$ [181]. The first term indicates mechanism (2.10) and (2.11). The second fourth-order term indicates the involvement of

$$\mathrm{H_2SeO_4} + \mathrm{H_2SeO_4} \xrightleftharpoons{\text{slow}} \mathrm{H_2Se_2O_7} + \mathrm{H_2O} \qquad (2.12)$$

2.7.3 Tellurium

2.7.3.1 TELLURIUM (IV): EXISTENCE OF CATIONIC $\mathrm{Te}^{\mathrm{IV}}(\mathrm{AQ})$?

Strong evidence for the formation of a cationic $\mathrm{Te}^{\mathrm{IV}}(\mathrm{aq})$ ion comes from measurements of the solubility of $\mathrm{TeO_2}$ in acidic solutions [182]. The solubility is found to increase linearly with acid concentration at low concentrations of acid. The following equilibrium is indicated from analysis of the data:

$$\mathrm{TeO_2}(c) + \mathrm{H_2O} + \mathrm{H}^+ \rightleftharpoons \mathrm{Te(OH)_3}^+(\mathrm{aq}) \qquad (2.13)$$

Values of $-\log Q$ for 2.13 vary from 1.5 to 2.2 [47]. Whether the ion should be represented as $\mathrm{Te(OH)_3}^+(\mathrm{aq})$ or $\mathrm{TeO(OH)}^+(\mathrm{aq})$ or $\mathrm{HTeO_2}^+(\mathrm{aq})$ is more uncertain. Further evidence for a cationic $\mathrm{Te}^{\mathrm{IV}}(\mathrm{aq})$ species in acidic solutions comes from attempts to partition $\mathrm{Te}^{\mathrm{IV}}$ between aqueous solutions and an extraction agent (dithizone or trioctylphosphine oxide) in $\mathrm{CCl_4}$ or hexane [183]. The distribution ratio parallels the acid concentration in the range 0.1–1.0 m. $\mathrm{Te(OH)_4}(\mathrm{aq})$ seems poorly characterised; however, the anions $\mathrm{TeO(OH)_3}^-(\mathrm{aq})$ and $\mathrm{TeO_2(OH)_2}^{2-}(\mathrm{aq})$ are believed to the present in $\mathrm{Te}^{\mathrm{IV}}(\mathrm{aq})$ solutions at higher pH values. The speciation plots of $\mathrm{Te}^{\mathrm{IV}}(\mathrm{aq})$ species show that $\mathrm{Te(OH)_3}^+(\mathrm{aq})$ is the form persisting in dilute ($< \mathrm{mM}$) solutions below pH 2. Above pH 4, $\mathrm{TeO(OH)_3}^-(\mathrm{aq})$ dominates and above pH 8 the dominant form is $\mathrm{TeO_2(OH)_2}^{2-}(\mathrm{aq})$. $\mathrm{Te(OH)_4}(\mathrm{aq})$ appears to reach a maximum ($\sim 13\%$ of the total tellurium) at around pH 3 [47].

The electrochemical reduction of $\mathrm{Te(OH)_3}^+(\mathrm{aq})$ has been studied at a glassy carbon surface using rotating electrode techniques [184]. By a combination of the electrochemical measurements with optical microscopic analysis of the electrodeposits it was found that the electrochemical behaviour is directly attributable to the type of deposit morphology being formed. $\mathrm{Te(OH)_3}^+(\mathrm{aq})$ is initially reduced by a four-electron reaction, producing a dendritic deposit of elemental tellurium. This can be further reduced to $\mathrm{H_2Te}$ at more cathodic potentials, with complete consumption of the dendritic tellurium. A comproportionation reaction between electrogenerated $\mathrm{H_2Te}$ and $\mathrm{Te(OH)_3}^+(\mathrm{aq})$ in the electrolyte then results in the deposition and subsequent growth of a smooth Te film on the electrode surface. The adsorption and colloidal behaviour of traces of $\mathrm{Te}^{\mathrm{IV}}$ in aqueous solutions has been investigated [185]. Studies on the electroxidation of gold telluride and of metallic tellurium have provided other investigations on the nature of tellurium species in aqueous solutions [186].

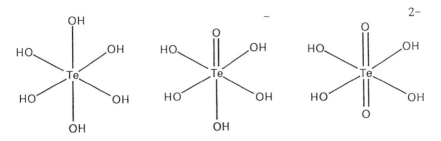

Figure 2.22. Structures for $Te(OH)_6$, $[TeO(OH)_5]^-$ and $[TeO_2(OH)_4]^{2-}$

2.7.3.2 TELLURIUM(VI): CHEMISTRY OF $Te(OH)_6$

The oxygen exchange behaviour of $Te(OH)_6$ is markedly different to that of either sulfate and selenate [163]. Whereas water exchange on both sulfate and selenate is acid catalysed, that on tellurate is base catalysed [187]. The explanation lies in the octahedral hexahydroxo structure (Fig. 2.22). A dissociative mechanism, involving the formation of '$TeO(OH)_4$', is suggested for exchange on octahedral $Te(OH)_6$ from the lack of a deuterium isotope effect on the rate. The $[H^+]$ dependence of the tellurate exchange indicates that $[TeO_2(OH)_4]^{2-}$ and $[TeO(OH)_5]^-$ (Fig. 2.22) are involved [187, 188]:

$$Te(OH)_6 + OH^- \rightleftharpoons [TeO(OH)_5]^- + H_2O \qquad (2.14)$$

$$[TeO(OH)_5]^- + OH^- \rightleftharpoons [TeO_2(OH)_4]^{2-} + H_2O \qquad (2.15)$$

$$[TeO(OH)_5]^- \rightleftharpoons TeO(OH)_4 + OH^- \qquad (2.16)$$

Optical spectroscopic measurements at 240 nm on 1.27 mM Te^{VI}(aq) solutions in 0.1 m $KClO_4$ show that $Te(OH)_6$ is the form persisting until pH 7 when $[TeO(OH)_5]^-$ ($-\log Q_{11} = 7.68$) and then $[TeO_2(OH)_4]^{2-}$ (pH 11) ($-\log Q_{12} = 18.68$) and eventually $[TeO_3(OH)_3]^{3-}$ (pH > 13) take over [189].

2.8 GROUP 17 ELEMENTS: FLUORINE, CHLORINE, BROMINE AND IODINE

2.8.1 Halide ion Hydration

Hydration around the chloride ion has been studied at the second-order difference level by Powell and Enderby in order to obtain sharper resolution in radial function plots. Data taken from a 2.0 m aqueous solution of $NiCl_2$ shows that each Cl^- ion is surrounded by 6.4 water molecules on average with distances of

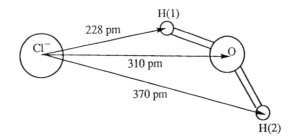

Figure 2.23. Orientation of a hydrating water molecule around a Cl^- ion

Cl—$H(1) = 228 \pm 30$ pm, Cl—$O = 310$ pm and Cl—$H(2) = 370$ pm [190, 191]. The findings are consistent with an orientation of the hydrating water molecule as shown in Fig. 2.23. A saturating hydration number of ~ 6 for Cl^- seems to be in agreement with diffraction and scattering data from a range of other chloride salts [192].

2.8.2 Halide Oxo Anions

Water exchange on the various oxo anions of group 17, as studied by the ^{18}O-labelling method, have revealed interesting effects down the group [163]. The remarkable inertness of perchlorate ($t_{1/2} > 100$ years for 9.0 M $HClO_4$ at 25 °C) and perbromate (no exchange observable on 1.0 M $HBrO_4$ after 14 hours at 120 °C) can be compared with that on periodate ($k = 4.5 \times 10^3 \, s^{-1}$, 25 °C for 1.0 M periodic acid). As with tellurate above, the difference stems from the structure H_5IO_6 of periodic acid. Here the mechanism is believed to be dissociative involving the formation of H_3IO_5. Similarly iodate, IO_3^-, is much more labile than either BrO_3^- or ClO_3^- and moreover is uniquely catalysed by a single H^+ ion. The mechanism is believed to involve protonation to HIO_3 which then undergoes a concerted reaction with H_2O [193]:

$$IO_3^- + H^+ \rightleftharpoons HIO_3 \tag{2.17}$$

$$HIO_3 + H_2O^* \left\{ \begin{array}{c} H-O \diagdown \begin{array}{c} O----H \\ I \diagdown \cdots O^* \diagup \\ O\cdots-H \end{array} \end{array} \right\} \longrightarrow HIO_2O^* + H_2O \tag{2.18}$$

kinetic
isotope
effect

as implied by a deuterium isotope effect. This pathway seems unavailable to both ClO_3^- and BrO_3^-.

2.9   GROUP 18 ELEMENTS: ARGON, KRYPTON AND XENON

The steady increase in solubility in water down the group reflects not only a greater polarisation of the stable inert gas electron configuration but also a significant degree of hydration.

Argon is an excellent blanketing gas for air-free aqueous solution work. This stems not only from the fact that it is heavier than either dioxygen or dinitrogen but also from the greater solubility in water reflecting a not insignificant hydration sphere. This has prompted a number of studies into the nature of the hydration sphere around argon atoms as an example of the contribution from instantaneous dipole–dipole attractive forces.

The first-order difference method of neutron diffraction and isotopic substitution has been applied to argon dissolved under pressure in D_2O [194]. The results show that argon atoms possess a well-defined primary hydration shell composed of 16 ± 2 water molecules in the range 280–540 pm. Also evident is a weaker second shell which extends to 800 pm. The results are in broad agreement with those obtained from Monte Carlo and MD computer simulation of models in which an argon atom interacts with a number of water molecules via pairwise additive forces. Similar hydration spheres have been detected around krypton and xenon [195]. Conversely, the influence of dissolved inert gas atoms on the structure and dynamics of water itself has been the subject of a number of studies both experimental [195, 196] and theoretical [197]. The generation of supercritical aqueous solutions is an area of growing interest [198].

REFERENCES

[1]   Bell, R. A., Christoph, G. G., Fronczek, F. R., and Marsh, R. E. *Science*, **190**, 151 (1975).

[2]   (a) Newton, M. D., and Ehrenson, S., *J. Am. Chem. Soc.*, **93**, 4971 (1971).
(b) Newton, M. D., *J. Chem. Phys.*, **67**, 5535 (1977).

[3]   Fornili, S. L., Migliore, M., and Palazzo, M. A., *Chem. Phys. Lett.*, **125**, 419 (1986).

[4]   Erdez-Gruz, T., *Transport Phenomena in Aqueous Solution*, Hilger, London, 327 (1974).

[5]   Corongiu, G., Kelterbaum, R., and Kochanski, E., *J. Phys. Chem.*, **99**, 8038 (1995).

[6]   Andaloro, G., Palazzo, M. A., Migliore, M., and Fornili, S. L., *Chem. Phys. Lett.*, **149**, 201 (1988).

[7]   Helper, L. G., and Wooley, E. M., in *Water, A Comprehensive Treatise* (ed. F. Franks) Vol. 3, Plenum, New York, 1972, Ch. 3.

[8]   Ikuta, S., *J. Comput. Chem.*, **5**, 374 (1984).

[9]   (a) Ruben, H. W., Templeton, D. H., Rosenstein, R. D., and Olovsson, I., *J. Am. Chem. Soc.*, **83**, 820 (1961).
(b) Levy, H. A., and Lisensky, G. C., *Acta Cryst.*, **B34**, 3502 (1978).

[10] Dickens, B., and Brown, W. E., *Acta Cryst.*, **B28**, 3056 (1972).

[11] (a) Newsome, J. R., Neilson, G. W., and Enderby, J. E., *J. Phys. Solid State Phys.*, **13**, L923 (1980).

(b) Radnai, T., Palinkas, G., Szasz, G. I., and Heinzinger, K., *Z. Naturforsch.*, **A36**, 1076 (1981).

(c) Tamura, Y., Yamaguchi, T., Okada, I., and Ohtaki, H., *Z. Naturforsch.*, **A42**, 367 (1987).

[12] (a) Narten, A. H., Vaslow, F., and Levy, H. A., *J. Chem. Phys.*, **58**, 5017 (1973).

(b) Licheri, G., Piccaluga, G., and Pinna, G., *Chem. Phys. Lett.*, **35**, 119 (1979).

[13] Palinkas, G., Radnai, T., and Hajdu, H., *Z. Naturforsch.*, **A35**, 107 (1980).

[14] Kameda, Y., and Uemura, O., *Bull. Chem. Soc. Japan*, **66**, 384 (1993).

[15] (a) Enderby, J. E., *Chem. Soc. Revs.*, 159 (1995).

(b) Enderby, J. E., and Neilson, G. W., *Adv. Inorg. Chem.*, **34**, 195 (1989).

(c) Enderby, J. E., 'The solvation of ions', *J. Phys. Cond. Matter*, F87–90 (1991).

(d) Enderby, J. E., Cummings, S., Herdman, G. J., Neilson, G. W., Salmon, P. S., and Skipper, N., 'Diffraction and the study of aqua ions', *J. Phys. Chem.*, **91**, 5851 (1987).

[16] Szasz, G. I., Heinzinger, K., and Riede, W. O., *Z. Naturforsch*, **A36**, 1067 (1981).

[17] Vorgin, J., Knapp, P. S., Flint, W. L., Anton, A., Highberger, G., and Malinkowski, E. R., *J. Chem. Phys.*, **54**, 178 (1971).

[18] (a) Pye, C. C., Rudolph, W., and Poirier, R. A., *J. Phys. Chem.* **100**, 601 (1996).

(b) Rudolph, W., Brooker, M. H., and Pye, C. C., *J. Phys. Chem.*, **99**, 3793 (1995).

[19] (a) Palinkas, G., Piccaluga, G., Pinna, G., and Magini, M., *Chem. Phys. Lett.*, **98**, 157 (1980).

(b) Ohtomo, N. and Arakawa, K., *Bull. Chem. Soc. Japan*, **53**, 1789 (1980) (and refs. therein).

[20] Noromoto, N., *Mineral. J. (Sapporo)*, **2**, 1 (1956).

[21] (a) Chandresekhar, J., Spellmeyer, D. C., and Jorgensen, W. L., *J. Am. Chem. Soc.*, **106**, 903 (1984).

(b) Mezei, M., and Beveridge, D. L., *J. Chem. Phys.*, **74**, 6902 (1981).

(c) Palinkas, G., Riede, W. O., and Heinzinger, K., *Z. Naturforsch*, **A32**, 1137 (1977).

(d) Limtakrul, J. P., and Rode, B. M., *Monatsch. Chem.*, **116**, 1377 (1985).

[22] Waizumi, K., Masuda, H., and Fukushima, N., *Inorg. Chim. Acta*, **209**, 207 (1993).

[23] Heinje, G., Luck, W. A. P., and Heinzinger, K., *J. Phys. Chem.*, **91**, 331 (1987).

[24] Neilson, G. W., and Skipper, N., *Chem. Phys. Lett.*, **114**, 35 (1985) (and refs therein).

[25] (a) Clementi, E., and Barsotti, R., *Chem. Phys. Lett.*, **59**, 21 (1978).

(b) Bounds, D., *Mol. Phys.*, **54**, 1335 (1985).

(c) Migliore, M., Fornili, S. L., Spohr, E., Palinkas, G., and Heinzinger, K., *Z. Naturforsch*, **A41**, 826 (1986).

[26] (a) Ohtomo, N., and Arakawa, K., *Bull. Chem. Soc. Japan*, **52**, 2755 (1979).

(b) Bertagnolli, H., Weidner, J.-U., and Zimmermann, H. W., *Ber. Bunsenges. Phys. Chem.*, **78**, 2 (1974).

[27] (a) Heinzinger, K., and Vogel, P. C., *Z. Naturforsch*, **A31**, 463 (1976).

(b) Szasz, G. I., and Heinzinger, K., *Z. Naturforsch*, **A38**, 214 (1983).

(c) Tamura, Y., Ohtaki, H., and Okada, I., *Z. Naturforsch*, **A46**, 1083 (1991).

[28] Wong, C. Y., and Woolins, J. D., *Coord. Chem. Revs.*, **130**, 243 (1994).

[29] Yamaguchi, T., Ohtaki, H., Spohr, E., Palinkas, G., Heinzinger, K., and Probst, M., *Z. Naturforsch.*, **A41**, 1175 (1986).

[30] Alei, M., and Jackson, J. A., *J. Chem. Phys.*, **41**, 3402 (1964).

[31] Connick, R. E., and Fiat, D. N., *J. Chem. Phys.*, **39**, 1349 (1963).

[32] Frahm, J., and Fuldner, H. H., *Ber. Bunsenges. Phys. Chem.*, **84**, 173 (1980).

[33] Pittet, P.-A., Elbaze, G., Helm, L., and Merbach, A. E., *Inorg. Chem.*, **29**, 1936 (1990).

[34] Probst, M. M., Spohr, E., and Heinzinger, K., *Chem. Phys. Lett.*, **161**, 405 (1989).
[35] Dance, I. G., and Freeman, H. C., *Acta Cryst.*, **B25**, 304 (1969).
[36] Dijakovic, V., Edenharter, A., Nowacki, W., and Ribar, B., *Z. Kristallorg. Kristalgeom. Kristallphys. Kristallchem.*, **114**, 314 (1976).
[37] Lutz, H. D., Lange, N., Maneva, M., and Georgiev, M., *Z. Anorg. Allg. Chem.*, **594**, 77 (1991).
[38] Maneva, M., and Georgiev, M., *Polyhedron*, **12**, 1821 (1993).
[39] Bell, N. A., *Adv. Inorg. Chem. Radiochem.*, **14**, 255 (1972).
[40] Bruno, J., *J. Chem. Soc. Dalton Trans.*, 2431 (1987).
[41] Inamo, M., Ishihara, K., Funahashi, S., Ducummon, Y., Merbach, A. E., and Tanaka, M., *Inorg. Chem.*, **30**, 1580 (1991).
[42] Kellersohn, T., Belaplane, R. G., and Olovsson, I., *Acta Cryst.*, **B50**, 316 (1994).
[43] Bock, C. W., and Glusker, J. P., *Inorg. Chem.*, **32**, 1242 (1993).
[44] Kalihana, H., and Sillen, L. G., *Acta Chem. Scand.*, **10**, 985 (1956).
[45] Carell, B., and Olin, A., *Acta Chem. Scand.*, **15**, 1875 (1961).
[46] Schwarzenbach, G., and Wenger, H., *Helv. Chem. Acta*, **52**, 644 (1969).
[47] Baes, C. F., and Mesmer, R. E., *The Hydrolysis of Cations*, Krieger, Florida, 1976.
[48] Brown, P. L., Ellis, J., and Sylva, R. N., *J. Chem. Soc. Dalton Trans.*, 2001 (1983).
[49] Bertin, F., Thomas-David, G., and Merlin, J. C., *Bull. Soc. Chim. Fr.*, 2393 (1967).
[50] Kalihana, H., and Maeda, M., *Bull. Chem. Soc. Japan*, **43**, 109 (1970).
[51] Lanza, E., and Carpeni, G., *Electrochim. Acta*, **13**, 519 (1968).
[52] Bertin, F., and Deroault, J., *Can. J. Chem.*, **57**, 913 (1979).
[53] Akitt, J. W., and Duncan, R. H., *J. Chem. Soc. Faraday Trans. I*, 2212 (1980).
[54] Faure, R., Bertin, F., Loiseleur, T., and Thomas-David, G., *Acta Cryst.*, **B30**, 462 (1974).
[55] (a) Bechtler, A., Breitschwerdt, K. G., and Tamm, K., *J. Chem. Phys.*, **52**, 2975 (1970).
(b) Eigen, M., and Tamm, K., *Z. Electrochem.*, **66**, 93, 107 (1962).
[56] Neely, J. W., Ph.D Thesis, University of California, Berkeley, UCRL-20580, 1971.
[57] Swaddle, T. W., *Inorg. Chem.*, **22**, 2663 (1983).
[58] Baldwin, W. G., and Stranks, D. R., *Aust. J. Chem.*, **21**, 2161 (1968).
[59] Lee, M. A., Winter, N. W., and Casey, W. H., *J. Phys. Chem.*, **98**, 8641 (1994).
[60] Crea, J., and Lincoln, S. F., *J. Chem. Soc. Dalton Trans.*, 2075 (1973).
[61] Ferraris, G., Jones, D. W., and Yerkess, J., *J. Chem. Soc. Dalton Trans.*, 816 (1973).
[62] Braibanti, A., Tiripicchio, A., Manotti-Lanfred, A. M., and Bigoli, F., *Acta Cryst.*, **B25**, 354 (1969).
[63] Flack, H., *Acta Cryst.*, **B29**, 656 (1973).
[64] Ledesert, M., and Monier, J. C., *Acta Cryst.*, **B37**, 652 (1981).
[65] Stephen, G. W., and MacGillavray, G. H., *Acta Cryst.*, **B28**, 1031 (1972).
[66] Neuman, M. A., Steiner, E. C., van Remoortere, F. P., and Boer, F. P., *Inorg. Chem.*, **14**, 734 (1975).
[67] Bock, C. W., Kaufman, A., and Glusker, J. P., *Inorg. Chem.*, **33**, 419 (1994).
[68] Matwiyoff, N. A., and Taube, H., *J. Am. Chem. Soc.*, **90**, 2796 (1968).
[69] Bol. W. A., Gerrits, G. J. A., and von Panthaleon van Eck, C. L., *J. Appl. Crystallogr.*, **3**, 486 (1970).
[70] Waizumi, K., Tamura, Y., Masuda, H., and Ohtaki, H., *Z. Naturforsch.*, **A46**, 307 (1991) (and refs therein).
[71] Szasz, G. I., Dietz, W., Heinzinger, K., Palinkas, G., and Radnai, T., *Chem. Phys. Lett.*, **92**, 338 (1982).
[72] Hostetler, P. B., *Am. J. Sci.*, **261**, 238 (1963).
[73] Lewis, D., *Acta Chem. Scand.*, **17**, 5 (1963).

[74]   Neely, J. W., and Connick, R. E., *J. Am. Chem. Soc.*, **92**, 3476 (1970).
[75]   Hewish, N. A., Neilson, G. W., and Enderby, J. E., *Nature, Lond.*, **297**, 138 (1982).
[76]   Yamaguchi, T., Hayashi, S., and Ohtaki, H., *Inorg. Chem.*, **28**, 2434 (1989) (and refs therein).
[77]   Putkas, D., Rotter, H. W., Thiele, G., Broderson, K., and Pezzei, G., *Z. Anorg. Allg. Chem.*, **595**, 193 (1991).
[78]   (a) Leclaire, A., and Borel, M. M., *Acta Cryst.*, **B33**, 2938 (1977).
       (b) Agron, P. A., and Busing, W. R., *Acta Cryst.*, **C42**, 141 (1986).
[79]   Mazzarella, L., Kovacs, A. L., De Santis, P., and Liquori, A. M., *Acta Cryst.*, **22**, 65 (1967).
[80]   Probst, M. M., Radnai, T., Heinzinger, K., Bopp, P., and Rode, B. M., *J. Phys. Chem.*, **89**, 753 (1985).
[81]   (a) Camaniti, R., Licheri, G., Piccaluga, G., and Pinna, G., *Chem. Phys. Lett.*, **47**, 275 (1977).
       (b) Licheri, G., Piccaluga, G., and Pinna, G., *J. Chem. Phys.*, **64**, 2437 (1976).
[82]   Atkinson, G., Emara, M. M., and Ferdandez-Prini, R., *J. Phys. Chem.*, **78**, 1913 (1974).
[83]   Eigen, M., *Pure Appl. Chem.*, **6**, 97 (1963).
[84]   Neilson, G. W., and Broadbent, R. D., *Chem. Phys. Lett.*, **167**, 429 (1990) (and refs therein).
[85]   Albright, J. N., *J. Chem. Phys.*, **56**, 3783 (1972).
[86]   Caminiti, R., Musiini, A., Paschina, G., and Pinna, G., *J. Appl. Crystallogr.*, **15**, 482 (1982).
[87]   Spohr, E., Palinkas, G., Heinzinger, K., Bopp, P., and Probst, M. M., *J. Phys. Chem.*, **92**, 6754 (1988).
[88]   Swift, T. J., and Sayre, W. G., *J. Chem. Phys.*, **44**, 3567 (1966).
[89]   Persson, I., Sandstrom, M., Yokohama, H., and Chaudhry, M., *Z. Naturforsch.*, **A50**, 21 (1995).
[90]   Marcus, Y., *Ion Solvation*, Wiley Interscience, Chichester, 1985, p. 306.
[91]   Buhl, J. C., Englehardt, G., and Felsche, J., *Zeolites*, **9**, 40 (1989).
[92]   Watenabe, K., and Sato, S., S., *Bull. Chem. Soc. Japan*, **67**, 379 (1994).
[93]   Chaudhuri, M. K., and Das, B., *J. Chem. Soc. Dalton Trans.*, 243 (1988).
[94]   Farmer, J. B., *Adv. Inorg. Chem. Radiochem.*, **25**, 187 (1982).
[95]   Ingri, N., *Acta Chem. Scand.*, **16**, 439 (1962) and **17**, 581 (1963).
[96]   Mesmer, R. E., Baes, C. F., and Sweeton, F. H., *Inorg. Chem.*, **11**, 537 (1972).
[97]   Ishihara, K., Nagasawa, A., Unemoto, K., Ito, H., and Saito, K., *Inorg. Chem.*, **33**, 3811 (1994).
[98]   (a) Ginderow, D., and Cesbron, F., *Acta Cryst.*, **B35**, 2499 (179).
       (b) Hermansson, K., *Acta Cryst.*, **C39**, 925 (1983).
[99]   Kriutsov, N. V., Shirokova, G. N., and Rosolovskii, V. Ya., *Russ. J. Inorg. Chem.*, **18**, 503 (1973).
[100] Adams, D. M., and Hills, D. J., *J. Chem. Soc. Dalton Trans.*, 782 (1978).
[101] Fiat, D., and Connick, R. E., *J. Am. Chem. Soc.*, **90**, 608 (1968).
[102] Baldwin, H. W., and Taube, H., *J. Chem. Phys.*, **33**, 206 (1960).
[103] Hugi-Cleary, D., Helm, L., and Merbach, A. E., *Helv. Chim. Acta*, **68**, 545 (1985).
[104] (a) Miceli, J., and Stuehr, J., *J. Am. Chem. Soc.*, **90**, 6967 (1968).
       (b) Kalidas, C., Knoche, W., and Papadopoulos, D., *Ber. Bunsenges. Phys. Chem.*, **75**, 196 (1971).
       (c) Rauh, H., and Konche, W., *Ber. Bunsenges. Phys. Chem.*, **83**, 518 (1979).
       (d) Secco, F., and Venturini, M., *Inorg. Chem.*, **14**, 978 (1975).
       (e) Matusek, M., and Strehlow, H., *Ber. Bunsenges. Phys. Chem.*, **73**, 982 (1969).

(f) Kuehn, C., and Knoche, W., *Trans. Faraday Soc.*, **67**, 2101 (1971).

[105] Akitt, J. W., Greenwood, N. N., Khandelwal, B. L., and Lester, G. D., *J. Chem. Soc. Dalton Trans.*, 604 (1972).

[106] (a) Akitt, J. W., and Farthing, A., *J. Chem. Soc. Dalton Trans.*, 1233, 1606, 1609, 1617, 1624 (1981).

(b) Akitt, J. W., and Farthing, A., *J. Magn. Reson.*, **32**, 345 (1978).

[107] (a) Akitt, J. W., *Prog. Nucl. Magn. Reson.*, **21**, 1 (1989).

(b) Akitt, J. W., Elders, J.M., Fontaine, X. L. R., and Kundu, A. K., *J. Chem. Soc. Dalton Trans.*, 1987 (1989).

(c) Akitt, J. W., in *Multinuclear NMR* (ed. J. Mason) Plenum, New York, p. 482 (1987).

(d) Schonter, S., Gorz, H., Bertram, R., Muller, D., and Gessner, W., *Z. Anorg. Allg. Chem.*, **476**, 188, 195 (1981), **483**, 153 (1981) and **502**, 113, 132 (1983).

[108] Akitt, J. W., and Elders, J. M., *Bull. Soc. Chim. Fr.*, 10 (1986).

[109] Akitt, J. W., and Milic, N. B., *J. Chem. Soc. Dalton Trans.*, 981 (1984).

[110] Johansson, G., *Arkiv. Kemi*, **20**, 305, 321 (1963).

[111] Akitt, J. W., and Elders, J. M., *J. Chem. Soc. Dalton Trans.*, 1347 (1988).

[112] (a) Henry, M., Jolivet, J.-P., and Livage, J., *Structure and Bonding*, **77**, 153–206 (1992).

(b) Brinker, C. J., and Scherer, G. W., *Sol-Gel Science*, Academic Press, New York, 1989.

[113] Akitt, J. W., Elders, J. M., and Letellier, P., *J. Chem. Soc. Faraday Trans I*, 1725 (1987).

[114] Beattie, J. K., Best, S. P., Skelton, B. W., and White, A. H., *J. Chem. Soc. Dalton Trans.*, 2105 (1981).

[115] (a) Armstrong, R. S., Beattie, J. K., Best, S. P., Braithwaite, G. P., Delfavero, P., Skelton, B. W., and White, A. H., *Aust. J. Chem.*, **43**, 393 (1990).

(b) Best, S. P., and Forsyth, J. B., *J. Chem. Soc. Dalton Trans.*, 395 (1990).

[116] (a) Johansson, G., 'Structures of metal complexes in solution', *Adv. Inorg. Chem.*, **39**, 178 (1992).

(b) Caminiti, R., and Paschina, G., *Chem. Phys. Lett.*, **82**, 487 (1981).

[117] Akitt, J. W., and Kettle, D., *Mag. Res. Chem.*, **27**, 377 (1989).

[118] Hugi-Cleary, D., Helm, L., and Merbach, A. E., *J. Am. Chem. Soc.*, **109**, 4444 (1987).

[119] Brown, P. L., *J. Chem. Soc. Dalton Trans.*, 399 (1989).

[120] (a) Brendler, E., Thomas, B., and Schonherr, S., *Monatsch. Chem.*, **123**, 285 (1992).

(b) Campisi, A., and Tregloan, P., *Inorg. Chim. Acta*, **100**, 251 (1978).

(c) Permutter-Hayman, B., Secco, F., Tapuhi, E., and Venturini, M., *J. Chem. Soc. Dalton Trans.*, 1124 (1980).

(d) Yamada, S., Iwanaga, A., Funahashi, S., and Tanaka, M., *Inorg. Chem.*, **23**, 3528 (1984).

[121] Dechter, J. P., *Prog. Inorg. Chem.*, **29**, 285 (1982).

[122] (a) Bradley, S. M., Kydd, R. A., and Yamdagni, R., *J. Chem. Soc. Dalton Trans.*, 413 (1990).

(b) Bradley, S. M., and Kydd, R. A., *J. Chem. Soc. Dalton Trans.*, 2407 (1993).

(c) Bradley, S. M., Lehr, C. R., and Kydd, R. A., *J. Chem. Soc. Dalton Trans.*, 2415 (1993).

[123] Bradley, S. M., Kydd, R. A., and Fyfe, C. A., *Inorg. Chem.*, **31**, 1181 (1992).

[124] Kudyska, J., Buckmaster, H. A., Kawano, K., Bradley, S. M., and Kydd, R. A., *J. Chem. Phys.*, **99**, 3329 (1993).

[125] Fratiello, A., *Prog. Inorg. Chem.*, **17**, 57 (1972).

[126] (a) Cannon, T. H., and Richards, R. E., *Trans. Farad. Soc.*, **62**, 1378 (1966).
       (b) Fratiello, A., Lee, R. M., Nishida, V. M., and Schuster, R. E., *J. Chem. Phys.*, **48**, 3705 (1968).
[127] Brown, P. L., Eillis, J., and Sylva, R. N., *J. Chem. Soc. Dalton Trans.*, 1911 (1982).
[128] Caminiti, R., Johansson, G., and Toth, I., *Acta Chem. Scand.*, **A40**, 435 (1986).
[129] Biedermann, G., and Ferri, D., *Acta Chem. Scand.*, **36**, 611 (1982).
[130] (a) Thompson, L. C. A., and Pacer, R., *J. Inorg. Nucl. Chem.*, **25**, 1041 (1963).
       (b) Akselrud, N. V., *Dokl. Akad. Nauk SSSR*, **132**, 1067 (1960).
[131] Ivanov-Emin, B. N., Nisel'son, L. A., and Greksa, Yu., *Zh. Neorg. Khim.*, **5**, 1966 (1960).
[132] Glass, G. E., Schwabacher, W. B., and Tobias, R. S., *Inorg. Chem.*, **7**, 2471 (1968).
[133] Berglund, J., Werndrup, P., and Elding, L. I., *J. Chem. Soc. Dalton Trans.*, 1435 (1994).
[134] Bell, R. P., and George, J. H. B., *Trans. Faraday Soc.*, **46**, 619 (1953).
[135] (a) Anbar, M., *Quart. Revs.*, **22**, 578 (1968).
       (b) Cercek, B., Ebert, M., and Swallow, A. J., *J. Chem. Soc. A*, 612 (1966).
[136] Symons, M. C. R., and Yandell, J. K., *J. Chem. Soc. A*, 760 (1971).
[137] (a) Burchill, C. E., and Hickling, G. G., *Can. J. Chem.*, **48**, 2466 (1970).
       (b) Burchill, C. E., and Wolodarsky, W. H., *Can. J. Chem.*, **48**, 2955 (1970).
[138] Stranks, D. R., and Yandell, J. K., *J. Phys. Chem.*, **73**, 840 (1969).
[139] Glaser, J., and Johansson, G., *Acta Chem. Scand.*, **A36**, 125 (1982).
[140] Glaser, J., and Johansson, G., *Acta Chem. Scand.*, **A35**, 639 (1981).
[141] (a) Biedermann, G., and Spiro, T. G., *Chem. Sci.*, **1**, 155 (1971).
       (b) Biedermann, G., *Ark. Kemi.*, **5**, 441 (1953).
[142] Sutin, N., *Ann. Rev. Phys. Chem.*, **17**, 119 (1960).
[143] (a) Blixt, J., Glaser, J., and Solymosi, P., *Inorg. Chem.*, **31**, 5288 (1992).
       (b) Banyai, I., and Glaser, J., *J. Am. Chem. Soc.*, **112**, 4703 (1990) and **111** 3186 (1989).
       (c) Henriksson, U., and Glaser, J., *Acta Chem. Scand.*, **A39**, 355 (1985).
[144] Blixt, J., Glaser, J., Mink, J., Persson, I., Persson, P., and Sandstrom, M., *J. Am. Chem. Soc.*, **117**, 5089 (1995).
[145] Spiro, T. G., *Inorg. Chem.*, **6**, 569 (1967) and **4**, 731, 1290 (1965).
[146] Satchell, D. P. N., Satchell, R. S., and Wassef, W. N., *J. Chem. Soc. Perkin. Trans. II*, 1199 (1992).
[147] Satchell, D. P. N., and Satchell. R. S., *J. Chem. Soc. Perkin. Trans. II*, 751 (1992).
[148] (a) Fadnis, A. G., and Bhatnagar, J., *J. Indian. Chem. Soc.*, **71**, 669 (1994).
       (b) Fadnis, A. G., and Shrivastava, S. K., *Ann. Quim.*, **A82**, 5 (1986).
[149] Agrawal, A., Nahar, S., Mishra, S. K., and Sharma, P. D., *J. Phys. Org. Chem.*, **6**, 179 (1993).
[150] (a) Rao, I., Mishira, S. K., and Sharma, P. D., *Ind. J. Chem.*, **A30**, 773 (1990).
       (b) Gupta, P., Sharma, P. D., and Gupta, Y. K., *J. Chem. Soc. Dalton Trans.*, 1867 (1984).
[151] Gupta, Y. K., Dumar, D., Jain, S., and Gupta, K. S., *J. Chem. Soc. Dalton Trans.*, 1915 (1990).
[152] (a) Glasser, L. J. D., *Chem. Br.*, 33 (1982).
       (b) Falcone Jr, J. S. (ed.), *Soluble Silicates*, ACS Symposium Series 194, Washington, DC, 1981.
[153] Englehardt, G., and Hoebbel, D., *J. Chem. Soc. Chem. Commun.*, 514 (1984).
[154] Ingri, N., *Acta Chem. Scand.*, **17**, 597 (1963).
[155] Yamaguchi, T., Lindquist, O., Claeson, T., and Boyce, J. B., *Chem. Phys. Lett.*, **93**, 528 (1982).

[156] Tobias, R. S., *Acta Chem. Scand.*, **12**, 198 (1958).
[157] (a) Grimvall, S., *Acta Chem. Scand.*, **27**, 1447 (1973).
(b) Davies, C. G., Donaldson, J. D., Laughlin, D. R., Howie, R. A., and Beddoes, R., *J. Chem. Soc. Dalton Trans.*, 2241 (1975).
[158] Johansson, G., and Ohtaki, H., *Acta Chem. Scand.*, **27**, 643 (1973).
[159] Taylor, M. J., and Coddington, J. M., *Polyhedron*, **11**, 1537 (1992).
[160] Hong, S. H., and Olin, A., *Acta Chem. Scand.*, **27**, 2304 (1973) and **28**, 233 (1974).
[161] (a) Spiro, T. G., Templeton, D. H., and Zalkin, A., *Inorg. Chem.*, **8**, 816 (1969).
(b) Olin, A., and Sonderquist, R., *Acta Chem. Scand.*, **26**, 3505 (1972).
[162] Johansson, G., and Olin, A., *Acta Chem. Scand.*, **22**, 3197 (1968).
[163] Gamsjager, H., and Murmann, R. K., *Adv. Inorg. Bioinorg. Mech.*, **2**, 317 (1983) (and refs. therein).
[164] Okumura, A., and Okazaki, N., *Bull. Chem. Soc. Japan*, **46**, 2937 (1973).
[165] Copenhafer, W. C., and Rieger, P. H., *J. Am. Chem. Soc.*, **100**, 3776 (1978).
[166] Okumura, A., Yamamoto, N., and Okazaki, N., *Bull. Chem. Soc. Japan*, **46**, 3633 (1973).
[167] Mishra, S. K., and Gupta, Y. K., *Ind. J. Chem.*, **6**, 757 (1968).
[168] Gayer, K. H., and Garrett, A. B., *J. Chem. Soc.*, **74**, 2353 (1952).
[169] (a) Lefebvre, J., and Maria, H., *Compt. Rend.*, **256**, 3121 (1963).
(b) Gate, S. H., and Richardson, E., *J. Inorg. Nucl. Chem.*, **23**, 257 (1961).
[170] Frank, W., Reiss, G. J., and Schneider, J., *Angew. Chem. Int. Ed. Engl. Trans.*, **34**, 2416 (1995).
[171] Shannon, R. D., *Acta Cryst.*, **A32**, 751 (1976).
[172] Sundvall, B., *Acta Chem. Scand.*, **A34**, 93 (1980).
[173] (a) Lazarini, B., *Acta Chem. Scand.*, **A35**, 448 (1979) and *Acta Cryst.*, **C8**, 69 (1979).
(b) Sundvall, B., *Acta Chem. Scand.*, **A33**, 219 (1979).
[174] (a) Asato, E., Katsura, K., Mikuriya, M., Fujii, T., and Reedijk, J., *Inorg. Chem.*, **32**, 5322 (1993).
(b) Asato, E., Katsura, K., Mikuriya, M., Turpeinen, U., Mutikainen, I., and Reedijk, J., *Inorg. Chem.*, **34**, 2447 (1995).
[175] Ford-Smith, M. H., and Hebeeb, J. J., *J. Chem. Soc. Dalton Trans.*, 461 (1973).
[176] Shibahara, T., Hashimoto, K., and Sakane, G., *J. Inorg. Biochem.*, **43**, 280 (1991) (and refs. therein).
[177] Saysell, D. M., and Sykes, A. G., *Inorg. Chem.* **35**, 5536 (1996) (and refs. therein).
[178] Cannon, W. R., Pettitt, B. M., and McCammon, J. A., *J. Phys. Chem.*, **98**, 6225 (1994).
[179] Radmer, R., *Inorg. Chem.*, **11**, 1162 (1972).
[180] Hoering, T. C., and Kennedy, J. W., *J. Am. Chem. Soc.*, **79**, 56 (1957).
[181] Okumura, A., and Okazaki, N., *Bull. Chem. Soc. Japan*, **46**, 1080 (1973).
[182] Nabivanets, B. I., and Kapantsyan, E. E., *Russ. J. Inorg. Chem.*, **13**, 946 (1968).
[183] (a) Mabuchi, H., and Okada, I., *Bull. Chem. Soc. Japan*, **38**, 1378 (1965).
(b) Sekine, T., Iwaka, H., Sakairi, M., and Shimada, F., *Bull. Chem. Soc. Japan*, **41**, 1 (1968).
[184] Dennison, S., and Webster, S., *J. Electroanal. Chem.*, **314**, 207 (1991).
[185] Maruyama, Y., and Yamaashi, Y., *J. Radioanal. Nucl. Chem.*, **91**, 67 (1985).
[186] (a) Jayasekera, S., Avraamides, J., and Ritchie, I. M., *Electrochim. Acta*, **41**, 879 (1996).
(b) Jayasekera, S., Ritchie, I. M., and Avraamides, J., *Aust. J. Chem.*, **47**, 1953 (1994).
[187] Wernli, B., PhD Thesis, Montyanuniversitat Leoben, Austria, 1980.

[188]   Luz, Z., and Pecht, I., *J. Am. Chem. Soc.*, **88**, 1152 (1966).
[189]   Earley, J. E., Fortnum, D., Wojcicki, A., and Edwards, J. O., *J. Am. Chem. Soc.*, **81**, 1295 (1959).
[190]   (a) Powell, D. H., Neilson, G. W., and Enderby, J. E., *J. Phys. Condens. Matter*, **5**, 5723 (1993) and *New. J. Chem.*, **19**, 27 (1995).
        (b) Powell, D. H., Barnes, A. C., Enderby, J. E., Neilson, G. W., and Salmon, P. S., *Faraday Disc. Chem. Soc.*, **85**, 137 (1988).
        (c) Cummings, S., Enderby, J. E., Neilson, G. W., Newsome, J. R., Howe, R. A., Howells, W. S., and Soper, A. K., *Nature, Lond.*, **287**, 714 (1980).
[191]   Powell, D. H., Neilson, G. W., and Enderby, J. E., *J. Phys. Condens. Matter*, **1**, 8721 (1989).
[192]   Ohtaki, H., and Radnai, T., *Chem. Revs.*, **93**, 1157 (1993).
[193]   von Felten, H., Gamsjager, H., and Baertschi, P., *J. Chem. Soc. Dalton Trans.*, 1683 (1976) (and refs. therein).
[194]   Broadbent, R. D., and Neilson, G. W., *J. Chem. Phys.*, **100**, 7543 (1994).
[195]   Tanaka, H., and Nakanishi, K., *J. Chem. Phys.*, **95**, 3719 (1991).
[196]   Haselmeier, R., Holz, M., Marbach, W., and Weingartner, H., *J. Phys. Chem.*, **99**, 2243 (1995).
[197]   Bushuev, Y. G., *Zhurnal Obshchei Khimii*, **64**, 1931 (1994).
[198]   (a) Pfund, D. M., Darab, J. G., Fulton, J. L., and Ma, Y. J., *J. Phys. Chem.*, **98**, 13102 (1994).
        (b) Cummings, P. T., Chialvo, A. A., and Cochran, H. D., *Chem. Eng. Sci.*, **49**, 2735 (1994).

Chapter 3

Group 3 Elements: Scandium, Yttrium, the Lanthanides and Actinides

The elements scandium, yttrium and the lanthanides (Ln) can be treated together as a group due to the similarity of their chemical properties. For example, all these elements are highly electropositive and readily form a stable trivalent aqua ion in solution ($E^\theta (Ln^{3+}/Ln) \sim -2.2\,V$). With the exception of the elements cerium, samarium, europium and ytterbium there is no aqueous redox chemistry of relevance and the trivalent state is the dominant form in solution due primarily to a relatively low value of the sum of the first three ionization potentials which is easily compensated for by the hydration energy of the resulting Ln^{3+} ion. The effective ionic radius of Sc^{3+} (74.5 pm) seems to be right on the limit for favourable formation of a hexahydrated ion and there are now strong indications (see Fig. 3.1) that it may prefer a heptahydrated arrangement. On the other hand, the radius of Y^{3+} (ca. 90 pm) places it roughly midway along the lanthanide series and in discussion it may truly be considered alongside this group of elements. Reflecting this, yttrium is found to occur naturally amongst the lanthanides and moreover many of its compounds are isomorphous with those of the lanthanides. Yttrium and the lanthanides are more commonly found with coordination numbers higher than six.

The older name of this group, the rare earth elements (REE), is perhaps a misnomer in that several of the elements are more abundant that lead and bismuth (not normally considered rare) and several more are more abundant than, for example, iodine. However, the name 'earth' is quite appropriate since the elements have properties certainly more akin to those of the alkaline earth elements, e.g. the formation of basic oxides and hydroxides, and a requirement for chelating hard base donor ligands for appreciable stability constants in complex-forming reactions. The strongest bonds are those that are highly polar and electrostatic in nature, reflective of the hard acid properties. Water-exchange

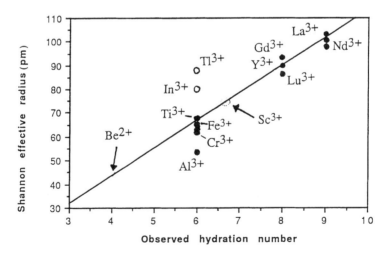

Figure 3.1. Plot of Shannon effective radius for six-coordination on M^{3+} ions vs the observed hydration number (the anomalous behaviour for In^{3+} and Tl^{3+} may be reflective of considerable covalency in their M—L bonds)

rates are fast and coordination numbers extremely variable, reflecting, for the lanthanides, a lack of appreciable crystal field splitting of the 4f orbitals. The reasons for this have been well highlighted elsewhere. Coordination numbers are thus largely dependent upon size and, to a certain degree, electrostatic factors, e.g. steric demands of the ligand donors in solution and crystal packing effects in the solid.

New mineral sources for, in particular, the heavier and rarer lanthanides, including Y, are continually sought, not at least because of their potential future large-scale use commercially in high T_c superconducting ceramics, fluorescent ceramics (Eu), oxidising ceramics (Ce) (e.g. for Pt metal dispersed catalytic converters), new magnetic materials (e.g. Nd, Eu, Gd) and in the lighting industry. One can add to this a resurgence in lanthanide coordination chemistry stemming from the use of certain lanthanide chelates (particularly of Gd^{3+} and Eu^{3+}) as contrast agents for magnetic resonance imaging. This is currently a billion dollar growth industry in the United States and in Japan. The long fluorescence lifetimes characteristic of many lanthanide compounds are also leading to increased applications in radioimmunoassay.

3.1 SCANDIUM

3.1.1 Chemistry of the Sc^{3+} (aq) Ion

3.1.1.1 HYDRATION NUMBER OF PUTATIVE $[Sc(OH_2)_x]^{3+}$

Solutions of Sc^{3+}(aq) are readily formed upon dissolution of simple Sc^{III} salts or of the oxide Sc_2O_3 in non-complexing acids. Sc^{3+}(aq) is frequently referred to as being hexa-coordinated by water [1]. Indeed, there is brief reference in the literature to the existence of an alum. However, closer analysis of the literature has revealed little direct evidence for a hexahydrated coordination sphere around Sc^{3+}(aq) despite much indirect evidence. For example, ^{45}Sc NMR ($I = 7/2$) linewidth studies of solutions of Sc^{3+} in various concentrations of aqueous HCl have been interpreted in terms of various octahedral chloro-aqua species [2]. Solid phases consisting of the octahedral $[Sc(OH)_6]^{3-}$ ion can be isolated from alkaline solutions of Sc^{III} and have been structurally characterised as in the case of $Na_3[Sc(OH)_6] \cdot 2H_2O$ [3]. X-ray structures show that octahedral six-coordination is quite common for Sc^{3+} with O-donor ligands, although not exclusively so [4]. One of the difficulties in determining the precise hydration number stems from the extensive complexation of Sc^{3+}(aq) in solution by anions, including often weakly coordinating ones such as NO_3^-. Sc^{III} is also extensively complexed by F^- in aqueous solution, reflecting its hard base nature and thus compatibility with the hard acid Sc^{3+} [5]. Although both Raman and infra-red spectra of aqueous Sc^{3+} solutions are well documented in the Russian literature [6], it is the result from more recent Raman studies performed in Japan coupled with X-ray diffraction measurements [7] that appear to have finally shed light on the hydration number. The Raman spectra from aqueous acidic solutions of Sc^{3+} were compared with those from corresponding solutions of Al^{3+} (aq). The acidic pH values used ($\geqslant 1.9$) prevented interference from hydrolysis products (see later). Despite an apparent similarity in the Raman spectra for the two tripositive ions there is a big difference in the depolarization ratio of the lower frequency bands around $300\,cm^{-1}$. Whereas the Raman band at $330\,cm^{-1}$ for Al^{3+} (aq) is depolarized the corresponding band at $300\,cm^{-1}$ for Sc^{3+} (aq) is polarized. Al^{3+} is well established as being hexa-coordinated by water and the Raman band at $330\,cm^{-1}$ can be assigned to the v_5 mode of a YX_6 type molecule. Two possibilities are inferred, either (a) the coordination number of Sc^{3+} (aq) is different from six or (b) coordination by both water and the counter-anions to Sc^{3+} is involved. This latter possibility was seemingly ruled out by observation of the same $300\,cm^{-1}$ Raman band from aqueous solutions of both $ScCl_3$ and $Sc(ClO_4)_3$. Furthermore, the X-ray diffraction studies have appeared to confirm that the inner coordination sphere of Sc^{3+} in these solutions consists of only water molecules. Further evidence of a deviation from octahedral six-coordination for Sc^{3+} was found from observation of four Raman bands from glassy solutions of $ScCl_3$ as opposed to the expected three bands for O_h symmetry as found for glassy solutions of $AlCl_3$. Similar results were found for glassy aqueous solutions of $Sc(ClO_4)_3$. The observation of more than one Raman band for all-aquated octahedral $[M(OH_2)_6]^{m+}$ ions is rather rare because of their inherent weakness. Thus it appears that the evidence is overwhelming for a coordination number for Sc^{3+} (aq) different from six. Certainly there is no crystal field driving force for six-coordination in the absence of any d electrons and therefore size is probably the dominant influence. The Shannon radius for six-coordination on Sc^{3+} ($74.5\,pm$) lies midway between that of Al^{3+} (ca. $54\,pm$), the first row transition metals (ca. $64\,pm$) and the heavier lanthanides (ca. $90\,pm$), which are believed to have a hydration number of eight (see later). Figure 3.1 suggests a preferred hydration number of around seven for Sc^{3+} (aq). Figure 3.2 further emphasises this in an apparent linear empirical relationship found by Kanno [8] between the frequency of the v_1 Raman band [8] and $1/r^2_{M-O}$ [9] for aqua metal ions with the same hydration number. Entry of the estimated average Sc—O distance from the X-ray diffraction studies ($\sim 218\,pm$) showed that Sc^{3+} deviates significantly from the line for the heavier lanthanides suggestive of a coordination number less than eight. Figure 3.2 also shows evidence, for Ln^{3+} (aq) ions, of the characteristic 'gadolinium break' believed to stem from a change in hydration number from nine (elements up to Sm, Eu) to eight (Gd and beyond including Y). It is not possible, however, to deduce the precise hydration geometry of putative $[Sc(OH_2)_7]^{3+}$ from the Raman data. There are three possibilities for the heptahydrated geometry, a capped octahedron, capped trigonal prism and a pentagonal bipyramid, all of which might be expected to have similar stability.

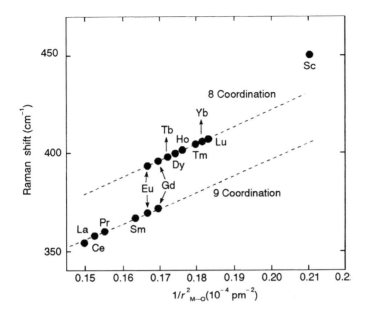

Figure 3.2. Correlation of Raman band v_1 and $1/r^2_{M-O}$ for aqua metal ions

Finally in 1989 Ouchi's group in Japan successfully isolated crystals of the aqua dimer $[(H_2O)_5Sc(\mu\text{-}OH)_2Sc(OH_2)_5](C_6H_5SO_3)_4\cdot4H_2O$ from dissolved solutions of Sc_2O_3 in 6.0 M HCl followed by evaporation and dissolution in aqueous benzenesulfonic acid [10]. The X-ray structure confirmed a coordination number of seven around Sc^{III} in the dimer in a distorted pentagonal bipyramidal arrangement (Fig. 3.3).

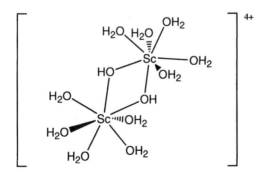

Figure 3.3. Structure of the dimeric unit in $[(H_2O)_5Sc(\mu\text{-}OH)_2Sc(OH_2)_5]$-$(C_6H_5SO_3)_4\cdot4H_2O$ (showing the distorted pentagonal bipyramidal geometry)

The Chemistry of Aqua Ions

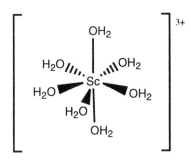

Figure 3.4. Proposed pentagonal bipyramidal geometry for hydrated Sc^{3+}(aq)

The Sc atoms are separated by a distance of 340 pm. The average Sc—O distance for the axial bonds in the pentagonal bipyramidal geometry are shorter than those equatorial (214.6 pm vs 222.7 pm) with the Sc—OH bonds even shorter (207.2 pm). Interestingly the average Sc—O distance over all seven directions (~ 216 pm) correlates with the average Sc—O value in Sc^{3+}(aq) determined by X-ray diffraction (~ 218 pm) [7]. This is the first and so far only purely aqua-derived complex known to exhibit seven-coordination. It is thus likely putative $[Sc(OH_2)_7]^{3+}$ is also pentagonal bipyramidal (Fig. 3.4). Indeed, the large basis set SCF calculations discussed in the previous chapter to estimate the binding energy of a seventh H_2O ligand indicate favourable low energy for the heptahydrated ion in the case of both Sc^{3+} and its neighbouring ion Ti^{3+}. Nonetheless, it is likely that very small crystal packing and perhaps steric effects can quickly facilitate a preference for six-coordination at Sc^{III} as seen in many compounds. The radius for Sc^{3+} probably sits on the borderline between a favoured coordination number of six and perhaps seven, and the presence of small amounts of octahedral $[Sc(OH_2)_6]^{3+}$ in equilibrium with $[Sc(OH_2)_7]^{3+}$ is a possibility that perhaps cannot be ruled out within the interpretation of the Raman and X-ray diffraction data [7]. Indeed, six-coordination is well established for Sc^{III} in a range of O-donor compounds. As well as the evidence for the formation of an alum, Sc^{3+} is also octahedrally coordinated to four waters and two pts$^-$ anions in the salt $Sc(pts)_3 \cdot 6H_2O$ [11]. In this regard the Sc^{3+} salt differs from the isomorphous series of pts$^-$ salts formed by Y^{3+}, Sm^{3+}, Gd^{3+}, Dy^{3+}, Ho^{3+}, Er^{3+} and Yb^{3+} wherein coordination to six water molecules and two pts$^-$ anions is observed. For these latter cations the coordination number of eight matches the hydration number believed to be relevant in solution. Sc^{3+}(aq) appears to adopt six-coordination in its series of chloro aqua complexes (see below).

The continuing controversy surrounding the precise hydration number of Sc^{3+}(aq) is apparent on the basis of its omission from the review of metal ion hydration numbers for trivalent metal ions by Ohtaki and Radnai [12] and the fact that the X-ray diffraction data on Sc^{3+}(aq) seems not to have been independently published.

3.1.1.2 HYDROLYSIS OF Sc^{3+}(AQ)

Hydrolysis of Sc^{3+}(aq) is extensive and there seems well-documented evidence for the formation of $ScOH^{2+}$(aq), the aqua dimer $Sc_2(OH)_2^{4+}$ (aq) and $Sc_3(OH)_5^{4+}$ (aq) [13–16]. Several groups have reported on the hydrolytic behaviour of Sc^{3+}(aq) at different temperatures and ionic strengths. The early results of Kilpatrick and Pokras [13] were later interpreted by Aveston [15] in terms of the above three hydrolysis products. Finally, Brown, Ellis and Sylva in 1983 [16] reported results from an extensive potentiometric study in 0.1 M KNO_3 which gave rise to the following hydrolysis constants: $ScOH^{2+}$(aq) ($pK_{11} = 4.840 \pm 0.008$), $Sc_2(OH)_2^{4+}$(aq) ($pK_{22} = 6.096 \pm 0.004$) and $Sc_3(OH)_5^{4+}$(aq) ($pK_{35} = 17.567 \pm 0.006$). The distribution of hydrolysis products is shown in Fig. 3.5. No evidence was found for the existence of $Sc_3(OH)_4^{5+}$ (aq), as had been earlier proposed by Aveston. The results reported above were, however, broadly

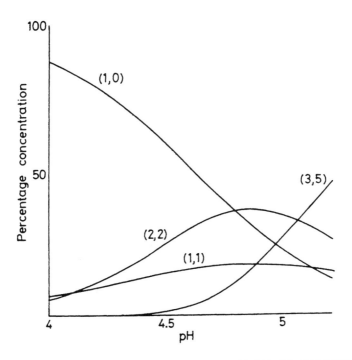

Figure 3.5. Distribution of hydrolysis products of Sc^{3+}(aq) in 0.1 M KNO_3

in agreement with values for the hydrolysis constants of these species reported by other workers in varying $NaClO_4$ solutions. In dilute solutions ($\sim 10^{-5}$ m) formation of the additional monomeric species $Sc(OH)_2^+$ (aq) ($pK_{12} = 9.7$) is also proposed prior to precipitation of the hydroxide around pH 8. The results broadly show that significant hydrolysis of Sc^{3+} (aq) quickly takes place above pH 3 and in concentrated solutions (~ 0.1 m) dimeric $Sc_2(OH)_2^{4+}$ (aq) is the dominant form in the pH range 4–5 with trimeric $Sc_3(OH)_5^{4+}$ (aq) dominant around pH 6. It has not been established whether the trimer is cyclic or linear. The trihydroxide is only weakly amphoteric but its solubility was nonetheless found to be proportional to the $[OH^-]$, suggesting equilibration to give some $Sc(OH)_4^-$ (aq) [3].

Finally, in view of the above pK values one must comment on the ready isolation of the pentagonal bipyramidal aqua dimer $Sc_2(OH)_2^{4+}$ (aq) from acidic solutions of Sc^{3+} (aq) under certain conditions [10]. Cole *et al.* have apparently studied the dimerisation of $ScOH^{2+}$ (aq) to give $Sc_2(OH)_2^{4+}$ (aq) by relaxation methods and have reported a rate constant (25° C) of 2×10^6 $M^{-1} s^{-1}$ [17]. With Sc^{3+} (aq) on the borderline between whether to be hexa-or hepta-hydrated it could be that such favourable and rapid formation of $Sc_2(OH)_2^{4+}$ (aq) from Sc^{3+} (aq) is facilitated by the adoption of the stable hepta-coordinated aqua-hydroxo geometry in the dimer. However, a more likely possibility is that hydrolysed Sc^{III} species, including complexes such as $[Sc_2(\mu\text{-}OH)_2(OH_2)_{8-n}Cl_n]Cl_{4-n}$, are actually formed during the evaporation of the $Sc_2O_3/6$ M HCl mixture which then rapidly aquate upon dissolution in the benzenesulfonic acid [10]. If so, aquation reactions of aqua chloro Sc^{III} complexes (see below) might be expected to be excellent candidates for the extreme A mechanism. Further work in this area should be encouraged.

3.1.1.3 COMPLEX FORMATION ON Sc^{3+} (AQ)

Complex formation on Sc^{3+} (aq) is extremely rapid requiring relaxation techniques. Solutions of Sc^{3+} (aq) in varying HCl solutions have been discussed as consisting of the following 'octahedral' species: $[ScCl(OH_2)_5]^{2+}$ (4.0 M HCl), $[ScCl_2(OH_2)_4]^+$ (7.0 M HCl) and $[ScCl_4(OH_2)_2]^-$ (11.5 M HCl) [2]. Formation constants (log K) determined for the first two Cl^- complexations are respectively 1.95 (log K_1), and 1.572 (log K_2), showing that Cl^- is not extensively complexed [18]. The formation of $ScCl_4^-$ (aq) in concentrated aqueous HCl has been confirmed in independent studies [19]. Using ultrasonic relaxation the retention of Cl^- ions within the inner coordination shell of Sc^{3+} was also found to be a feature of solutions in aqueous methanol which differentiated it from the Ln^{3+} ions [20]. Sc^{3+} is an extremely hard acid and the much higher formation constants, 7.08 (log K_1), 5.81 (log K_2), 4.48 (log K_3) and 2.85 (log K_4), relevant for F^-, are reflective of its hard base nature [5]. Although only ScF_4^- (aq) has been identified in aqueous solution, the $[ScF_6]^{3-}$ ion is established in isolated salts.

Few kinetic studies of complexation have been reported. Using ultrasonic relaxation Geier found a rate constant of 4.8×10^7 M^{-1} s^{-1} for anation of $Sc^{3+}(aq)$ by murexide at a pH of 4.0 at 12 °C [21]. This pH is, however, in the range where hydrolysis of Sc^{3+} is beginning to become significant (Fig. 2.4) and thus some concern might be appropriate. Nonetheless, this result illustrates the high kinetic lability of $Sc^{3+}(aq)$ similar to, for example, the heavier $Ln^{3+}(aq)$ ions (see later), $[Mn(OH_2)_6]^{2+}$ (Section 7.1) and $[Zn(OH_2)_6]^{2+}$ (Section 12.1). Little dependence of rate on the incoming ligand is thus anticipated for such a labile metal ion. In the absence of measurements thus far of the water-exchange process on $Sc^{3+}(aq)$, studies on a number of $Sc^{III}O_6$ complexes are supportive of the existence of an associative A or I_a mechanism for substitution. Exchange of trimethylphosphate (TMP) on $[Sc(TMP)_6]^{3+}$ has been studied by variable temperature and pressure ^{45}Sc NMR [22]. The second-order rate constant (25 °C), from studies in neat TMP, was ~ 140 M^{-1} s^{-1}, the much lower rate constant probably reflective of the more bulky nature of TMP as ligand. An I_a mechanism approaching limiting A was suggested from both the highly negative $\Delta S^{\ddagger}$ value (-60 J K^{-1} mol^{-1}) and $\Delta V^{\ddagger}$ value of ~ -20 cm^3 mol^{-1}, the latter being only slightly smaller than the $\Delta V°$ (-23.0) found for the equilibration reaction of $[Nd(TMP)_6]^{3+}$ with TMP to give $[Nd(TMP)_7]^{3+}$ [23]. A highly negative $\Delta S^{\ddagger}$ value (-105 J K^{-1} mol^{-1}) is also relevant for the acetylacetone-dependent step of the exchange on Sc(acac)$_3$ consistent with an I_a mechanism [4d]. The existence of a similar associative mechanism for $Sc^{3+}(aq)$ might be considered likely but should be tempered with caution given the uncertainty over the hydration number. At the same time, however, it might also be added that eight-coordination is not unknown for Sc^{III} [4].

3.2 YTTRIUM AND THE LANTHANIDES

Yttrium is best discussed alongside the lanthanide elements given the similarity in the ionic radius of Y^{3+} to that of Ho^{3+} and the isomorphicity of many Y^{III} and Ln^{III} coordination compounds. The oxides M_2O_3 are quite basic and solutions of $Y^{3+}(aq)$ and the various $Ln^{3+}(aq)$ ions are readily obtained by dissolution of the appropriate oxide in the required non-complexing acid. Upon crystallisation various hydrated M^{3+} salts can be isolated which, for different counter-anions, are found to markedly differ in hydration/coordination number around the M^{3+} centre.

3.2.1 Spectroscopic and Magnetic Properties of the $Ln^{3+}(aq)$ Ions

In the Ln^{3+} ions the 4f orbitals are very efficiently shielded from the external ligand field such that f–f splitting is minimal (only ~ 200 cm^{-1}). This unquenching of the orbital degeneracy results in well-defined and discrete electronic energy levels (J states) which arise from full manifestation of coupled total spin (S) and orbital angular momentum (L). Thus electronic spectra for the Ln^{3+}(aq) are unique to each element consisting of a whole series of orbitally forbidden transitions from the ground state ($4f^n$) ($n \neq 7$ or 14) to the various J states, the lack of crystal field splitting leading to their characteristic narrow linewidth (Fig. 3.6). The characteristic line spectrum of Ho^{3+} in Ho_2O_3 oxide filters is often used for calibrating ultraviolet–visible spectrophotometers. Little orbital mixing is ap-

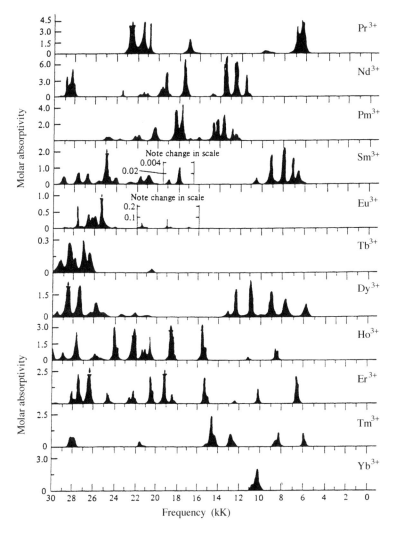

Figure 3.6. Electronic spectra for the Ln^{3+}(aq) ions

Table 3.1. Magnetic properties (at 300 K) for lanthanide (aq) ions

Ln	Number of f electrons	Ground state	μ(expt)	$\mu(SO)^a$	$\mu(L-S)^b$
Ce^{3+}	1	$^2F_{5/2}$	2.28	1.73	2.54
Pr^{3+}	2	$^3H_4$	3.40	2.83	3.58
Nd^{3+}	3	$^4I_{9/2}$	3.50	3.87	3.62
Sm^{3+}	5	$^6H_{5/2}$	1.58	5.90	1.60^c
Eu^{3+}	6	$^7F_0$	3.42	6.92	3.61^c
Sm^{2+}	6	$^7F_0$	3.57	6.92	3.61^c
Gd^{3+}	7	$^8S_{3/2}$	7.91	7.94	7.94
Eu^{2+}	7	$^8S_{3/2}$	7.91	7.94	7.94
Tb^{3+}	8	$^7F_6$	9.50	6.92	9.72
Dy^{3+}	9	$^6J_{15/2}$	10.40	5.90	10.63
Ho^{3+}	10	$^5I_8$	10.40	4.89	10.60
Er^{3+}	11	$^4I_{15/2}$	9.40	3.87	9.57
Tm^{3+}	12	$^3H_6$	7.10	2.83	7.63
Yb^{3+}	13	$^2F_{1/2}$	4.86	1.73	4.50

^{a} SO = spin only.
^{b} Using $g[J(J + 1)]$ except where noted.
^{c} Calculation includes mixing of ground and higher energy terms.

parent and selection rules for the 4f–4f transitions are strongly obeyed, leading to very low extinction coefficients, typically $< 2 \, M^{-1} cm^{-1}$. The colours exhibited and number of spectral bands obtained is therefore solely a function of the $4f^n$ configuration and by and large ligand/geometry effects are minimal. There are some exceptions, however (e.g. Nd^{3+}, see later). The spectra for Ce^{3+} $(4f^1)$, $Gd^{3+}(4f^7)$ and $Lu^{3+}(4f^{14})$ show intense bands in the u.v. region from orbitally allowed 4f–5d transitions. The ground states are well defined and the observed magnetic moments can easily be calculated from simple $L–S$ coupling, (Table 3.1). In the ions Sm^{3+} and Eu^{3+} there is some mixing-in of the higher J states into the ground state, which raises the observed moment. A number of the lanthanides, particularly Eu^{3+} and Gd^{3+}, are sought for their fluorescent properties, this stemming from the long radiative lifetimes characteristic of their excited states. This property is a further reflection of the highly shielded nature of the 4f orbitals which effectively suppresses many of the quenching mechanisms.

3.2.2 Hydration Numbers of Y^{3+}(aq) and the Ln^{3+}(aq) Ions

Considerable evidence has now been gathered pointing towards a change in the solution hydration number for the Ln^{3+}(aq) ions from nine to eight around Sm, Eu and Gd, the so-called 'gadolinium break'. One of the difficulties that has surrounded this premise has been the variability of coordination numbers found for the Ln^{3+} ions in different compounds (e.g. both six- and nine-coordination is known in solid compounds covering the entire series) and the general problems inherent in estimating hydration numbers for highly labile metal ions (water-exchange rates vary for the $Ln^{3+}(aq)$ ions in the range 10^9 s^{-1} (lighter elements) to 10^7 s^{-1} (heavier elements)). The decrease in both the water-exchange rate and the Ln—OH_2 distance in the structures of many solid compounds is a direct consequence of the so-called 'lanthanide contraction' which results in a steady, almost monotonic, decrease in ionic radius for Ln^{3+} of nearly 20 pm along the series. This arises from the significant build-up in the effective nuclear charge due to the poor shielding of the nucleus by the filling 4f orbitals resulting in valence shell contraction. The smooth nature of the contraction (Fig. 3.7) reflects the lack of crystal field splitting of the highly shielded 4f orbitals with only the faintest evidence of a double cusp shape maximising at $La^{3+}(4f^0)$, $Gd^{3+}(4f^7)$ and $Lu^{3+}(4f^{14})$.

Various X-ray structures on hydrated lanthanide salts have illustrated the problems inherent in assessing the relevant hydration number for the $Ln^{3+}(aq)$ ions. Although the apparent change in solution hydration number for the $Ln^{3+}(aq)$ ions roughly halfway along the series (see later) can be seen as a direct consequence of the decreasing ionic radius, it appears that in the crystal structures of many classes of solid compound other factors can overide the effects of the lanthanide contraction, leading to retention of a constant coordination geometry over the entire series. For example, X-ray crystal structures of hydrated bromate and ethylsulfate salts show evidence of a hydration number of nine, in a tricapped trigonal prismatic arrangement, for Ln^{3+} in the case of Pr, Nd, Ho

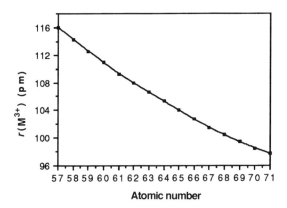

Figure 3.7. The lanthanide contraction (Ln^{3+} radii used are Shannon values [24] for eight-coordination). (Reproduced by permission of The Royal Society of Chemistry)

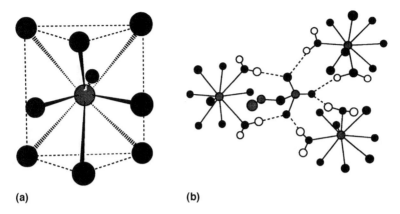

(a)   (b)

Figure 3.8. (a) The $[Nd(OH_2)_9]^{3+}$ unit in $Nd(BrO_3)_3 \cdot 9H_2O$ (b) The $[Ho(OH_2)_9]^{3+}$ unit in $Ho(C_2H_5SO_3)_3 \cdot 9H_2O$

and Yb (Fig. 3.8) [25], whereas corresponding X-ray crystal structures for a series of $M(ClO_4)_3 \cdot 6H_2O$ salts (M = La, Tb and Er) showed evidence of regular octahedral hexaaqua coordination [26]. On the other hand, La^{3+} is nine-coordinated (tricapped trigonal prism) in the salt $La_2(SO_4)_3 \cdot 9H_2O$ (six prismatic La—OH_2 bonds with three face-capping La—O bonds to SO_4^{2-} ions) [27].

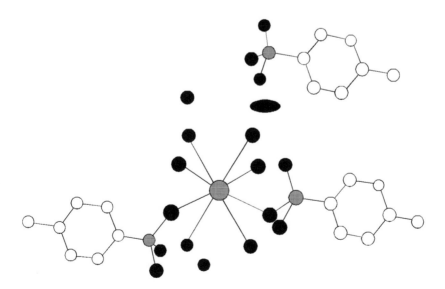

Figure 3.9. Square antiprismatic $[Y(OH_2)_6(pts)_2]^+$ unit in $Y(pts)_3 \cdot 9H_2O$

Finally, in the series of isomorphous pts$^-$ salts, M(pts)$_3$·9H$_2$O (M = Y, Sm, Gd, Dy, Ho, Er and Yb) eight-coordination to six water ligands and two oxygens of pts$^-$ anions is present in a square antiprismatic geometry (Fig. 3.9) [23]. It is clear from these findings that crystal packing and perhaps even the intimate hydrogen-bonding network can drastically influence the preferred hydration number and coordination geometry around Ln^{3+} in crystalline hydrates. It was as a result of both ^{17}O NMR and structural evidence (e.g. Fig. 3.8) that it was presumed at one time that a constant hydration number (nine) was preserved along the entire series also in solution [24, 28–30]. Moreover, around the same time an apparent constant solvation number (x) of eight had already been established for the [Ln(DMF)$_x$]$^{3+}$ ions [28, 31]. The hydration number of nine was based upon conclusions reached by Karracher in 1968, on the basis of f–f spectra, of nine-coordination for Nd^{3+}(aq) [32]. It was found that certain of the f–f transitions of Nd^{3+}, particularly the transitions: ^{4}I$_{9/2}$ (ground state) to ^{2}H$_{9/2}$, ^{4}F$_{5/2}$ ($\sim$ 800 nm), ^{4}I$_{9/2}$ to ^{4}S$_{3/2}$, ^{4}F$_{7/2}$ ($\sim$ 750 nm), ^{4}I$_{9/2}$ to ^{4}G$_{5/2}$, ^{2}G$_{7/2}$

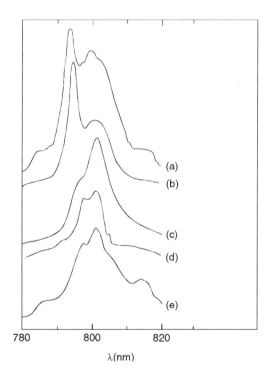

Figure 3.10. Nd^{3+} 'hypersensitive' f–f spectra: (a) solid Nd(BrO$_3$)$_3$·9H$_2$O, (b) 0.0535 M Nd^{3+}(aq), (c) 0.0535 M Nd^{3+}(aq) in 11.5 M HCl, (d) solid NdCl$_3$·6H$_2$O and (e) solid Nd$_2$(SO$_4$)$_3$·8H$_2$O) [32]. (Reprinted with permission from D. G. Karracher, *Inorg. Chem.*, **7**, 473, 1968. Copyright 1968 American Chemical Society)

($\sim 580\,\text{nm}$) and $^4I_{9/2}$ to $^2K_{13/2}$, $^4G_{7/2}$ and $^2G_{9/2}$ ($\sim 520\,\text{nm}$), were, unlike those from most other Ln^{3+} ions, surprisingly sensitive to changes to the coordination geometry (termed 'hypersensitive'). Thus spectral characteristic of both eight- and nine-coordinate Nd^{3+} could be distinguished (Fig. 3.10). Comparisons of corresponding spectra obtained from crystalline samples of $Nd(BrO_3)_3 \cdot 9H_2O$, $NdCl_3 \cdot 6H_2O$ and $Nd_2(SO_4)_3 \cdot 8H_2O$ showed that $Nd^{3+}(aq)$ was almost certainly present as $[Nd(OH_2)_9]^{3+}$, as in the bromate salt. In the other two salts Nd^{3+} was eight-coordinated. The results also showed a rapid change to eight-coordination upon addition of Cl^- ions to the $Nd^{3+}(aq)$ solutions [32]. On the other hand, corresponding spectra obtained for the $^5I_8$ to $^5G_6$, $^5F_1$ transition for Ho^{3+} and the $^4I_{15/2}$ to $^2H_{11/2}$ transition for Er^{3+} (also hypersensitive) were largely unchanged for solutions obtained at different HCl concentrations and moreover on moving from $11.0\,\text{M}$ $HClO_4$ to $11.0\,\text{M}$ HCl, indicative of a constant coordination (hydration) number of eight for these two Ln^{3+} ions.

Strong indications of a change in hydration number from nine to eight around Sm to Gd stemmed from the studies by Spedding and coworkers in the mid 1960s

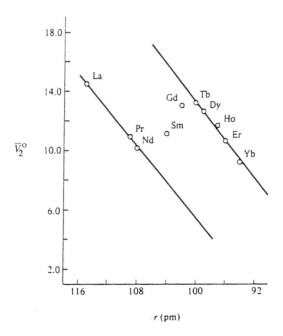

Figure 3.11. The trend in partial molar volumes for Ln^{3+} at infinite dilution (V°_2) with Pauling ionic radius r (pm) [33]. (Reprinted with permission from F. H. Spedding, M. J. Pikal and B. O. Ayers, *J. Phys. Chem.*, **70**, 2440, 1966. Copyright 1966 American Chemical Society)

which involved measurements of partial molar volumes [33], heat capacities [34], molar entropies [34] and viscosities [35] along the entire Ln^{3+} series. The 'gadolinium break' is most clearly illustrated by the trend in partial molar volumes (Fig. 3.11). Partial molar volumes should decrease smoothly in accordance with the decrease in effective ionic radius. The discontinuity and the increase in V_2° with decreasing radius around Sm to Gd was rationalised by a decrease in hydration number, resulting in less breakdown in water structure and a corresponding decrease in $\Delta V_{H_2O}^{\circ}$. The hydration numbers of nine (for Nd^{3+}(aq)) and eight (for Gd^{3+}(aq), Tb^{3+}(aq), Er^{3+}(aq) and also Y^{3+}(aq)) are also consistent with the results from independent X-ray and neutron diffraction studies on aqueous solutions of their chloride and perchlorate salts [9, 36]. In addition, hydration numbers of nine (Nd^{3+}), between eight and nine (Sm^{3+} and Eu^{3+}) and eight (Gd^{3+}, Tb^{3+}, Dy^{3+}, Er^{3+}, Tm^{3+}, and Lu^{3+}) appear to be consistent with the results from EXAFS experiments [37]. Finally first-order difference neutron scattering studies appear to have unequivocally established a hydration number of nine for Nd^{3+}, eight for Dy^{3+}(aq) and Yb^{3+}(aq), and between eight and nine for Sm^{3+} [38]. The uniform coordination number of eight observed in the X-ray structures of the series of isomorphous pts$^-$ salts (M = Y, Sm, Gd, Dy, Ho, Er and Yb) [23] is therefore of interest as reflecting the same hydration number in solution for these ions and moreover suggests that the preferred geometry for the $[Ln(OH_2)_8]^{3+}$ ions might be square antiprismatic (Fig. 3.12).

The change in preferred hydration number reflects the lanthanide contraction (Fig. 3.6), resulting in a radius for Sm^{3+} and Eu^{3+} that is too small to fit comfortably in the cavity provided by the nine-coordinate tricapped trigonal prismatic hydration sphere. Thus there is a switch to the slightly more compacted square antiprismatic geometry. It is believed that these two geometries play an intimate role in the water-exchange mechanism on the ions Gd^{3+}(aq) through to Yb^{3+}(aq).

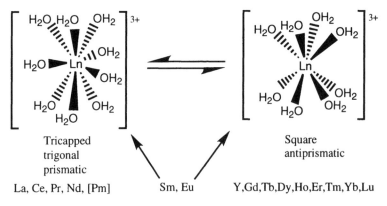

Figure 3.12. Possible hydration geometries for Ln^{3+}(aq) ions

3.2.3 Hydrolysis of $[Y(OH_2)_8]^{3+}$ and the Ln^{3+}(aq) Ions

The pattern of hydrolysis of Y^{3+}(aq) and the Ln^{3+}(aq) ions appears to parallel that of Sc^{3+}(aq) in so far as the available hydrolysis data are consistent with the formation of the species MOH^{2+}(aq), $M_2(OH)_2^{4+}$(aq) and $M_3(OH)_5^{4+}$(aq). Baes and Mesmer [39] tabulated much of the earlier potentiometric and solubility data which showed evidence across the series of a steady decrease in pK_{11}, pK_{22} and pK_{35} from respectively ~ 8.5, ~ 17 and 38 (La) to ~ 7.6, 13.6, 29 (Er), consistent with an increase in acidity of Ln^{3+} as the charge/size ratio increases due to the lanthanide contraction in radius (Fig. 3.6). Y^{3+}(aq) is quoted as having a pK_{11} value of 7.7, consistent with a radius similar to that of Ho^{3+} and Er^{3+}. However, despite general agreement from a number of groups [40] as to these magnitudes of pK_{xy} values, more recent studies on La^{3+}, Eu^{3+} and Lu^{3+}, based upon solvent extraction into xylene with thenoyl trifluoroacetone, appear to be suggestive of pK_{11} values rather between 3.8 (La) and 3.1 (Lu) [41]. Significant adsorption problems encountered on walls of containers in the range $4 < pH < 6$ (believed due to hydrolysis) were cited as further evidence of significant hydrolysis for the three Ln^{3+}(aq) ions studied at pH values well below 6. Little attention was paid to the possible formation of polynuclear species. As in the case of scandium, the lanthanide trihydroxides are only weakly amphoteric but dissolve in excess base proportional to the $[OH^-]$, giving evidence for the formation of $Ln(OH)_4^-$(aq) [42]. Values for pK_{14} are in the range 36.5(Y) to 32.7(Yb) [39].

3.2.4 Water Exchange and Complex Formation on Y^{3+}(aq) and the Ln^{3+}(aq) Ions

Merbach and coworkers have carried out an extensive and thorough study of the water-exchange process on the ions Pr^{3+}(aq) to Yb^{3+}(aq) under conditions of variable temperature and pressure using ^{17}O NMR at high fields (up to 14.1T) [29, 43]. The measurements were made using a Swift–Connick line-broadening approach wherein the chemical shift and both the transverse (T_2) and longitudinal (T_1) relaxation times for the bound water solvent resonance become modified by the presence of a paramagnetic metal aqua ion in the fast exchange region in solution. The relevant theory has been well discussed (ref. [44] and Chapter 1) and in its simplest form can be expressed by

$$\left(\frac{1}{T_2} - \frac{1}{T_1}\right) = \frac{(\Delta\omega_s)^2 \tau_m}{P_m} \tag{3.1}$$

wherein the exchange rate constant for the bound water is measured in terms of τ_m (the mean lifetime of a water molecule bound to Ln^{3+}), $\Delta\omega_s$ is the chemical shift between the free and the free and bound coalesced water resonance and P_m is the partial molar fraction of water bound to Ln^{3+}. The values obtained by the Lausanne group [29, 43] are viewed as the most reliable and comprehensive, the

Table 3.2. Rate constants and activation parameters for the water-exchange process on $[Ln(OH_2)_8]^{3+}$, Ln = Pr to Yb [29, 43, 46]

Ln^{3+}	$10^{-7} k_{ex}{}^a$ (s^{-1})	$\Delta H^{\ddagger}$ (kJ mol^{-1})	$\Delta S^{\ddagger}$ (J K^{-1} mol^{-1})	$\Delta V^{\ddagger b}$ (cm^3 mol^{-1})	$10^2 \, \Delta\beta^b$ cm^3 mol^{-1} mPa^{-1}
Pr	$\geqslant 40$	12.9 ± 2.0	-36.3 ± 6.0	—	
Nd	$\geqslant 50$	7.0 ± 2.0	-54.4 ± 7.0	—	
Gd	83.0 ± 1.0	12.0 ± 1.3	-30.9 ± 4.1	-3.3 ± 0.2	Not quoted
Tb	55.8 ± 1.3	12.1 ± 0.5	-36.9 ± 1.6	-5.7 ± 0.5	$+0.3 \pm 1.6$
Dy	43.4 ± 1.0	16.6 ± 0.5	-24.0 ± 1.5	-6.0 ± 0.4	$+0.4 \pm 1.4$
Ho	21.4 ± 0.4	16.4 ± 0.4	-30.5 ± 1.3	-6.6 ± 0.4	$+0.6 \pm 1.3$
Er	13.3 ± 0.2	18.4 ± 0.3	-27.8 ± 1.1	-6.9 ± 0.4	$+0.3 \pm 1.2$
Tm	9.1 ± 0.2	22.7 ± 0.6	-16.4 ± 1.9	-6.0 ± 0.8	$+1.3 \pm 3.0$
Yb	4.7 ± 0.2	23.3 ± 0.9	-21.0 ± 3.3	—	

^{a} Rate constants are reported at 298 K.
^{b} Values of $\Delta V^{\ddagger}$ and $\Delta\beta^{\ddagger}$ were determined at the following temperatures: Gd (298.2 K), Tb (269.1 K), Dy (268.5 K), Ho (268.8 K), Er (268.8 K) and Tm (269.1 K).

earlier values reported by Reuben and Fiat [45], ~ one order of magnitude lower, being at a much lower field (1.4 T) and thus viewed as giving only lower limiting values. Rate constants and activation parameters were later adjusted [43, 46] to take account of the evidence for octaaqua coordination for Gd^{3+} to Yb^{3+}. Values for Pr^{3+} and Nd^{3+} required very high field NMR measurements but are only lower limits. Table 3.2 summarises the avaliable data.

A gradual decrease in the rate constant for water exchange (Fig. 3.13) Gd^{3+} to Yb^{3+} parallels the lanthanide contraction as expected. Activation volumes,

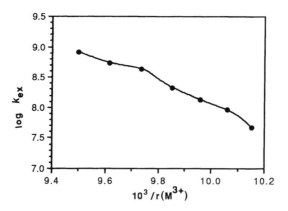

Figure 3.13. Logarithm of k_{ex} values for water exchange on Ln^{3+}(aq) vs reciprocal ionic radius [24] for $[Ln(OH_2)_8]^{3+}$ (Ln = Gd to Yb)

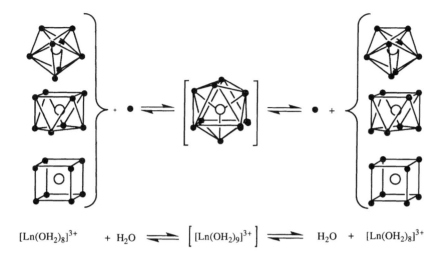

$$[Ln(OH_2)_8]^{3+} \quad + H_2O \quad \rightleftharpoons \quad \left[[Ln(OH_2)_9]^{3+} \right] \quad \rightleftharpoons \quad H_2O \quad + \quad [Ln(OH_2)_8]^{3+}$$

Figure 3.14. Mechanism for water exchange on the octaaqua lanthanide ions Gd^{3+} to Yb^{3+}

from variable high-pressure studies, are also of interest. From Tb^{3+} to Tm^{3+} appreciably negative values $\sim -6\,cm^3\,mol^{-1}$ are found which have been discussed in terms of an associative process in which a nine-coordinated tricapped trigonal prismatic transition state is formed (Fig. 3.14).

In a separate study the 295 nm absorption band of $Ce^{3+}(aq)$ was found to be hypersensitive (Fig. 3.15) and a variable-pressure spectrophotometric study

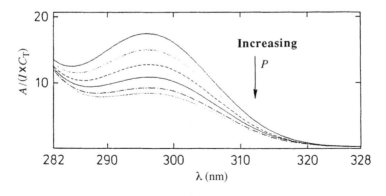

Figure 3.15. UV spectrum of $Ce^{3+}(aq)$ as a function of pressure in the range 2.2–200.4 mPa [47]. (Reproduced by permission of Verlag Helvetica Chimica Acta AG)

allowed a measure of the reaction volume for the following equilibrium as $+10.9\,\text{cm}^3\,\text{mol}^{-1}$ [47]:

$$[Ce(OH_2)_9]^{3+} \underset{\Delta V^\circ = +10.9\,\text{cm}^3\,\text{mol}^{-1}}{\rightleftharpoons} [Ce(OH_2)_8]^{3+} + H_2O \qquad (3.2)$$

Calculated values for the loss of the molar volume of water in this type of equilibrium based on a semi-empirical model are $\sim +12.9\,\text{cm}^3\,\text{mol}^{-1}$ [48]. The observed $\Delta V^\ddagger_{ex}$ ($\sim -6\,\text{cm}^3\,\text{mol}^{-1}$) for the water-exchange process on the $[Ln(OH_2)_8]^{3+}$ ions from Gd^{3+} to Yb^{3+} [43, 46a], in being lower than the limiting value, might initially be viewed as reflective of the highly solvated interchange process that has been discussed as relevant to water exchange on the d-block ions in Chapter 1. Here, though, given the clear evidence of the equilibrium between the eight- and nine-coordinate $Ln^{3+}(aq)$ species, one might be more confident in concluding that the negative $\Delta V^\ddagger_{ex}$ values observed in this case are probably reflective of an associatively activated interchange process with a degree of incoming ligand participation.

The use of very high field ^{17}O NMR measurements at 14.1 T has permitted estimations of the lower limits for the water-exchange rate constants on $Pr^{3+}(aq)$ and $Nd^{3+}(aq)$ [46b]. Indications are that the water-exchange rate may maximise around Sm^{3+}, consistent with there being no preference for either of the two geometries (CN of ~ 8.5 from neutron diffraction studies). Indeed, the activation enthalpy for its near neighbour Nd^{3+} is the lowest of the series. However, kinetic data for complexation of $Ln^{3+}(aq)$ by sulfate is available for comparison over the entire series [49]. Complexation reactions on the $Ln^{3+}(aq)$ ions have been analysed by an Eigen–Wilkins approach:

$$[Ln(OH_2)_n] + L \underset{K_{os}}{\rightleftharpoons} [Ln(OH_2)_n\text{---}L] \underset{k_I}{\rightleftharpoons} [Ln(OH_2)_{n-1}L] + H_2O \quad (3.3)$$

wherein estimation of the outer-sphere ion-pair association constant K_{os}, e.g. via the electrostatic models of Bjerrum [50] or Fuoss [51], allows a comparison of the interchange rate constant $k_I(s^{-1})$ with the water-exchange rate constant k_{ex} for $Ln^{3+}(aq)$. In the region from Gd^{3+} onwards k_I (sulfate) parallels closely k_{ex} in both magnitude and in the smooth steady decrease (Fig. 3.16). Of further interest, however, is that the k_I value for sulfate indeed does appear to maximise in the region around Sm to Gd with lower rate constants appearing to be relevant at the beginning of the series. A similar maximum in k_I in the same region around Sm is seen in the case of acetate [52]. It is recalled that the region Sm to Gd is where there is an apparent change in hydration number for $Ln^{3+}(aq)$ from nine to eight. Thus one possible conclusion is a disfavouring of the associative process for the elements before Sm (due to a requirement for a ten-coordinated transition state). Alternatively, the rate fall-off towards the beginning of the series, e.g. for $Ce^{3+}(aq)$, might also be a reflection of an enforced change to a more dissociative

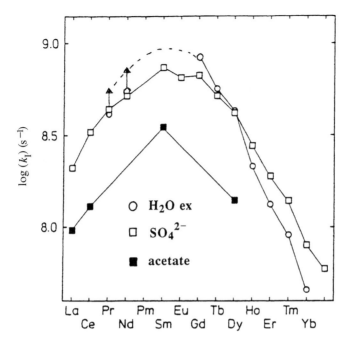

Figure 3.16. Interchange rate constants for complex formation on Ln^{3+}(aq) ions, a comparison with $k_{ex}(H_2O)$

mode of activation (eight-coordinate transition state) (Fig. 3.14).An independent confirmation of this in terms of measured water-exchange rate constants for, for example, Ce^{3+}(aq) would be of interest. One could indeed speculate that k_{ex} for Ce^{3+}(aq), in view of the possible changeover in mechanism, might be smaller than that for Gd^{3+}(aq) and possibly of the same magnitude as those for Dy^{3+}(aq) and Ho^{3+}(aq). However, evidence from a number of sources points towards a very low activation energy for coordination equilibria in Ce^{3+}(aq) solutions [47, 53], which is reflective of an extremely low activation energy for the interchange of the eight- and nine-coordinate geometries for all the lighter Ln^{3+}(aq) ions and a water-exchange rate at or approaching the diffusion controlled limit. The apparent maximum in the interchange rate constant for sulfate complexation at around Sm to Gd, also virtually at the diffusion controlled limit, presumably reflects them as having essentially identical energies for the eight- and nine-coordinate species. The move towards a smooth 'interchange' process around Sm to Gd is also reflected in the smaller value of $\Delta V^{\ddagger}$ ($-3.3\,\text{cm}^3\,\text{mol}^{-1}$) for water exchange on Gd^{3+}(aq) versus the later members. Thereafter the lanthanide contraction slowly leads to a disfavouring of the coordination geometry increase and a slowing down of the water-exchange rate.

Merbach and coworkers have recently performed a number of molecular dynamic simulation studies of the hydration sphere around Nd^{3+}(aq), Sm^{3+}(aq) and Yb^{3+}(aq) [54]. These involved the use of *ab initio* three-body potential functions that take into account the effects of water polarisation in the first hydration shell. The simulated radial functions give rise to hydration numbers respectively of 9.0 (Nd^{3+}, tricapped trigonal prismatic), 8.5 (Sm^{3+}, equilibrium between tricapped trigonal prismatic and square antiprismatic) and 8.0 (Yb^{3+}, square antiprismatic), in keeping with the results from neutron diffraction. Furthermore, the fast rates of water exchange have allowed simulation of the microscopic mechanisms of the water-exchange process between the first and second hydration shells. Residence times for the water molecules in the first hydration shell were computed as 1577, 170 and 410 ps for the above three Ln^{3+}(aq) ions respectively. Whilst reflecting somewhat smaller residence times than those experimentally measured using NMR the calculated transition state structures, square antiprismatic for Nd^{3+}(aq) and tricapped trigonal prismatic for Yb^{3+}(aq), supported the proposal of an apparent changeover in water-exchange mechanism. The maximum rate of water exchange was found for Sm^{3+}(aq) which was consistent with a small energy difference between the ground state and transition state configurations.

Kinetic data for complex formation on Ln^{3+}(aq) have also been reported for a number of other ligands, e.g. oxalate and murexide [55]. Here only overall k_f values are available and so a similar close comparison with water-exchange rate constants and moreover of the rate constants for different ligands themselves is not possible. For oxalate and murexide, for example, k_f values are found to be essentially constant for the elements up to Sm. Thereafter, however, rate constants drop off sharply, suggestive again of an influence from the change in hydration number of Ln^{3+}(aq). For such highly labile metal ions largely ligand-independent complex formation rate constants at each Ln^{3+}(aq) ion might be expected since ligand discrimination becomes less important (irrespective of the mechanisms) for fast rates of complexation. This is particularly well illustrated in the case of Mn^{2+}(aq) (Chapter 7) and for both Zn^{2+}(aq) and Cd^{2+}(aq) (Chapter 12). For the Ln^{3+}(aq) ions this seems also to be the case for ligands such as sulfate (Fig. 3.15), wherein available k_1 values for Gd^{3+} to Yb^{3+} are found to largely overlap with $k_{ex}(H_2O)$ despite the indications of an associative process. In this regard, further kinetic data allowing a comparison of k_1 with k_{ex} would be useful to provide confirmation.

Lanthanum-139 NMR has been successfully used to study complex formation on La^{3+}(aq) and its dissociation from complexes in aqueous solution. Dissociation (aquation) rates are slow enough in certain systems to be characterised by observation of two separate signals from La^{3+}(aq) and the La^{III} complex in the presence of excess La^{3+}(aq). Kinetic information on the dissociation process is easily extractable from the measured linewidth of the two signals as a function of temperature. Merbach and coworkers studied the aquation of $[LaL(aq)]^+$

(L = 2,6-dicarboxy-4-hydroxy-pyridine) by this method and obtained $k_{aq}(25\,°C) = 207 \pm 51\,s^{-1}$, $\Delta H^{\ddagger} = 38.3 \pm 3.1\,kJ\,mol^{-1}$ and $\Delta S^{\ddagger} = -72 \pm 8\,J\,K^{-1}$ mol^{-1} [56]. Detellier used ^{139}La NMR to study the interactions of La^{3+}(aq) with the anions Cl$^-$, NO$_3^-$ and ClO$_4^-$. The chemical shift and transverse relaxation rate data were interpreted in terms of a 1:1 contact ion pair model. Formation constants determined for the contact ion pair $\{LaX\}^{2+}$ were $0.45 \pm 0.05\,M^{-1}$ (NO$_3^-$), $0.15 \pm 0.09\,M^{-1}$ (Cl$^-$) and $0.03 \pm 0.01\,M^{-1}$ (ClO$_4^-$) [57]. It was concluded that very fast exchange of water in the first coordination sphere was responsible for the time fluctuation of the electric field gradient at the ^{139}La nucleus.

3.2.5 Redox Reactions of the Lanthanide (aq) Ions

3.2.5.1 TETRAVALENT LANTHANIDES: THE Ce^{IV} (AQ) ION

In minerals cerium is found with mobility, indicating its presence in the tetravalent state. However, the trivalent state remains the thermodynamically stable one in solution for cerium as with of the lanthanides and Ce^{IV}, despite forming an f^0 configuration, is strongly oxidising. It is well established as a convenient oxidising reagent in analytical and organic chemistry. Ce^{IV}(aq) is more extensively complexed than Ce^{III}(aq) [58] reflective of the higher positive charge, and this manifests itself in large variations in the Ce^{IV}/Ce^{III} potential (Table 3.3) for different media [59, 60]. An interesting consequence of this shift in E^{θ} is that whereas in HClO$_4$ media Ce^{IV} will oxidise [Mn(OH$_2$)$_6$]$^{2+}$ to MnO$_4^-$, the reverse is true in H$_2$SO$_4$ media. In all cases the E^{θ} value suggests that Ce^{IV} can spontaneously oxidise H$_2$O to O$_2$ (E^{θ} (O$_2$,4H$^+$/2H$_2$O) = + 1.23 V).

However, kinetic control, usually as a result of anion complexation, can lead to kinetically stable Ce^{IV}(aq) solutions showing little deterioration over many months. Thus Ce^{IV}(aq) is usually employed as an analytical volumetric oxidant in H$_2$SO$_4$. In HClO$_4$ solutions the Ce^{IV}(aq) ion is essentially uncomplexed and solutions are now only metastable, decaying to Ce^{III}(aq) within days. Table 3.3

Table 3.3. Reduction potential for the Ce^{IV}/Ce^{III} couple (V vs NHE) in various media [59, 60]

Molarity	HClO$_4$	HNO$_3$	H$_2$SO$_4$	HCl
1.0	1.70	1.61	1.44	1.47
2.0	1.71	1.62	1.44	
4.0	1.75	1.61	1.43	
6.0	1.81	1.56	1.42	
8.0	1.87			

shows that the highest E^θ values are obtained in perchlorate media. Ce^{IV}(aq) is also strongly hydrolysed and highly acidic solutions (> 1.0 M H^+) are required to prevent extensive hydrolytic polymerisation and ultimately precipitation of CeO_2(aq) ('$Ce(OH)_4$'). The presence of variable amounts of hydrolysis products probably also contributes to the wide variation in the E^θ value for the Ce^{IV}/Ce^{III} couple in the different media. Crystalline CeO_2 adopts the fluorite structure (eight-coordinated) and is finding growing use as the supporting ceramic material in catalytic convertors. CeO_2 is extremely hard and its use as a glass abrasive has been well documented.

Preparation of Ce^{IV}(aq) solutions have normally involved precipitation of '$Ce(OH)_4$' with aqueous ammonia from solutions of Ce^{IV} salts, such as $(NH_4)_2[Ce(NO_3)_6]$ [61]. Here if the ceric hydroxide is allowed to precipitate over a period of 48 hours and then leached with H_2O it readily redissolves to generate solutions of Ce^{IV}(aq) in the required acid of choice. This method suffers from potentially introducing hydrolysis products. An alternative route to generate largely uncomplexed Ce^{IV}(aq) is via simple dissolution of the nitrato salt in the required acid, e.g. $HClO_4$. However, G. Frederick Smith Co. Ltd in the United States, for many years suppliers of ceric salts, have introduced a stock solution of '$H_2Ce(ClO_4)_6/6$ M $HClO_4$' available for general purpose use. The 'enforced' complexation of Ce^{IV} by ClO_4^- ions in this solution is presumably responsible for the kinetic control of the reduction by H_2O and thus a reasonably long shelf life.

Solutions of Ce^{IV}(aq) are bright yellow to orange in color (conveniently self-indicating) due to a favourable O to Ce^{IV} charge transfer absorption in the u.v. region which appreciably 'tails' into the visible. Ce^{III}(aq) solutions are colourless. The spectrum of Ce^{IV}(aq) differs in the various media due to hydrolysis and/or extensive anion complexation. In 1 M $HClO_4$, Ce^{IV}(aq) is reported to have a peak around 295 nm. However, the ε value seems to vary in different solutions from 507 M^{-1} cm^{-1} (prepared via solutions of the nitrato salt) [61] to as high as ~ 2000 M^{-1} cm^{-1} (prepared via precipitated '$Ce(OH)_4$'), the latter due probably to the presence of hydrolysis products. In 0.5 M H_2SO_4 solution, Ce^{IV}(aq) is reported to absorb with a maximum at 316 nm ($\varepsilon = 5580$ M^{-1} cm^{-1}) [62], the higher ε value here being due to extensive complexation by SO_4^{2-} ions.

3.2.5.2   HYDROLYSIS OF Ce^{IV}(AQ)

Considerably uncertainty surrounds the values of the hydrolysis constants for Ce^{IV}(aq). It is generally agreed, however, that Ce^{4+} itself is already extensively hydrolysed even in 1 M H^+. Since Ce^{3+} is essentially unhydrolysed under these conditions use has been made of the E^θ value for the Ce^{IV}/Ce^{III} couple as a function of $[H^+]$ to probe the extent of hydrolysis in Ce^{4+}(aq). Baes and Mesmer [63] have summarised the early E^θ work. A typical cell employed for these studies is shown below:

$$Pt, H_2/H^+, ClO_4^- //H^+, Ce^{3+}, Ce^{IV}, ClO_4^-/Pt \qquad (3.4)$$

It is generally agreed that formation of monomeric $CeOH^{3+}(aq)$ and $Ce(OH)_2{}^{2+}(aq)$ is relevant. Their effect on the observed E^θ value can be expressed by

$$E^\theta_{obs} = E^{\theta'} + \left(\frac{RT}{F}\right) \ln \left[1 + \left(\frac{K_{11}}{[H^+]}\right) + \left(\frac{K_{12}}{[H^+]^2}\right) \right] \tag{3.5}$$

Values for pK_{11} have been reported ranging between -0.7 and 0.7. The formation of $Ce(OH)_2{}^{2+}(aq)$ is, however, frequently overlooked and Baes and Mesmer, in summarising the available data, estimated probable values for pK_{11} and pK_{12} as respectively -1.1 and -0.3. On this basis they calculated the $E^\theta(Ce^{IV}/Ce^{III})$ value for the 1.3 M $HClO_4$ solutions studied by Sherrill, King and Spooner [64] as 1.774 V, which agreed well with the experimental value of Baker, Newton and Kahn of 1.777 V in 2 M $HClO_4$ [65]. Hardwick and Robinson found that solutions of $Ce^{IV}(aq)$ in 2 M $HClO_4$ did not obey Beers law as the Ce^{IV} concentration was raised [66]. They concluded that polynuclear hydrolysis products were present and interpreted their results in terms of the additional formation of dimeric $Ce_2(OH)_2{}^{6+}(aq)$. Formation of the dimeric species appears much less extensive in more dilute solutions [61]. Subsequent work by King and Pondow led to a proposal of an additional dimeric species $Ce_2(OH)_3{}^{5+}(aq)$ [67]. Values of pK_{22} and pK_{23} have been estimated as 3.6 and 4.1 respectively. In both nitrate and sulfate media the more extensive anion complexation suppresses hydrolysis to a limited extent and in nitrate media there have been claims for the formation of $Ce_2(OH)_4{}^{4+}(aq)$ ($pK_{24} = 2.29$ and $Ce_6(OH)_{12}{}^{12+}(aq)$ ($pK_{612} = 1.98$) [68]. The effective positive charges on these species is probably much lower than those stated due to the presence of complexed nitrate and this is almost certainly responsible for the supression of the hydrolysis in these solutions.

The extensive hydrolytic tendency of $Ce^{4+}(aq)$ has meant that there are no precise measurements of the hydration number. The hydration number for $Ce^{3+}(aq)$ is believed to be nine so a reduction to eight on the basis of an expected radius reduction, 97 pm, Ce^{IV} vs 114 pm, Ce^{3+} [24], might not be unreasonable. Moreover, eight-coordination is present in the fluorite structure of CeO_2. Eight-coordination is also present in $Ce(acac)_4$ [69] but now in a dodecahedral arrangement. Higher coordination numbers (e.g. 12 in $[Ce(NO_3)_6]^{2-}$) are, however, well established with, in this case, an effective Ce—O distance (~ 114 pm) not unlike the value in typical Ce^{III} compounds. In the absence therefore of definitive data some caution as to the predicted hydration number of $Ce^{IV}(aq)$ might be appropriate.

3.2.5.3 Oxidations by Ce^{IV} (aq)

A wide range of oxidations have been carried out by $Ce^{IV}(aq)$ solutions in the various acid media, ranging from simple oxidations of organic compounds to use in assay procedures for certain natural products and potential pharmaceuticals.

Ce^{IV}(aq) also catalyses the polymerisation of acrylamide, acrylonitrile and other related compounds. The use of Pt metal compounds as efficient catalysts of many Ce^{IV} oxidations including that of water is well documented. The use of RuO_2 in this regard has also been well documented as leading to oxidation of H_2O to O_2 by Ce^{IV}(aq) with as much as 73% efficiency [70]. These studies can be seen as related to the use of CeO_2 as an oxidative support ceramic for dispersed Pt metal based catalytic converters. A detailed survey of the use of Ce^{IV} as an oxidant is outside the scope of this book. The subject has, however, been well reviewed [71]. Two redox studies are worthy of mention.

In the rather complex oxidation of Hg_2^{2+}(aq) (Chapter 12) by Ce^{IV}(aq) in $HClO_4$ media, the rate-determining step was found to be Hg independent, suggestive of Ce^{IV}(aq) decay, via reaction with H_2O, to generate a more reactive oxidant, conceivably a 'Ce^{III}-peroxo species', which then reacts rapidly with Hg^I [72]. The nature of this intermediate has not been fully substantiated. This, however, prompted a detailed investigation of the spontaneous decay of Ce^{IV}(aq) solutions in $HClO_4$ which showed evidence of a two-step process ($t_{1/2} \sim 100$ min) to generate an intermediate which then decayed to Ce^{3+}(aq) in a slower process (factor of ~ 2) [72]. The rate constant of the initial step was found to be very similar to that for the rate-determining step in the oxidation of Hg_2^{2+}(aq). This reaction is efficiently catalysed by light, glass surfaces and also metallic mercury.

The oxidation of the aquamolybdenum(V) dimer $[Mo_2O_2(\mu\text{-}O)_2(OH_2)_6]^{2+}$ (Chapter 6) by Ce^{IV}(aq) at $\mu = 2$ M ClO_4^- was interpreted as proceeding solely via the monohydroxy species $CeOH^{3+}$(aq) [73]. This led to a kinetic estimation of the acid dissociation constant for Ce^{4+}(aq) under these conditions as 0.46 M. This study, however, assumed that only Ce^{4+}(aq) and $CeOH^{3+}$(aq) were relevant and ignored possible formation of $Ce(OH)_2^{2+}$(aq). The observed Ce^{IV} dependence did not imply the formation and involvement of polynuclear hydrolysis products. The rate constant for the reaction of the Ce^{IV}(OH) species with $Mo_2O_4^{2+}$(aq) was 2.73×10^4 $M^{-1}s^{-1}$. An inner-sphere mechanism, involving hydroxide bridge formation from Mo^V to Ce^{IV}, was proposed to account for the preferential reaction of Ce^{IV}(OH). The rate constant was of an order of magnitude sufficient to suggest substitution of the water ligands on Mo^V (see Section 6.4).

3.2.5.4   OTHER TETRAVALENT LANTHANIDES

Ce^{IV} is the only lanthanide known to form a stable tetravalent aqua ion in solution. Figure 3.17 shows, however, that both Pr and Tb have the propensity to form the tetravalent state. This correlated directly with the value of the fourth ionization potential Tb^{IV} stabilised by the $4f^7$ configuration and Pr by being presumably close to an empty $4f^0$. Studies have shown that tetravalent Pr and Tb can be stabilised in strongly alkaline media and can be electrolytically generated as hydroxy species in 1 M KOH [74]. Under these conditions the two species are characterised by intense absorption peaks (O to M^{IV} charge transfer) respectively

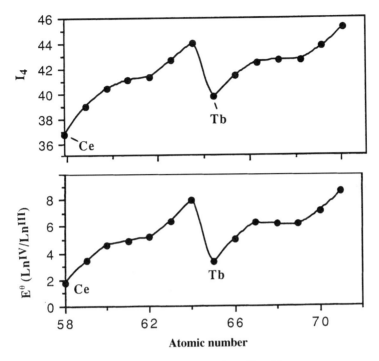

Figure 3.17. Correlation of E^θ (Ln^{IV}/Ln^{III}) with I_4

at 283 nm (Pr^{IV}) and 365 nm ($\varepsilon \sim 1000\,M^{-1}\,cm^{-1}, Tb^{IV}$). The use of alkaline solutions shifts the E^θ to more negative values, helping to promote formation of the tetravalent state. Under these conditions the Ce^{IV}/Ce^{III} couple has an E^θ value of only $+0.05$ V, sufficient to promote air oxidation of Ce^{III} to Ce^{IV}. Both Pr^{IV} and Tb^{IV} are also reported to be stable in alkaline carbonate solutions. PrO_2 (dark brown) and TbO_2 (dark red), like CeO_2, adopt the fluorite structure also known for ZrO_2. On contact with acidic solution both PrO_2 and TbO_2 rapidly evolve O_2 to produce solutions of Pr^{III}(aq) and Tb^{III}(aq).

3.2.5.5 Divalent lanthanides: the Sm^{2+}(aq), Eu^{2+}(aq) and Yb^{2+}(aq) ions

Figure 3.18 shows that these three elements have the propensity to form the divalent state in compounds, the E^θ (Ln^{III}/Ln^{II}) value correlating with the values of the third ionisation potential. The formations of Eu^{2+}(aq) (E^θ (Ln^{III}/Ln^{II}) $= -0.35$ V) and Yb^{2+}(aq) (-1.1 V) are promoted by the stable $4f^7$ and $4f^{14}$ configurations and Sm^{2+}(aq), the most highly reducing of the three aqua ions (-1.5 V), by presumably an approach to $4f^7$. The divalent tendency of Eu

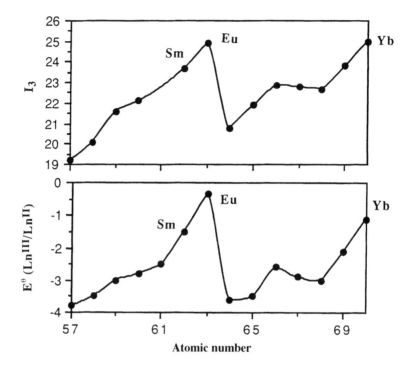

Figure 3.18. Correlation of E^θ (Ln^{III}/Ln^{II}) with I_3

and Yb is exemplified in lower heats of atomization and sublimation and in lower melting points for these elements compared to the other elements of the series. In particular, a depletion (or enrichment) of Eu, relative to the rest, is a commonly observed geochemical phenomenon in lanthanide deposits stemming from a tendency of europium to have mobility within divalent alkaline earth bearing minerals such as feldspars. Metallic Eu and Yb dissolve in liquid NH_3 to give blue solutions containing Ln^{2+} and $e^-(NH_3)_x$ (cf. the alkali and alkaline earths). However, Na reduction of $Sm(ClO_4)_3$ is required to generate similar solutions of Sm^{2+}. Crown ether ligands such as 18-crown-6 have also been shown to stabilise Eu^{2+} and Yb^{2+} with respect to their trivalent counterparts [75].

Solutions of Eu^{2+} can be easily generated via amalgamated zinc reduction of Eu^{3+} solutions in the acid of choice. Alternatively, freshly precipitated $EuCO_3$ or $Eu(OH)_2$(aq) can be used. The more highly reducing Yb^{2+}(aq) and Sm^{2+}(aq) ions require alkali–metal amalgam or electrolytic reduction. Pulse radiolysis has also been employed [76]. Solutions of Eu^{2+}(aq) (pale yellow-green) are reported to be stable in the absence of O_2 for more than 10 days. Yb^{2+} (aq) (pale green) and Sm^{2+}(aq) (red) are much more unstable and decay to the trivalent state within hours. In a study of Yb^{2+}(aq), prepared by mercury cathode electrolysis, the half-life was found to range between 0.5 and 2.5 hours in a number of $HClO_4$–$LiClO_4$ mixtures [H^+] varying from 0.01 to 0.9 M) [77]. These rates of decay were, however, much slower than those found for solutions of Yb^{2+}(aq) prepared under high dilution 10^{-5}–10^{-6} M via pulse radiolysis. Moreover, the decay of Yb^{2+}(aq) here was found to carry both a dependence on Yb^{2+} and Yb^{3+} which was still present in vast excess [76]. It was concluded that oxidising impurities, present in the Yb^{3+}, were now at a level in the pulse radiolytically generated solutions as to be probably responsible for the apparent dependence on Yb^{3+} and the faster rate of decay ($t_{1/2} = 7.7$ ms). Electrolytically generated Sm^{2+}(aq) is reported to decay with a half-life of <1 hour [78]. Short-lived solutions ($t_{1/2} < 1$ minute) of Tm^{2+}(aq), Er^{2+}(aq) and Gd^{2+}(aq) have also been generated using pulse radiolytic methods [79]. Ho^{2+}(aq) has, however, been reported to persist in solution with a ($t_{1/2}$ of ~ 1 hour (~ 28 min in the air) following γ irradiation of Ho_2O_3 and subsequent dissolution in dilute acid [80].

Solutions of the Ln^{2+}(aq) ions are best standardised by titrimetry, e.g. reduction of excess Fe^{3+} and titration of the Fe^{2+} generated (usually in SO_4^{2-} media) although the electronic spectra for all three aqua ions have been recorded (Fig. 3.19) [81]. the spectra of Eu^{2+}(aq) and Yb^{2+}(aq) are characterised by two broad and quite intense bands. These arise due to transitions from the ground state $4f^7$ and $4f^{14}$ configurations of the $4f^65d^1(^7F)$ and $4f^{13}5d^1(^2F)$ excited states respectively, which are each split into two from coupling to other d configurations. the two bands are respectively at 248 ($\varepsilon = \sim 2100\ M^{-1}\ cm^{-1}$) and 321 nm

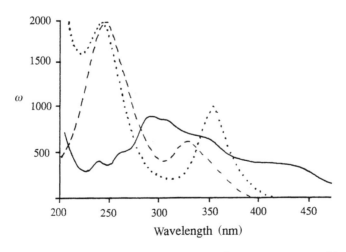

Figure 3.19. Electronic spectra of Sm^{2+}(aq) (—), Eu^{2+}(aq) (– – –) and Yb^{2+}(aq) (......)

($\sim$ 580) (Eu^{2+}) and 246 ($\sim$ 2000) and 352 nm ($\sim$ 1000) (Yb^{2+}). The intensity of the bands arises from the orbitally allowed nature of these spin-allowed 4f–5d transitions. The spectrum of Sm^{2+}(aq) is more complex, consisting of around six bands with medium intensity ($\varepsilon \sim$ 400–800), also due to 4f–5d transitions.

Hydration numbers for the Ln^{2+} ions have not been determined. An increase to nine, or even ten, for Sm^{2+}(aq) and Eu^{2+} would not be unreasonable given the expansion in ionic radius (cf. 125 pm for Eu^{2+} vs 107 pm for Eu^{3+} for eight-coordination) [24]. Similarly, one could propose a hydration number of nine for Yb^{2+}. Addition of NaOH solution to Eu^{2+}(aq) precipitates a yellow hydrated hydroxide which, on drying at 100 °C, assumes the formula $Eu(OH)_2 \cdot 2H_2O$. The ready formation of $Eu(OH)_2$(aq) when metallic Eu is treated with 10 M NaOH [82] is a further reflection of its divalent tendency.

3.2.5.6 REDUCTIONS BY LN^{2+}(AQ) IONS

Ln^{2+}(aq) ions, particularly Eu^{2+}(aq), have been extensively used as reducing ions in redox reactions. Similarities to Cr^{2+}(aq) (Chapter 6) in both reduction potential and mechanism have been noted. Both inner-sphere and outer-sphere mechanisms are believed to operate. For example, the reduction of $[Co(en)_3]^{3+}$ by Eu^{2+}(aq) operates by an outer-sphere pathway [83] whereas the strong dependence upon X in the rate of reduction of $[Co(NH_3)_5X]^{2+}$ species by Eu^{2+}(aq) (variations in $k(M^{-1}s^{-1})$ from 0.05 (X = N-bonded NCS) to 2.6×10^4 (X = F^-)) implies an inner-sphere mechanism [84,85]. The $6 \times 10^4 \times$ faster rate of reaction of the S-bonded NCS complex versus the N-bonded form [86] emphasis the inner-sphere nature of the reduction. Similar behaviour has been found for Yb^{2+}(aq) with reduction of both $[Co(NH_3)_6]^{3+}$ and $[Co(en)_3]^{3+}$ occurring by an outer-sphere path and the reaction with $[Cr(NH_3)_5X]^{2+}$ appearing to be an inner-sphere mechanism [77]. The reduction of Fe^{III}(aq) by Eu^{2+}(aq) involves both $[Fe(OH_2)_6]^{3+}$ and $[Fe(OH_2)_5OH]^{2+}$, with the latter reacting faster. An inner-sphere reaction is suggested [87]. A number of outer-sphere reductions by Eu^{2+}(aq) have been successfully modelled by the Marcus approach [88]. In the outer-sphere reduction of $[Co(NH_3)_5(py)]^{3+}$ by Eu^{2+}(aq) the rate is faster than that observed with Cr^{2+}(aq) as reductant [98], the opposite being true for the case of inner-sphere processes [85].

Eu^{2+}(aq) has also proved to be a very convenient reducing agent for generating high concentrations of certain low valent oligomeric aqua ion species, e.g. those of molybednum (see Chapter 6), for ^{17}O NMR investigations [90]. The paramagnetic Eu^{2+}(aq)/Eu^{3+}(aq) ions do not interfere since water-substitution (exchange) rates on both are close to the diffusion-controlled limit resulting in no shift or broadening of the bound ^{17}O signals observed. For these studies concentrated solutions of Eu^{2+}(aq) as high as 1.0 M in 2 M CF_3SO_3H can be conveniently prepared by amalgamated zinc reduction of Eu^{3+}, the latter solution prepared by dissolution of Eu_2O_3 in the acid. This approach allows syringe injection of small volumes ($< 0.5\,cm^3$) of Eu^{2+}(aq) so as to avoid significant dilution for the NMR studies.

3.3   THE ACTINIDES

Extremely high hydration numbers, probably > 9, are believed to be relevant for the large early actinide ions such as Ac^{3+}(aq) and Th^{4+}(aq). For example, ^1H NMR measurements give the hydration number n of $[Th(OH_2)_n]^{4+}$ as around 9.1–10.0 (see Section 1.2.3). The elements uranium to americium have aquated cations representing four oxidation states: *trans*-AnO_2^{2+}(aq),* *trans*-AnO_2^+(aq), An^{4+}(aq) and An^{3+}(aq). Interestingly, the redox potentials linking the four plutonium species are such that all four aquated ions are able to coexist together in solution without any detectable comproportionation or disproportionation, the E^θ values for each couple being close to $+1.00$ V. For the elements on either side of plutonium differing redox properties are apparent. For uranium the *trans*-UO_2^{2+}(aq) ion is the stable form whereas for americium it is the Am^{3+}(aq) ion. This is now recognised as due to the so-called 'actinide contraction' caused by the poor shielding of the increasing nuclear charge by the 5f orbitals on moving from left to right along the series. This causes a decrease in the effective ionic radius reflected by a corresponding increase in the valence shell ionisation potential. The result is a shift in the redox potentials for the various oxidation state couples to more positive values, thus favouring the lower valent ions. This phenomenon has been harnessed in the industrial separation of plutonium from uranium from spent nuclear fuel in the Purex process. Here use of a specific reducing medium in aqueous nitric acid leads to efficient partitioning of the two dissolved actinides, the uranium (as *trans*-$[U^{VI}O_2(OH_2)_2(NO_3)_2]$) extractable into a non-polar organic phase with the plutonium (reduced to the $4+$ state as $Pu^{IV}(OH_2)_n(NO_3)_4$) remaining in the aqueous phase. The Shannon ionic radius for six-coordination on U^{6+} (73 pm) is comparable to that for scandium, suggesting that it should be capable of adopting six- or seven-coordination. Indeed, pentagonal bipyramidal coordination has been established in several X-ray structures of complexes containing the aqua dihydroxy dimer unit $(UO_2)_2(\mu\text{-}OH)_2^{2+}$ (Fig. 3.20) [91–93]. The eight-coordination present in the nitrate complex is probably due to the fact that bidentate nitrate ions require less space (smaller bite angle) than two water molecules, as is apparent in the accommodation of six nitrate ions in twelve-coordinate structures such as $[Ce(NO_3)_6]^{2-}$. This is reinforced in the structure of the dimer $[(H_2O)_3$-$(UO_2)(\mu\text{-}OH)_2(UO_2)(NO_3)_2]$ (Fig. 3.21), containing both seven- and eight-

---

*Throughout the book the abbreviation (aq) is used either as a shorthanded notation for $(H_2O)_n$ or when n is not known for certain, as in the case of the actinide aqua ions.

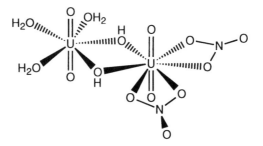

Figure 3.20. Structure of *trans*-$[(UO_2)_2(\mu\text{-}OH)_2(OH_2)_6]^{2+}$

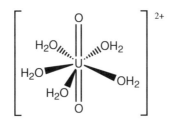

Figure 3.21. Structure of $[(H_2O)_3\,(\textit{trans-}UO_2)(\mu\text{-}OH)_2\,(\textit{trans-}UO_2)(NO_3)_2]$

coordinated U^{VI} centres [92]. In the aqua dimer, three distinct U—O bond distances are found: U—O(yl) 170, 173 ppm, U—O(μ-OH) 227, 231 ppm and U—O(OH$_2$) 224 and 247 pm [91]. The structure of the *trans*-UO_2^{2+}(aq) ion is therefore assumed to be the pentagonal bipyramidal hydration arrangement (Fig. 3.22).

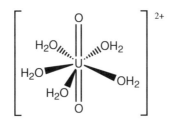

Figure 3.22. Structure of *trans*-$[UO_2(OH_2)_5]^{2+}$

The rate of water exchange on $trans$-$[UO_2(OH_2)_5]^{2+}$ and on the aqua dimer, together with the rate of formation and decomposition of the aqua dimer, has been studied using ^{17}O NMR [94,95]. Water exchange on the $U{=}O$ groups is extremely slow and carries a reciprocal $[H^+]$ dependence, $k_{obs} = k[H^+]^{-1}$. At 25 °C, $I = 3.79$ M and $k = 9.9 \times 10^{-9} s^{-1}$. However, the exchange is found to be efficiently catalysed by amounts of pentavalent $trans$-UO_2^+(aq) which exchanges much faster. As a result the study of the $trans$-UO_2^{2+}(aq) exchange is best carried out in the presence of added Cl_2 which oxidises any $trans$-UO_2^+(aq) present to $trans$-UO_2^{2+}(aq). In 0.8 M HCl the half-life for the exchange is reported to be ~ 95 days [95].

The slow exchange of the terminal oxygens of $trans$-UO_2^{2+}(aq) also contrasts with the much higher lability (14 orders of magnitude) shown by the five equatorial water ligands, $k = 7.6 \times 10^5 s^{-1}$, as studied by variable temperature ^{17}O linewidth measurements [94]. The rate constant for the water ligands on the aqua dimer is comparable, $k = 2.14 \times 10^5 s^{-1}$, the difference in the rates exemplified in the differing U—O bond lengths for the two types of oxygen atom. Catalysis by the pentavalent $trans$-dioxo aqua ion is also a feature of the water-exchange process on the yl oxygens of $trans$-NpO_2^{2+}(aq) and $trans$-PuO_2^{2+}(aq) [97].

Baes and Mesmer have evaluated hydrolysis constants for various uranium aqueous species. U^{4+}(aq) is estimated to have $pK_{11} = 0.65$. Hydrolysis of $trans$-$[UO_2(OH_2)_5]^{2+}$ gives $trans$-$UO_2(OH)^+$(aq) ($pK_{11} = 5.8$), ($trans$-$UO_2)_2(\mu$-$OH)_2^{2+}$(aq) ($pK_{22} = 5.62$) and ($trans$-$UO_2)_3(\mu$-$OH)_5]^+$ ($pK_{35} = 15.63$) [98]. The strong tendency for $trans$-$UO_2(OH)^+$(aq) to dimerise to give ($trans$-$UO_2)_2(\mu$-$OH)_2^{2+}$(aq) (Fig. 3.20) is exemplified by the ratio K_{22}/K_{11}^2 of $\sim 10^6$.

REFERENCES

[1] Cotton, F. A. and Wilkinson, G., *Advanced Inorganic Chemistry*, 5th edn. Wiley Interscience, New York, 1988, p. 973.

[2] (a) Yarasov, V. P., Kirakosyan, G. A., Trots, S. V., Buslaev, Yu. A., and Panyushin, V. T., *Koord. Khim.*, **9**, 205 (1983).
(b) Kirakosyan, G. A. and Tarasov, V. P., *Koord. Khim.*, **8**, 261 (1982).

[3] Ivanov–Emin, B. N., Borzova, L. D., Zaitsev, B. E., and Fisher, M. M., *Russ. J. Inorg. Chem.*, **13**, 1156 (1986).

[4] (a) Anderson, T. J., Neuman, M. A., and Melson, G. A., *Inorg. Chem.*, **12**, 927 (1973) and **13**, 1884 (1974).
(b) Hansson, E., *Acta Chem. Scand.*, **26**, 1337 (1972) and **27**, 2841 (1973).
(c) Guseinova, M. K., Antsyshkina, A. S., and Porai Koshits, M. A., *J. Struct. Chem. Engl. Transl.*, **9**, 926 (1968).
(d) Hatakeyama, Y., Kido, H., Harada, M., Tomiyasu, H., and Fukutomi, H., *Inorg. Chem.*, **27**, 992 (1988).

[5] Kury, J., Paul, A. D., Hepler, L. G., and Connick, R. E., *J. Am. Chem. Soc.*, **81**, 4185, (1959).

[6]   Sipachev, V. A., and Grigo'ev, A. I., *J. Struct. Chem. Engl. Trans.*, **10**, 710 (1969).
[7]   Kanno, H., Yamaguchi, T., and Ohtaki, H., *J. Phys. Chem.*, **93**, 1695 (1989).
[8]   Kanno, H., *J. Phys. Chem.*, **92**, 4232 (1988).
[9]   Habenschuss, A., and Spedding, F. H., *J. Chem. Phys.*, **70**, 2797, 3758 (1979).
[10] Matsumoto, F., Ohki, Y., Suzuki, Y., and Ouchi, A., *Bull. Chem. Soc. Japan*, **62**, 2081 (1989).
[11] Ohki, Y., Suzuki, Y., Takeuchi, T., and Ouchi, A., *Bull. Chem. Soc. Japan*, **61**, 393 (1988).
[12] Ohtaki, H., and Radnai, T., *Chem. Revs.*, **93**, 1157 (1993).
[13] Kilpatrick, M., and Pokras, L., *J. Electrochem. Soc.*, **100**, 85 (1953) and **101**, 39 (1954).
[14] Biedermann, G., Kilpatrick, M., Pokras, L., and Sillen, L.G., *Acta Chem. Scand.*, **10**, 1327 (1956).
[15] Aveston, J., *J. Chem. Soc. A*, 1599 (1966).
[16] Brown, P. L., Ellis, J., and Sylva, R. N., *J. Chem. Soc. Dalton Trans.*, 35 (1983).
[17] Cole, D. L., Rich, L. D., Owen, J. D., and Eyring, E. M., *Inorg. Chem.*, **8**, 682 (1969).
[18] Meyer, R. J., Wassjuchnow, A., Drapier, N., and Bodlander, E., *Z. Anorg. Allg. Chem.*, **86**, 257 (1914).
[19] Kraus, K., Nelson, F., and Smith, G.W., *J. Phys. Chem.*, **59**, 11 (1954).
[20] Silber, H., and Mioduski, T., *Inorg. Chem.*, **23**, 1577 (1984).
[21] Geier, G., *Ber. Bunsenges Phys. Chem.*, **69**, 617 (1965).
[22] Helm, L., Ammann, C., and Merbach, A.E., *Z. Phys. Chem.*, **155**, 145 (1987).
[23] Pisaniello, D. L., Nichols, P. J., Ducommun, Y., and Merbach, A. E., *Helv. Chim. Acta*, **65**, 1025 (1982).
[24] Shannon, R. D., *Acta Cryst.*, **A32**, 751 (1976).
[25] (a) Helmholz, L., *J. Am. Chem. Soc.*, **61**, 1544 (1939).
      (b) Hubbard, C. R., Quicksall, C. O., and Jacobson, R. A., *Acta Cryst.*, **B30**, 2613 (1974).
      (c) Albertson, J., and Elding, I., *Acta Cryst.*, **B34**, 1460 (1977).
[26] Glaser, J., and Johansson, G., *Acta Chem, Scand.*, **A35**, 639 (1981).
[27] Sherry, E. G., *J. Solid State Chem.*, **19**, 271 (1976).
[28] Lincoln, S. F., *Adv. Inorg. Bioinorg. Mech.*, **4**, 217 (1983).
[29] Cossy, C., Helm, L. and Merbach, A. E., *Inorg. Chem.*, **27**, 1973 (1988).
[30] (a) Sikka, S. K., *Acta Cryst.*, **A25**, 621 (1969).
      (b) Paiva Santos, C. O., Castellano, E. E., Machado, L. C., and Vincentini, G., *Inorg. Chim. Acta*, **110**, 83 (1985).
      (c) Harrowfield, J. McB., Kepert, D.L., Patrick, J.M., and White, A.H., *Aust. J. Chem.*, **36**, 483 (1983).
[31] Pisanello, D. L., Helm, L., Meier, P., and Merbach, A. E., *J. Am. Chem. Soc.*, **105**, 4528 (1983).
[32] Karracher, D. G., *Inorg. Chem.*, **7**, 473 (1968).
[33] Spedding, F. H., Pikal, M. J., and Ayers, B. O., *J. Phys. Chem.*, **70**, 2440 (1966).
[34] Spedding, F. H., and Jones, K. C., *J. Phys. Chem.*, **70**, 2450 (1966).
[35] Spedding, F. H., and Pikal, M. J., *J. Phys. Chem.*, **70**, 2430 (1966).
[36] (a) Helm, L., and Merbach, A. E., *Eur. J. Solid State Inorg. Chem.*, **28**, 245 (1991).
      (b) Habenshuss, H., and Spedding, F. H., *J. Chem. Phys.*, **73**, 442 (1980).
      (c) Ryss, A. I., Lesocitskaya, M. K., and Shapovalov, I. M., *Chem. Abst.*, **89**, 95116 (1978).
      (d) Yamaguchi, T., Tanaka, S., Wakita, H., Misawa, M., Okada, I., Soper, A.K., and Howells, S. W., *Z. Naturforsch.*, **A46**, 84, (1991).
      (e) Johansson, G., Ninitso, L., and Wakita, H., *Acta Chem. Scand.*, **A39**, 359 (1985).

[37] Yamaguchi, Y., Nomura, M., Wakita, H., and Ohtaki, H., *J. Chem. Phys.*, **89**, 5153, (1988).
[38] (a) Cossy, C., Barnes, A. C., and Enderby, J. E., *J. Chem. Phys.*, **90**, 3254 (1989).
(b) Cossy, C., Helm, L., Powell, D. H., and Merbach, A. E., *New. J. Chem.*, **19**, 27 (1995).
[39] Baes, C. F., and Mesmer, R. E., *The Hydrolysis of Cations*, Krieger, Florida, 1986, pp. 129–38.
[40] (a) Biedermann, G., and Ciavatta, L., *Ark. Kemi*, **22**, 253 (1964).
(b) Amaya, T. H., Kakihana, H., and Maeda, M., *Bull. Chem. Soc. Japan*, **46**, 1720, 2889 (1973).
(c) Burkov, K. A., Lilich, L. S., Ngo, N.D., and Smirnov, A. Yu., *Zh. Neorg. Khim.*, **18**, 1513 (1973) and *Russ. J. Inorg. Chem.*, **18**, 797 (1973).
(d) Usherenko, L. N., and Skorik, N. A., *Russ. J. Inorg. Chem.*, **17**, 1533 (1972).
[41] Mohapatra, P. K., and Khopkar, P. K., *Polyhedron*, **8**, 2071 (1989).
[42] Ivanov–Emin, B. N., Siforova, E. N. Fischer, M. M., and Kampos, V. M., *Russ J. Inorg. Chem.*, **11**, 258, 1054 (1966).
[43] Cossy, C., Helm, L., and Merbach, A. E., *Inorg. Chem.*, **28**, 2699 (1989).
[44] Swift, T. J., and Connick, R. E., *J. Chem. Phys.*, **37**, 307 (1962).
[45] Reuben, J., and Fiat, D., *J. Chem. Phys.*, **51**, 4918 (1969).
[46] (a) Micskei, K., Powell, D. H., Helm, L., Brucher, E., and Merbach, A.E., *Mag. Res. Chem.*, **31**, 1011 (1993).
(b) Powell, D. H., and Merbach, A. E., *Mag. Res. Chem.*, **32**, 739 (1994).
[47] Laurenczy, G., and Merbach, A. E., *Helv. Chim. Acta*, **71**, 1971 (1988).
[48] Swaddle, T.W., *Adv. Inorg. Bioinorg. Mech.*, **2**, 95 (1983).
[49] Fay, D., Litchinski, D., and Purdie, N., *J. Phys. Chem.*, **73**, 544 (1969).
[50] Bjerrum, N., *K. Dan. Vidensk. Selsk. Skr. Naturvidensk. Math. Afd.*, **7**, 9 (1926).
[51] Fuoss, R. M., *J. Am. Chem. Soc.*, **80**, 5059 (1958).
[52] Garza, V. L., and Purdie, N., *J. Phys. Chem.*, **74**, 275 (1970).
[53] Okada, K., Maizu, Y., Kobayashi, H., Tanaka, K., and Marumo, F., *Mol. Phys.*, **54**, 1293 (1985).
[54] Kowall, T., Foglia, F., Helm, L., and Merbach, A. E., *Chem. Eur. J.*, **2**, 285 (1996), *J. Am. Chem. Soc.*, **117**, 3790 (1995) and *J. Phys. Chem.*, **99**, 13078 (1995).
[55] (a) Farrow, M. M., Purdie, N., and Eyring, E. M., *Inorg. Chem.*, **13**, 2024 (1974).
(b) Graffeo, A. J., and Bear, J. L., *J. Inorg. Nucl. Chem.*, **30**, 1577 (1968).
[56] Ducummon, Y., Helm, L., Laurenczy, G., and Merbach, A.E., *Mag. Res. Chem.*, **26**, 1023 (1988).
[57] Chen, Z., and Detellier, C., *J. Solution Chem.*, **21**, 941 (1992).
[58] McAuley, A., *Coord. Chem. Revs.*, **5**, 245 (1970).
[59] Wadsworth, E., Duke, F. R., and Goetz, C. A., *Anal. Chem.*, **29**, 1824 (1957).
[60] Maverick, A. W., and Yao, Q., *Inorg. Chem.*, **32**, 5626 (1993).
[61] Smith, G. F., and Fly, W. H., *Anal. Chem.*, **21**, 1233 (1949).
[62] Medalia, A. I., and Byrne, B. J., *Anal. Chem.*, **23**, 453 (1951).
[63] Baes, C. F., and Mesmer, R. E., *The Hydrolysis of Cations*, Krieger, Florida, 1986. pp. 138–146.
[64] Sherrill, M. S., King, C. B., and Spooner, R. C., *J. Am. Chem. Soc.*, **65**, 170 (1943).
[65] Baker, F. B., Newton, T. W., and Kahn, M., *J. Phys. Chem.*, **64**, 109 (1960).
[66] Hardwick, T.J., and Robinson, E., *Can. J. Chem.*, **29**, 818 (1951).
[67] King, E. L., and Pandow, M. L., *J. Am. Chem. Soc.*, **74**, 5296 (1952).
[68] Danesi, P. R., Magani, M., Margherita, S., and D'Alessandro, G., *Energ. Nucl. (Milan)*, **15**, 335 (1968).

[69] Mihara, T., Tomiyasu, H., and Jung, W. S., *Polyhedron*, **13**, 1747 (1994) (and refs. therein).
[70] (a) Mills, A., *J. Chem. Soc. Dalton Trans.*, 1213 (1982).
       (b) Kiwi, J., Gratzel, M., and Blondeel, G., *J. Chem. Soc. Dalton Trans.*, 2215 (1983).
[71] See, for example (a) Molanda, G. A., 'Applications of lanthanide reagents in organic synthesis', *Chem. Revs.*, **92**, 29–68 (1992). (b) Ruck, K., and Kunz, H., 'Ceric ammonium nitrate—a versatile oxidising reagent' *J. Prakt. Chem. Chem. Zeit.*, **336**, 470 (1994).
[72] Granica Meza, J. M., and Spiro, M., *Inorg. Chim. Acta*, **184**, 53 (1991).
[73] Chappelle, G. A., Macstay, A., Pittenger, S. T., Ohashi, K., and Hicks, K. W., *Inorg. Chem.*, **18**, 2768 (1984).
[74] Hobart, D. E., Samhoun, K., Young, J. P., Norvell, V. E., Mamantov, G., and Peterson, J. R., *Inorg. Nucl. Chem. Lett.*, **16**, 321 (1980).
[75] Yamana, H., Mitsugashira, T., Shiokawa, Y., and Suzuki, S., *Bull. Chem. Soc. Japan*, **55**, 2615 (1982).
[76] Farragi, M., and Tendler, Y., *J. Chem. Phys.*, **56**, 3287 (1972).
[77] Christensen, R. J., Espenson, J. H., and Butcher, A. B., *Inorg. Chem.*, **12**, 564 (1973).
[78] Johnson, D. A., *Adv. Inorg. Chem. Radiochem.*, **20**, 1 (1977).
[79] Gordon, S., Sullivan, J. C., and Mulac, W. A., *Proc. Sympos. Rad. Chem. Keszthely*, 763 (1976).
[80] Apers, D. J., DeBlock, R., and Capron, P. C., *J. Inorg. Nucl. Chem.*, **36**, 1441 (1974).
[81] (a) Butement, F. D. S., *Trans. Faraday Soc.*, **44**, 617 (1948).
       (b) Johnson, K. E., McKenzie, J. R., and Sandoe, J. N., *J. Chem. Soc. A.*, 2644 (1968).
[82] Baernighausen, H., *Z. Anorg. Allg. Chem.*, **342**, 233 (1966).
[83] Doyle, J., and Sykes, A. G., *J. Chem. Soc. A.*, 2836 (1968).
[84] Adin, A., and Sykes, A. G., *J. Chem. Soc. A.*, 354 (1968).
[85] Wilkins, R. G., *The Kinetics and Mechanisms of Reactions of Transition Metal Complexes*, VCH, Weinheim, 1991, p. 271.
[86] Adegite, A., and Kuku, T. A., *J. Chem. Soc. Dalton Trans.*, 158 (1976).
[87] Caryle, D. W., and Espenson, J. H., *J. Am. Chem. Soc.*, **90**, 2273 (1968).
[88] Poon, C. K., and Tang, T. W., *Inorg. Chem.*, **23**, 2130 (1984).
[89] Dockal. E.R., and Gould, E.S., *J. Am. Chem. Soc.*, **94**, 6673 (1972).
[90] Richens, D. T., Helm, L., Pittet, P.-A., and Merbach, A.E., *Inorg. Chim. Acta*, **132**, 85 (1987).
[91] Aberg, M., *Acta Chem. Scand.*, **23**, 791 (1969).
[92] Navaza, A., Villain, F., and Charpin, P., *Polyhedron*, **3**, 143 (1984).
[93] Perrin, A., *Acta Cryst.*, **B32**, 1658 (1976).
[94] Jung, W.-S., Harada, M., Tomiyasu, H., and Fukutomi, H., *Bull. Chem. Soc. Japan*, **61**, 3895 (1988), **60**, 489 (1987), **59**, 3761 (1986), **58**, 938 (1985) and **57**, 2317 (1984).
[95] Kato, Y., Suzuki, F., Fukutomi, H., Tomiyasu, H., and Gordon, G., *Bull. Tokyo, Inst. Tech.*, **96**, 133 (1970).
[96] Rabideau, S. W., *J. Phys. Chem.*, **67**, 2655 (1963).
[97] Masters, B. J., and Rabideau, S. W., *Inorg. Chem.*, **2**, 1 (1963).
[98] Baes, C. F., and Mesmer, R. E., *The Hydrolysis of Cations*, Krieger, Florida, 1986, pp. 180–182.

Chapter 4

Group 4 Elements: Titanium, Zirconium and Hafnium

4.1 TITANIUM

The aqueous chemistry of titanium is basically limited to the oxidation states III and IV [1]. The IV state is the stable one in the presence of air with the III state readily oxidised to it. The III state is, however, stable under a nitrogen or argon atmosphere and moreover is characterised as possessing the only true aqua ion of Group 4 in the form of octahedral hexaaqua $[Ti(OH_2)_6]^{3+}$.

4.1.1   Titanium(II) (d^2)

Ti^{2+}(aq) is said to persist for short times in ice-cold hydrochloric acid solutions of TiO but has no relevant solution chemistry [2]. Complexation of Ti^{II} by Cl^- is probably an important stabilising factor. The redox potential for reduction of Ti^{3+} to Ti^{2+} has been quoted as -0.37 V [3].

4.1.2   Titanium(III) (d^1): Chemistry of the $[Ti(OH_2)_6]^{3+}$ Ion

4.1.2.1   PREPARATION OF SOLUTIONS OF $[Ti(OH_2)_6]^{3+}$

The pale reddish-purple hexaaqua ion is the principal species present in dilute HCl or H_2SO_4 solutions of Ti^{III} added either as $TiCl_3$ or the oxide Ti_2O_3 [4]. Solutions of $[Ti(OH_2)_6]^{3+}$ have also been obtained via air-free dissolution of pure titanium metal in dilute, e.g. H_2SO_4, solution or by reduction of solutions of Ti^{IV} (aq) (see later) with zinc [5]. Complexation by Cl^- or SO_4^{2-} in dilute solutions is not extensive and a number of kinetic studies of complexation reactions have been carried out in Cl^- media with due allowance given for the formation of species such as $[Ti(OH_2)_5Cl]^{2+}$ and *cis*- and *trans*-$[Ti(OH_2)_4Cl_2]^+$. Simple dissolution of freshly precipitated Cl^- free Ti_2O_3 in dilute triflic acid (CF_3SO_3H) followed by ice-cold air-free ion-exchange chromatography (see Section 1.5) using a DOWEX 50W X2 resin column and elution with dilute CF_3SO_3H would seem to be the method of choice to prepare pure solutions of $[Ti(OH_2)_6]^{3+}$ free of complexing ligands. Unfortunately, the use of the more desirable ClO_4^- anion is not possible owing to its ready oxidation of Ti^{III} to Ti^{IV} $(t_{1/2} = 2$ hours, 25 °C [6]. Merbach and coworkers have, however, devised an alternative method to prepare Cl^--free solutions of $[Ti(OH_2)_6]^{3+}$ via preparation of solid anhydrous $Ti(CF_3SO_3)_3$ [7]. In a typical glove box/vacuum line procedure 3.0 g of pure $TiCl_3$ was mixed with 9.6 g of freshly degassed anhydrous CF_3SO_3H. Degassed doubly distilled water (5.5 g) was then vacuum distilled on to the mixture and, after homogenisation, distilled off again along with the HCl produced by the reaction. A second portion of degassed water (5.5 g) is again distilled on to and then off the reaction mixture to eliminate the residual HCl. The greyish-blue salt was then dried by heating to 359 K under vacuum $(10^{-4}$ mm Hg) wherein microanalyses gave rise to a residual formula $Ti(CF_3SO_3)_{2.96}Cl_{0.04}$. This salt then readily dissolves in water/CF_3SO_3H to generate pure solutions of $[Ti(OH_2)_6]^{3+}$ quantitatively at whatever stoichiometry is desired, and at the same time allowing for the introduction of controlled amounts of enriched H_2O^{17}. As a result such solutions were subsequently used to measure the exchange rate of coordinated water by ^{17}O NMR line broadening under conditions of variable temperature and pressure [7] (see below). The hexaaqua ion persists in the solid state structures of $TiCl_3 \cdot 6H_2O$ [8] and the caesium alum $CsTi(SO_4)_2 \cdot 12H_2O$ [9]. The Ti—O distance in the alum is quoted as 203 pm. The X-ray crystal structure of the pts$^-$ salt $[Ti(OH_2)_6](pts)_3 \cdot 3H_2O$ has also been reported [10]. Here the geometry at the Ti is essentially octahedral with Ti—O bond lengths ranging from 201.8(5) to 204.6(6) pm.

4.1.2.2 ELECTRONIC SPECTRUM OF $[Ti(OH_2)_6]^{3+}$

Historically Ti^{III} complexes and $[Ti(OH_2)_6]^{3+}$ in particular have been connected with the development of ligand field theory in providing an example of the simplest situation, namely an octahedral d^1 species. As a result its electronic spectrum (Fig. 4.1) characteristically appears in almost every comprehensive textbook on inorganic chemistry dealing with the electronic spectra of transition metal complexes, the ion allowing observation of the simplest case of the transition of one single triply degenerate t_{2g} electron ($^2T_{2g}$ ground state) into one of the doubly degenerate e_g orbitals (^2E_g excited state). Thus a single electronic transition ($^2T_{2g} \rightarrow {}^2E_g$) should be observable. Ilse and Hartmann in 1951 first appreciated that the rather broad asymmetric band observed in the visible spectrum of solutions of $[Ti(OH_2)_6]^{3+}$ around 20100 cm^{-1} (Fig. 4.1) was indeed due to the $^2T_{2g} \rightarrow {}^2E_g$ transition and this proved to be the first correct assignment of a ligand field band [11]. It was subsequently appreciated that the energy of the

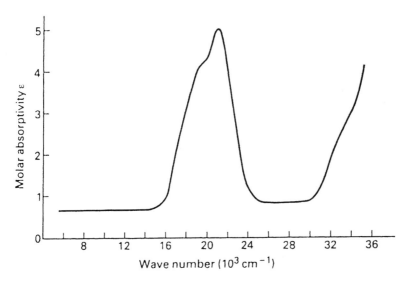

Figure 4.1. Electronic spectrum of $[Ti(OH_2)_6]^{3+}$

$^2T_{2g} \rightarrow {}^2E_g$ transition was a direct measure of the strength of the ligand field at Ti^{III} and this paved the way for the setting up of the first spectrochemical series of ligands as a result of comparisons of the electronic spectra from a whole family of Ti^{III} complexes having six identical ligands. Subsequently in 1962 the observed splitting of the 'single' band (Fig. 4.1) was rationalised in terms of a Jahn–Teller splitting of the $(e_g)^1$ excited state [12], it being eventually appreciated that an electronic transition from one of the t_{2g} orbitals to the $d_{x^2-y^2}$ orbital should not necessarily take place with the same energy as a corresponding transition to the d_{z^2} orbital. For $[Ti(OH_2)_6]^{3+}$ the splitting of the ^2E_g state amounts to 3100 cm^{-1}, the two resolved bands having energies (ε, M^{-1} cm^{-1}) of 20100 cm^{-1} (5.7) and 17000 cm^{-1} (3.9) respectively [13] (only the hexacyano and hexafluoro complexes show a greater splitting in the ^2E_g state). In the reflectance spectrum of the caesium alum [9c] the band energies are respectively 19900 cm^{-1} ($\sim$4.2) and 18000 cm^{-1} ($\sim$2.2), splitting 1900 cm^{-1}, suggestive of the expected slight distortion commonly observed in alums from regular octahedral towards trigonal prismatic geometry. The geometrical and hydration structure and electronic states of $[Ti(OH_2)_6]^{3+}$ have recently been re-examined by *ab initio* calculations and ESR/electron spin echo studies [14].

4.1.2.3 HYDROLYSIS OF $[Ti(OH_2)_6]^{3+}$

Baes and Mesmer have recalculated a value of pK_{11} for the formation of $[Ti(OH_2)_5OH]^{2+}$ from $[Ti(OH_2)_6]^{3+}$ as (2.2 ± 0.3) based on an accumulation of earlier data [15]. Dimeric $Ti_2(OH)_2^{4+}$ (aq) is also formed in appreciable amounts ($pK_{22} = (3.6 \pm 0.5)$) [16]. Hydrolysis of Ti^{III} is observed to lead to a marked increase in reactivity with respect to oxidation to Ti^{IV}, a factor that plays a key role in certain redox reactions of Ti^{3+} (see below).

4.1.2.4 SUBSTITUTION OF WATER ON $[Ti(OH_2)_6]^{3+}$

A number of kinetic studies involving 1:1 complex forming reactions on $[Ti(OH_2)_6]^{3+}$ have been carried out by Diebler and coworkers employing the temperature-jump technique [17]. Table 4.1 summarises rate constants so far obtained with a number of ligands. Included is the recently determined value for water exchange as measured by Merbach and coworkers using a Swift–Connick ^{17}O NMR treatment of the linewidth of the free water in the presence of added $[Ti(OH_2)_6]^{3+}$ and CF_3SO_3H [7]. Even at temperatures as low as $-24\,^{\circ}C$ the exchange is too fast to permit observation of a bound water ^{17}O NMR resonance. Variation in formation rate constants spanning >3 orders of magnitude was argued as supportive of an associative interchange mechanism [16]. A trend was also observed with increasing rate constants paralleling an increase in basicity of the ligand. Such a mechanism is consistent with the expected behaviour of an octahedral t_{2g}^1 metal ion. The lack of any involvement from the hydrolysed

Table 4.1. Complex formation rate constants for reaction of $[Ti(OH_2)_6]^{3+}$ with various incoming ligands

Ligand	pK_a	$k\,(M^{-1}\,s^{-1})$	Conditions[a]	Ref.
$ClCH_2CO_2H$	—	6.7×10^2	15, 0.5 M	[17c]
CH_3CO_2H	—	9.7×10^2	15, 0.5 M	[17c]
H_2O	−1.74	8.6×10^{3b}	12, 2–3 m	[7]
NCS^-	−1.84	8.0×10^3	8–9, 1.5 M	[17a]
$HC_2O_4^-$	1.23	3.9×10^5	10, 1.0 M	[17b]
$Cl_2CHCO_2^-$	1.25	1.1×10^5	15, 0.5 M	[17c]
$HO_2CCH_2CO_2^-$	2.43	4.2×10^5	15, 0.5 M	[17d]
$ClCH_2CO_2^-$	2.46	2.1×10^5	15, 0.5 M	[17c]
$HO_2CCH(CH_3)CO_2^-$	2.62	3.2×10^5	15, 0.5 M	[17c]
$CH_3CO_2^-$	4.47	1.8×10^6	15, 0.5 M	[17d]

[a]Temperature in °C, M = molarity, m = molality.
[b]This rate constants refers to the exchange of one of six-coordinated waters and is thus converted to second-order units $(6k_{ex}\,(12\,°C)\,/\,55.56)$; $k_{ex}\,(25\,°C) = (1.81 \pm 0.03) \times 10^5\,s^{-1}$, $\Delta H^{\ddagger} = (43.4 \pm 0.7)\,kJ\,mol^{-1}$, $\Delta S^{\ddagger} = (+1.2 \pm 2.2)$ $J\,K^{-1}\,mol^{-1}$, $\Delta V^{\ddagger}(-12.1 \pm 0.4)\,cm^3\,mol^{-1}$, $\Delta\beta^{\ddagger} = (-1.9 \pm 0.4)\,cm^3\,mol^{-1}\,MPa^{-1}$.

species $TiOH^{2+}$ is also indicative of the tendency towards a dominant associative pathway, as will be seen in other examples subsequently.

In the water-exchange study, plots of $\ln(k_p/k_0)$ showed slight curvature (non-zero $\Delta\beta^{\ddagger}$ but the resulting $\Delta V^{\ddagger}$ is independent of temperature and has the most negative value $(-12.1\,cm^3\,mol^{-1})$ so far observed for a hexaaqua ion [7]. Moreover, it is extremely close to the limiting value of $-13.5\,cm^3\,mol^{-1}$ proposed by Swaddle as representing an extreme A (S_N2) mechanism [18]. It may be concluded that substitution on $[Ti(OH_2)_6]^{3+}$ is strongly associative, tending perhaps towards limiting A.

$[Ti(OH_2)_6]^{3+}$ forms one of the most stable heptaaqua complex of all the di- and trivalent first row metal ions on the basis of gas phase energies (see Section 1.4) and indeed seven-coordination around Ti^{3+} is present in the oxalato complex $[Ti(C_2O_4)_3(OH_2)]^{3-}$ [19].

4.1.2.5 REDOX REACTIONS INVOLVING $[Ti(OH_2)_6]^{3+}$

In contrast to the substitution reactions, rate laws governing redox reactions involving Ti^{3+} as reductant carry a dominant $[H^+]^{-1}$ term implying involvement of $TiOH^{2+}$ as reactive species. This is perfectly consistent with oxidation to Ti^{IV}, a more extensively hydrolysed species, in view of the higher charge. The involvement of $TiOH^{2+}$ has been shown not to correlate necessarily with the presence of an inner-sphere mechanism, e.g. via OH^- bridge formation, and indeed $TiOH^{2+}$ has been shown to exhibit outer-sphere Marcus behaviour [20]. Nonetheless, if suitable bridging substituents on the 'oxidant' are present, inner-sphere mechanisms are often observed for reduction by Ti^{3+} [21] and the formation of discrete binuclear intermediates has been detected in a number of cases [22]. Indeed, in a number of examples substitution into the Ti^{III} coordination sphere has been found to be rate determining [23]. The oxidation of $[Ti(OH_2)_6]^{3+}$ by molecular oxygen also carries a dominating $[H^+]^{-1}$ term such that $TiOH^{2+}$ is the sole reactant [24]. At 25 °C the bimolecular rate constant for reaction of $TiOH^{2+}$ with O_2 is $(4.25 \pm 0.13) \, M^{-1} s^{-1}$ based on a value of pK_{11} for $[Ti(OH_2)_6]^{3+}$ of 2.2.

4.1.3   Titanium(IV) (d⁰)

4.1.3.1   PUTATIVE TiO^{2+} (AQ): FACT OR FICTION?

Two papers published in the mid 1980s have sought to further investigate the nature of colourless Ti^{IV} in non-complexing acidic aqueous media [25, 26]. In early work regarding Ti^{IV}(aq), the principal species in acidic aqueous solution, usually in perchloric acid, was reported as containing TiO^{2+}(aq) [27]. In fact a close examination of the hard evidence shows that the putative 'titanyl' ion has only been poorly characterised in solution. There seems general agreement that a mononuclear dipositive cationic species is relevant on the basis of ion-exchange [28], potentiometric titration [29], electromigration [28b] and kinetic studies of electron transfer and complexation reactions [24, 27]. Conceivably under the conditions studied the relevant species could consist of an equilibrium mixture of one or more of the following species: $Ti(OH)_2^{2+}$(aq), $TiCl(OH)^{2+}$(aq) and $TiCl_2^{2+}$(aq) in addition to the putative TiO^{2+}(aq) ion. Furthermore, in contrast to vanadyl $(V=O^{2+})$, the $Ti=O^{2+}$ group is only known discretely in a few complexes such as TiO (porphyrin) [30], TiO(phthalocyanine) [31], $[TiOCl_4]^{2-}$ [32] and $[TiOF_5]^{3-}$ [33]. However, no real definitive evidence for the $Ti=O^{2+}$ group persisting in aqueous solution appeared to exist. Indeed, many 'TiO^{2+}' containing compounds tend rather to have oxygen-bridged chain structures rather than possessing a terminal oxo group, a good example being the structure of '$TiOSO_4$'. However recent studies [25, 26] have shown that TiO^{2+}(aq) is probably present in aqueous solutions but existing in equilibrium with the dihydroxy ion, $Ti(OH)_2^{2+}$(aq).

4.1.3.2   HYDROLYTIC (PROTIC) EQUILIBRIA INVOLVING Ti^{IV}(AQ) SPECIES IN NON-COMPLEXING AQUEOUS SOLUTION

Ciavatta, Ferri and Riccio [25] reinvestigated the hydrolysis of Ti^{IV}(aq) in 3.0 M chloride media at 25 °C by measuring Ti^{IV}/Ti^{III} ratios in solutions of Ti^{IV} chloride in the presence of H_2(g) at different $[H^+]$. $Ti(OH)_2^{2+}$(aq) (or 'TiO^{2+}') was found to predominate in the concentration ranges $0.5 \, M < [H^+] < 2.0 M$ and $1.5 \times 10^{-3} M < [Ti^{IV}] < 0.05 M$. From the equilibrium data the following reduction potential was evaluated:

$$Ti(OH)_2^{2+} + 2H^+ + e^- \rightleftharpoons Ti^{3+} + 2H_2O, \qquad E° = (7.7 \pm 0.6) \times 10^{-3} V \qquad (4.1)$$

Variation in the redox potential of the Ti^{IV}/Ti^{III} couple as a function of $[H^+]$ confirmed the consumption of two protons. The potentiometric data in the acidity range $0.3 M < [H^+] < 12 M$ were explained on the basis of the following processes:

$$Ti^{4+} + e^- \rightleftharpoons Ti^{3+}, \qquad\qquad E^\theta = (0.202 \pm 0.002)V \qquad (4.2)$$

$$Ti^{4+} + H_2O \rightleftharpoons Ti(OH)^{3+} + H^+, \qquad pK_{11} = -0.3 \pm 0.01 \qquad (4.3)$$

$$Ti^{4+} + 2H_2O \rightleftharpoons Ti(OH)_2^{2+} + 2H^+, \qquad pK_{12} = 1.38 \pm 0.05 \qquad (4.4)$$

Further hydrolysis to give $Ti(OH)_3^+(aq)$ ($pK_{13} \geqslant 2.3$) and $Ti(OH)_4(aq)$ ($pK_{14} = 4.8$) has been reported based on the solubility behaviour of hydrous TiO_2 [28]. In earlier reports the existence of polynuclear Ti^{IV} species were not considered. However a recent extensive ^{17}O NMR study by Comba and Merbach [26] in ClO_4^- media under conditions of variable temperature, $[H^+]$ and $[Ti^{IV}]$ in conjunction with light scattering and Raman measurements has detected not only the presence of a number of polynuclear species as relevant under certain conditions but also the conditions under which the putative $TiO^{2+}(aq)$ ion exists. In mixed water-methanol solutions of acidified Ti^{IV} perchlorate at $\sim 196 K$ characteristic ^{17}O NMR chemical shifts of a terminal $Ti{=}O$ group are observed around 1000–1100ppm. Shifts also observed in the ranges 700–800 ppm and 0–100 ppm were assigned respectively to bridging oxo (OH^-) and terminal H_2O (OH^-) ligands of various polynuclear/mononuclear Ti^{IV} species. The data showed that at low $[H^+]$ $\sim 0.9 m$ and relatively high $[Ti^{IV}]$ $> 0.05 m$ (m-molality), polynuclear species are formed extensively and the data were analysed on the basis of a system containing two trimers ($Ti_3O_4^{4+}(A)$, $Ti_3O_3^\alpha O_2^\beta H_3^{5+}(B)$) and a tetramer ($Ti_4O_4^\alpha O_2^\beta H_4^{8+}(C)$) in addition to mononuclear species TiO^{2+} or $Ti(OH)_2^{2+}$ (with O^α and O^β referring to different bridging oxygens as detected by ^{17}O NMR). Formation constants for the various species were evaluated at $25°C$, $\mu = 4.0 m$, as K_3^3 (A)$=(0.38 \pm 0.06)$, K_3^2 (B)$=(1.64 \pm 0.06)m^{-1}$ and K_4^1 (C)$=(2.31 \pm 0.03) m^{-3}$. Oxygen-exchange rate constants ($25°C$) were determined from the temperature dependence of the line widths of the ^{17}O NMR signals and had the following values; oxo or OH^- bridges, $k_{ex} = (100 \pm 50) s^{-1}$; terminal H_2O (OH^-), $k_{ex} = (3400 \pm 200) s^{-1}$ (lower limit) and $Ti{=}O$ of TiO^{2+} (68% aqueous methanol), $k_{ex} \sim (16\,000 \pm 5000) s^{-1}$. This remarkably fast rate of exchange on the $Ti{=}O$ oxygen of TiO^{2+} (nine orders of magnitude faster than the oxygen of $V{=}O^{2+}$) appears to be due to a favourable acid-catalysed mechanism involving ready protonation to $Ti{-}OH^{3+}$ followed by rapid proton transfer (exchange of the oxygen of $V{=}O^{2+}$ is known to be base catalysed, see later). In turn this is almost certainly responsible for its lack of detection in pure acidic aqueous solution by ^{17}O NMR under normal ambient conditions. Whether protonation induces exclusive formation of $Ti(OH)_2^{2+}$ is not in fact clear from these studies and an equilibrium involving both species seems to be the best representation (in none of the ^{17}O NMR experiments was definitive evidence for formation of $Ti(OH)_2^{2+}$ shown). An explanation for the ready protonation of the $Ti{=}O$ group could be a higher electron density placed on oxygen as a result of greater oxygen character in σ- and π molecular orbitals within the $Ti{=}O$ bond owing to the high energy of the titanium 3d orbitals, for example, vs those on vanadium.

4.1.3.3 SUBSTITUTION REACTIONS INVOLVING $Ti^{IV}(AQ)$

Formation rate constants and activation parameters for reactions of $Ti^{IV}(aq)$ with a number of incoming ligands, NCS^-, HF and pyrophosphate, have been determined using temperature jump techniques [27c]. Rate constants and activation enthalpies appear to be largely independent of the nature of the incoming ligand, $k_f \sim 10^3\,M^{-1}\,s^{-1}$, $\Delta H \sim 45\text{kJ mol}^{-1}$ for all, suggestive of an I_d process which was believed to reflect the influence of the presumed *trans* labilising $Ti{=}O$ group (cf. the behaviour of $V{=}O^{2+}$ and dimeric $Mo_2O_4^{2+}$ ($Mo{=}O$) species, see later).

The reaction of $Ti^{IV}(aq)$ and 'TiO^{2+}'compounds with H_2O_2 has received much attention [24, 34]. The intense orange colour of peroxo $Ti^{IV}(aq)$, shown now unambiquously to contain the ion $Ti(O_2)^{2+}$, has long been used for the spectrophotometric determination of titanium [35]. Second order rate constants for the reaction of $Ti^{IV}(aq)$ with H_2O_2 have been determined at $25\,^\circ C$. For the acid independent step, $k_f \sim 120M^{-1}\,s^{-1}$ [34a]. Terms in the rate law involving $[H^+]$ and $[H^+]^2$ were also detected. In contrast to the complexation reactions described above all reactions between H_2O_2 and 'TiO^{2+}' species, to give peroxo Ti^{IV} complexes (all 1:1), appear to be associatively activated on the basis of negative activation volumes determined by high pressure stopped flow [34c]. Oxygen-18 labelling studies have confirmed the simple substitution by O_2^{2-} on $Ti^{IV}(aq)$ without cleavage of the O—O bond [36].

4.2 ZIRCONIUM AND HAFNIUM

Much of the early studies regarding the comparative aqueous chemistry of these two elements concerned the problems of separation. The two elements have extremely similar chemical properties stemming from their virtually identical atomic (ionic) radii (145 and 86 pm for Zr and Zr^{4+}; 144 and 85 pm for Hf and Hf^{4+}). The aqueous chemistry itself is confined to the chemistry of the IV state species in the acid range > 0.1 M. Earlier reports of transiently stable aqua Zr^{III}

species do not seem to have been substantiated by further studies. The aqueous chemistry of zirconium and hafnium has been the subject of several reviews [37].

4.2.1   Zirconium(III) and Hafnium(III) (d¹)

Baker and Janus [38] reported on the aqueous solution properties of dissolved ZrI_3 in water in 1964. On the basis of spectrophotometric changes in the u.v.–visible region an aqua iodo complex, $[Zr(OH_2)_{6-n}I_n]^{(3-n)+}$ $(n < 3)$, was presumed initially formed in a rapid process (seconds) followed by slower aquation (few minutes) to '$[Zr(OH_2)_6]^{3+}$' which slowly oxidised over a 40 minute period to $Zr^{IV}(aq)$. Formation of putative $[Zr(OH_2)_6]^{3+}$ was correlated with the transient appearance of a band maximum at 24 400 cm^{-1} which was tentatively assigned to the $^2T_{2g} \rightarrow {^2E_g}$ transition, cf. $[Ti(OH_2)_6]^{3+}$. The existence of these species have not, however, been substantiated by further investigations. Little is known of the corresponding solution chemistry of $Hf^{III}(aq)$.

4.2.2   Zirconium(IV) and Hafnium(IV) (d⁰)

4.2.2.1   MONONUCLEAR $ZR^{IV}(AQ)$ AND $HF^{IV}(AQ)$ SPECIES

It was once thought that the simple mononuclear ions ZrO^{2+} and HfO^{2+} were present in the non-complexing acidic solutions of Zr^{IV} and Hf^{IV} based upon the stoichiometry present in basic salts such as $MOCl_2 \cdot 8H_2O$. From the results of a number of potentiometric [39], complexometric [40], spectrophotometric [41], X-ray scattering [42] and ultracentrifugation [43] studies it is now clear that hydrolytic polymerisation occurs extensively even in acidic solutions > 0.1 M $[H^+]$ to produce OH^- bridged polynuclear chains and/or clusters. An important early study was that of Zielen and Connick [41]. They showed that at low $[Zr^{IV}] < 10^{-4}$ M, hydrolysis of $Zr^{4+}(aq)$ is suppressed in 1.0 M $[H^+]$. Noren [39] conducted a study of the effect of hydrolysis on the stability of $ZrF^{3+}(aq)$ and $HfF^{3+}(aq)$ in 4.0 M $NaClO_4$ and, on the basis of their findings, Baes and Mesmer estimated values for pK_{11} of the $M^{4+}(aq)/MOH^{3+}(aq)$ pair as respectively -0.3 (Zr) and 0.25 (Hf). Estimates of formation constants for the further hydrolysis products, $M(OH)_2^{2+}(aq)$, $M(OH)_3^+(aq)$, $M(OH)_4(aq)$ and $M(OH)_5^-(aq)$ (M = Zr or Hf), were based upon less reliable data regarding the solubility properties of the hydrous dioxide, $MO_2 \cdot nH_2O$, under different conditions [44]. These estimates are listed in Table 4.2.

4.2.2.2   POLYNUCLEAR $ZR^{IV}(AQ)$ AND $HF^{IV}(AQ)$ SPECIES: THE TETRAMERIC $[M_4(OH)_8(OH_2)_{16}]^{8+}$ IONS

Zielen and Connick [41] studied the effect of polymerisation on the stability of a Zr^{IV} complex with 2-thenoyltrifluoroacetone and concluded that at 5×10^{-4} M

Table. 4.2. Formation constants (pK_{xy}) for various hydrolytic forms of zirconium and hafnium in acidic perchlorate
solution[a]

| | pK_{xy} | |
Species	Zr	Hf
Mononuclear		
$M(OH)^{3+}$	− 0.3	0.25
$M(OH)_2^{2+}$	(1.7)	(2.4)
$M(OH)_3^{+}$	(5.1)	(6.0)
$M(OH)_4$	(9.7)	(10.7)
$M(OH)_5^{-}$	(16.0)	(17.2)
Polynuclear		
$M_3(OH)_4^{8+}$	(0.6)	
$M_3(OH)_5^{7+}$	3.7	
$M_4(OH)_8^{8+}$	6.0	

[a]All data are calculated from the values obtained by Zielen and Connick
[41]. Values in parentheses are estimates based on solubility properties of
hydrous MO_2.

$< [M^{IV}] < 0.02$ M and between 1.0 and 2.0 M $[H^+]$ a trinuclear species and
a tetranuclear species were the dominant forms present. The trinuclear species was
formulated as $Zr_3(OH)_4^{8+}$(aq) and the tetranuclear species as $Zr_4(OH)_8^{8+}$(aq).
McVey [40] has since proposed that the trinuclear species should be better
represented as $Zr_3(OH)_5^{7+}$(aq). In 1956 Clearfield and Vaughan [45] showed in
an X-ray structural study that solid $ZrOCl_2 \cdot 8H_2O$ actually consisted of the
tetranuclear structure shown in Fig. 4.2 based on a square plane of zirconium
atoms each doubly bridged by hydroxy ligands and each coordinated to four
terminal water ligands, making eight-coordination in all at each metal centre.

The $[Zr_4(OH)_8(OH_2)_{16}]^{8+}$ structure itself has been since refined and is best
viewed as distorted dodecahedral at each zirconium [46]. It was subsequently
shown by X-ray scattering studies on solutions of $MOCl_2$ (M = Zr or Hf) that this
tetranuclear structure was indeed the major form representative in solution in
addition to smaller amounts of the trinuclear form [42]. The only difference in the
solution structure from that of solid $ZrOCl_2 \cdot 8H_2O$ was the position of the X^-
anions, e.g. Cl^- or Br^- being located between terminal water ligands on adjacent
metal atoms of the tetramer (Fig. 4.3). ^1H NMR studies of Zr^{IV} perchlorate in
water–acetone solution at − 70 °C [47] gave rise to a hydration number of four
at zirconium consistent with the tetranuclear structure if only shifts involving
the coordinated waters ligands are relevant. Earlier infra-red and Raman studies
on hydrated $ZrOCl_2$, which wrongly assigned a band at $1000\,cm^{-1}$ to the
stretching vibration of a Zr=O group, were shown by more recent Raman
studies [48] to be consistent with an assignment of the band rather to a

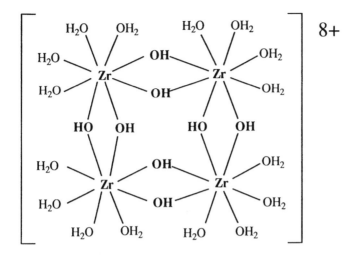

Figure 4.2. Structure of the μ-hydroxy bridged tetranuclear cation, $[Zr_4(OH)_8(OH_2)_{16}]^{8+}$, present in solid $ZrOCl_2 \cdot 8H_2O$

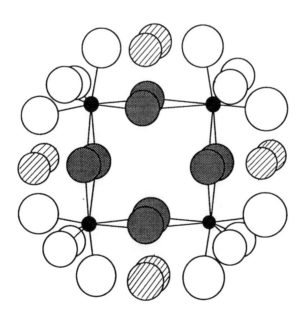

Figure 4.3. Aqueous solution structure of $ZrOCl_2$ obtained via X-ray scattering studies showing all eight positions for anion binding

Zr—OH—Zr bridging group (observed shift in the band to lower frequency on deuteration). Aberg and Glaser [49] have criticised the water–acetone results of Fratiello, Vidulich and Mako [47] following a careful reinvestigation of the ^1H NMR properties of 2.0 M aqueous solutions of $[Zr_4(OH)_8(OH_2)_{16}](ClO_4)_8$ in addition to reporting the results of a ^{17}O NMR study. A broad ^{17}O NMR resonance at 180 ppm from bulk water was assigned to the coordinated H_2O ligands on the basis of an enrichment observed with the addition of 1.8 atom % H_2O^{17} under conditions wherein the OH^- bridged core structure remained intact. Two kinetically distinct H_2O's may also be relevant from complexation studies with NO_3^-. At least one if not both are slow exchanging on the NMR timescale. It may be noted from Fig. 4.3 that two of the H_2O point in the directions of the OH^- bridges while the other two do not. One might envisage a greater lability in the former H_2O ligands.

XRD work has shown that the corresponding tetrameric Hf^{IV}(aq) species $Hf_4(OH)_8^{8+}$(aq) is the present in aqueous solution of Hf^{IV} salts [42]. Solutions of 'HfOCl$_2$' 0.5–2.0 m have the tetramer formulated as '$Hf_4(OH)_8(OH_2)_{12}$', indicative of replacement of four of the terminal water ligands by Cl^- ions. However, the fully aquated non–complexed $[Hf_4(OH)_8(OH_2)_{16}]^{8+}$ ion is apparently present in 2.0 m solutions of the bromide complex. The Hf—Hf distance, 357 pm, is not, surprising, in view of the comparative size of the two metals, identical to the Zr—Zr distance in the corresponding tetrameric Zr^{IV} species.

Livage and coworkers have attempted a rationale of the formation of mono-meric $Ti(OH)_2^{2+}$ ('TiO^{2+}') (aq) species for Ti^{IV}, as distinct from the formation of tetrameric OH-bridged species for Zr^{IV} and Hf^{IV}, on the basis of the partial charge model (Section 1.3.3.2). Calculations reveal that the precursor species $[Ti(OH)(OH_2)_5]^{3+}$ has a positive partial charge on the OH group; hence the species behaves as an acid and further deprotonates to give monomeric $Ti(OH)_2^{2+}$ (aq). At this point a very small negative charge does result on the OH group (-0.01) but not apparently sufficient to induce nucleophilicity leading to bridged species.

The situation for the larger Zr^{4+} (and Hf^{4+}) ions is different. Firstly, there is a higher coordination number (eight) at the metal which means that there are more waters competing for the acidity of the metal. As a result the precursor monomers $[M(OH)(OH_2)_7]^{3+}$ and $[M(OH)_2(OH_2)_6]^{2+}$ are much less acidic and lead to retention of an appreciably negative partial charge on the OH groups. This promotes condensation to form the OH-bridged tetramers. The propensity for Zr^{IV} to form di-μ-OH bridged units, based upon a square antiprismatic arrangement of Zr^{IV} ions, is exemplified by the existence of such units in the structures of both crystalline and amorphous ZrO_2 and HfO_2. Indeed, in crystalline ZrO_2 (Fig. 4.4), one can readily identify the same di-μ-OH bridged tetrameric units as present in the aquated tetramer [50].

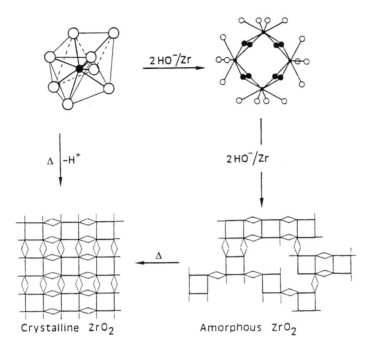

Figure 4.4. Basis structural aqueous chemistry involved in the interconversion between crystalline and amorphous ZrO_2

4.2.2.3 REACTIONS OF $M_4(OH)_8^{8+}$ (AQ) (M = ZR OR HF)

Few studies on the solution reactivity of $[Zr_4(OH)_8(OH_2)_{16}]^{8+}$ have been reported. A kinetic study of the $[H^+]$ promoted depolymerisation of $Zr_4(OH)_8^{8+}$ (aq) to mononuclear Zr^{IV} in 2.0 M $Na(H)ClO_4$ was consistent with a rate law, rate $= k_H[H^+][Zr_4]$, with $k_H = 9.5 \times 10^{-4}\,M^{-1}s^{-1}$ at 25 °C. General acid catalysis of the depolymerisation reation was also found to occur with second-order rate constants, k_{HX} ($M^{-1}s^{-1}$, 25 °C), following the order:

$$H_3PO_4 > H_2C_2O_4 > H_4P_2O_7 > HF > H_2SO_4$$
$$> CCl_3CO_2H > CH_2ClCO_2H\ [51]$$

An interesting tetranuclear peroxy complex of Zr^{IV} having the formula $Zr_4(O_2)_2(OH)_4^{8+}$ (aq) has been reported [52]. In this species simple substitution of two of the di-μ-hydroxy bridges in $Zr_4(OH)_8^{8+}$ (aq) by a side-on bonded O_2^{2-} group appears to have occurred. Further studies on both the structural nature of this species and on other aspects of the chemistry of the tetramer would be of interest.

At $[H^+] < 0.2\,M$ solutions of $MOCl_2$ show evidence of increasing polymerisation and polydispersion with Zr showing the greater tendency [43]. Solutions $<0.1\,M$ in $[H^+]$ kept at $100\,°C$ for one hour showed evidence of polymers containing up to 20–40 metal ions. Table 4.2 summarises the currently available data regarding formation constants for the various hydrolytic species of Zr^{IV} and Hf^{IV} in acidic aqueous solution.

REFERENCES

[1]   (a) McAuliffe, C. A., and Barratt, D. S., *Comprehensive Coordination Chemistry* (eds. G. Wilkinson, R. D. Gillard and J. A. McCleverty), Vol. 3, Pergamon, Oxford, 1987, p. 330.
      (b) Clark, R. J. H., *The Chemistry of Titanium and Vanadium*, Elsevier, Amsterdam (1968).
[2]   Cotton, F. A., and Wilkinson, G., *Advanced Inorganic Chemistry*, 5th edn, Wiley, New York, p. 661 (1988).
[3]   deBethune, A J., and Loud, N. A. S., *Standard Electrode Potentials and Temperature Coefficients at 25 °C*, C. A. Hampel, Stokie, Illinois (1964).
[4]   Hartmann, H., and Schlafer, H. L., *Z. Phys. Chem.*, **197**, 116 (1951).
[5]   McQuillan, A. D., and McQuillan, M. K., *Titanium*, Butterworths, London, 1956.
[6]   Rabideau, S. W., and Quinney, P. R., *J. Am. Chem. Soc.*,**76**, 3800 (1954).
[7]   Hugi, A. D., Helm, L., and Merbach, A. E., *Inorg. Chem.*, **26**, 1763 (1987).
[8]   Furman, S. C., and Garner, C. S., *J. Am. Chem. Soc.*, **73**, 4528 (1951).
[9]   (a) Lipson, H., *Proc. Roy. Soc. London, A*, **151**, 347 (1935).
      (b) Spangenberg, K., *Neues Jb. Mineral. Geol. Palaentolog., Abst. A*, 99 (1949).
      (c) Hartmann, H., and Schlafer, H. L., *Z. Naturforsch, Teil A*, **6**, 754 (1951).
[10] Aquino, M. A. S., Clegg, W., Liu, Q. T., and Sykes, A. G., *Acta Cryst.*, **C51**, 560 (1995).
[11] Ilse, F. E., and Hartmann, H., *Z. Phys. Chem.*, **197**, 239 (1951).
[12] Liehr, A. D., *Prog. Inorg. Chem.*, **3**, 281 (1962).
[13] Hartmann, H., Schlafer, H. L., and Hansen, K. H., *Z. Anorg. Allg. Chem.*, **284**, 153 (1956).
[14] Tachikawa, H., Ichikawa, T., and Yoshida, H., *J. Am. Chem. Soc.*, **112**, 977, 982 (1990).
[15] Baes, C. F., and Mesmer, R. E., *The Hydrolysis of Cations*, Krieger, Florida, 1986, p. 148.
[16] (a) Paris, M. R., and Gregoire, C., *Anal. Chim. Acta*, **42**, 439 (1968).
      (b) Krentizien, H., and Brito, F., *Ion*, **30**(342), 14 (1970).
[17] (a) Diebler, H., *Z. Phys. Chem. (Munich)*, **68**, 64 (1969).
      (b) Chaudhuri, P., and Diebler, H., *J. Chem. Soc. Dalton Trans*, 596 (1977).
      (c) Chaudhuri, P., and Diebler, H., *Z. Phys. Chem. (Munich)*, **139**, 191 (1984).
      (d) Chaudhuri, P., and Diebler, H., *J. Chem. Soc. Dalton Trans*, 1693 (1986).
[18] Swaddle, T. W., *Adv. Inorg. Bioinorg. Mech.*, **2**, 95 (1983).
[19] Eve, D. A., *Inorg. Chim. Acta.*, **174**, 205 (1990).
[20] (a) McAuley, A., Olubuyide, O., Spencer, L., and West, P. R., *Inorg. Chem.*,**22**, 2594 (1984).
      (b) Ogino, H., Kikkawa, M., Shimura, M., and Tanaka, N., *J. Chem. Soc. Dalton Trans.*, 894 (1981).

(c) Chalilpoyil, P., Davies, K. M., and Earley, J. E., *Inorg. Chem.*, **16**, 3344 (1977).
(d) Bakac, A., Marcec, R., and Orhanovic, M., *Inorg. Chem.*, **16**, 3133 (1977).
(e) Davies, K. M., and Earley, J. E. *Inorg. Chem.*, **17**, 3350 (1978).
[21] (a) Berrie, B. H., and Earley, J. E., *Inorg. Chem.*, **23**, 774 (1984).
(b) Ram, M. S., Martin, A. H., and Gould, E. S., *Inorg. Chem*, **22**, 1103 (1983).
(c) Ellis, J. D., and Sykes, A. G., *J. Chem. Soc., Dalton Trans.*, 2553 (1973).
(d) Bakac, A., and Orhanovic, M., *Inorg. Chim. Acta*, **21**, 173 (1977).
(e) Thompson, G. A. K., and Sykes, A. G., *Inorg. Chem.*, **15**, 638 (1976)
(f) Hery, M., and Wieghardt, K., *Inorg. Chem.*, **17**, 1130 (1978).
[22] (a) Marcec, R., Orhanovic, M., Wray, J. A., and Cannon, R. D., *J. Chem. Soc. Dalton Trans.*, 663 (1984).
(b) Birk, J. P., and Logan, T. P., *Inorg. Chem.*, **12**, 580 (1973).
(c) Bose, R. N., and Earley, J. E., *J. Chem. Soc. Chem. Commun.*, 50 (1983).
[23] See for example, Bose, R. N., Cornelius, R. D., and Mullen, A. C., *Inorg. Chem.*, **26**, 1414 (1987).
[24] Rotzinger, F. P., and Gratzel, M., *Inorg. Chem.*, **26**, 3704 (1987).
[25] Ciavatta, L., Ferri, D., and Riccio, G., *Polyhedron*, **4**, 15 (1985).
[26] Comba, P., and Merbach, A. E., *Inorg. Chem.*, **26**, 1315 (1987).
[27] (a) Ellis, J. D., and Sykes, A. G., *J. Chem. Soc. Dalton Trans.*, 537 (1973).
(b) Ellis, J. D., Thompson, G. A. K., and Sykes, A. G., *Inorg. Chem.*, **15**, 3172 (1976).
(c) Thompson, G. A. K., Taylor, R. S., and Sykes, A. G., *Inorg. Chem.*, **16**, 2880 (1977).
[28] (a) Beukenkamp, F. L., and Herrington, K. D., *J. Am. Chem. Soc.*, **82**, 3025 (1960).
(b) Nabivanets, B. I., *Russ. J. Inorg. Chem. (Engl. Transl.)*, **7**, 210 (1962).
[29] Caglioti, V., Ciavatta, L., and Liberti, A., *J. Inorg. Nucl. Chem.*, **15**, 115 (1960).
[30] Dwyer, P. N., Puppe, L., Buchler, J. W., and Scheidt, W. R., *Inorg. Chem.*, **14**, 1782 (1975).
[31] (a) Taube, R., *Z. Chem.*, **3**, 194 (1963).
(b) Block, B. P., and Meloni, E. G., *Inorg. Chem.*, **4**, 111 (1965).
[32] Feltz, A., *Z. Chem.*, **7**, 158 (1967).
[33] Dehnicke, K., Pausewang, G., and Rudorff, W., *Z. Anorg. Allg. Chem.*, **366**, 64 (1969).
[34] (a) Orhanovic, M., and Wilkins, R. G., *J.Am. Chem. Soc.*, **89**, 278 (1967).
(b) Thompson, R. C., *Inorg. Chem.*, **23**, 1794 (1984).
(c) Inamo, M., Funahashi, S., and Tanaka, M., *Inorg. Chem.*, **22**, 3734 (1983) and **24**, 2475 (1985).
[35] Kolthoff, I. M., and Elving, P. J., *Treatise on Analytical Chemistry Part II*, Vol. 5, Interscience, New York, 1961.
[36] (a) Latour, J.-M., Galland, B., and Marchon, J.-C., *J. Chem. Soc. Chem. Commun.*, 570 (1979).
(b) Inamo, M., Funahashi, S., and Tanaka, M., *Bull. Chem. Soc. Japan.*, **59**, 2629 (1986).
[37] (a) Larsen, E. M., *Adv. Inorg. Chem. Radiochem.*, **13**, 1 (1970).
(b) Solovkin, A. S., and Tsvetkova, S. V., *Russ. Chem. Rev. (Engl.)*, **31**, 655 (1962).
(c) Clearfield, A., *Rev. Pure Appl. Chem.*, **14**, 91 (1964).
[38] Baker, W. A., and Janus, A. R., *J. Inorg. Nucl. Chem.*, **26**, 2087 (1964).
[39] Noren, B., *Acta Chem. Scand.*, **27**, 1369 (1973).
[40] McVey, W. H., USAEC Report HW-21487, 1951.
[41] Zielen, A. J., and Connick, R. E., *J. Am. Chem. Soc.*, **78**, 5785 (1956).
[42] (a) Muha, G. M., and Vaughan, P. A., *J. Chem. Phys.*, **33**, 194 (1960).
(b) Aberg, M., *Acta Chem. Scand. A*, **31**, 171 (1977).

[43] (a) Kraus, K. A., and Johnson, J. A., *J. Am. Chem. Soc.*, **75**, 5769 (1953).
      (b) Johnson, J. A., Kraus, K. A., and Holmberg, R.W., *J. Am. Chem. Soc.*, **78**, 26 (1956).
      (c) Johnson, J. A., and Kraus, K. A., *J. Am. Chem. Soc.*, **78**, 3937 (1956).
[44] (a) Bilinski, H., Branica, M., and Sillen, L. G., *Acta Chem. Scand.*, **20**, 853 (1966).
      (b) Sheka, I. A., and Pevzner, T. V., *Russ. J. Inorg. Chem. (Engl. Transl.)*, **5**, 1119 (1960).
[45] Clearfield, A., and Vaughan, P. A., *Acta Cryst.*, **9**, 155 (1956).
[46] Mak, T. C. W., *Can. J. Chem.*, **46**, 3493 (1968).
[47] Fratiello, A., Vidulich, G. A., and Mako, F., *Inorg. Chem.*, **12**, 470 (1973).
[48] Burkov, K. A., Koznevnikova, G. V., Lilich, L. S., and Myund, L. A., *Russ. J. Inorg. Chem. (Engl. Transl.)*, **27**, 804 (1982).
[49] Aberg, M., and Glaser, J., *Inorg. Chim. Acta*, **206**, 53 (1993)
[50] (a) Clearfield, A., *Pure Appl. Chem.*, **14**, 91 (1964).
      (b) Fryer, J. R., Hutchinson, J. L., and Paterson, R., *J. Colloid. Interface Sci.*, **34**, 238 (1970).
[51] Devia, D. H., and Sykes, A. G., *Inorg. Chem.*, **20**, 910 (1981).
[52] Thompson, R. C., *Inorg. Chem.*, **25**, 3542 (1985).

Chapter 5

Group 5 Elements: Vanadium, Niobium and Tantalum

5.1   VANADIUM

Vanadium possesses a rich aqueous chemistry with 'aqua ions' characterising four oxidation states II–V. V^V has a complex aqueous chemistry with the species present dependent strongly upon the pH and vanadium concentration. The chemistry of the II–IV oxidation states, on the other hand, is dominated by well-characterised mononuclear cationic species stable over a reasonably wide pH range with simple hexaaqua ions known for the II and III states. No aqueous chemistry exists for oxidation states below II. The aqueous chemistry of vanadium has recently been the subject of a number of excellent reviews primarily from the viewpoint of a link to its biological role in certain plants, fungi and sea creatures [1].

5.1.1   Vanadium(II) (d^3): Chemistry of the $[V(OH_2)_6]^{2+}$ Ion

5.1.1.1   PREPARATION OF $[V(OH_2)_6]^{2+}$

Solutions of violet $[V(OH_2)_6]^{2+}$ have long been prepared by exhaustive reduction of acidic aqueous solutions (usually HCl) of V^{IV} (see later) either electrolytically using a mercury pool electrode, or by using amalgamated zinc. Because of the ready oxidation of V^{II} by ClO_4^- ions, solutions of $[V(OH_2)_6]^{2+}$ were generated for many years via dissolution of $VCl_2 \cdot 2H_2O$ in dilute HCl, complexation by Cl^- ions not being extensive [2]. Some preparations have employed reduction of V^{IV} (aq) solutions in $HClO_4$ with the V^{2+} (aq) rapidly used before the onset of oxidation [3,4]. Nonetheless the best method would seem to be that eliminating the use of Cl^- or ClO_4^- altogether. In this regard the use of triflate, $CF_3SO_3^-$, as a counter-anion would seem desirable as in the case of Ti^{3+}(aq). In a recent report Larkworthy and coworkers showed that vanadium metal readily dissolves in refluxing aqueous solutions of CF_3SO_3H under a nitrogen atmosphere to produce violet solutions from which crystals of $[V(OH_2)_6](CF_3SO_3)_2$ (unhydrated) can be readily isolated [5]. This compound is proving a useful 'lead-in' towards the synthesis of a whole range of simple hexa-coordinated V^{II} compounds.

In a typical air-free preparation vanadium turnings (5 g) are heated under reflux with a mixture of CF_3SO_3H (20 cm³) and water (20 cm³). A light blue-green colour initially develops which turns to dark blue on heating for 12 hours. On cooling a purple colour develops from which purple crystals begin to separate. Following heating, to redissolve the crystals, and decanting, to remove undissolved vanadium metal, the solution deposited purple needles of $[V(OH_2)_6]$-$(CF_3SO_3)_2$ on recooling. These could be filtered off and washed with anhydrous diethyl ether. An approximate 20% yield was reported based upon the weight of recovered vanadium metal. Using this crystalline material other V^{II} compounds can be readily made, including $[V(OH_2)_6]Br_2$, $[V(OH_2)_6](BF_4)_2$, $VCl_2(OH_2)_4$ and $[V(en)_3](CF_3SO_3)_2$. The salt readily dissolves in acetone, methanol, ethanol, ethyl acetate and 1,2-dimethoxyethane, allowing the likely synthesis of a number of non-aqueous solvates. $[V(OH_2)_6](CF_3SO_3)_2$ obeys the Curie–Weiss law having a temperature-independent μ_{eff} (3.86 BM at 90 K, 3.83 BM at 295 K) close to the expected spin–only value for an octahedral d^3 ion ($^4A_{2g}$ ground state).

5.1.1.2   ELECTRONIC SPECTRUM OF $[V(OH_2)_6]^{2+}$

A recent study by Taube and coworkers reported electronic spectral data for V^{II} in a number of solvents. For $[V(OH_2)_6]^{2+}$, (Fig. 5.1) , peak maxima (energy, cm⁻¹, ε, M⁻¹ cm⁻¹) are as follows: 843 nm (11 862 cm⁻¹, 4.5) ($^4A_{2g} \rightarrow {}^4T_{2g}$), 556 nm (17 986, 4.3) ($^4A_{2g} \rightarrow {}^4T_{1g}(F)$) and $\sim$390 nm (shoulder) (25 641cm⁻¹)-

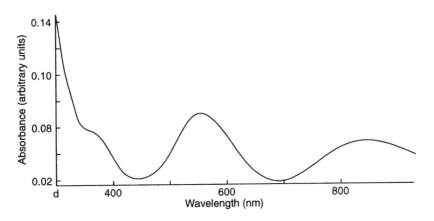

Figure 5.1. Electronic spectrum of $[V(OH_2)_6]^{2+}$

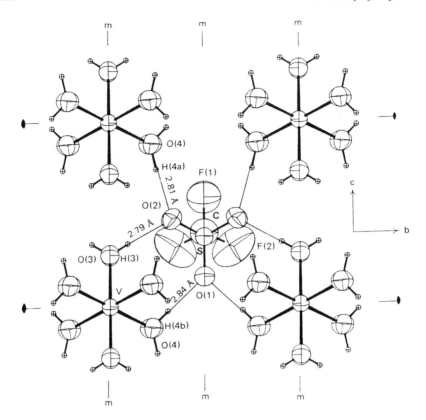

Figure 5.2. Structure of the hydrogen-bonded $V(OH_2)_6$ units in $[V(OH_2)_6](CF_3SO_3)_2$ [5]. (Reprinted from *Inorganica Chimica Acta*, **169**, D. G. Holt, L. F. Larkworthy, D. C. Povey, G. W. Smith and G. J. Leigh, p. 201, 1990 with kind permission from Elsevier Science S. A., Lausanne, Switzerland)

$(^4A_{2g} \rightarrow {}^4T_{1g}(P))$ [6]. This may be compared with the solid reflectance spectrum of $[V(OH_2)_6](CF_3SO_3)_2$, with bands at 847 nm (11 806) cm^{-1}, broad), 575 nm (17 400 cm^{-1}) and 376 nm (26 600 cm^{-1}). The agreement in energy between these bands confirms the retention of the octahedral hexaaqua geometry in both the solid and solution structures of V^{2+}. The X-ray structure of $[V(OH_2)_6](CF_3SO_3)_2$ (Fig. 5.2) shows the hydrogen-bonding network to the $CF_3SO_3^-$ group in addition to the regular octahedral geometry of water ligands around each V^{2+}[5]. The mean V—OH$_2$ distance is 211.9(1) pm. $[V(OH_2)_6]^{2+}$ also occurs in the Tutton salts, $M_2[V(OH_2)_6](SO_4)_2$ ($M = NH_4^+$, K^+, Rb^+ and Cs^+). In the NH_4^+ salt the mean V—OH$_2$ distance was reported as 215 pm [7].

5.1.1.3 SUBSTITUTION REACTIONS ON $[V(OH_2)_6]^{2+}$

Water exchange on solutions of $[V(OH_2)_6]Cl_2$, generated in HCl solutions containing H_2O^{17}, have been followed by measurement of the ^{17}O NMR linewidth of the free-water signal as a function of temperature and pressure (cf. $Ti^{3+}(aq)$). The kinetic parameters, $k_{ex}(25\,°C) = 87 \pm 4\,s^{-1}$, $\Delta H^{\ddagger} = 61.8 \pm 0.7\,kJ\,mol^{-1}$, $\Delta S^{\ddagger} = -0.4 \pm 1.0\,J\,K^{-1}\,mol^{-1}$ and $\Delta V^{\ddagger} = -4.1 \pm 0.1\,cm^3\,mol^{-1}$[8], have been interpreted in terms of an associative interchange I_a process. The full volume profile obtained from a variable-pressure study of the complexation of V^{2+} with NCS^- [9] (Fig. 5.3) shows that the reaction can be broken down to reveal an interchange step largely in keeping with the characteristics of the water-exchange process. The reaction was analysed by assuming the Eigen–Wilkins mechanism involving an ion pair preassociation of the charged reactants followed by ligand interchange. Thus, two additive contributions to the observed $\Delta V_f^{\ddagger}$ value ($-2.1\,cm^3\,mol^{-1}$) can be envisaged: $\Delta V_{os}^{\ddagger}$ (ion pair preassociation) and $\Delta V^{\ddagger}$ (interchange of ligands). $\Delta V_{os}^{\ddagger}$ was estimated from electrostatics as $+3.2\,cm^3\,mol^{-1}$, leaving a value of $\Delta V^{\ddagger}$ of $-5.3\,cm^3\,mol^{-1}$, close to the value for water exchange. Thus, as with all divalent ions the controlling process in a water ligand substitution reaction is the water-exchange (dissociation) process itself. The lower charge and greater d-electron population vs $[Ti(OH_2)_6]^{3+}$ (d^1) probably accounts for the smaller negative volume observed here for $[V(OH_2)_6]^{2+}$, indicative of a smaller contribution from factors responsible for a volume contraction to the transition state (see Section 1.4). Unfortunately, there is a surprisingly dearth of kinetic studies regarding substitution reactions on $[V(OH_2)_6]^{2+}$ [10]. However, redox processes involving $[V(OH_2)_6]^{2+}$ seem to be

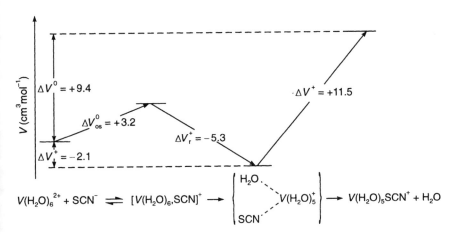

Figure 5.3. Full volume profile for the reaction of NCS^- with $[V(OH_2)_6]^{2+}$

largely substitution controlled and these have provided some insights into the relative rates for different entering 'ligands'.

5.1.1.4 REDOX REACTIONS INVOLVING $[V(OH_2)_6]^{2+}$

Whilst few substitution studies involving $[V(OH_2)_6]^{2+}$ have been reported much attention has been devoted to redox reactions in which $[V(OH_2)_6]^{2+}$ acts as a reductant . $[V(OH_2)_6]^{2+}$ is quite a powerful reducing agent, as exemplified by the standard reduction potential of the $V^{3+/2+}$ couple (-0.256 V). Despite the favourable thermodynamic driving force many redox reactions have been found to proceed rather slowly, consistent with a substitution controlled inner-sphere process [11]. A number of kinetic studies have focused on the reduction of a large number of Co^{III} complexes by $[V(OH_2)_6]^{2+}$ [3, 12]. In many cases reaction rates are within an order of magnitude of that characterising substitution at V^{II} in support of an inner-sphere process. However, a number of reactions clearly proceed via outer-sphere pathways. Reactions that are outer sphere have usually smaller values of $\Delta H^{\ddagger}$ than those characterising substitution-controlled inner-sphere processes. Examples of some typical reactions are presented in Table 5.1 [12].

Inner-sphere reactions tend to have rate constants occurring within a narrow range close to typical substitution rates on $[V(OH_2)_6]^{2+}$, implying that the

Table 5.1. Rate parameters for redox reactions involving $[V(OH_2)_6]^{2+}$ at 25 °C [12]

Reactant	k ($M^{-1}s^{-1}$)	$\Delta H^{\ddagger}$ (kJ mol^{-1})	$\Delta S^{\ddagger}$ (JK^{-1}mol^{-1})	Mechanism
$[Cr(OH_2)_5SCN]^{2+}$	8.0	54	-46	Inner sphere
VO^{2+}(aq)	1.6	51	-71	
$[Co(NH_3)_5SCN]^{2+}$	30.0	69	$+25$	
$[Co(NH_3)_5C_2O_4]^+$	45.0	51	-42	
$[Co(NH_3)_5N_3]^{2+}$	13.0	49	-54	
$[Co(NH_3)_5SO_4]^+$	25.5	49	-54	
$[Co(en)_3N_3SCN]^{2+}$	32.7	—	—	
$[Ru(OH_2)_5Cl]^{2+}$	1.9×10^3	—	—	Outer sphere
$[Fe(OH_2)_6]^{3+}$	1.8×10^4	—	—	
$[Co(NH_3)_6]^{3+}$	0.004	38	-167	
$[Co(NH_3)_5Cl]^{2+}$	10	31	-121	
$[Co(NH_3)_5OH_2]^{3+}$	0.53	34	-134	
Substitution				
H_2O^a	87	62	-0.4	
NCS^-	24	67	$+5$	

^{a} First-order rate constant in s^{-1}.

rate-determining step is controlled by bridging ligand entry into the V^{2+} co-ordination sphere. The small spread of rates nonetheless provides further support for the general premise that substitution reactions on divalent ions, including V^{2+}, show little dependence on the nature of the entering ligand and, despite the observed negative $\Delta V^{\ddagger}$, are largely controlled by water dissociation from the V^{2+} coordination sphere (see arguments presented in Section 1.4.4). In some cases, however, the redox mechanism is difficult to assign since rates for reactions that are clearly outer sphere are frequently close to those characterising substitution-controlled processes. The presence of potential bridging ligands would favour the inner-sphere route although at the same time one might expect the t_{2g}^3 configuration of V^{2+} to favour an outer-sphere process. Thus it is not a surprise that both processes are found to operate. The apparent 'outer-sphere' nature of the reaction of $V^{2+}(aq)$ with $[Co(NH_3)_5Cl]^{2+}$ (cf. $Cr^{2+}(aq)$ later) is of interest, however, and may reflect the poor affinity of V^{2+} for Cl^- as the incoming ligand.

The reaction of $[V(OH_2)_6]^{2+}$ with molecular oxygen gives V^{IV} directly in a two-electron process since V^{IV} reacts at a much slower observed rate [13]. The following reactions are believed to be relevant at $25\,°C$ [14]:

$$V^{2+} + O_2 \longrightarrow VO_2^{2+}, \quad k_1 = 2 \times 10^3 \text{ M}^{-1}\text{s}^{-1} \tag{5.1}$$

$$VO_2^{2+} + H_2O \longrightarrow VO^{2+} + H_2O_2, \quad k_2 \sim 10^2 \text{ s}^{-1} \tag{5.2}$$

$$VO_2^{2+} + V^{2+} \rightleftharpoons VOOV^{4+}, k_3 = 3.7 \times 10^3 \text{ M}^{-1}\text{s}^{-1}, k_{-3} = 20\,\text{s}^{-1} \tag{5.3}$$

$$VOOV^{4+} \longrightarrow 2VO^{2+}, \quad k_4 = 35\,\text{s}^{-1} \tag{5.4}$$

Reaction (5.2) suggests that the final product of reaction (5.1) should be written as $V^{IV}(O_2^{2-})^{2+}$. Reactions (5.3) and (5.4) are relevant if excess V^{2+} is present. The values of k_1 and k_3 would appear to suggest outer-sphere processes (Table 5.1) although inner-sphere products result. H_2O_2 also oxidises $[V(OH_2)_6]^{2+}$ to V^{IV} in a two-electron process, presumably involving an intermediate such as $VOOV^{4+}$ which then undergoes reaction (5.4) [13]. However, with either V^V, I_2 or Br_2 as oxidant the product of reaction with $V^{2+}(aq)$ is V^{III}.

Oxidation of $[V(OH_2)_6]^{2+}$ by a number of alkyl radicals, to give $[V(OH_2)_6]^{3+}$ and the corresponding alkane, has been recently studied [15]. It is believed that a transient seven-coordinate intermediate $[(H_2O)_6VR]^2$ is first produced which then decays yielding $[V(OH_2)_5OH]^{2+}$ and thus $[V(OH_2)_6]^{3+}$ in addition to RH (Fig. 5.4). The VR^{2+} compounds are much more reactive than the ones resulting from corresponding reactions on Cr^{II} and this is believed to reflect the formation of the seven-coordinate radical coupled 'colligated' intermediate (Fig. 5.4) [16]. Second-order rate constants are typically $\sim 10^5$ M^{-1}s^{-1} for a series of primary radicals far in excess of the normal substitution rate on $[V(OH_2)_6]^{2+}$. Conceivably therefore a similar 'colligated' reaction could occur to explain the fast rate of reaction with the 'diradical' triplet dioxygen, reaction (5.1) above.

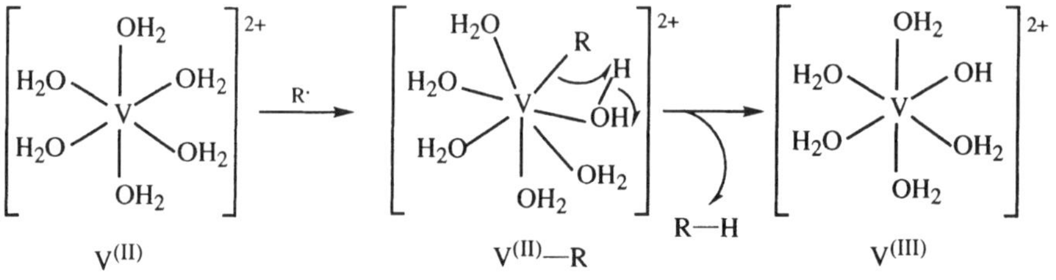

Figure 5.4. Mechanism of reaction of $[V(OH_2)_6]^{2+}$ with $R^{\cdot}$

5.1.2   Vanadium(III) (d²): Chemistry of the $[V(OH_2)_6]^{3+}$ Ion

5.1.2.1   PREPARATION OF $[V(OH_2)_6]^{3+}$

The $[V(OH_2)_6]^{3+}$ ion is present in dulute acidic aqueous solutions of VCl_3 or $V_2(SO_4)_3$ since coordination by these anions on V^{3+} is not extensive. As with $[Ti(OH_2)_6]^{3+}$ the best method to prepare pure solutions of $[V(OH_2)_6]^{3+}$ is via the anhydrous triflate, $V(CF_3SO_3)_3$. This is prepared from VCl_3 and triflic acid by a procedure identical to that reported for the preparation of $Ti(CF_3SO_3)_3$ (Chapter 4, Ref. [7]). Required solutions of $[V(OH_2)_6]^{3+}$ are then prepared by dissolution of $V(CF_3SO_3)_3$ in aqueous triflic acid.

5.1.2.2   ELECTRONIC AND MOLECULAR STRUCTURE OF $[V(OH_2)_6]^{3+}$

Solutions of $[V(OH_2)_6]^{3+}$ show two electronic absorption maxima at 400 nm ($25\,000$ cm^{-1}, $\varepsilon = 8.35$ M^{-1} cm^{-1}) and 580 nm ($17\,250$ cm^{-1}, 5.6 M^{-1} cm^{-1}) assigned to the transitions $^3T_{1g}(F) \rightarrow {}^3T_{1g}(P)$ and $^3T_{1g}(F) \rightarrow {}^3T_{2g}$ respectively [17]. The higher energy transition to the $^3A_{2g}$ state is usually obscured in solution by $L \rightarrow V^{3+}$ charge transfer absorptions, but it has been observed occurring at 290 nm ($34\,500$ cm^{-1}) for V^{3+} doped into the octahedral lattice of Al_2O_3 wherein the charge transfer absorptions occur at higher energy [18]. The structure of the $[V(OH_2)_6]^{3+}$ ion has been examined by X-ray diffraction studies on the caesium alum [19] and by neutron diffraction on the salt $[V(OH_2)_6][H_5O_2](CF_3SO_3)_4$ [20]. The latter compound was crystallised from a 2.0M triflic acid solution containing the trinuclear complex $V_3(\mu_3\text{-}O)(\mu\text{-}CH_3CO_2)_6(CH_3CO_2H)_2(thf)$. In the triflate salt the water protons were disposed as to cause a reduction in symmetry of the VO_6 unit from O_h to D_{3d}, (Fig. 5.5). From infra-red and single-crystal Raman studies on the caesium alum, Best, Armstrong and Beattie assigned the following vibrations: 525 cm^{-1} (symm. V—OH_2 stretch), 532 cm^{-1} (asymm. V—OH_2 stretch) and 607 cm^{-1} (asymm. V—OH_2 bend) [21]. Low-temperature electronic spectral studies on the ammonium alum by Hitchmann *et al.* have led to a determination of V—OH_2 bonding parameters. It was shown

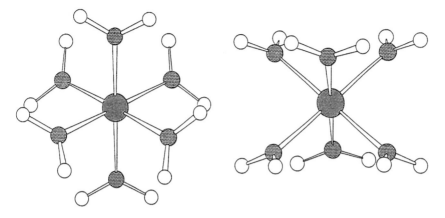

Figure 5.5. Structure of the $[V(OH_2)_6]^{3+}$ unit in $[V(OH_2)_6][H_5O_2][H_5O_2](CF_3SO_3)_4$ [20]

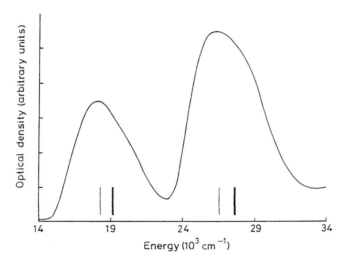

Figure 5.6. Single-crystal electronic spectrum of $(NH_4)[V(OH_2)_6](SO_4)_2 \cdot 6H_2O$ recorded at $10\,K$ [22]

that π-bonding in the plane of each water ligand is weaker than that perpendicular to it [22]. The electronic spectrum of a crystal of the ammonium alum determined at $10\,K$ (Fig. 5.6) shows the appearance of higher energy shoulders on the two visible transitions. This has been assigned to a splitting of the $^3T_{2g}$ and $^3T_{1g}$ excited states to give pairs of $^3A_{1g}$ and ^3E_g and $^3A_{2g}$ and ^3E_g states respectively as a result of the lowering of symmetry of the $V(OH_2)_6$ unit to approximately D_{3d} [22].

5.1.2.3   HYDROLYSIS OF $[V(OH_2)_6]^{3+}$

$[V(OH_2)_6]^{3+}$ hydrolysis leads to mononuclear $[V(OH_2)_5(OH)]^{2+}$ ($pK_{11} = 2.26$) and dimeric $[V_2(OH)_2(OH_2)_8]^{4+}$ ($pK_{22} = 3.8$) [23]. The di-μ-hydroxo formulation for the aqua dimer was based upon the observation of no change in magnetic susceptibility (per V) upon dimerisation [24]. A brown μ-oxo dimer is also believed to result from the reaction of $[V(OH_2)_6]^{2+}$ with $[VO(OH_2)_5]^{2+}$. Despite the $V_2(OH)_2{}^{4+}$(aq) formulation most, if not all, structurally characterised V^{III} dimeric complexes possess a single μ-oxo bridge and there have been recent claims of the existence of only one aqua dimer, namely the single μ-oxo species, following resonance Raman studies [25]. At pH 1.5 the aqua dimer is characterised by an absorption maximum at 415 nm which increases in intensity and shifts to slightly lower energy as the pH is raised (Fig. 5.7). At pH 2.2

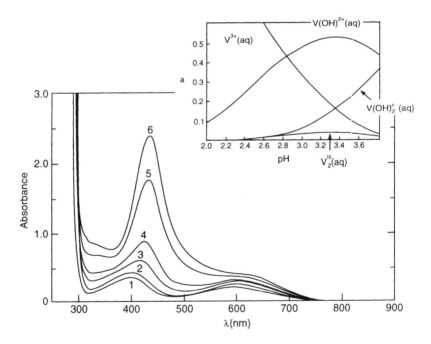

Figure 5.7. Electronic spectrum of aqueous $V_2(SO_4)_3$ as a function of pH: (1) 0.8, (2) 1.1, (3) 1.5, (4) 1.7, (5) 2.0, (6) 2.2. (Inset: distribution curves for V^{III}(aq) species (2.9×10^{-3} M) as a function of pH [25].) (Reprinted with permission from K. Kanamori, Y. Ookuto, K. Ino, K. Kawai and H. Michabata, *Inorg. Chem.*, **30**, 3832, 1991. Copyright 1991 American Chemical Society)

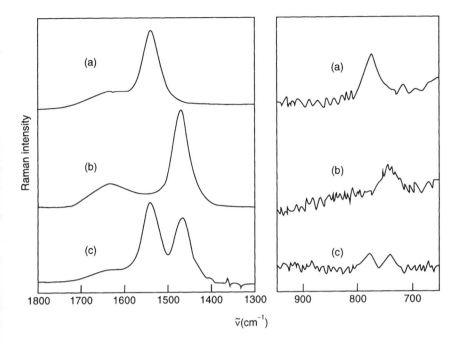

$\tilde{v}(cm^{-1})$

Figure 5.8. Raman spectra of VCl_3 in $H_2{}^{16}O/H_2{}^{18}O$ at pH 2.0: (a) $H_2{}^{16}O$ (b) $H_2{}^{18}O$ (c) 1:1 mixture of $H_2{}^{16}O$ and $H_2{}^{18}O$ [25]. (Reprinted with permission from K. Kanamori, Y. Ookubo, K. Ino, K. Kawai and H. Michabata, *Inorg. Chem.*, **30**, 3832, 1991. Copyright 1991 American Chemical Society)

it is located at 425 nm. Raman bands at 1533 and 774 cm^{-1} in solutions of VCl_3 at pH 2.0 (Fig. 5.8) were assigned to stretching vibrations of a μ-oxo dimer, similar bands being found in a number of single μ-oxo V^{III} complexes. Isotope shifts observed on substituting ^{18}O for ^{16}O were found to be similar to those observed in a number of Mo—O—Mo species. The presence of only two bands due to ^{16}O and ^{18}O labelling indicated isotopically pure species pointing towards the single μ-oxo formulation [25]. A speciation diagram for 2.9×10^{-3} M V^{III}(aq) as a function of pH [24] is shown in Fig. 5.7.

5.1.2.4 SUBSTITUTION OF WATER ON $[V(OH_2)_6]^{3+}$

From variable temperature and pressure high-field ^{17}O NMR studies the following parameters have been determined for $[V(OH_2)_6]^{3+}$: $k_{ex}(25\,°C) = (5.0 \pm 0.3) \times 10^2 s^{-1}$, $\Delta H^{\ddagger} = (49.4 \pm 0.8) kJ\,mol^{-1}$, $\Delta S^{\ddagger} = (-27.8 \pm 2.2) J\,k^{-1}\,mol^{-1}$, $\Delta V^{\ddagger} = (-8.9 \pm 0.4)\,cm^3\,mol^{-1}$ and $\Delta \beta^{\ddagger} = (-1.1 \pm 0.3)\,cm^3\,mol^{-1}\,MPa^{-1}$ [26]. These values, particularly the significant dependence of $\Delta V^{\ddagger}$ on pressure, are in

Table 5.2. Rate constants for complexation reaction on $[V(OH_2)_6]^{3+}$ at 298 K [27, 28]

Incoming ligand	pK_a	$k_f/(M^{-1}s^{-1})$	Ref.
Cl^-		< 3	[27a]
Br^-		< 10	[27a]
5-Nitrosalicylic acid		3.5	[27b]
Salicyclic acid		4.9	[27c]
NCS^-	− 1.84	114, 104	[28a, b, c]
H_2O	− 1.74	54^a	[26]
$HC_2O_4^-$	1.3	1.3×10^2	[27a]
5-Nitrosalicylate(1-)	2.24	1.23×10^3	[27b]
Salicylate(1-)	2.80	1.4×10^3	[27c]
Malonate(1-)	2.83	3.8×10^3	[27d]
4-Aminosalicylate(1-)	3.6	7×10^3	[27e]
1-Cysteine		1.12×10^3 (pH 8.0)	[27f]

^{a} Second-order rate constant computed as $6k_{ex}/55.56$.

accordance with the expected associative interchange mechanism occurring on expanded low d-electron population trivalent ions (see the previous chapter and Section 1.4). Correspondingly, rate constants for complex forming reactions on $[V(OH_2)_6]^{3+}$ (298 K) cover a significantly wide range (Table 5.2) [27,28]. A feature of substitution reactions on $[V(OH_2)_6]^{3+}$, as found on $[Ti(OH_2)_6]^{3+}$, is also the absence of a pathway involving hydrolysis products such as $[V(OH_2)_5(OH)]^{2+}$ [27, 28]. Such species are known to promote dissociatively activated pathways to substitution and their insignificance here is believed to reflect the dominance of the associative pathway promoted by the low t_{2g} orbital population and relatively large ionic radius of these early transition metal ions. In one case, however, the preferential reaction of $[V(OH_2)_5(OH)]^{2+}$ with hydrazoic acid, HN_3, over $[V(OH_2)_6]^{3+}$ reacting with N_3^- (the proton ambiguity problem) is inferred from the measurement of $\Delta S^{\ddagger}$ for the reverse hydrolysis reaction of the product $[V(OH_2)_5N_3]^{2+}$ [29].

5.1.2.5   REDOX REACTIONS INVOLVING $[V(OH_2)_6]^{3+}$

Few redox reactions involving $[V(OH_2)_6]^{3+}$ have been reported. Strong reductants are required to form V^{II} so that only reaction, with, for example, Cr^{2+}(aq) have been studied in any detail. Rate laws are characterised by $[H^+]^{-1}$ terms which would tend to imply involvement of $[V(OH_2)_5(OH)]^{2+}$, possibly in an inner-sphere process. In the reaction between $[V(OH_2)_6]^{3+}$ and O_2 the rate equation is first order in both reactants [30]. A two-electron process giving V^V and HO_2^- has been suggested, rapid reaction of V^V with $[V(OH_2)_6]^{3+}$ being responsible for generating the final product $[VO(OH_2)_5]^{2+}$. The existence of binuclear intermediates has not been substantiated (cf. the reaction of $[V(OH_2)_6]^{2+}$. Solutions of $[V(OH_2)_6]^{3+}$ cannot be prepared in non-complexing ClO_4^- media owing to the slow accompanying redox reaction to give $[VO(OH_2)_5]^{2+}$. At elevated temperatures ($> 45\,°C$) reactions are fast enough to enable full kinetic studies. Indeed, under these conditions $[V(OH_2)_6]^{3+}$ and $[V(OH_2)_6]^{2+}$ react with ClO_4^- at comparable rates which has tended to complicate study of the latter [31]. Kinetic studies of the oxidation of $[V(OH_2)_6]^{3+}$ with Fe^{III}, Co^{III}, Np^V and Tl^{III}, giving $[VO(OH_2)_5]^{2+}$, have been reported [32]. The reaction with Fe^{III} is catalysed by Cu^{II}, rate laws again being characterised by the presence of $[H^+]^{-1}$ terms [33]. Despite reaction of $[V(OH_2)_6]^{3+}$ with Fe^{III} occurring 10 times slower than with Co^{III} the rates of both reactions remain comparable enough to that of substitution to suggest the possibility of inner-sphere processes [34]. The reaction with Tl^{III} is believed to occur via stepwise one-electron processes involving formation of unstable Tl^{II}.

Large concentrations of vanadium (up to $\sim 10\,000$ ppm), present almost certainly as $[V(OH_2)_6]^{3+}$ with some $[VO(OH_2)_5]^{2+}$ (see below), are accumulated within the cells of certain marine creatures called tunicates or 'sea squirts' [1]. Significant concentrations of SO_4^{2-} ions are also present. The precise role of these high vanadium concentrations remains a mystery. It has been proposed that the vanadium is concentrated as part of the organism's immune response function, perhaps acting as a kind of protection against surface fouling via generation of reactive peroxo V^V species.

5.1.3   Vanadium(IV) (d¹): Chemistry of the $[VO(OH_2)_5]^{2+}$ Ion

In contrast to Ti^{IV}, the $V{=}O^{2+}$ group dominates the solution chemistry of V^{IV} and is present in the sky-blue mononuclear ion $[VO(OH_2)_5]^{2+}$. V^{IV} is the stable oxidation state for vanadium in aqueous solution in the air. Surprisingly, however, much still remains to be understood regarding the solution dynamics of $[VO(OH_2)_5]^{2+}$ due to the problem of separating base-catalysed processes involved with exchange at the terminal $V{=}O$ group from those involved with substitution at the water ligands. There also exits the possibility of both intra- and intermolecular ligand-exchange processes.

5.1.3.1   PREPARATION OF $[VO(OH_2)_5]^{2+}$

Aqueous solutions containing sky-blue $[VO(OH_2)_5]^{2+}$ can readily be prepared via mild reduction (e.g. with SO_2) [35] of acidified solutions of V^V (see below). The interaction of V_2O_5 with ethanolic HCl gives rise to solutions containing $[VOCl_5]^{3-}$ from which $[VO(OH_2)_5]^{2+}$ can be obtained via simple aquation of coordinated Cl^- ions [36]. The dark blue amphoteric oxide VO_2 also dissolves in non-complexing acids to give $[VO(OH_2)_5]^{2+}$. However, the most convenient method is via simple aquation of commercial vanadyl sulphate 'VOSO$_4$' in, for example, 2.0 M $HClO_4$ or triflic acid followed by DOWEX 50W X2 cation-exchange chromatography (dilution to $[H^+]$ 0.05 M, loading and elution with 2.0 M of the required acid). Solutions containing up to 0.2 M in VO^{2+} can be readily prepared via elution from a saturated short column (length 5 × height) [37]. From solutions in $HClO_4$ the blue deliquescent salt $[VO(OH_2)_5](ClO_4)_2$ has been isolated [38].

5.1.3.2 ELECTRONIC AND MOLECULAR STRUCTURE OF $[VO(OH_2)_5]^{2+}$

A number of X-ray structural studies have been carried out on the hydrated sulphates, $VOSO_4 \cdot nH_2O$ ($n = 3, 5$ or 6). In the structure of $VOSO_4 \cdot 6H_2O$ discrete $[VO(OH_2)_5]^{2+}$ and $[VO(OH_2)_4]^{2+}$ (square pyramidal) ions are both present [39]. For the lower hydrates coordination of sulphate occurs in the equatorial position *cis* to the V=O group, leading to three-dimensional networks linked by sulphate bridges [40].

The electronic spectrum of $[VO(OH_2)_5]^{2+}$ shows a broad band centred around 760 nm (13 200 cm^{-1}, $\varepsilon = 17.2$ M^{-1}cm^{-1}) with a distinct shoulder on the high energy side [37]. In addition strong absorptions are seen occurring below 500 nm. The single unpaired electron resides in the d_{xy} orbital (b_2 symmetry) and three transitions are assigned [41]:

$$b_2(xy) \longrightarrow e(xz, yz), \quad 680\text{--}910 \text{ nm}$$
$$b_2(xy) \longrightarrow b_1(x^2-y^2), \quad ca. \ 670 \text{ nm}$$
$$b_2(xy) \longrightarrow a_1(z^2), \quad 313\text{--}476 \text{ nm}$$

The triply bonded V=O group is explained by two components: a σ component $(sp_\sigma(O) \rightarrow (4s + 3d_{z^2} (V))$ and a π component $(p_x, p_y(O) \rightarrow 3d_{xz}, 3d_{yz} (V))$. The four equatorial waters are described via the σ-bonded component $(sp_\sigma(O) \rightarrow 4p_x, 4p_y, 3d_{x^2-y^2} (V))$. The weakly bonded axial water is described via an interaction with the vanadium $4p_z$ orbital. $[VO(OH_2)_5]^{2+}$ gives the characteristic eight-line ESR pattern, characteristic of VO^{2+} compounds, with g values in the range 1.95–2.00, the eight lines arising from strong coupling of the single d_{xy} electron to the ^{51}V nucleus (100%, $I = 7/2$) [41].

The multiple bonding in the V=O group is evident in the short V—O distance e.g. 159.4 pm in $VOSO_4 \cdot 5H_2O$ [40]. A characteristic V—O stretching vibration is observed in the i.r. spectrum in the region 950–1000 cm^{-1}, the exact frequency being very sensitive to whether a ligand is present in the axial position. The vanadium atom is usually found lying between 10 and 30 pm above the plane defined by the equatorial ligands and many compounds are known to have square pyramidal geometry with no axial ligand. These tend to be characterised by having V—O stretching frequencies around the upper limit of 1000 cm^{-1}. Bond lengths to the axial ligand, when present, are correspondingly long, V—OH_2(a) = 228.4 pm in $VOSO_4 \cdot 5H_2O$, with the bonds in the equatorial

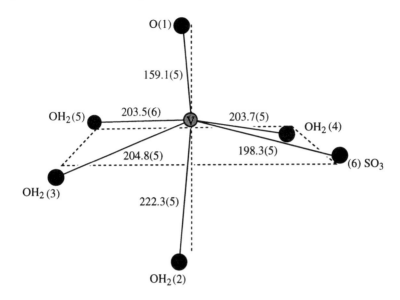

Figure 5.9. Structure of the VO^{2+} unit in $VOSO_4 \cdot 5H_2O$

positions intermediate, $V\!-\!OH_2(e) = 200\text{--}205.6$ pm (Fig. 5.9) [40]. Average $V\!-\!O$ bond distances in $[VO(OH_2)_5]^{2+}$, investigated in solution by electron spin echo modulation (ESEM) and electron nuclear double resonance (ENDOR) techniques [42], are consistent with the above range of values.

5.1.3.3 HYDROLYSIS OF $[VO(OH_2)_5]^{2+}$

$[VO(OH_2)_5]^{2+}$ undergoes hydrolysis above pH 4 to give mononuclear $[VO(OH_2)_4(OH)]^+$ ($pK_{11} = 5.67$) and dinuclear $(VO)_2(OH)_2^{2+}(aq)$ ($pK_{22} = 6.67$) [43]. Between pH 6 and 10 anions such as $(VO)_2(OH)_5^-(aq)$ are formed. At vanadium concentrations $> 10^{-2}$ M above pH 4 grey hydrous VO_2 precipitates but redissolves above pH 10 to give clear solutions from which brown–black salts such as $K_{12}[V_{18}O_{42}] \cdot 24H_2O$ can be isolated [44]. The retention of the $V\!=\!O$ group in this cluster anion (Fig. 5.10), highlights the remarkable stability of this entity over a large range of pH and vanadium concentrations (compare Ti^{IV}).

5.1.3.4 WATER EXCHANGE AND SUBSTITUTION ON $[VO(OH_2)_5]^{2+}$

Oxygen-18 labelling studies conducted by Johnson and Murmann [45] have examined oxygen exchange on the $V\!=\!O$ group and have concluded that the species $[VO(OH_2)_4(OH)]^+$ is involved in addition to $[VO(OH_2)_5]^{2+}$. A two-

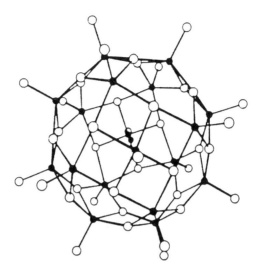

Figure 5.10. Structure of the $[V_{18}O_{42}]^{12-}$ anion in $K_{12}[V_{18}O_{42}] \cdot 24H_2O$ [44]. (Reprinted with permission from G. K. Johnson and E. O. Schlemper, *J. Am. Chem. Soc.*, **100**, 3645, 1978. Copyright 1978 American Chemical Society)

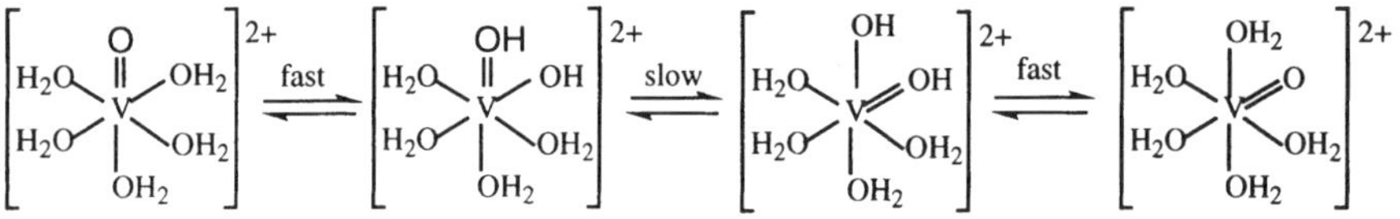

Figure 5.11. Mechanism of exchange at the terminal V=O group of $[VO(OH_2)_5]^{2+}$

term rate law has been found:

$$\text{Rate} = (k_0 + k_{OH} K_a [\text{H}^+]^{-1})[\text{VO}^{2+}] \tag{5.5}$$

At $0\,^\circ C$ and $I = 2.5$ M, $k_0 = 2.4 \times 10^{-5}\,\text{s}^{-1}$ and $k_{OH} = 1.3\,\text{s}^{-1}$, with K_a given as 4×10^{-7} M. A mechanism has been suggested (Fig. 5.11) involving rapid proton transfer and an internal electronic rearrangement involving the V=O group [46]. Within the scheme the V=O group is transformed into a more labile terminal water. The highly favoured base-catalysed pathway involving $VO(OH)^+(aq)$ contrasts with the acid-catalysed path involved in the corresponding exchange at the more labile Ti=O group in $TiO^2(aq)$; this difference in mechanism is believed to be responsible for the $\sim 10^9$ times difference in magnitude of exchange rate.

The influence of the VO^{2+} group can be seen on the relative labilities of the two types of water ligand as measured by ^{17}O NMR [47, 48]. The weakly bonded axial water is extremely labile (k_{ex} (a) $> 5 \times 10^8 \, s^{-1}$) with those in the equatorial positions somewhat more inert (k_{ex} (e) $= 5 \times 10^2 \, s^{-1}$), the exchange rate constants correlating with the respective V—O lengths. One of the difficulties, however, with the equatorial water-exchange rate is knowing whether it represents intermolecular bulk water exchange or contains a component arising from intramolecular exchange via the highly labile axial site. This is a problem not easily resolved although Saito and Sasaki have proposed the existence of the intramolecular exchange process on the basis of 10^3 times slower rates of substitution at a single equatorial water in a number of VO^{2+} compounds wherein the axial and all the other equatorial sites were taken up by a chelating ligand (Fig. 5.12). Rate constants *ca.* $0.1–1.0 \, s^{-1}$ so obtained for direct equatorial substitution on VO^{2+} imply a $> 10^8$ times difference in lability between axial and equatorial sites.

Calculations based on the data from a number of independent substitution studies have shown that a narrow range of bimolecular rate constants, of the order *ca.* $10^3 \, M^{-1} \, s^{-1}$, is obtained if consideration of a significant contribution from $VO(OH)^+$(aq) is made [50]. This has been used to suggest the presence of a dissociative mechanism. Unfortunately the contribution involving $VO(OH)^+$(aq) is difficult to assess since the precise nature of the $VO(OH)^+$(aq) species itself is uncertain, the studies indicating the base-catalysed exchange at the V=O group could be of a comparable rate to that of direct equatorial ligand exchange. Finally, a pressure-independent volume of activation ($+ 1.9 \, cm^3 \, mol^{-1}$) was recently determined for water exchange on $[VO(OH_2)_5]^{2+}$ presumed at an equatorial site [48]. No allowance was made for involvement of $VO(OH)^+$(aq). The small positive value could, however, be consistent with the proposed dissociative mechanism if consideration is given of initial substitution occurring at the more labile, and expanded, axial site. Here the configuration

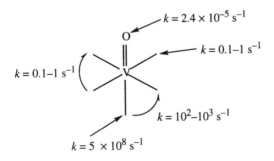

Figure 5.12. Oxygen exchange rate constants for different sites on $[VO(OH_2)_5]^{2+}$

might be similar to that of the transition state and thus could explain the small activation volume occurring even for a predominantly dissociative process (see also the discussion of water exchange on Cu^{2+}(aq), Section 11.1.2).

5.1.3.5 REDOX REACTIONS INVOLVING $[VO(OH_2)_5]^{2+}$

$[VO(OH_2)_5]^{2+}$ can be both oxidised and reduced but requires reasonably powerful reagents. Reduction of $[VO(OH_2)_5]^{2+}$ has been carried out with a number of one-electron reagents, V^{2+}(aq), Cr^{2+}(aq), Eu^{2+}(aq) and Cu^{+}(aq) [11,12]. Studies indicate the formation of oxo-(or hydroxo)-bridged binuclear intermediates via inner-sphere processes [12]. The reaction with $[V[OH_2)_6]^{2+}$, as already described, gives the stable dinuclear ion $[(H_2O)_5VOV(OH_2)_5]^{4+}$ (μ-oxo bridge). The standard reduction potential for the VO^{2+}/V^{3+} couple at 25 °C has been quoted as $+0.366$ V vs NHE.

A number of reactions involving oxidation of $[VO(OH_2)_5]^{2+}$ have been shown to involve $[VO(OH_2)_4(OH)]^{+}$ in addition to $[VO(OH_2)_5]^{2+}$, the hydroxy form being more reactive (by a factor of 10^5 in rate) [5]. Such behaviour is common in reflecting the more 'hydrolysed' nature of the product, V^V. The creation of a hydroxy ligand *cis* to the V=O group may well be the precursor for generation of the final *cis*-dioxo VO_2^{+} product (see below). A roughly linear relationship holds between $\log k(Fe^{2+})$ and $\log k(VO(OH)^+)$ for reduction of a number of macrocyclic Ni^{III} and polypyridine M^{III} complexes (M = Fe, Ru and Os) [51]. The standard potential for the VO_2^{+}/VO^{2+} couple at 25 °C is $+1.00$ V vs NHE.

5.1.4   Vanadium (V) (d⁰)

The aqueous chemistry of V^V is complex, dominated for vanadium concentrations $> 10^{-2}$ M in the pH range 2–13 by a range of polynuclear anionic species [1, 52]. Altogether twelve contributing species have been identified. At either extreme of pH, mononuclear species are present: yellow *cis*-dioxo $[VO_2(OH_2)_4]^{+}$ (pH < 2) and colourless tetrahedral $[VO_4]^{3-}$ (pH > 13). For very high vanadium con-centrations (> 1.0 M) brick-red amphoteric V_2O_5 precipitates between pH 5 and 9.

5.1.4.1   PREPARATION AND PROPERTIES OF *cis*-$[VO_2(OH_2)_4]^{+}$

The only true aqua ion for V^V exists as yellow *cis*-dioxo $[VO_2(OH_2)_4]^{+}$ below pH 2. Preparation is simply achieved via acidification of solutions of salts such as NH_4VO_3 or $Na_6[V_{10}O_{28}]\cdot 18H_2O$. In some cases V_2O_5 has been used as the source of V^V [53]. In a preparation described by Rahmoeller and Murmann [54], 3.5 g of $Na_6[V_{10}O_{28}]\cdot 18H_2O$, dissolved in 75 cm³ of water, was passed through a column of DOWEX 50W X8 resin in its H^+ form (2.5-fold capacity). The column was rinsed with $50 \, cm^3$ of water and the combined eluents acidified with $\sim 15 \, cm^3$ of concentrated 70% $HClO_4$. Following dilution, solutions of $[VO_2(OH_2)_4]^+$ (~ 0.2 M in vanadium) could be readily obtained in 1.0 M $HClO_4$. Standardisation of VO_2^+ solutions is usually performed via titration with Fe^{II}. The *cis*-dioxo formulation is believed to arise in order to maximise d_π–p_π bonding. From structural studies conducted on a number of *cis*-VO_2^+ complexes containing EDTA and oxalate ligands the O—V—O angle is in the range 104–107° [55].

5.1.4.2 HYDROLYSIS OF *cis*-$[VO_2(OH_2)_4]^+$

At concentrations below 2×10^{-5} M only mononuclear species occur, namely $[VO(OH)_3]$ (pH 3–4, $pK_{11} = 3.3$) [56], $[VO_2(OH)_2]^-$ (pH 4–8, $pK_{12} = 7.3$) [56] $[VO_3(OH)]^{2-}$ (pH 8–13) [57] and $[VO_4]^{3-}$ (pH > 13) [58], a change from octahedral to tetrahedral coordination at the vanadium occurring as the hydrolysis progresses [59]. For vanadium concentrations > 10^{-2} M polynuclear species are formed between pH 2 and 13. The predominant species in the pH range 2–7 is the decavanadate ion $[V_{10}O_{28}]^{6-}$ [60], together with a range of protonated derivatives. The pK value ($25 \, °C$) for the overall hydrolysis reaction represented by the following equation is ~ 10.7 [61]:

$$10 \, VO_2^+ + 8 \, H_2O \longrightarrow [H_2V_{10}O_{28}]^{4-} + 14 \, H^+ \qquad (5.6)$$

The pK values for the further two deprotonations to give $[V_{10}O_{28}]^{6-}$ are respectively 4.34 and 6.94 [59]. The ready assembly of the $[V_{10}O_{28}]^{6-}$ unit, occurring upon either acidification of solutions of $[VO_4]^{3-}$ or upon raising the pH of $[VO_2(OH_2)_4]^+$ solutions, is clearly a remarkable process. A number of mechanisms for its assembly and subsequent destruction have been suggested following measurement of the exchange processes involving the various oxygen sites [62]. The seven structurally distinct oxygen atoms have been identified by ^{17}O NMR (Fig. 5.13) and exchange at similar, but probably not identical, rates [16, 64]. The exchange half-life has been found to be sensitive to the vanadium concentration which supports a mechanism involving concerted breakdown to a half-opened intermediate [64] rather than extensive breakdown to produce smaller V_4 and V_3 units. In the pH range 7–9 a trimer, probably $[V_3O_{10}]^{5-}$ [57], and tetramer $[V_4O_{12}]^{4-}$ [65] are believed to form. Between pH 9 and 13 dimeric forms $[V_2O_7]^{4-}$ and $[HV_2O_7]^{3-}$ are eventually produced [58]. Similarities to the behaviour of P^V have been noted and indeed the toxicity of V^V may well relate to its disruption of biological phosphorylation/dephosphorylation processes. Indeed, V^V is established as an inhibitor of certain phosphorylase enzymes [66]. V^V is, however, well established as an essential component of a number of mammalian haloperoxidases [1].

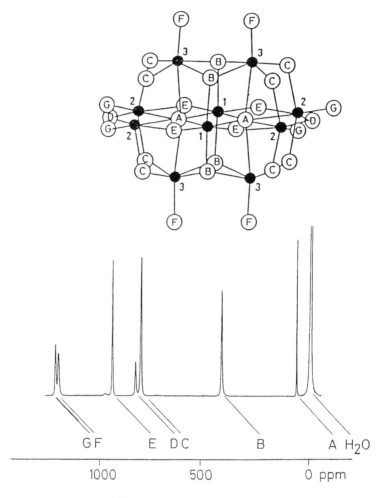

Figure 5.13. Assignment of ^{17}O NMR signals to the structure of the decavanadate ion, $[V_{10}O_{28}]^{6-}$ [64]. (Reproduced by permission of Verlag Helvetica Chimica Acta AG)

5.1.4.3   SUBSTITUTION REACTIONS INVOLVING cis-$[VO_2(OH_2)_4]^+$

Rate constants for terminal water exchange on cis-$[VO_2(OH_2)_4]^+$ are too fast to be measured by ^{18}O labelling techniques. The rate constant for exchange at the cis-$V{=}O$ group has, however, been estimated as ~ 4.7 s^{-1} (0 °C) by an ^{18}O labelling method involving competition from rapid complexation and reduction by NCS$^-$ to produce slow exchanging $[VO(NCS)_5]^{3-}$ [54]. Second-order rate constants for complex formation with a number of amino carboxylate chelates on $[VO_2(OH_2)_4]^+$ are contained within a fairly narrow range, 10^4–10^6 M^{-1} s^{-1},

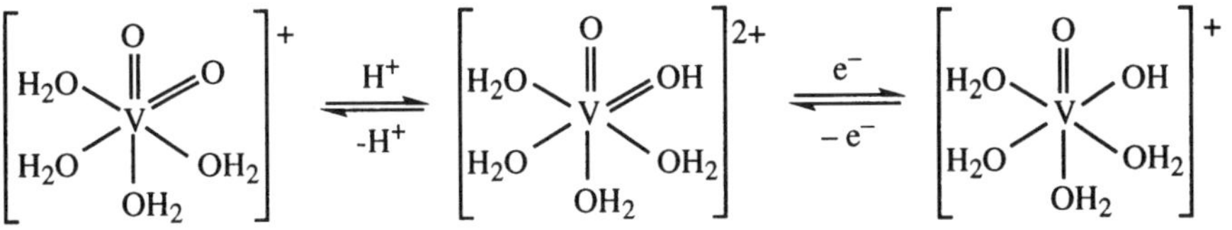

Figure 5.14. Mechanism for one-electron reduction of $[VO_2(OH_2)_4]^+$

implying an I_d process [49]. Moreover, the differences observed have been attributed to differences in ion pair association constants, K_{os}. A number of complex forming reactions have also been performed on the anions, $VO_2(OH)_2^-$(aq) and $VO_3(OH)^{2-}$(aq), in the pH range 8–10 [49]. Solutions containing the anions were readily prepared by adjusting aqueous solutions of $Na_6[V_{10}O_{28}]\cdot18H_2O$ to pH 9.0 and allowing to stand until the bright yellow colour of decavanadate fades. Determination of V^V is normal carried out via the formation of brightly coloured peroxo complexes on treatment with H_2O_2 [67].

5.1.4.4 REDOX REACTIONS INVOLVING *cis*-$[VO_2(OH_2)_4]^+$

A number of reductions of *cis*-$[VO_2(OH_2)_4]^+$ show acid catalysis involving no rate saturation at high $[H^+]$. A protonation equilibrium [68] (Fig. 5.14) is indicated wherein the two forms of V^V react separately. The $[VO(OH)(OH_2)_4]^{2+}$ form is set up for one-electron reduction to the corresponding V^{IV} species, $VO(OH)^+$(aq), and as such an extended Marcus treatment has been successfully applied to outer-sphere reactions of the two forms and a self-exchange rate constant for the $[VO(OH)(OH_2)_4]^{2+/+}$ couple was estimated as $\sim 1 \times 10^{-3}$ M^{-1} s^{-1} [51]. Reductants used for the $[VO_2(OH_2)_4]^+$ reactions included a number of trisbipyridyl and tris-1,10-phenanthroline Os^{II} complexes as well as $[V(OH_2)_6]^{2+}$.

5.2 NIOBIUM AND TANTALUM

The chemistry of these two elements is reminiscent of the heavier Group 4 members in being very similar [69,70]. However, the one single feature that characterises the coordination chemistry of these two elements is the propensity for forming polynuclear oxo anionic species in the highest oxidation state V, cf. vanadium, and for forming metal–metal bonded cluster species in the lower oxidation states, II–IV. Indeed, the chemistry of the lower oxidation state species has only developed very recently in the case of niobium and is even less developed in the case of tantalum. One of the principal reasons behind this is the ease of oxidation to the V state, with the tantalum species generally more reactive than those of niobium. Nonetheless, considerable progress in opening up the aqueous chemistry of these elements is being made with the recent preparation of the first authentic niobium 'aqua ion', albeit as a mixed-valence trinuclear cluster species. In discussing the historical developments in the aqueous chemistry of these elements it will therefore be pertinent to discuss firstly the chemistry of V state representative species.

5.2.1 Niobium(V) and Tantalum(V) (d^0)

5.2.1.1 CHEMISTRY OF ISOPOLYNIOBATES

The hexaniobate anion $[H_xNb_6O_{19}]^{(8-x)-}$ dominates in aqueous Nb^V solutions throughout the pH range above 7.0. Aqueous solutions are generally prepared via dissolution of freshly precipitated Nb_2O_5 in alkali. The structural unit has been characterised in salts such as $Na_7H[Nb_6O_{19}]\cdot15H_2O$ (Fig. 5.15) [71]. The conservation of the same structural unit in solution has been verified by a number of ultracentrifugation, light scattering, Raman [72] and ^{17}O NMR studies (Fig. 5.15) [73]. Successive protonation constants (log K) for $[Nb_6O_{19}]^{8-}$ in 1 M KCl are respectively 12.3, 10.94 and 9.71 [74]. In strong alkali (1–12 M KOH), degradation into tetrameric and monomeric species is reported [72]. The decaniobate anion, analogous to that of vanadium, has been characterised

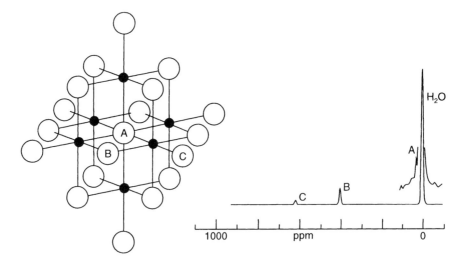

Figure 5.15. Structure of the hexaniobate anion $Nb_6O_{19}{}^{8-}$ and ^{17}O NMR spectrum of a 5 m solution (natural abundance) in water [73]. (Reprinted with permission from M. Filowitz, R. K. C. Ho, W. G. Klemperer and W. Shum, *Inorg. Chem.*, **18**, 93, 1979. Copyright 1979 American Chemical Society)

in salts such as $(NMe_4)_6[Nb_{10}O_{28}] \cdot 6H_2O$ [75]. The corresponding hexatantalate anion, $[Ta_6O_{19}]^{8-}$, has been structurally characterised in the salt $K_4Na_2H_2$ $[Ta_6O_{19}] \cdot 2H_2O$ [76] and in solution by ultracentrifugation, light scattering, Raman [72] and ^{17}O NMR measurements [73]. Oxygen exchange with free water on $[Ta_6O_{19}]^{8-}$ occurs more slowly than on the niobium analogue, a feature to be repeated in the properties of subsequent second and third row members. Unlike on decavanadate, however, oxygen exchange on both $[Nb_6O_{19}]^{8-}$ and $[Ta_6O_{19}]^{8-}$ occurs much faster at the terminal oxo groups than at the bridging ones. Protonation constants (log K) for $[Ta_6O_{19}]^{8-}$, respectively 12.68, 10.81 and 9.28 [74], are similar to those for hexaniobate.

Despite early reports of species such as $Nb(OH)_4^+$ (aq) existing at high dilution [77] no corresponding cationic chemistry exists for either element in the V state in acidic solution, acidification leading in all cases to quantitative precipitation of the hydrated pentoxide. Certain ions such as F^-, Cl^- [78] and oxalate [79] are, however, highly effective in solubilising Nb^V in acidic media, wherein a number of anionic 'aqua' complexes have been characterised. Their use in dissolution procedures for the often highly refractory minerals that contain niobium and tantalum have been well documented. Highly coloured complexes such as $NbO(OH)_2(C_6H_7O_6)(C_6H_7O_6 =$ ascorbate$^-$) have been used for the colorimetric determination of niobium [80] and could be considered as deriving from the putative 'aqua ion' $[NbO(OH_2)_5]^{3+}$, as could the oxalato complexes $[NbO(OH_2)_2(C_2O_4)_2]^-$ [81] and $[NbO(OH)(OH_2)(C_2O_4)_2]^{2-}$ [82]. Ta_2O_5 also dissolves in oxalic acid but here the species are less well characterised owing to the formation of polynuclear species [83]. An example is the 'adamantane-like' tetranuclear anion, present in $(NEt_4)_4[Ta_4O_6F_{12}]$ [84], which is formed under conditions where niobium gives mononuclear and dinuclear species. Finally, the recent successful isolation of the tetrathianiobate anion, $[NbS_4]^{3-}$, by Lee and Holm [85], following on from the earlier synthesis and characterisation of the μ-S cluster species $[Nb_6S_{17}]^{2-}$ [86], may ultimately lead, on reduction, to the synthesis and characterisation of a number of low oxidation state oxo-sulphido niobium cluster aqua species (cf. the section on molybdenum and tungsten, Section 6.2.3). Recently a trinuclear μ-S $Nb^{III,III,IV}$ complex with stoichiometry $[Nb_3S_7(C_5Me_5)_3]$ was isolated from the reaction of $(C_5Me_5)NbBH_4$ with elemental sulphur [87] and could represent the first authentic sample of triangular μ-S cluster of the lower states. The structure of this complex has not been determined. Nonetheless, it is within the lower oxidation states for niobium that considerable progress is being made in developing a limited 'cationic' aqueous chemistry.

5.2.2 Trinuclear Niobium(III,IV,IV) Cluster Species

Niobium has a propensity for forming triangular cluster complexes and, in particular, where the formal oxidation state over the three niobium atoms is 3.67, $Nb^{III,IV,IV}$. The arrangement of only four d electrons about the three Nb—Nb

Table 5.3. Structurally characterised triniobium cluster complexes

Complex	Nb—Nb (pm)	Formal oxidation state	Nb—Nb bond order	Ref.
$[Nb_3(\mu_2\text{-}O)_2(OH)_2(O_2CH)_3Cp^*_3]$	314	IV, IV, IV	1/3	[88]
$[Nb_3(\mu_3\text{-}O)_2(O_2CCH_3)_6(THF)_3]^+$	283	III, IV, IV	2/3	[89]
$[Nb_3(\mu_3\text{-}O)_2(O_2CCMe_3)_6(THF)_3]^+$	284	III, IV, IV	2/3	[90]
$[Nb_3(\mu_3\text{-}O)_2(\mu_2\text{-}SO_4)_6(OH_2)_3]^{5-}$	287	III, IV, IV	2/3	[91]
$[Nb_3(\mu_3\text{-}S)(\mu_2-O)_3(NCS)_9]^{6-}$	276	III, IV, IV	2/3	[92]
Nb_3Cl_8	281	III, III, II	2/3	[93]
$[Nb_3Cl_{10}(PEt_3)_3]^-$	298	III, III, III	1	[94]

bonds means that these triangular Nb_3 species are electron deficient. The properties of a number of such examples are shown in Table 5.3. A wide range of bridging and capping groups is seen ranging from Cl^-, O^{2-}, OH^- and S^{2-} to SO_4^{2-} and RCO_2^- (R = H, Me, CMe_3).

In each case an average Nb—Nb separation of 282 ± 5 pm correlates with the presence of the formal 3.67 oxidation state and a formal bond order of 2/3. The trinuclear niobium complexes $[Nb_3Cl_{10}(PEt_3)_3](Nb^{III,III,III})$ and $[Nb_3(\mu\text{-}O)_2\text{-}(\mu\text{-}OH)_2(\mu\text{-}O_2CH)_3(Cp^*)_3]$ ($Nb^{IV,IV,IV}$) have significantly longer Nb—Nb separations of 314 and 298 pm respectively. The structure of the complex $[Nb_3(\mu_3\text{-}S)(\mu\text{-}O)_3(NCS)_9]^{6-}$ (Fig. 5.16), prepared by Cotton *et al.*, [92] is interesting in that it contains the same core structure as present in a number of well-character-

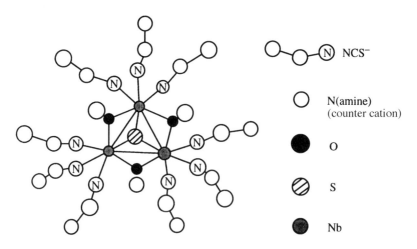

Figure 5.16. Structure of the anion $[Nb_3(\mu_3\text{-}S)(\mu_2\text{-}O)_3(NCS)_9]^{6-}$

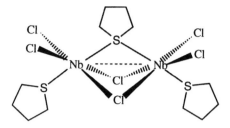

Figure 5.17. Structure of Nb_2Cl_6 (tetrahydrothiophen)$_3$

ised trinuclear oxo-sulphido aqua ion cluster of Mo^{IV} and W^{IV} (see Section 6.3).
Cotton's complex was prepared via hydrolysis of the dimeric $Nb^{III,III}$ complex
Nb_2Cl_6 (tetrahydrothiophen)$_3$ [95] (Fig. 5.17) in aqueous HCl followed by
DOWEX 50W X2 cation-exchange purification and treatment with NCS^- ions.
Two sources exist therefore for the origin of the μ_3-S group. Cotton and
coworkers reported that the species elutable from the DOWEX column was
green in colour and postulated that it might contain the aqua ion $[Nb_3(\mu_3\text{-}O)(\mu_2\text{-}O)_3(OH_2)_9]^{3+}$ (assumed to be $Nb^{III,IV,IV}$ as in the final complex). Since the most
likely source of S^{2-} in the acidic media employed is from NCS^- it would seem
likely that S^{2-} enters the trinuclear core during or following the complexation
reaction with NCS^-. However, more recent results from our own laboratory
have shed further light on the possible structure for the green aqua species.

5.2.3 Cationic 'Aqua Ions' of Niobium

5.2.3.1 CHARACTERISATION OF $[Nb_3(\mu_3\text{-}Cl)(\mu\text{-}O)_3(OH_2)_9]^{4+}$

A further lead-in compound has proved to be the complex '$NbCl_3$(1,2-
dimethoxyethane)' [96] marketed by Aldrich since 1989 as a Lewis acid catalyst
for promoting certain C—C bond forming reactions. It is made by air-free
reduction of sublimed anhydrous niobium pentachloride with tri-n-butyltin
hydride in 1,2-dimethoxyethane solvent. The '$NbCl_3$(1,2-DME)' complex separ-
ates as an air-sensitive 'brick-red' precipitate that can be readily filtered off,
washed with 1,2-DME and pentane and dried under vacuum. $NbCl_3$(1,2-DME)
seems not to have been structurally characterised since it appears to only dissolve
in solvents with which it reacts. It is a $12e^-$ compound and hence doubts are cast
over the mononuclear formulation (Fig. 5.18), given the tendency for Nb^{III} to
form metal–metal bonded chloride-bridged dimers and trimers. Nonetheless, this
complex dissolves readily in air-free HCl solutions (4–12 M) accompanied by
spontaneous hydrolytic decomposition to produce a dark brownish-green species which, on dilution to 0.5 M in $[H^+]$, can be chromatographed on a DOWEX
50W X2 cation-exchange column. Elution with 2–4 M solutions of HCl yields

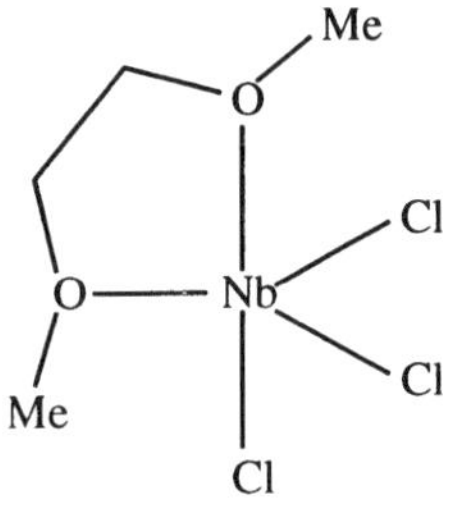

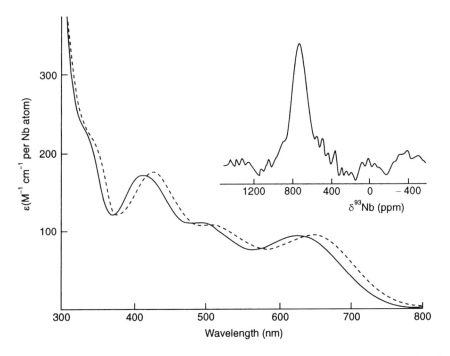

Figure 5.18. Proposed structure for $NbCl_3(1,2\text{-DME})$

Figure 5.19. Electronic spectrum of the green niobium aqua ion in (------) 4M HCl and (——) 4 M Hpts (Inset ^{93}Nb NMR spectrum in 4M Hpts, ppm vs $NbCl_6^-$)

air-sensitive green solutions [97] which exhibit the electronic spectrum in Fig. 5.19. Band maxima (4 M HCl) are observed at 427 nm ($\varepsilon = 170$ M^{-1} cm^{-1} per Nb) and 634 nm (80) with shoulders at ~ 330 and ~ 500 nm. The green species can also be eluted with 4 M Hpts wherein the band maxima shift to higher energy, 414 and 622 nm (Fig. 5.19) (negligible change in ε value), implying H_2O substitution for Cl$^-$ within the chromophore of a niobium aqua species. Niobium was determined gravimetrically by oxidation and quantitative hydrolysis to the hydrated pentoxide, filtration and ignition to anhydrous Nb_2O_5. A single rather broad ^{93}Nb NMR resonance is seen (Fig. 5.19 inset) ~ 800 ppm downfield from the reference line of $NbCl_6{}^-$. An improved method for generating highly concentrated solutions of the green species has been developed. This involves evaporation of the HCl eluates to dryness on a vacuum line at 10^{-3} torr, leaving a green residue, presumed to be a chloro aqua complex, which is then dissolved in solutions of the required acid. In this way solutions of the green ion > 0.1 M in Nb have been obtained in 3 M CF_3SO_3H/H_2O^{17} for the purpose of obtaining the ^{17}O NMR spectrum shown in Fig. 5.20 [97]. In these solutions Mn^{2+} (~ 0.1 M) is added in order to relax out the large signal due to free water at 0 ppm. Four resonances are observed which are assigned as follows: 34 and -15 ppm (coordinated water), 165 ppm ($CF_3SO_3{}^-$, natural abundance) and 305 ppm

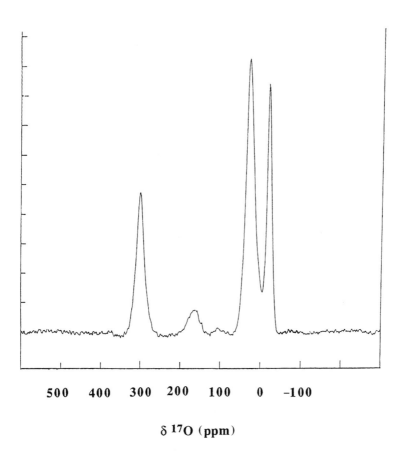

500   400   300   200   100   0   -100

$\delta\ ^{17}O$ (ppm)

Figure 5.20. 40.56 MHz ^{17}O NMR spectrum for the green niobium aqua ion in 3 M CF_3SO_3H (5 atom % ^{17}O enrichment, chemical shifts relative to free $H_2O = 0$ ppm

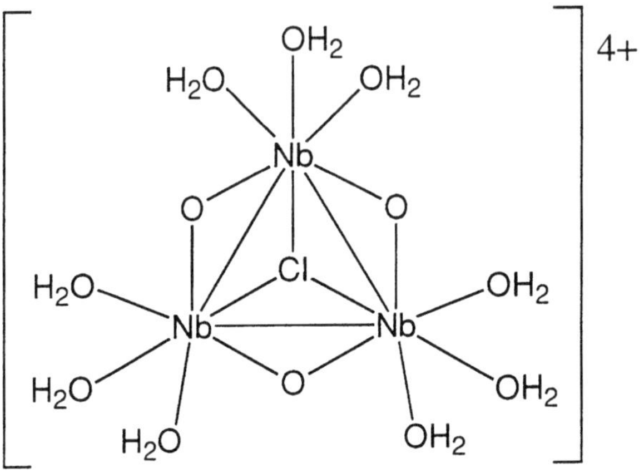

Figure 5.21. Proposed structure for the green Nb(aq) cation

(bridging O^{2-} (OH^-)). The two-coordinate water resonances are in a 2:1 ratio consistent with the two non-equivalent terminal sites that would be present in a triangular structure such as shown in Fig. 5.16. Such ^{17}O NMR features characterise the triangular M^{IV} ions $[M_3X_4(OH_2)_9]^{4+}$ (M = Mo or W; X = O or S, Chapter 1, Fig. 1.16). Here ^{17}O NMR seems able to distinguish between μ_3-O (capping) and μ_2-O (bridging) groups (see also Section 6.2.3). The single resonance is in a ratio of 1:2:1 with respect to the two terminal waters, indicating an assignment to a bridging μ-O(OH) group. The absence of the expected capping μ_3-O group is of interest, pointing against the core structure $Nb_3(\mu_3$-O$)(\mu_2$-O$)_3^{3+}$ proposed by Cotton *et al.* [92]. In the absence of a source of sulfur the only other candidate for a capping group, given its presence in other triangular Nb complexes, is chloride. Thus the structure $[Nb_3(\mu_3$-Cl$)(\mu$-O$)_3(OH_2)_9]^{4+}$ is proposed for the green aqua ion (Fig. 5.21). The ^{93}Nb NMR spectrum of the ion (Fig. 5.19 inset) implies equivalent Nb atoms on the NMR timescale.

5.2.3.2 $[Nb_3(\mu_3$-S$)(\mu$-O$)_3(OH_2)_9]^{3+}$?

A second cationic species more olive-green in colour, along with the above green cation, is obtained following pretreatment of $NbCl_3$(1,2-DME) with aqueous Na_2S prior to hydrolysis in HCl. It may be separated from the green ion above by cation-exchange chromatography and has been assigned to the μ_3-sulfido species $[Nb_3(\mu_3$-S$)(\mu$-O$)_3(OH_2)_9]^{3+}$ [97]. The olive-green colour arises from an intense band in the u.v. region around 360 nm tailing into the visible and is believed to be characteristic of the presence of the triangular $M_3(\mu_3$-S$)^{n+}$ (aq) core. The trinuclear Mo(aq) clusters $[Mo_3(\mu_3$-S$)(\mu$-O$)_3(OH_2)_9]^{4+}$ and $[Mo_3(\mu_3$-S$)(\mu$-S$)_3(OH_2)_9]^{4+}$ both possess intense band maxima in this region (see Section 6.2.3). Thus it appears that the origin of the μ_3-S group in Cotton's NCS$^-$ complex (Fig. 5.16) is not the THT ligand in the $Nb_2Cl_6(THT)_3$ complex but NCS$^-$ via sulfur abstraction in the presence of methanolic HCl. Further work in this area is continuing.

5.2.3.3 OTHER DERIVATIVES OF $NB_3(AQ)$?

The solid chloro-aqua residue proved to have alcohol solubility leading to potential routes to a range of derivative complexes of the $Nb_3(\mu_3\text{-Cl})(\mu\text{-O})_3{}^{4+}$ core. In one such attempt dissolution of the residue in ethanol (or methanol) was followed by the addition of an excess of the salt $K^+[HBpz_3]^-$ (pz = 1-pyrazolyl). In each case a reddish-brown precipitate resulted which, upon extraction with CH_2Cl_2 and slow evaporation, afforded pinkish-red crystals of a triangular Nb_3 complex, the crystal structure of which is shown in Fig. 5.22 [98].

The presence of the unusual capping borate ligand presumably derives from the hydrolysis of one of the hydridotris(1-pyrazolyl) borate ligands in the

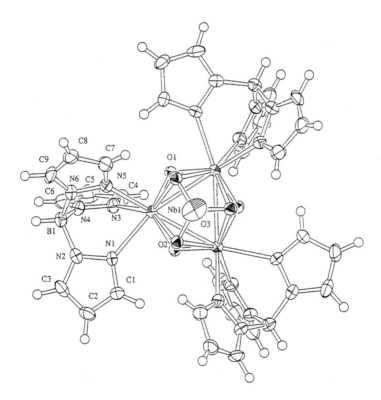

Figure 5.22. Structure of the triangular Nb_3 complex, $[Nb_3(\mu_3\text{-BO}_3(OH))(\mu\text{-O})_3(HBpz_3)_3]$ shown looking down the terminal B—O bond of the capping borate [98]

presence of the aqua niobium cluster. Assignment of a formal oxidation state for the $HBpz_3^-$ complex is not straightforward on the basis of the observed Nb—Nb separation of 284 pm. Triangular Nb clusters having such Nb—Nb distances range from the formal oxidation state $Nb^{IV,IV,V}$ (two cluster d electrons) to $Nb^{III,IV,IV}$ (four cluster d electrons) as in the green aquated cation. Both can be accommodated in the triangular structure (Fig. 5.22) for the $HBpz_3^-$ complex depending upon whether the bridging oxygens are oxo or hydroxo. Unfortunately this cannot be deduced from the crystal data as is often the case with heavy atom clusters. One feature in support of the higher formal oxidation state and presence of μ-O groups is the surprising air stability of the $HBpz_3^-$ complex. Furthermore, Nb^V is known to form extremely strong mineral composites with borate.

5.2.3.4 WATER-EXCHANGE KINETIC MEASUREMENTS ON $[NB_3(\mu_3-CL)(\mu-O)_3(OH_2)_9]^{4+}$

The two water ligands on $[Nb_3(\mu_3-Cl)(\mu-O)_3(OH_2)_9]^{4+}$ have markedly differing exchange rates. Each has been followed ^{17}O NMR measurements following enrichment up to 5 atom % in the isotope. Exchange at the waters (labelled d) that are *trans* to the bridging μ-O groups have been followed by observing the temperature dependence of the linewidth $(1/T_2)$ of the bound water resonances (Fig. 5.20) in solutions of $Nb_3(aq)$ (0.05 M) in 2.5 M CF_3SO_3H. Above 30 °C the linewidth of the d waters increases whereas that for the c waters, *trans* to the capping group, continues to show quadrupolar narrowing. By assuming a similar quadrupolar relaxation energy E_Q for the two water ligands, the exchange rate constant can be directly extracted from the observed linewidth (Hz) according to

$$\frac{1}{T_2 \text{ (obs)}} = \frac{1}{T_{2Q}} + k_{ex} \tag{5.7}$$

From data points where k_{ex} makes at least a 20% contribution to $1/T_2$ the following kinetic parameters are obtained: $k_{ex}(d)(25\,°C) = (3.52 \pm 0.41) \times 10^2\,s^{-1}$, $\Delta H_{ex}^{\ddagger} = 36.8 \pm 2.2$ kJ mol^{-1} and $\Delta S_{ex}^{\ddagger} = -72.4 \pm 6.9$ J K^{-1} mol^{-1}. Exchange at the more inert c waters was followed by monitoring the increase in height of the resonance at -24 ppm, following addition of a sample of isotopically enriched water plus Mn^{2+} (aq) to a solution of Nb_3(aq) in 2.5 M CF_3SO_3H. The resonance of the fully exchanged d waters serves as a convenient internal reference line. At 25 °C, following fitting of the growth in the peak height to a standard exponential function, $k_{ex}(c) = (8.3 \pm 1.7) \times 10^{-3}s^{-1}$. The 4×10^4-fold difference in exchange rate between the two water ligands is in keeping with their respective assignments in exactly paralleling the behaviour of the corresponding two water ligands on the trinuclear $Mo_3O_4{}^{4+}$(aq) and $W_3O_4{}^{4+}$(aq) ions, which also have the same basic M_3X_{13} cluster core (see Section 6.2.3).

The low $\Delta H^{\ddagger}$ and negative $\Delta S^{\ddagger}$ values coupled with the apparent absence of an $[H^+]^{-1}$ dependence might be tentatively argued as evidence for an associative mechanism on the Nb_3(aq) cluster ion (cf. water exchange on Ti^{3+}(aq) and V^{3+}(aq)). Indeed, such a mechanism might be favoured since the Nb_3 cluster core is somewhat electron deficient (bond order of 2/3), the larger size of the Nb atoms also favouring a greater approach of the entering group. This behaviour would contrast with the more dissociatively activated pathways believed to operate on the 'electron-complete' d^6 cluster ions of Mo^{IV} and W^{IV} (see Section 6.2.3). Further studies on the Nb_3(aq) analogue systems are continuing.

The analogous dimeric Ta^{III} complex Ta_2Cl_6 (tetrahydrothiophen)$_3$ [99] is known, leading to speculation that corresponding triangular Ta(aq) species may be formed. The formal reduction potential for the $Nb^{V/III}$ couple has been estimated to be $-0.275\,V$ vs NHE in 3 M H_2SO_4 [100]. That for the $Ta^{V/III}$ couple is estimated as $\sim -0.4V$, similar to that for $Cr^{III/II}$ (Chapter 6).

5.2.4 Hexanuclear Niobium and Tantalum Cluster Ions

In the formal oxidation states between $+2$ and $+3$ both elements have a propensity for forming hexanuclear cluster species [101] wherein the metal atoms are placed at the corner of an octahedron, similar to the arrangement present in the $[M_6O_{19}]^{8-}$ anions. The basic unit consists of the cation $M_6X_{12}{}^{n+}$ (M = Nb or Ta; X = Cl or Br; $n = 2,3$, or 4) with six additional ligands, usually X^- or H_2O, coordinated at the terminal positions (one per metal). Salts containing the anions $[M_6X_{18}]^{m-}$ ($m = 2,3$ or 4) have all been characterised. For niobium, $m = 4$, replacement of up to four of the terminal Cl^- ligands with H_2O can be achieved, with the best characterised species being represented by $[(M_6X_{12})]X_2(OH_2)_4]\cdot 4H_2O$. However, in the case of tantalum replacement of all of the terminal Cl^- ligands by H_2O has been achieved and the 'hexaaqua' ion, $[Ta_6Cl_{12}(OH_2)_6]^{3+}$, has been structurally characterised within the double salt $Me_4N[Ta_6Cl_{12}$-$(OH_2)_6]Br_4$ [102] (Fig. 5.23).

The preparation involves treatment of $[Ta_6Cl_{12}]Cl_2\cdot 8H_2O$ in methanol with Me_4NOH followed by filtration and treatment with concentrated HCl or HBr. A red precipitate of $[Ta_6Cl_{12}(OH)_4]\cdot 10H_2O$ is initially formed [103], which redissolves with excess of the acid. On standing overnight a colour change from red ($Ta_6Cl_{12}{}^{2+}$) to olive-green ($Ta_6Cl_{12}{}^{3+}$) occurs and eventually brown crystals of the above double salt separate. Salt of the anion $[Ta_6Cl_{12}(OH)_6]^{2-}$ [104] as well as a labile triflate derivative [105] have also been reported. Recently a number of preparative routes for the corresponding cationic hexaniobium aqua cluster $[Nb_6Cl_{12}(OH_2)_6]^{2+}$ have been described [106] and this should open up a number of interesting comparative studies between the two heavy group 5 elements. As with the triangular species these hexanuclear clusters are electron

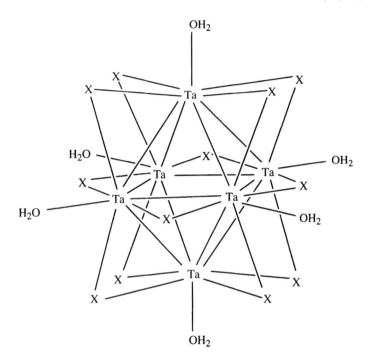

Figure 5.23. Structure of the $[(Ta_6X_{12})(OH_2)_6]^{3+}$ unit in the salt $Me_4N[(Ta_6X_{12})\cdot(OH_2)_6]Br_4$ (X = Cl)

deficient with the M—M distance (280 pm) consistent with the formal bond order of 2/3 at each M—M bond and an overall $16e^-$ count for the M_6 unit.

A number of redox studies were reported on the $\{M_6X_{12}\}^{n+}$ species (M = Nb and Ta) as far back as the late 1960 s [107]. These served to demonstrate the rich and sometimes complex redox chemistry of these seemingly 'simple' clusters. Both one- and two-electron redox processes were characterised. The existence now of the hexaaqua ion $[Ta_6X_{12}(OH_2)_6]^{3+}$ (X = Cl) should lead on to interesting further studies (redox and water ligand substitution) on these electron-deficient aqueous metal cluster species.

REFERENCES

[1]   (a) Butler, A., and Carrano, C. J., *Coord. Chem Revs.* **109**, 61 (1991).
      (b) Wever, R., and Kustin, K., *Adv. Inorg. Chem.*, **35**, 81 (1990).
      (c) Kustin, K., and McCara, I. C., *Comm. Inorg. Chem.*, **2**, 1 (1982).
      (d) Kustin, K., Mcleod, G., Gilbert, T. R., and Briggs, L. R., *Struct. Bond.*, **53**, 139 (1983).
      (e) Smith, M. J., *Experientia*, **45**, 452 (1989).

[2] (a) Kranz, M., *Inorganic Syntheses.*, **7**, 94 (1963).
      (b) Larkworthy, L. F., Patel, K. C., and Phillips, D. J., *J. Chem. Soc. A*, 1095 (1970).
      (c) Brauer, G., *Handbook of Preparative Inorganic Chemistry*, Vol. 2, 2nd edn, Academic Press, London, 1965, p. 337.
[3] Green, M.,Taylor. R. S., and Sykes, A. G., *J. Chem. Soc. A.*, 509 (1971).
[4] Orhanovic, M., Po, H. N., and Sutin, N., *J. Am. Chem. Soc.*, **90**, 7224 (1968).
[5] Holt, D. G., Larkworthy, L. F., Povey, D. C., Smith, G. W., and Leigh, G. J., *Inorg. Chim. Acta*, **169**, 201 (1990).
[6] Dobson, J. C., Sano, M., and Taube, H., *Inorg. Chem.*, **30**, 456 (1991).
[7] Montgomery, H., Chastain, R. V., Natt, J. J., Wotkowska, A. M., and Lingafelter, E. C., *Acta., Cryst.*, **22**, 775 (1967).
[8] Ducummon, Y., Zbinden, D., and Merbach, A. E., *Helv. Chim. Acta*, **65**, 1385 (1982).
[9] Nichols, P. J., Docummon, Y., and Merbach, A. E., *Inorg. Chem.*, **22**, 3993 (1983).
[10] Vilas Boas, L., and Costa Pessoa, J., *Comprehensive Coordination Chemistry* (eds. G. Wilkinson, R. D. Gillard and J. A. McClevetry), Vol. 3, Pergamon, London, 1987, pp. 453–583.
[11] Sutin, N., *Acc. Chem. Res.*, **1**, 225 (1968).
[12] (a) Cannon, R. D., *Electron Transfer Reactions*, Butterworths, London, 1980.
      (b) Sykes, A. G., and Green, M., *J. Chem. Soc. A*, 3221 (1970).
      (c) Davies, K. M., and Espenson, J. H., *J. Am. Chem. Soc.*, **91**, 3093 (1969).
[13] Swinehart, J. H., *Inorg. Chem.*, **4**, 1069 (1965).
[14] Rush, J. D., and Bielski, B. H. J., *Inorg. Chem.*, **24**, 4282 (1985).
[15] (a) Espenson, J. H., Bakac, A., and Kim, J.-Y., *Inorg. Chem.*, **30**, 4830 (1991).
      (b) Bakac, A., and Espenson, J. H., *Inorg. Chem.*, **28**, 3901 (1989).
      (c) Kelley, D. G., Espenson, J. H., and Bakac., *Inorg. Chem.*, **29**, 4996 (1990).
[16] Ingold, C. K., *Structure and Mechanism in Organic Chemistry*, 2nd edn, Cornell University Press, Ithaca, New York, 1969, pp. 4–7.
[17] Clark, R. J. H., *The Chemistry of Titanium and Vanadium*, Elsevier, Amsterdam, 1968.
[18] McClure, D. J., *J. Chem. Phys.*, **36**, 2757 (1962).
[19] (a) Hartmann, H., and Schlafer, H. L., *Z. Naturforsch., Teil A*, **6**, 754 (1951).
      (b) Beattie, J. K., Best, S. P., Skelton, B. W., and White, A. H., *J. Chem. Soc. Dalton Trans.*, 2105 (1981).
[20] Cotton, F. A., Fair, C. K., Lewis, G. E., Mott, G. N., Ross, F. K., Schultz, A. J., and Williams, J. M., *J. Am. Chem. Soc.*, **106**, 5319 (1984).
[21] Best, S. P., Armstrong, R. S., and Beattie, J. K., *Inorg. Chem.*, **19**, 1958 (1980).
[22] Hitchmann, M. A., McDonald, R. G., Smith, P. W., and Stranger, R., *J. Chem. Soc. Dalton Trans.*, 1393 (1988).
[23] Baes, C. K., and Mesmer, R. E., *The Hydrolysis of Cations*, Krieger, Florida, 1986, p. 202.
[24] Pajdowski, L., and Jezowska-Trzebiatowska. B., *J. Inorg. Nucl. Chem.*, **28**, 443 (1966).
[25] Kanamori, K., Ookubo, Y., Ino, K., Kawai, K., and Michabata, H., *Inorg. Chem.*, **30**, 3832 (1991).
[26] Hugi, A. D., Helm, L., and Merbach, A. E., *Helv. Chim. Acta*, **68**, 508 (1985).
[27] (a) Patel, R. C., and Diebler, H., *Ber. Bunsenges. Phys. Chem.*, **76**, 1035 (1972).
      (b) Perlmutter-Hayman, B., and Tapuhi, E., *J. Coord. Chem.*, **10**, 219 (1980).
      (c) Perlmutter-Hayman, B., and Tapuhi, E., *J. Coord. Chem.*, **9**, 177 (1979).
      (d) Mathur, P. N., and Fukutomi, H., *J. Inorg. Nucl. Chem.*, **43**, 2869 (1981).

(e) Perlmutter-Hayman, B., and Tapuhi, E., *Inorg. Chem.*, **18**, 2872 (1979).

(f) Konstantos, J., Kalatzis, G., Vrachnou-Astra, E., and Katakis, D., *J. Chem. Soc. Dalton Trans.*, 2461 (1985) and *Inorg. Chim Acta*, **151**, 191 (1988).

[28] (a) Baker, B. R., Sutin, N., and Welch, T. J., *Inorg. Chem.*, **6**, 1948 (1967).

(b) Kruse, W., and Thusius, D., *Inorg. Chem.*, **7**, 464 (1968).

(c) Sauvageat, P.-Y., Ducummon, Y., and Merbach, A. E., *Helv. Chim. Acta*, **72**, 1801 (1989).

[29] Espenson, J. H., and Pladziewicz, J. R., *Inorg. Chem.*, **9**, 1380 (1970).

[30] Ramsey, J. B., Sugimoto, R., and de Vorkin, H., *J. Am. Chem. Soc.*, **63**, 3480 (1941).

[31] King, W. B., and Garner, C. S., *J. Phys. Chem.*, **58**, 29 (1954).

[32] Sykes, A. G., *Kinetics of Inorganic Reactions.*, Pergamon, New York, 1970, pp. 175–212.

[33] Higginson, W. C. E., and Sykes, A. G., *J. Chem. Soc. A*, 2841 (1962).

[34] Higginson, W. C. E., and Rosseinsky, D. R., *J. Chem. Soc.*, 31 (1960).

[35] Steele, M. C., and Hall, F. M., *Anal. Chim. Acta*, **9**, 384 (1953).

[36] Selbin, J., *Coord. Chem. Rev.*, **1**, 293 (1966) and *Angew. Chem. Int. Ed. (Eng.)*, **5**, 712 (1966).

[37] See, for example, Richens, D. T., Harmer, M. A., and Sykes, A. G., *J. Chem. Soc. Dalton Trans.*, 2099 (1984).

[38] Selbin, J., and Holmes Jr, L. H., *J. Inorg. Nucl. Chem.*, **24**, 1111 (1962).

[39] (a) Kierkegaard, P., and Longo, J. M. *Acta Chem. Scand.*, **19**, 1906 (1965).

(b) Tachez, M., and Theobold, F., *Acta Cryst.*, **B36**, 1757 (1980).

[40] Ballhausen, C. J., Djurinskii, B. F., and Watson, K. J., *J. Am. Chem. Soc.*, **901**, 3305 (1968).

[41] Ballhausen, C. J., and Gray, H. B., *Inorg. Chem.*, **1**, 111 (1962).

[42] Kevan L., *J. Phys. Chem.*, **88**, 327 (1984).

[43] (a) Henry, R. P., Mitchell, P. C. H., and Prue, J. E., *J. Chem. Soc. Dalton Trans.*, 1156 (1973).

(b) Lutz, B., and Wendt, H., *Ber. Bunsenges Phys. Chem.*, **74**, 372 (1970).

(c) Rossotti, F. J. C., and Rossotti, H. S., *Acta Chem. Scand.*, **9**, 1177 (1955).

[44] Johnson, G. K., and Schlemper, E. O., *J. Am. Chem. Soc.*, **100**, 3645 (1978).

[45] Johnson, M. D., and Murmann, R. K., *Inorg. Chem.*, **22**, 1068 (1983).

[46] Gamsjager, H., and Murmann, R. K., *Adv. Inorg. Bioinorg. Mech.*, **2**, 317 (1983).

[47] (a) Reuben, J., and Fiat, D., *Inorg. Chem.*, **6**, 579 (1967).

(b) Wuthrich, K., and Connick, R. E., *Inorg. Chem.*, **7**, 1377 (1968).

[48] Kuroiwa, Y., Harada, M., Tomiyasu, M., and Fukotomi, H., *Inorg. Chim Acta*, **146**, 7 (1988).

[49] Saito, K., and Sasaki, Y., *Adv. Inorg. Bioinorg. Mech.*, **1**, 186 (1982).

[50] Che, T. M., and Kustin, K., *Inorg. Chem.*, **19**, 2275 (1980).

[51] (a) Macartney, D. H., McAuley, A., and Olubuyide, O. A., *Inorg. Chem.*, **24**, 307 (1985).

(b) Macartney. D. H., *Inorg. Chem.*, **25**, 2222 (1986).

[52] (a) Tytko, K. H., and Mehmke, J., *Z. Anorg. Allg. Chem.*, **503**, 67 (1983).

(b) Pope, M. T., and Dale, B. W., *Chem. Soc. Quart. Revs.*, **22**, 527 (1968).

[53] (a) Yamada, S., Funahashi, S., and Tanaka, M., *J. Inorg. Nucl. Chem.*, **37**, 835 (1975).

(b) Kustin, K., and Toppen, D. L., *Inorg. Chem.*, **12**, 1404 (1973).

[54] Rahmoeller, K. M., and Murmann, R. K., *Inorg. Chem.*, **22**, 1072 (1983).

[55] (a) Scheidt, W. R., Tsai, C., and Hoard J. L., *J. Am. Chem. Soc.*, **93**, 3867 (1971).

(b) Scheidt, W. R., Collins, D. M., and Hoard, J. L., *J. Am. Chem. Soc.*, **93**, 3873 (1971).

[56] Dyrssen, D., and Sekine, T., *J. Inorg. Nucl. Chem.*, **26**, 981 (1964).
[57] Ingri, N., and Brito. F., *Acta Chem. Scand.*, **13**, 1971 (1959).
[58] Newmann, L., LaFleur, W. J., Brousaides, F. J., and Ross, A. M., *J. Am. Chem. Soc.*, **80**, 4491 (1958).
[59] Schwarzenbach, G., and Geier, G., *Helv. Chim. Acta*, **46**, 906 (1963).
[60] (a) Pope, M. T., *Heteropoly and Isopolyoxometallates*, Springer, Berlin, 1983.
(b) Evans Jr, H. T., Swallow, A. G., and Barnes, W. H., *J. Am. Chem. Soc.*, **86**, 4209 (1964).
(c) Swallow, A. G., Ahmed, F.R., and Barnes, W. H., *Acta Cryst.*, **21**, 397 (1966).
[61] (a) Baes, C. K., and Mesmer, R. E., *The Hydrolysis of Cations*, Krieger, Florida, 1986, p. 209.
(b) Rossotti, F. J. C., and Rossotti, H., *Acta Chem. Scand.*, **10**, 957 (1956).
[62] (a) Howarth, O. W., and Richards, R. E., *J. Chem. Soc.*, 864 (1965).
(b) English, A. D., Jesson, J. P., Klemperer, W. G., Mamonneas, T., Messerle, L., Shum, W., and Tramontano, A., *J. Am. Chem. Soc.*, **97**, 4785 (1975).
(c) Klemperer, W. G., and Shum, W., *J. Am. Chem. Soc.*, **99**, 3544 (1977).
[63] Murmann, R. K., and Geise, K. C., *Inorg. Chem.*, **17**, 1160 (1978).
[64] Comba, P., and Merbach, A. E., *Helv. Chim. Acta*, **71**, 1406 (1988).
[65] (a) Brito, F., *An. Quim.*, **62B**, 123, 197 (1966).
(b) Brito, F., *Acta Chem. Scand.*, **21**, 1968 (1967).
[66] Cantley Jr, L. C., Josephson, L., Warner, R., Yanagisawa, M., Lechene, C., and Guidotti, G., *J. Biol. Chem.*, **252**, 7421 (1977).
[67] (a) Griffith, W. P., and Lesniak, P. J. B., *J. Chem. Soc. A*, 1066 (1969).
(b) Thompson, R. C., *Inorg. Chem.*, **21**, 859 (1982) and **22**, 1072 (1983).
[68] Littler, J. S., and Waters, W. A., *J. Chem. Soc.*, 1299 (1959).
[69] (a) Cotton, F. A., and Wilkinson, G., *Advanced Inorganic Chemistry*, 5th edn, Wiley, New York, 1988, p. 787.
(b) Fairbrother, F., *The Chemistry of Niobium and Tantalum*, Elsevier, New York, 1967.
(c) Hubert-Pfalzgraf, L. G., Postel, M., and Riess, J. G., *Comprehensive Coordination Chemistry* (eds. G. Wilkinson, R. D. Gillard and J. A. McCleverty), Vol. 3, Pergamon, London, 1987, p. 585.
[70] Berg, R. W., *Coord. Chem. Revs.*, **113**, 1 (1992).
[71] Goiffon, A., Philippot, E., and Maurin, M., *Rev. Chim. Miner.*, **17**, 466 (1980).
[72] Goiffon, A., and Spinner, R., *Rev. Chim. Miner.*, **11**, 262 (1974) and *Bull. Soc. Chim. Fr.*, 2435 (1975).
[73] Filowitz, M., Ho, R. K. C., Klemperer, W. G., and Shum, W., *Inorg. Chem.*, **18**, 93 (1979).
[74] Goiffon, A., Granger. R., Bockel, C., and Spinner, R., *Rev. Chim Miner.*, **10**, 487 (1973).
[75] Graeber, E. J., and Morosin, B., *Acta Cryst.*, **B33**, 2136 (1977).
[76] Thiele A., and Fuchs, J., Private communication in Hubert-Pfalzgraf, L. G., Postel, M., and Riess, J. G., *Comprehensive Coordination Chemistry* (eds. G. Wilkinson, R. D. Gillard and J. A. McCleverty), Vol. 3, Pergamon, London, 1987, p. 585.
[77] Babko, A. K., Gridchina, G. I. and Nabivanets, B. I., *Russ. J., Inorg. Chem.*, **7**, 66 (1963).
[78] Kanzelmeyer, J. H., Ryan, J., and Freund, H., *J. Am. Chem. Soc.*, **78**, 3020 (1956).
[79] Britton, H. T. S., and Robinson, R. A., *J. Chem. Soc.*, 419 (1933).
[80] Gupta, R. N., and Sen, B. K. *J. Inorg. Nucl. Chem.*, **37**, 1548 (1975).
[81] Kojik-Prodic, R., Liminga, R., and Scavnicar, S., *Acta Cryst.*, **B29**, 864 (1973).

[82] Galesic, M., Markovic, B., Herceg, M., and Sljukic, M., *J. Less Comm. Met.*, **25**, 234 (1971).
[83] Brnicevic, N., and Djordjevic, C., *J. Less Comm. Met.*, **21**, 469 (1970).
[84] Sala-pala, J., Guerchais, J. E., and Edwards, A. J., *Angew. Chem. Int. Ed. (Eng.)*, **21**, 870 (1982).
[85] Lee, S. C., and Holm, R. H. *J. Am. Chem. Soc.*, **112**, 9654 (1990).
[86] Sola, J., Do, Y., Berg, J. M., and Holm, R. H., *J. Am. Chem. Soc.*, **105**, 7784 (1983).
[87] Brunner, H., Klement, U., Wachter, J., Tsunoda, M., Leblanc, J.-C., and Moise, C., *Inorg. Chem.*, **29**, 584 (1990).
[88] Kalinnikov, V. T., Pasynskii, A. A., Larin, G. M., Novotortsev, V. M., Struchkov, Yu, T., Gusev, A. I., and Kirillova, N. I., *J. Organo. Met. Chem.*, **74**, 91 (1974).
[89] Cotton, F. A., Diebold, M. P., and Roth, W. J., *Inorg. Chem.*, **27**, 2347 (1988).
[90] Cotton, F. A., Duraj, S. A., and Roth, W. J., *J. Am. Chem. Soc.*, **106**, 3527 (1984).
[91] Bino, A., *Inorg. Chem.*, **21**, 1917 (1982).
[92] Cotton, F. A., Diebold, M. P., Llusar, R., and Roth, W. J., *J. Chem. Soc., Chem. Commum.*, 1276 (1986).
[93] Simon, A., and von Schnering, H. G. *Z. Anorg. Allg. Chem.*, **339**, 155 (1965).
[94] Cotton, F. A., Diebold, M. P., Feng, X., and Roth, W. J., *Inorg. Chem.*, **27**, 3414 (1988).
[95] Maas, E. T., and McCarley, R. E., *Inorg. Chem.*, **12**, 1096 (1973).
[96] Roskamp, E. J., and Pedesen, S. F., *J. Am. Chem. Soc.*, **109**, 6551 (1987).
[97] Minhas, S., and Richens, D. T. *J. Chem. Soc. Dalton Trans.*, 703 (1996).
[98] Minhos, S., Devlin, A., Richens, D. T., Benyei, A., and Lightfoot, P. *Acta Cryst. C* (1997) (in press).
[99] Templeton, J. L., and McCarley, R. E., *Inorg. Chem.*, **17**, 2293 (1978).
[100] Gupta, R. N., and Sen, B. K., *J. Inorg. Nucl. Chem.*, **37**, 1044 (1975).
[101] (a) Hughes, B. G., Meyer, J. L., Fleming, P. B., and McCarley, R. E., *Inorg. Chem.*, **9**, 1343 (1970).
      (b) Converse, J. G., and McCarley, R. E., *Inorg. Chem.*, **9**, 1361 (1970).
      (c) Spreckelmeyer, B., *Z. Anorg. Allg. Chem.*, **358**, 148 (1968).
      (d) McCarley, R. E., Hughes, B. G., Cotton, F. A., and Zimmermann, R., *Inorg. Chem.*, **4**, 1691 (1965).
      (e) Kuhn, P. J., and McCarley, R. E., *Inorg. Chem.*, **4**, 1482 (1965).
[102] Brnicevic, N., Ruzic-Toros, Z., and Kojic-prodic, B., *J. Chem. Soc. Dalton Trans.*, 455 (1985).
[103] Brnicevic, N., and Schafer, H., *Z. Anorg. Allg. Chem.*, **441**, 219 (1978).
[104] Brnicevic, N., Mesaric, S., and Schafer, H., *Croat. Chem. Acta*, **57**, 529 (1984).
[105] Kennedy, V. O., Stern, C. L., and Shriver, D. F., *Inorg. Chem.*, **33**, 5967 (1994).
[106] (a) Brnicevic, N., Kodic-Prodic, B., Luic, M., Kashta, A., Planinic, P., and McCarley, R. E., *Croat. Chem. Acta*, **68**, 861 (1995).
      (b) Brnicevic, N., Plannic, P., McCarley, R. E., Antolic, S., Luic, M., and Kodic-Prodic, B., *J. Chem. Soc. Dalton Trans.*, 1441 (1995).
[107] (a) Espenson, J. H., and Boone, D. J., *Inorg. Chem.*, **7**, 636 (1968).
      (b) Espenson, J. H., *Inorg. Chem.*, **7**, 631 (1968).

Chapter 6

Group 6 Elements: Chromium, Molybdenum and Tungsten

The elements of Group 6 collectively possess the richest cationic aqueous chemistry of any group of transition elements. Whilst cationic aqueous species are known for molybdenum in five oxidation states, the highest number for any transition element, the slow attainment of solution equilibria for Cr^{III} aqueous species has led to a detailed understanding of hydrolysis processes probably unrivalled by any other transition metal oxidation state. The aqueous chemistry of tungsten is much more similar to molybdenum than chromium, continuing the trend established for the second and third row members in the preceding two groups. Finally, the aqueous chemistry of Cr/Mo/W is dominated by the formation of polynuclear and cluster species remeniscent of the V/Nb/Ta group but notably unlike Mn/Tc/Re (next chapter).

6.1 CHROMIUM

The aqueous chemistry of stable chromium species is limited to oxidation states II, III and VI and cationic species to the first two oxidation states. Complexes of Cr^{III} being representative of some of the oldest known and most stable coordination compounds [1]. The existence of the now-extensive series of slowly equilibrating hydroxy-bridged polymeric cations was first postulated by Bjerrum in 1908 [2]. $Cr^{II}(aq)$ is highly air sensitive and as such comparative knowledge of its aqueous chemistry is limited. $Cr^{IV}(aq)$ has a fleeting existence in aqueous solution contrasting with the stability of CrO_2, e.g. as used in magnetic tapes. $Cr^{V}(aq)$ is even more short lived in solution. The transient nature of $Cr^{IV}(aq)$ and $Cr^{V}(aq)$ parallels the behaviour of the corresponding mononuclear species for Mo and W. The known stability of $Mo(W)^{IV}(aq)$ and $Mo(W)^{V}(aq)$ in solution arises as a result of the formation of M—M bonded polynuclear structures (see below). The aqueous chemistry of Cr^{VI} is basically that of a number of simple oxo anions.

6.1.1 Chromium(II) (d^4): Chemistry of the $[Cr(OH_2)_6]^{2+}$ Ion

6.1.1.1 PREPARATION OF SOLUTIONS OF $[CR(OH_2)_6]^{2+}$

Beautiful air-sensitive bright blue solutions containing the $[Cr(OH_2)_6]^{2+}$ ion are readily prepared by two common methods: amalgamated zinc reduction of acidified solutions of Cr^{VI} or Cr^{III} [1,3] or via dissolution of electrolytic grade chromium metal [4] in the appropriate acid under air-free conditions. Using the latter method the blue crystalline hydrated chloride, $CrCl_2 \cdot 4H_2O$, is readily isolated from concentrated HCl solutions [5]. Dissolution of $CrCl_2 \cdot 4H_2O$, or other Cr^{II} salts, in air-free dilute aqueous solutions of HCl or other non-oxidising weakly coordinating acids liberates the blue $[Cr(OH_2)_6]^{2+}$ ion. The ion is well characterised in the solid state in the series of Tutton salts, $M_2SO_4 \cdot CrSO_4 \cdot nH_2O$ ($M^+ = NH_4^+, Na^+, K^+, Rb^+$ and Cs^+; $n = 6$) [6]. The Cr—$O(OH_2)$ distances obtained from a room temperature X-ray diffraction study on the ammonium salt are respectively 212.2 and 205.2 pm (equatorial) and 237.2 pm (axial), showing the expected Jahn–Teller octahedral distortion of the high-spin d^4 configuration, (Fig. 6.1). $(ND_4)_2Cr(SO_4)_2 \cdot 6D_2O$ has also been the subject of low-temperature neutron diffraction studies [7]. EXAFS measurements on solutions of $[Cr(OH_2)_6]^{2+}$ show the Jahn–Teller distorted hexaaqua structure with Cr—OH_2 (equatorial) of 199 pm and Cr—OH_2 (axial) of 230 pm [8]. The presence of

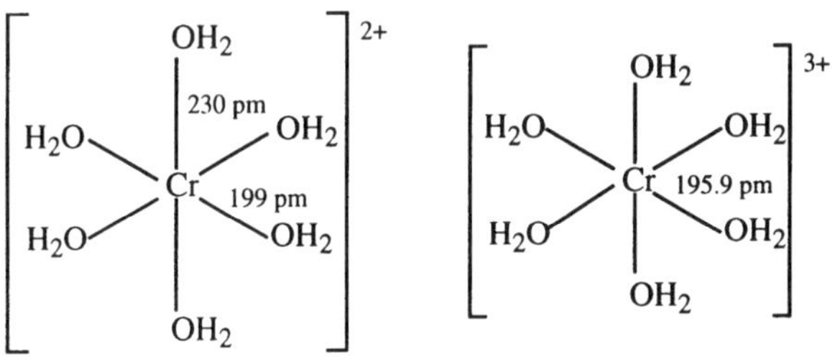

$$\left[\begin{array}{c} \mathrm{OH_2} \\ \overset{230\ \mathrm{pm}}{\underset{199\ \mathrm{pm}}{\mathrm{Cr}}} \\ \mathrm{OH_2} \end{array}\ \substack{\mathrm{H_2O} \\ \mathrm{H_2O}}\ \substack{\mathrm{OH_2} \\ \mathrm{OH_2}} \right]^{2+} \qquad \left[\begin{array}{c} \mathrm{OH_2} \\ \underset{195.9\ \mathrm{pm}}{\mathrm{Cr}} \\ \mathrm{OH_2} \end{array}\ \substack{\mathrm{H_2O} \\ \mathrm{H_2O}}\ \substack{\mathrm{OH_2} \\ \mathrm{OH_2}} \right]^{3+}$$

Figure 6.1. Structures of $[\mathrm{Cr(OH_2)_6}]^{2+}$ (from solution EXAFS) and of $[\mathrm{Cr(OH_2)_6}]^{3+}$

four short $\mathrm{Cr—OH_2}$ bonds (average distance 208 pm) and two long $\mathrm{Cr—Cl}$ bonds (at 276 pm) has been detected in the crystal structure of $\mathrm{CrCl_2 \cdot 4H_2O}$ [9]. $\mathrm{CrSO_4 \cdot 5H_2O}$ is isomorphous with $\mathrm{CuSO_4 \cdot 5H_2O}$ and possesses a similar Jahn–Teller distorted octahedral geometry with four short equatorial $\mathrm{M—OH_2}$ bonds ($\sim 205\,\mathrm{pm}$) and two long bonds (242–245 pm) to axial sulphates [10]. The crystalline salt $[\mathrm{Cr(OH_2)_6}]\mathrm{SiF_6}$ is unusual in having all $\mathrm{Cr—O(OH_2)}$ bonds identical [11]. Here the Jahn–Teller distortion is suppressed as a result of strong hydrogen bonding to the $\mathrm{SiF_6}^{2-}$ ions together with crystal packing effects. A similar effect is apparent in the corresponding $\mathrm{Cu^{II}}$ salt. However, $[\mathrm{Cr(OH_2)_6}]\mathrm{SiF_6}$ is unstable and a simple recrystallisation process readily results in formation of the Jahn–Teller distorted tetraaqua salt $\mathrm{Cr(OH_2)_4SiF_6}$ wherein two of the water ligands are replaced by longer bonds (239.6 pm) to F atoms from $\mathrm{SiF_6}^{2-}$ [12].

6.1.1.2 SPECTROSCOPIC AND MAGNETIC PROPERTIES

Mononuclear $\mathrm{Cr^{II}}$ species are magnetically dilute, possessing largely temperature-independent moments close to the spin-only value of 4.90 BM for high spin $\mathrm{d^4}$. The blue colour of $[\mathrm{Cr(OH_2)_6}]^{2+}$ arises from a characteristic absorption band centred around 714 nm with in addition a shoulder at *ca.* 1000–1050 nm [13]. The two bands are as a result of the splitting of the high spin $\mathrm{d^4}$ doublet ground state term $^5E_{2g}$ into $^5B_{1g}(x^2 - y^2)$ (higher) and $^5A_{1g}(z^2)$ (lower) terms due to the tetragonal distortion. The two visible–NIR bands have been assigned to the spin-allowed transitions $^5B_{1g}$–$^5B_{2g}$ $(x^2 - y^2 - xy)$ ($\sim 1000\,\mathrm{nm}$) and $^5B_{1g}$–^5E_g $(x^2 - y^2 - xz, yz)$. The same two bands are also evident in reflectance spectra from various hydrated $\mathrm{Cr^{II}}$ salts. In the case of the violet dihydrate $\mathrm{Cs_2SO_4 \cdot CrSO_4 \cdot 2H_2O}$, the anomalously low magnetic moment and temperature dependency (0.88 BM (300 K), 0.48 BM (90 K)) and shift in the two bands to higher energy ($\sim 550\,\mathrm{nm}$ and $\sim 800\,\mathrm{nm}$) is suggestive of antiferromagnetic coupling arising from formation of a sulphate-bridged dimer complex [10] analogous to the structures found in the well-known dimeric carboxylates such as $\mathrm{Cr_2(O_2CMe)_4(OH_2)_2}$.

The Jahn–Teller distorted geometry with two weakly bonded axial water ligands is almost certainly responsible for the extremely fast water-exchange rate on $[Cr(OH_2)_6]^{2+}$ ($k_{ex} \sim 10^9 \, s^{-1}$) [8, 14], similar to that for $[Cu(OH_2)_6]^{2+}$ (see Section 11.1.2). The extreme air sensitivity has precluded detailed study of its solution chemistry. The strongly reducing nature of $[Cr(OH_2)_6]^{2+}$ is exemplified by the Cr^{3+}/Cr^{2+} reduction potential (values quoted ranging from -0.398 to -0.454 V). The most recently reported pK_{11} value for $[Cr(OH_2)_6]^{2+}$ (5.3) [15] seems somewhat low for a divalent transition metal ion (see subsequent chapters) and may need re-evaluation. A number of earlier determinations quote the pK_{11} value more in the range 8.7–11.

Whilst the solution chemistry of $[Cr(OH_2)_6]^{2+}$ is limited in itself its primary importance and use stems from its tendency to participate in inner-sphere redox reactions resulting in the formation of a variety of Cr^{III}(aq) species [16] (see below). The favoured inner-sphere path arises from the σ character of its donor d orbital and retention of the electrons forming what will be the stable t_{2g}^3 configuration of the Cr^{III} product. Despite the often favourable driving force the rates of many $[Cr(OH_2)_6]^{2+}$ redox reactions are still many orders of magnitude less than typical ligand substitution rates on Cr^{II}, due partly to the reorganisation in bond lengths required by the change from tetragonally distorted octahedral (Cr^{II}) to regular octahedral (Cr^{III}). The reduction of a variety of $[Co(NH_3)_5L]^{n+}$ complexes by $[Cr(OH_2)_6]^{2+}$ has been well documented as providing good evidence for the favourability of the inner-sphere pathway [17]. With highly effective bridging ligands L on Co^{III} such as the halides, N_3^- and S-bonded SCN^- bimolecular rate constants are usually of magnitude 10^5–10^6 $M^{-1} \, s^{-1}$, with L transferred across to the Cr^{III} product. When $L = NH_3$, H_2O, OAc^- or pyridine correspondingly much smaller rate constants between 10^{-1} and 10^{-5} $M^{-1} \, s^{-1}$ are observed.

6.1.1.3 THE REACTION OF $[Cr(OH_2)_6]^{2+}$ WITH DIOXYGEN [18]

The reaction of $[Cr(OH)_2)_6]^{2+}$ with O_2 was first examined more than 80 years ago [19]. However this study identified the presence of two intermediates both more highly oxidising than Cr^{VI}(aq) (see later). The simple 1:1 reaction of Cr^{2+}(aq) with O_2 forms the O_2 adduct aqua complex, $[Cr(OH_2)_5(O_2)]^{2+}$. The bimolecular rate constant determined by pulse radiolysis and flash photolysis, $1.6 \times 10^8 \, M^{-1} \, s^{-1}$, is one of the largest for a reaction with O_2 [20]. The equilibrium constant ($6 \times 10^{11} \, M^{-1}$), determined kinetically from the forward and back rate constants, reflects the solution stability of the O_2 'adduct'. Indeed, $[Cr(OH_2)_5(O_2)]^{2+}$ is stable for several hours in the absence of further $[Cr(OH_2)_6]^{2+}$ and O_2 [20, 21]. Resonance Raman measurements (O—O stretching band at 1165 cm^{-1}) indicate that $[Cr(OH_2)_5(O_2)]^{2+}$ is probably best formulated as the η_1-superoxo aqua Cr^{III} complex $[Cr^{III}(OH_2)_5(O_2^-)]^{2+}$ [22]. $[Cr^{III}(OH_2)_5(O_2^-)]^{2+}$ is characterised by two intense absorptions at 290 nm

($\varepsilon = 3100\,M^{-1}\,cm^{-1}$) and 245 nm ($7000\,M^{-1}\,cm^{-1}$), the latter band being assigned to a transition within the O_2^- group. Further reaction of $[Cr^{III}(OH_2)_5(O_2^-)]^{2+}$ with $[Cr(OH_2)_6]^{2+}$ gives rise first to $[(H_2O)_5CrOOCr(OH_2)_5]^{4+}$ and ultimately the dihydroxy-bridged Cr^{III} dimer $[(H_2O)_4Cr(\mu\text{-}OH)_2Cr(OH_2)_4]^{4+}$ [23]. For this reason solutions of $[Cr^{III}(OH_2)_5(O_2^-)]^{2+}$ are best generated in the presence of an excess of O_2. In this way solutions $\sim 10^{-3}\,M$ can be generated. $[Cr^{III}(OH_2)_5(O_2^-)]^{2+}$ reacts with one-electron donors to give the hydroperoxo aqua ion $[Cr^{III}(OH_2)_5(HO_2^-)]^{2+}$. The best yields are obtained from the reaction of the superoxo complex with $[Ru(NH_3)_6]^{2+}$ ($k = 9.5 \times 10^5\,M^{-1}\,s^{-1}$) [22]. This is because $[Ru(NH_3)_6]^{2+}$ does not undergo further reactions. $[Cr^{III}(OH_2)_5(HO_2^-)]^{2+}$ somewhat surprisingly has a reactivity profile very similar to that of H_2O_2 [24]. Reduction to $[Cr(OH_2)_6]^{3+}$ is thus brought about by the same set of reactants that reduce H_2O_2, namely V^{2+}(aq), VO^{2+}(aq), Ti^{3+}(aq), Fe^{2+}(aq) and Co(cyclam)$^{2+}$. Likewise the lack of reactivity of $[Cr^{III}(OH_2)_5(HO_2^-)]^{2+}$ with outer-sphere reductants such as $[Ru(NH_3)_6]^{2+}$ and Co^{2+} complexes of cage-type macrocycles is a characteristic of H_2O_2 [24]. Indeed, despite the presence of the more 'electrophilic' Cr^{3+} centre, gas phase ionization energies, solution reduction potentials and estimated acidity constants all point to aqueous H_2O_2 as being a better one-electron oxidant than $Cr(HO_2)^{2+}$ (aq).

In the presence of CH_3OH the reaction of O_2 with $[Cr(OH_2)_6]^{2+}$ has been shown to give rise to a short-lived Cr^{IV}(aq) species (see Section 6.1.3). The reaction of $[Cr(OH_2)_6]^{2+}$ with alkyl halides gives rise to a whole range of $[RCr(OH_2)_5]^{2+}$ species [25] (see later). The reaction is believed to proceed in two stages:

$$[Cr(OH_2)_6]^{2+} + R\text{---}X \longrightarrow [Cr(OH_2)_5R]^{2+} + X^{\cdot} + H_2O \qquad (6.1)$$

$$[Cr(OH_2)_6]^{2+} + X^{\cdot} \longrightarrow [Cr(OH_2)_5X]^{2+} + H_2O \qquad (6.2)$$

A high-pressure kinetic investigation of the reaction of $[Cr(OH_2)_6]^{2+}$ with a range of alkyl radicals has given rise to an essentially constant $\Delta V^{\ddagger}$ of $(+4.3 \pm 1.0)\,cm^3\,mol^{-1}$, implying an I_d process perhaps common to substitution reactions generally on $[Cr(OH_2)_6]^{2+}$ [26]. Reaction of $[Cr(OH_2)_6]^{2+}$ with $H^{\cdot}$ radicals generated via pulse radiolysis or flash photolysis gives rise to $[CrH(OH_2)_5]^{2+}$ ($k = 1.5 \times 10^9\,s^{-1}$) [27].

6.1.2   Chromium(III) (d³): Chemistry of the $[Cr(OH_2)_6]^{3+}$ Ion

6.1.2.1   PREPARATION AND PROPERTIES OF $[Cr(OH_2)_6]^{3+}$

The blue-violet $[Cr(OH_2)_6]^{3+}$ ion is found in the chrome alum, $MCr(SO_4)_2 \cdot 12H_2O$ [28], and in crystals of the salt $H_2Cr_4(SO_4)_7 \cdot 24H_2O$ [29]. The hydrous nitrate also contains $[Cr(OH_2)_6]^{3+}$ ions whereas green $CrCl_3 \cdot 6H_2O$ contains appreciable amounts of chloroaqua species such as $[Cr(OH_2)_5Cl]^{2+}$ and *cis*- and *trans*-$[Cr(OH_2)_4Cl_2]^+$ [30]. Solutions containing $[Cr(OH_2)_6]^{3+}$ are best prepared via reduction of acidified aqueous solutions of Cr^{VI} with, for example, amalgamated zinc and separation from hydrolytic polymers and Zn^{2+} ions via cation-exchange chromatography. Reduction of Cr^{VI} solutions with H_2O_2 has also been used. Alternatively, oxidation of $[Cr(OH_2)_6]^{2+}$ solutions can be used. Reduction of *o*-benzoquinone with $[Cr(OH_2)_6]^{2+}$ yields an unstable dimeric form of Cr^{III}, once claimed to be $[(H_2O)_5CrOCr(OH_2)_5]^{4+}$ [31] but now known to be semiquinone complex [32]. Acidification of this complex yields $[Cr(OH_2)_6]^{3+}$ [33]. Redox reactions involving oxidation of $[Cr(OH_2)_6]^{2+}$ can generally be used in the absence of ligand transfer to the inert Cr^{III} product.

$[Cr(OH_2)_6]^{3+}$ has the expected regular octahedral structure for a d^3 ion. In the alum the $Cr-OH_2$ bond length is 196.1 pm (Fig. 6.1) [28]. The ion is magnetically dilute and possesses a temperature-independent moment close to the spin-only value for three unpaired electrons, 3.89 BM. Solutions of $[Cr(OH_2)_6]^{3+}$ show two visible absorption bands at 575 nm ($\varepsilon = 13.2\,M^{-1}\,cm^{-1}$) and 408 nm ($15.5\,M^{-1}\,cm^{-1}$) assigned to the $^4A_{2g} \rightarrow {}^4T_{2g}$ and $^4A_{2g} \rightarrow {}^4T_{1g}(F)$ transitions. The third transition, $^4A_{2g} \rightarrow {}^4T_{1g}(P)$, is obscured by LMCT bands below 300 nm. Spin-forbidden $^4A_{2g} \rightarrow {}^2T_{2g}$, ^2E_g (650 nm) and $^4A_{2g} \rightarrow {}^2A_{1g}$ (350 nm) transitions are often observable. The inertness of $[Cr(OH_2)_6)]^{3+}$ and ready preparation in stable concentrated solutions has allowed studies by vibrational spectroscopy [34]. In solid samples of the alum, $Cr-OH_2$ vibrations were observed at $555\,cm^{-1}$ (v_{as} stretch) and $329\,cm^{-1}$ (v_{as} bend).

6.1.2.2 WATER EXCHANGE AND COMPLEX FORMATION ON $(Cr(OH_2)_6]^{3+}$

The rate of water exchange on the $[Cr(OH_2)_6]^{3+}$ ion has been extensively studied and its rate constant was one of the earliest to be determined [35, 36]. Oxygen-18 labelling techniques have been employed using stop-quench methods (NMR methods cannot be used owing to the paramagnetism). Recent improvements in the technique have allowed derivation of the full rate law [36]:

$$k_{ex} = k_1 + k_{OH}[H^+]^{-1} \qquad (6.3)$$

At 25 °C, $k_1 = 2.4 \times 10^{-6}\,s^{-1}$, $\Delta H_1^{\ddagger} = (108.6 \pm 2.7)$ kJ mol^{-1}, $\Delta S_1^{\ddagger} = (+11.6 \pm 8.6)$ JK^{-1} mol^{-1} and $\Delta V_1^{\ddagger} = -9.6\,cm^3\,mol^{-1}$; $k_{OH} = 1.8 \times 10^{-4}\,s^{-1}$, $\Delta H_{OH}^{\ddagger} = (111 \pm 2.5)$ kJ mol^{-1}, $\Delta S_{OH}^{\ddagger} = (+55.6 \pm 8.1)$ JK^{-1} mol^{-1}, $\Delta V_{OH}^{\ddagger} = +2.7\,cm^3\,mol^{-1}$. A significantly wide range of rate constants for complex forming reactions on $[Cr(OH_2)_6]^{3+}$, coupled with the negative activation volume, would suggest an appreciable role played by the incoming ligand in complex formation/water-exchange reactions on the hexaaqua ion (Table 6.1) [37, 38]. On the other hand, rate constants for complex formation on $[Cr(OH_2)_5OH]^{2+}$ cover a much narrower range of values indicating, along with the positive $\Delta V^{\ddagger}$, a much more dissociative mode of reaction in line with comparable behaviour observed on other tripositive aqua and the corresponding dipositive hydroxy aqua ions (see Section 1.4).

Table 6.1. Rate constants for complex forming reactions on
$[Cr(OH_2)_6]^{3+}$ and $[Cr(OH_2)_5OH]^{2+}$ at 25 °C, $\mu = 1.0\,M$

Incoming ligand	$10^7\,k_1$ $(M^{-1}s^{-1})$	$10^5\,k_{OH}$ $(M^{-1}s^{-1})$	Ref.
H_2O	24^a	18^a	[36]
NCS^-	18	9.7	[37]
SCN^-	0.1	2.1	
Cl^-	0.29	4.2	[38]
Br^-	0.09	2.7	
I^-	0.008	0.46	
NO_3^-	7.1	15.0	
SO_4^{2-}	110	61	

^{a} Units of s^{-1}.

As mentioned in Section 1.4, the more dissociative mode of reaction observed on hydroxoaqua ions may be a consequence of the lower 2 + charge (lower δ + charge on the protons of the water ligands of the primary shell to attract the incoming nucleophile), as from a specific electronic effect from the hydroxy ligand. The lack of a pressure dependence on $\Delta V^{\ddagger}$ implies complete retention of the solvation shell during interchange.

From neutron diffraction studies and molecular dynamics simulations a well-defined secondary shell of around 13 water molecules has been confirmed to reside around $[Cr(OH_2)_6]^{3+}$ in aqueous solutions. Merbach and coworkers have recently employed a modified Swift–Connick approach (see Section 1.4) using variable high field ^{17}O NMR to follow the very fast exchange between a secondary hydration shell and bulk water on a paramagnetic metal ion for the first time using $Cr^{3+}(aq)$. The measured kinetic parameters were $k_{ex}^{298} = (7.8 \pm 0.2) \times 10^9\,s^{-1}$, $\Delta H^{\ddagger} = (21.3 \pm 1.1)\,kJ\,mol^{-1}$ and $\Delta S^{\ddagger} = (+16.2 \pm 3.7)\,J\,K^{-1}\,mol^{-1}$, corresponding to a lifetime of 128 ps for a water molecule in the second hydration shell. This value compares very well with a corresponding value of 144 ps calculated using a molecular dynamics simulation and provide further evidence of the usefulness and increasing power of this approach in modelling the structure and dynamics of metal ions within aqueous solutions.

6.1.2.3   HYDROLYSIS OF $[Cr(OH_2)_6]^{3+}$

From the kinetics of complex formation/water exchange it is not possible to determine a value for pK_{11} for formation of $[Cr(OH_2)_5OH]^{2+}$. However, pK values representing the formation of a whole series of mononuclear [39–42] and polynuclear [39–41, 43–49] hydrolysis products have been measured by independent equilibration studies due to the slow attainment of equilibrium in reactions involving Cr^{III}. More is known regarding the hydrolysis behaviour of $[Cr(OH_2)_6]^{3+}$ than arguably any other transition metal ion [39–50]. The following overall hydrolysis constants have been reported: $[Cr(OH_2)_5OH]^{2+}$ ($pK_{11} = 4.29 \pm 0.03$); $[Cr(OH_2)_4(OH)_2]^+$ ($pK_{12} = 9.7 \pm 0.1$) [39–41]; $[Cr(OH_2)_3(OH)_3]$ ($pK_{13} = 18$); $[Cr(OH_2)_2(OH)_4]^-$ ($pK_{14} = 27.4$) [41]. In addition, blue-green dimeric $[(H_2O)_4Cr(OH)_2Cr(OH_2)_4]^{4+}$ ($pK_{22} = 5.06 \pm 0.1$) [39, 43–46], green trimeric $[Cr_3(OH)_4(OH_2)_{9 \text{ or } 10}]^{5+}$ ($pK_{34} = 8.15 \pm 0.15$) [39, 47] and green tetrameric $Cr_4(OH)_6^{6+}$ (aq) [39, 47] forms have been characterised.

6.1.2.4 CHARACTERISATION OF CHROMIUM(III) HYDROLYTIC OLIGOMERIC IONS

The dihydroxy-bridged dimer species (Fig. 6.2) is the major product following the O_2 oxidation of $[Cr(OH_2)_6]^{2+}$ [23] and is readily separated from other Cr^{III} species by cation-exchange chromatography [39–49]. The green trimer, olive-green tetramer and higher oligomers can also be isolated following O_2 oxidation of high concentrations (> 0.5 M) of $[Cr(OH_2)_6]^{2+}$ prepared via dissolution of chromium metal in $HClO_4$ solution. Alternative procedures have involved refluxing aqueous solutions of $Cr(NO_3)_3$ or $Cr(ClO_4)_3$, generated via H_2O_2 reduction of Cr^{VI} in the appropriate acid at pH ~ 3 for several hours [44] or simple addition of OH^- to Cr^{III} solutions [39]. In the former the dimer is the major product.

For the purpose of separating the oligomers various cation-exchange elution methods have been developed. Finholt, Thompson and Connick [47] have used columns of Dowex 50W X2 resin (200–400 mesh) to separate the various oligomers from the monomer by displacement elution with aqueous 0.05 M $Th(ClO_4)_4$. Stunzi and Marty [39] developed methods based on columns of Sephadex SP C25 resin (30 cm × 4 cm) wherein elution was carried out successfully with aqueous solutions of $NaClO_4$ (0.5–4.0 M) containing 0.005–0.04 M $HClO_4$. Elution was performed as follows: 0.5–1.0 M $NaClO_4$, 0.005–0.01 M $HClO_4$ (monomer), 2.0 M $NaClO_4$, 0.02–0.04 M $HClO_4$ (dimer) and 4.0 M $NaClO_4$ (trimer and tetramer). These methods proved superior to simple elution with $HClO_4$ and moreover enabled high concentrations of the various oligomers

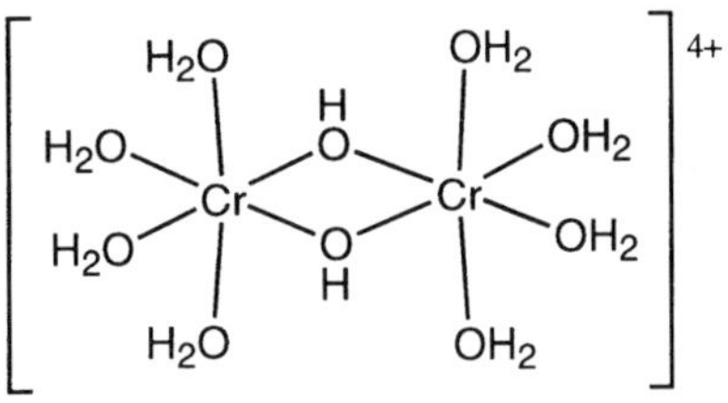

Figure 6.2. Structure of the di-μ-hydroxo bridged dimer, $[(H_2O)_4Cr(\mu\text{-}OH)_2\text{-}Cr(OH_2)_4]^{4+}$

to be obtained at relatively low background acid concentrations. This was a desirable feature which allowed characterisation of several oligomers via determination of the number of OH^- ligands and charge per Cr atom. For the trimer, titration of OH^-/Cr was carried out as follows. The free H^+ content of trimer solutions was found via dilution with water and pH measurement using a glass electrode. Chromium was determined via oxidation to Cr^{VI}. The trimer solution was then made alkaline by adding a known amount of OH^- and oxidised by heating with an excess of 30% H_2O_2. The OH^- content was then determined by titration with standard HCl and by difference the original number of OH^-/Cr could be determined. The charge/Cr was found by loading a known amount of trimer solution on to a column containing excess Dowex 50W X2 resin in its H^+ form. The column was then washed with water and the background and displaced H^+ were collected and titrated with standard NaOH. The free $[H^+]$ is again determined by pH measurement on a diluted sample. The charge/Cr was then determined from

$$\text{Charge/Cr} = \frac{\text{total charge} - \text{free } [H^+]}{Cr} \qquad (6.4)$$

In this way the $Cr_3(OH)_4{}^{5+}$ unit of the trimer was identified [47]. The kinetics and equilibria defining $[H^+]$-assisted conversion of the dihydroxy-bridged dimer to monohydroxy-bridged $[(H_2O)_5Cr(OH)Cr(OH_2)_5]^{5+}$ (Fig. 6.3) [44] and ultimately $[Cr(OH_2)_6]^{3+}$ have been recently investigated [46, 51]. The kinetics of the reverse process, the dimerisation of Cr^{3+}, has also been studied [51], Both $CrOH^{2+}$ (aq) and $Cr(OH)_2{}^+$ (aq) are believed to be involved. Room temperature magnetic moments (per chromium) for the two dimers and trimer are respectively 3.74 BM (doubly hydroxy-bridged), 3.48 BM (single hydroxy-bridged) [44] and 3.37 BM (trimer) [47], showing little evidence of strong spin interactions, the lower value for the single bridged dimer possibly indicative of a more favourable Cr–OH–Cr bond angle, promoting slightly stronger antiferromagnetic coupling. Roughly equal concentrations of the two forms are at equilibrium at 25 °C in 2.0 M $HClO_4$ [44]. The scheme of reactions (Fig. 6.4) for 25 °C, $I = 1.0$ M $NaClO_4$ has been identified.

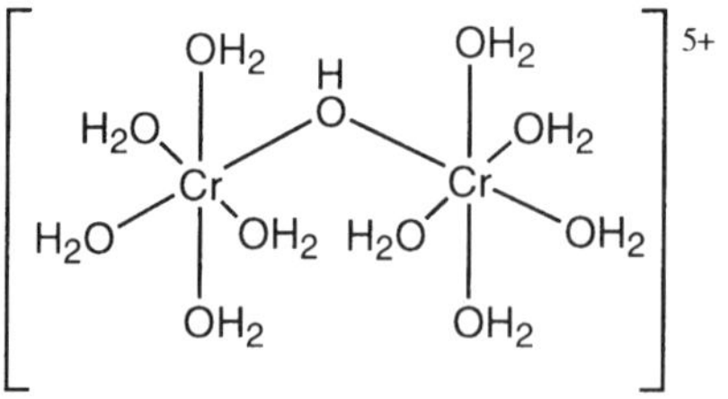

Figure 6.3. Structure of the mono-μ-hydroxo bridged dimer, $[(H_2O)_5Cr(\mu\text{-}OH)\text{-}Cr(OH_2)_5]^{5+}$

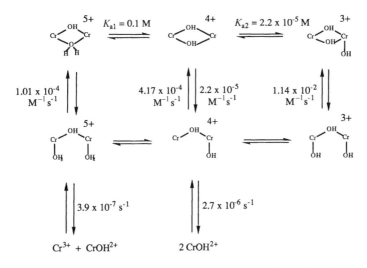

Figure 6.4. Scheme of reactions of aqua chromium dimers in solution

The crystal structure of $[(H_2O)_4Cr(\mu\text{-}OH)_2Cr(OH_2)_4](1,3,5\text{-}Me_3C_6H_2SO_3)_4 \cdot 4H_2O$ (Fig. 6.5) has revealed a Cr—Cr distance of 300.6 pm [45]. No significant lengthening of the Cr—OH$_2$ bonds *trans* to the bridging OH$^-$ groups was observed. Some properties of the various hydrolytic oligomers of Cr^{III} are shown in Table 6.2. The $-J/k$ value of 16 °C has been used as evidence in favour of a triangular structure $[Cr_3(OH)_4(OH_2)_9]^{5+}$ (Fig. 6.6), for the trimeric ion [47], the value of 22 °C found by fitting the linear alternative (Fig. 6.7), seemingly rather large when compared to values in the range 6–10 °C for double hydroxy-bridged dimers. A further trimeric structure, $[Cr_3(OH)_4(OH_2)_{10}]^{5+}$ (Fig. 6.8), is also possible, similar to the unit present in the structure of $[Cr_3(OH)_4(NH_3)_{10}]Br_5$ (Fig. 6.9) [52] (see under Mo later). In the structure of $[Cr_3(OH)_5(tacn)_3]I_4 \cdot 5H_2O$, (Fig. 6.10) [53], the remaining coordinated water is deprotonated and hydrogen bonded to one of the bridging OH$^-$ groups.

Marty and coworkers developed an improved method for generating high yields of the tetramer via addition of OH$^-$ to solutions of the dimer [39, 40]. In a typical experiment the pH of a dimer solution was adjusted to 3.8. After 3 days at 25 °C $\sim$ 50% of the total chromium was recoverable as the tetramer following cation exchange (very little trimer was produced). This led on to suggestions that the predominant route to formation of the tetramer is via the condensation of two dimer units. However, similar treatment of equimolar amounts of trimer and monomer also gave high yields of tetramer. As a result a number of 'cyclic' structures for the tetramer have been proposed (Fig. 6.11). Structure A is also known in the blue-violet 'rhodoso complex' $[Cr_4(OH)_6(NH_3)_{12}]Cl_6$ [54]. Pfeiffer's compound $[Cr_4(OH)_6(en)_6]X_6(X = N_3^-, BPh_4^-, NO_3^-, ClO_4^-$ and 0.5

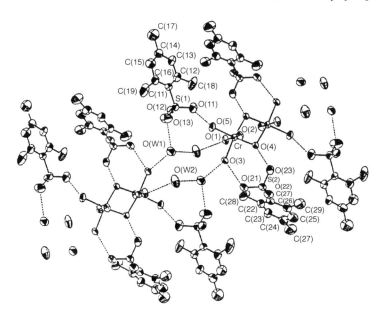

Figure 6.5. Structure of the cationic repeat unit in $[(H_2O)_4Cr(\mu\text{-}OH)_2Cr(OH_2)_4](1,3,5\text{-}Me_3C_6H_2SO_3)_4\cdot4H_2O$ [45]. (Reprinted with permission from L. Spiccia, H. S. Evans, W. Marty and R. Giovanoli, *Inorg. Chem.*, **26**, 474, 1987. Copuright 1987 American Chemical Society)

Table 6.2. Some properties of hydrolytic polymeric ions of Cr^{III} [39, 44, 47]

Species	$pK_{xy}{}^a$	μ_{eff}/Cr (BM)	$-J/k$ (°C)	$4A_{2g}{\to}4T_{2g}$	$4A_{2g}{\to}4T_{1g}(F)$
				nm($\varepsilon(M^{-1}cm^{-1})$)	
$[Cr_2(OH)(OH_2)_{10}]^{5+}$	—	3.48	16.0	592(16.5)	420(20.8)
$[Cr_2(OH)_2(OH_2)_8]^{4+}$	3.68	3.74	7.5	582(17.4)	417(20.4)
	6.04				
$[Cr_3(OH)_4(OH_2)_{9\ or\ 10}]^{5+}$	4.35	3.37	16.0	584(19.2)	425(30.5)
	5.63				
	6.01				
$[Cr_4(OH)_6(OH_2)_{12}]^{6+}$	2.55	—	—	580(15.6)	426(30.3)
	5.08				

$^a pK_{xy}$ defined as in Section 1.3.

$SO_4{}^{2-}$) also contains the rhodoso structure A [55]. Structures B and C were proposed on the basis of formation of tetramer from trimer and monomer solutions. Structure B has been favoured by Stunzi and Marty [39] in order to explain the particular low first pK_a (2.55) of the tetramer via formation of

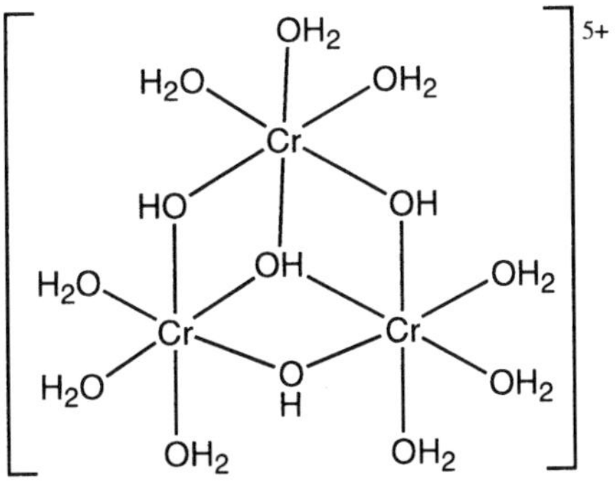

Figure 6.6. Cyclic Structure for Trimeric $Cr^{III}(aq)$

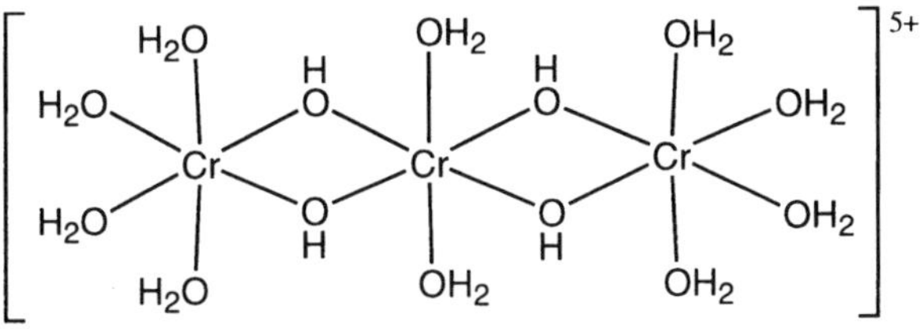

Figure 6.7. Linear Structure for Trimeric $Cr^{III}(aq)$

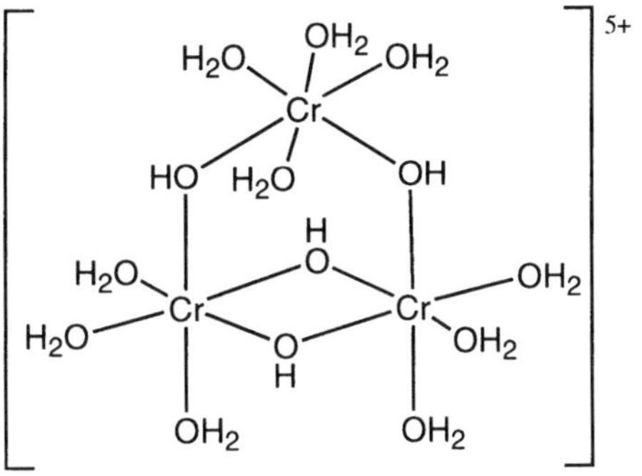

Figure 6.8. Alternative Cyclic Structure for Trimeric $Cr^{III}(aq)$

structure D, but its existence has been challenged. Monsted, Monsted and Springborg [56] have moreover challenged the necessitated formation of 'cyclic' tetramers overall by postulating stabilisation of other more classical 'linear' structures E and F (Fig. 6.12) via intramolecular hydrogen bonding analogous to that occurring in dimeric species. Structure F is moreover established in the violet-blue tetrameric compound $\Lambda\Lambda\Lambda\Lambda$-$[Cr_4(OH)_6(en)_6]X_6$ ($X = Br^-, I^-, NO_3^-$) [57].

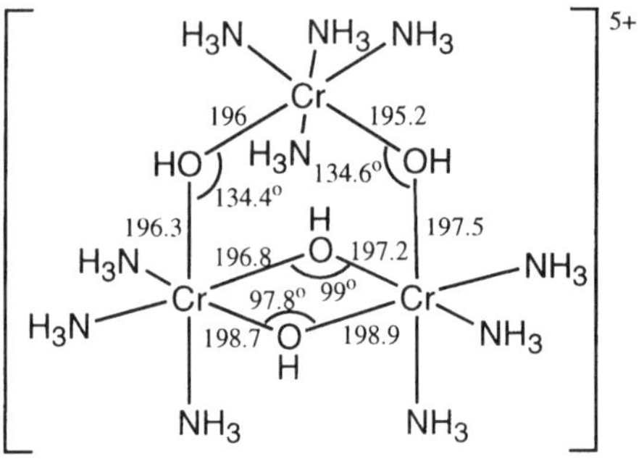

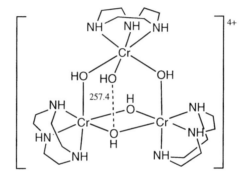

Figure 6.9. Cyclic Structure of $[Cr_3(\mu\text{-OH})_4(NH_3)_{10}]^{5+}$

Figure 6.10. Structure of the cationic unit present in the complex $[Cr_3(OH)_5(tacn)_3]I_4 \cdot 5H_2O$

Further work by Marty and coworkers has established that active forms of Cr^{III} 'hydroxide' can be isolated which retain the oligomeric structure present in the original aqueous Cr^{III} solution. Acidification of an active dimer hydroxide '$Cr_2(\mu\text{-OH})_2(OH)_4(aq)$' was used to obtain highly concentrated dimer solutions from which crystals of the $1,3,5\text{-Me}_3C_6H_2SO_3{}^-$ salt above suitable for X-ray diffraction were subsequently isolated [45]. Spiccia and Marty [50] have studied the ageing process occurring to $Cr(OH)_3(OH_2)_3$ following precipitation and its effect on the distribution of oligomeric solution products upon reacidification. Acidification one minute after precipitation allowed 99.4% of the chromium to be recovered as monomer, $[Cr(OH_2)_6]^{3+}$. At times thereafter the amount of monomer decreased at the expense of oligomer formation. The distribution was pH dependent. After 16 hours buffered at pH 5.06, 25 °C and $I = 1.0 \, \text{M} \, (NaClO_4)$, acidification showed that 19% of the chromium was present in the form of lower oligomers, i.e. dimer, trimer and tetramer, whereas 27% was present as higher oligomers ($n > 4$). After 72 hours the amount of higher oligomers rose to 54% at

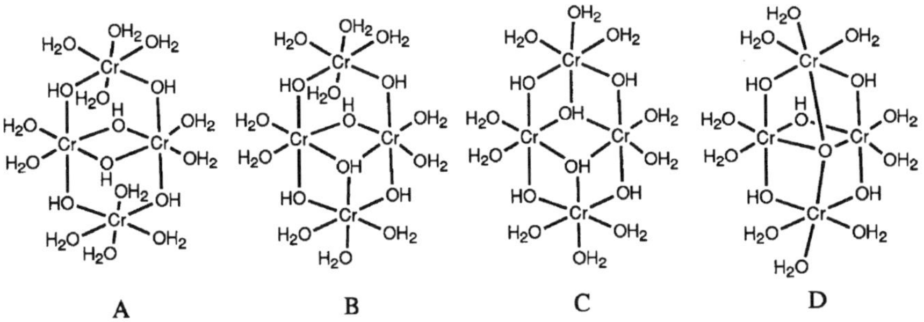

Figure 6.11. Proposed 'cyclic' structures for tetrameric $Cr_4(OH)_6)(aq)^{6+}$ [39]

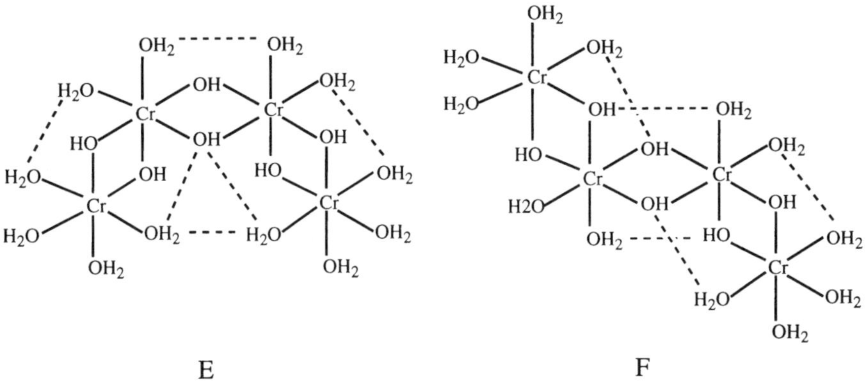

Figure 6.12. Alternative 'classical' structures for tetrameric $Cr_4(OH)_6{}^{6+}(aq)$. Stabilising hydrogen bonds are shown by (- - - - -) [56]

the expense of lower oligomers (13%) and monomer (33%). Between pH 6 and 8.5, however, the amount of monomer recovered never fell below 87%, even after 72 hours. The species were separated and identified following cation-exchange column chromatography using Sephadex SP C25 as described above [39]. The higher oligomers were eluted with saturated solutions of potassium oxalate in 0.2 M NaOH. Stability constants for the overall formation of hydrolytic dimer, trimer and tetramer from $[Cr(OH_2)_6]^{3+}$ have been reported [58].

Kinetic studies of the water-exchange process on the aqua dimer have been performed using ^{18}O labelling studies at $I = 1.0$ M [59]. Three distinct pathways for exchange were identified. Two of the rate constants have been assigned to

Table 6.3. Kinetic parameters for water exchange on $Cr_2(\mu\text{-}OH)_2^{4+}$(aq) [59]

	$Cr_2(\mu\text{-}OH)_2^{4+}$(aq)	$Cr_2(\mu\text{-}OH)_2(OH)^{3+}$(aq)
Trans to μ-OH		
$k(298)(\mathrm{s}^{-1})$	3.6×10^{-4}	$2.6 \times 10^{-6\,a}$
$\Delta H^{\ddagger}(\mathrm{kJ\,mol}^{-1})$	81 ± 8	157 ± 2
$\Delta S^{\ddagger}(\mathrm{J\,K}^{-1}\,\mathrm{mol}^{-1})$	-40 ± 26	174 ± 8
$k(298)/K_a^{\,b}(\mathrm{s}^{-1})$		1.2×10^{-2}
Cis to μ-OH		
$k(298)\,(\mathrm{s}^{-1})$	6.6×10^{-5}	$1 \times 10^{-6\,a}$
$\Delta H^{\ddagger}(\mathrm{kJ\,mol}^{-1})$	97 ± 10	114 ± 4
$\Delta S^{\ddagger}(\mathrm{J\,K}^{-1}\,\mathrm{mol}^{-1})$	0 ± 32	23 ± 14
$k(298)/K_a^{\,b}(\mathrm{s}^{-1})$		4.8×10^{-3}
Cleavage of OH *bridge*		
$k(298)(\mathrm{s}^{-1})$	1.1×10^{-5}	$3.1 \times 10^{-7\,a}$
$k(298)/\,K_a^{\,b}/(\mathrm{s}^{-1})$	.	1.5×10^{-3}

^{a} Units of $\mathrm{M}^{-1}\,\mathrm{s}^{-1}$.
^{b} $K_a(298)$ measured as 2.09×10^{-4} M.

exchange of the coordinated water molecules occupying positions *trans* (more labile) and *cis* to the bridging OH groups. Each show a dependence on $[\mathrm{H}^+]$ of the type $k = k_0 + k_{\mathrm{OH}}[\mathrm{H}^+]^{-1}$, consistent with the additional involvement of monodeprotonated forms. The third is believed to involve an OH bridge cleavage and reformation reaction. Some of the relevant kinetic parameters are listed in Table 6.3.

Interestingly, a definitive correlation was found to exist between the rate constants for the various water exchange, OH bridge cleavage and complex formation processes on Cr^{III} aqua oligomers with the OH/Cr ratio. The finding in these studies of differing labilisation of the waters *cis* and *trans* to the bridging OH groups also gives strong support for the electronic nature of the so-called 'conjugate base effect'.

Extensive studies of the hydrolysis of $[Cr(OH_2)_6]^{3+}$ have also been reported in the Chinese literature by Luo *et al.* [60]. Linked hydroxo- or oxo-bridged chromium atoms (Fig. 6.13) are believed to play a key role in the hardening and fixing process that occurs upon heating in the so-called 'chrome tannage' process [61] used in the manufacture of leather from skin hide.

6.1.2.5 AQUA IONS OF CHROMIUM (III) OF GENERAL FORMULA $[CrL(OH_2)_5]^{2+}$

The tendency of inert Cr^{III} to capture ligands present in, or transferred to, its first coordination sphere during redox reactions involving oxidation of

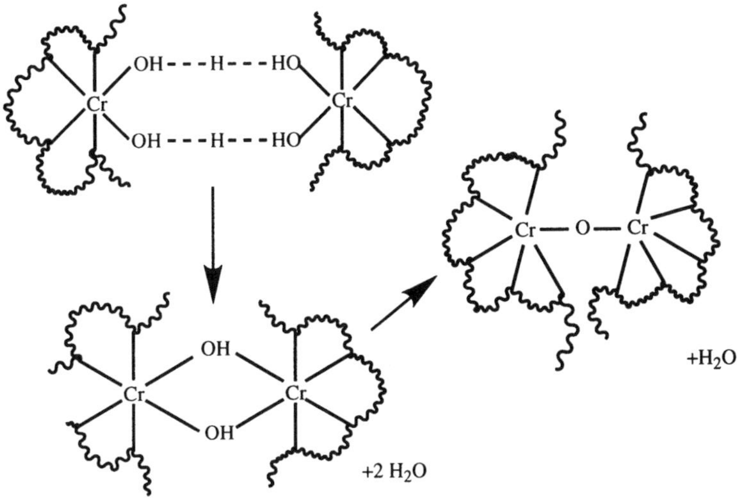

Figure 6.13. Formation of hydroxo- and oxo-bridged species (olation and oxolation) in the chrome tannage hardening process

$[Cr(OH_2)_6]^{2+}$ has been exploited to prepare a whole series of Cr^{III} aqua ions of general formula $[CrL(OH_2)_5]^{2+}$ and a brief consideration of these is appropriate.

$[Cr(SH)(OH_2)_5]^{2+}$ [62, 63]   This green ion is prepared via anaerobic oxidation of $[Cr(OH_2)_6]^{2+}$ with PbS, Ag_2S, S_n^{2-} or $S_2O_3^{2-}$. The highest yields (10–20%) are obtained with S_n^{2-}. Protonation to give $[Cr(SH_2)(OH_2)_5]^{2+}$ does not readily occur, in agreement with the expected low pK_a ($\ll 0$) of coordinated H_2S. However, kinetic evidence for the transient existence of the protonated form has been found from studies of its anation reaction with NCS^- and aquation reaction to give $[Cr(OH_2)_6]^{3+}$ [63, 64]. The aquation processes are easily monitorable by observation of the disappearance of the intense S—Cr charge transfer band at 258 nm ($\varepsilon = 6520\,M^{-1}\,cm^{-1}$) [64]. The favourable path for aquation is via $[Cr(SH_2)(OH_2)_5]^{3+}$. Infra-red bands in brown-green solid $[Cr(SH)(OH_2)_5]SO_4$ are observed at $2560\,cm^{-1}$ (v_{SH}) and $340\,cm^{-1}$ (v_{MS}) [63]. A labilising effect from the SH^- ligand vs, for example, OH^- in $[Cr(OH)(OH_2)_5]^{2+}$ has been detected with regard to the rates of substitution at the five remaining water ligands. Rate constants (25 °C) for 1:1 NCS^- anation are respectively $8.6 \times 10^{-1}\,M^{-1}\,s^{-1}$($CrSH^{2+}$) and $4.9 \times 10^{-5}\,M^{-1}\,s^{-1}$($CrOH^{2+}$). A ratio $> 10^3$ may be relevant between corresponding reactions on $CrSH^{3+}$(aq) and Cr^{3+}(aq). Anaerobic oxidation with I_2 yields the yellow-brown dimer $[(H_2O)_5CrSSCr(OH_2)_5]^{4+}$. Here the species $[CrS(OH_2)_5]^+$ is the kinetically active form. Many compounds of general formula $[Cr(SR)(OH_2)_5]^{2+}$ (R = alkyl or aryl) have now been characterised [65, 66].

$[CrR(OH_2)_5]^{2+}$ These alkyl Cr^{III} species form the second major class of substituted pentaaqua ions prepared in this case via reaction of $[Cr(OH_2)_6]^{2+}$ with RCl. Recent reviews have appeared [25]. Although no $[CrR(OH_2)_5]^{2+}$ species has been structurally characterised clear evidence of a *trans* labilisation from the alkyl group has been detected in related species. The lability vs $[Cr(OH_2)_2]^{3+}$ is shown by the following half-times respectively (25 °C) for 1:1 complexation with NCS^-: 42 s ($[(H_2O)_5CrCH_2Cl]^{2+}$), 128 s ($[(H_2O)_5CrCH$-$Cl_2]^{2+}$) and 9.3×10^6 s ($[Cr(OH_2)_6]^{3+}$) [67]. The labilising is a kinetic effect since the equilibrium constants for all three species are essentially identical. Reaction of $[CrR(OH_2)_5]^{2+}$ with H_3O^+ gives $[Cr(OH_2)_6]^{3+}$ and R—H. Interestingly $[Cr(OH_2)_6]^{3+}$ and not $[Cr(OH_2)_5X]^{2+}$, in addition to RX and X^-, are rapidly formed on treatment of $[CrR(OH_2)_5]^{2+}$ with X_2 (X = Cl or Br), suggesting that X^- forms only after generation of $[Cr(OH_2)_6]^{3+}$ [25]. With alkyl groups carrying β-H or halogen groups treatment with H_3O^+ leads to β elimination, giving $[Cr(OH_2)_6]^{3+}$ and the corresponding alkene.

6.1.3 Chromium(IV)(d^2): Characterisation of a Cr^{IV} Aqua Ion

Bakac, Espenson and coworkers have recently re-examined the O_2 oxidation of $[Cr(OH_2)_6]^{2+}$ under certain conditions. They initially found that when $[Cr(OH_2)_6]^{2+}$ is injected into acidic aqueous solutions containing O_2 at very low Cr/O_2 ratios < 0.05:1, $Cr^{III}(O_2^-)^{2+}$(aq) is formed quantitatively [22]. Stopped-flow mixing of such solutions of $Cr(O_2^-)^{2+}$(aq), the μ-peroxo ion, $CrOOCr^{4+}$(aq) [68], or Tl^{3+}(aq) with an equal concentration of $[Cr(OH_2)_6]^{2+}$ (0.1–0.5 mM) in acidic $HClO_4$ solution (0.1–1.0 M) has been shown to generate a new chromium aqua species [69]. The same species is also generated via mixing of $[Cr(OH_2)_6]^{2+}$ and O_2 in a 1:1 ratio (stopped flow) or a 1:5 ratio (syringe injection). Yields are low (15–50%) because of competing oxidation of $[Cr(OH_2)_6]^{2+}$. The weak absorption in the visible region (Fig. 6.14) [70] and short life time ($t_{1/2} = 0.50$ min at 25 °C, $I = 1.0$ M) have made direct spectral characterisation of the new chromium aqua species very difficult. However, the presence of Cr^{IV} has been deduced on the basis of its stoichiometric reactivity with Ph_3P to generate $OPPh_3$ and $[Cr(OH_2)_6]^{2+}$ [70]:

$$CrO^{2+}(aq) + Ph_3P \xrightarrow[k = (2.1 \pm 0.2) \times 10^3 M^{-1} s^{-1}]{} Ph_3PO + [Cr(OH_2)_6]^{2+} \quad (6.5)$$

It was found subsequently that the addition of methanol to solutions of Cr^{IV} produced via mixing $[Cr(OH_2)_6]^{2+}$ and O_2 generated the spectrum of $Cr^{III}(O_2^-)^{2+}$(aq). The rate of appearance of $Cr(O_2^-)^{2+}$(aq) was first order in

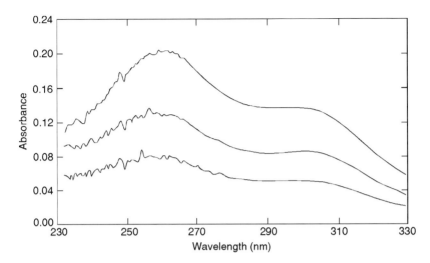

Figure 6.14. Difference spectrum of $Cr^{IV}O^{2+}(aq)$ formed by stopped-flow mixing of 0.3 mM $Cr^{2+}(aq)$ and 0.26 mM O_2 in 1.0 M $HClO_4$. Time interval 20 s (extrapolated ε value at 260 nm $= (5 \pm 1) \times 10^3 \, M^{-1} \, cm^{-1}$) [70]. (Reprinted with permission from S. L. Scott, A. Bakac and J. H. Espenson, *J. Am. Chem. Soc.*, **114**, 4205 (1992). Copyright 1992 American Chemical Society)

methanol and zero order in O_2. However, one mole of O_2 is required in the overall process:

$$CrO^{2+}(aq) + CH_3OH + O_2 \xrightarrow[k = (52.2 \pm 1.4 M^{-1} s^{-1})]{} Cr(O_2^-)^{2+}(aq) + HCHO + H_2O$$

(6.6)

An intercept in the methanol-dependent plots signified a methanol-independent decomposition of Cr^{IV} ($k = 0.033 \, s^{-1}$) and consistent with this, if the methanol addition was delayed by several minutes, $Cr(O_2^-)^{2+}(aq)$ is not formed. A significant kinetic isotope effect found on replacing CH_3OH with CD_3OH ($k_H/k_D = 3.46$) supports C—H bond breaking and formation of HCHO. A general hydride transfer $2e^-$ mechanism has been proposed. Two overall steps are probably involved:

$$CrO^{2+}(aq) + CH_3OH \xrightarrow{k} HCHO + Cr^{2+}(aq) + H_2O \qquad (6.7)$$

$$Cr^{2+}(aq) + O_2 \xrightarrow{fast} CrO_2^{2+}(aq) \qquad (6.8)$$

A structure similar to that of $VO^{2+}(aq)$ (Fig. 6.15) has been proposed for $Cr^{IV}(aq)$ based upon its 1:1 reaction with Ph_3P to give Ph_3PO.

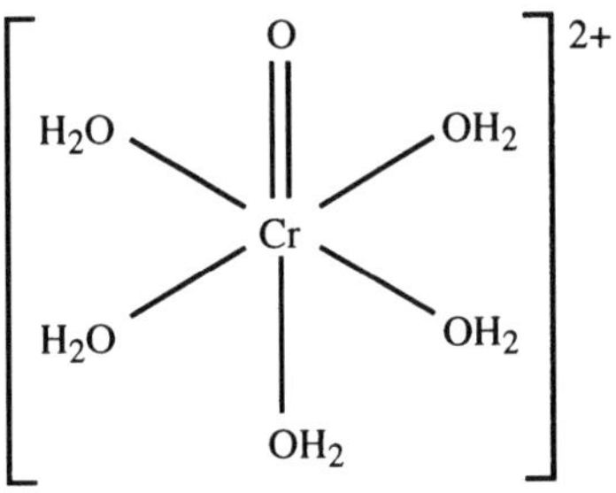

$$\left[\begin{array}{c} \overset{\displaystyle O}{\underset{\displaystyle OH_2}{\overset{\displaystyle \|}{\underset{\displaystyle |}{Cr}}}} \end{array}\right]^{2+}$$

Figure 6.15. Proposed structure for the Cr^{IV} aqua ion, $[CrO(OH_2)_5]^{2+}$

$Cr^{III}(O_2^-)^{2+}$(aq) has been observed as a product of the Cr^{VI} oxidation of a range of alcohols in the presence of O_2, leading to the proposal of $Cr^{IV}=O^{2+}$(aq) as a necessitated intermediate. On the basis of its oxidation to a stable $Cr^V=O$ species a $Cr^{IV}=O^{2+}$ complex with the ligand 2-ethyl-2-hydroxybutanoate has been characterised as an intermediate following reaction of Cr^{VI} with As^{III} in solutions buffered with the ligand at $\sim$ pH 3 [71]. Preparative routes to, and subsequent reactions of, CrO^{2+}(aq) are summarised in Scheme 6.1. Despite a number of stable compounds containing the $Cr^V = O^{3+}$ group no aqua ions containing Cr^V have any appreciable lifetime.

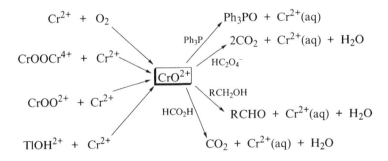

Scheme 6.1. Syntheitc routes to, and reactions of, CrO^{2+}(aq)

6.1.4. Chromium(VI) (d^0): The Anions $[CrO_4]^{2-}$ and $[Cr_2O_7]^{2-}$ [72]

Cr^{VI} is a powerful oxidising agent in acidic solution as judged by the following standard potential:

$$Cr_2O_7^{2-} + 6e^- + 14 H^+ \rightleftharpoons 2 Cr^{3+} + 7 H_2O, \qquad E^0 = +1.33 \text{ V} \qquad (6.9)$$

No cationic chemistry is relevant, the aqueous chemistry being restricted basically to that of the oxo anions, $Cr_2O_7^{2-}$ and CrO_4^{2-}, the orange dimeric form persisting in acidic solution. Oxygen exchange occurring from water on $Cr^{VI}(aq)$ has been measured between pH 7 and 12 using oxygen-18 labelling techniques [73]. The following rate law has been established:

$$\text{Rate} = k_1[CrO_4^{2-}] + k_2[HCrO_4^-] + k_3[H^+][HCrO_4^-]$$

$$+ k_4[HCrO_4^-][CrO_4^{2-}] + k_5[HCrO_4^-]^2 \qquad (6.10)$$

At $25\,°C, I = 1.0\,M, k_1 = 3.2 \times 10^{-7}\,s^{-1}, k_2 = 2.3 \times 10^{-3}\,s^{-1}, k_3 = 7.3 \times 10^5\,M^{-1}s^{-1}$, $k_4 \sim 10^{-3}\,M^{-1}s^{-1}$ and $k_5 = 9.0\,M^{-1}s^{-1}$. Perhaps not surprisingly the most favourable pathway is k_3. It has been suggested that protonation helps to facilitate exchange by polarizing the Cr=O bond. The rate of attainment of the

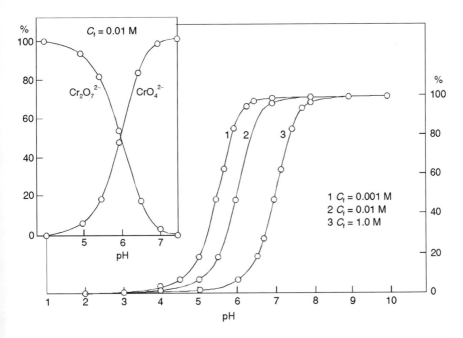

Figure 6.16. Distribution curves for $Cr^{VI}(aq)$ species as a function of pH [72]. (Reprinted from *Coordination Chemistry Reviews*, **109**, M. Cieslak-Golonka, p. 223, 1991, with kind permission from Elsevier Science S. A., Lausanne, Switzerland)

following equilibria has been measured by flow and relaxation techniques [74]:

$$Cr_2O_7^{2-} + H_2O \rightleftharpoons 2HCrO_4^-$$ (6.11)

$$HCrO_4^- \rightleftharpoons CrO_4^{2-} + H^+$$ (6.12)

The first step can be either acid or base catalysed. The existence of $HCrO_4^-$ as a long-lived intermediate species has been questioned on the basis of Raman measurements [75, 76]. Recent studies on the nucleophilic catalysis of the hydrolysis of $Cr_2O_7^{2-}$ by certain buffer bases have, however, been interpreted as involving the formation of $HCrO_4^-$ [77]. Distribution curves showing the species $[Cr_2O_7]^{2-}$ and $[CrO_4]^{2-}$ based upon available data are illustrated in Fig. 6.16. The highly negative $\Delta V^{\ddagger}$ values for base-catalysed hydrolysis of $Cr_2O_7^{2-}$ by OH^-, NH_3, H_2O and lutidine are in agreement with an associatively activated addition process [78].

6.2 MOLYBDENUM AND TUNGSTEN

Molybdenum has a diversity of chemistry arguably unrivalled by any other metallic element. A wealth of stable compounds, mononuclear and polynuclear and involving a wide range of donor atoms and coordination geometries, exists in all oxidation states from zero to the maximum $+6$ state. In addition the various oxidation states are characterised by a rich redox chemistry. It is no surprise therefore to find that molybdenum is the only second row transition element known to serve at the active site of a number of redox metalloenzymes. Moreover, many of these enzymes catalyse crucial steps involved with carbon, nitrogen and sulfur metabolism in bacteria, plants and higher animals and as a result they are essential for maintainance of a stable healthy organism [79]. A molybdenum–iron–sulfur cluster, of stoichiometry $MoFe_7S_8$, is established at the active site of a principal group of bacterial nitrogenases found in the root nodules of many leguminous plants. Here the molybdenum centre plays an important role in the mediation of electrons and protons to what appears to be an iron-bound N_2 [80]. Recently one tungsten-containing metalloenzyme has been identified.

The aqueous chemistry of molybdenum is also rich and diverse with every aspect represented from classical Werner coordination compounds (e.g. $[Mo(NH_3)_6]^{3+}$ and $[Mo(OH_2)_6]^{3+}$), organometallic compounds (e.g. $[(\pi$-$C_6H_6)Mo(OH_2)_3]$), isopolyanion aggregates (e.g. $Mo_8O_{26}^{4-}$), metal–metal bonded clusters (e.g. $[Mo_3O_4(OH_2)_9]^{4+}$) and multiple metal–metal bonded compounds (e.g. $[(H_2O)_4Mo\equiv Mo(OH_2)_4]^{4+}$) [81]. The propensity for both molybdenum and tungsten to form metal–metal bonded species, particularly in the lower oxidation states, continues the trend established in the heavier Group 5 members and is somewhat responsible for the rich diversity of aqueous compounds known. Additional factors such as a relatively low electronegativity, particularly in the oxidation states below VI, coupled with an appropriate d-electron count in the various oxidation states, results in molybdenum being the only d-block element having cationic oxo-aqua ion species which characterise five different oxidation states, II to VI [82]. A number of related mixed-valence species are also known. Within the transition metal series only vanadium (four) comes anywhere near matching this (evidence perhaps of the diagonal relationship). In many aspects the aqueous chemistry of tungsten parallels that of molybdenum although cationic aqua ions of tungsten are at present restricted to the IV and V states. Even here the species found for tungsten are inherently more unstable than the corresponding ones for molybdenum, primarily because of ready oxidation to the W^{VI} state. However, a number of these have been studied and have offered a comparison of chemical properties between the two elements. Some definitive trends are emerging. It is therefore opportune to review the aqueous chemistry of the heavier Group 6 members together as we have done for the previous two Groups. Detailed methods describing the synthesis of many of the molybdenum aqua ions have been reviewed. There are no aqua ions for either element representing oxidation states below II.

6.2.1 Molybdenum(II) and Tungsten(II)(d^4)

6.2.1.1 QUADRUPLY-BONDED Mo^{II}_2 SPECIES: THE AQUA DIMER ION $[Mo_2(OH_2)_8]^{4+}$

The chemistry of these species was developed during the 1960s, the first species to be characterised being the dimolybdenum tetracarboxylates analogous to $Cr_2(O_2CCH_3)_4 \cdot 2H_2O$ [83]. However, in contrast to Cr^{II}, quadruply-bonded dimers for Mo^{II} exist without the need for bridging ligands, a consequence of the much more effective overlap between the 4d orbitals on molybdenum. As a result the 'aqua ion' of Mo^{II} exists not as $[Mo(OH_2)_6]^{2+}$ (cf. Cr) but as the quadruply-bonded dimer $[(H_2O)_4 Mo\equiv Mo(OH_2)_4]^{4+}$ (Fig. 6.17), which was first characterised by Bowen and Taube in 1974 [84]. It can be viewed as the prototypal quadruply-bonded Mo^{II}_2 species from which all others are derived.

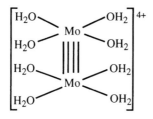

Figure 6.17. The structure of quadruply-bonded $Mo^{II}_2(OH_2)_8]^{4+}$

6.2.1.2   Preparation of $[Mo_2(OH_2)_8]^{4+}$

There are two established preparative routes to quadruply-bonded Mo^{II}_2 species. The most established is the refluxing of $Mo(CO)_6$ in mixtures of acetic acid and acetic anhydride to first make the yellow tetraacetate, $Mo_2(O_2CCH_3)_4$ [85]. This reaction also produces Mo^{IV}_3 clusters (see Section 6.2.3). The tetraacetate is then converted first to the red octachloride $[Mo_2Cl_8]^{4-}$ by treatment with 8.0 M HCl followed by KCl to precipitate the K^+ salt [86] and then finally to the pinkish-red tetrasulfate $[Mo_2(SO_4)_4]^{4-}$ by treatment with excess K_2SO_4 [87]. $[Mo_2(SO_4)_4]^{4-}$ is the 'lead-in' to the preparation of the pinkish-red aqua ion via treatment with a Ba^{2+} salt of an appropriate non-complexing acid, usually $Ba(CF_3SO_3)_2$. All of these reactions (Fig. 6.18) are complete within a few minutes at room temperature although stirring is normally continued for a few hours. Solutions of the aqua ion have been used as obtained after filtration of the precipitated $BaSO_4$ but can be further purified by ice-cold air-free ion-exchange chromatography. Perchlorate ions cannot be used owing to rapid oxidation of the Mo^{II} ion. A second route involving reduction directly from Mo^{VI} has also been devised. The key step appears to involve the generation of dimeric Mo^{III} species such as $[Mo_2Cl_9]^{3-}$ containing $Mo\equiv Mo$ bonds. In a preparative procedure reported by Bino and Gibson in 1980 [88] a solution of molybdenum trioxide (2.0 g) in 50 cm³ of HCl (12.0 M) was reduced electrolytically using a platinum cathode and a carbon anode separated in the cell by a porous clay pot.

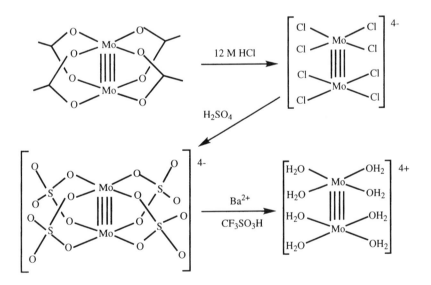

Figure 6.18.   Preparative route to $[Mo_2(OH_2)_8]^{4+}$

Use of a mercury pool cathode for this reduction has also been widely documented. The product of this reduction is Mo^{III} ($[MoCl_6]^{3-}$ with some $[MoCl_5(OH_2)]^{2-}$) [89]. This solution is then evaporated to near dryness on a rotary evaporator, leading to a condensation reaction to produce dimeric $[Mo_2Cl_9]^{3-}$ [90]. At this point $80\,cm^3$ of HCl (0.6 M) is added and the solution passed down an ice-cold Jones reductor (Zn/Hg) column ($15\,cm \times 1\,cm$) into a solution of sodium acetate (5.0 g) in $30\,cm^3$ H_2O. Yields of precipitated yellow tetraacetate as high as 80% (2.37 g in the present case) have been reported by this route, far in excess of those normally obtained starting from $Mo(CO)_6$. The aqua ion is then prepared by the route as shown in Fig. 6.18. A 60% yield of Mo_2^{4+} (aq) has also been obtained following the mercury pool electrolytic reduction of Mo^{VI}(aq) in a mixture of HCl (0.5 M) and Me_4NCl (3.5 M), the $[Mo_2Cl_9]^{3-}$ ion presumably formed as an intermediate under these conditions.

The d^4 configuration is appropriate for quadruple-bond formation in allowing the filling of each of the one σ (d_{z^2}–d_{z^2}) two π ($d_{xz,yz}$–$d_{xz,yz}$) and one δ (d_{xy}–d_{xy}) bonding molecular orbitals. The remaining $d_{x^2-y^2}$ orbital is believed to be involved primarily with bonding to the terminal H_2O ligands. However, an involvement in the δ interaction has also been proposed [83]. Although the $[Mo_2(OH_2)_8]^{4+}$ ion has not as yet been crystallised it is assumed to have the same eclipsed structure as adopted by the octachloro complex (Fig. 6.17). It is also possible, but not confirmed, that two weakly bonded axial H_2O ligands are present. The strength of the δ interaction can be assessed by the energy of the δ–δ^* transition, this being a more reliable indicator than bond length. The prominent appearance of the δ–δ^* transition in the visible region around 500 nm is responsible for the pinkish-red colour of both the aqua ion and octachloro complex. In the tetraacetate the yellow colour arises from a shift in the δ–δ^* transition to higher energy (450 nm), reflecting the ability of the acetate bridges to bring the metal centres closer together and thus resulting in greater δ overlap. However, when monodentate ligands are present the wavelength of the δ–δ^* transition is invariably around 500 ($\pm$ 20) nm, its invariance with the nature of the terminal ligands confirming the assignment. The record for the shortest $Mo\equiv Mo$ bond is held by the bridged bidentate complex with 2-hydroxy-6-methylpyridine ($Mo\equiv Mo$ 207 pm). EXAFS measurements indicate a Mo—Mo separation for the aqua ion as 212 pm [91]. Mo_2^{4+} species also have characteristic low field ^{95}Mo NMR resonances occurring around 3000–4000 ppm from MoO_4^{2-}, the highest frequencies by far of any molybdenum-containing species [92]. As a result the ^{95}Mo chemical shift scale for all molybdenum compounds now covers some 7000 ppm [93]. In 1.0 M CF_3SO_3H, $[Mo_2(OH_2)_8]^{4+}$ resonates at 4056 ppm [92].

Quadruply-bonded W^{II}_2 compounds are fewer owing to the tendency towards oxidation of the W^{II} units. It appears that no equivalent aqua ion for tungsten exists, the W_2^{4+} unit being apparently too reducing to exist in the presence of H_3O^+. Correspondingly, $W_2(O_2CCH_3)_4$ and $[W_2Cl_8]^{4-}$ cannot be made via the same routes as for the molybdenum species and require synthesis under non-aqueous conditions. The dark blue $[W_2Cl_8]^{4-}$ unit has been identified in the crystalline compound $Na_4(TMEDA)_4[W_2Cl_8]$ obtained following Na/Hg reduction of $(WCl_4)_x$ in THF followed by treatment with TMEDA [94]. The blue colour arises from the appearance of the δ–δ* transition occurring now at 600 nm. The much weaker δ overlap is apparent in the much longer bond lengths of W_2^{4+} compounds, typically 10 pm longer than their Mo_2^{4+} counterparts. The reason is believed to relate to the poorly expanded nature of the 5d orbitals on W in the spacial direction appropriate for δ overlap, which cannot compensate for the greater repulsion from the more electron dense cores of the tungsten atoms. Thus ready oxidation to triply-bonded W_2^{III} species is seen (loss of the δ bond), which in contrast gives evidence of stronger W—W bonds than in the case of Mo in a number of cases, a good example being $[W_2Cl_9]^{3-}$. Quadruply-bonded W_2^{II} compounds are also characterised by extreme downfield [183]W NMR chemical shifts vs W^{VI} (WO_4^{2-}) [82].

6.2.1.3 REACTIONS OF $[Mo_2(OH_2)_8]^{4+}$ AND RELATED SPECIES

The results of only a few studies in solution on the aqua ion $[Mo_2(OH_2)_8]^{4+}$ have appeared. A kinetic study of substitution at the aqua ligands by NCS^- and oxalate was reported by Finholt, Leupin and Sykes in 1983 [95]. Consistent with the evidence of rapid ligand replacement reactions in the steps in Fig. 6.18, the studies required the use of stopped-flow techniques. For NCS^- as the incoming ligand, Mo_2^{4+}(aq) in tenfold excess, equilibrium kinetics were relevant and the rate law $k_{eq} = k_1[Mo_2^{4+}] + k_{-1}$, for reactions studied in $I = 0.1$ M ($NaCF_3SO_3$). Rate constants (25 °C) obtained were $k_1 = 590 \, M^{-1} s^{-1}$ and $k_{-1} = 0.21 \, s^{-1}$. Activation parameters for k_1, $\Delta H^{\ddagger} = 57.6 \, kJ \, mol^{-1}$ and $\Delta S^{\ddagger} = +3.0 \, J \, K^{-1} \, mol^{-1}$ are similar to those characterising substitution by NCS^- on both $[VO(OH_2)_5]^{2+}$ and TiO^{2+}(aq), suggesting a common mechanism involving initial coordination at the weakly bonded axial position followed by slower movement into the equatorial position (Fig. 6.19). A much slower reaction occurs with $HC_2O_4^-$ as the incoming ligand, $k_1 = 0.49 \, M^{-1} s^{-1}$, implying here a different rate-determining process perhaps involving carboxylate bridge formation. The respective $\Delta H^{\ddagger}$ values characterising the dissociation steps (57.2 kJ mol^{-1}, oxalate; 16.3 kJ mol^{-1}, NCS^-) probably reflect the greater kinetic stability of the μ-carboxylato product [95]. A kinetic study of water exchange on the diamagnetic $[Mo_2(OH_2)_8]^{4+}$ ion using ^{17}O NMR should be feasable and would be of interest with regard to the differing behaviour observed for NCS^- and $HC_2O_4^-$ as the substituting ligand.

In H_2SO_4 solution Mo_2^{4+}(aq) oxidises in air with effective loss of one δ-bonding electron, giving $[Mo_2(SO_4)_4]^{3-}$, which has been isolated as its blue K^+ salt [96]. The δ–δ* transition is shifted to lower energy (573 nm) concurrent with an observed increase in bond length (now 216 pm) and decrease in bond

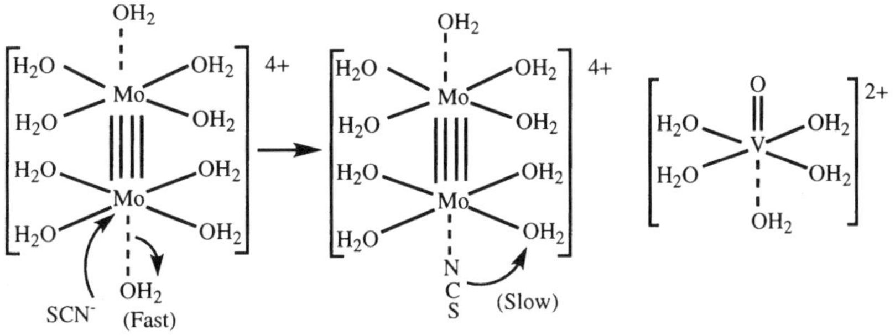

Figure 6.19. The mechanism of substitution at Mo_2^{4+}(aq) and its similarity to that on VO^{2+}(aq)

order to 3.5 [97]. In H_3PO_4 solution oxidation to the triply-bonded Mo^{III}_2 complex $[Mo_2(HPO_4)_4]^{2-}$ (Mo—Mo bond length of 223 pm) occurs (loss of the δ bond) [98]. Irradiation of Mo_2^{4+}(aq) at 254 nm results in oxidation to dimeric Mo^{III}_2(aq) [99] (see next section). In the presence of strong π-accepting ligands (such as CO, RNC and NO) loss of the π-electron density at Mo results in complete fission of the M—M bond and the formation of mononuclear Mo^{II} products.

6.2.2 Molybdenum(III) and Tungsten(III) (d^3)

6.2.2.1 SYNTHESIS AND PROPERTIES OF THE $[Mo(OH_2)_6]^{3+}$ ION

Pale yellow $[Mo(OH_2)_6]^{3+}$ represents the only true mononuclear aqua ion for the heavier Group 6 members. It was first reported by Bowen and Taube in 1971 [100] and later in more purified form by Sasaki and Sykes in 1975 [101]. The earlier samples were contaminated with amounts of the yellow Mo^V dimer cation $[Mo_2O_4(OH_2)_6]^{2+}$ (see Section 6.2.4) to which $[Mo(OH_2)_6]^{3+}$ is readily air oxidised. Preparations have employed air-free acid-catalysed aquation of red $[MoCl_6]^{3-}$ or $[MoCl_5(OH_2)]^{2-}$ in non-complexing acidic solution, usually 0.5 M Hpts or CF_3SO_3H. The aquation process normally requires a 24 hour period. However, a more convenient compound has proved to be $Na_3[Mo(HCO_2)_6]$, which aquates to the aqua ion in 0.5 M H^+ over a period of a few minutes. $Na_3[Mo(HCO_2)_6]$ is easily prepared as a pale yellow air-sensitive solid by dissolving $K_3[MoCl_6]$ in aqueous formic acid followed by treatment with an excess of sodium formate [102]. When dry, $Na_3[Mo(HCO_2)_6]$ is only mildly air sensitive and can be handled for short periods in the air. Following dilution to 0.1 M $[H^+]$ separation and purification of the aqua ion can be carried out using ice-cold air-free cation-exchange chromatography (usually Dowex

50 W X2 resin 200–400 mesh is used). Techniques for the purification, handling and manipulation of air-sensitive aqua species like $[Mo(OH_2)_6]^{3+}$ were described in Section 1.5. Following washing of the column with 0.5 M $[H^+]$, which removes any $Mo^V(aq)$ $(Mo_2O_4{}^{2+})$ present, elution of pure solutions of $[Mo(OH_2)_6]^{3+}$ can be carried out using 1.0 or 2.0 M solutions of the desired acid (Hpts or CF_3SO_3H). Sporadic but complete oxidation of Mo^{3+}(aq) by $ClO_4{}^-$ ions [103] means that use of $HClO_4$ as the supporting medium is not possible. An alternative preparative route has involved the anhydrous triflate $Mo(CF_3SO_3)_3$. This compound is made by refluxing $Mo(CO)_6$ in anhydrous CF_3SO_3H for 3–4 hours according to

$$Mo(CO)_6 + 3CF_3SO_3H \longrightarrow Mo(CF_3SO_3)_3 + 6CO + \tfrac{3}{2}H_2 \qquad (6.13)$$

wherein the pale yellow triflate precipitates as an air-sensitive off-white powder which can be removed by filtration and washed with dry diethyl ether and dried under vacuum [104]. $Mo(CF_3SO_3)_3$ dissolves readily in aqueous CF_3SO_3H to generate solutions of $[Mo(OH_2)_6]^{3+\cdot}$ However, its primary use may well be in the general preparation of $[MoL_6]^{3+}$ solvates (e.g. L = DMF, THF, MeCN, NH_3), several of which have now been characterised. $MoCl_3(THF)_3$ has proved a versatile lead-in to Mo^{III}–amine chemistry [105], the recent characterisation of fac-$[Mo(NH_3)_3(CF_3SO_3)_3]$ [106] and $[Mo(NH_3)_6](CF_3SO_3)_3$ from reaction in THF/NH_3 being good examples. The air-sensitive pale yellow caesium alum, prepared by treating solutions of $Na_3[Mo(HCO_2)_6]$ in H_2SO_4 with CsCl, has been characterised [102, 107]. $Na_3[Mo(HCO_2)_6]$, like $Mo(CF_3SO_3)_3$, has the advantage of allowing rapid generation of high concentrations of $[Mo(OH_2)_6]^{3+}$ (> 0.5 M) via acid-catalysed aquation before the onset of hydrolytic polymerisation (see below). Figure 6.20 shows the electronic spectrum of a solution of $CsMo(SO_4)_2 \cdot 12H_2O$ (~ 0.03 M) in 2.0 M CF_3SO_3H which matches well that of the aqua ion freshly generated in solution following cation-exchange chromatography [108]. The two bands observed at 386 nm ($\varepsilon = 13.3$ M^{-1} cm^{-1}) and 320 nm (19.0 M^{-1} cm^{-1}) are assigned respectively to the $^4A_{2g} \rightarrow {}^4T_{2g}$ (10 Dq) and $^4A_{2g} \rightarrow {}^4T_{1g}$(F) transitions. The $^4A_{2g} \rightarrow {}^4T_{1g}$(P) transition is obscured by charge transfer absorption below 250 nm. The Racah parameter B for the alum is calculated to be 476 cm^{-1}. The $[Mo(OH_2)_6]^{3+}$ unit in the alum is shown in Fig. 6.21 and is identical to that found in the alums for V^{3+} and Cr^{3+}. The unit cell packing diagram (Fig. 6.22) shows the face-centred cubic arrangement of octahedral $[Mo(OH_2)_6]^{3+}$ units and the hydrogen-bonded network (thin lines). The average $Mo-O(OH_2)$ distance is 209 pm. As often found in alum structures for low d-electron population M^{3+} ions, the water ligands are planarly ligated with respect to the $Mo-O(OH_2)$ bond (β structure) (see Section 1.2). However, the β-alum structure is also found in the case of the Ga^{3+} and In^{3+} alums (closed shell), suggesting that intermolecular hydrogen bonding may also be important in affecting the orientation of the O—H bonds [109].

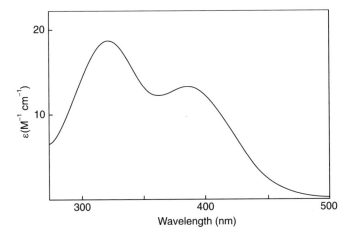

Figure 6.20. Electronic spectrum of $[Mo(OH_2)_6]^{3+}$ from $CsMo(SO_4)_2 \cdot 12H_2O$ dissolved in $2.0\,M\ CF_3SO_3H$

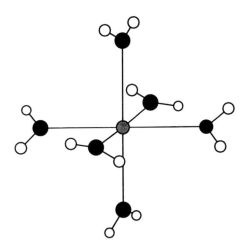

Figure 6.21. Structure of the $[Mo(OH_2)_6]^{3+}$ unit in $CsMo(SO_4)_2 \cdot 12H_2O$

$CsMo(SO_4)_2 \cdot 12\,H_2O$ is unstable, decomposing under N_2 in a sealed tube at RT within a few weeks, due presumably to slow oxidation of Mo^{3+} by SO_4^{2-}. The absorption minimum below 300 nm (Fig. 6.20) is *extremely* sensitive to oxidation, the presence of $< 1\%$ of Mo^V causing an increase of 10% in absorption at the 250 nm minimum.

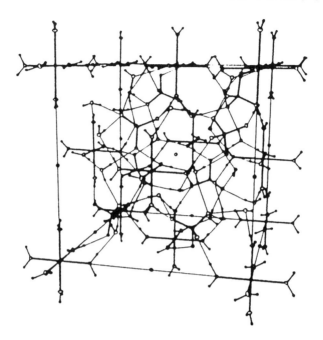

Figure 6.22. Packing within the unit cell of $CsMo(SO_4)_2 \cdot 12H_2O$ [107]. (Reprinted with permission from M. Brorson and M. Gajhede, *Inorg. Chem.*, **26**, 2109, 1987. Copyright 1987 American Chemical Society)

6.2.2.2   COMPLEX FORMATION REACTIONS ON $[Mo(OH_2)_6]^{3+}$

The rates of ligand replacement reactions occurring on Mo^{3+} are unique for a second row transition element in being faster than those occurring on its lighter group partner Cr^{3+}. Rate constants for substitution by Cl^- and NCS^- on $[Mo(OH_2)_6]^{3+}$ are some 10^5 times larger than those correspondingly on $[Cr(OH_2)_6]^{3+}$, which has led to the suggestion of a limiting S_N2 or associative (A) mechanism [101, 110]. Additional support for this mechanism comes from the following observations: (a) there is no participation from $[Mo(OH_2)_5OH]^{2+}$ in substitution reactions in contrast to reactions on Cr^{3+}, (b) a significantly large k_{NCS}/k_{Cl} ratio (59) is observed ($I = 1.0$ M pts$^-$), (c) the volume of activation for the NCS^- anation reaction is highly negative ($-11.4 \, cm^3 \, mol^{-1}$) [99]. The relevant data are summarised in Tables 6.4 and 6.5.

The range of incoming ligands studied with regard to Mo^{3+} is unfortunately limited but measured rate constants cover some 5 orders of magnitude. For some incoming ligands, e.g. the halides, reactions are difficult to follow reliably owing to accompanying oxidation to $Mo^V(aq)$, which generates absorbance changes of

Table 6.4. Kinetic data for substitution reactions on $[Mo(OH_2)_6]^{3+}$

Incoming ligand	k_f $(M^{-1}s^{-1})$	$\Delta H^{\ddagger}$ $(kJ\,mol^{-1})$	$\Delta S^{\ddagger}$ $(J\,K^{-1}\,mol^{-1})$	$\Delta V^{\ddagger}$ $(cm^3\,mol^{-1})$	Ref.
NCS^{-a}	0.268	68.1 ± 1.7	-26.7 ± 5.4		[101]
	0.317^b	67.2 ± 2.7	-29.2 ± 9.4	-11.4 ± 0.5	[110]
Cl^{-a}	0.0046	98.2 ± 2.5	$+40.1 \pm 8.8$	—	[101]
$HC_2O_4^{-a}$	0.49	—	—	—	[111]
$[Co(C_2O_4)_3]^{3-c}$	0.34	—	—	—	[108]
O_2^c	180	—	—	—	[112]
MoO_2^{2+c}	42	—	—	—	[112]

$^a I = 1.00\,M\,(pts^-)$.
$^b I = 1.00\,M\,(CF_3SO_3^-)$.
$^c I = 2.00\,M\,(pts^-)$.
The data here are for the substitution controlled rate determining step of an inner-sphere redox reaction.

Table 6.5. Comparison of kinetic data for NCS^- and Cl^- complexation on trivalent hexaaqua metal ions

Metal ion	k_{Cl} $(M^{-1}s^{-1})$	k_{NCS} $(M^{-1}s^{-1})$	k_{H_2O} (s^{-1})	k_{NCS}/k_{Cl}
Ti^{3+}	Not studied	8×10^3	1.8×10^5	not known
V^{3+}	$\leqslant 3$	1.1×10^2	5.0×10^2	$\geqslant 36$
Cr^{3+}	2.9×10^{-8}	1.8×10^{-6}	2.4×10^{-6}	62
Mo^{3+}	4.6×10^{-3}	0.27–0.32	Not known	59–69
Fe^{3+}	9.4	1.27×10^2	8.2×10^3	13.5
Co^{3+}	$\leqslant 2$	86.5	Not known	$\geqslant 43$

similar magnitude to the substitution process. It is proposed that the A mechanism is promoted by a combination of the low d-electron population and the larger radius of Mo^{3+}, leading to ready formation of a seven-coordinate transition state. The magnitude of the rate constant for water exchange on $[Mo(OH_2)_6]^{3+}$ (estimated between 0.1 and $1.0\,s^{-1}$ at 25 °C) is unfortunately not in a range amenable to study by established methods, being too slow for line-broadening NMR studies and too fast to be followed by oxygen isotopic labelling. If one applies the Eigen–Wilkins model (ion pair preassociation followed by interchange) to the NCS^- anation reaction,

$$k_{NCS} = k_I K_{os}; \qquad \Delta V^{\ddagger}_{NCS} = \Delta V^{\ddagger}_I + \Delta V^{\ddagger o}_{os} \qquad (6.14)$$

ΔV^o_{os} can be estimated as $+5.3\,cm^3\,mol^{-1}$ from the Fuoss equation. This allows a value for $\Delta V^{\ddagger}_I$ to be estimated from $\Delta V^{\ddagger}_{NCS}(-11.4)$ as $\sim -17\,cm^3\,mol^{-1}$ [110].

This value should be similar to that for $\Delta \bar{V}^{\circ}_{H_2O}$ and this may be compared with the value or water exchange on $[Ti(OH_2)_6]^{3+}$ (t^1_{2g}) (-12.1), for which a limiting A mechanism has been suggested. Taken together there seems strong evidence for a further example of a limiting A (S_N2) mechanism for substitution on $[Mo(OH_2)_6]^{3+}$.

6.2.2.3 REDOX REACTIONS INVOLVING $[Mo(OH_2)_6]^{3+}$

There is little reduction chemistry of Mo^{3+}(aq) in the absence of Cl^- ligands [88]. However, the ready oxidation of $[Mo(OH_2)_6]^{3+}$ to Mo^V and eventually Mo^{VI} has led to a number of kinetic studies with a variety of coreagents [108, 113]. Both inner-sphere and outer-sphere processes have been detected. With Mo^{3+} in excess the final product is invariably Mo^V as dimeric $[Mo_2O_4(OH_2)_6]^{2+}$. A key study has been that carried out with the series of oxidants, $[IrCl_6]^{2-}$, $[Co(C_2O_4)_3]^{3-}$ and $[VO(OH_2)_5]^{2+}$, having varying reduction potentials [108]. The reactions were followed by monitoring the appearance of the dimeric Mo^V product, giving the stoichiometry of the observed rate law

$$\frac{d(Mo_2O_4{}^{2+}(aq))}{dt} = 2k_{Ox}[Mo^{3+}][Ox] + 4k_d[Mo^{3+}]^2 + 2k_m[Mo^{3+}] \quad (6.15)$$

for reaction in pts$^-$ media. With the strong oxidant $[IrCl_6]^{2-}$ ($E^\theta = +0.89$ V), a rapid outer-sphere reaction, $k_{Ox} = k_1 + k_2$ $[H^+]^{-1}$, is observed. At 25 °C, $k_1 = 3.4 \times 10^4$ $M^{-1}s^{-1}$ and $k_2 = 2.9 \times 10^4$ s^{-1}. The $[H^+]^{-1}$ path is typical of the involvement of hydrolysis products (here $MoOH^{2+}$(aq)) as resembling more closely the oxidation product, Mo^V. However, with the weaker oxidant VO^{2+}(aq) ($E^\theta = +0.36$ V) the dominant term is k_d, zero order in [oxidant] and second order in $[Mo^{3+}]$. Here the slow step is hydrolytic polymerisation to give the more reactive dimer species $[Mo_2(\mu\text{-}OH)_2(OH_2)_8]^{4+}$, which then oxidises rapidly to dimeric Mo^V ($[Mo_2O_4(OH_2)_6]^{2+}$). Interestingly, if the anion is changed to $CF_3SO_3{}^-$ both zero-order terms represented by k_d and k_m disappear and a first-order dependence of these terms on the anion pts$^-$ has been found. It appears that pts$^-$ may promote oxidation of Mo^{3+}, possibly through charge neutralisation via outer-sphere ion association and/or coordination. It may also play a role in facilitating the formation of the more reactive OH-bridged dimer. Solutions (mM) of $[Mo(OH_2)_6]^{3+}$ in Hpts, even under air-free conditions, slowly become greenish-brown in colour due to hydrolytic polymerisation, giving $Mo_2(OH)_2{}^{4+}$(aq) and higher forms, and there is growing evidence that pts$^-$ may not be the innocent counter-anion previously supposed, a finding to be expanded upon further in Section 6.2.3. With an oxidant of intermediate strength, $[Co(C_2O_4)_3]^{3-}$ [$E^\theta = +0.58$ V), and a potential bridging ligand, $C_2O_4{}^{2-}$, a substitution controlled inner-sphere process is observed in addition to the two zero-order pathways, the value of k_{Co} being very close to that for substitution by $HC_2O_4^-$ (Table 6.4).

The reaction of $[Mo(OH_2)_6]^{3+}$ with O_2 takes place in three distinct stages, the first two being first order in $[Mo^{3+}]$ and the last independent of both $[Mo^{3+}]$ and $[O_2]$ [112]. The final product with excess Mo^{3+} present is Mo^V. A scheme of reactions is indicated involving successive formation of superoxo $Mo^{IV}(aq)$ and peroxo $Mo^{IV}_2(aq)$ intermediates prior to eventual formation of $[Mo_2O_4(OH_2)_6]^{2+}$. Figure 6.23 summarises the various pathways for oxidation of $[Mo(OH_2)_6]^{3+}$ to Mo^V.

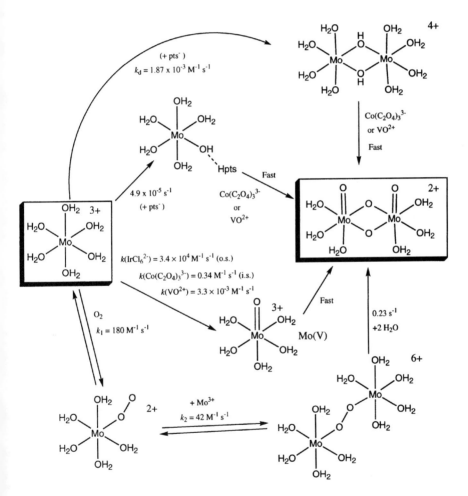

Figure 6.23. Redox reactions involving oxidation of $[Mo(OH_2)_6]^{3+}$ to Mo^V ($[Mo_2O_4(OH_2)_6]^{2+}$)

The reaction of $[Mo(OH_2)_6]^{3+}$ with aqueous HNO_3 has been briefly reported in the context of mimicking the molybdoenzyme nitrate reductase [114]. However, whereas the product of the enzyme reaction is NO_2^-, here the product is NO with some N_2O along with Mo^V as $[Mo_2O_4(OH_2)_6]^{2+}$. This arises because any NO_2^- produced competes successfully with NO_3^- for reduction by Mo^{3+}. $[Mo(OH_2)_6]^{3+}$ also reacts with DMSO in aqueous Hpts solution to give Me_2S and $[Mo_2O_4(OH_2)_6]^{2+}$, presumably via an oxygen atom abstraction reaction to give initially $[Mo^VO(OH_2)_5]^{2+}$, which then rapidly dimerises to $[Mo_2O_4(OH_2)_6]^{2+}$ (Fig. 6.23).

Two other forms of $Mo^{III}(aq)$ are known analogous to those established for Cr^{III}, a dimeric form $[Mo_2(\mu\text{-}OH)_2(OH_2)_8]^{4+}$ and a trimeric form $[Mo_3(OH)_4(OH_2)_{9\,or\,10}]^{5+}$. The dimeric form can be obtained directly by reduction of acidified solutions of $Mo^{VI}(aq)$ with, for example, Zn/Hg or electrochemically (Hg pool). This is because dimeric $Mo^V(aq)$ is generated as an intermediate under these conditions which then gets further reduced. Alternatively, reduction of freshly purified $[Mo_2O_4(OH_2)_6]^{2+}$ can be employed. Trimeric $Mo^{III}(aq)$, however, can only be obtained via reduction from $Mo^{IV}(aq)$, which also possesses a trimeric structure (see Section 6.2.3).

6.2.2.4   PREPARATION AND PROPERTIES OF $[Mo_2(\mu\text{-}OH)_2(OH_2)_8]^{4+}$

Solutions of the blue-green dimer, first reported by Ardon and Pernick [115], are most conveniently prepared by passing deoxygenated solutions of Na_2MoO_4 (10^{-2} M) in 1.0 M HCl ($100\,cm^3$) slowly down a Jones reductor (Zn/Hg) column (25×1 cm). The solution that emerges at the bottom is then collected, diluted to 0.5 M in $[H^+]$ and transferred to an ice-water jacketted column of deoxygenated Dowex 50W X2 resin (30×1 cm) and purified as described for $[Mo(OH_2)_6]^{3+}$ (Section 1.5). Amounts of $[Mo_2O_4(OH_2)_6]^{2+}$ are easily removed by washing the column with 0.5 M Hpts. Solutions of $[Mo_2(\mu\text{-}OH)_2(OH_2)_8]^{4+}$ can then be obtained by elution with 2.0 M Hpts and, for concentrations > 0.01 M, require use within 24 hours to avoid the onset of further hydrolytic polymerisation, which gives amounts of the trimeric and higher oligomeric forms. Freshly prepared solutions are characterised by absorption maxima at 360 nm ($\varepsilon = 910\,M^{-1}\,cm^{-1}$ per Mo_2), 572 nm ($96\,M^{-1}\,cm^{-1}$) and 624 nm ($110\,M^{-1}\,cm^{-1}$). Electronic spectra for monomeric, dimeric and trimeric forms of $Mo^{III}(aq)$ are shown in Fig. 6.24. Concentrations of Mo^{III} are normally determined by adding an excess of Fe^{III} under air-free conditions and titrating the Fe^{II} generated with $Ce^{IV}(aq)$ in 1.0 M H_2SO_4 using ferroin ($[Fe(phen)_3]^{2+}$) as redox indicator. The $[Mo_2(\mu\text{-}OH)_2(OH_2)_8]^{4+}$ structure has been verified by ^{17}O-labelling NMR studies (Fig. 6.25) [116].

Consistent with Fig. 6.23, a separate kinetic study of the oxidation of $[Mo_2(\mu\text{-}OH)_2(OH_2)_8]^{4+}$ by $[Co(C_2O_4)_3]^{3-}$, to give $[Mo_2O_4(OH_2)_6]^{2+}$, required stopped-flow monitoring at $25\,°C$ [117]. A rate law, $k_{obs} = Kk_{et}$

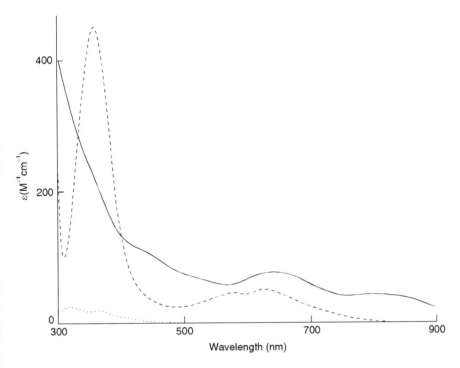

Figure 6.24. Electronic spectra for monomeric (⋯⋯), dimeric (– – –) and trimeric (——) forms of Mo^{III}(aq) (2.0 M Hpts)

$[Mo_2^{III}]/(1 + K[Mo_2^{III}])$ is relevant with K and k_{et} (25 °C) respectively 5090 M^{-1} and 1.8 s^{-1} for reaction in 2.0 M Hpts. The reaction of $Mo_2(OH)_2^{4+}$ (aq) with O_2, to give Mo(V), also occurs much faster (factor of $\sim 10^2$) than the corresponding reaction of O_2 with $[Mo(OH_2)_6]^{3+}$. The facile redox interconversion between Mo^{III} and Mo^V here stems from their similar dimeric structures. Indeed polarograms obtained from solutions of $Mo_2(OH)_2^{4+}$ (aq) exhibit a 4e$^-$ oxidation wave to Mo^V ($E° = -0.35$ V) [118]. In contrast, freshly prepared solutions of $[Mo(OH_2)_6]^{3+}$ (at mM concentrations) exhibit no such oxidation wave to Mo^V on the same polarographic timescale, reflecting the need for a structural change. As will prove apparent, the rates governing redox interconversions within the Mo^{n+} (aq) species are often a consequence of the structural changes involved (see Section 6.2.3). A further dimeric cationic form of Mo^{III}(aq), yellow $[Mo_2Cl_4(OH_2)_4]^{2+}$ ($\lambda_{max} = 430$ nm), is obtained following electrochemical oxidation of solutions of Mo_2^{4+} in HCl [98]. It can also be obtained upon treatment of $[Mo_2(HPO_4)_4]^{2+}$ with HCl. The species is believed to contain a $Mo\equiv Mo$ bond analogous to that present in $[Mo_2(HPO_4)_4]^{2-}$.

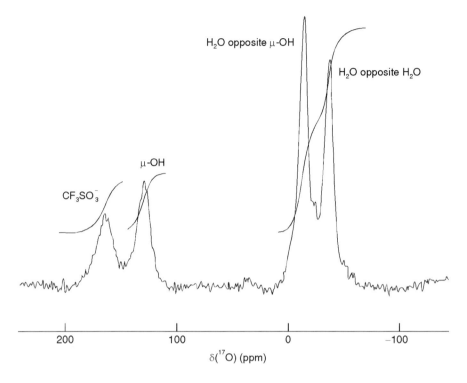

Figure 6.25. 54.24 MHz ^{17}O NMR spectrum for a solution (5 atom % ^{17}O enriched) of $Mo_2^{III}(aq)$ (0.01 M) in 1.0 M CF_3SO_3H (contains 0.1 M Mn^{2+})

Mononuclear W^{III} compounds are rare in the absence of strong π-acceptor ligands owing to ready oxidation and a propensity for strong $W\equiv W$ bond formation through effective σ and π overlap of the 5d orbitals (better than 4d) due to relativistic expansion in the appropriate directions. As a result, however, there is no evidence of a stable $[W(OH_2)_6]^{3+}$ or indeed of any other $W^{III}(aq)$ species (probably too reducing towards H_3O^+). The only cationic aqua species containing W^{III} are trinuclear and as such are discussed in the next section.

6.2.3 Molybdenum(IV) and Tungsten(IV) (d²)

6.2.3.1 CHEMISTRY OF THE $[M_3O_4(OH_2)_9]^{4+}$ IONS (M = Mo OR W)
AND OF RELATED TRINUCLEAR SPECIES

The existence of a Mo^{IV} aqua ion was first demonstrated by Souchay, Cadiot and Duhameaux in 1966 [119]. Later, efforts to establish the nuclearity of the aqua ion produced proposals of both mononuclear and dinuclear structures. These early conclusions were based upon electrochemical, kinetic [120], chromato-graphic [121] and cryoscopic [122] measurements. As time went by a number of trinuclear Mo^{IV} complexes containing a $Mo_3(\mu_3\text{-}O)(\mu\text{-}O)_3^{4+}$ core unit with various terminal ligands were identified, many obtained via simple treatment of the aqua ion with the complexing ligand under mild conditions [123]. Finally, in 1980 Murmann and coworkers showed conclusively by isotope labelling with ^{18}O that the aqua ion had the molecular structure shown in Fig. 6.26 [124]. The success of the isotope labelling method reflected the extreme inertness of the μ-oxo groups in the trinuclear core towards exchange. Since then the trinuclear structure has been verified in solution by ^{17}O NMR on a ^{17}O-labelled sample [116] (see page 36) and finally by an X-ray crystal structure of the salt $[Mo_3O_4(OH_2)_9][pts]_4 \cdot 13H_2O$ (Fig. 6.26) [125].

The Mo—Mo (248 pm) and Mo—O (core O and OH_2) distances were found to be in close agreement with predictions by EXAFS [91] and with those in other

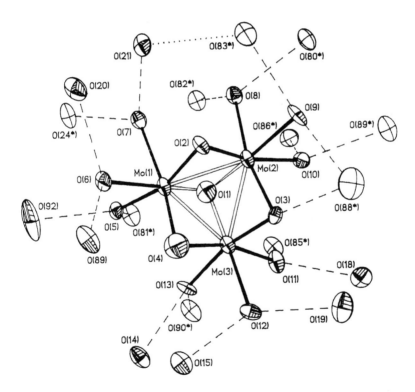

Figure 6.26. X-ray crystal structure of $[Mo_3O_4(OH_2)_9][pts]_4 \cdot 13H_2O$ showing the trinuclear core structure [125]. (Reprinted with permission from D. T. Richens, L. Helm, P.-A. Pittet, A. E. Merbach, F. Nicolo and G. Chapuis, *Inorg. Chem.*, **28**, 1394, 1989. Copyright 1989 American Chemical Society)

$Mo_3O_4^{4+}$-containing structures [123]. The burgundy-red colour of the aqua ion is distinctive and stems from a visible transition at 505 nm) ($\varepsilon = 217\,M^{-1}\,cm^{-1}$ per Mo_3) which is assigned to a transition within the MOs of the triangular M—M and Mo—O—Mo bonded framework [126, 127]. A further maximum appears in the u.v. region at 300 nm ($890\,M^{-1}\,cm^{-1}$). The construction of the triangular M_3 unit is easily appreciated on the basis of a total of six d electrons forming three M—M bonds.

6.2.3.2 PREPARATIVE ROUTES TO THE $[Mo_3O_4(OH_2)_9]^{4+}$ ION AND RELATED SPECIES

The preparative method of choice remains a comproportionation reaction between Mo^{VI}(aq) (added as either MoO_3 or $Na_2[MoO_4]$) or Mo^V(aq) (as $[Mo_2O_4(OH_2)_6]^{2+}$) and Mo^{III}(aq) (total concentration $\sim 3 \times 10^{-2}\,M$) in 2.0 M HCl (100 cm³) with heating at $\sim 90\,°C$ for 2 hours to cause assembly of the trinuclear core. In principle any form of Mo^{III}(aq) will suffice and in the early work the air-stable salt $K_3[MoCl_6]$ was the reactant of choice if available [120, 125]. Other methods devised have involved heating samples of $Mo_2(OH)_2^{4+}$(aq), generated via the Jones reductor, with Mo^{VI}(aq) in the correct molar ratio 1:1:

$$Mo^{VI}(aq) + Mo^{III}_2(aq) \longrightarrow Mo^{IV}_3(aq) \qquad (6.16)$$

this method having the virtue of using $Na_2[MoO_4]$ as the only lead-in required [128]. $[Mo_2(OH_2)_8]^{4+}$ has also been used as the reductant. Because of the inertness of the core μ-O groups towards exchange complete ^{17}O labelling of the aqua ion for the NMR measurements required the use of enriched precursor species. Here the readily exchangeable $[Mo_2O_4(OH_2)_6]^{2+}$ (Section 6.2.4) and $[Mo(OH_2)_6]^{3+}$ ions were employed [116]. Following the 2 hour heating period the crude Mo^{IV} solution is cooled and diluted to 0.5 M [H⁺], usually with Hpts, allowed to stand at RT for 24 hours to allow aquation of coordinated Cl⁻ ions (if relevant), and then loaded on to a column of Dowex 50 W X 2 resin in the H⁺ form. A second method of preparing ^{17}O-labelled samples of $Mo_3O_4^{4+}$(aq) is via simple acid hydrolysis (Hpts or HCl) of the air-sensitive green salt $K_2[Mo^{IV}Cl_6]$, analogous to the method used in the synthesis of the corresponding $W_3O_4^{4+}$(aq) ion from $K_2[WCl_6]$ (see later). Solid samples of $K_2[MoCl_6]$ are conveniently prepared *in situ* by simple treatment of powdered solid $K_3[MoCl_6]$ with liquid Br_2:

$$K_3[Mo^{III}Cl_6] + \tfrac{1}{2}Br_2(l) \longrightarrow K_2[Mo^{IV}Cl_6] + KBr \qquad (6.17)$$
$$\text{(red)} \qquad\qquad\qquad \text{(green)}$$

followed by evaporation of the Br_2 with a stream of Ar or N_2 gas. The contamination with KBr is not a problem. Here the trimer assembles from its mononuclear precursor as H_2O replaces coordinated Cl⁻, allowing ready introduction of the ^{17}O label.

A typical size of column would be 20 cm × 1 cm for chromatographic puri-fication of ~ 30 mmol of Mo^{IV}(aq). Amounts of $[Mo_2O_4(OH_2)_6]^{2+}$ and $Mo_3O_4Cl_x^{(4-x)+}$(aq) species are readily removed by washing the column with 0.5 M and then 1.0 M Hpts (check for Cl^- with Ag^+(aq)). Pure solutions of $[Mo_3O_4(OH_2)_9]^{4+}$ can then be eluted with a 2.0–4.0 M solution of any strong acid: Hpts, CF_3SO_3H, CH_3SO_3H or HBF_4. Concentrations as high as 0.15 M in $Mo_3O_4^{4+}$(aq) can be obtained from a saturated Dowex 50W X2 column by elution with 4.0 M Hpts. Use of CF_3SO_3H gives more dilute solutions. $HClO_4$ can also in principle be used, but here Mo^{IV}(aq) solutions are more unstable due to oxidation to Mo^V by ClO_4^- ions, the precise lifetime appearing to depend upon the presence of trace metal ions, e.g. Fe^{3+}(aq) which are efficient catalysts for the reaction. All manipulations with $[Mo_3O_4(OH_2)_9]^{4+}$ should be perform-ed under air-free conditions although air oxidation is quite slow, $< 10\%$ per day consistent with the need for a structural change to Mo^V(aq).

An interesting finding has been the high acidity of the water ligands on $[Mo_3O_4(OH_2)_9]^{4+}$ [129]. Measurement of the acidity constant (K_{aM}) can be performed by stepwise dilution of 2.0 M $[H^+]$ solutions of $[Mo_3O_4(OH_2)_9]^{4+}$ into solutions of the Li^+ salt (2.0 M) at 0 °C and measurement of the electronic spectrum in the range 350–600 nm at different temperatures. Figure 6.27 shows absorbance changes for a series of measurements at 15 °C. A plot of ε_{obs}^{505} vs pH (assumed $= -\log_{10}[H^+]_{adjusted}$) is shown in the inset to Fig. 6.27. From measure-ments in the range 5–35 °C, K_{aM} (25 °C) $= (0.43 \pm 0.04)$ M, $\Delta H_a^\circ = +3.9$ kJ mol^{-1}, $\varepsilon^{505}(Mo_3O_4^{4+}) = (2.17 \pm 3.6)$ M^{-1} cm^{-1}, $\varepsilon^{505}(Mo_3O_4(OH)^{3+}) = (106 \pm 1.1)$ M^{-1} cm^{-1}, $I = 2.0$ M, Lipts. In 2.0 M Hpts, usually the standard condition for estimation, ε^{505} is 189 M^{-1} cm^{-1} containing $\sim 5\%$ of the mono-hydroxy form. As a result air-free solutions of $Mo_3O_4^{4+}$(aq) 'age' over a period of several months due to hydrolytic polymerisation and require column purifica-tion. Solutions in HCl are, however, more stable due to coordination by Cl^- which reduces both the cationic charge at the metal and the sites available for hydrolysis. Such solutions are thus recommended for long-term storage.

$[Mo_3O_4(OH_2)_9]^{4+}$ is in fact a prototypal species since routes have now been devised to allow the synthesis of a whole range of related chalcogenide-sub-stituted trimeric ions $[Mo_3O_nX_{4-n}(OH_2)_9]^{4+}$ ($n = 0$–4; X = S, Se and possibly Te) [130, 131]. Although X-ray structures of a number of derivative complexes of the $Mo_3O_nX_{4-n}^{4+}$ cores have been determined the only structurally character-ised aqua ion is $[Mo_3S_4(OH_2)_9][pts]_4 \cdot 9H_2O$ [132]. The Mo—Mo separation is 274 pm, significantly longer than typical Mo—Mo separations in $Mo_3O_4^{4+}$ compounds (250 pm). This is clearly a consequence of the significantly larger radius of S^{2-} vs O^{2-}. A further increase of ~ 6 pm in Mo—Mo occurs on going to the μ-Se derivatives [131]. The subsequent studies carried out on these derivatives have led to interesting insights not only into the precise mechanism of assembly of the trinuclear unit but also as to the role played by the core X atoms in affecting both the electronic interactions within the cluster and the lability

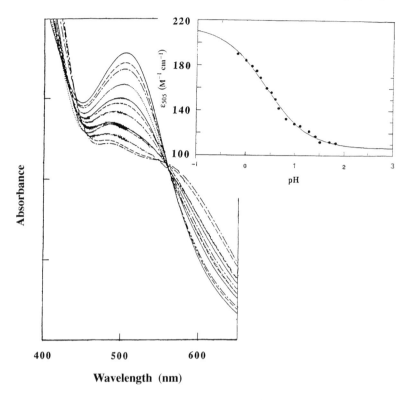

Figure 6.27. Electronic spectral changes for $Mo_3O_4{}^{4+}$(aq) in the $[H^+]$ range 0.01–2.00 M, $I = 2.00$ M (Lipts), 15 °C. (Inset: plot of ε_{obs}^{505} vs pH ($-\log[H^+]_{adjusted}$))

(acidity) of the terminal water ligands. The versatile lead-in compounds have proved to be the chalogenide-bridged Mo^V dimers ions $[Mo_2O_2(\mu\text{-}O)_n$ $(\mu\text{-}X)_{2-n}(OH_2)_6]^{2+}$ ($n = 0,1$ or 2; $X = S$ or Se) (Section 6.2.4), which are then treated with a reducing agent and heated at 90 °C for 2 hours. The reducing agents have normally been Mo^{III}, usually as added $K_3[MoCl_6]$, or $BH_4{}^-$. Electrochemical reduction using a carbon 'cloth' or mercury 'pool' cathode has also been employed. The synthetic reaction is a comproportionation reaction between Mo_2^V and Mo^{III} (added or generated *in situ*) to give a Mo_3^{IV} product. Some of the typical synthetic routes are shown in Fig. 6.28. The standard method of choice appears now to be $BH_4{}^-$ reduction of the appropriate Mo_2^V aqua ion, usually in 1.0 M HCl followed by heating in the presence of air and purification by Dowex cation exchange, as demonstated recently in the preparation of the series of μ-Se trimers [131]. In all cases where S or Se atoms are present at least one

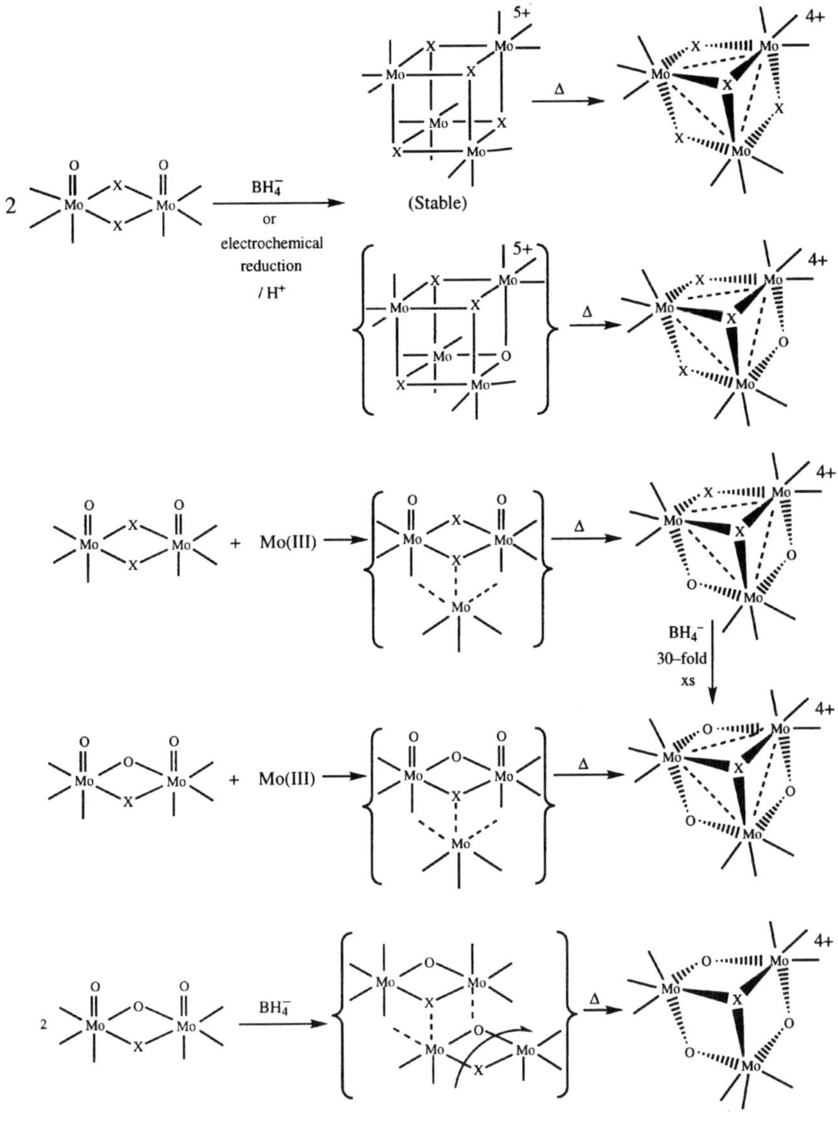

Figure 6.28. Synthetic routes to $[Mo_3O_nX_{4-n}(OH_2)_9]^{4+}$ ions (X = S or Se)

always ends up located in the capping μ_3 position, as exemplified in the preparation of $Mo_3(\mu_3\text{-X})(\mu\text{-O})_3^{4+}$(aq) from the dimer $Mo_2^VO_2(\mu\text{-O})(\mu\text{-}X)^{2+}$(aq). A mechanism involving nucleophilic attack of a $\mu\text{-S}^{2-}$ or $\mu\text{-Se}^{2-}$ group at the incoming Mo centre has been proposed to account for this. The conversion of $Mo_3(\mu_3\text{-}X)(\mu\text{-}X)(\mu\text{-}O)_2^{4+}$ (aq) to $Mo_3(\mu_3\text{-}X)(\mu\text{-}O)_3^{4+}$ (aq) in the presence of excess BH_4^- explains why $Mo_3(\mu_3\text{-}X)(\mu\text{-}X)(\mu\text{-}O)_2^{4+}$ (aq) is not observed as a product following the BH_4^- reduction of $Mo^V_2O_2(\mu\text{-}O)(\mu\text{-}X)^{2+}$ (aq) [131].

The isolation and characterisation of the cuboidal $[Mo_4X_4(OH_2)_{12}]^{5+}$ ions (X = S or Se) suggested that cubes formed first from two Mo^V dimers upon reduction and protonation [131, 133]. The cubes then oxidise upon heating in air with extrusion of one Mo to yield the trimers. For this reason the trimers have been termed 'incomplete cubes'. However, two other possible assembly mechanisms are also shown in Fig. 6.28 and are consistent with the major products obtained. The cube form is not known for all μ-O species and for stability requires at least three chalcogenide atoms to be present [130]. The cubes also require formally at least three Mo^{III} to be present, the $Mo_4S_4^{5+}$ (aq) ion being formally $Mo^{III}_3Mo^{IV}$ (see Section 6.2.3.4). The driving force for break-up of the cube structure, with extrusion of one Mo atom, is likely to be the formation of the M—M bonded triangle when the d-electron count reaches six at three of the Mo's. Properties of the various $[Mo_3O_nX_{4-n}(OH_2)_9]^{4+}$ ions are given in Table 6.6.

The most interesting aspect of these 'incomplete cuboidal' and cuboidal cluster species has been with regard to their solution chemical behaviour, the rates of mechanism of terminal water substitution and redox properties. Most attention has focused on reactions of the trimers, initially starting with the prototypal $[Mo_3O_4(OH_2)_9]^{4+}$ ion [129] but extending more recently to reactions of all the $[Mo_3O_nX_{4-n}(OH_2)_9]^{4+}$ ions [134, 135] where a much deeper insight into the reactivity has been gained.

Table 6.6. Some Solution properties of various $[Mo_3O_nX_{4-n}(OH_2)_9]^{4+}$ ions ($I = 2.00\,M\,[H^+]$)

	X = S					X = Se		
n	λ_{max} (nm)	ε $(M^{-1}cm^-)$	^{95}Mo (ppm)	NMR $(\Delta\nu_{1/2}(Hz))$	$K_{aM}{}^d$ (M)	λ_{max} (nm)	ε^e $(M^{-1}cm^{-1})$	$K_{aM}{}^d$ (M)
4	505	(189)	1003	$(2200)^b$	0.39			
3 (μ_3-X)	512	(153)	1070	$(960)^c$	0.6	525	(184)	0.44
2 (μ_3-X)	572	(202)	—			634	(398)	
1 (μ_3-X)	590	(280)	—			658	(347)	
0	603	(351)	2100	$(1400)^c$	0.22	648	(263)	0.32

^{a} Recorded in 2.00 M Hpts at 25 °C.
^{b} Recorded in 2.00 M Hpts.
^{c} Recorded in 32.596 MHz on 2.00 M natural abundance solutions at 25 °C in 2.00 M $HClO_4$.
^{d} Obtained from kinetic studies of complex formation with NCS^- at $I = 2.00$ M, 25 °C.
^{e} Recorded in 2.00 M $HClO_4$.

6.2.3.3 COMPLEX FORMATION AND WATER EXCHANGE ON THE $[Mo_3O_nX_{4-n}(OH_2)_9]^{4+}$ IONS (X = S OR SE)

Complex formation by NCS^- has been studied on most of the species for comparison. The range of ligands studied is, however, somewhat limited; Cl^- was also studied on $[Mo_3S_4(OH_2)_9]^{4+}$ [136] and additionally $HC_2O_4^-$ in the case of $[Mo_3O_4(OH_2)_9]^{4+}$ [129]. However, of most relevance has been a ^{17}O NMR study of water exchange on the $[Mo_3S_4(OH_2)_9]^{4+}$ ion, chosen because of its redox stability towards the desirable ClO_4^- medium [136]. For the early work on $[Mo_3O_4(OH_2)_9]^{4+}$ labels for the four types of oxygen atom were defined as follows: (a) μ_3-O, (b) μ-O, (c) H_2O opposite μ_3-O and (d) H_2O opposite μ-O [127] (see Fig. 6.29). The same labels for the two H_2O ligands have been retained for all the trimers. In addition Sykes has labelled the metal centres respectively as (e) M bonded to μ_3-O and two μ-O, (f) M bonded to μ_3-X and two μ-O, (g) M bonded to μ_3-X, one μ-X, and one μ-O and (h) M bonded to μ_3-X and two μ-X [124]. These labels will be used in the following discussion. The water-exchange studies have shown that the c- and d-H_2O are kinetically distinct (factor of $\sim 10^5$ difference) with the d-H_2O more labile [125, 136]. The greater lability towards exchange at the d-H_2O is due to activation via a monohydroxy form whereas exchange at the c-H_2O is not similarly activated. For the d-H_2O exchange the $[H^+]$ dependence

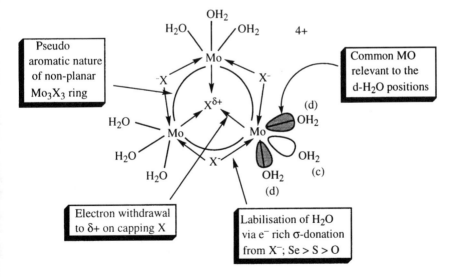

Figure 6.29. Summary of reactivity patterns on $[Mo_3O_nX_{4-n}(OH_2)_9]^{4+}$ ions (X = S, Se)

of k_{ex} is of the form given in

$$k_{ex}(d) = \frac{k_1[H^+] + k_{OH}K_{aM}}{[H^+] + K_{aM}} \qquad (6.18)$$

consistent with the mechanism shown in the following steps for the case of $Mo_3S_4^{4+}$(aq) [136]:

$$[Mo_3S_4(OH_2)_9]^{4+} \overset{K_{aM}}{\rightleftharpoons} [Mo_3S_4(OH_2)_8(OH)]^{3+} + H^+ \qquad (6.19)$$

$$[Mo_3S_4(OH_2)_9]^{4+} + H_2O^{17} \overset{k_1}{\rightleftharpoons} [Mo_3S_4(OH_2)_8(d\text{-}O^{17}H_2)]^{4+} + H_2O \quad (6.20)$$

$$[Mo_3S_4(OH_2)_8(OH)]^{3+} + H_2O^{17} \overset{k_{OH}}{\rightleftharpoons} [Mo_3S_4(OH_2)_7(OH)(d\text{-}O^{17}H_2)]^{3+}$$
$$+ H_2O \quad (6.21)$$

At 25 °C, k_1 accounts for less than 1% of the reaction, the dominant pathway for exchange being via the monohydroxy species (k_{OH}). In the case of $[Mo_3S_4(OH_2)_9]^{4+}$, k_{OH} (25 °C) = $(7.5 \pm 2.1) \times 10^3$ s^{-1}, $\Delta H^{\ddagger} = (83.0 \pm 4.0)$ kJ mol^{-1}, $\Delta S^{\ddagger} = (107 \pm 15)$ J K^{-1} mol^{-1} and $K_{aM} = (0.18 \pm 0.07)$ M for $I = 2.00$ M, ClO_4^-. For the case of 1:1 complexation with anions Y^- (NCS^-, Cl^- or $HC_2O_4^-$), equilibration kinetics is observed under psuedo first-order conditions and the rate law

$$k_{obs} = k_f([Y^-]/n \text{ or } [Mo_3X_4^{4+}]) + k_b \qquad (6.22)$$

is relevant with k_f and k_b representing the forward and backward rate constants respectively. A feature of equation (6.22) is the statistical factor n detected from comparisons of pseudo first-order kinetics with either Y^- or trimer in excess [137]. Rate constants k_f obtained from studies with Y^- in excess require to be divided by n in order to match those obtained with trimer in excess and arise due to the presence, in the trimers, of more than one identical and independently reacting metal centre which become statistically relevant with excess Y^- present, e.g. for $Mo_3S_4^{4+}$(aq), $n = 3$. In some of the mixed O,X-bridged trimers, different Mo centres f, g or h are present and n values of 1 and 2 have been detected [134]. The $[H^+]$ dependence of k_f and k_b are both of the form

$$k_f = \frac{k_1[H^+] + k_{OH}K_{aM}}{[H^+] + K_{aM}}, \qquad k_b = \frac{k_{-1}[H^+] + k_{-OH}K_{aMY}}{[H^+] + K_{aMY}} \qquad (6.23)$$

identical to that found for k_{ex} (d) (equation (6.18)) and thought to represent 1:1 complex formation at a d-H_2O:

$$[Mo_3S_4(OH_2)_9]^{4+} \rightleftharpoons [Mo_3S_4(OH_2)_8(OH)]^{3+} + H^+ \ (K_{aM}) \qquad (6.24)$$

$$[Mo_3S_4(OH_2)_9]^{4+} + Y^- \rightleftharpoons [Mo_3S_4(OH_2)_8Y]^{3+} + H_2O \ (k_1, k_{-1}) \quad (6.25)$$

Table 6.7. Comperisons of rate constants (250 °C) for equilibration of NCS^- with $[Mo_3O_nX_{4-n}(OH_2)_9]^{4+}$ ions (X = S or Se) in 2.00 M $HClO_4{}^+$ (taken from ref. [134c]

		Mo–S			Mo–Se		
n	M type	k_f $(M^{-1}s^{-1})$	k_b (s^{-1})	K_{eq} (M^{-1})	k_f $(M^{-1}s^{-1})$	k_b (s^{-1})	K_{eq} (M^{-1})
4	e	2.13	2.2×10^{-3}	970	2.13	2.2×10^{-3}	970
3	f	0.37	1.6×10^{-4}	2310	0.19^a	1.25×10^{-4}	1520
2	f	Not determined			2.8	6.2×10^{-4}	4520
	g	7.7	2.7×10^{-3}	2850	13.5	5.9×10^{-3}	2290
1	g	23.0	1.4×10^{-2}	1640	52.0	3.0×10^{-2}	1730
	h	82.0	2.4×10^{-2}	3420	131.0	9.3×10^{-2}	1410
0	h	212.0	9.2×10^{-2}	2300	480.0	2.0×10^{-1}	2400

^{a} Recorded in 2.00 M Hpts.

$$[Mo_3S_4(OH_2)_8(OH)]^{3+} + Y^- \leftrightarrows [Mo_3S_4(OH_2)_7(OH)Y]^{2+} + H_2O \ (k_{OH}, k_{-OH})$$
(6.26)

$$[Mo_3S_4(OH_2)_8Y]^{3+} \rightleftharpoons [Mo_3S_4(OH_2)_7(OH)Y]^{2+} + H^+ (K_{aMY})$$ (6.27)

A number of definitive trends have emerged from comparative studies in 2.0 M $HClO_4$ (25 °C) on the different trimers (Table 6.7). Replacement of μ_3-O by firstly μ_3-S and then μ_3-Se on $[Mo_3O_4(OH_2)_9]^{4+}$ results in a retardation in the rate of substitution, presumed at a d-H_2O, by factors of 6 and 11 respectively [134]. However, replacement of μ-O by μ-S and then μ-Se results in labilisation, with a factor of ~ 10(S) and ~ 20(Se) for each μ-O replaced [134, 135]. The trend O < S < Se suggests an influence from the increasing electron-rich σ-donating ability of the μ-X group. The opposite trend for the μ_3-X group suggests rather an electron-withdrawing role (Se > S > O). Water-exchange studies on both the $Mo_3O_4{}^{4+}$ (aq) and $Mo_3S_4{}^{4+}$ (aq) ions have shown that both the d- and c-H_2O positions are labilised when μ-O is replaced by μ-S. Of interest, however, is the site of deprotonation in the monohydroxy form because of the specific labilisation of the d-H_2O. In the case of $[Mo_3O_4(OH_2)_9]^{4+}$, ^{17}O NMR studies show an upfield shift in the resonance of the d-H_2O, but not the c-H_2O, upon decreasing the $[H^+]$ the shift being in the range appropriate for a deprotonation with $K_{aM} \sim 0.4$ M (Fig. 6.27) [125]. If one can extrapolate this behaviour as typical of all the trimers then the site for deprotonation would appear to be at a d-H_2O. One could suppose that the d-H_2O positions influence each other via a common MO which in turn has little or no involvement in the bonding of the c-H_2O. However, corresponding measurements on $Mo_3S_4{}^{4+}$ (aq) have failed to reveal a shift in the ^{17}O resonances of either H_2O ligand over a similar $[H^+]$ range [136] and this parallels an apparent insensitivity of the u.v.–visible chromophore of $Mo_3S_4^{4+}$ (aq) to $[H^+]$ change [135], this despite the detection of a similarly large K_{aM} of ~ 0.2 M for $Mo_3S_4^{4+}$ (aq) from complex formation and water-exchange kinetics (equations (6.18) to (6.27)). Significant changes to both the u.v.–visible chromophore and ^{95}Mo NMR signals of the trimers (Table 6.6) are, however, apparent on replacing μ-O with μ-S (Se) but not in the case of the μ_3-capping group. It is now clear that the electronic structure of these clusters is profoundly influenced by the nature of the μ-X ligand, leading to a proposal of possible 'quasi-aromatic' behaviour in the non-planar Mo_3X_3 ring when X = S or Se (σ donation perhaps encouraged by π acceptance back into empty low-lying 3d and 4d orbitals respectively) [127]. Moreover, when X = S or Se the MOs involved in the Mo—X—Mo bonding framework will be at a significantly different energy than those involved in bonds to the terminal H_2O ligands, a factor that might explain the insensitivity of the u.v.–visible chromophore of, for example, $Mo_3S_4^{4+}$ (aq) to the presence of an OH^- group as distinct from H_2O at a terminal position.

In discussing the mechanism of substitution on aqua metal ions a fairly reliable indicator has proved to be the ratio of anation rate constants k_{NCS^-}/k_{Cl^-} (Section 6.2.2) [101]. For reaction on $Mo_3S_4(OH)^{3+}$ (aq), replacement at a d-H_2O (Table 6.8), ratio is close to unity (1.4) and an I_d mechanism is assigned. Reactions involving conjugate base forms are frequently dissociative in nature and the assignment gains further support from the fairly high $\Delta H^{\ddagger}$ value (83 kJ mol^{-1}) and large positive $\Delta S^{\ddagger}$ (> 100 J K^{-1} mol^{-1}) for the water-exchange process. For reaction on $Mo_3S_4^{4+}$ (aq), itself, however, the anation rate is well in excess of that for water exchange and here an associative process is tentatively suggested despite $Mo_3S_4^{4+}$ (aq) seemingly unable to distinguish between NCS^- and Cl^- as the incoming ligand. A similar changeover in the substitution mechanism occurs in the case of Fe^{3+} (aq) (I_a) and its conjugate base $FeOH^{2+}$ (aq) (I_d) (see Section 8.1.2). The overall reactivity patterns relevant to the $Mo_3X_4^{4+}$ (aq) ions are summarised schematically in Fig. 6.29.

Table 6.8. Comparison of kinetic data for water exchange and complex formation on the $[Mo_3S_4(OH_2)_9]^{4+}$ ion, 25 °C, $I = 2.00$ M, $LiClO_4$ (from ref. [136])

Ligand X$^-$	$k_1(Mo_3S_4^{4+})$ (M^{-1}s^{-1})	$k_{OH}(Mo_3S_4(OH)^{3+}$ (M^{-1}s^{-1})	$k_1{}^a$ (s^{-1})	K_{aM} (M)
NCS$^-$	108 ± 4	1120 ± 55	~ 5600	0.22 ± 0.03
Cl$^-$	91 ± 1	790 ± 10	~ 3950	0.23
d-H_2O	$< 10^b$		$7500(k_{OH})$	0.18 ± 0.07

^{a} The Eigen–Wilkins model, $k_{OH}(X^-) = K_{OS}k_1$, has been assumed for reaction on $[Mo_3S_4(OH_2)_8(OH)]^{3+}$ to enable interchange rate constants (k_1) to be compared with those for water exchange, k_{OH}(d-H_2O). The kinetic data indicate that K_{OS} can be assumed to have a value ~ 0.2 M^{-1}.
^{b} Units of s^{-1}.

6.2.3.4 CUBOIDAL AND MIXED-METAL CUBOIDAL MO AND W AQUA IONS

The entire series of cube ions $[Mo_4S_4(OH_2)_{12}]^{4+/5+/6+}$ has now been synthesised. The 5 + cube (formally $Mo_3^{III}Mo^{IV}$) is the stable form. Air-sensitive $[Mo_4S_4(OH_2)_{12}]^{4+}$ (Mo_4^{III}) has been characterised following electrochemical reduction (carbon cloth cathode) of the 5 + cube [138]. A further more high yielding route has been reported from the reduction of $[Mo_3S_4(OH_2)_9]^{4+}$ with metallic Pb [139]. The red 6 + cube (formally $Mo_2^{III}Mo_2^{IV}$) has been prepared by oxidation of the 5 + cube with a 2:1 excess of *cis*-$[VO_2(OH_2)_4]^+$ in Hpts solution [140]. Redox potentials linking the three cubes have been determined:

$$[Mo_4S_4(OH_2)_{12}]^{6+} \overset{0.86\,V}{\rightleftharpoons} [Mo_4S_4(OH_2)_{12}]^{5+} \overset{0.21\,V}{\rightleftharpoons} [Mo_4S_4(OH_2)_{12}]^{4+}$$

(6.28)

The pattern of substitution lability at the water ligands of these cube ions is somewhat intriguing. The water ligands of the 6 + cube, even allowing for the presence of a labilising conjugate-base pathway, and the statistical factor are more labile (substitution believed at Mo^{IV}) than at the single Mo^{IV} centre of the 5 + cube. In the latter a conjugate-base pathway is not seen. Likewise the water ligands of the 4 + cube (Mo^{III}_4) are more labile than at the three presumed Mo^{III} centres of the 5 + cube. Distinct Mo^{III} and Mo^{IV} centres appear to be relevant from the observed biphasic kinetics on the 5 + cube. The likely origin of these conflicting effects might be viewed as arising from the actual lack of discrete localised oxidation states in these cluster species with the degree of delocalisation possibly varying from cube to cube (perhaps with the 5 + cube somewhat different than the other two, given its stability). Nonetheless, in several of the mixed-metal cubes (see below) substitution behaviour reminiscent of that of localised oxidation states is apparent and it is clear that a fuller understanding of the nature of these intriguing kinetic effects is required. Substitution rates (NCS^- for H_2O) on $Mo_4S_4^{5+}$(aq) are slower than on $Mo_3S_4^{4+}$(aq), consistent with the presence of H_2O ligands only *trans* to μ_3-S^{2-} [141].

The electron-rich nature of the μ-S^{2-} groups on $[Mo_3S_4(OH_2)_9]^{4+}$ has been used to coordinate a fourth metal ion, other than Mo, to make heterometallic cube aqua ions, $[Mo_3MS_4(OH_2)_{10}]^{n+}$ ($n = 4$ or 5; M = Fe, Co, Ni, Cu, Sn and In) [142]. With the odd-electron species formed by Co and Cu [142], and also in the case of many of the main group elements [143] (see also Chapter 2), dimeric species result which can be linked by edges (Co, Cu) or corners (Hg, In, Tl, Sn, Pb, Sb and Bi) (Fig. 6.30). Interestingly in the case of Co and Cu, formation of the dimer seems to be promoted by the presence of weakly coordinating anions such as ClO_4^- or pts$^-$. In the presence of Cl^- ions monomer cubes are formed. Preparative methods have involved treating $[Mo_3S_4(OH_2)_9]^{4+}$ with the metal (in the cases of Fe, Co, Cu, Sn, In, Sb, Bi and Hg) or with M^{2+}(aq) in the presence of $NaBH_4$ (Ni, Pb, Sn, In, Tl, Pb, Sb and Bi). In addition, the Cu-bound ion

$[Mo_3CuS_4(OH_2)_{10}]^{4+}$ can be prepared by heterometal exchange via simple treatment of the hetero Fe and Ni complexes with Cu^{2+} (aq) [144]. The hetero-metal 'M' is tetrahedrally coordinated in the case of Fe, Co, Ni and Cu but octahedrally coordinated in the case of Cr, Sn, In Tl, Pb, Sb, Bi and Hg (Fig. 6.30).

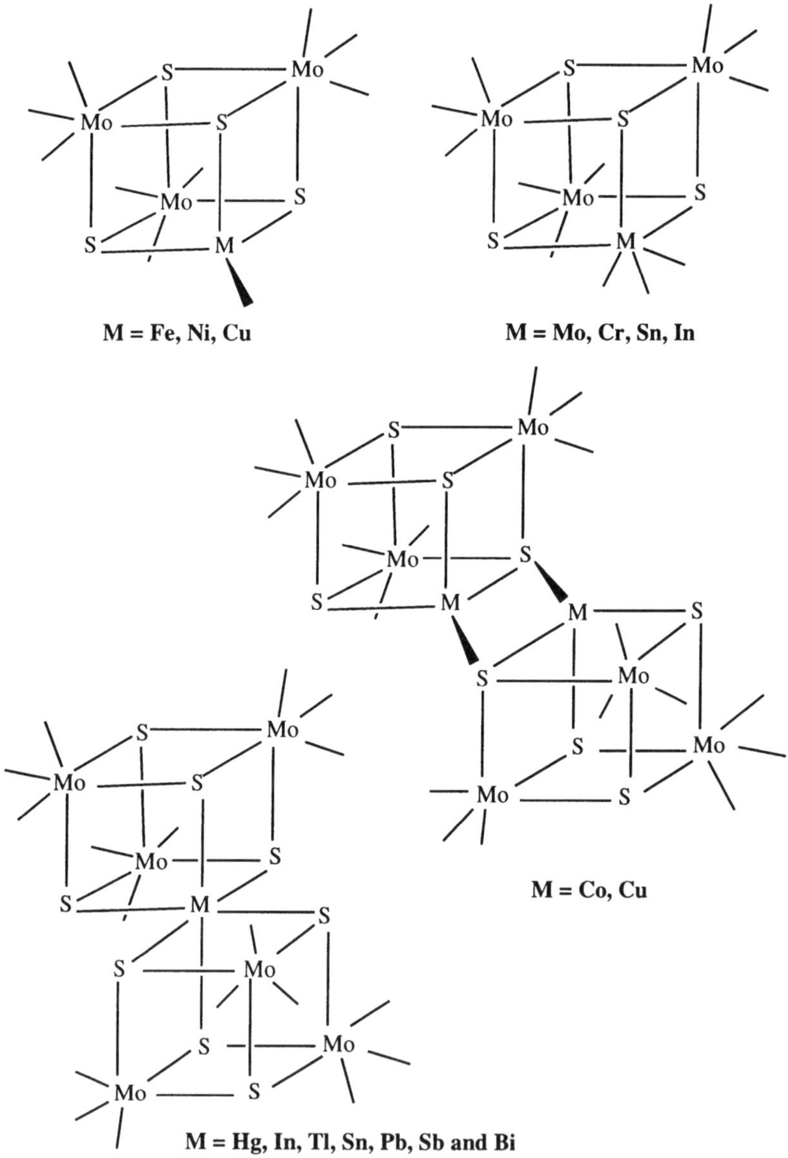

M = Fe, Ni, Cu   M = Mo, Cr, Sn, In

M = Co, Cu

M = Hg, In, Tl, Sn, Pb, Sb and Bi

Figure 6.30. Structures of heterometallic cube aqua ions of molybdenum

The synthesis of these species has allowed studies on the substitution behaviour of the $(\mu\text{-S})_3M(OH_2)_n$ moiety and, in the case of $M = Ni$, significantly slower substitution rates for the H_2O ligand are found on comparison with $[Ni(OH_2)_6]^{2+}$ [145]. Redox studies on, for example, $[Mo_3NiS_4(OH_2)_{10}]^{4+}$ have shown that both inner- and outer-sphere pathways towards oxidation can operate.

6.2.3.5 THE $[Mo_3CrS_4(OH_2)_{12}]^{4+}$ CLUSTER AQUA ION [146]

This mixed Group 6 element cluster is prepared by the reaction of $[Mo_3S_4(OH_2)_9]^{4+}$ with a tenfold excess of $[Cr(OH_2)_6]^{2+}$, three equivalents being required for the stoichiometric reaction:

$$Mo_3S_4{}^{4+} + 3Cr^{2+} \longrightarrow Mo_3CrS_4{}^{4+} + 2Cr^{3+} \qquad (6.29)$$

The ion is formally $Mo^{III}{}_3Cr^{III}$. Substitution of water by NCS^- requires stopped-flow monitoring (contrasting with the slower reaction on $[Mo_4S_4(OH_2)_{12}]^{4+}$) and is assigned to reaction at the single Cr^{III} centre labilised by the presence of the three bridging μ-S ligands (cf. the labilisation at $Cr^{III}(aq)$ in the presence of SH^- and H_2S ligands; Section 6.1.2.5). The presence of a conjugate-base pathway is moreover reminiscent of substitution reactions on $[Cr(OH_2)_6]^{3+}$ (Section 6.1.2). In $[Mo_4S_4(OH_2)_{12}]^{4+}$ substitution at Mo^{III} has no $[H^+]^{-1}$ dependent pathway which is reminiscent of reactions on $[Mo(OH_2)_6]^{3+}$ (Section 6.2.2). Thus it appears that this cluster behaves as if formally localised Mo^{III} and Cr^{III} centres are present. Aerial oxidation of $Mo_3CrS_4{}^{4+}(aq)$ results in cube breakdown and formation of $Mo_3S_4{}^{4+}(aq)$ and $Cr^{3+}(aq)$.

6.2.3.6 PREPARATION AND PROPERTIES OF $[W_3O_4(OH_2)_9]^{4+}$ AND RELATED SPECIES

The corresponding series of $[W_3O_nX_{4-n}(OH_2)_9]^{4+}$ ions have also been identified and comparisons of properties made with the series for Mo $(X = S)$. The prototypal orange $[W_3O_4(OH_2)_9]^{4+}$ ion was first characterised by Segawa and Sasaki in 1985 [147] following acid hydrolysis (HCl or Hpts) of the purple-red salt $K_2[WCl_6]$. About 40% conversion to $W_3O_4{}^{4+}(aq)$ can be achieved on hydrolysis of 2.0 g of salt with 50 cm^3 of acid solution. The comproportionation reaction, so successful in the case of Mo, does not work for W. The only other promising route to the $[W_3O_4(OH_2)_9]^{4+}$ ion appears to be from slow aerial oxidation of acidic solutions of the trinuclear $W^{III,III,IV}$ complex $[W_3(\mu_3\text{-}O)(\mu\text{-}O_2CCH_3)_6(OH_2)_3]^{2+}$. Solutions of $[W_3O_4(OH_2)_9]^{4+}$ can be purified by Dowex 50W X2 cation-exchange column chromatography as in the case of Mo^{IV}. In 2.0 M Hpts solution $W_3O_4{}^{4+}(aq)$ is characterised by a λ_{max} at 455 nm $(\varepsilon = 375 \text{ M}^{-1} \text{cm}^{-1}$ per $W_3)$ with 'shoulders' at ~ 500 nm $(\sim 300 \text{ M}^{-1}\text{cm}^{-1})$ and ~ 300 nm $(\sim 750 \text{ M}^{-1}\text{cm}^{-1})$ [147]. The electronic spectrum of

$[W_3O_4(OH_2)_9]^{4+}$ is shown for comparison with that for $[Mo_3O_4(OH_2)]^{4+}$ in Fig. 6.31. The preparative method for $W_3O_4^{4+}$ (aq) allows ready introduction of a ^{17}O label [148] and Fig. 6.31 also shows comparative ^{17}O NMR spectra showing the four types of oxygen atom a to d. A feature is the sensitivity towards the nature of M (Mo or W) shown by the ^{17}O chemical shifts of the bridging core oxygens a and b but not shown by the coordinated H_2O c and d. The crystal structures of a number of derivatives of the $W_3O_4^{4+}$ core have been reported [147] but no crystal structure exists as yet for the aqua ion. The W—W separation is typically ~ 253 pm, slightly, but not significantly, longer than typical Mo—Mo separations in $Mo_3O_4^{4+}$ compounds (249 pm). The complete series of $[W_3O_nX_{4-n}(OH_2)_9]^{4+}$ ions (X = S) have been characterised and several derivatives have been structurally determined [149]. The X-ray crystal structure of the only aqua ion thus far is that of $[W_3S_4(OH_2)_9][pts]_4 \cdot 7H_2O$ [150]. The W—W distance here (272 pm), as in the case of Mo, is significantly longer than that in typical $W_3O_4^{4+}$ compounds. It is significant that S vs O causes a larger change to the M—M separation than W vs Mo.

Rate constants for terminal H_2O replacement on the series of $[W_3O_nX_{4-n}(OH_2)_9]^{4+}$ ions X = S) have been determined and comparisons have been made with the molydenum series [151]. The same basic I_d mechanism is believed to apply, with substitution (presumed at a d-H_2O) on the $W_3X_4(OH)^{3+}$ (aq) form dominant. Effects observed upon replacing μ_3-O and μ-O groups with S are the same as those observed on the Mo species. However, in each case the rate constants for tungsten are approximately one-tenth those on the corresponding molybdenum species. This is a phenomenon frequently observed on comparing the two metals, particularly when a dissociative process is relevant. It is believed to reflect relativistic expansion in the 5d orbitals for W which leads to greater overlap and stronger covalent bonding with terminal H_2O.

5.2.3.7 REDOX REACTIONS INVOLVING $[M_3X_4(OH_2)_9]^{4+}$ IONS (M = Mo OR W; X = O OR S)

Reduction   As mentioned earlier, a third form of Mo^{III}(aq) (green) (Fig. 6.24) can be obtained on reduction of $[Mo_3O_4(OH_2)_9]^{4+}$. This can be achieved either electrochemically (Hg pool cathode) [152] or chemically with Zn/Hg [152], Cr^{2+}(aq) [153] or Eu^{2+}(aq) [154]. The reversibility of the Mo^{IV}/Mo^{III} reaction implies a cyclic trinuclear Mo^{III} product. Murmann and coworkers have moreover demonstrated, using ^{18}O labelling, that the four core oxygen atoms are retained during the redox cycle [155]. Electrochemical measurements have further verified that all four oxo groups of $Mo_3O_4^{4+}$(aq) become protonated upon reduction and the structure $Mo_3(OH)_4^{5+}$(aq) is relevant [152]. From studies in 2.0 M CF_3SO_3H, Paffett and Anson obtained electrochemical evidence for two forms of Mo_3^{III}(aq) following rapid reduction of $Mo_3O_4^{4+}$ through a zinc column, the second form building up over a period of ~ 30 hours [152c]. Each

309

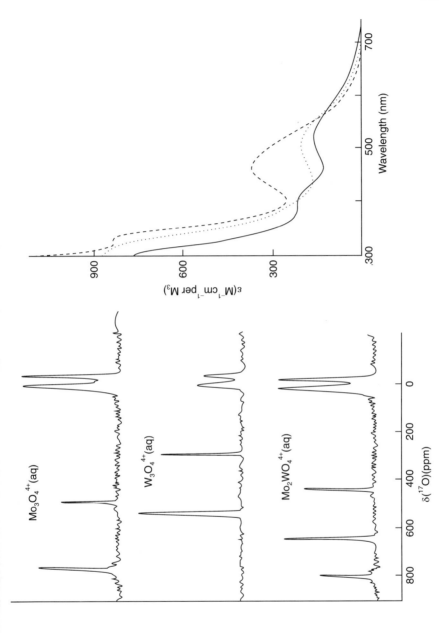

Figure 6.31. Electronic and ^{17}O NMR spectra for $[M_3O_4(OH_2)_9]^{4+}$ ions, $M_3 = Mo_3$ (.....), W_3 (— — —) and Mo_2W (——)

had different formal potentials for oxidation back to $Mo_3O_4{}^{4+}$ (aq). Two forms of Mo_3^{III}(aq) are also evident from ^{17}O NMR monitoring of the Eu^{2+} (aq) reduction of ^{17}O-enriched $Mo_3O_4{}^{4+}$ [154]. The results are shown in Fig. 6.32. The Mo_3^{IIIA} form (Fig. 6.33) has the basic 'Mo_3X_4' core with a capping μ_3-OH group ($d^{17}O = 209$ ppm) and three μ-OH groups (355 ppm) and the formula $[Mo_3(OH)_4(OH_2)_9]^{5+}$. The Mo_3^{IIIB} form has lost the capping group and has only μ-OH groups (single .resonance at 232 ppm) and the formula $[Mo_3(OH)_4(OH_2)_{10}]^{5+}$. A cyclic voltammogram of 5.0 mM Mo_3^{III}(aq) generated in 2.0 M Hpts [152b] is also shown in Fig. 5.32 for comparison. Both forms of Mo_3^{III}(aq) are produced within minut i Hpts solution whereas in CF_3SO_3H, Mo_3^{IIIA} forms first which then slowly converts to an equilibrium mixture of the two forms. Formation of Mo_3^{IIIB} (di-μ-OH structure) appears to be promoted by pts$^-$ as counter-anion (cf. $Mo^{3+}-Mo_2(OH)_2{}^{4+}$ conversion earlier). The two forms also differ in their pathways for reoxidation to $Mo_3O_4{}^{4+}$ (aq). Whereas Mo_3^{IIIA}(aq) is oxidised in a single $3e^-$ step to $Mo_3O_4{}^{4+}$(aq) ($E^\theta \sim -0.1$ V), reflecting their similar structures, Mo_3^{IIIB}(aq) is oxidised in two steps, firstly to a mixed-valence $Mo_3^{III,III,IV}$(aq) intermediate ($E^\theta = -0.17$ V, 2.0 M Hpts) and then finally to $Mo_3O_4{}^{4+}$ (aq) ($E^\theta = +0.05$ V, irreversible). Only one ^{17}O resonance is seen for the $Mo_3^{III,III,IV}$(aq) intermediate (403 ppm) (Fig. 5.32), suggesting a similar μ-OH-bridged structure to that of Mo_3^{IIIB}. Figure 6.32 shows both oxidation pathways taking place concurrently with some $Mo_3O_4{}^{4+}$(aq), from the direct $3e^-$ oxidation of Mo_3^{IIIA}, appearing along with the $Mo_3^{III,III,IV}$(aq) intermediate, spectrum (c). The similar structures for Mo_3^{IIIB}(aq) and $Mo_3^{III,III,IV}$(aq) are also reflected in the magnitude of their self-exchange rate constant, $\log k_{11}$ $(M^{-1} s^{-1}) = (4.2 \pm 0.6)$ [156]. The redox interconversions are summarised in Fig. 6.33. $[Mo_3^{III,III,IV}(\mu\text{-}OH)_4(OH_2)_{10}]^{6+}$ has a characteristic broad maximum at 1050 nm ($\varepsilon = 300$ M^{-1} cm^{-1} per trimer (Fig. 6.34). The band profile analysis indicates assignment to an intervalence charge-transfer transition within a class IIA mixed-valence system [152, 154]. Small but significant solvent shifts in the band maximum have been detected consistent with this assignment. The existence of a stable $Mo_3^{III,III,IV}$ form (eight-cluster d electrons) had been earlier suggested on the basis of Fenske–Hall type calculations which predict an available low-lying empty MO largely non-bonding in character with respect to the M_3X_4 framework [157].

Corresponding studies on the series of $[Mo_3O_nS_{4-n}(OH_2)_9]^{4+}$ ions have shown that for ready reduction the presence of protonatable μ-O groups is required, the case of reduction decreasing with the introduction of μ-S. In the case of the ion $[Mo_3(\mu_3\text{-S})(\mu\text{-O})_3(OH_2)_9]^{4+}$ formation of the equivalent mixed valence $Mo_3^{III,III,IV}$(aq) ion requires reduction in 8.0 M HCl, the more negative E^θ values and higher acidity required reflecting the reluctance of the μ_3-S group to protonate (Fig. 6.33) [130].

$[W_3O_4(OH_2)_9]^{4+}$ can also be reduced to a $W_3^{III,III,IV}$(aq) form but the reduction potential is about 0.9 V more negative that in the case of the Mo analogue

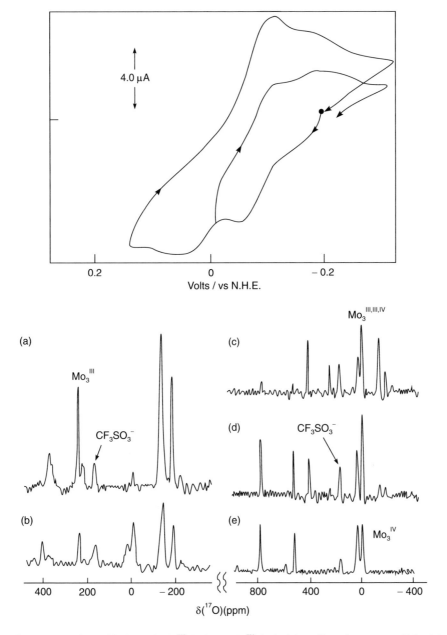

Figure 6.32. The oxidation of $Mo^{III}_3(aq)$ to $Mo^{IV}_3(aq)$: (a) cyclic voltammetry (5.0 mM Mo^{III}_3 solution, HMDE, 0.1 V s^{-1}); (b) ^{17}O NMR studies of the air oxidation of 5 atom % ^{17}O-enriched Mo^{III}_3 (54.24 MHz, 0.01 M Mo_3, 1.2 M Hpts, I = 2.0 M ($CF_3SO_3^-$)

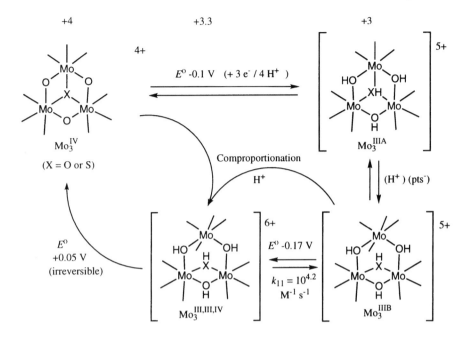

Figure 6.33. Redox interconversions involving trinuclear aqua ions of Mo^{IV} and Mo^{III} (X = 0, 5, 5e)

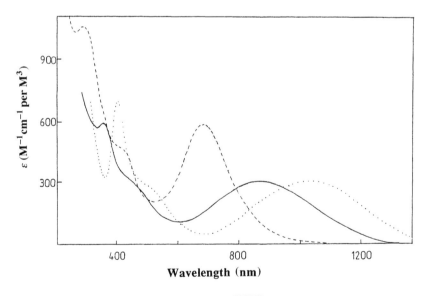

Figure 6.34. Electronic spectra for trinuclear $M_3^{III,III,IV}$(aq) ions $M_3 = Mo_3(\cdots)$, $W_3 (---)$ and $Mo_2W(—)$

and for chemical reduction requires Zn/Hg [148]. The ^{17}O NMR spectrum shows a single μ-OH resonance at 250 ppm consistent with the $[W_3(\mu\text{-}OH)_4(OH_2)_{10}]^{6+}$ structure. The $W_3^{III,III,IV}$(aq) ion reduces both O_2 an H_3O^+ and amounts of reductant (Zn/Hg) are required to be present in the sample tubes to enable NMR measurement. Reduction to W^{III} does not seem possible and there is little evidence for a stable W^{III}(aq) ion of any kind.* The electronic spectrum of $W_3^{III,III,IV}$(aq) in 2.0 M Hpts is similar to that of $Mo_3^{III,III,IV}$(aq), with the bands slighly more intense, sharp and occurring at higher energy (Fig. 6.34). The intervalence band occurs now at 680 nm and is suggestive of slightly greater delocalisation in the W case, indicative perhaps of the tendency towards stronger W—W single bonds.

Successful preparation of a mixed Mo—W trimer of formula $[Mo_2WO_4(OH_2)_9]^{4+}$ has been reported and its properties studied [158]. Synthesis was achieved via acid hydrolysis of 2:1 mixtures of $K_2[MoCl_6]$ and $K_2[WCl_6]$ in HCl followed by cation-exchange purification. Attempts to make the mixed ion $[MoW_2O_4(OH_2)_9]^{4+}$ by using 1:2 Mo/W ratios of salts produce the Mo_2W ion contaminated with amounts of $[W_3O_4(OH_2)_9]^{4+}$. The preferential formation of the Mo_2W ion is believed to stem from kinetic factors, with ligand replacement and hence incorporation of Mo units in the assembling trinuclear core occurring more rapidly than in the case of W. Use of an excess of W^{IV} salt is preferred since it has proved easier to separate the Mo_2W ion on the column from $W_3O_4^{4+}$(aq) than from $Mo_3O_4^{4+}$(aq). Properties of the $[Mo_2WO_4(OH_2)_9]^{4+}$ ion are compared with those of the Mo_3 and W_3 counterparts in Fig. 6.31. Enrichment of the ion with ^{17}O is straightforward and its ^{17}O NMR spectrum shows the expected three resonances for the core oxygens: 442 ppm (μ_3-O), 646 ppm (Mo—O—W) and 815 ppm (Mo—O—Mo) [158]. The ^{95}Mo NMR spectrum shows a resonance at 1189 ppm (from MoO_4^{2-}) which may be compared with that for $Mo_3O_4^{4+}$ (1003 ppm). Cyclic voltammetry shows that the first reduction wave of $Mo_2WO_4^{4+}$ (aq) occurs with $E^\theta \sim 100$ mV more negative than in the case of $Mo_3O_4^{4+}$(aq) in 2.0 M Hpts. Taken together these results imply a higher electron density at Mo and bond polarization $W^{\delta+}\text{-}Mo^{\delta-}$ in the Mo_2W ion. Mixed-valence $Mo_2W^{III,III,IV}$(aq) and $Mo_2W^{III,III,III}$(aq) ions have been identified, the former generated using Eu^{2+}(aq), the latter requiring Zn/Hg. This tends to imply that the mixed-valence ion is $Mo_2^{III}W^{IV}$(aq). The electronic spectrum of $Mo_2^{III}W^{IV}$(aq) (Fig. 6.34) is remarkably similar to that of the $Mo_3^{III,III,IV}$(aq) and $W_3^{III,III,IV}$(aq) ions and lies between them in energy, suggesting that there is significant delocalisation over the M_3 unit in defining the chromophore of these species. The appearance of only one μ-OH

---

*Cyclic voltammograms from aqueous solutions of the trinuclear $W^{III,III,IV}$ complex $[W_3(\mu_3\text{-}O)(\mu\text{-}O_2CCH_3)_6(OH_6)]^{2+}$ shows a reversible 1e$^-$ reduction process at -0.92 vs SCE (0.1 M CF_3SO_3H, glassy carbon electrode) to what appears to be a W^{III} aqua complex. However, efforts to isolate this extremely sensitive species have thus far failed.

resonance in each of the mixed-valence ions and of only one ^{95}Mo resonance (996 ppm) in the case of Mo$_3^{III,III,IV}$(aq) implies considerable electronic delocalisation on the NMR timescale. The rate constants for 1:1 complexation of Mo$_2$WO$_4^{4+}$(aq) with NCS$^-$ are consistent with substitution occurring at the more labile Mo atoms (a statistical factor of 2 is detected). In the case of the [M$_3$S$_4$(OH$_2$)$_9$]$^{4+}$ ions the successful synthesis and separation of both Mo$_2$W (major) and MoW$_2$ (minor) mixed derivatives have been reported [159]. Interestingly there is little change to the binding energies of Mo (3d$_{1/2}$ and 3d$_{5/2}$) and W(4f$_{5/2}$ and 4f$_{7/2}$) (from XPS spectra) on replacing Mo with W and W with Mo over the entire series of M$_3$S$_4^{4+}$(aq) ions, in contrast to the properties of the μ-O bridged species. This is suggestive of a buffering effect from μ-S towards changes in electron density at the metal centres. Here ^{95}Mo NMR measurements would be of interest.

Oxidation   A number of kinetic studies have appeared describing oxidation of [Mo$_3$O$_4$(OH$_2$)$_9$]$^{4+}$ to Mo^V and Mo^{VI}. With appropriate reagents either can be the major product. Even with strong oxidants ($E^\theta > 0.9$ V) reactions are slow and can be monitored by conventional means, reflecting the mismatch between the structures of the Mo^{IV}, Mo^V and Mo^{VI}(aq) species. In each case reaction rate laws possess strong [H$^+$]$^{-1}$ dependences, implying involvement of the Mo$_3$O$_4$(OH)$^{3+}$(aq) species. Inner-sphere processes seem to predominate but one reaction with [Fe(phen)$_3$]$^{3+}$ is presumed to be an outer-sphere process [160]. A particularly interesting observation is the facilitation of the Fe^{3+}(aq) oxidation of Mo$_3$O$_4^{4+}$(aq) by the addition of NCS$^-$, presumably via an inner-sphere bridged Fe^{III}–NCS–Mo^{IV} species [161]. The rate constant (25 °C) for oxidation by [IrCl$_6$]$^{2-}$ ($k = 4.5$ M^{-1} s^{-1}) is very close to typical values for substitution on Mo$_3$O$_4$(OH)$^{3+}$(aq) (e.g. NCS$^-$$k_{OH}$ (25 °C) = 4.8 M^{-1} s^{-1}) and an inner-sphere process is indicated here also [160]. Table 6.9 summarises some of the relevant kinetic data.

6.2.4   Molybdenum(V) and Tungsten(V) (d^1): The Dinuclear [M$_2$O$_2$(μ-O)$_2$(OH$_2$)$_6$]$^{2+}$ Ions and Related Derivatives

6.2.4.1 Preparation and properties of [Mo$_2$O$_2$(μ-O)$_2$(OH$_2$)$_6$]$^{2+}$

The yellow-orange aqua Mo^V dimer ion, [Mo$_2$O$_2$(μ-O)$_2$(OH$_2$)$_6$]$^{2+}$, was first reported by Ardon and Pernick in 1973 [163]. The first preparations involved reactions of Klason's salt, (NH$_4$)$_2$[MoOCl$_5$], prepared by mild reduction of Mo^{VI} in 12 M HCl followed by addition of NH$_4$Cl. The salt retains its green colour and paramagnetism when dissolved in 12 M HCl. However, on dilution to < 2.0 M [H$^+$] a change to a yellow-orange colour occurs, with loss of the diamagnetism, from which a yellow-orange cation could be separated and purified by Dowex cation-exchange chromatography. Retention on the column

Table 6.9. Kinetic data for oxidation of $[Mo_3O_4(OH_2)_9]^{4+}$ by various reagents[a]

Oxidant	Rate law[b]	Parameters[c]	Ref.
$[IrCl_6]^{2-}$ [g]	$kK_{aM}[Red]/([H^+] + K_{aM})$	$k = 4.5\,M^{-1}\,s^{-1}$, $K_{aM}{}^d = 0.42\,M$	[160]
$V^V(aq)$ [h]	$(kK_{aM} + k'K_{aM}[H^+]^{-1})$ $[Ox]^2/([H^+] + K_{aM})$	$k = 2.6 \times 10^3\,M^{-2}\,s^{-1}$, $k' = 830\,M^{-1}\,s^{-1}$, $K_{aM} = 0.19\,M$	[161]
$BrO_3{}^-$ [i]	$kKK_{aM}[Ox]/$ $(K_{aM} + [H] + KK_{aM}[Ox]$	$k = 0.29\,s^{-1}$, $K = 150\,M^{-1}$, $K_{aM} = 0.18\,M$	[161]
H_5IO_6 [j]	Same	$k = 44\,s^{-1}$, $K = 70\,M^{-1}$, $K_{aM} = 0.19\,M$	[161]
$Fe(NCS)^{2+}$ [k]	$(k_1 + k_2 K_{Mo}[NCS^-])[Red]/$ $(1 + K_{Fe}[NCS^-])$ $(1 + K_{Mo}[NCS^-])[H]^2$	$k_1 = 0.19\,M^2\,s^{-1}$, $k_2 = 0.14\,M^2\,s^{-1}$, $K_{Mo} = 300^e$, $K_{Fe} = 138^f$	[161]

[a] Reactions monitored at 25 °C at 505 nm except in the case of $Fe(NCS)^{2+}$ (460 nm) and $[IrCl_6]^{2-}$ (300 nm).
[b] Rate laws describe $-d(\ln[Mo_3O_4{}^{4+}])/dt$.
[c] K values pertain to the $Mo_3O_4{}^{4+}$–Ox association quotient.
[d] K_{aM} is the acid dissociation constant for $Mo_3O_4{}^{4+}(aq)$.
[e] Value obtained by extrapolation of data in ref. [118].
[f] K value reported in ref. [162].
[g] $I = 2.0\,M$ (Lipts).
[h] $I = 1.2\,M$ (Napts).
[i] $I = 2.0\,M$ (Napts).
[j] $I = 1.2\,M$ (Napts).
[k] $I = 0.1\,M$ (Napts).

required loading at $[H^+] < 0.2\,M$ which suggested a charge $< 3+$. Elution of the ion could be carried out with a $> 0.5\,M$ solution of any strong acid, including $HClO_4$. The dinuclear $Mo_2O_4{}^{2+}$ structure was implied on the basis of redox and cryoscopic behaviour and charge/Mo determinations. The reversible formation of the green colour of $[MoOCl_5]^{2-}$ upon saturating a solution of $Mo_2O_4{}^{2+}(aq)$ with HCl gas suggested retention of the Mo=O group in the dimer. Finally, the $[Mo_2O_2(\mu\text{-}O)_2(OH_2)_6]^{2+}$ formulation (Fig. 6.35) was indicated via the ready formation of structural characterised derivatives with the same $Mo_2O_4{}^{2+}$ core upon simple treatment of the aqua ion with complexing ligands such as, for example, EDTA, $C_2O_4{}^{2-}$ and cysteinate under mild conditions. In solutions $\sim 6.0\,M$ in HCl a paramagnetic single-bridged dimer, $[Mo_2O_2(\mu\text{-}O)Cl_8]^{4-}$ (Fig. 6.35), is also believed to exist.

A convenient method of direct synthesis of the $[Mo_2O_4(OH_2)_6]^{2+}$ ion is via reduction of solutions of Mo^{VI} (Na_2MoO_4) in 2.0 M HCl with hydrazine for 2 hours at 50 °C followed by filtration, dilution to 0.1 M $[H^+]$ and ion-exchange purification. From a saturated column, elution of the ion with 2.0 M $[H^+]$ can give solutions $\sim 0.2\,M$ in $Mo_2O_4{}^{2+}(aq)$.

Solutions of $[Mo_2O_4(OH_2)_6]^{2+}$ possess peak maxima at 384 nm ($\varepsilon = 103\,M^{-1}\,cm^{-1}$ per Mo_2), 295 nm ($3550\,M^{-1}\,cm^{-1}$) and 254 nm

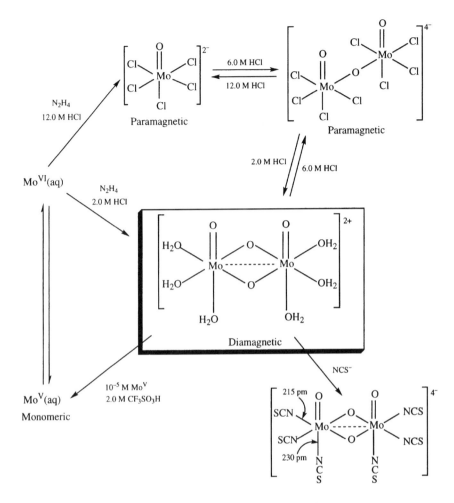

Figure 6.35. Scheme of reactions involving $Mo^V(aq)$ species

$(4120\,M^{-1}\,cm^{-1})$. No crystal structure exists for the aqua ion itself although the crystal structure of $(pyH)_4[Mo_2O_4(NCS)_6]\cdot H_2O$ indicates non-equivalent iso-thiocyanate ligands, those Mo—NCS bonds *trans* to the Mo=O groups being longer (230 pm) than those *trans* to μ-O (215 pm) [164]. The diamagnetism probably arises through coupling of the d^1 Mo^V centres via superexchange through the μ-O groups and/or direct Mo—Mo interaction (Mo—Mo = 256 pm, EXAFS) [91]. A hydrolysis constant (pK_a) of > 2 is implied by lack of changes to the electronic spectrum of $Mo_2O_4^{2+}$ in the $[H^+]$ range

0.01–1.0 M. Above pH 2 a change to a deeper orange colour is noticed prior to precipitation of '$Mo(OH)_5$'. A soluble polymeric form of $Mo^V(aq)$ has been reported [165]. Monomeric Mo^V exists under conditions of high dilution $(1 \times 10^{-5} M)$ from reversible cyclic voltammograms from solutions of Mo^{VI} ($[MoO_2(OH_2)_4]^{2+}$, Section 6.2.5) in 2.0 M CF_3SO_3H [166]. The structure of the ion is therefore probably $[MoO_2(OH_2)_4]^+$. At higher concentrations rapid dimerisation to $[Mo_2O_4(OH_2)_6]^{2+}$ occurs ($k_d \sim 10^3 M^{-1}s^{-1}$). The hydrolysis of mononuclear Mo^V at high dilution in ClO_4^- media has also been studied [167].

6.2.4.2 COMPLEX FORMATION AND WATER EXCHANGE ON $[Mo_2O_2(\mu\text{-}O)_2(OH_2)_6]^{2+}$

An ^{18}O labelling study by Murman [168] revealed fast exchange at the $Mo{=}O$ groups ($t_{1/2} = 4$ min, 0 °C) but much slower exchange at the μ-O groups ($t_{1/2} = 100$ hours, 40 °C, confirmed by Raman analysis). Further analysis of the μ-O exchange revealed a dependence of the form $k_{ex} = k$ $[H^+]^2$, implying a mechanism wherein double protonation of a μ-O group breaks the Mo—O—Mo bridge, giving a fast exchanging water molecule. Mo^{VI}, added as $Na_2[MoO_4]$, also promotes exchange at the μ-O atoms, dependence $k_{ex} = k_o + k_1[Mo^{VI}]$. In 0.3 M HCl, 40 °C, $k_o = 4.0 \times 10^{-6} s^{-1}$, $\Delta H^{\ddagger}_o = (108.3 \pm 6.6)$ kJ mol^{-1}, $\Delta S^{\ddagger}_o = (1.7 \pm 0.8)$ J K^{-1} mol^{-1} and $k_1 = 4.6 \times 10^{-2} M^{-1}s^{-1}$. The large value of $\Delta H^{\ddagger}_o$ reflects the Mo—O bond-breaking process. The exchange process has also been studied by ^{17}O NMR [116]. ^{17}O NMR spectra obtained (25 °C) following treatment of a solution of $[Mo_2O_4(OH_2)_6]^{2+}$ in 1.0 M CF_3SO_3H with $H_2^{17}O$ (to a final ^{17}O enrichment of 5 atom %) are shown in Fig. 6.36. Rapid appearance of a peak at 964 ppm ($Mo{=}O$), within the time taken to mix and recording one spectrum (5 min), is followed by the slow appearance of a second peak at 582 ppm (μ-O) which reaches a maximum height after 10 days. The $t_{1/2}$ of $\sim$ 2 days under these conditions matches that predicted by Murmann. The H_2O ligands exchange on the fast timescale ($k_{ex} > 10^3 s^{-1}$, 25 °C) and thus their ^{17}O resonances (expected ± 50 ppm from H_2O) are unobservable. The broad resonance at 159 ppm in Fig. 6.36 is from the natural abundance non-exchanging $CF_3SO_3^-$ anion. Fast exchange of the H_2O ligands was predicted from complex formation studies which showed fast equilibration requiring the use of relaxation techniques.

A kinetic study of the 1:1 equilibration of NCS^- with $[Mo_2O_4(OH_2)_6]^{2+}$, using the temperature-jump method, revealed an $[H^+]$ independent process with $k_f = (2.9 \pm 0.1) \times 10^4 M^{-1}s^{-1}$, $\Delta S^{\ddagger}_f = (47.2 \pm 3.8)$ kJ mol^{-1}, $\Delta S^{\ddagger}_f = (-1.25 \pm 13.0)$ J K^{-1} mol^{-1} and $k_b = (120 \pm 10)s^{-1}$, $\Delta H^{\ddagger}_b = (57.3 \pm 10.5)$ kJ mol^{-1}, $\Delta S^{\ddagger}_b = (-12.5 \pm 33.4)$ J K^{-1} mol^{-1}, $I = 2.0$ M, $LiClO_4$ [169]. The kinetically determined K_{eq} (240 M^{-1}) compared well with the value (250 $\pm$ 25 M^{-1}) obtained spectrophotometrically. The rate constants and activation parameters for

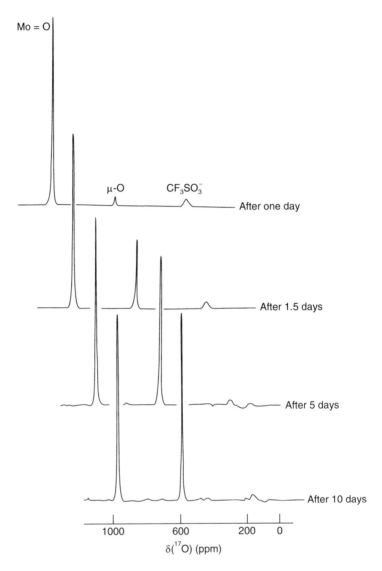

Figure 6.36. ^{17}O NMR spectra for a solution of $[Mo_2O_2(\mu-O)_2(OH_2)_6]^{2+}(0.05M)$ in $1.0\,M\ CF_3SO_3H/0.1\,M\ Mn(CF_3SO_3)_2$, $25\,°C$, equilibrated with 5 atom % H_2O^{17} over a period of one week

the 1:1 NCS$^-$ equilibration process support a mechanism similar to that occurring on Mo_2^{4+} (aq) and VO^{2+} (aq) wherein initial rapid complexation at the position *trans* to the Mo=O group is followed by isomerisation to the equatorial positions (see Fig. 6.19). Rate constants are, however, somewhat faster than those occurring on VO^{2+}(aq), implying additional labilisation of the equatorial positions stemming from the presence of the μ-O groups. The faster exchange occurring at the M=O groups of $Mo_2O_4^{2+}$(aq) vs that on VO^{2+}(aq) also implies a labilisation from the μ-O groups in the dimer structure.

Substitution by S and Se occurs readily into the bridging μ-O positions of $Mo_2O_4^{2+}$(aq) with both $Mo_2O_2(\mu\text{-O})(\mu\text{-X})^{2+}$ and $Mo_2O_2(\mu\text{-X})_2^{2+}$ units structurally characterised [130a]. Substitution of each μ-O by μ-S results in a downfield shift in the ^{95}Mo NMR resonance of ~ 170 ppm (Fig. 6.37), implying effective donation of σ-electron density from μ-S. Each Mo^V dimer can be conveniently isolated as the 1-cysteinato complex, $[Mo_2O_2(\mu\text{-X})_2(1\text{-cys}$-teinate)$_2]^{2-}$, either by reduction of Mo^{VI} in HCl with N_2H_4 followed by the addition of either $Na_2S \cdot 9H_2O$ (acidic pH) (mixed μ-O μ-S) [130] or H_2S (neutral pH) (both μ-S) [170] and then 1-cysteine, or, alternatively, in the case of Se, reduction of Mo^{VI} directly with HSe^- (both μ-Se) and then addition of 1-cysteine [131]. Acid catalysed aquation of the appropriate 1-cysteinate complex then gives the corresponding Mo_2^V aqua ion. As mentioned earlier, these species have proved useful 'lead-ins' to the synthesis of a range of mixed O,S (Se) Mo^{IV} trimer ions (see Section 6.2.3). Although a number of Mo^V species are known containing Mo=S groups, including some $Mo_2S_2(\mu\text{-S})^{2+}$ dimeric complexes, no aqua ion exists with this unit due to rapid replacement of Mo=S by Mo=O in the presence of H_3O^+. At least one Mo=S group is believed to be present at the active site of a number of molybdo-oxygen transferase enzymes such as xanthine oxidase [79].

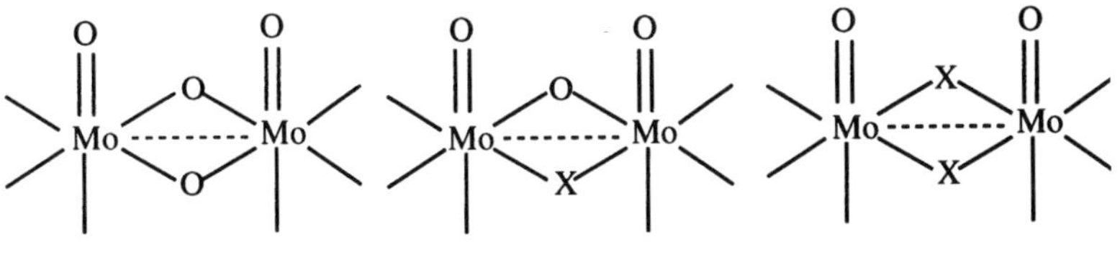

		X = S	X = S, Se
δ^{95}Mo (ppm from Mo^{VI})	544	722	974 (S)
Mo–Mo (pm) (typical)	255	272	281 (S) 283 (Se)

Figure 6.37. Known dimeric units for Mo^V(aq)

6.2.4.3	REDOX REACTIONS INVOLVING $[Mo_2O_2(\mu\text{-}O)_2(OH_2)_6]^{2+}$ AND
RELATED SPECIES

Replacement of μ-O by μ-S introduces a degree of stability towards oxidation to Mo^{VI}. The μ-Se analogues are intrinsically unstable thermodynamically due to ready loss of Se [131]. Solutions of $[Mo_2O_4(OH_2)_6]^{2+}$ oxidise only slowly in the air but eventually colourless Mo^{VI} is produced, sometimes contaminated with an insoluble blue $Mo^{V,VI}$ mixed-valence polymer.

Reduction of $[Mo_2O_4(OH_2)_6]^{2+}$ yields the blue-green Mo_2^{III} ion $[Mo_2(\mu\text{-}OH)_2(OH_2)_8]^{4+}$ in a 4e- step and is a convenient route to its preparation. The Mo_2^{III} species resulting from reduction of the $[Mo_2O_2(\mu\text{-}O)_n(\mu\text{-}X)_{2-n}(OH_2)_6]^{2+}$ ions (X = S, Se) seem not to have been fully characterised. Most kinetic studies have, however, concentrated on the $Mo^{V}-Mo^{VI}$ oxidation process. Consistent with the structural mismatch, and the need for bond breaking, oxidation of $Mo^{V}(aq)$ is rather sluggish and requires quite powerful reagents for study.

The kinetics of oxidation of $[Mo_2O_4(OH_2)_6]^{2+}$ with MnO_4^- [170], $[IrCl_6]^{2-}$ and $[Fe(phen)_3]^{3+}$ [171] and $VO_2^+(aq)$ [172] have been reported. With $[IrCl_6]^{2-}$, conventional monitoring at 487 nm (loss of $[IrCl_6]^{2-}$) with Mo_2^V in excess, the rate law was of the following form containing a term (k_1) zero order in oxidant:

$$-\frac{d([IrCl_6]^{2-})}{2dt} = k_1[Mo_2O_4{}^{2+}][H^+]^{-1}$$

$$+ (k_2 + k_3[H^+]^{-1})[Mo_2O_4{}^{2+}][IrCl_6{}^{2-}] \qquad (6.30)$$

At 25 °C, $I = 2.00$ M, $LiClO_4$, $k_1 = (7.95 \pm 0.13) \times 10^{-6}$ M s^{-1}, $k_2 = (0.114 \pm 0.012)$ M^{-1} s^{-1} and $k_3 = (0.052 \pm 0.003)$ s^{-1} [171]. A feature of both pathways is the strong $[H^+]^{-1}$ dependence. A similar rate law was found in the case of oxidation by $[Fe(phen)_3]^{3+}$, $k_1 = 3.09 \times 10^{-6}$ M s^{-1}, $k_2 = 31$ M^{-1} s^{-1}. The agreement in k_1 values supports a common process. Both oxidant-dependent processes are probably outer-sphere. With oxidants such as the μ-superoxo complex $[Co_2(NH_3)_{10}(\mu\text{-}O_2{}^-)]^{5+}$ only the k_1 pathway $(k_1 = 4.3 \times 10^{-6}$ M s$^{-1})$ is observed. The oxidant-independent process is believed to involve cleavage of one μ-O bridge to give '$Mo_2O_2(\mu\text{-}O)^{4+}(aq)$' which is then rapidly oxidised. Bridge cleavage to give monomeric $Mo^{V}(aq)$ is not suggested on the basis of the slow rate of oxidation of the dimer by NO_3^-. Nonetheless, the $[H^+]^{-1}$ dependence for k_1 has not been adequately explained and remains a somewhat surprising observation in view of the $[H^+]^2$ dependence observed for the μ-O water exchange process [168]. With the strong oxidant MnO_4^-, a rate law, $(k_1 + k_2K_p[H^+])[Mo_2O_4{}^{2+}][MnO_4^-]$, is observed [170]. Here the oxidant-dependent term dominates $(k_1 = (5.4 \pm 0.4) \times 10^3$ M^{-1} s^{-1}, $k_2 = (2.4 \pm 0.2) \times 10^2$ M^{-2} s$^{-1})$ with the $[H^+]$ dependence implying two parallel oxidation pathways for $Mo_2O_4{}^{2+}(aq)$ involving MnO_4^- and $HMnO_4$, $(K_p (MnO_4^- + H^+ \rightleftharpoons HMnO_4) = 2.99 \times 10^3$ M$^{-1})$.

5.2.4.4 PREPARATION AND PROPERTIES OF $[W_2^V O_2(\mu\text{-}O)_2(OH_2)_6]^{2+}$

Support for the existence of a dimeric W^V aqua ion first arose from knowledge of the $W_2O_2(\mu\text{-}O)_2^{2+}$ unit in complexes such $[W_2O_4(EDTA)]^{2-}$ [173]. The corresponding pale-yellow $[W_2O_2(\mu\text{-}O)_2(OH_2)_6]^{2+}$ ion was eventually isolated and characterised in the late 1980s by Sykes' group [174]. As found in the case of $[W_3O_4(OH_2)_9]^{4+}$, simple reduction from W^{VI} does not work to generate W^V. Instead air-free acid hydrolysis of monomeric W^V complexes, such as blue $[W^VOCl_5]^{2-}$ (best), has to be used. A useful high yield route through to $[WOCl_5]^{2-}$ involves treatment of the oxalato complex $[W^VO_2(C_2O_4)_2]^{3-}$ with concentrated HCl in the presence of NH_4Cl. Solutions of yellow $[W_2O_4(OH_2)_6]^{2+}$ are then prepared by treatment of $[WOCl_5]^{2-}$ with, for example, aqueous Hpts, filtration to remove a blue $W^{V,VI}$ oxide precipitate, and then purification by Dowex cation-exchange column chromatography as for $[Mo_2O_4(OH_2)_6]^{2+}$. Freshly columned solutions of $[W_2O_4(OH_2)_6]^{2+}$ in 2.0 M Hpts possess absorption maxima at 430 nm ($\varepsilon = 193$ M^{-1} cm^{-1} per W_2) and 340 (278 M^{-1} cm^{-1}) (Fig. 6.38) [174]. The electronic spectrum of $[Mo_2O_4(OH_2)_6]^{2+}$ is shown for comparison. Natural abundance aqueous CF_3SO_3H solutions of $[W_2O_4(OH_2)_6]^{2+}$, following treatment with ^{17}O, show a single ^{17}O NMR resonance at 443 ppm (W—O—W), slow to exchange, which may be compared

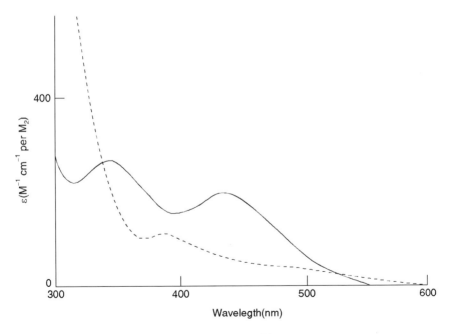

Figure 6.38. Electronic spectra of $[M_2O_4(OH_2)_6]^{2+}$ ions $M_2 = Mo_2(\text{---})$, $W_2(—)$.

to the Mo—O—Mo resonance in $[Mo_2O_4(OH_2)_6]^{2+}$ (582 ppm). The upfield shift in the μ-O resonance seems to be a general phenomenon observed on replacing Mo with W.

Earlier attempted routes to make $W_2O_4^{2+}$(aq), via acid aquation of the $[W_2O_4(EDTA)]^{2-}$ complex, produced a species with a similar electronic spectrum to that prepared from $[WOCl_5]^{2-}$, but recomplexation with EDTA on lowering the $[H^+]$ meant that the aqua ion could not be separated and purified from the free EDTA using ion-exchange chromatography [175].

Manipulations involving $[W_2O_4(OH_2)_6]^{2+}$ require rigorous air-free conditions as the aqua ion is acutely air sensitive. Solutions exposed to air show a 50% increase in absorbance at 430 nm after one hour and deposit a blue precipitate after several hours. The ion is sensitive to oxidation by ClO_4^- and for most solution studies Hpts has been used. The use of freshly columned solutions of $W_2O_4^{2+}$(aq) is thus to be recommended.

6.2.4.5   Complex formation studies on $[W_2O_2(\mu\text{-O})_2(OH_2)_6]^{2+}$

Early indications of slower rates of ligand replacement on the W^V ion were evident from reactions involving HCl. Whereas 10^{-3} M solutions of $Mo_2O_4^{2+}$(aq) produce green $[MoOCl_5]^{2-}$ instantaneously upon treatment with 10.0 M HCl, corresponding treatment of solutions of $W_2O_4^{2+}$(aq) require at least 10 seconds for development of the blue colour of $[WOCl_5]^{2-}$. Detailed kinetic studies have been performed regarding complexation with NCS^-. An $[H^+]$-independent equilibration process is observed as found with $Mo_2O_4^{2+}$(aq). Here, however, the reaction was slow enough to be followed by stopped-flow methods [174]. The rate law

$$k_{eq} = k_1([W_2O_4^{2+}]_{xs} \text{ or } [NCS^-]_{xs}/2) + k_{-1} \qquad (6.31)$$

was relevant with a statistical factor of 2 for reaction on the two identical and independent W sites. At 25 °C, $k_1 = 2.52 \times 10^3$ M^{-1} s^{-1} which may be compared with a k_1 value of 2.9×10^4 M^{-1} s^{-1} for the corresponding reaction on $Mo_2O_4^{2+}$(aq). Consistent with this it has been shown that the rate constant for aquation of $[W_2O_4(EDTA)]^{2-}$ (1.6×10^{-4} s^{-1}) is some 60 times smaller than that for $[Mo_2O_4(EDTA)]^{2-}$ (6.9×10^{-3} s^{-1}) [175]. The greater inertness of the W ion is as a result of the relativistic expansion effect, resulting in greater 5d orbital participation into the W—L MOs, leading to stronger and thus more inert W—L bonds. The slightly higher $\Delta H^{\ddagger}$ value (55.6 kJ mol^{-1}) for NCS^- substitution on $W_2O_4^{2+}$(aq) versus that on $Mo_2O_4^{2+}$ (47.2 kJ mol^{-1}) would be in keeping with the mechanism.

6.2.4.6   Redox reactions involving $[W_2O_2(\mu\text{-O})_2(OH_2)_6]^{2+}$

Kinetic studies involving oxidation of $[W_2O_4(OH_2)_6]^{2+}$ with $[IrCl_6]^{2-}$, BrO_3^-, and $[PtCl_6]^{2-}$ have been described [176]. The rate laws show no evidence of the zero-order pathway that characterised corresponding oxidation of $Mo_2O_4^{2+}$ (aq). With amounts of oxidant insufficient to cause oxidation to W^{VI}, deep blue coloured products result characterised by intense broad absorptions above 700 nm extending into the near i.r. region. Mixed-valence W^V-W^{VI} polymeric products are believed to be involved. For oxidation by $[IrCl_6]^{2-}$, the rate law $-d[Ir^{IV}]/dt = 2k_{Ir} [W_2O_4^{2+}][IrCl_6^{2-}]$ was relevant. Comparisons with Mo show that oxidation of $W_2O_4^{2+}$ (aq) ($k_{Ir} = 6.6 \times 10^4 M^{-1} s^{-1}$) occurs some 5 orders of magnitude faster than corresponding oxidation of $Mo_2O_4^{2+}$ (aq) (oxidant-dependent pathway). Surprisingly the $W_2O_4^{2+}$ (aq) oxidation shows no $[H^+]$ dependence in contrast to strong $[H^+]^{-1}$ dependences for oxidation of $Mo_2O_4^{2+}$ (aq). The reasons for this are not clear. The stronger reducing property of W^V is also evident from a similar 10^5 times faster rate of oxidation, by $[IrCl_6]^{2-}$, of $[W_2O_4(EDTA)]^{2-}$ vs $[Mo_2O_4(EDTA)]^{2-}$ [175]. With BrO_3^- and $[PtCl_6]^{4-}$ as oxidants $W_2O_4^{2+}$ (aq) likewise behaves as a much stronger reductant than $Mo_2O_4^{2+}$ (aq). The rate constant (25 °C) for oxidation of $[W_2O_4(OH_2)_6]^{2+}$ by $[IrCl_6]^{2-}$ is ~ 26 times larger than that observed for substitution of H_2O by NCS^- and, as in the case of $[Mo_2O_4(OH_2)_6]^{2+}$, the reaction is most likely outer-sphere.

The preparation and properties of the mixed Mo^V-W^V aqua ion $[MoWO_4(OH_2)_6]^{2+}$ has been reported along with a crystal structure of the complex $[MoWO_4(EDTA)]^{2-}$ prepared by neutralisation of solutions containing EDTA and the salts $[MoOCl_5]^{2-}$ and $[WOCl_5]^{2-}$ (1:1 ratio). The aqua ion has been obtained via acid aquation of the EDTA complex. The Mo—W distance in $[MoWO_4(EDTA)]^{2-}$ (255 pm) is identical to that found for $[W_2O_4(EDTA)]^{2-}$. The ^{183}W NMR shift for the Mo—W complex (549 ppm from WO_4^{2-}) can be compared with that for $[W_2O_4(EDTA)]^{2-}$ (798 ppm). Corresponding consideration of ^{95}Mo NMR shifts for $[MoWO_4(EDTA)]^{2-}$ (877 ppm) and $[Mo_2O_4(EDTA)]^{2-}$ (612 ppm) are indicative, as in the case of the M^{IV} trimer ions (Section 6.2.3), of a bond polarization $Mo^{\delta-}$—$W^{\delta+}$ in the mixed Mo—W derivatives. A review of the properties of mixed Mo—W species has appeared [177].

6.2.5 Molybdenum(VI) and Tungsten(VI) (d^0)

6.2.5.1 Cationic Mo^{VI} aqua ions

Cationic aqua ions exist in acidic aqueous solution for Mo^{VI} but not for W^{VI}, although, for the latter, soluble transiently stable blue mixed-valence $W^{V,VI}$(aq) species are known to result from oxidation of $[W_2O_4(OH_2)_6]^{2+}$. In $HClO_4$ solutions $\geqslant 6.0 M$, the principal species is believed to be monomeric cis-$[MoO_2(OH_2)_4]^{2+}$ [178]. Between 0.2 and 3.0 M $[H^+]$, a monomer–dimer equilibrium exists and the kinetics of this process have been investigated by Sykes' group in $I = 3.0 M$, $LiClO_4$ solution using the temperature-jump tech-

324   *The Chemistry of Aqua Ions*

nique [179]. In this range of acidity, a unipositive cationic monomer is relevant and may be written as $HMoO_3^+$(aq) or more probably as *cis*-$[MoO_2(OH)(OH_2)_3]^+$. The principal pathway for the equilibrium appears to be

$$2[MoO_2(OH)(OH_2)_3]^+ \underset{k_{-1}}{\overset{k_1}{\rightleftharpoons}} [Mo_2O_4(\mu\text{-}O)(OH_2)_6]^{2+} + H_2O \quad (6.32)$$

with a minor contribution from

$$[MoO_2(OH)(OH_2)_3]^+ + [MoO_2(OH_2)_4]^{2+} \underset{k_{-2}}{\overset{k_1}{\rightleftharpoons}}$$

$$[Mo_2O_4(\mu\text{-}OH)(OH_2)_6]^{3+} + H_2O \quad (6.33)$$

At 25 °C, $k_1 = (1.71 \pm 0.1) \times 10^5\,M^{-1}s^{-1}$, $k_{-1} = (3.2 \pm 0.2) \times 10^3\,s^{-1}$, $k_2 = (0.3 \pm 0.3) \times 10^5\,M^{-1}s^{-1}$ and $k_{-2} = (30 \pm 20)\,s^{-1}$. Krumenacker had earlier obtained independent evidence for dimeric 2+ and 3+ cationic species in this acidity range [180]. Three dimeric species have been proposed following spectrophotometric studies by Cruywagen and coworkers in $HClO_4$ solution (0.5 – 3.0 M) [181]. Evidence for dinuclear Mo^{VI}(aq) species (Mo—O—Mo bridge) has also been gained from Raman studies in 3.0 M $HClO_4$. Bands were assigned respectively as 953 and 920 cm^{-1} (*v* of *cis*-MoO_2), 825 (*v* Mo—O—Mo) and 378 ($\delta\,MoO_2$) [182]. The principal species under these conditions was thus deduced to be $[Mo_2O_4(\mu\text{-}O)(OH_2)_6]^{2+}$. Additional evidence for the species $[Mo_2O_5(OH_2)_6]^{2+}$, $[MoO_2X_2(H_2O)_2]$ and *cis*-$[MoO_2X_4]^{2-}$ (X = Cl or Br) has been gained from ^{95}Mo NMR and vibrational spectroscopic studies of Mo^{VI}(aq) in aqueous HX solutions [183]. The ion $[Mo_2O_5(OH_2)_6]^{2+}$ (Fig. 6.39) resonates at -63 ppm from $[MoO_4]^{2-}$ and appears to predominate in solutions up to 6.0 M in $[H^+]$. The $Mo_2O_5^{2+}$ core is also established in the complex

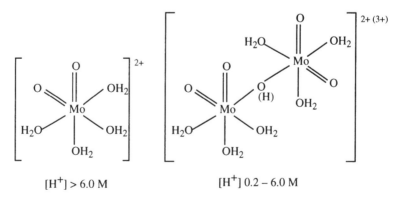

[H⁺] > 6.0 M   [H⁺] 0.2 – 6.0 M

Figure 6.39. Structures of Mo^{VI}(aq) cations in strongly acidic solution

$[Mo_2O_5(C_2O_4)_2(OH_2)_2]^{2-}$ At pH 2 the principal species appears to be monomeric. It is often written as $Mo(OH)_6$ but is more probably $[MoO_2(OH)_2(OH_2)_2]$.

6.2.5.2 ISOPOLYMOLYBDATES

Between pH 2 and 4 condensation of these monomeric units occurs to give remarkably two dominating isopolymolybdate anions $[Mo_7O_{24}]^{6-}$ and $[Mo_8O_{26}]^{4-}$ (Fig. 6.40). The $[Mo_8O_{26}]^{4-}$ ion can coexist in two isomeric forms α and β. In the α form two distorted tetrahedra are located above and below a crown of six octahedra formed by edge sharing. The β form has only octahedral Mo present [184]. Formation constant data are best evaluated if one considers small amounts of a protonated tetrahedral dimeric form, $[HMo_2O_7]^-$ (cf. $Cr_2O_7^{2-}$), in additional to protonated heptamolybdates $[HMo_7O_{24}]^{5-}$ and $[H_2Mo_7O_{24}]^{4-}$. The results of further ^{17}O and ^{95}Mo NMR investigations have suggested the formation of $[H_3Mo_8O_{28}]^{5-}$ as an intermediate in the interconversion between $[Mo_7O_{24}]^{6-}$ and the β form of $[Mo_8O_{26}]^{4-}$ [185]. Indeed,

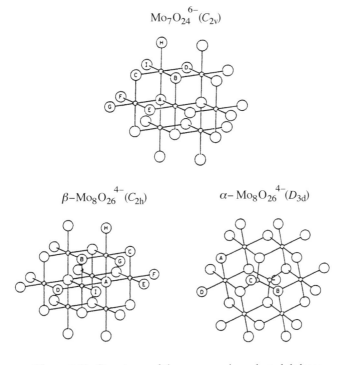

$Mo_7O_{24}^{6-}$ (C_{2v})

$\beta-Mo_8O_{26}^{4-}$ (C_{2h}) $\alpha-Mo_8O_{26}^{4-}$ (D_{3d})

Figure 6.40. Structures of the common isopolymolybdates

indications are that this species is a kinetic intermediate in the rapid oxygen exchange occurring between the isopolymolybdates and solvent water in the pH range 3–4. For the specific condition $[H^+]/Mo^{VI} = 1.8$ ultracentrifugation and EMF studies have shown that very large isopolymolybdate structures can exist, such as $[Mo_{36}O_{112}(OH_2)_{18}]^{8-}$ [183, 186]. Figure 6.41 shows distribution curves for Mo^{VI} species as a function of pH [81]. Above pH 4 the predominant form is tetrahedral $[MoO_4]^{2-}$. The equilibration reaction between $[MoO_4]^{2-}$

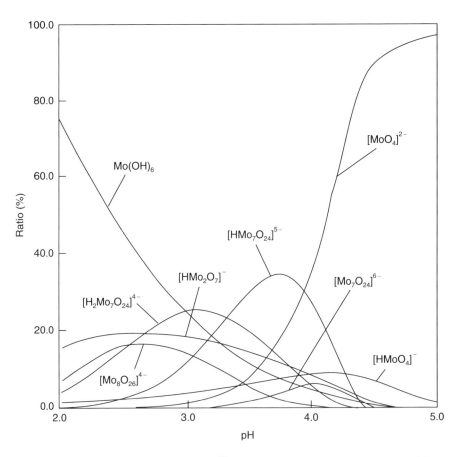

Figure 6.41. Distribution curves for Mo^{VI} species as a function of pH. Condsitions: $Mo^{VI} 5 \times 10^{-4}$ M, 25 °C,I = 1.0 M, NaCl) [81]. (Reprinted from A. G. Sykes, in *Comprehensive Coordination Chemistry* (eds. G. Wilkinson, R. D. Gillard and J. A. McCleverty), Vol. 5, 1987, p. 1258, with kind permission from Elsevier Science Ltd. The Boulevard, Hangford Hane, Kindlington 0 × 5 1GB, UK)

Group 6 Elements: Chromium, Molybdenum and Tungsten

Table 6.10. Thermodynamic data for the equilibration of $[MoO_4]^{2-}$ with H^+, 25 °C, $I = 3.0$ M, $NaClO_4$ [187]

Reaction	$\log Q$	ΔH° (kJ mol^{-1})	ΔS° (J K^{-1} mol^{-1})
$[MoO_4]^{2-} + H^+ \Leftrightarrow [HMoO_4]^-$	3.89	59 ± 30	272 ± 96
$[MoO_4]^{2-} + 2H^+ \Leftrightarrow cis\text{-}[MoO_2(OH)_2(OH_2)_2]$	7.50		
$7[MoO_4]^{2-} + 8H^+ \Leftrightarrow [Mo_7O_{24}]^{6-} + 4H_2O$	57.74	-234.0 ± 0.8	317 ± 4
$7[MoO_4]^{2-} + 9H^+ \Leftrightarrow [HMo_7O_{24}]^{5-}$	62.14		
$7[MoO_4]^{2-} + 10H^+ \Leftrightarrow [H_2Mo_7O_{24}]^{4-}$	65.68		
$7[MoO_4]^{2-} + 11H^+ \Leftrightarrow [H_3Mo_7O_{24}]^{3-}$	68.21		
$[Mo_7O_{24}]^{6-} + H^+ \Leftrightarrow [HMo_7O_{24}]^{5-}$		10.9 ± 0.1	121 ± 4
$[HMo_7O_{24}]^{5-} + H^+ \Leftrightarrow [H_2Mo_7O_{24}]^{4-}$		3.3 ± 2.1	79 ± 4
$[H_2Mo_7O_{24}]^{4-} + H^+ \Leftrightarrow [H_3Mo_7O_{24}]^{3-}$		-2.5 ± 0.5	42 ± 17

and H^+ has received detailed study and some of the findings are listed in Table 6.10 [187].

The mechanism of aggregation is complex but arises almost certainly from the increase in coordination number at Mo from four to six. Protonation of the $Mo=O$ groups of $[MoO_4]^{2-}$ increases the positive charge at Mo which becomes susceptible to nucleophilic attack by not only H_2O, giving $[MoO_2(OH)_2(OH_2)_2]$, but also the $O^{2-}(OH^-)$ groups of other Mo^{VI} moieties, leading to aggregation as the coordination number increases to six. The specific oligomeric structures that form probably arise from a balance between cation/anion size and electrostatic forces.

Oxygen-17 NMR spectroscopy has been successfully used to assign the different oxygen environments in heptamolybdate and the α and β form of $[Mo_8O_{26}]^{4-}$ [188]. The surface structure of isopolymolybdates has been likened to that of a metal oxide and, as such, considerable interest has arisen from detection of some degree of heterogeneous catalytic activity shown by a number of solid derivatives. These simpler 'surface analogues' have the advantage of possessing a well-defined molecular geometry and an extensive organometallic chemistry is rapidly developing [184].

Oxygen exchange between tetrahedral $[MoO_4]^{2-}$ and solvent water has been followed by ^{18}O labelling at pH > 11 [189]. The rate law is of the form: rate $= k_o[MoO_4{}^{2-}] + k_{OH}[MoO_4{}^{2-}][OH^-]$. At 25 °C, $I = 1.0$ M, $NaClO_4$, $k_o = 0.33$ s^{-1}, $\Delta H_o^\ddagger = (62.8 \pm 2.5)$ kJ mol^{-1}, $\Delta S_o^\ddagger = (-43 \pm 9)$ J K^{-1} mol^{-1}, and $k_{OH} = 2.22$ M^{-1} s^{-1}, $\Delta H_{OH}^\ddagger = (70.3 \pm 5.0)$ kJ mol^{-1}, $\Delta S_{OH}^\ddagger = (-2.5 \pm 13.0)$ J K^{-1} mol^{-1}. At lower pH values faster exchange occurs as the protonated forms and isopolyanions become involved [185].

6.2.5.3 ISOPOLYTUNGSTATES

Protonation of tetrahedral $[WO_4]^{2-}$ yields a number of isopolytungstates, the principal ones of which are defined as

$$7\,[WO_4]^{2-} + 8\,H^+ \xrightleftharpoons{\text{pH7}} [W_7O_{24}]^{6-} + 4\,H_2O$$

$$\textit{para}\text{-tungstate A} \qquad (6.34)$$

$$12\,[WO_4]^{2-} + 14\,H^+ \xrightleftharpoons{\text{pH 6}} [W_{12}O_{41}]^{10-} + 7\,H_2O$$

$$\textit{para}\text{-tungstate B} \qquad (6.35)$$

$$12\,[WO_4]^{2-} + 18\,H^+ \xrightleftharpoons{\text{pH 4}} [W_{12}O_{39}]^{6-} + 9\,H_2O$$

$$\textit{meta}\text{-tungstate} \qquad (6.36)$$

All isopolytungstates contain WO_6 octahedra but delicate balanced equilibria are present with the isolated structures dependent strongly upon the counter cation present in the solution. In contrast to the rapid attainment of equilibrium for molybdenum, formation of the isopolytungstates often takes several days to reach equilibrium. Much of the complexity has, however, been resolved by the combined use in solution of ^{17}O and ^{183}W NMR spectroscopy [188, 190, 191]. Some of the representative structures are shown in Fig. 6.42.

The $[W_6O_{19}]^{2-}$ ion is analogous to that found in solution for Nb^V and Ta^V but is only known in the solid state for molybdenum. The *para*-tungstates A and B are so named since they are interconvertable depending upon the tungsten concentration. *Para*-tungstate A, $[W_7O_{24}]^{6-}$, has been shown by ^{17}O and ^{183}W

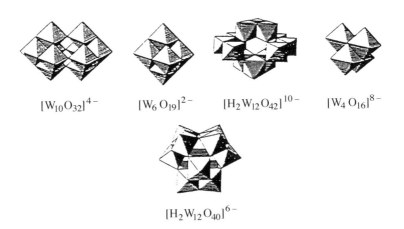

$$[W_{10}O_{32}]^{4-} \qquad [W_6O_{19}]^{2-} \qquad [H_2W_{12}O_{42}]^{10-} \qquad [W_4O_{16}]^{8-}$$

$$[H_2W_{12}O_{40}]^{6-}$$

Figure 6.42. Representative structures of isopolytungstates

NMR to have the same structure as heptamolybdate. By studying the structures of many isopolymetallates, Klemperer has been able to correlate the ^{17}O chemical shifts with the type of oxygen atom present. Increased metal to oxygen bonding correlated with large downfield shifts. Thus in essence the more metal atoms that coordinate to a given oxygen atom the further upfield is its ^{17}O NMR resonance [188] (see Chapter 5, Fig. 5.15).

Oxygen exchange at tetrahedral $[WO_4]^{2-}$ above pH 11 possesses the same rate law as found for $[MoO_4]^{2-}$. At 25 °C the value of $k_o = 0.44 \, s^{-1}$, very similar to that found for $[MoO_4]^{2-}$. However, this agreement is fortuitous since the spontaneous exchange on $[WO_4]^{2-}$ is characterised by a lower $\Delta H^{\ddagger}$ value $((29.25 \pm 3.22) \, kJ \, mol^{-1})$ compensated by a highly negative $\Delta S^{\ddagger}$ value $((-153.5 \pm 11.3) \, J \, K^{-1} \, mol^{-1})$ [189]. A linear isokinetic relationship $\Delta H^{\ddagger}$ vs $\Delta S^{\ddagger}$ is found on considering the k_o pathway for the series $[MoO_4]^{2-}$, $[WO_4]^{2-}$ and $[ReO_4]^{-}$, implying a common rate-determining process which is suggested to be solvent assisted oxygen dissociation (Fig. 6.43). The 10^6 times faster spontaneous exchange occurring on both $[MoO_4]^{2-}$ and $[WO_4]^{2-}$ vs that on $[CrO_4]^{2-}$ reflects the larger radius of the two heavier members as well as greater expansion in the acceptor d orbitals. A similar trend in activation parameters $Mo \rightarrow W$ has been observed to characterise terminal water exchange on the trinuclear bioxo-capped acetato complexes, $[M^{IV}_3(\mu_3\text{-}O)_2(\mu\text{-}O_2CCH_3)_6(OH_2)_3]^{2+}$ [192].

6.2.5.4 OXOTHIOMOLYBDATES

The complete series of thiomolybdate $[MoO_nS_{4-n}]^{2-}$ ions is known. In $[MoOS_3]^{2-}$ the $Mo{=}O$ distance of 176 pm is significantly longer than in other $Mo{=}O$ compounds, a further consequence of the σ-donating ability of S^{2-}. The $Mo{=}S$ distance, however, is typical of other $Mo{=}S$ compounds (218 pm). A number of kinetic studies regarding the interconversion between the various oxothiomolybdate ions have been reported [193, 194]. For the replacement of

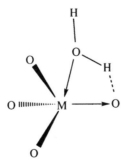

Figure 6.43. Mechanism of spontaneous oxygen exchange with solvent water on tetrahedral $[MO_4]^{2-}$ ions (M = Mo, W)

oxo by sulphido, rate constants decrease as the number of Mo=S groups increases. The reaction of H_2S with $[MoO_4]^{2-}$, giving $[MoO_3S]^{2-}$, is best considered as reaction of HS^- with $[HMoO_4]^-$, $k = 1.3 \times 10^6 \, M^{-1} s^{-1}$, pH 8–10, 0.25 M NH_3/NH_4^+ buffer, $I = 0.5$ M, NaCl [193]. The value of k here is similar in magnitude to that generally found for substitution reactions involving $[HMoO_4]^-$. Thiomolybdates, most probably $[MoS_4]^{2-}$, may be involved in the take-up of molybdenum into the cells of higher organisms for assimilation into enzymes and in ruminant animals responsible for the promotion of copper deficiency syndrome via the immobilisation of the element within various $Cu_xMo_yS_z$ species [195].

REFERENCES

[1]   (a) Rollinson, C. L., in Comprehensive Inorganic Chemistry (eds. J. C. Bailar Jr, H. J. Emeleus, R. S. Nyholm and A. F. Trotman-Dickenson), Vol. 3, Pergamon, Oxford, 1973, pp. 623–770.
      (b) Gmelins Handbuch der Anorganische Chemie, Teil 52C, Chromium, Springer-Verlag, Berlin, 1965.
[2]   Bjerrum, N., PhD Thesis, University of Copenhagen, 1908.
[3]   Hein, F., and Herzog, S. in Brauer, G. (ed) Handbook of Preparative Inorganic Chemistry, 2nd edn, Vol. 2, Academic, New York, 1965, p. 1361.
[4]   (a) Kranz, M., and Duczmal, W., Rosz. Chem., 47, 1823 (1973).
      (b) Holah, D. G., and Fackler Jr, J. P., Inorg. Syn., 10, 26 (1967).
[5]   (a) Larkworthy, L. F., Coord. Chem. Rev., 37, 91 (1981), 45, 105 (1982) and 57, 189 (1984).
      (b) Colton, R., Coord. Chem. Rev., 58, 245 (1984) and 62, 85 (1985).
[6]   (a) Cotton, F. A., Daniels, L. M., Murillo, C. A., and Quesada, J. F., Inorg. Chem., 32, 4861 (1993).
      (b) Earnshaw, A., Larkworthy, L. F., Patel, K. C., and Beech, G., J. Chem. Soc. A., 1334 (1969).
[7]   (a) Figgis, B. N., Kucharski, E. S., and Forsyth, J. B., Acta Cryst., C47, 419 (1991).
      (b) Delfs, C. D., Figgis, B. N., Forsyth, J. B., Kucharski, E. S., and Reynolds, P. A., Proc. Roy. Soc. Lond., Ser. A, 436, 417 (1992).
[8]   (a) Sham, T. K., Hastings, J. B., and Perlmann, M. L., Chem. Phys. Lett., 83, 391 (1981).
      (b) Micskei, K., and Nagypal, I., J. Chem. Soc. Dalton Trans., 1301 (1990).
[9]   von Schnering, H. G., and Brand. B.-H., Z. Anorg. Allg. Chem., 402, 159 (1973).
[10] (a) Hitchmann, M. A., Lichon, M., McDonald, R. G., Smith, P. W., Stranger, R., Skelton, B. W., and White, A. H., J. Chem. Soc. Dalton Trans., 1817 (1987).
      (b) One, K., J. Phys. Soc. Japan, 12, 1231 (1957).
[11] Cotton, F. A., Falvello, L. R., Murillo, C. A., and Quesada, J. F., J. Solid State Chem., 96, 192 (1992).
[12] Cotton, F. A., Daniels, L. M., and Murillo, C. A., Inorg. Chem., 32, 4868 (1993).
[13] Fackler Jr, J. P., and Holah, D. G., Inorg. Chem., 4, 954 (1965).
[14] Hunt, J. P., and Friedman, H. L., Prog. Inorg. Chem., 30, 359 (1983).
[15] Nagypal, I. J. Chem. Soc. Dalton Trans., 1335 (1983).

[16] Deutsch, E., Root, M. J., and Nasco, D. L., *Adv. Inorg. Bioinorg. Mech.*, **1**, 269 (1982).
[17] (a) Cannon, R. D., *Electron Transfer Reactions*, Butterworths, London, 1980.
(b) Wilkins, R. G., *Kinetics and Mechanisms of Reactions of Transition Metal Complexes*, 2nd edn, V.C.H., Weinheim, Germany, 1991, pp. 270–1.
[18] Bakac, A., and Espenson, J. H., *Acc. Chem. Res.*, **26**, 519 (1993).
[19] Picard, J., *Ber.*, **56**, 2477 (1913).
[20] (a) Ilan, Y. A., Czapski, G., and Ardon, M., *Isr. J. Chem.*, **13**, 15 (1975).
(b) Sellers, R. M., and Simic, M. G., *J. Am. Chem. Soc.*, **98**, 6145 (1976).
[21] Bruhn, S. L., Bakac, A., and Espenson, J. H., *Inorg. Chem.*, **25**, 535 (1986).
[22] (a) Bakac, A., Scott, S. L., Espenson, J. H., and Rodgers, K. R., *J. Am. Chem. Soc.*, **117**, 6493 (1995).
(b) Brynildson, M. E., Bakac, A., and Espenson, J. H., *Inorg. Chem.*, **27**, 2592 (1988).
(c) Brynildson, M. E., Bakac, A., and Espenson, J. H., *J. Am. Chem. Soc.*, **109**, 4579 (1987).
[23] Kolaczkowski, R. W., and Plane, R. A., *Inorg. Chem.*, **3**, 322 (1964).
[24] Wang, W.-D., Bakac, A., and Espenson, J. H., *Inorg. Chem.*, **32**, 5034 (1993).
[25] Espenson, J. H., *Acc. Chem. Res.*, **25**, 222 (1992), *Prog. Inorg. Chem.*, **30**, 189 (1983) and *Adv. Inorg. Bioinorg. Mech.*, **1**,1 (1982).
[26] van Eldik, R., Gaede, W., Cohen, H., and Meyerstein, D., *Inorg. Chem.*, **31**, 3695 (1992).
[27] Cohen, H., and Meyerstein, D., *J. Chem. Soc. Dalton Trans.*, 2449 (1974). Ryan, D. A., and Espenson, J. H., *Inorg. Chem.*, **20**, 4401 (1981).
[28] Best, S. P., and Forsyth, J. B., *J. Chem. Soc. Dalton Trans.*, 1721 (1991).
[29] Gustaffson, T., Lundgren, J. O., and Olovsson, I., *Acta Cryst.*, **B36**, 1323 (1980).
[30] (a) Magini, M., *J. Chem. Phys.*, **73**, 2499 (1980).
(b) Epple, M., and Massa, W., *Z. Anorg. Allg. Chem.*, **444**, 47 (1978).
[31] Holwerda, R. A., and Petersen, J. S., *Inorg. Chem.*, **19**, 2279 (1980).
[32] Scott, S. L., Bakac, A., and Espenson, J. H., *J. Am. Chem. Soc.*, **114**, 4605 (1992).
[33] Johnson, R. F., and Holwerda, R. A., *Inorg. Chem.*, **22**, 2942 (1983) and **24**, 3181 (1985).
[34] Best, S. P., Armstrong, R. A., and Beattie, J. K., *Inorg. Chem.*, **19**, 1958 (1980).
[35] (a) Hunt, J. P., and Plane, R. A., *J. Am. Chem. Soc.*, **76**, 5960 (1954).
(b) Stranks, D. R., and Swaddle, T. W., *J. Am. Chem. Soc.*, **93**, 2783 (1971).
[36] Hu, F.-C., Krouse, H. R., and Swaddle, T. W., *Inorg. Chem.*, **24**, 267 (1985).
[37] Arnau, C., Ferrer, M., Martinez, M., and Sanchez, A., *J. Chem. Soc. Dalton Trans.*, 1839 (1986).
[38] Abdullah, M. A., Barrett, J., and O'Brien, P., *J. Chem. Soc. Dalton Trans.*, 1647 (1984).
[39] Stunzi, H., and Marty, W., *Inorg. Chem.*, **22**, 2145 (1983).
[40] Rotzinger, F. P., Stunzi, H., and Marty, W., *Inorg. Chem.*, **25**, 489 (1986).
[41] Baes, C. F., and Mesmer, R. E., *The Hydrolysis of Cations*, 2nd edn., Kreiger, Malibar, Florida, 1986, pp. 211–20.
[42] Swaddle, T. W., and Kong, P. C., *Can. J. Chem.*, **48**, 3223 (1970).
[43] Ardon, M., and Linenberg, A., *J. Phys Chem.*, **65**, 1443 (1961).
[44] Thompson, M. E., and Connick, R. E., *Inorg. Chem.*, **20**, 2279 (1981).
[45] Spiccia, L., Evans, H. S., Marty, W., and Giovanoli, R., *Inorg. Chem.*, **26**, 474 (1987).
[46] Spiccia, L., and Marty, W., *Polyhedron*, **10**, 619 (1991).
[47] Finholt, J. E., Thompson, M. E., and Connick, R. E., *Inorg. Chem.*, **20**, 4151 (1981).

[48] Stunzi, H., Rotzinger, F. P., and Marty, W., *Inorg. Chem.*, **2160** (1984).
[49] (a) Laswick, J. A., and Plane, R. A., *J. Am. Chem. Soc.*, **81**, 3564 (1959).
     (b) Finholt, J. E., PhD Thesis, Lawrence Radiation Laboratory Report UCRL-8879, University of California, Berkeley, 1960.
[50] Spiccia, L., and Marty, W., *Inorg. Chem.*, **25**, 266 (1986).
[51] Spiccia, L., *Polyhedron*, **10**, 1865 (1991).
[52] Andersen, P., Damuus, T., Pedersen, E., and Pedersen, A., *Acta Chem. Scand.*, **38**, 359 (1984).
[53] Wieghardt, K., Schmidt, W., Endres, H., and Wolfe, C. R., *Chem. Ber.*, **112**, 2837 (1979).
[54] Gudel, H. H., Hauser, H., and Furrer, A., *Inorg. Chem.*, **18**, 2730 (1979)
[55] Flood, M. T., Marsh, R. E. and Gray, H. B., *J. Am. Chem. Soc.*, **91**, 193 (1969).
[56] Monsted, O., Monsted, L., and Springborg, J., *Inorg. Chem.*, **24**, 3496 (1985).
[57] (a) Andersen, P., and Berg, T., *Acta Chem. Scand.*, **32**, 989 (1978).
     (b) Damuus, T., and Pedersen, E., *Inorg. Chem.*, **23**, 695 (1984).
[58] Stunzi, H., Spiccia, L., Rotzinger, F. P., and Marty, W., *Inorg. Chem.*, **28**, 66 (1989).
[59] Crimp, S. J., Spiccia, L., Krouse, H. R., and Swaddle, T. W., *Inorg. Chem.*, **33**, 465 (1994).
[60] See, for example, Luo, Q., Shen, M., Ding, Y., Tu, Q., and Dai, A., *Hauxue Xuebao*, **44**, 568 (1986) (and refs. therein).
[61] Gustavson, K. H., *The Chemistry of the Tanning Process*, Academic, New York, 1956.
[62] Ardon, M., and Taube, H., *J. Am. Chem. Soc.*, **89**, 3661 (1967).
[63] Ramasami, T., and Sykes, A. G., *Inorg. Chem.*, **15**, 1010 (1976).
[64] Ramasami, T., Taylor, R. S., and Sykes, A. G., *J. Chem. Soc. Chem. Commun.*, 383 (1976) and *Inorg. Chem.*, **16**, 1931 (1977).
[65] Asher, L. E., and Deutsch, E., *Inorg. Chem.*, **12**, 1774 (1973).
[66] Asher, L. E., and Deutsch, E., *Inorg. Chem.*, **11**, 2927 (1972) and **15**, 1531 (1976).
[67] Bakac, A., Espenson, J. H., and Miller, L., *Inorg. Chem.*, **21**, 1557 (1982).
[68] Adams, C. A., Crook, J. R., Bockhoff, F., and King, E. L., *J.Am. Chem. Soc.*, **90**, 5761 (1968).
[69] Scott, S. L., Bakac, A., and Espenson, J. H., *J. Am. Chem. Soc.*, **113**, 7787 (1991).
[70] Scott, S. L., Bakac, A., and Espenson, J. H., *J. Am. Chem. Soc.*, **114**, 4205 (1992).
[71] Ghosh, M. C., and Gould, E. S., *Inorg. Chem.*, **30**, 491 (1991).
[72] Cieslak-Golonka, M., *Coord. Chem. Revs.*, **109**, 223 (1991).
[73] Okumara, A., Kitani, M., Toyomi, Y. and Oxazaki, N., *Bull. Chem. Soc Japan*, **53**, 3143 (1980).
[74] Gamsjager, H., and Murmann, R. K., *Adv. Inorg. Bioinorg. Mech.*, **2**, 317 (1983).
[75] Michel, G., and Machiroux, R., *J. Raman Spectrosc.*, **14**, 22 (1983).
[76] Michel, G., and Cahay, R., *J. Raman Spectrosc.*, **17**, 89 (1986).
[77] Brasch, N., Buckingham, D. A., and Clark, C. R., *Inorg. Chem.*, **33**, 2683 (1994) and *Aust. J. Chem.*, **47**, 2283 (1994).
[78] Moore, P., Ducummon, Y., Newman, K. E., and Merbach, A. E., *Helv. Chim. Acta*, **66**, 2445 (1983).
[79] Spiro, G. (ed.), *Molybdenum Enzymes*, Wiley, New York, 1985.
[80] (a) Sellman, D., *Angew. Chem. Int. Ed. (Engl.)*, **32**, 64 (1993).
     (b) Georgiadis, M. M. Komiya, H., Charabarti, P., Woo, D. Kornuf, J. J., and Rees, D. C., *Science*, **257**, 1653 (1992).
[81] Sykes, A. G., in *Comprehensive Coordination Chemistry*, (eds. G. Wilkinson, R. D. Gillard and J. A. McCleverty), Vol. 5, Pergamon, London, 1987, pp. 1229–61.

[82]   (a) Richens, D. T., and Sykes, A. G., *Comm. Inorg. Chem.*, **1** 141 (1981).
       (b) Richens, D. T., and Sykes, A. G., *Inorg. Syn.*, **23**, 130–9 (1985).
[83]   Cotton, F. A., and Walton, R. A., *Multiple Bonds Between Metal Atoms*, Wiley, New York, 1982.
[84]   Bowen, A. R., and Taube, H., *Inorg. Chem.*, **13**, 2245 (1974).
[85]   Brignole, A. B., and Cotton, F. A., *Inorg. Syn.*, **13**, 88 (1972).
[86]   Brencic, J. V., and Cotton, F. A., *Inorg. Chem.*, **9**, 351 (1970).
[87]   Cotton, F. A., Bertram, F. A., Pedersen, E., and Webb, T. E., *Inorg. Chem.*, **14**, 391 (1975).
[88]   Bino, A., and Gibson, D., *J. Am. Chem. Soc.*, **102**, 4277 (1980).
[89]   Lohmann, K. H., and Young, R. C., *Inorg. Syn.*, **4**, 97 (1953).
[90]   Lewis, J., Nyholm, R. S., and Smith, P. W., *J. Chem. Soc. A*, 57 (1969).
[91]   Cramer, S. P., Eidem, P. K., Paffett, M. T., Winkler, J. R., Dori, Z., and Gray, H. B., *J. Am. Chem. Soc.*, **105**, 799 (1983).
[92]   Gheller, S. F., Hambley, T. W., Brownlee, R. C., O'Connor, M. J., Snow, M. R., and Wedd, A. G., *J. Am. Chem. Soc.*, **105**, 1527 (1983).
[93]   Minelli, M., Enemark, J. H., Brownlee, R. C., O'Connor, M. J., and Wedd, A. G., *Coord. Chem. Revs.*, **68**, 169–278 (1985).
[94]   Cotton, F. A., Mott, G. N., Schrock, R. A., and Sturgeoff, L. G., *J. Am. Chem. Soc.*, **104**, 6781 (1982).
[95]   Finholt, J. E., Leupin, P., and Sykes, A. G. *Inorg. Chem.*, **22**, 3315 (1983).
[96]   Cotton, F. A., Frenz, B. A., and Webb, T. E., *J. Am. Chem. Soc.*, **95**, 4431 (1973).
[97]   Pernick, A., and Ardon, M., *J. Am. Chem. Soc.*, **97**, 1255 (1975).
[98]   Bino, A., *Inorg. Chem.*, **20**, 623 (1981).
[99]   Trogler, W. R., Erwin, D. K., Geoffroy, G. L., and Gray, H. B., *J. Am. Chem. Soc.*, **100**, 1160 (1978).
[100] Bowen, A. R., and Taube, H., *J. Am. Chem. Soc.*, **93**, 3287 (1971).
[101] Sasaki, Y., and Sykes, A. G., *J. Chem. Soc. Dalton Trans.*, 1048 (1975).
[102] Brorson, M., and Schaffer, C., *Acta Chem. Scand.*, **A40**, 358 (1986).
[103] Hills, E. F., and Sykes, A. G., *Polyhedron*, **5**, 511 (1986).
[104] Mayer, J. M., and Abbott, E. H., *Inorg. Chem.*, **22**, 2774 (1983).
[105] Hyldtoft, J., Larsen, S., and Monsted, O., *Acta Chem. Scand.*, **A43**, 842 (1989).
[106] Jacobsen, C. J. H., Villadsen, J., and Weihe, H., *Inorg. Chem.*, **32**, 5396 (1993).
[107] Brorson, M., and Gajhede, M., *Inorg. Chem.*, **26**, 2109 (1987).
[108] Richens, D. T., Harmer, M. A., and Sykes, A. G., *J. Chem. Soc. Dalton Trans.*, 2099 (1984).
[109] Beattie, J. K., Best, S. P., Skelton, B. W., and White, A. H., *J. Chem. Soc. Dalton Trans.*, 2105 (1981).
[110] Richens, D. T., Ducommun, Y., and Merbach, A. E., *J. Am. Chem. Soc.*, **109**, 603 (1987).
[111] Kelly, H. M., Richens, D. T., and Sykes, A. G., *J. Chem. Soc. Dalton Trans.*, 1229 (1984).
[112] Hills, E. F., Norman, P. R., Ramasami, T., Richens, D. T., and Sykes, A. G., *J. Chem. Soc. Dalton Trans.*, 157 (1986).
[113] Diebler, H., and Millan, C., *Polyhedron*, **5**, 539 (1986).
[114] Ketchum, P. A., Taylor, R. C., and Young, D. C., *Nature, Lond.*, **259**, 203 (1976).
[115] Ardon, M., and Pernick, A., *Inorg. Chem.*, **13**, 2276 (1974).
[116] Richens, D. T., Helm, L., Pittet, P.-A., and Merbach, A. E., *Inorg. Chim. Acta*, **132**, 85 (1987).
[117] Harmer, M. A., and Sykes, A. G., *Inorg. Chem.*, **20**, 3963 (1981).
[118] Chalilpoyil, P., and Anson, F. C., *Inorg. Chem.*, **17**, 2418 (1978).

[119] Souchay, P., Cadiot, M. and Duhameaux, M., *CR. Hebd. Seances Acad. Sci.*, **262**, 1524 (1966).
[120] Ojo, J. F., Sasaki, Y., Taylor, R. S., and Sykes, A. G., *Inorg. Chem.*, **15**, 1006 (1976).
[121] Ardon, M., Bino, A., and Yahaw, G., *J. Am. Chem. Soc.*, **98**, 2338 (1976).
[122] Cramer, S. P., and Gray, H. B., *J. Am. Chem. Soc.*, **101**, 2770 (1979).
[123] See, for example, Bino, A., Cotton, F. A., and Dori, Z., *J. Am. Chem. Soc.*, **101**, 3842 (1979).
[124] (a) Murmann, R. K., and Shelton, M. E., *J. Am. Chem. Soc.*, **102**, 3984 (1980).
       (b) Schlemper, E. O., Hussain, M. S., and Murmann, R. K., *Acta Cryst. Sec. C.*, *Cryst. Struct. Commun.*, **11**, 89 (1982).
[125] Richens, D. T., Helm, L., Pittet, P.-A., Merbach, A. E., Nicolo, F., and Chapuis, G., *Inorg. Chem.*, **28**, 1394 (1989).
[126] (a) Bursten, B. E., Cotton, F. A., Hall, M. B., and Najjar, R. C., *Inorg. Chem.*, **21**, 302 (1982).
       (b) Cotton, F. A., and Feng. X., *Inorg. Chem.*, **30**, 3666 (1991).
       (c) Cotton, F. A., *Polyhedron*, **5**, 3 (1986).
[127] (a) Wendan, C., Qianer, Z., Jinshun, H., and Jia-xi, L., *Polyhedron*, **8**, 2785 (1989).
       (b) Li, J., Liu, C.-W., and Jia-xi, L., *Polyhedron*, **13**, 1841 (1994).
[128] Cotton, F. A., Marler, D. O., and Schwotzer, W., *Inorg. Chem.*, **23**, 3671 (1984).
[129] Ooi, B.-L., and Sykes, A. G., *Inorg. Chem.*, **27**, 310 (1988).
[130] (a) Shibahara, T., *Coord. Chem. Revs.*, **123**, 73 (1993).
       (b) Martinez, M., Ooi, B.-L., and Sykes, A. G., *J. Am. Chem. Soc.*, **109**, 4615 (1987).
       (c) Shibahara, T., Akashi, H., Nagahata, S., Hattori, H., and Kuroya, H., *Inorg. Chem.*, **28**, 362 (1989).
       (d) Kathirgamanathan, P., Martinez, M., and Sykes, A. G., *Polyhedron*, **5**, 505 (1986).
[131] Nasreldin, M., Henkel, G., Kampmann, G., Krebs, B., Lamprecht, G. J., Routledge, C. A., and Sykes, A. G., *J. Chem. Soc. Dalton Trans.*, 737 (1993).
[132] Akashi, H., Shibahara, T., and Kuroya, H., *Polyhedron*, **9**, 1671 (1990).
[133] Akashi, H., Shibahara, T., Narabara, T., Tsuru, H., and Kuroya, H., *Chem. Lett.*, 129 (1989).
[134] (a) Ooi, B.-L., Martinez, M., and Sykes, A. G., *J. Chem. Soc. Chem. Commun.*, 1324 (1988).
       (b) Ooi, B.-L., Martinez, M., Shibahara, T., and Sykes, A. G., *J. Chem. Soc. Dalton Trans.*, 2239 (1988).
       (c) Lamprecht, G. J., Martinez, M., Nasreldin, M., Routledge, C. A., Al-Shatti, N., and Sykes, A. G., *J. Chem. Soc. Dalton Trans.*, 747 (1993).
[135] Ooi, B.-L., and Sykes, A. G., *Inorg. Chem.*, **28**, 3799 (1989).
[136] Richens, D. T., Pittet, P.-A., Merbach, A. E., Humanes, M., Lamprecht, G. J., Ooi, B.-L., and Sykes, A. G., *J. Chem. Soc. Dalton Trans.*, 2305 (1993).
[137] Kathirgamanathan, P., Soares, A. B., Richens, D. T., and Sykes, A. G., *Inorg. Chem.*, **24**, 2950 (1985).
[138] Ooi, B.-L., Sharp, C., and Sykes, A. G., *J. Am. Chem. Soc.*, **111**, 125 (1989).
[139] Brorson, M., Hyldtoft, J., Jacobsen, C. J. M., and Olesen, K. G., *Inorg. Chim. Acta*, **232**, 171 (1995).
[140] Hong, M. C., Li. Y.-J., Lu, J.-X, Nasreldin, M., and Sykes, A. G., *J. Chem. Soc. Dalton Trans.*, 2613 (1993).
[141] Li. Y.-J., Nasreldin, M., Humanes, M., and Sykes, A. G., *Inorg. Chem.*, **31**, 3011 (1992).
[142] (a) Shibahara, T., Akashi, H., and Kuroya, H., *J. Am. Chem. Soc.*, **108**, 1342 (1986).
       (b) Shibahara, T., Akashi, H., and Kuroya, H., *J. Inorg. Biochem.*, **36**, 178 (1989).

(c) Shibahara, T., and Kuroya, H., *J. Coord. Chem.*, **18**, 233 (1988).

(d) Shibahara, T., Akashi, H., and Kuroya, H., *J. Am. Chem. Soc.*, **110**, 3313 (1988).

(e) Akashi, H., and Shibahara, T., *Inorg. Chem.*, **28**, 2906 (1989).

(f) Sakane, G., and Shibahara, T., *Inorg. Chem.*, **32**, 777 (1993).

[143] Shibahara, T., Akashi, H., Yamasaki, M., and Hashimoto, K., *Chem. Lett.*, 689 (1991).

[144] (a) Shibahara, T., Asano, T., and Sakane, G., *Polyhedron*, **10**, 2351 (1991).

(b) Shibahara, T., Hashimoto, K., and Sakane, G., *J. Inorg. Biochem.*, **43**, 280 (1991).

(c) Saysell, D. M., and Sykes, A. G., *Inorg. Chem.*, **35**, 5536 (1996) (in press).

[145] Dimmock, P. W., Lamprecht, G. J., and Sykes, A. G., *J. Chem. Soc. Dalton Trans.*, 955 (1991).

[146] Routledge, C. A., Humanes, M., Li, Y.-J., and Sykes, A. G., *J. Chem. Soc. Dalton Trans.*, 1275, 2023 (1994).

[147] Segawa, M., and Sasaki, Y., *J. Am. Chem. Soc.*, **107**, 5565 (1985).

[148] Patel, A., McMahon, M. R., and Richens, D. T., *Inorg. Chim. Acta*, **163**, 79 (1989).

[149] (a) Dori, Z., Cotton, F. A., Llusar, R., and Schwotzer, W., *Polyhedron*, **5**, 907 (1986).

(b) Shibahara, T., Kohda, K., Ohtsuji, A., Yasuda, K., and Kuroya, H., *J. Am. Chem. Soc.*, **108**, 2757 (1986).

(c) Fedin, V. P., Sokolov, M. N., Virovets, A. V., Podberezskaya, N. V., and Fedorov, V. Y., *Polyhedron*, **11**, 2973 (1992).

[150] Shibahara, T., Takeuchi, A., Ohtsuji, A., Kohda, K., and Kuroya, H., *Inorg. Chim. Acta*, **122**, L45 (1987).

[151] (a) Ooi, B.-L., Petrou, A., and Sykes, A. G., *Inorg. Chem.*, **27**, 3626 (1988).

(b) Nasreldin, M., Olatunji, A., Dimmock, P. W., and Sykes, A. G., *J. Chem. Soc. Dalton. Trans.*, 1765 (1990).

(c) Li, Y.-J., Routledge, C. A., and Sykes, A. G., *Inorg. Chem.*, **30**, 5043 (1991).

(d) Routledge, C. A., and Sykes, A. G., *J. Chem. Soc. Dalton Trans.*, 325 (1992).

[152] (a) Richens, D. T., and Sykes, A. G., *Inorg. Chim. Acta*, **54**, L3 (1981).

(b) Richens, D. T., and Sykes, A. G., *Inorg. Chem.*, **21**, 418 (1982).

(c) Paffett, M. T., and Anson, F. C., *Inorg. Chem.*, **22**, 1347 (1983).

[153] Hills, E. F., and Sykes, A. G., *J. Chem. Soc. Dalton Trans.*, 1397 (1987).

[154] Richens, D. T., and Guille-Photin, C. G., *J. Chem. Soc. Dalton Trans.*, 407 (1990).

[155] Rodgers, K. R., Murmann, R. K., Schlemper, E. O., and Shelton, M. E., *Inorg. Chem.*, **24**, 1313 (1985).

[156] Ghosh, S. P., and Gould, E. S., *Inorg. Chem.*, **32**, 864 (1993).

[157] (a) Muller, A., Jostes, R., and Cotton, F. A., *Angew. Chem. Intl. Ed. (Eng)*, **19**, 875 (1980).

(b) Cotton, F. A., Shang, M., and Sun, Z. S., *J. Am. Chem. Soc.*, **113**, 3007 (1991).

[158] (a) Patel, A., Siddiqui, S., Richens, D. T., Harman, M. E., and Hursthouse, M. B., *J. Chem. Soc. Dalton Trans.*, 767 (1993).

(b) Patel, A., and Richens, D. T., *J. Chem. Soc., Chem. Commun.*, 274 (1990).

[159] Shibahara, T., and Yamasaki, M., *Inorg. Chem.*, **30**, 1688 (1991).

[160] Harmer, M. A., Richens, D. T., Soares, A. B., Thornton, A. T., and Sykes, A. G., *Inorg. Chem.*, **20**, 4155 (1981).

[161] Ghosh, S. P., and Gould, E. S., *Inorg. Chem.*, **30**, 3662 (1991).

[162] Carlyle, D. W., and Espenson, J. H., *J. Am. Chem. Soc.*, **91**, 599 (1969).

[163] Ardon, M., and Pernick, A., *Inorg. Chem.*, **12**, 2484 (1973).

[164] Cayley, G. R., and Sykes, A. G., *Inorg. Chem.*, **15**, 2882 (1976).

[165] Armstrong, F. A., and Sykes, A. G., *Polyhedron*, **1**, 109 (1982).

[166]   Paffett, M. T., and Anson, F. C., *Inorg. Chem.*, **20**, 3967 (1981).
[167]   Nabivanets, B. I., and Gorina, D. O., *Zh. Neorg. Khim.*, **29**, 1738 (1984).
[168]   Murmann, R. K., *Inorg. Chem.*, **19** 1765 (1980).
[169]   Sasaki, Y., Taylor, R. S., and Sykes, A. G., *J. Chem. Soc. Dalton Trans.*, 396 (1975).
[170]   McAllister, R., Hicks, K. W., Hurless, M. A., Thurston Pittenger, S., and Gedridge, R. W., *Inorg. Chem.*, **21**, 4098 (1982).
[171]   Cayley, G. R., Taylor, R. S., Wharton, R. K., and Sykes, A. G., *Inorg. Chem.*, **16**, 1377 (1977).
[172]   Martire, D. O., Feliz, M. R., and Capparelli, A. L., *Polyhedron*, **10**, 359 (1991).
[173]   (a) Novak, J., and Podlaha, J., *J. Inorg. Nucl. Chem.*, **36**, 1031 (1974).
        (b) Khahil, S., Sheldrick, B., Soares, A. B. and Sykes, A. G., *Inorg. Chim. Acta*, **25**, L83 (1977).
[174]   Sharp, C., Hills, E. F., and Sykes, A. G., *J. Chem. Soc. Dalton Trans.*, 2293 (1987).
[175]   Soares, A. B., Taylor, R. C., and Sykes, A. G., *J. Chem. Soc. Dalton Trans.*, 1101 (1980).
[176]   Sharp, C., and Sykes, A. G., *Inorg. Chem.*, **27**, 501 (1988).
[177]   Wang, B., Sasaki, Y., Nagasawa, A., and Ito, T., *J. Coord. Chem.*, **18**, 45 (1988).
[178]   Burclova, J., Prasilova, J., and Bines, P., *J. Inorg. Nucl. Chem.*, **35**, 909 (1973).
[179]   Ojo, J. F., Taylor, R. S., and Sykes, A. G., *J. Chem. Soc. Dalton Trans.*, 500 (1975).
[180]   Krumenacker, L., *Ann. Chim.*, **7**, 425 (1972).
[181]   Cruywagen, J. J., Heyns, J. B. B., and Rohwer, E. F. C. H., *J. Inorg. Nucl. Chem.*, **40**, 53 (1978).
[182]   Himeno, S., and Hasegawa, M., *Inorg. Chim. Acta*, **83**, L5 (1984).
[183]   Coddington, J. M., and Taylor, M. J., *J. Chem. Soc. Dalton Trans.*, 41 (1990).
[184]   Pope, M. T., *Heteropoly and Isopoly Oxometallates*, Springler-Verlag, Berlin, 1983.
[185]   Howarth, O. W., Kelly, P., and Pettersson, L., *J. Chem. Soc. Dalton Trans.*, 81 (1990).
[186]   Krebs, B., and Paulat-Boeschen, I., *Acta Cryst. Sec. B.*, **38**, 1710 (1982).
[187]   Baes, C. F., and Mesmer, R. E., *The Hydrolysis of Cations*, 2nd edn, Krieger, Malibar, Florida, 1986, pp. 255–7 (and refs. therein).
[188]   Klemperer, W. G., *Angew, Chem. Int. Ed. (Engl)*, **17**, 246 (1978).
[189]   von Felton, H., Wernli, B., Gamsjager, H., and Baertschi, P., *J. Chem. Soc. Dalton Trans.*, 496 (1978).
[190]   Maksimovskaya, R. I., and Burtseva, K. G., *Polyhedron*, **4**, 1559 (1985).
[191]   Errington, R. J., Kerlogue, M. D., and Richards, D. G., *J. Chem. Soc., Chem. Commun.*, 649 (1993).
[192]   Powell, G., and Richens, D. T., *Inorg. Chem.*, **32**, 4021 (1993).
[193]   Harmer, M. A., and Sykes, A. G., *Inorg. Chem.*, **19**, 2881 (1980).
[194]   (a) Clarke, N. J., Laurie, S. H., Blandamer, M. J., Burgess, J., and Hakin, A., *Inorg. Chim. Acta*, **130**, 79 (1987).
        (b) Brule, J. E., Hayden, Y. T., Callahan, K. P., and Edwards, J. O., *Gazz. Chim. Ital.*, **118**, 93 (1988).
[195]   Mills, C. F., *Phil. Trans. Roy. Soc. Lond., Ser. B.*, **288**, 51 (1979).

Chapter 7

Group 7 Elements: Manganese, Technetium and Rhenium

In contrast to those of Group 6, the elements of Group 7 possess a much less extensive cationic aqueous chemistry [1]. Only Mn^{II} and Mn^{III} form an authentic hexaaqua ion with no cationic chemistry known above Mn^{III}. The tendency towards more stable (less reducing) higher oxidation states for technetium and rhenium, coupled with a strong tendency towards metal–metal bond formation, leads to a propensity for these elements to form polynuclear species or extensively hydrolysed/deprotonated aqua species. Despite a similar propensity for the heavier Group 6 elements the lack of a correspondingly rich cationic aqueous chemistry for technetium and rhenium is somewhat surprising, but might be a reflection of the need to go to a higher oxidation state to achieve the same electron count, e.g. as that present in the stable M—M bonded aqua ions for Mo^{IV} and Mo^V (see Chapter 6). However, a rich chemistry of dimeric oxo-bridged terminal nitrido aqua ion complexes for $Tc^{VI}(d^1)$, analogous to that of the $Mo_2O_4^{2+}$ (aq) ion, has now emerged largely as a result of the terminal N^{3-} group supplying the extra negative charge (vs. O^{2-}) to effectively compensate for the higher positive charge at Tc^{VI} (vs the isoelectronic d^1 Mo^V). Thus the nitrido Tc^{VI} aqua dimer $[Tc_2(N)_2(\mu\text{-}O)_2(OH_2)_6]^{2+}$ is the direct isoelectronic analogue of the Mo^V dimer $[Mo_2(O)_2(\mu\text{-}O)_2(OH_2)_6]^{2+}$ and has corresponding stability in solution. The effective neutralisation of the positive charge on the Tc^{VI} centre in turn stabilises it towards disproportionation. A corresponding chemistry for rhenium may also exist. In the absence of the nitrido group both M^{VI}(aq) and M^V(aq) readily disproportionate to give the insoluble hydrous dioxide, $MO_2 \cdot xH_2O$, and the tetrahedral $M^{VII}O_4^-$ ion. The corresponding $Tc_2^{VI}(O)_2(\mu\text{-}O)_2^{4+}$ species is, for example, unknown. The high acidic nature of the M^{VI}=O unit is exemplified in the formation of the brown insoluble amide complex $[ReO(NH_2)_4]_n$ upon treatment of $[Re^{VI}OCl_4]$ with liquid ammonia at $-33\,°C$. For $Re^{III}(d^4)$ a number of quadruply-bonded dimeric species such as $[Re_2Cl_8]^{2-}$ and $[Re_2(NCS)_8]^{2-}$ are well characterised, the former being the first crystallographically characterised quadruply-bonded dimer [2], cf. Mo_2^{II} and W_2^{II} species. However, despite the formation of cationic species such as $[Re_2(O_2CPr^n)_4(OH_2)_2]^{2+}$ there is no evidence that the aqua dimer $[Re_2(OH_2)_{10}]^{6+}$ exists, possibly because of the higher positive charge. Clear similarities between Re_2^{III} and Mo_2^{II} and W_2^{II} chemistry, however, remain. Finally, there have been repeated claims for the existence of aqua ions such as $[Tc(OH_2)_6]^{3+}$, or a hydroxo derivative, as well as $Tc^{IV}(O)OH^+$(aq) and even dimeric $Tc_2^{IV}(O)_2(OH)_3^+$(aq) in aqueous perchlorate solutions, but it appears none have been fully characterised.

7.1 MANGANESE

Much recent interest in the aqueous chemistry of manganese has been stimulated by the knowledge of its involvement (Mn_4 'cluster') at the oxygen evolving site of chloroplasts [3] and its role in a number of metalloenzymes such as bacterial catalase [4], superoxide dismutase [5] and ribonucleotide reductase [6]. How manganese might enter the biological matrix is unclear but the prime candidate is probably the $[Mn(OH_2)_6]^{2+}$ ion, the only relatively stable and soluble species likely to be present under physiological conditions. Overall the chemistry of Mn^{2+} resembles more that of Ca^{2+} and Mg^{2+} rather than that of its neighbouring transition elements, and indeed Mn^{2+} has been used as a biological probe (ESR activity) for these alkaline earth ions. It also seems likely that manganese as Mn^{2+} is employed simply as a Lewis acid in several of the enzymes [7].

7.1.1 Manganese(II) (d^5): Chemistry of the $[Mn(OH_2)_6]^{2+}$ Ion

The pale pink $[Mn(OH_2)_6]^{2+}$ is the species immediately present upon dissolution of all simple Mn^{2+} salts. The extremely pale colour results from the spin-forbidden nature of all d–d transitions from the singlet $^6A_{1g}$ $(t_{2g}^3 e_g^2)$ ground state. The u.v.–visible spectrum (Fig. 7.1) is the classic textbook illustration of spin-forbidden transitions, in this case to spin quartet states. The $[Mn(OH_2)_6]^{2+}$ formulation has been deduced on the basis of the u.v.–visible spectrum [8] and its characteristic six-line EPR spectrum (coupling to the $I = 5/2$ ^{55}Mn nucleus). $[Mn(OH_2)_6]^{2+}$ has been shown a highly efficient paramagnetic relaxation agent at 0.1 M concentrations for removing ^{17}O NMR signals of bulk water (see page 35) [9]. Correspondingly no coordinated water signals can be observed.* No X-ray structure exists for the aqua ion but an average $Mn\!-\!OH_2$ distance of 218 pm has been established from EXAFS studies [10].

7.1.1.1 HYDROLYSIS OF $[Mn(OH_2)_6]^{2+}$

Hydrolysis occurs but only in alkaline solution and only just before precipitation of the hydrous $Mn(OH)_2$. Hydrolysed solutions of $Mn^{2+}(aq)$ contain $MnOH^+$ ($pK_{11} = 10.59$), $Mn_2(OH)^{3+}$ ($pK_{21} = 9.87$) and $Mn_2(OH)_3^+$ ($pK_{23} = 25.47$) [11], the latter dominating in solution just before precipitation of the hydroxide. $Mn(OH)_2$ is itself a well-established crystalline compound having a similar hexagonal lattice structure to that of $Mg(OH)_2$. It also occurs naturally (as pyrochroite) and is mildly amphoteric. In sufficiently strong alkali and at low metal ion concentrations, tetrahedral $[Mn(OH)_4]^{2-}$ is formed ($pK_{14} = 48.3$) [12].

---

*The relative ion pair associating (coordinating) ability of a number of 'innocent' counter-anions used in the study of aqua ions can be seen by the properties of their natural abundance ^{17}O NMR resonances in the presence of 0.1 M Mn^{2+} (ClO_4^-, loss of resolved coupling to $^{35,37}Cl$ nuclei but resonance easily seen ($\Delta v < 100\,Hz$); $CF_3SO_3^-$, resonance seen but highly broadened ($\Delta v > 500\,Hz$); pts$^-$, resonance completely broadened into the baseline). A trend in ion-pair associating (coordinating) ability pts$^-$ $> CF_3SO_3^- \gg ClO_4^-$ can be deduced.

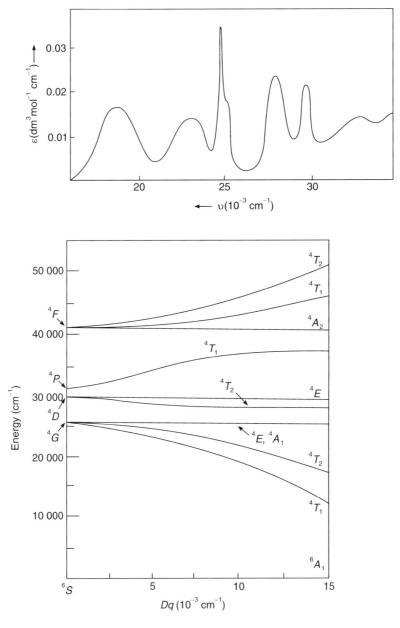

Figure 7.1. Electronic spectrum for the $[Mn(OH_2)_6]^{2+}$ ion

7.1.1.2 WATER EXCHANGE AND COMPLEX FORMATION ON $[Mn(OH_2)_6]^{2+}$

$[Mn(OH_2)_6]^{2+}$ is extremely labile ($k_{ex}(25\,°C) = 2.15 \times 10^7\,s^{-1}$) [13], resulting from its high spin d^5 configuration, low charge and absence of LFSE. Mn—L bond lengths are quite long and variable, as are coordination numbers which are dominated by the steric and electronic effects of the ligands themselves. An I_a mechanism for water exchange has been suggested from the activation volume ($-5.4\,cm^3\,mol^{-1}$) [14]. However, as discussed in Chapter 1, the manifestation of the negative $\Delta V^{\ddagger}$ probably arises from a combination of factors largely deriving from the expanded nature of the hydration sphere. Thus the extra space created by the long Mn—OH_2 bonds (218 pm in the Tutton salt) might be a factor that allows a significant approach of the entering group rather than the tendency for the long Mn—OH_2 bonds to favour the expected dissociative process. At the same time the expanded coordination sphere might lead to an appreciable volume contraction to the five-coordinate transition state. Indeed, the DMF solvent-exchange process on the bulkier $[Mn(DMF)_6]^{2+}$ is duly characterised by a positive activation volume [15].

Precise mechanistic assignments for substitution reactions on Mn^{2+}(aq) are somewhat difficult owing to the typical behaviour, for such a labile metal ion, of a general lack of a dependence of rate with the nature of the entering ligand. Variable pressure kinetic studies of complex formation with bipy and terpy on $[Mn(OH_2)_6]^{2+}$ have nonetheless been interpreted as indicating an I_a process [16, 17].

Mn^{2+} is a class A metal ion with a preference for forming complexes with ligands having donor atoms promoting a high degree of ionic character in the Mn—L bonds. Thus O coordination is favoured over N coordination. It is not an exaggeration to say that detailed complex formation studies on Mn^{2+} have been largely neglected owing to the extreme lability and the difficulty of knowing the exact nature of the coordination polyhedra involved. Complex formation with a number of amino acids has, however, been reported, in which O donation via carboxyl groups, often resulting in polymeric species [18], seems relevant. Coordination via the NH_2 groups is only believed to be promoted in alkaline solution [19]. It is therefore likely that biological coordination of Mn^{2+} to amino acids occurs via carboxylate groups. However, there seems to be a rich non-aqueous chemistry of Mn^{II} with pyridinyl, imidazolyl and RS^- groups [20], which suggest that such coordination cannot be ruled out if a hydrophobic environment is relevant to the binding site. Interestingly, RS^- binding seems to be favoured more for Mn^{II} than for Mn^{III}.

7.1.1.3 REDOX CHEMISTRY OF $[Mn(OH_2)_6]^{2+}$

$[Mn(OH_2)_6]^{2+}$ has little accessible redox chemistry and indeed the $t_{2g}^3 e_g^2$ configuration leads to a marked stability towards oxidation or reduction. Of all the transition metal M^{2+} ions, $[Mn(OH_2)_6]^{2+}$ is slowest to react with $e^-(aq)$ ($k = 7.7 \times 10^7\,M^{-1}s^{-1}$), with $Mn^+(aq)$ being highly reactive [21]. The introduction of strongly π-acceptor ligands such as CNR^- can, however, stabilise Mn^+ [22]. Treatment of $[Mn(OH_2)_6]^{2+}$ in acid solution with $[MnO_4]^-$ is a convenient route through to $[Mn(OH_2)_6]^{3+}$(see below). The application of Marcus theory to the $[Mn(OH_2)_6]^{3+/2+}$ self-exchange process has given rise to a wide range of rate constants ($10^{-3}–10^{-9}\,M^{-1}s^{-1}$) [23]. A semi-classical approach has given a value of $10^{-4\pm1}\,M^{-1}s^{-1}$. Nonetheless, the slow nature of the exchange is consistent with the expected large change in Mn—O bond length accompanying the transfer of a σ^* electron.

7.1.2   Manganese(III) (d⁴): Chemistry of the $[Mn(OH_2)_6]^{3+}$ Ion

Green solutions containing the $[Mn(OH_2)_6]^{3+}$ ion [23, 24] can be obtained in strongly acidic solution ($[H^+] \geqslant 3.0\,M$) by treating $[Mn(OH_2)_6]^{2+}$ with $[MnO_4]^-$. In the presence of excess Mn^{2+}, solutions of $[Mn(OH_2)_6]^{3+}$ are stable for several days at room temperature. Lowering the $[H^+]$ and $[Mn^{2+}]$ promotes disproportionation to give Mn^{2+} and MnO_2 according to

$$2\,Mn^{3+} + 2\,H_2O \xrightarrow{\log K = 9} Mn^{2+} + MnO_2(s) + 4\,H^+ \qquad (7.1)$$

The $[Mn(OH_2)_6]^{3+}$ ion has also been characterised in the alum $CsMn(SO_4)_2 \cdot 12\,H_2O$ [25]. The average Mn—$O(OH_2)$ distance in the alum is 199.1(6) pm. The $[Mn(OH_2)_6]^{3+}$ ion was found to be in a three fold symmetry environment, consistent with an expected dynamic Jahn–Teller distortion of the high spin d^4 configuration at the 295 K measuring temperature. No exceptionally large thermal parameters were, however, found. The ion is strongly acidic with a reported pK_{11} value of 0.7 [26]. The green aqueous solutions of Mn^{3+} reported tend to have higher ε values than those found for Mn^{3+} in the alum and may be indicative of the presence of variable amounts of hydrolysed species such as $Mn_2(OH)_2^{4+}(aq)$ and/or higher oligomers, and, as such, may cast some doubt on the validity of the pK_{11} value [27].

The hydroxide '$Mn(OH)_3$' is not believed to exist; it is formulated instead as the oxyhydroxide MnO(OH) which is known to exist in two crystalline forms. Groutite, α-MnO(OH), has the goethite (α-FeO(OH)) structure. In strongly alkaline solution the hydroxo ion $[Mn(OH)_6]^{3-}$ is said to exist and is known in a number of alkai metal salts. This ion is believed to be present, along with the Mn^V ion, $[MnO_4]^{3-}$, in the blue solutions of 'Mn^{IV}' in strongly alkaline solutions ($\geqslant 8.0\,M$) [28]. The $[Mn(OH)_6]^{3-}$ ion is reported to have a peak maximum around 470 nm typical of other octahedral $Mn^{III}O_6$ complexes.

$[Mn(OH_2)_6]^{3+}$ ion is highly oxidising (E^θ ($Mn^{3+/2+}$) = + 1.5 V, acidic solutions) in addition to its tendency to disproportionate and as a result its simple aqueous chemistry is poorly developed [24]. Even in the presence of Mn^{2+}/H^+

(for stability), solutions of $[Mn(OH_2)_6]^{3+}$ eventually become cloudy and darken due to the formation of insoluble MnO_2 promoting the disproportionation reaction. In alkaline solution, however, certain hydroxo-acids and polyol compounds can effectively suppress the disproportionation reaction leading to stable monomeric aqueous Mn^{III} and Mn^{IV} species (see below).

The rate constant for water exchange has not been measured although estimates of the magnitude do exist (see Section 1.4). Redox reactions involving ligand reductants such as $C_2O_4^{2-}$, Br^- and HN_3 are probably substitution controlled [24]. Complexation with carboxylate ligands leads to aggregation to form polynuclear species ranging from the well-known trinuclear basic acetate type [29] to higher aggregates [30] containing both Mn^{III} and Mn^{IV}. These species are of interest in providing likely routes towards the biological aggregation of the metal, possibly resulting in the formation of Mn-cluster species relevant to the photosynthetic oxygen-evolving site.

7.1.3 Manganese(IV) (d^3)

A polynuclear Mn^{IV}-containing species is believed to be present at the oxygen-evolving site of chloroplasts, but further details are still sketchy [1, 31]. The aqueous chemistry of Mn^{IV} is extremely limited, being dominated by formation of the highly insoluble black MnO_2. The Mn^{4+}(aq) ion may have a transient existence at high dilution in strongly acidic solutions before hydrolysis occurs to deposit the black hydrous dioxide. In strongly alkaline solution Mn^{IV} is soluble but unstable towards disproportionation, giving Mn^{III} and Mn^V. However, soluble red-brown peroxo Mn^{IV} complexes are known, from which salts such as $K_2H_2[MnO(O_2)_3]$ have been isolated [32]. Neutral aqueous solutions of these salts slowly evolve O_2 accompanied by precipitation of MnO_2. The stabilisation of Mn^{IV} in complexes requires extremely hard class A donors. In strongly alkaline solution ($\geqslant 0.5$ M $[OH^-]$) soluble and stable mononuclear d^3 Mn^{IV} complexes, EPR active, can be formed in the presence of a number of polyol ligands, one such example being provided by the tris(sorbitolate) complex (Fig. 7.2) [33, 34]. The complex exhibits the typical $S = 3/2$ EPR signal and the room temperature magnetic moment of 3.84 BM expected for a tris-chelated mononuclear d^3 Mn^{IV} complex [33]. The $^4A_{2g}$–$^4T_{2g}$ transition is observable as a shoulder at ~ 500 nm on the intense RO–Mn^{IV} charge transfer absorption occurring at 360 nm ($\varepsilon = 1000$ M^{-1} cm^{-1}). In these alkaline solutions (0.5 M polyol ligand, 0.5 M OH^-, 1–5 mM Mn) a rich aqueous redox chemistry of soluble mononuclear and binuclear purple-brown Mn^{III} and reddish-brown Mn^{IV} species is evident from spectroscopic and electrochemical studies [34]. The binuclear species form as the pH is lowered, detectable by a degree of spin pairing, as evident from solution measurements of the magnetic moment per Mn atom.

The tris-3,5-di-butylcatecholato complex $[Mn(cat)_3]^{2-}$ [35], the red trisiodate complex $[Mn(IO_3)_3]^{2-}$ [36] and trisperiodate complex $Na_7H_4[Mn(IO_6)_3]$ [37]

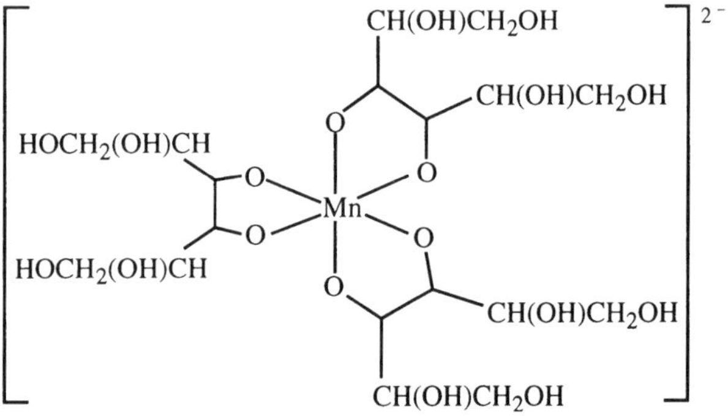

CH(OH)CH₂OH

$$\left[\begin{array}{c} \text{CH(OH)CH}_2\text{OH} \\ \text{CH(OH)CH}_2\text{OH} \\ \text{HOCH}_2\text{(OH)CH} \quad \text{O} \\ \text{O} \quad \text{Mn} \quad \text{O} \\ \text{O} \\ \text{HOCH}_2\text{(OH)CH} \quad \text{O} \\ \text{O} \\ \text{CH(OH)CH}_2\text{OH} \\ \text{CH(OH)CH}_2\text{OH} \end{array}\right]^{2-}$$

Figure 7.2. Proposed structure for the tris-chelated mononuclear $Mn^{IV}(\text{sorbitolato})_3]^{2-}$ anion

are other examples of octahedral $Mn^{IV}O_6$ complexes. Octahedral Mn^{IV} has also been stabilised as the central metal ion in Anderson-type polyoxometallate structures. The synthesis and characterisation of clusters containing both Mn^{III} and Mn^{IV} have been of considerable interest as possible models for the Mn cluster present at the water evolving site of chloroplasts. Figure 7.3 shows representative

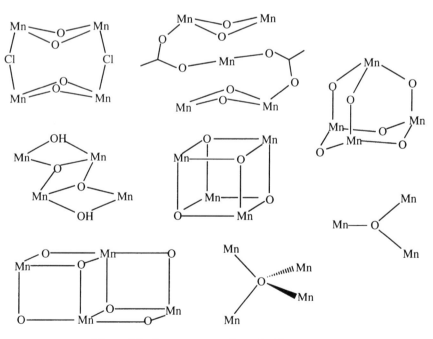

Figure 7.3. Core structures for poly Mn aggregates

core structures found in a number of poly Mn aggregates. As is the case in most areas of manganese chemistry, the exact core structure one obtains derives from a number of factors, not least the electronic and steric properties of the surrounding ligands.

7.1.4 Manganese(VII), (VI) and (V): Chemistry of the Tetrahedral $[MnO_4]^{-/2-/3-}$ Ions

The only stable aqueous species at pH 7.0 is the intensely purple tetrahedral permanganate(VII) anion $[MnO_4]^-$. Permanganate has found wide use as a bleaching agent, preservative, antiseptic and astringent as well as being a powerful reagent for a number of organic oxidations. Because of its very intense colour (self-indicating) it has also found use as an oxidative titrant in laboratory analytical chemistry. The latter application stems from a good example of kinetic control which allows acidic $[MnO_4]^-$ solutions to be stable for many weeks despite the strong thermodynamic driving force towards oxidation of H_2O

$$MnO_4^- + 8H^+ + 5e^- \underset{}{\overset{+1.51\ V}{\rightleftharpoons}} Mn^{2+}(aq) + 4H_2O \tag{7.2}$$

$$MnO_4^- + 2H_2O + 3e^- \underset{}{\overset{+1.23V}{\rightleftharpoons}} MnO_2 + 4OH^- \tag{7.3}$$

The corresponding Mn^{VI} and Mn^V tetrahedral anions, green $[MnO_4]^{2-}$ [38] and blue $[MnO_4]^{3-}$ [39], are only stable respectively in $> 1.0\,M\,OH^-$ and $> 8.0\,M$ OH^- solution, rapid disportionation occurring otherwise to give MnO_2 (alkaline solution) or Mn^{2+} (neutral or acid solution) in addition to $[MnO_4]^-$. The redox stability of the various aqueous species of manganese in both acid and alkaline solution can be conveniently illustrated in the classical oxidation state diagram (Fig. 7.4).

Oxygen exchange with water on both $[MnO_4]^{2-}$ and $[MnO_4]^-$ has been studied with half-lives respectively of: $[MnO_4]^{2-}$ ($\sim 1\,min$), $[MnO_4]^-$ (200 hours) (pH 7.0, 25 °C) [40]. The rate law for exchange on $[MnO_4]^-$ carries both $[H^+]$ and $[H^+]^{-1}$ pathways. These are believed to reflect different mechanisms, the former involved with protonation of an oxo ligand to a hydroxo followed by proton transfer and exchange with an entering water ligand and the latter involving attack by an OH^- ion followed by proton transfer. Dehydration of aqueous solutions of $HMnO_4$ has been shown to give rise to crystals of the complex $(H_3O^+)[Mn^{IV}(Mn^{VII}O_4)_6]\cdot nH_2O$ [41]. The Mn—O distance in the $[Mn^{VII}O_4]^-$ moieties (163 pm) is, as expected, slightly shorter than in the crystalline salts containing the $[Mn^{VI}O_4]^{2-}$ ion (av. 166 pm). In concentrated H_2SO_4 solution the cation $Mn^{VII}O_3^+(aq)$ is said to have existence [39 b].

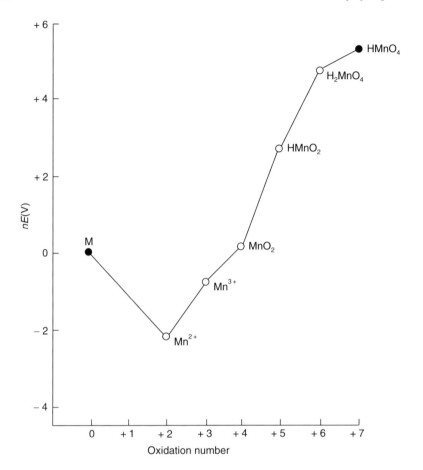

Figure 7.4. Oxidation state diagram for manganese species

7.2 TECHNETIUM AND RHENIUM

A cationic aqueous chemistry is now well established for nitrido Tc^{VI} species with some optimism of a similar chemistry for Re^{VI} and possibly Re^{V}. Other areas of potential promise include a number of dinuclear di-μ-oxo and di-μ-sulphido Re^{IV} species, trinuclear halo-bridged Re^{III} clusters and metal–metal bonded Re^{II} and Re^{III} dimers. The $+7$ oxidation state is much less oxidising for these elements, especially for rhenium, and as a result compounds of the lower states are somewhat reducing and frequently oxygen sensitive. There is also a marked tendency to form metal–metal bonded species and clusters, as in the case of the Group 5 and 6 elements. Furthermore, stable complexes of the lower oxidation states, wherein cationic aqua species are most likely to be found, require the presence of one or more π-acceptor ligands. Thus 'simple' complexes with only water-derived ligands are rare and no purely water-derived cationic species seem to have been characterised. The only 'true' examples of soluble water-derived species are the oxo anion analogues of permanganate, $[Tc^{VII}(Re^{VII})O_4]^-$. Recent interest in Tc coordination and aqueous chemistry has, however, been stimulated by the use of a number of ^{99m}Tc coordination compounds as radiotracers and more recently as radiopharmaceuticals in body tissue imaging. This has led to a rapid development of a number of areas of technetium aqueous chemistry. The ^{99m}Tc isotope has a much shorter half-life (6.05 h) than ^{99}Tc (2.15×10^5 y) and as such is the isotope of choice for radiopharmaceutical applications. Most chemical studies are normally carried out using the much longer lived ^{99}Tc isotope which is a fairly mild β emitter and compounds are easily handled in a clean, glass glove box. ^{99}Tc was first prepared in gram quantities via n,γ reactions on ^{99}Mo, although today tonne quantities are available from fission products of uranium nuclear fuel. The interesting result from this is that there may well be more man-made technetium around on the earth's surface today than there is naturally occurring quantities of rhenium! The major bulk use of rhenium is for the preparation of Pt/Re catalysts used in the production of high octane unleaded petrol.

7.2.1 Technetium(I) and Rhenium(I) (d⁶)

A number of interesting cationic aqua carbonyl Re^I compounds have recently been characterised [42]. Ag$^+$ catalysed aquation of $(Et_4N)_2[ReBr_3(CO)_3]$, prepared via treatment of $ReBr(CO)_5$ with Et_4NBr, yields the triaqua cation $[Re(OH_2)_3(CO)_3]^+$(I), the lability of the Br$^-$ ions presumably arising from the presence of a *trans* effect from the CO ligands. $[Re(OH_2)_3(CO)_3]^+$ is reported to be very stable in aqueous solution with no decomposition detectable in the air for several weeks. Controlled titration with OH$^-$ up to 1.34 OH$^-$/Re yields a solution from which the novel trinuclear $[Re_3(\mu_3\text{-}OH)(\mu\text{-}OH)_3(CO)_9]^-$ anion (II) was isolated (cf. the M_3X_{13} clusters of Group 6, Section 6.2.3). The slow consumption of OH$^-$ was consistent with a slow substitution/oligomerisation process. Very rapid addition of OH$^-$, however, allows formation of the dimeric ion $[Re_2(\mu\text{-}OH)_3(CO)_6]^-$ (III). Surpris-

(I)   (II)   (II)

ingly no evidence for the formation of monomeric $[Re(OH)_3(CO)_3]^{2-}$ can be found. By extrapolation a *facial* coordination geometry for (I) is proposed.

These compounds are interesting 'hybrids' between classical Werner coordination compounds and low oxidation state organometallics in the context of exploring the feasibility of adapting established organometallic/homogeneous catalytic reactions into an aqueous environment. For example, $[Re(OH_2)_3(CO)_3]^+$ might be considered to be a strong $12e^-$ Lewis acid [43] and thus show a tendency to expand its coordination number. Aqua ligand substitution rates appear to be quite slow which would be consistent, in a classical coordination chemical sense, with a high LFSE resulting from a low-spin t_{2g}^6 configuration and possible dissociatively activated path. Alternatively, the $12e^-$ Lewis acid formulation [42, 43] might lead to a favourable associative path being relevant. Interesting mechanistic studies should doubtless result from further studies on these 'hybrid' systems. Since the cationic Tc^I compound $[Tc(CO)_3(NCCH_3)_3]^+$ is known the existence of similar carbonyl aqua technetium hybrid complexes seems likely. Other organo aqua metal ions are discussed in Chapter 6 (Cr) and in Chapter 8 (Ru, Os).

7.2.2 Technetium(II) and Rhenium(II) (d^5)

Here the chemistry is dominated by metal–metal bonded species and/or mononuclear species stabilised by π-acceptor ligands such as CN^- or CNR^- or PR_3 (this also being true for several of the higher states) [1]. Solvated cations such as $[Re_2(NCCH_3)_8]^{4+}$ and $[Re_2(NCCH_3)_{10}]^{4+}$ have been characterised in salts and offer some hope of preparing corresponding aqua species. The X-ray structure of $[Re_2(NCCH_3)_{10}](Mo_6O_{19})_2$(Re—Re = 226 pm) [44] reveals a staggered arrangement of CH_3CN ligands as expected for a triply-bonded $s^2p^4\delta^2\delta^{*2}$ species. Ready loss of the axial CH_3CN ligands is observed. The chemistry of Tc^{II} is somewhat poorly developed.

7.2.3 Technetium(III) and Rhenium(III) (d^4)

A number of mononuclear six-coordinate compounds are known such as $[Tc(Re)(NCS)_6]^{3-}$ [45, 46], $[Tc(thiourea)_6]Cl_3$ [47] and $ReCl_3$ (thiourea)$_3$ [48], the latter is prepared by treatment of $[ReO_4]^-$ with thiourea in 2.0 M HCl. Six-coordination is common although higher coordination numbers are not uncommon. In the absence as yet of detailed kinetic studies one could speculate on the existence of an associative mechanism for substitution on six-coordinate low-spin t_{2g}^4 $Tc(Re)^{III}$ compounds.

7.2.3.1 EXISTENCE OF A Tc^{3+} (AQ) ION?

Aqua ions such as $[Tc(OH_2)_6]^{3+}$ have not been characterised for either element although polarographic studies [49–51] indicate that in dilute solutions clean

$4e^-$ reduction of Tc^{VII} to Tc^{III} can be achieved under mildly acidic conditions (pH 1.0–2.0). Stable solutions of Tc^{III} seem only to be obtained at $\leqslant mM$ concentrations in SO_4^{2-}, PO_4^{3-} or Cl^- media and as such their detailed constitution remains unknown. The presence of 'complexed' species is likely. Studies in SO_4^{2-} media also indicate a pH-dependent quasi-reversible one electron reoxidation to Tc^{IV} (Fig. 7.5) [51]. The final Tc^{IV} product has been proposed [51] as consisting of the complex $[Tc(OH)_2(SO_4)_2]^{2-}$ analogous to that prepared independently by Spitsyn *et al.* [52] (similar u.v.–visible spectrum). However, related studies rather imply the formation of dinuclear di-μ-oxo species

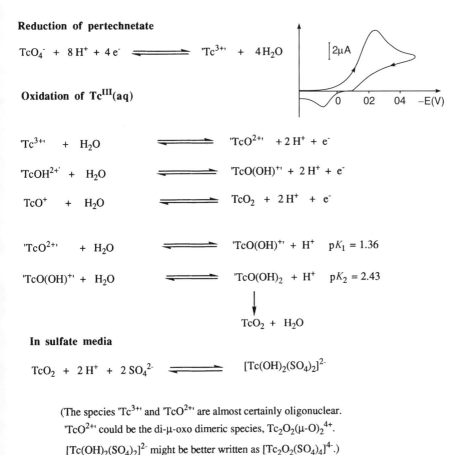

Reduction of pertechnetate

$$TcO_4^- + 8H^+ + 4e^- \rightleftharpoons \text{'}Tc^{3+\text{'}} + 4H_2O$$

Oxidation of Tc^{III}(aq)

$$\text{'}Tc^{3+\text{'}} + H_2O \rightleftharpoons \text{'}TcO^{2+\text{'}} + 2H^+ + e^-$$

$$\text{'}TcOH^{2+\text{'}} + H_2O \rightleftharpoons \text{'}TcO(OH)^{+\text{'}} + 2H^+ + e^-$$

$$TcO^+ + H_2O \rightleftharpoons TcO_2 + 2H^+ + e^-$$

$$\text{'}TcO^{2+\text{'}} + H_2O \rightleftharpoons \text{'}TcO(OH)^{+\text{'}} + H^+ \quad pK_1 = 1.36$$

$$\text{'}TcO(OH)^{+\text{'}} + H_2O \rightleftharpoons \text{'}TcO(OH)_2 + H^+ \quad pK_2 = 2.43$$

$$\downarrow$$

$$TcO_2 + H_2O$$

In sulfate media

$$TcO_2 + 2H^+ + 2SO_4^{2-} \rightleftharpoons [Tc(OH)_2(SO_4)_2]^{2-}$$

(The species $\text{'}Tc^{3+\text{'}}$ and $\text{'}TcO^{2+\text{'}}$ are almost certainly oligonuclear.
$\text{'}TcO^{2+\text{'}}$ could be the di-μ-oxo dimeric species, $Tc_2O_2(\mu\text{-O})_2^{4+}$.
$[Tc(OH)_2(SO_4)_2]^{2-}$ might be better written as $[Tc_2O_2(SO_4)_4]^{4-}$.)

Figure 7.5. Redox interconversions of Tc(aq) species in SO_4^{2-} and Cl^- media. Inset: cyclic voltammogram (HMDE) of 0.93 mM $[TcO_4]^-$ in 0.5 M Na_2SO_4–$NaHSO_4$ (pH 2.0), scan rate 8.4 V min^{-1}

(see Section 6.2.3). Hg pool coulometric reduction of mM Tc^{VII} in 0.5 M $HClO_4$ also yields Tc^{III}, but here solutions are more unstable and eventually deposit a black precipitate (TcO_2) owing to spontaneous oxidation of Tc^{III} by H^+ [49]. Coulometric reduction of Tc^{VII} in 0.1 M PO_4^{3-}, pH 7 is quoted as yielding pink Tc^{IV} and finally green Tc^{III}. Solutions of Tc^{III} in Cl^- or SO_4^{2-} media, generated at pH $\sim$ 1.5, are said to be pale yellow. In alkaline solution, the product of the electrochemical reduction of Tc^{VII} is TcO_2 [52]. However, voltammetric studies of the $1e^-$ electrochemical reduction of $[TcBr_6]^{2-}$ indicate ready loss of the six Br^- ions, suggesting formation of a Tc^{III} aqua ion or a mixed aqua hydroxo complex [53]:

$$\left.\begin{array}{c} [TcBr_6]^{2-} \\ \text{or} \\ [TcBr_5(OH_2)]^- \end{array}\right\} + e - \xrightarrow{H_2O} \left\{\begin{array}{c} `[Tc(OH)(OH_2)_5]^{2+}` \\ \| \\ `[Tc(OH)_2(OH_2)_4]^+` + H^+ \end{array}\right. + 6Br^- + H^+ \quad (7.4)$$

Despite suggestions of a possible route to Tc^{3+}(aq), neither of the proposed mononuclear products, $[TcOH(OH_2)_5]^{2+}$ and $[Tc(OH)_2(OH_2)_4]^+$, has been substantiated, presumably because of an instability at the higher preparative concentrations [49]. Corresponding aqueous reduction of $[TcCl_6]^{2-}$ is reported, however, to give rise to the stable mononuclear Tc^{III}(aq) complex $[TcCl_4(OH_2)_2]^-$.

7.2.3.2 QUADRUPLY-BONDED Re^{III} DIMERS

A much more stable NCS^- complex for Re^{III} is the quadruply-bonded dimer ion $[Re_2(NCS)_8]^{2-}$, one of a whole family of Re_2^{II} dimers [2]. There is no aqua ion representative of the Re_2^{6+} unit although further work in this area should be encouraged. For example, treatment of $[Re_2Cl_8]^{2-}$ with Na_2SO_4 gives $Na_2[Re_2(SO_4)_4]\cdot8H_2O$. The structure confirms the presence of two weakly bonded axial H_2O ligands ($Re-OH_2 = 228$ pm) [54]. Treatment of $[Re_2(SO_4)_4]^{2-}$ with refluxing HCl regenerates the $[Re_2Cl_8]^{2-}$ ion. The product obtained upon treatment of $[Re_2(SO_4)_4]^{2-}$ with $Ba(CF_3SO_3)_2$ in aqueous CF_3SO_3H would be of interest, cf. the preparation of $[Mo_2(OH_2)_8]^{4+}$ from $[Mo_2(SO_4)_4]^{4-}$ (Section 6.2.1). Both formally Tc^{II} and Tc^{III} are present in the dimer $[Tc_2Cl_8]^{3-}$ but again no aqua ions have been characterised.

7.2.3.3 TRINUCLEAR Re^{III} CLUSTERS

Re^{III} also forms an extensive number of trinuclear complexes containing formally the Re_3^{9+} core $Re_3(\mu_3-X)_3Y_6L_3$, where X can be Cl^-, Br^- or I^-, $Y = Cl^-$, Br^-, I^-, CN^-, NCS^- or CNR^- and $L = PR_3$, Ph_3PO, DMF, HMPA, THF, NCS^- or H_2O [2, 55]. The terminal ligands are, as expected, more easy to replace than the bridging ones. The novel preparation of the THF complex involved the co-condensation of rhenium atoms, generated via a positive hearth electron gun furnace, and reactive halocarbons such as 1,2-dibromo (or dichloro) ethane. Extraction of the reaction matrix with THF gave $Re_3X_9(THF)_3$ (X = Cl or Br) in yields of $\sim 90\%$ [56]. This complex could provide a useful lead-in to the aqueous chemistry. The diaqua complex $Cs[Re_3(\mu\text{-}Cl)_3Br_7(OH_2)_2]$ has been structurally characterised [57] but the neutral triaqua complex has yet to be isolated.

7.2.3.4 Hexanuclear μ_3-sulfido clusters of Re^{III}

A number of solid phases are known which appear to contain the discrete octahedral $Re^{III}{}_6S_8{}^{2+}$ core as in $M^I{}_4[\{Re_6S_8\}S_2(S\text{—}S)]$ (M^I = Na or K), prepared by the reaction of Re, $KReO_4$ or ReS_2 with a large excess of $M^I{}_2CO_3$ and sulfur at a temperature of 750–800 °C [58]. The Re—Re distance in the $Re_6S_8{}^{2+}$ unit is between 259 and 263 pm. The alkaline earth salts $Ba_2(Sr_2)$ $[Re_6S_{11}]$ contain the $[\{Re_6S_8\}S_3]^{4-}$ unit. Interest in these compounds stems not only because of their similarity to so-called Chevrel phases, MMo_6Y_8(Y = S, Se or Te), but also because of the possible existence of aqua derivatives (cf. $[Mo_4S_4(OH_2)_{12}]^{5+}$ in Section 6.2.3). Only recently has the derivative chemistry of these cluster species been investigated with the structural characterisation of the compound $[Re_6(\mu_3\text{-}S)_4(\mu_3\text{-}Cl)_4Cl_6]$ [59]. This would imply that the hypothetical $Re_6S_8{}^{2+}$ aqua ion would most likely have the formula '$[Re_6S_8(OH_2)_6]^{2+}$' (one H_2O ligand on each Re). Surprisingly the existence of $[Re_6S_8(CN)_6]^{4-}$ (preparation via the addition of NCS^- to the above μ_3-Cl complex) has not been confirmed (cf. Section 6.2.3). The corresponding Tc cluster chemistry is less well developed.

7.2.4 Technetium(IV) and Rhenium(IV) (d^3)

As mentioned above, Tc^{IV}(aq) species have been detected during polarographic reduction of TcO_4^- [51,52]. Dilute 0.1–1.0×10^{-3} M solutions of ReO_4^- in 4 M $HClO_4$ give well-defined polarograms consistent with an irreversible $3e^-$ reduction to give soluble Re^{IV}(aq) [60]. At higher Re concentrations hydrous ReO_2 invariably precipitates so an already extensively hydrolysed product is probably present within the electrical double layer of the dropping mercury electrode. In HCl the product is octahedral $[ReCl_6]^{2-}$. Hydrolysis of $[ReCl_6]^{2-}$ was reported as giving $[Re(OH)_3(OH_2)_3]^+$ prior to precipitation of the dioxide [61], but this has been disputed [62]. Fusion of $K_2[ReCl_6]$ in a molten KSCN/KCN mixture gives the cubiodal cluster $[Re_4(\mu_3\text{-}S)_4(CN)_{12}]^{4-}$ (Re—Re = 275.5 pm) [63]. Reaction with KSeCN gives the corresponding μ_3-Se cube. The reaction of $NaReO_4$ with H_2S in the presence of NaCN gives the di-μ-sulfido dimer $[Re_2(\mu\text{-}S)_2(CN)_8]^{4-}$ (Re—Re = 260 pm) [64]. A further example of the $Re_2(\mu\text{-}S)_2{}^{4+}$ unit is provided by $[Re_2(\mu\text{-}S)_2(S_2CNBu_2)_4]$ [65].

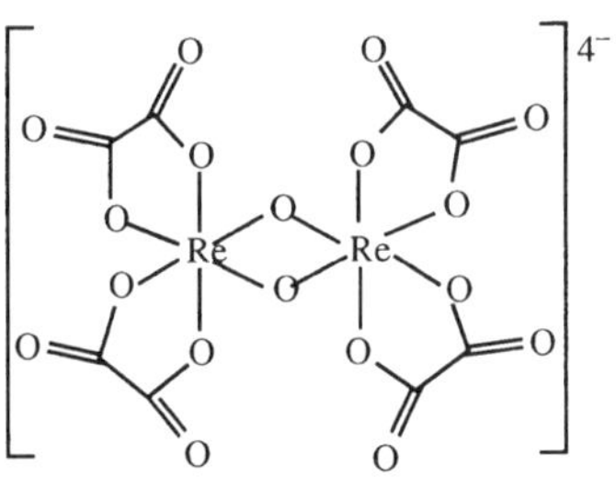

Figure 7.6. Structure of
$[(C_2O_4)_2Re(\mu\text{-}O)_2Re(C_2O_4)_2]^{4-}$

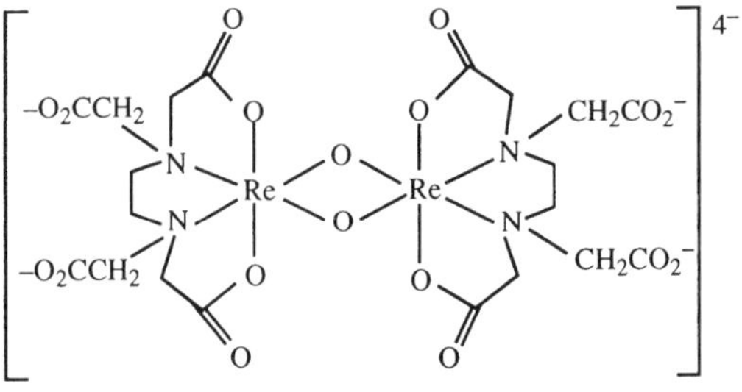

Figure 7.7. Structure of $[(edta)Re(\mu\text{-}O)_2Re(edta)]^{4-}$

ReO_2 is reported to dissolve in a mixture of oxalic acid and potassium oxalate to give a range of products, one of which has been structurally characterised as $K_4[Re_2(\mu\text{-}O)_2(C_2O_4)_4]$ (Re—Re = 236.2 pm (Fig. 7.6) [66]. $[Re_2(\mu\text{-}O)_2$ $(edta)_2]^{4-}$ (Fig. 7.7) has been prepared by two methods: (a) reduction of $Cs_2[Re-OCl_5]/Na_2[H_2edta]$ in acetate buffer with zinc or (b) treatment of $[ReCl_6]^{2-}$ in acetate buffer with $Na_2[H_2edta]$ [67]. In each case a black precipitate of hydrous ReO_2 accompanies complex formation. The crystal structure of the Ba^{2+} salt reveals an Re—Re distance similar to that in the oxalato complex (236.2 pm), implying multiple Re—Re bonding [67]. Wieghardt and coworkers have developed a rich aqueous chemistry of dinuclear Re species using the ligand L = 1,4,7-triazacyclononane including some Re^{IV}_2 species [68]. $[L_2Re^{IV}_2Cl_2(\mu\text{-}O)_2]I_2$ (Re—Re = 237.6 pm) was prepared either by a disproportionation reaction involving Re^V (see below) or via oxidation of Re^{III} (Fig. 7.8). Indications are that these compounds could be useful lead-ins to other $Re^{IV}_2(\mu\text{-}O)_2{}^{4+}$(aq) species and routes to the related di-μ-sulfido(selenido) species are being explored, as are routes to $Re_4(\mu_3\text{-}S)_4{}^{8+}$ cubanes. There appears a similarly extensive chemistry emerging for di-μ-oxo Tc^{IV} species such as $[Tc_2(\mu\text{-}O)_2(H_2edta)_2]$ [69] and

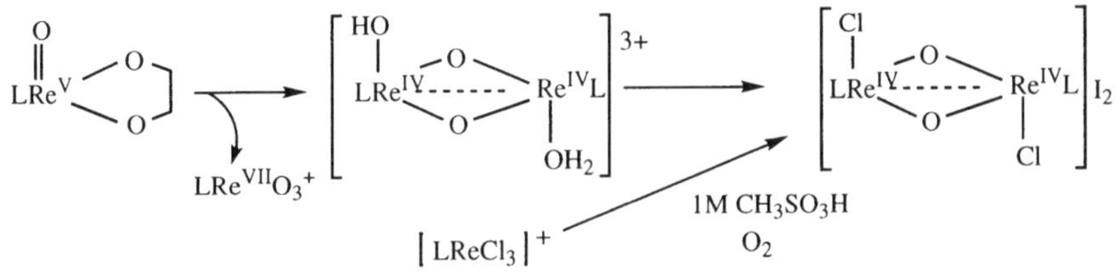

Figure 7.8. Preparative routes to di-μ-oxo Re^{IV} species

$[Tc_2(\mu\text{-}O)_2(tcta)_2]^{2-/3-}$ (TCTA—$N,N'N''$-1,4,7-triazacyclononanetriacetic acid) [70], the latter, being a mixed-valence Tc^{IV}–Tc^{III} species. The Tc—Tc distance (233 pm) in the H_2edta complex implies multiple Tc—Tc bonding.

7.2.4.1 THE EXISTENCE OF PUTATIVE Tc^{IV} (AQ)

Several reports [71–73] have claimed the existence of cationic Tc^{IV}(aq) species such as $Tc_2O_2(OH)_3^+$ (aq) and even monomeric TcO^{2+}(aq), $TcO(OH)^+$(aq) and $Tc(OH)_2^{2+}$(aq). In one study the soluble Tc^{IV}(aq) species are prepared following the reduction of TcO_4^- with hydrazine hydrate in perchloric acid solution [71]. The stability of these species proved apparently sufficient for measurements of hydrolysis constants between various monomeric species to be carried out using an electrophoretic method [71] and for speciation studies to be conducted on the basis of the pH dependence of the electronic spectrum [72]. A monomer–dimer equilibrium forming $Tc_2O_2(OH)_3^+$(aq) (presumably di-μ-oxo) has also been described [72]. However, despite these reports truly authentic cationic Tc^{IV}(aq) species have not been characterised. The use of hydrazine as reductant (above) [71] makes one suspect that the soluble Tc^{IV}(aq) species reported above might in fact be $Tc^{VI}\equiv N$(aq) species (see Section 7.2.6 below). It is very possible that at somewhat high dilution, partly hydrolysed aquated cationic species, such as the di-μ-oxo ion $[Tc_2O_2(OH)_n(OH_2)_{8-n}]^{(4-n)+}$, may well exist. However, at the higher isolable concentrations it appears that extensive hydrolysis leads rapidly to precipitation of the hydrous dioxide.

7.2.5 Technetium(V) and Rhenium(V) (d^2)

Three basic core 'oxo' structures are known in complexes, $Tc(Re)O^{3+}$, *trans*-ReO_2^+ and $Tc(Re)_2O_2(\mu\text{-}O)^{4+}$ [1]. However, no aqua cations are established in any case. The principal reason for this is that in the absence of π-accepting ligands such as phosphines, the M^V state, especially in the case of $Re=O^{3+}$, is unstable with respect to disproportionation to give Re^{IV} and Re^{VII} [48, 74]:

$$3\,Re^V \rightarrow 2\,Re^{IV} + Re^{VII} \tag{7.5}$$

the Re^{IV} then almost invariably precipitating as the highly insoluble hydrous ReO_2, further promoting the disproportionation reaction. $[ReOCl_5]^{2-}$, for example, is only stable in concentrated HCl and is invariably contaminated with $[ReCl_6]^{2-}$ and Re_2O_7. Aquation of $[ReOCl_5]^{2-}$ to give $[ReOCl_4(OH_2)]^-$ has been characterised. Terminal nitrido $Re^V=N^{2+}$ species are known but require a number of π-acceptor ligands, e.g. phosphines for stability, and there are no aqua species to speak of. Phosphine complexes play a major role in defining much of the chemistry of Re^V [1]. One of the most stable and versatile lead-in compounds is $ReOCl_3(PPh_3)_2$. A well-established route to $Re\equiv N^{2+}$ species, from $Re=O^{3+}$, is via treatment with ethanolic hydrazine. A rare example of

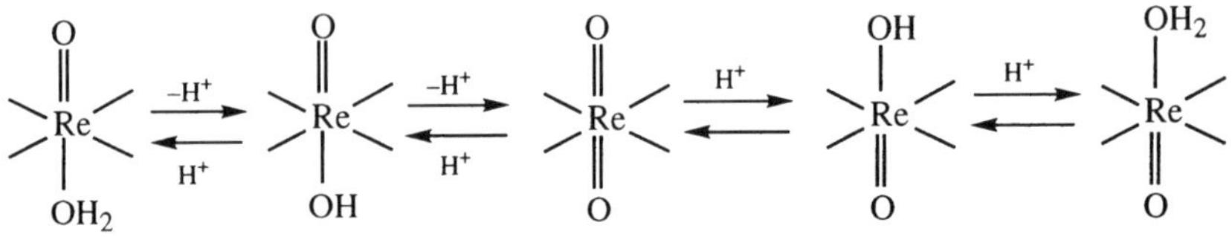

$$
\underset{OH_2}{>\!Re\!<}\ \overset{-H^+}{\underset{H^+}{\rightleftarrows}}\ \underset{OH}{>\!Re\!<}\ \overset{-H^+}{\underset{H^+}{\rightleftarrows}}\ \underset{O}{>\!Re\!<}\ \overset{H^+}{\leftarrow}\ \underset{O}{>\!Re\!<}\ \overset{H^+}{\leftarrow}\ \underset{O}{>\!Re\!<}
$$

Figure 7.9. Mechanism of the proton-assisted oxygen exchange in trans-ReO_2 compounds

a cationic Re^V complex is trans-$[Re(NMe)(H_2NMe)_4Cl](ClO_4)_2$ prepared by air oxidation of an aqueous solution of $[ReCl_6]^{2-}$ in the presence of $MeNH_2$ [75]. Nitrido species are more extensively characterised for both Tc^{VI} and Re^{VI} (see below).

A number of trans-dioxo $Re^VO_2L_4{}^+$ compounds are known with $L_4 = en_2$, $(CN^-)_4$, py_4 and also tetraaza macrocycles such as cyclam [1, 76–78]. The trans-$O\!=\!Re\!=\!O^+$ group in these complexes stabilises Re^V in solution to a marked extent. Protonation occurs in acidic solution to give trans-$ReO(OH)^{2+}$ and trans-$ReO(OH_2)^{3+}$ groups respectively [76], the latter probably best formulated as containing trans-$Re(OH)_2{}^{3+}$ on the basis of recent oxygen-exchange studies using ^{17}O NMR (one ^{17}O NMR signal seen from equivalent oxygen environments) [77, 78]. Rapid proton exchange in acid solution results in spontaneous ^{17}O enrichment of both oxo groups in trans-$ReO_2{}^+$ moieties owing to the equilibria in Fig. 7.9. Over a period of hours trans-$[ReO(OH)L_4]^{2+}$ species convert to give dimeric $[Re_2O_3L_n]$ complexes. This reaction has been followed in aqueous solution ($L = CN^-$) using ^{17}O NMR [77].

7.2.6 Technetium(VI) and Rhenium(VI) (d^1)

7.2.6.1 CHEMISTRY OF THE NITRIDO DIMER ION $[Tc^{IV}{}_2N_2(\mu\text{-}O)_2(OH_2)_6]^{2+}$

Progress in this area owes much to the pioneering work of Baldas and coworkers at the Australian Radiation Laboratory [79–83] in the late 1980s and early 1990s. The work was simulated by the need to understand further the solution chemistry of species such as $[^{99m}TcNCl_4]^-$, one of the first compounds to be screened as a radiopharmaceutical and today used frequently as a lead-in to the preparation of other Tc^{VI} radiopharmaceutical compounds. Paramagnetic $Cs_2[TcNCl_5]$ [84], when dissolved in water, results in the formation of a brown precipitate, formulated as '$TcN(OH)_3$' (analogous to '$MoO(OH)_3$'), which readily dissolved in solutions of weakly complexing strong acids such as Hpts or CF_3SO_3H to give a diamagnetic yellow solution [79, 80]. In '$TcN(OH)_3$' the $Tc\!\equiv\!N$ group appears in the infra-red spectrum as a strong band at $1046\,cm^{-1}$ (cf. $Mo\!=\!O$ in $MoO(OH)_3$ occurs at $945\,cm^{-1}$). However, the presence of a band

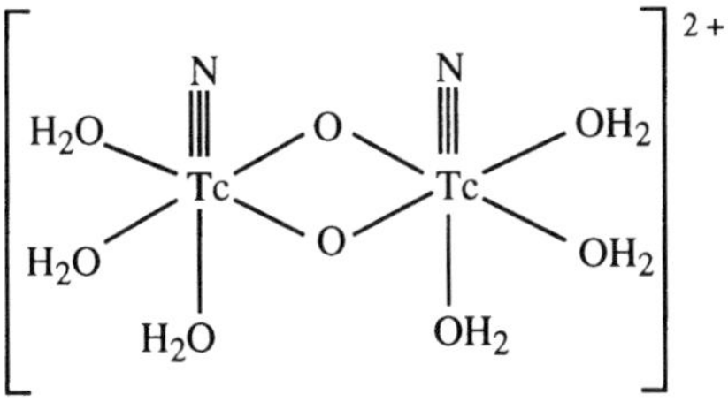

Figure 7.10. Structure of the nitrido aqua Tc^{VI} dimer $[Tc_2N_2(\mu\text{-}O)_2(OH_2)_6]^{2+}$

at 734 cm^{-1}, assignable to ν_{asym} (Tc—O—Tc), confirms the presence of a μ-oxo-bridged species [80]. The yellow solution was formulated as containing the di-μ-oxo-bridged cation, $[Tc_2N_2(\mu\text{-}O)_2(OH_2)_6]^{2+}$ (Fig. 7.10), due to reaction with aqueous $Na[S_2CNEt_2]$, giving the structurally characterised di-μ-oxo-bridged dimer $[Tc_2N_2(\mu\text{-}O)_2(S_2CNEt_2)_2]$ [81].

The yellow species in 0.5 M Hpts exhibited an electrophoretic migration characteristic of a 2+ cation. The yellow nitrido Tc^{VI} aqua dimer is the isoelectronic analogue of the Mo^V oxo-aqua dimer $[Mo_2O_2(\mu\text{-}O)_2(OH_2)_6]^{2+}$ (see Chapter 6). It was eventually shown that good yields of the aqua-nitrido dimer could be obtained directly from dissolution of $[TcNCl_5]^{2-}$ in aqueous Hpts or CF_3SO_3H followed by cation-exchange purification [83]. The electronic spectrum of solutions of $[Tc_2N_2(\mu\text{-}O)_2(OH_2)_6]^{2+}$ in 1.0 M CF_3SO_3H (Fig. 7.11) has λ_{max} values at 251 nm ($\varepsilon = 2760$ M^{-1} cm^{-1} per Tc_2 unit). 295 nm (2580 M^{-1} cm^{-1}) and a shoulder at 331 nm (~ 1930 M^{-1} cm^{-1}) [81]. The features are generally reminiscent of those of $[Mo_2^VO_2(\mu\text{-}O)_2(OH_2)_6]^{2+}$, with the shoulder at 331 nm believed to correspond to the less intense band for $Mo_2O_4^{2+}$(aq) at 384 nm ($\varepsilon = 103$ M^{-1} cm^{-1}) (page 315). In $\geqslant 7.0$ M CF_3SO_3H the electronic spectrum changes to reveal an intense peak at 474 nm ($\varepsilon > 3500$ M^{-1} cm^{-1}), which has been assigned to the single μ-oxo-bridged cation, $[Tc_2N_2(\mu\text{-}O)(OH_2)_8]^{4+}$ (Fig. 7.12) [83]. EPR signals from paramagnetic Tc^{VI} species can be observed upon rapid freezing of solutions of $[TcNCl_5]^{2-}$ in 2.0 M Hpts but disappear if solutions are kept at room temperature for 20 min prior to freezing down. Similarly, no signals are seen upon rapid freezing of $[TcNCl_5]^{2-}$ solutions in 0.25 M Hpts. In each case the EPR signals disappear due to formation of the spin-paired diamagnetic di-μ-oxo dimer. EPR signals, presumably from monomeric or partially spin-coupled single oxo-bridged dimer, species, have been reported from $[TcNCl_5]^{2-}$ dissolved in 10 M H_2SO_4 and 17.5 M H_3PO_4. In 10 M H_2SO_4 the strong band observed at 485 nm ($\varepsilon > 6700$ M^{-1} cm^{-1}) has been assigned to the species $[Tc_2N_2(\mu\text{-}O)(SO_4)_4]^{4-}$. The behaviour appears consistent with an equilibrium between mono-μ-oxo dimer, di-μ-oxo dimer and monomer. However, only in $\geqslant 7$ M HCl do significant

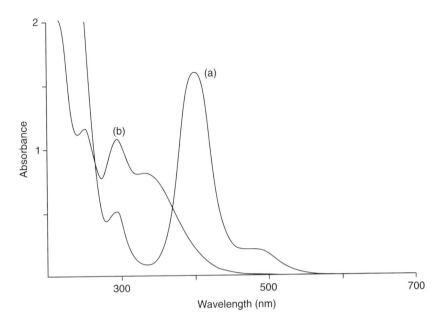

Figure 7.11. Electronic spectra of (a) $[TcNCl_5]^{2-}$ in 7.5 M HCl and (b) $[Tc_2N_2(\mu\text{-O})_2\text{-}(OH_2)_6]^{2+}$ in 1.0 M CF_3SO_3H

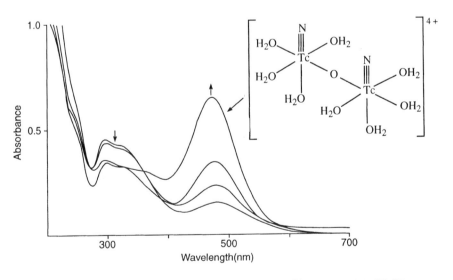

Figure 7.12. Formation of $[Tc_2N_2(\mu\text{-O})(OH_2)_8]^{4+}$ in 7.5 M CF_3SO_3H

amounts of monomeric species prevail, λ_{max} 300 nm ($\varepsilon = 4780\,\text{M}^{-1}\,\text{cm}^{-1}$). Much of this chemistry is reminiscent of $Mo^V(aq)$ although the mono-μ-oxo dimer seems more prevalent for $Tc^{VI}(aq)$ in weakly complexing media. One reason for this could lie in the much slower rate of exchange at the μ-oxo groups on $Tc_2^{VI}(aq)$, as exemplified by the slow rate of interconversion of the two Tc^{VI} aqua dimers (days in $\geqslant 7\,\text{M}\,H^+$ as opposed to hours for μ-oxo exchange on $Mo_2O_4^{2+}(aq)$) under the same conditions). No kinetic studies of H_2O or μ-oxo ligand exchange have so far been carried out on any of these cationic nitrido-aqua Tc^{VI} species. Rapid exchange at H_2O is to be expected due to strong *trans* labilising effects from the $Tc\equiv N$ groups as well as the μ-oxo groups.

Nitrido rhenium(VI) complexes such as $Ph_4As[ReNCl_4]$ [85] and $K_2[ReNCl_5]$ are known, but despite this no corresponding nitrido aqua chemistry of Re^{VI} has so for been established.

7.2.7 Technetium(VII) and Rhenium(VII) (d⁰)

7.2.7.1 THE TETRAHEDRAL IONS, TcO_4^- AND ReO_4^-

These are the only examples of purely 'water-derived' species for the heavier Group 7 elements and the only stable oxo anions in solution [1, 48]. Both $HTcO_4$ and $HReO_4$ are strong acids, like $HMnO_4$, and are obtained when the heptoxides are dissolved in H_2O. Careful evaporation of such Tc^{VII} solutions produce red crystals of $HTcO_4$. In the case of Re^{VII}, yellow crystals of $Re_2O_7(OH_2)_2$ result. The X-ray structure shows that both H_2O molecules are bound to one of the Re^{VII} atoms (Fig. 7.13) [86]. Interesting adducts have been obtained with dioxane such as $Re_2O_7(OH_2)_2(\text{dioxane})$ [87]. Upon addition of water, polymeric OH-bridged adduct species such as '$Re_2O_6(\mu\text{-OH})_2(\text{dioxane})_3$' have been characterised [48, 87].

ReO_4^- exchanges oxygen with H_2O under acid catalysis. The rate law

$$\text{Rate} = k_0[ReO_4^-] + k_1[ReO_4^-][H^+]^2 + k_2[ReO_4^-][OH^-] \quad (7.6)$$

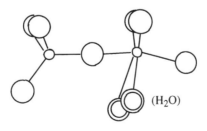

(H₂O)

Figure 7.13. Structure of $Re_2O_7(OH_2)_2$

has been established by ^{18}O-labelling studies with $k_0 = 7.7 \times 10^{-7}\,\text{s}^{-1}$, $k_1 = 1.9 \times 10^3\,\text{M}^{-2}\text{s}^{-1}$ and $k_2 = 8.6 \times 10^{-3}\,\text{M}^{-1}\text{s}^{-1}$ in 0.1 M LiCl at 25 °C [88]. At pH < 4 only the $[\text{H}^+]^2$ term dominates and under these conditions the rate is faster than that occurring on MnO_4^- [40, 88] and TcO_4^- [89]. A mechanism has been proposed wherein ReO_4^- rapidly equilibrates to form $\text{ReO}_3(\text{OH}_2)^+$ followed by slow dissociative exchange between $\text{ReO}_3(\text{OH}_2)^+$ and H_2O assisted by hydrogen bonding to the solvent. Studies in methanol and in aqueous methanol have established that the rate of exchange is independent of $[\text{H}_2\text{O}]$ [90]. ReO_4^- is a significantly weaker oxidant than TcO_4^-:

$$\text{TcO}_4^- + 4\,\text{H}^+ + 3\,\text{e}^- \xrightleftharpoons{\;+0.7\text{V}\;} \text{TcO}_2 + 2\,\text{H}_2\text{O} \qquad (7.7)$$

$$\text{ReO}_4^- + 4\,\text{H}^+ + 3\,\text{e}^- \xrightleftharpoons{\;+0.51\text{V}\;} \text{ReO}_2 + 2\,\text{H}_2\text{O} \qquad (7.8)$$

but both are much weaker than MnO_4^-. This provides another typical example of the large difference apparent between the chemistry of the first row member (Mn) and the heavier members (Tc and Re) of Group 7. Solutions of ReO_4^- are colourless due to a shift in the intense O^{2-} to Re^{VII} CT transition into the u.v. region (Fig. 7.14) [91]. The red colour of concentrated HTcO_4 solutions has been attributed to some distortion of the regular tetrahedral symmetry, causing

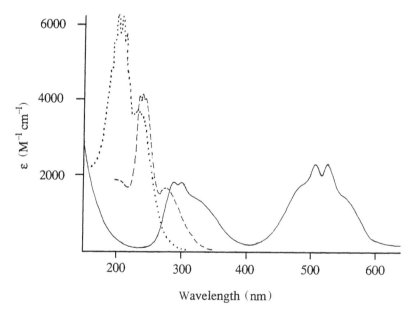

Figure 7.14. Electronic spectra of MnO_4^- (—), TcO_4^- (– – –) and ReO_4^- (···) ions in aqueous solution

intensity enhancement of the band on the edge of the visible spectral region. The lowest transition in Fig. 7.14 has been correlated with the effective ionic radius of M^{7+} [91]. More importantly, though, the band energies reflect the trend in oxidising property of M^{7+}.

Pulse radiolytic reduction (e^-(aq)) of both TcO_4^- and ReO_4^- has been studied [92, 93]. In each case the products are presumed to be the M^{VI} species MO_4^{2-}. As with $[MnO_4]^{2-}$, these anions are reasonable stable in alkaline solutions but rapidly disproportionate at lower pH values in a second-order process to give the hydrous dioxide and MO_4^-. The reduction itself proceeds for each at diffusion controlled rates ($\sim 10^{10} M^{-1} s^{-1}$ for reduction at pH 13). TcO_4^{2-} is reported to have a broad maximum at ~ 500 nm and a peak at 340 nm. For ReO_4^{2-} the maximum is reported at 290 nm [93]. In neutral solution the disproportionation reaction proceeds with second-order rate constants of $1.3 \times 10^8 M^{-1} s^{-1}$ (Tc) and $1.7 \times 10^9 M^{-1} s^{-1}$ (Re).

Cationic species $LReO_3^+$ (L = 1,4,7-triazacyclononane) have been characterised by Wieghardt and coworkers [68]. The well-characterised nitrido and sulfido-oxo rhenates(VII) ions, ReO_3N^{2-} and $ReO_{4-n}S_n^-$ [1], are also of interest in offering, via controlled reduction, potential entries into lower oxidation state aqueous species of Re such as $Re_2^{VI}N_2(\mu\text{-O})_2^{2+}$(aq) (cf. Tc) and $Re_2^{IV}(\mu\text{-S})_2^{4+}$(aq). Further work in these areas is needed.

REFERENCES

[1] Chiswell, B., McKenzie, E. D., and Lindoy, L. F. (manganese), Conner, K. A., and Walton, R. A. (technetium and rhenium), in *Comprehensive Coordination Chemistry* (eds.) G. Wilkinson, R. D. Gillard, and J. A. McCleverty, Vol. 6, Pergamon, London, 1987, pp. 1–216.

[2] Cotton, F. A., and Walton, R. A., *Multiple Bonds Between Metal Atoms*, Wiley, New York, 1982.

[3] (a) Vincent , J. B., and Christou, G., *Adv. Inorg. Chem.*, **33**, 197 (1989).
    (b) Brudvig, G. V., and Crabtree, R. H., *Prog. Inorg. Chem.*, **37**, 99 (1989).

[4] Beyer, W. F., and Fridovich, I., *Biochemistry*, **64**, 6460 (1985).

[5] Michelson, A. M., McCord, J. M., and Fridovich, I. (eds.), *Superoxide and Superoxide Dismutases*, Academic Press, New York, 1977.

[6] Willing, A., Follmann, H., and Auling, G., *Eur. J. Biochem.*, **170**, 603 (1988).

[7] See, for example, Hughes, M. N., *The Inorganic Chemistry of Biological Processes*, 2nd edn, Wiley, New York, pp. 116–18 (and refs. therein).

[8] Hunt, J. P., and Friedman, H. L., *Prog. Inorg. Chem.*, **30**, 377 (1983).

[9] Hugi-Cleary, D., Helm, L., and Merbach, A. E., *J. Am. Chem. Soc.*, **109**, 4444 (1987).

[10] Ref. [1], p. 35.

[11] Fontano, S., and Brito, F., *Inorg. Chim. Acta*, **2**, 179 (1968).

[12] Baes, C. J., and Mesmer, R. E., *The Hydrolysis of Cations*, Krieger, Malibar, Florida, 1986, p. 222 (and refs. therein).

[13] Swift, T. J., and Connick, R. E., *J. Chem. Phys.*, **37**, 307 (1962) and **41**, 2553 (1964).

[14] (a) Merbach, A. E., in *High Pressure Inorganic Chemistry: Kinetics and Mechanisms* (ed. R. van Eldik), Elsevier, London, 1986.

(b) Ducummon, Y., Newman, K. E., and Merbach, A. E., *Inorg. Chem.*, **19**, 3696 (1980).
[15] Moore, P., and Fielding, L., *J. Chem. Soc. Chem. Commun.*, 49 (1988).
[16] van Eldik, R., Asano, T., and LeNoble, W. J., *Chem. Revs.*, **89**, 549 (1989).
[17] (a) Mohr, R., Mietta, L. A., Ducummon, Y., and van Eldik, R., *Inorg. Chem.*, **24**, 757 (1985).
(b) Doss, R., and van Eldik, R., *Inorg. Chem.*, **21**, 4108 (1982).
(c) Mohr, R., and van Eldik, R., *Inorg. Chem.*, **24**, 3396 (1985).
[18] Ciunik, Z., and Glowiak, T., *Acta Cryst. Sec. B.*, **37**, 693 (1981).
[19] (a) Reddy, D., Sethuram, B., and Rao, T. N., *Indian J. Chem.*, **20**, 150 (1981).
(b) Berezina, L. P., Pozigun, A. I., Misyura, A. V., and Samoilenko, V. G., *Zh. Obshch. Khim.*, **49**, 1595 (1979).
[20] Henkel, G., Greuve, K., and Krebs, B., *Angew. Chem. Int. Ed. Engl. Trans.*, **24**, 117 (1985).
[21] Buxton, G. V., and Sellers, R. M., *Coord. Chem. Revs.*, **25**, 195 (1977).
[22] Stebler, M., Nielson, R. M., Siems, W. F., Hunt, J. P., Dodgen, H. W., and Wherland, S., *Inorg. Chem.*, **27**, 2893 (1988).
[23] Macartney, D. H., and Sutin, N., *Inorg. Chem.*, **24**, 3403 (1985).
[24] Davies, G., *Coord. Chem. Revs.*, **4**, 199 (1969).
[25] Beattie, J. K., Best, S. P., Skelton, B. W., and White, A. H., *J. Chem. Soc. Dalton. Trans.*, 2105 (1981).
[26] Diebler, H., and Sutin, N., *J. Phys. Chem.*, **68**, 174 (1964).
[27] Siskos, P. A., Peterson, N. C., and Huie, R. E., *Inorg. Chem.*, **23**, 1134 (1984).
[28] Lott, K. A. K., and Symons, M. C. R., *J. Chem. Soc.*, 829 (1959).
[29] Vincent, J. B., Chang, H. R., Folting, K., Huffman, J. C., Christou, G., and Hendrickson, D. N., *J. Am. Chem. Soc.*, **109**, 5703 (1987).
[30] (a) Kulawiec, R. J., Crabtree, R. H., Brudvig, G. V., and Schulte, G. K., *Inorg. Chem.*, **27**, 1309 (1988).
(b) Christmas, C., Vincent, J. B., Chang, H. R., Huffman, J. C., Christou, G., and Hendrickson, D. N., *J. Am. Chem. Soc.*, **110**, 823 (1988).
(c) Christou, G., *Acc. Chem. Res.*, **22**, 328 (1989).
[31] (a) Brudvig, G. W., *New J. Chem.*, **11**, 103 (1987).
(b) Beck, W. J., *J. Am. Chem. Soc.*, **108**, 4002 (1986).
[32] Scholder, R., and Kolb, A., *Z. Anorg. Allg. Chem.*, **260**, 31, 231 (1949).
[33] Richens, D. T., and Sawyer, D. T., *J. Am. Chem. Soc.*, **101**, 3681 (1979).
[34] Richens, D. T., Smith, C. G., and Sawyer, D. T., *Inorg. Chem.*, **18**, 706 (1979).
[35] (a) Hartman, J. A. R., Foxman, B. M., and Cooper, S. R., *J. Chem. Soc. Chem. Commun.*, 583 (1982).
(b) Chin. D. -H., Sawyer, D. T., Schaefer, W. P., and Simmons, C. J., *Inorg. Chem.*, **22**, 752 (1983).
[36] Pavlov, V. L., and Melezhik, A. V., *Russ. J. Inorg. Chem.*, **20**, 378, 532 (1975).
[37] Brown, I. D., *Can. J. Chem.*, **47**, 3779 (1969).
[38] Clark, R. J. H., Dines, T. J., and Doherty, J. M., *Inorg. Chem.*, **24**, 3088 (1985).
[39] (a) Waters, W. A., *Quart. Revs.*, **12**, 277 (1958).
(b) Carrington, A., and Symons, M. C. R., *Chem. Revs.*, **63**, 443 (1963)
[40] (a) Heckner, K. H., and Landsberg, R., *J. Inorg. Nucl. Chem.*, **29**, 413 (1967).
(b) Gamsjager, H. and Murmann, R. K., in *Advances in Inorganic and Bioinorganic Mechanics* (ed. A. G. Sykes), Vol. 2, Academic Press New York, 1983, p. 350.
[41] Krebs, B., and Hasse, K.-D., *Angew. Chem. Int. Ed. Engl. Trans.*, **13**, 603 (1974).
[42] Alberto, R., Egli, A., Abram, U., Hegetschweiler, K., and Schubiger, P. A., *Abstracts*

of the 30th International Conference on Coordination Chemistry, Kyoto, Japan, 1994, p. 31.

[43] Beck, W., and Sunkel, K., *Chem. Revs.*, **88**, 1405 (1988).
[44] Bernstein, S. N., and Dunbar, K. R., *Angew. Chem. Int. Ed. (Engl.)*, **31**, 1360 (1992).
[45] Trop, H. S., Davison, A., and Jones, A. G., *Inorg. Chim. Acta*, **54**, L61 (1981).
[46] Trop, H. S., Davison, A., Jones, A. G., Davis, M. A., Szaldo, D. J., and Lippard, S. J., *Inorg. Chem.*, **19**, 1105 (1980).
[47] Abrams, M. J., Davison, A., Faggiani, R., Jones, A. G., and Lock, C. J. L., *Inorg. Chem.*, **23**, 3284 (1984).
[48] Rouschias, G., *Chem. Revs.*, **74**, 531–66 (1974).
[49] Rulfs, C. L., Pacer, R., and Anderson, A., *J. Electroanal. Chem.*, **15**, 61 (1967).
[50] Russell, C. D., and Cash, A. G., *J. Electroanal. Chem.*, **92**, 85 (1978).
[51] Grassi, J., Devynck, J., and Tremillon, B., *Anal. Chim. Acta*, **107**, 47 (1979).
[52] Spitsyn, V. I., Kuzina, A. F., Oblova, A. A., Glinkina, M. I., and Stepovaya, L. I., *J. Radioanal. Chem.*, **30**, 561 (1976).
[53] Huber, E. W., Heinemann, W. R., and Deutsch, E., *Inorg. Chem.*, **26**, 3718 (1987).
[54] Cotton, F. A., Frenz, B. A., and Shive, L. W., *Inorg. Chem.*, **14**, 649 (1975).
[55] Bursten, B. E., Cotton, F. A., Green, J. C., Seddon, E. A., and Stanley, G. G., *J. Am. Chem. Soc.*, **102**, 955 (1980).
[56] Brown, P. R., Cloke, F. G. N., Green, M. L. H., and Tovey, R. C., *J. Chem. Soc., Chem. Commun.*, 519 (1982).
[57] Elder, M., Gainsford, G. J., Papps, M. D., and Penfold, B. R., *J. Chem. Soc., Chem. Commun.*, 731 (1969).
[58] (a) Chen, S., and Robinson, W. R., *J. Chem. Soc., Chem. Commun.*, 879 (1978).
      (b) Bronger, W., and Spangenberg, M., *J. Less-Common Metals*, **76**, 73 (1980).
[59] Federov, V. E., Mishchenko, A. V., Kolesov, B. A., Gubin, S. P., Slovokhotov, Yu. L., and Struchkov, Yu. T., *Izv. Akad. Nauk. SSSR, Ser. Khim.*, **9**, 2159 (1984).
[60] Lingane, J. J., *J. Am. Chem. Soc.*, **64**, 1001, (1942).
[61] Rulfs, C. L., and Meyer, R. J., *J. Am. Chem. Soc.*, **77**, 4505 (1955).
[62] Pavlova, M., Jordanov, N., and Popova, N., *J. Inorg. Nucl. Chem.*, **36**, 3945 (1974).
[63] Laing, M., Kiernan, P. M., and Griffith, W. P., *J. Chem. Soc., Chem. Commun.*, 221 (1977).
[64] (a) Griffith, W. P., Kiernan, P. M., and Bregeault, J. -M., *J. Chem. Soc. Dalton Trans.*, 1411 (1978).
      (b) Laing, M., Bregeault, J. -M., and Griffith, W. P., *Inorg. Chim. Acta*, **26**, L27 (1978).
[65] Wei, L., Halbert, T. R., Murray, III, R. R., and Stiefel, E. I., *J. Am. Chem. Soc.*, **112**, 6431 (1990).
[66] Lis, T., *Acta. Crystallogr. Sec. B.*, **31**, 1594 (1975).
[67] Ikari, S., Ito, T., McFarlane, W., Nasreldin, M., Ooi, B.-L., Sasaki, Y., and Sykes, A. G., *J. Chem. Soc. Dalton Trans.*, 2621 (1993).
[68] Bohm, G., Wieghardt, K., Nuber, B., and Weiss, J., *Inorg. Chem.*, **30**, 3464 (1991).
[69] Burgi, H. B., Anderegg, G., and Blauenstein, P., *Inorg. Chem.*, **20**, 3829 (1981).
[70] Linder, K. E., Dewan, J. C., and Davison, A., *Inorg. Chem.*, **28**, 3820 (1989).
[71] Gorski, B., and Koch, H., *J. Inorg. Nucl. Chem.*, **31**, 3565 (1969).
[72] Sundrehagen, E., *Int. J. Appl. Radiat. Isotop.*, **30**, 739 (1979).
[73] (a) Eckelman, W. C., and Levenson, S. M., *Int. J. Appl. Radiat. Isotop.*, **28**, 67 (1967).
      (b) Schwochau, K., and Koch, H., *Z. Anorg. Allg. Chem.*, **318**, 198 (1962).
      (c) Ianovici, E., Lerch, P., Proso, Z., and Maddock, A. G., *J. Radioanal. Chem.*, **46**, 11 (1978).

[74] Kostromin, A. I., Evgen'ev, M. I., and Novikova, L. A., *Zh. Fiz. Khim.*, **49**, 509 (1975).

[75] Shandles, R. S., Murmann, R. K., and Schlemper, E. O., *Inorg. Chem.*, **13**, 1373 (1974).

[76] Lawrance, G. A., and Sangster, D. F., *Polyhedron*, **5**, 1553 (1986).

[77] Roodt, A., Leipoldt, J. G., Helm, L., and Merbach, A. E., *Inorg. Chem.*, **31**, 2864 (1992).

[78] Roodt, A., Leipoldt, J. G., Helm, L., and Merbach, A. E., *Inorg. Chem.*, **33**, 140 (1994).

[79] Baldas, J., Boas, J. F., and Bonnyman, J., *Aust. J. Chem.*, **42**, 639 (1989).

[80] Baldas, J., Boas, J. F., and Bonnyman, J., Colmanet, S. F., and Williams, G.A., *Inorg. Chim. Acta*, **179**, 151 (1991).

[81] Baldas, J., Boas, J. F., and Bonnyman, J., Colmanet, S. F., and Williams, G. A., *J. Chem. Soc., Chem. Commun.*, 1163 (1990).

[82] Baldas, J., Boas, J. F., Colmanet, S. F., and Williams, G. A., *J. Chem. Soc. Dalton Trans.*, 2845 (1992).

[83] Baldas, J., Boas, J. F., Ivanov, Z., and James, B. D., *Inorg. Chim. Acta*, **204**, 199 (1993).

[84] Baldas, J., Colmanet, S. F., and Williams, G. A., *Inorg. Chim. Acta*, **179**, 189 (1991).

[85] Liese, W., Dehricke, K., Rogers, R. D., Shakir, R., and Atwood, J. L., *J. Chem. Soc. Dalton Trans.*, 1061 (1981).

[86] Beyer, H., Glemser, O., Krebs, B., and Wanger, G., *Z. Anorg. Allg. Chem.*, **376**, 87 (1970).

[87] Fischer, D., and Krebs, B., *Z. Anorg. Allg. Chem.*, **491**, 73 (1982).

[88] Murmann, R. K., *J. Phys. Chem*, **71**, 974 (1967).

[89] Wiechen, A., Herr, W., Hess, B., and Pieper, H. H., *Angew. Chem. Int. Ed. (Engl.)*, **6**, 1003 (1967).

[90] Murmann, R. K., *J. Am. Chem. Soc.*, **93**, 4184 (1971).

[91] Carrington, A., and Symons, M. C. R., *Chem. Revs.*, **63**, 443 (1963). Gray, H. B., *Coord. Chem. Revs.*, **1**, 2 (1966).

[92] Libson, K., Sullivan, J. C., Mulac, W. A., Gordon, S., and Deutsch, E., *Inorg. Chem.*, **28**, 375 (1989).

[93] Lawrance, G. A., and Sangster, D., *Polyhedron*, **4**, 1095 (1985).

Chapter 8

Group 8 Elements: Iron, Ruthenium and Osmium

Iron is the most abundant transition metal element within the solar system and on the earth. By chance it has also proved to be a key element in all stages in the evolution of life from primitive bacteria and algae to higher forms including ourselves. It is also, by far, the most abundant transition metal present in higher mammals with some 3–4g of the element present in the normal human body alone. Approximately 70 % is present in the hemoglobin and myoglobin oxygen transport proteins with only 0.7 % present in other intracellular proteins and enzymes. The remaining $\sim 29\%$ is stored. Like most biologically important elements, maintaining the required levels of iron in the body is crucial and its loss can be traced certain forms of anaemia. Prescribed iron is usually administered in the form of simple Fe^{II}(aq) salts. This reflects the fact that most biological sources of iron, e.g. from meats and vegetables, contain the element in a highly complexed and thus less accessible form. Elaborate iron storage and transport systems [1] have been developed for the simple reason that iron as Fe^{III}(aq)(via the oxidation of Fe^{II}) is extensively hydrolysed at the physiological pH (6–8) and would be precipitated as the highly insoluble hydrous 'Fe(OH)$_3$' (solubility product 10^{-36}). Thus uncomplexed Fe^{III}(aq) must be rapidly picked up, transported and stored to prevent hydrolytic loss. Indeed, the mechanism of iron storage makes use of its hydrolytic tendency but in a controlled manner. The importance of binding and transporting Fe^{III} is exemplified in the stability constant (log $\beta_{11} \sim 52$) for binding Fe^{III} by the sequestering agent enterobactin. This is one of the highest known stability constants and certainly the highest for any Fe^{III} chelating agent. Fe^{IV} and Fe^V, probably containing Fe=O groups, have been implicated as reactive intermediates in oxygen–activating heme proteins, such as cytochrome P450, but there are no stable aqueous forms for these oxidation states. Fe^{VI} is, however, stable as the tetrahedral oxo anion [FeO$_4$]$^{2-}$ in strongly alkaline solution, becoming spontaneously reduced by water as the pH is decreased towards neutrality.

In contrast, ruthenium and osmium have no known biological activity and indeed are highly toxic metals. Ruthenium forms [M(OH$_2$)$_6$]$^{3+/2+}$ ions but these are low spin in contrast to the iron analogues and this markedly affects the resulting reaction properties. Thus [Ru(OH$_2$)$_6$]$^{3+/2+}$ ions are noticeably more inert than their iron counterparts (by a factor of 10^8) with the divalent ion $\sim 10^4$ times more labile than its trivalent analogue. The expected increase in stability of the higher oxidation states is seen on progressing down the group. Thus [Ru(OH$_2$)$_6$]$^{2+}$ is more easily oxidised in acidic media than [Fe(OH$_2$)$_6$]$^{2+}$, reflecting a 560 mV negative shift in the reduction potential. Ru^{IV}(aq) is also perfectly stable in acidic solution as a cationic oxo-bridged tetramer. Like [FeO$_4$]$^{2-}$, [RuO$_4$]$^{2-}$ (or rather [RuO$_3$(OH)$_2$]$^{2-}$) can be stabilised in alkaline solution but it is much less oxidising. Finally, OsO$_4$ and RuO$_4$ are the only examples of authentic M^{VIII} species and both are soluble and stable in acidic aqueous solution in the absence of reducing agents.

8.1 IRON

8.1.1 Iron(II) (d^6): Chemistry of the [Fe(OH$_2$)$_6$]$^{2+}$ Ion

8.1.1.1 PREPARATION AND PROPERTIES OF [FE(OH$_2$)$_6$]$^{2+}$

The extremely pale bluish-green hexaaqua ion [Fe(OH$_2$)$_6$]$^{2+}$ is readily liberated upon simple dissolution of Fe^{II} salts in water and requires alkaline pH before hydrolysis occurs. The hexaaqua formulation has been established in solution by XRD [1], neutron scattering [2] and also EXAFS [3], as well as by ^{17}O [4] and ^{1}H [5] NMR measurements. The ion is well established in Tutton-type salts such as (NH$_4$)$_2$Fe(SO$_4$)$_2$·6H$_2$O (originally known as Mohr's salt) [6], the fully

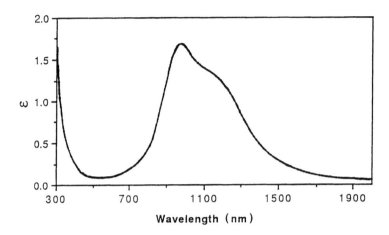

Figure 8.1. Electronic spectrum of $[\mathrm{Fe(OH_2)_6}]^{2+}$

deuterated form of which having been the subject of a number of low temperature neutron diffraction studies [7]. The $\mathrm{FeO_6}$ geometry is close to regular octahedral with average $\mathrm{Fe-OH_2}$ distances around 213 pm . Spectroscopic and magnetic properties are consistent with the high-spin $t_{2g}^4 e_g^2$ configuration. The room temperature magnetic moment is around 5.4 BM due to an orbital contribution from the t_{2g}^4 ground state, the highly mixed nature of the t_{2g} orbitals of both high-spin $\mathrm{Fe^{2+}}$ and $\mathrm{Co^{2+}}$ (Section 9.1.1) being consistent with spin and charge density calculations [8]. The electronic spectrum of $[\mathrm{Fe(OH_2)_6}]^{2+}$ is shown in Fig. 8.1. Two d–d transitions from the triply degenerate $^5T_{2g}$ ground state to the Jahn–Teller split ^5E_g excited state can be seen in the near infra-red region at 1205 nm ($\varepsilon = 1.4\,\mathrm{M^{-1}cm^{-1}}$) and 962 nm ($1.7\,\mathrm{M^{-1}cm^{-1}}$) [9]. This absorption tails into the red part of the visible spectrum and this, combined with the rising absorption in the blue at 400 nm (probably charge transfer), is the origin of the characteristic pale bluish-green colour.

Hydrolysis of $[\mathrm{Fe(OH_2)_6}]^{2+}$ is not extensive and only occurs at alkaline pH. The hydrolysis constants quoted by Baes and Mesmer are pK_{11} (9.5), pK_{12} (20.6), pK_{13} (31.0) and pK_{14} (46.0) [10,11]. Hydrous $\mathrm{Fe(OH)_2}$ precipitates at around pH 10. The formation of only mononuclear hydrolysis products provides a marked contrast with the behaviour of $\mathrm{Fe^{3+}}$ (aq) and also the neighbouring elements $\mathrm{Co^{2+}}$ and $\mathrm{Ni^{2+}}$. Solutions of $\mathrm{Fe^{2+}}$ (aq) become highly oxygen sensitive as the pH is raised due to a combination of a negative shift in the $\mathrm{Fe^{III/II}}$ reduction potential and resulting hydrolytic polymerisation of the $\mathrm{Fe^{III}}$ product (cf. E^θ $[\mathrm{Fe(OH_2)_6}]^{3+/2+}$ $= +0.77\,\mathrm{V}$, but E^θ ($\mathrm{Fe(OH)_3}$/

$Fe(OH)_2, OH^-) = -0.56 V$). The stable solid phase in equilibrium with Fe^{2+} (aq) in the presence of a reducing agent (1 atm of H_2) appears to be magnetite, Fe_3O_4 [12].

8.1.1.2 WATER EXCHANGE AND SUBSTITUTION REACTIONS ON $[Fe(OH_2)_6]^{2+}$

Early estimates for the water-exchange rate constant (25 °C) on $[Fe(OH_2)_6]^{2+}$ were based on ^{17}O NMR measurements ($3.2 \times 10^6 s^{-1}$) [13] and ultrasound absorption (SO_4^{2-} complexation, $1-2 \times 10^6 s^{-1}$) [14]. These values are in fact close to the remeasured value determined by Merbach and coworkers ($4.4 \times 10^6 s^{-1}$) [15] using a Swift–Connick treatment of the free water ^{17}O NMR resonance obtained from a solution of $Fe(ClO_4)_2$ as function of temperature. The activation parameters are $\Delta H_{ex}^{\ddagger} = 41.4$ kJ mol^{-1}, $\Delta S_{ex}^{\ddagger} = +21.2$ J K^{-1} mol^{-1} and $\Delta V_{ex}^{\ddagger} = +3.8$ cm^3 mol^{-1}. As discussed in Section 1.4, the change in sign of the activation volume from negative (Mn^{2+} (aq), -5.4) to positive (Fe^{2+} (aq), $+3.8$) reflects a decreasing importance of the factors largely responsible for volume contraction to the transition state within a basic reaction profile largely controlled by M—OH_2 bond dissociation across the entire divalent first row series. As is the case generally with dipositive transition metal ions, substitution reaction rate constants on $[Fe(OH_2)_6]^{2+}$ vary little as a function of the incoming ligand (Table 8.1).

8.1.1.3 REDOX REACTIONS INVOLVING $[Fe(OH_2)_6]^{2+}$

The rapid rate of reduction of oxidative titrants such as $[Cr_2O_7]^{2-}$ and Ce^{IV} (aq) by Fe^{2+} (aq) have led to the use of the ion, in solutions of Mohr's salt in dilute H_2SO_4, as a usefull primary redox standard. However, many reactions involving reductions by Fe^{2+} (aq) are quite slow and often several orders of magnitude slower than typical substitution reactions. For example, a large variation of rates

Table. 8.1. Rate constants (25 °C) for substitution reactions on $[Fe(OH_2)_6]^{2+}$

Incoming ligand	$10^{-6}k$ (M^{-1}s^{-1})	Method	Ref.
H_2O	4.4 [a]	^{17}O NMR	[15]
SO_4^{2-}	2.4	Ultrasound	[15]
NO	0.62	Stopped-flow	[16]
HF	0.93	^{19}F NMR	[17]
F^-	1.4	^{19}F NMR	[17]
bipy	0.16	Stopped-flow	[18]
TPTZ[b]	0.13	Stopped-flow	[19]

[a] Units of s^{-1}
[b] TPTZ = 2,4,6-tripyridyl-S-triazine

Table. 8.2. Rate constant $(25\,^\circ\mathrm{C})$ for reduction of $[Co(NH_3)X]^{2+}$ complexes by $[Fe(OH_2)_6]^{2+}$ [20]

X	$10^3 k_{11}$ $(\mathrm{M}^{-1}\,\mathrm{s}^{-1})$
F^-	6.6
Cl^-	1.4
Br^-	0.73
N_3^-	8.7
NCS^-	3.0
SCN^-	120
$HC_2O_4^-$	430

is seen for the reduction of a range of pentammine Co^{III} complexes by Fe^{2+} (aq) (Table 8.2). These reactions are likely to be inner sphere in nature on the grounds that the outer-sphere reaction is disfavoured since it is in this case uphill. Such a variation could arise if the rate of formation of the intermediate is rapid followed by the slow preorganisation or dissociation step. Co^{III} would prefer to receive an σ electron but the electronic structure of the Fe^{3+} product $(t_{2g}^3\,e_g^2)$ suggest that the lowest energy process for Fe^{2+} (aq) might rather involve donation of a $t_{2g}\pi$ electron. This might reflect a large activation barrier to σ–σ electron transfer within the rapidly formed intermediate and this may be largely responsible for the slow rate of redox.

8.1.1.4   THE $[Fe(OH_2)_6]^{3+/2+}$ SELF-EXCHANGE PROCESS

Central to an understanding of redox reactions involving $[Fe(OH_2)_6]^{2+}$ and $[Fe(OH_2)_6]^{3+}$ is the self-exchange reaction

$$[Fe(OH_2)_6]^{3+} + [^*Fe(OH_2)_6]^{2+} \overset{k_{11}}{\rightleftharpoons} [Fe(OH_2)_6]^{2+} + [^*Fe(OH_2)_6]^{3+} \quad (8.1)$$

The rate and mechanism of this seemingly 'simple' process has, however, been a source of considerable controversy since the original measurements of Silverman and Dodson in 1952 [21]. It has become apparent that the reaction actually occurs in two parallel pathways, the second involving the monohydroxo species of Fe^{III}, $[Fe(OH_2)_5OH]^{2+}$:

$$[Fe(OH_2)_6]^{3+} \overset{K_{11}}{\rightleftharpoons} [Fe(OH_2)_5OH]^{2+} + H^+ \quad (8.2)$$

$$[Fe(OH_2)_5OH]^{2+} + [^*Fe(OH_2)_6]^{2+} \overset{k_{11}OH}{\rightleftharpoons} [Fe(OH_2)_6]^{2+} + [^*Fe(OH_2)_5OH]^{2+}$$

$$(8.3)$$

which is likely to be of an inner-sphere nature, as is the case for the $Fe^{3+/2+}$ (aq) couple in the presence of complexing ions such as Cl^- or SO_4^{2-}. The mechanism of the $[Fe(OH_2)_6]^{3+/2+}$ exchange process itself is more difficult to pin down in view of the substitution lability on the exchange timescale. A bridged aqua ligand intermediate was proposed by Hupp and Weaver [22] and Ludi and Merbach and coworkers [23] to account for the perceived anomalously rapid rate of self-exchange in the hexaaqua iron couple in homogeneous solution, the electrochemical rate estimate of Hupp and Weaver being 10^4 times slower than that found by the homogeneous solution measurements of Silverman and Dodson [21]. Sutin and coworkers [24] subsequently found agreement with Silverman and Dodson and with predictions from an updated Marcus approach. Swaddle has since argued that reactions involving the $[Fe(OH_2)_6]^{3+/2+}$ couple might be better represented as an example of a non-adiabatic electron transfer process, i.e. one involving extremely weak $Fe^{II}–Fe^{III}$ electronic coupling in the encounter complex, and that the question of non-adiabaticity may be central to an understanding of the rates and mechanisms of electron tranfer reactions involving both $[Fe(OH_2)_6]^{2+}$ and $[Fe(OH_2)_6]^{3+}$. Swaddle and coworkers have re-examined the $Fe^{3+/2+}$ (aq) self-exchanging system under variable high pressure using radiolabelling with ^{59}Fe at $+2°C$ [25]. The results could be accounted for quantitatively on the basis of Marcus–Hush theory (modified to account for the pressure dependence of the Fe—Fe separation and of likely cation–anion pairing) in terms of an adiabatic outer-sphere mechanism for the $[Fe(OH_2)_6]^{3+/2+}$ exchange ($\Delta V^{\ddagger} = -11.1 \pm 0.4 \, cm^3 \, mol^{-1}$) and an OH-bridged inner-sphere mechanism for the $[Fe(OH_2)_5OH]^{2+}/[Fe(OH_2)_6]^{2+}$ reaction ($\Delta V^{\ddagger} = +0.8 \pm 0.9 \, cm^3 \, mol^{-1}$). The self-exchange rate constant determined for the hexaaqua couple ($1.17 \, M^{-1} \, s^{-1}$) at ambient pressure (0.1 MPa) [24] is in good agreement with the earlier value of Silverman and Dodson. The rate constant for the combination of reactions (8.2) and (8.3), representing the pathway through $[Fe(OH_2)_5OH]^{2+}$, was $0.41 \, s^{-1}$. Thus, as Furholz and Haim [26], Taube and coworkers [27] and Sutin and coworkers [28] have suggested, the observed anomalies in the kinetics of cross-reactions involving the $[Fe(OH_2)_6]^{3+/2+}$ couple *vis-à-vis* the self-exchange rate may be due to a closer approach to full adiabacity in the latter.

8.1.2   Iron(III) (d^5): Chemistry of the $[Fe(OH_2)_6]^{3+}$ Ion

The $[Fe(OH_2)_6]^{3+}$ ion has only a weak violet colour due to the spin-forbidden nature of transitions from the $^6A_{1g}$ ground state (high-spin $t_{2g}^3 e_g^2$). It is established in simple hydrated Fe^{III} salts such as $Fe(ClO_4)_3 \cdot 10H_2O$ and $Fe(NO_3)_3 \cdot 6–9H_2O$ as well as in the NH_4^+, Rb^+ and Cs^+ alums, $M^I[Fe(OH_2)_6](XO_4)_2 \cdot 6H_2O$ ($X = S, Se$) [29, 30]. A neutron diffraction data study of both the caesium sulfate and selenate alums at 15 K has been reported [31]. Whereas the sulfate alum adopts the normal β modification with the waters bonded trigonal planar, the selenate alum

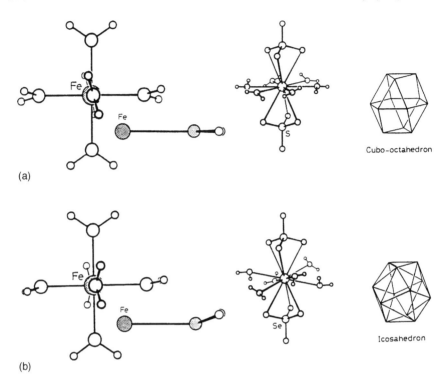

Figure 8.2. Coordination geometry about Fe^{3+} and Cs^+ in the respective (a) $Cs[Fe(OH_2)_6](SO_4)_2 \cdot 6H_2O$ and (b) $Cs[Fe(OH_2)_6](SeO_4)_2 \cdot 6H_2O$ alums

has α modification with trigonal pyramidal geometry at the oxygens, with the plane of the waters inclined to the Fe—O bond vector at an angle of 18°C (Fig. 8.2). These features are believed to be largely due to differing hydrogen bonding effects within the two lattices. the small difference in Fe—O distance (0.8 pm) between the sulfate (199.4 pm) and the selenate (200.2 pm) alums is not thought relevant, apart from being a consequence of the different water geometries. Indeed, at room temperature the average Fe—O distance in the two alums is identical within uncertainties and, furthermore, the $v_1(MO_6)$ vibration is found at an identical frequency for both at 80 K. The geometry around the Cs^+ ion in the respective alums is also different for the two alums, (Fig. 8.2). In the sulfate alum, twelve-coordinate cube-octahedral structure is seen with six 'equatorial' water molecules in addition to two 'axial' triply-bonded sulfates. In the selenate alum, however, the 'equatorial' water bonds twist to adopt an icosahedral geometry.

The hexaaqua formulation has been confirmed by both XRD [32] neutron scattering [33] and EXAFS [3] measurements. Typical Fe—OH_2 distances from solution XRD data are around 200 pm. In solution the hexaaqua ion itself is only stable in strongly acidic media ($[H^+] > 6.0$ M) owing to a strong tendency towards hydrolytic polymerisation. In studies of simple reactions on Fe^{3+} (aq) the use of low Fe^{III} concentrations and strongly acid media have proved to be essential in order to suppress this process.

The invariable presence of hydrolysis products and for $[Fe(OH_2)_6]^{3+}$ itself a strong charge transfer absorption below 350 nm have combined to prevent a detailed unequivocal assignment of the d–d electronic spectrum of $[Fe(OH_2)_6]^{3+}$. High-spin Fe^{3+} ($t_{2g}^3 e_g^2$) has the same electronic ground state $^6A_{1g}$ as high-spin Mn^{2+}, and as such the d–d transitions are likewise spin-forbidden and weak. A number of lower energy bands (Fig. 8.3) are observable and assignments have been made largely on the basis of a similarity to those for $[Mn(OH_2)_6]^{2+}$. These are respectively the bands at 794 nm ($\varepsilon \sim 0.1$ $M^{-1}cm^{-1}$) ($^6A_{1g}-^4T_{1g}(G)$), 541 nm (0.1 $M^{-1}cm^{-1}$) ($^6A_{1g}-^4T_{2g}(G)$) and 412 and 407 nm (~ 0.5 $M^{-1}cm^{-1}$) ($^6A_{1g}-^4A_{1g}$, $^4E_g(G)$) [34]. The intense H_2O to Fe^{3+} charge transfer band maxi-

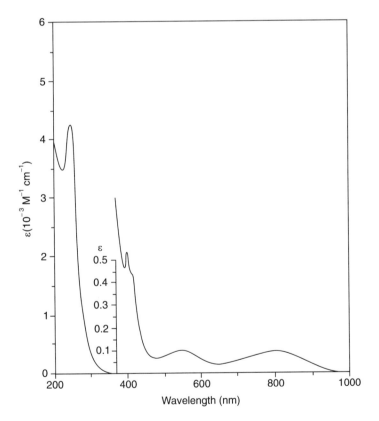

Figure 8.3. Electronic spectrum of freshly dissolved $Fe(ClO_4)_3 \cdot 6H_2O$ in 6.0 M $HClO_4$

mum, responsible for obscuring the remainder of the spin-forbidden d–d features, is seen at 240 nm (4250 $M^{-1}cm^{-1}$) [35]. The ε values for the d–d transitions within $[Fe(OH_2)_6]^{3+}$ are ~ 10 times larger than those for the corresponding transitions within $[Mn(OH_2)_6]^{2+}$, a feature possibly reflective of a somewhat greater covalency in the Fe—OH_2 bonds. A well-resolved crystal d–d spectrum for Fe^{3+} is, however, seen for the ion doped into beryl [36], where both octahedral and tetrahedral sites are occupied. Here the spin-forbidden bands have narrower linewidths and resolution into those representing both octahedral and tetrahedral Fe^{3+} ions has been possible. The octahedral Fe^{3+} ions have bands respectively at 704 nm ($^6A_{1g}$–$^4T_{1g}(G)$), 571 and 487 nm ($^6A_{1g}$–$^4T_{2g}(G)$), 423 nm ($^6A_{1g}$–$^4A_{1g}$, $^4E_g(G)$), 373 nm ($^6A_{1g}$–$^4T_{2g}(D)$) and 366 nm ($^6A_{1g}$–$^4E_g(D)$).

8.1.2.1 THE HYDROLYSIS OF $[Fe(OH_2)_6]^{3+}$

This topic has been extensively studied and reviewed such is its importance to the aqueous chemistry of Fe^{III} species. Sylva [37] compiled the first comprehensive review of the topic in 1972 while Flynn [38] and Schneider and coworkers [39] have more recently provided an update. Schneider in particular has highlighted the link between an understanding of Fe^{III} hydrolysis and that of biomineralisation processes, such as those occurring within the iron storage protein ferritin [40] (see below). Even in strongly acidic solutions (several molar in $[H^+]$) the pale purple colour characteristic of freshly dissolved Fe^{3+} (aq) salts eventually takes on a yellow coloration arising from species taking part in the slow and complex set of equilibria that are involved in the hydrolytic polymerisation process.

The hydrolysis of $[Fe(OH_2)_6]^{3+}$ in acidic solution is generally agreed [37–42] to involve the first two equilibra

$$[Fe(OH_2)_6]^{3+} + H_2O \rightleftharpoons [Fe(OH_2)_5(OH)]^{2+} + H_3O^+ \quad (8.4)$$

$$2[Fe(OH_2)_5(OH)]^{2+} \rightleftharpoons \begin{array}{l} [Fe_2(OH)_2(OH_2)_8]^{4+} + 2H_2O \\ (\text{or } [Fe_2O(OH_2)_{10}]^{4+} + H_2O \end{array} \quad (8.5)$$

$$\begin{array}{l} Fe_2(OH)_2^{4+}(aq) \\ (\text{or } Fe_2O^{4+}(aq)) \end{array} + FeOH^{2+}(aq) \xrightarrow{H_2O} \begin{array}{l} Fe_3(OH)_4^{5+}(aq) \\ (\text{or } Fe_3O(OH)_2^{5+}(aq) \end{array} + H^+ \quad (8.6)$$

and possibly a third [40–43] prior to condensation to high molecular weight species and eventual precipitation. In dilute solution $[Fe(OH_2)_5(OH)]^{2+}$ and the aqua dimer ($[Fe_2(OH)_2(OH_2)_8]^{4+}$ or $[Fe_2O(OH_2)_{10}]^{4+}$) have characteristic electronic spectra (Fig. 8.4) and solutions containing up to 10 % of the aqua dimer are reported to be stable if kept in supersaturation with respect to solid phases such as α-FeO(OH) (see below). Processes (8.4) and (8.5) are fully reversible and occur rapidly. A number of potentiometric studies [42, 43] required the trimeric species, $Fe_3(OH)_4^{5+}$ (aq), to be present for solutions ~ 0.1 M in Fe^{III}. In more dilute solutions the formation of an Fe_{12} species has been suggested [44]. Baes and Mesmer have estimated thermodynamic K_{xy} values for the various hydrolysis products as $pK_{11} = 2.19$ ($[Fe(OH_2)_5(OH)]^{2+}$, $pK_{22} = 2.96$ ($Fe_2(OH)_2^{4+}$(aq) or Fe_2O^{4+}(aq)) and $pK_{34} = 6.3$ ($Fe_3(OH)_4^{5+}$ (aq)) [41]. At lower Fe^{III} concentrations ($\sim 10^{-5}$ M) only mononuclear hydrolysis products are relevant. In addition to $FeOH^{2+}$(aq), one sees $[Fe(OH_2)_4(OH)_2]^+$ ($pK_{12} = 5.67$) and $[Fe(OH)_4]^-$ ($pK_{14} = 21.6$) in addition to the precipitation of hydrous $Fe(OH)_3$ around pH 7. The hydrolysis of $[Fe(OH_2)_6]^{3+}$ has been widely studied by different groups with regard to anion and ionic strength effects with somewhat differing findings. Brown and coworkers [43] evaluated hydrolysis constants for Fe^{3+} (aq) species in perchlorate, nitrate and chloride media at I = 1.0 M. Values of $-\log Q_{xy}$ (anion) were found to follow the trend $-\log Q_{xy}(Cl^-) > -\log Q_{xy}(NO_3^-) > -\log Q_{xy}(ClO_4^-)$, interpreted in terms of decreasing anion complexation of Fe^{3+} (aq) in this order. These results are listed in Table 8.3 along with the thermodynamic values for I = O evaluated by Baes and Mesmer. Danielo *et al.* [45] compared $-\log Q_{xy}$ values at different ionic strengths in both nitrate

Table. 8.3. Formation quotients ($-\log Q_{xy}$) for the hydrolysis products of Fe^{3+}(aq) in acidic 1.0 M perchlorate, nitrate and chloride media [43] along with estimated thermodynamic K_{xy} values [41]

Species	$-\log Q_{xy}$			pK_{xy}
	1.0 M Cl^-	1.0 M NO_3^-	1.0 M ClO_4^-	
$FeOH^{2+}$(aq)	3.21 ± 0.023	2.77 ± 0.014	2.73 ± 0.006	2.19
$Fe(OH)_2^+$(aq)	6.73 ± 0.029	6.61 ± 0.037	6.29 ± 0.009	5.67
$Fe_2(OH)_2^{4+}$(aq) (or Fe_2O^{4+}(aq))	4.09 ± 0.037	3.22 ± 0.011	3.20 ± 0.004	2.96
$Fe_3(OH)_4^{5+}$(aq)	7.58 ± 0.042	6.98 ± 0.035		6.30

Table. 8.4. Ionic strength dependence of $-\log Q_{xy}$ for hydrolysis of $[Fe(OH_2)_6]^{3+}$ in perchlorate and nitrate media [45, 46]

Ionic strength	Perchlorate			Nitrate
	$-\log Q_{11}$	$-\log Q_{12}$	$-\log Q_{22}$	$-\log Q_{11}$
0 ($\log K_{xy}$)	2.18 ± 0.01	5.6 ± 0.1	2.92	2.21 ± 0.04
0.05				2.50 ± 0.02
0.1	2.57 ± 0.01	5.9 ± 0.1	2.88	2.56 ± 0.02
0.5	2.73 ± 0.01	6.0 ± 0.1	2.80	2.59 ± 0.02
1.0	2.78 ± 0.01	5.9 ± 0.1	2.74	2.52 ± 0.04
3.0	3.09 ± 0.01	6.3 ± 0.1	2.88	—

(KNO_3) and perchlorate ($NaClO_4$) media. The results, listed in Table 8.4, extrapolate well at zero I to the values of Baes and Mesmer [41] and Biedermann [46], but appear to be in conflict with thoses of Brown and coworkers [43]. Here a significant dependence for $-\log Q_{xy}$ on I is seen in perchlorate, but only a slight dependence in nitrate, with the result that at $I = 1.0$ M, $-\log Q_{xy}$ is smaller in nitrate than in perchlorate media [45]. The somewhat surprising inference was that nitrate was a weaker interacting anion for Fe^{3+} than perchlorate. Several groups have moreover commented on the apparent ClO_4^- dependence of $-\log Q_{11}$, which has been interpreted in terms of a significant outer-sphere interaction of the anion with $[Fe(OH_2)_6]^{3+}$. Indeed, there is structural evidence for coordination of perchlorate within an oxo-bridged Fe^{III}_2 complex [47]. Daniele *et al.* [45] also found evidence in the nitrate data for the large oligomer $Fe_{12}(OH)_{34}^{2+}$ (aq) in addition to the aqua dimer. Van-Eldik and coworkers [48] have studied both the ionic strength and pressure dependence of $-\log Q_{11}$ in relation to complex formation studies [49]. Values of $\log Q_{11}$ were determined spectrophotometrically at 340 nm on dilute $Fe(ClO_4)_3 \cdot 6H_2O$ solutions in distilled water wherein $FeOH^{2+}$ (aq) is the only species having an appreciable absorbance. The dependence of $\log Q_{11}$ on I in perchlorate media was observed to take the form

$$\log Q_{11} = \log K_{11} - \frac{2.04\sqrt{I}}{1 + 2.97\sqrt{I}} \qquad (8.7)$$

consistent with the following values: $-\log Q_{11} = 2.31$ ($I = 0.005$ M), 2.65 (0.5 M), 2.70 (1.0 M) and 2.76 (3.0 M) with $\log K_{11}$ ($I = 0$) assumed as 2.19 [41] and ε_{FeOH}^{2+} (340 nm) calculated as 980 ± 40 M^{-1} cm^{-1} (see Fig. 8.4). The values of $-\log Q_{xy}$ compare well with those quoted by Daniele *et al.* above [45] at the lower ionic strengths and with the value of Brown and coworkers [43] at $I = 1.0$ M. Overall there seems to be substantial evidence in support of a significant ionic strength dependence on $\log Q_{11}$ for $[Fe(OH_2)_6]^{3+}$, certainly in perchlorate media.

The aqua dimer has been formulated as having either the di-μ-hydroxo or mono-μ-oxo formulation. Obtaining definitive information is difficult owing to the complex nature of hydrolysed Fe^{III} solutions when the dimer is present in appreciable quantities. The nature of such solutions is also strongly medium dependent (see below) and one cannot rule out the presence of condensed colloidal species. The two aqua dimer species may in fact be present in equilibrium [50]:

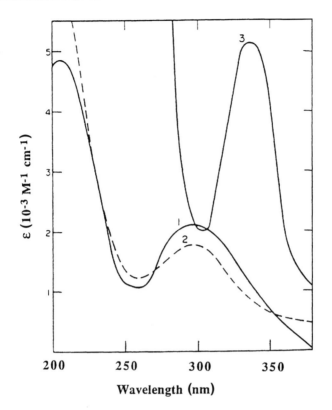

Figure 8.4. Electronic spectrum for $[Fe(OH_2)_5OH]^{2+}$ (1) shown along with the calculated spectrum for $[Fe(OH_2)_4(OH)_2]^+$ (2) and the aqua Fe^{III} dimer (3) in perchlorate solution

Certain pyridine dicarboxylic acid (dipicolinic acid) ligands* are capable of forming binuclear iron complexes containing either bridging moiety, viz. $[L_3(H_2O)Fe(OH)_2Fe(OH_2)L_3]$ [51] and $[L_{3'}(H_2O)_2FeOFe(OH_2)_2L_{3'}] \cdot 4H_2O$ [52], suggesting that the energy difference governing equilibrium (8.8) may be quite small. The di-μ-hydroxo formulation for the aqua dimer has been favoured on the basis of magnetic, infra-red and kinetic data [50, 53]. Attempts at investigating the dimer structure using X-ray absorption methods [54] proved not to be definitive [55]. Mossbauer studies on the dimer in solution have been interpreted in terms of the μ-oxo formulation [56]. However, a study of Fe^{3+} (aq) ion-exchanged on to a Nafion®+membrane gave rise to two quadrupole doublets which were

---

* L_3 = dipicolinic acid, $L_{3'}$ = 4-chlorodipicolinic acid. +-Nafion®-Na+ salt of a fluorinated sulfonated polymer marked by DuPont.

assigned to the species $\{Fe\!-\!O\!-\!Fe\}^{4+}$(Nafion) and $\{Fe\!-\!(H_3O_2)\!-\!Fe\}^{5+}$ (Nafion) [57], with the μ-oxo dimer the major species at higher temperatures. The presence of a μ-H_3O_2-bridging group has in fact been structurally characterised in a μ-oxo Fe_2^{III} complex with the ligand tris(2-pyridyl)amine (TPA) [46, 58]. There is also other evidence in support of $H_3O_2{}^-$ bridging groups as intermediates in the hydrolysis of a number of other aqua Fe^{III} complexes [59].

The electronic spectrum of the aqua dimer has been evaluated by an iterative treatment using absorbance data from dilute 10^{-3}–10^{-2} M solutions of $Fe(ClO_4)_3$ in the pH range 1.7–2.3 and calculated hydrolysis constants for the formation of the $FeOH^{2+}$ (aq), $Fe(OH)_2{}^+$ (aq) and the aqua dimer [60]. Figure 8.4 shows the calculated spectra of $Fe(OH)_2{}^+$ (aq) and the aqua dimer along with that of $FeOH^{2+}$ (aq). Features of the aqua dimer are a well-defined maximum at 335 nm ($\varepsilon \sim 5000$ M^{-1} cm^{-1}) and a shoulder at 240 nm ($\sim 12\,000$ M^{-1} cm^{-1}) on a rising absorption extending into the u.v. The ε value for $FeOH^{2+}$ (aq) at 340 nm (~ 1000 M^{-1} cm^{-1}) agrees well with that calculated by van-Eldik and coworkers [48]. The spectrum of $FeOH^{2+}$ (aq) seems well established on the basis of the well-defined isobestic points observed at 225 and 273 nm [61]. These disappear at higher Fe^{III} concentrations as the aqua dimer builds up. The spectrophotometric value of $-\log Q_{11}$ at $I = 0.1$ M (2.54) [60] is in keeping with the potentiometric values in perchlorate media quoted from Biedermann's work (2.57) [46] and the spectrophotometric values of van-Eldik and coworkers (2.52) [48].

The determination of the spectrum of the Fe^{III} aqua dimer in the presence of monomer species is certainly feasable since it has been shown that the dimer condenses to higher polymers relatively slowly under conditions when an excess of $[Fe(OH_2)_6]^{3+}$ and $FeOH^{2+}$ (aq) are present in dilute solution at pH $\leqslant 2$. This can be interpreted as due to a suppression of reaction (8.6), a possible pathway to condensed polynuclears. Under these conditions precipitation of the solid phase can take months [40b,63]. Solutions prepared at higher pH, where the dimer should be formed in higher amounts invariably contain apppreciable amounts of collioidal 'FeO(OH)' species in accordance with the data of Schindler and Biedermann [62] and Feitknecht and Michaelis [63]. These workers showed that the aqua dimer is thermodynamically unstable in perchlorate media with respect to mononuclear Fe^{3+} (aq) and the solid phase FeO(OH).

Kinetic studies with regard to the break-up of the aqua dimer back to monomeric Fe^{III}(aq) species:

$$Fe_2^{III}(aq) \xrightarrow{\ H^+/H_2O\ } 2\,Fe(OH)^{2+}(aq) \xrightarrow{\ 2H^+\ }{\rightleftharpoons} 2[Fe(OH_2)_6]^{3+} \qquad (8.9)$$

are consistent with the rate law

$$\text{Rate} = (k_1 + k_2[H^+])\,[Fe_2^{III}(aq)] \qquad (8.10)$$

At 25 °C, $k_1 = 0.4$ s^{-1} and $k_2 = 3.1$ M^{-1} s^{-1} [64]. These rate constants have been substantiated by several independent studies [65].

Several studies have reported on the hydrolysis of ferric sulfate solutions. In one study the similarity in the physical and chemical properties of hydrolysed $Fe_2(SO_4)_3$ solutions to those of Al^{3+} (aq) and Ga^{3+} (aq) suggested the likely formation of the tridecamer $[FeO_4(Fe(OH)_2(OH_2))_{12}]^{7+}$ (see Chapter 2) or a related species [66]. However, this species has not been definitively identified.

Finally an *ab initio* molecular dynamics simulation study of $[Fe(OH_2)_6]^{3+}$ and its monomeric hydrolysis products, $[Fe(OH_2)_5(OH)]^{2+}$ and *cis*- and *trans*-$[Fe(OH_2)_4(OH)_2]^+$, in aqueous solution has been reported [67]. Calculated Fe—O distances (205 and 417 pm) for the first and second hydration shells of water molecules around Fe^{3+} are in excellent agreement with the data from XRD [32] and neutron diffraction [33]. The hydration number of 6 was verified. The major effect of the hydrolysis was seen as an adjustment of the second hydration shell towards the donation of new hydrogen bonds to the OH group.

8.1.2.2 IRON(III) OXIDES AND OXYHYDROXIDES

The nature of the solid phases resulting from hydrolytic precipitation from Fe^{3+} (aq) is a topic that has been well studied and reviewed [38–40] such is its fundamental importance to corrosion, medicine, industry, soil science and environmental chemistry. Therefore only a brief overview will be given here. Solid phases such as goethite (α-FeO(OH)) and haematite (α-Fe_2O_3) are widely employed as pigments and are also the precursors of maghaemite (γ-Fe_2O_3) which is used in magnetic tapes. As mentioned above, the equilibria leading to the low molecular weight species $FeOH^{2+}$ (aq), $Fe(OH)_2^+$ (aq) and $Fe_2(OH)_2(or(O))^{4+}$ (aq) is established rapidly; these then interact with each other to produce species with increasing nuclearity. This process starts immediately and can continue over very long periods.

Addition of base to Fe^{3+} (aq) solutions rapidly forms the mono and dinuclear species which, with increasing OH/Fe^{3+} ratios, give rise to successively larger polynuclear species which age to either an amorphous precipitate or to a variety of crystalline products. If sufficient base is added so that OH/Fe exceeds three, an amorphous precipitate forms immediately (see below) which gradually transforms the iron oxides or oxohydroxides depending upon the conditions. Which crystalline iron hydroxide forms depends upon the OH/Fe ratio and on the anion present. Thus α-FeO(OH) (goethite) precipitates from nitrate solutions [69] whereas β-FeO(OH) (akaganeite) is precipitated when chloride ion is present [69, 70] and lepidocrocite (γ-FeO(OH)) is precipitated, along with goethite, from perchlorate solutions [63]. Heating acidic suspensions of FeO(OH) transforms it to α-Fe_2O_3. This form is also produced by heating Fe^{3+} (aq) solutions above $80\,^\circ$C. Techniques such as TEM, XRD, laser light scattering, magnetic measurements and Mossbauer spectroscopy have been used to follow the nucleation processes involved in the transformations of the polynuclears into crystalline products [39, 40, 71]. It was found that the ageing process to the crystallites did not involve conversion of hydroxo to oxo bridges, but rather an alteration in the polynuclear shape from initially 'needle-like' to shorter wider assemblages.

The amorphous red–brown precipitate that results upon the addition of sufficient base to give $OH/Fe^{3+} > 3$ is a poorly ordered ferric hydroxide that resembles the mineral ferrihydrite and shows some similarity to the oxyhydroxide core structure of the biomineral ferritin. Ferrihydrite appears to contain 15–25% water by weight. The formula does not seem to have been unequivocally established. Suggested formulae range from $5Fe_2O_3 \cdot 9H_2O$ through to $Fe_2O_3 \cdot FeO(OH)_2 \cdot 6H_2O$. Ferrihydrite is thermodynamically unstable with respect to the more stable crystalline oxides α-$FeO(OH)$ (goethite) and α-Fe_2O_3 (haematite):

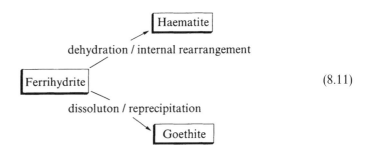

$$(8.11)$$

Formation of ferrihydrite is in fact suppressed when amounts of Fe^{2+} (aq) are present. If the Fe^{II}/Fe^{III} ratio equals 0.5 the product at high pH is magnetite (Fe_3O_4) which is then readily oxidised to maghaemite (γ-Fe_2O_3) at low pH [72]. A possible mechanistic scheme for the growth of Fe_3O_4 has been presented by Livage and coworkers [73]. This involves the condensation of two mixed Fe^{II}–Fe^{III} dimers $[Fe_2(OH)_5(OH_2)_5]$ at right angles to produce a compact tetrameric '$Fe_4O_4(OH)_2(OH_2)_{10}$' critical nucleus having optimised magnetic couplings. After further olation (formation of OH bridges) and oxolation (oxo bridge formation) reactions between the tetramers, electronic delocalisation can occur inside the resulting chains and planes. These chains and planes are then further associated through monomeric $[Fe(OH)_3(OH_2)_3]$ precursors. These octahedral monomers (with no CFSE requirement) are believed to be those eventually occupying the tetrahedral sites in the resulting cubic inverse spinel structure. The concept of a critical tetrameric nucleus, which further condenses with the aid of breakaway monomers, has been suggested [74] for the growth of Fe^{III} polynuclears from kinetic models based upon the TEM data of Murphy and coworkers [69b, 75]. Reference [39a] should be consulted for a recent overview of some the mechanisms surrounding the solid phase transformation of iron oxides and oxyhydroxides.

8.1.2.3 FERRITIN

Storage of iron in mammals occurs within a number of remarkable proteins possessing the capacity to bind large numbers of iron atoms within a small volume, the best known and understood of these being ferritin [76, 77]. Here the iron atoms are bound together in the form of a micellar core consisting of an oxohydroxophosphato polymeric Fe^{III} complex of composition $[(FeO(OH))_8\text{-}(FeO(PO_4)H_2]_n$. The iron content by weight within the micelle can be as high as 57%, each micelle having the capacity to store as many as 4500 iron atoms. Some success at modelling the ferritin core structure, via controlled hydrolysis of Fe^{3+} salts, has been achieved with Mwts up to 150 000 reported for the resulting 'FeO(OH)' polynuclear unit. A red hydrous polymer has been reported, consisting of spheres 2–4 nm in diameter containing approximately 100 Fe^{III} ions [38]. Single crystals of complex molecules containing from 6 up to 17 and 19 Fe atoms have been prepared with some degree of control [78]. This important work has demonstrated the ability of certain chelating ligands to allow the controlled assembly of a polynuclear iron oxyhydroxy core in a manner reminiscent of that occurring within ferritin.

8.1.2.4 WATER EXCHANGE AND SUBSTITUTION OF $[Fe(OH_2)_6]^{3+}$

Despite the common occurrence and importance of $[Fe(OH_2)_6]^{3+}$ it was the last of the air-stable first row hexaaqua ions to have its water-exchange rate determined. This has stemmed largely from the complex nature of the hydrolytic polymerisation process which is frequently unavoidable at workable Fe^{III} concentrations. In particular the ready formation of the hydrolytic dimer, even in fairly strong acidic solution, has proved a persistent problem. One striking feature emerges, however, from all the studies carried out, evidence of a strong inverse acid dependence on the rate indicating an important, and often dominating, contribution from the monohydroxo species, $[Fe(OH_2)_5OH]^{2+}$. The rate constant for water exchange is defined by the two-term expression involving a contribution from both $[Fe(OH_2)_6]^{3+}$ and $[Fe(OH_2)_5OH]^{2+}$ [79, 80]:

$$k_{ex} = k_{Fe^{3+}} + k_{FeOH^{2+}}K_{11}[H^+]^{-1} \tag{8.12}$$

The parameters obtained are listed in Table 8.5.

Figure 8.5 shows how the pressure dependence on the water exchange rate of Fe^{3+}(aq) changes from a pressure-accelerated to a pressure-deccelerated process in the $[H^+]$ region where first $[Fe(OH_2)_6]^{3+}$ ($\Delta V_{ex}^{\ddagger} = -5.4\,cm^3\,mol^{-1}$) and then $[Fe(OH_2)_5OH]^{2+}$ ($\Delta V_{ex}^{\ddagger} = +7.0\,cm^3\,mol^{-1}$) are the dominating species present. One requires a detailed examination of kinetic data regarding a range of incoming ligands in order to verify whether this feature is indeed reflective of a changeover in the activation process from one involving an importance of bond making ($[Fe(OH_2)_6]^{3+}$) to one involving predominantly bond breaking

Table. 8.5. Rate constants and activation parameters for water exchange on $[Fe(OH_2)_6]^{3+}$ and $[Fe(OH_2)_5OH]^{2+}$ [79, 80][a]

k		$\Delta H_{ex}^{\ddagger}$ (kJ mol^{-1})	$\Delta S_{ex}^{\ddagger}$ (J K^{-1} mol^{-1})	$\Delta V_{ex}^{\ddagger}$ (cm^3 mol^{-1})
$k_{Fe^{3+}}$	$(1.6 \pm 0.2) \times 10^2$	63.95 ± 2.51	$+12.12 \pm 6.69$	-5.4 ± 0.4
$k_{FeOH^{2+}}$	$(1.2 \pm 0.1) \times 10^5$	42.39 ± 1.46	$+5.27 \pm 3.97$	$+7.0 \pm 0.3$

[a] The temperature-dependent study was carried out at variable ionic strength in the range 0.38–5.4 m [79] and the pressure-dependent study at ionic strength 6.0 m [80]

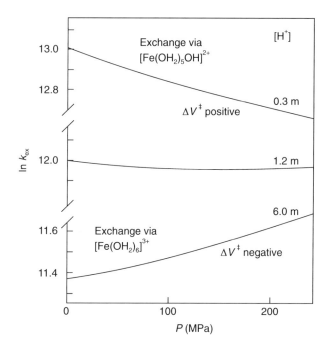

Figure 8.5. Acidity dependence of the plot of In k_{ex} vs pressure (MPa) for the water-exchange process on Fe^{3+}(aq)

$([Fe(OH_2)_5OH]^{2+})$. Unfortunately many of the potentially desirable ligand systems are those having ionizable protons themselves leading to the problem of proton ambiguity, a topic that has been reviewed and discussed by Jordan and colleagues with particular reference to substitution reactions on Fe^{3+} (aq) [79, 81]. Selected rate constants for a range of complexation reactions on both $[Fe(OH_2)_6]^{3+}$ and $[Fe(OH_2)_5OH]^{2+}$ have been compiled in Table 8.6, including

Table. 8.6. A selection of rate constants (25 °C) for substitution reactions on both $[Fe(OH_2)_6]^{3+}$ and $[Fe(OH_2)_5OH]^{2+}$

Incoming ligands	Method	$[Fe(OH_2)_6]^{3+}$ $10^{-1}k_f(M^{-1}s^{-1})$	$[Fe(OH_2)_5OH]^{2+}$ $10^{-4}k_f(M^{-1}s^{-1})$	Ref.
H_2O	^{17}O NMR	$16\,(s^{-1})$	$12\,(s^{-1})$	[79, 80]
Cl^-	T-J	0.48	0.55	[82]
	S-F	0.94		[83]
Br^-	T-J	0.16	0.28	[84]
NCS^-	S-F	9.0	0.50	[85]
	S-F	9.5^a	0.71^a	[86]
N_3^-	S-F	Not detected	38^a	[86]
HN_3	S-F	Not detected	0.50^a	[86]
$CCl_3CO_2^-$		6.3	0.78	[87]
$CHCl_2CO_2^-$		11.8	1.9	[87]
$CH_2ClCO_2^-$		150	4.1	[87]
CH_3CO_2H		2.7	$\leqslant 0.28$	[88]
C_6H_5OH		Not observed	0.15	[89]
$C_{10}H_{12}(=O)(OH)^b$	S-F	2.2	0.63	[90]
$CH_3C(O)NH(OH)^c$	S-F	0.48	0.57	[91]
		0.64	0.49	[92]
Desferrioxamine	S-F	0.38	0.36	[91]
		0.20	0.36	[93]

^{a} Units of molality (m).
^{b} 4-Isopropyltropolone.
^{c} Acetohydroxamic acid.

those for ionizable protic ligands where the proton ambiguity problem seems largely to have been combated. Table 8.7 also lists volumes of activation resulting from a selection of pressure-dependent kinetic studies. It may again surprise the reader to learn that most of the data in Table 8.6 has only been gathered since the mid 1970s.

A considerably larger spread of rate constants is relevant from Table 8.6 for reactions on $[Fe(OH_2)_6]^{3+}$ than for $[Fe(OH_2)_5OH]^{2+}$, in keeping with the conclusions from the pressure dependency of the water exchange. With the exception of N_3^- the rate constants for reactions on $[Fe(OH_2)_5OH]^{2+}$ fall in a narrow range around $0.5 \pm 0.3 \times 10^4$ $M^{-1}s^{-1}$, whereas those for $[Fe(OH_2)_6]^{3+}$ cover more than three orders of magnitude. These conclusions are largely borne out by the activation volumes in Table 8.7 which are negative for reactions on $[Fe(OH_2)_6]^{3+}$ but are positive for reactions on $[Fe(OH_2)_5OH]^{2+}$. In only one case (a temperature jump study of the NCS^- reaction [94]) was a positive activation volume found for substitution on $[Fe(OH_2)_6]^{3+}$, a result not apparently in keeping with the negative values found from two other independent stopped-flow kinetic studies. The behaviours observed for $[Fe(OH_2)_6]^{3+}$ and $[Fe(OH_2)_5OH]^{2+}$ offer further

Table. 8.7. Activation volumes for selected substitution reactions on $[Fe(OH_2)_6]^{3+}$ and $[Fe(OH_2)_5OH]^{2+}$

Incoming ligands	$\Delta V^{\ddagger}_{Fe^{3+}}$ (cm³ mol⁻¹)	$\Delta V^{\ddagger}_{FeOH^{2+}}$ (cm³ mol⁻¹)	I (m)	Ref.
H_2O	-5.4 ± 0.4	$+7.0 \pm 0.3$	6.0	[80]
NCS^-	-5.7 ± 0.3	$+9.0 \pm 0.4$	1.0	[86]
	-6.1 ± 1	$+8.5 \pm 1.2$	1.5	[92]
	$+4.3 \pm 0.6$	$+15.6 \pm 1.3$	1.0	[94]
		$(+5.4 - +16.5)$	0.1–1.5	[94]
N_3^-		$+12.9 \pm 1.5$	1.0	[86]
HN_3	Not observed	$+6.8 \pm 0.5$	1.0	[86]
Cl^-	-4.5 ± 1.1	$+7.8 \pm 1.0$	1.5	[82]
Br^-	-8 ± 4	—	2.0	[84]
$C_{10}H_{12}(=O)(OH)$	-8.7 ± 0.8	$+4.1 \pm 0.6$	1.0	[92]
$CH_3C(O)NH(OH)$	-10.0 ± 1.4	$+7.7 \pm 0.6$	1.0	[92]
	-6.3 ± 1.4	$+5.2 \pm 0.5$	2.0	[91]
Desferrioxamine	-4.7 ± 1.6	$+4.3 \pm 0.5$	2.0	[91]

support for the general conclusions reached in Section 1.5 of a greater participation from the entering ligands regarding the pathway to the transition state for M^{3+}(aq) ions. For dipositive ions (M^{2+} (aq) and MOH^{2+} (aq)) the mechanism is largely controlled by water dissociation. As is the case for a number of other monohydroxo species, the greater lability in the $[Fe(OH_2)_5OH]^{2+}$ ion probably stems not from the reduction in cationic charge but from promotion of the leaving ligand dissociation process through π-donation from the OH group (the conjugate base effect).

8.1.2.5 $[Mo_3FeS_4(OH_2)_{10}]^{4+}$

One of the now extensive series of mixed Mo—M cubane sulfur-bridged clusters (see Section 6.2.3.4) the Fe derivative (Fig. 8.6), can be prepared either by treatment of acidic (2.0 M $HClO_4$ or HCl) solutions of the $[Mo_3S_4(OH_2)_9]^{4+}$ ion with iron wire or by treatment of $[Mo_3S_4(OH_2)_9]^{4+}$ with Fe^{2+} (aq) in the presence of a reducing agent (BH_4^-) [95]. Electrochemical reduction of solutions of $[Mo_3S_4(OH_2)_9]^{4+}$ with Fe^{2+}(aq) at a carbon cloth cathode can also be used [96]. In each case purification is achieved by Dowex 50W X2 cation-exchange chromotography. The 4+ cube is found to elute with 2.0 M H^+ ahead of the $Mo_3S_4^{4+}$ (aq) ion. Mossbauer measurements indicate a spin-coupled Fe^{III} centre. Water ligand substitution at the Fe centre is the only significant process and is too fast to be followed by stopped flow. The estimated second-order rate constant (25°C) for Cl^- substitution ($> 2 \times 10^4$ $M^{-1}s^{-1}$) may be compared to typical

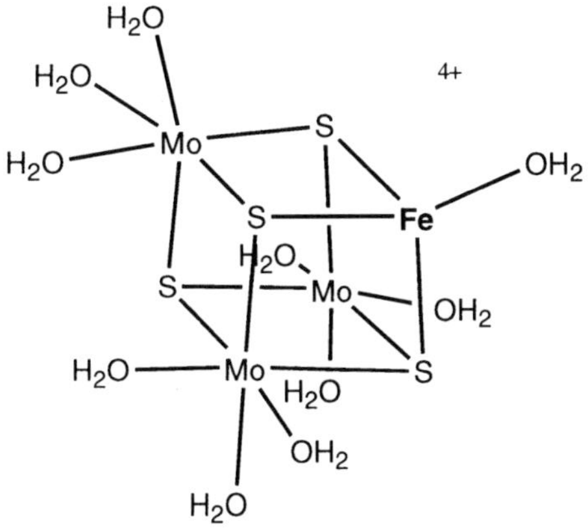

Figure 8.6. The $[Mo_3FeS_4(OH_2)_{10}]^{4+}$ aqua ion

values for substitution on Fe^{2+} (aq) (Table 8.1) and Fe^{3+} (aq) (Table 8.6). Complexation of Cl^- at the Fe centre in the cube is favoured ($\beta_{11} = 560 \, M^{-1}$) when compared with 1:1 equilibration of Cl^- on Fe^{3+} ($\beta_{11} = 3–5 \, M^{-1}$) and Fe^{2+} (aq) ($\beta_{11} \sim 0.5 \, M^{-1}$).

8.1.2.6 COMPLEXATION OF FE^{3+}(AQ) RELEVANT TO IRON TRANSPORT AND STORAGE

In humans and other higher mammals iron is transported as Fe^{III} within the glycoprotein transferrin [76, 97, 98]. While extremely high formation constants are relevant to the specific iron-binding sites (two are believed to be involved), the release of iron is thought to be triggered by the additional binding of anions (possibly phosphate or pyrophosphate) to the protein. In microbes iron is transported within so-called siderophores, relatively small Mwt (< 3000) molecules containing highly specific and powerful cavity binding sites for Fe^{III}, usually catecholates or hydroxamates. Enterobactin, a cyclic tripeptide ester of 1,2-dihydroxybenzoyl-l-serine, possesses the highest formation constants (log $\beta_{11} \sim 52$) of any known Fe^{III} chelating agent. Ferrichromes provide the alternative cyclic hydroxamate mode of binding (log $\beta_{11} \sim 30$). Within these siderophores the mechanism of iron release is less well understood. Reduction to more labile Fe^{II} may be involved or alternatively some change to the nature of the binding site, activated by an interaction with a cellular component. Further studies are required. Reduction of free Fe^{3+} (aq) by both l-cysteine [99] (giving the disulfide, l-cystine) and l-ascorbic acid [100] has been the subject of detailed kinetic studies. In the latter, as expected, both Fe^{3+} (aq) and $FeOH^{2+}$ (aq) take part in the reaction. A significant acceleration is seen in the presence of Cl^- ions, attributed to the higher redox activity of $FeCl^{2+}$ (aq) and $Fe(OH)Cl^+$ (aq) [100].

Consistent with the general redox behaviour of the $Fe^{3+/2+}$ (aq) and $FeOH^{2+/+}$ (aq) couples, an outer-sphere reaction was indicated for the Fe^{3+} (aq) species whereas reaction of $FeOH^{2+}$ (aq) proceeds by an inner-sphere path.

There are no stable aqua ions or oxo anions characterising either Fe^{IV} or Fe^V in aqueous media. Concievably transient Fe^{IV} (aq) and Fe^V (aq) species may be formed in the oxidation of alkaline Fe^{III} (aq) to $[FeO_4]^{2-}$ (below) and also during the reduction of $[FeO_4]^{2-}$, but none have been characterised. Further work in this area should be encouraged, however, given the recent developments in Cr^{IV} (aq) chemistry (Section 6.1.3). On the other hand, there seems to be mounting evidence in favour of the existence of reactive $Fe^{IV}=O$ and $Fe^V=O$ intermediate species in the catalytic cycles of iron-containing enzymes such as methane monooxygenase, cytocrome P450, catalase and horseradish peroxidase.

8.1.3   Iron(VI) (d^2): Chemistry of the tetrahedral $[FeO_4]^{2-}$ Ion

Tetrahedral $[FeO_4]^{2-}$, existing only in very strongly alkaline solutions, represents the highest valent stable compound of aqueous iron. Preparations have involved electrolysis of concentrated aqueous alkali using an iron anode [101] as well as KOCl oxidation of aqueous alkaline Fe^{III} [102]. In each case the Fe^{VI} (aq) is rapidly precipitated as $K_2[FeO_4]$ from the cold solutions by the addition of further amounts of KOH. Rapid isolation of the K^+ salt is essential to avoid excessive decomposition . The spectrum of the $[FeO_4]^{2-}$ ion in 9.0 M KOH is shown in Fig. 8.7 [103] and is consistent with the tetrahedral geometry present in

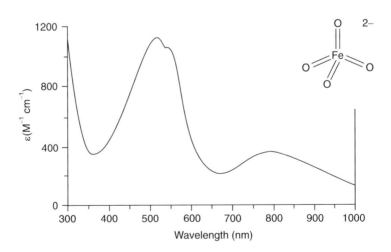

Figure 8.7. Electronic spectrum of $[FeO_4]^{2-}$ in aqueous 9.0 M KOH

the solid salts [104]. The low energy band at 794 nm ($\varepsilon = 400$ $M^{-1}cm^{-1}$) has been assigned to the $^3T_2 - ^3A_2$ transition of the tetrahedral d^2 ion. Magnetic moments are in the range 2.8–3.2 BM [102b, 105] consistent with two unpaired electrons. The aqueous stability of Fe^{VI}(aq) presumably arises from the stable tetrahedral geometry of the oxo anion.

The exchange of the $[FeO_4]^{2-}$ anion with solvent water has been investigated in strongly alkaline solution using ^{18}O labelling [106]. At 25°C, $k_{ex} = 1.6 \times 10^{-2} s^{-1}$, $\Delta H^{\ddagger} = 62$ kJ mol^{-1} and $\Delta S^{\ddagger} = -71.2$ J K^{-1} mol^{-1}. No dependence on the concentrations of OH^- ions or lower oxidation state species was found, the latter being confirmed by the lack of a retardation effect upon the addition of OCl^-.

Decomposition of $[FeO_4]^{2-}$ upon lowering the pH occurs rapidly with evolution of O_2 according to

$$2[FeO_4]^{2-} + 10H^+ \xrightarrow{H_2O} 2Fe^{3+}(aq) + 5H_2O + \tfrac{3}{2}O_2 \qquad (8.13)$$

The reduction potential for the half-reaction

$$[FeO_4]^{2-} + 8H^+ + 5e^- \longrightarrow Fe^{3+}(aq) + 4H_2O \qquad (8.14)$$

is $\sim +2.0$ V, making it a stronger oxidant than $[MnO_4]^-$. As a result it has found use as a powerful organic oxidant in aqueous alkaline solution.

8.2 RUTHENIUM

Next to vanadium and molybdenum, ruthenium has the richest aqueous chemistry of cationic aqua species with oxidation states $+2$, $+3$ and $+4$ all represented. In addition three oxo anions, formally $[RuO_4]^{0,1-,2-}$, are known with varying stability in aqueous solution. Only Ru^V is not represented. Like osmium (see later), ruthenium forms a highly oxidising and toxic tetraoxide which is a gas at RT. As such it is normally handled following *in situ* generation in solution from lower valent Ru salts. It has been used extensively as a powerful oxidising reagent in organic synthesis. RuO_4 is also the lead-in to the aqueous chemistry of the $+2$, $+3$ and $+4$ states via controlled chemical reduction in non-complexing acidic media. There are no representative aqua ions below the $+2$ state.

Commercial ruthenium 'trichloride', '$RuCl_3 \cdot nH_2O$', is marketed as a lead-in compound to ruthenium chemistry although its composition varies enormously from sample to sample. Despite the formula quoted commercial '$RuCl_3$', which is normally prepared via reduction of RuO_4 with HCl, can contain up to 50% Ru^{IV} chloro species [107] and some reports quote as much as 80% Ru^{IV} [108]. Some samples have also been reported to contain nitrogenous material [109]. Although '$RuCl_3$' has been used for a variety of syntheses a more definable lead-in compound is probably desirable. Useful alternative lead-ins have proved to be $Ru^{II}Cl_2(Me_2SO)_4$ [110] and $[Ru(OH_2)_6]$ (pts)$_2$ (see below). The aqueous chemistry of ruthenium receives comprehensive coverage in the book by Seddon and Seddon [111] and in separate review articles by Seddon [109] and Rard [112].

8.2.1 Ruthenium(II) (d^6): Chemistry of the $[Ru(OH_2)_6]^{2+}$ Ion

8.2.1.1 SYNTHESIS AND SPECTROSCOPIC PROPERTIES

Ludi and others have shown that crystalline $[Ru(OH_2)_6]$ (pts)$_2$ is a versatile alternative lead-in compound to ruthenium complex chemistry. The solid compound is less air sensitive than solutions of Ru^{2+} (aq) and can be stored for months when protected from air by an atmosphere of argon. The preparation involves firstly the generation of RuO_4 followed by its reduction to the $+2$ state in an appropriate non-complexing acidic media. In the first preparations by Kallen and Earley $K_2[RuCl_5(OH_2)]$ or RuI_3 were digested with concentrated H_2SO_4 (to remove halide ions) and then oxidised with $S_2O_8{}^{2-}$ using Ag^I as catalyst (see Chapter 11) [113]. $MnO_4{}^-$, $IO_4{}^-$ and ClO^- have also been used as chemical oxidants for the generation of RuO_4 from lower oxidation state salts. The volatile RuO_4 can then be swept out using a stream of nitrogen or argon gas and, if required, isolated in an dry ice–ethanol cold trap. The RuO_4 is then dissolved in 2.0 M HBF_4 and reduced to the $+2$ state using tin metal. Air-free cation-exchange chromotography is then used to separate dissolved Sn^{II} ions, e.g. $Sn_3(OH)_4{}^{2+}$ (aq) from Ru^{2+} (aq) by washing with 0.1 M HBF_4. Elution with 1.0 M HBF_4 gives solutions of $[Ru(OH_2)_6]^{2+}$. Complete removal of Sn^{II} (aq) is a problem and in two reported alternative procedures Ludi and coworkers have employed the use of metallic lead as reductant with Pb^{II} (aq) ions removed with H_2SO_4 prior to the cation-exchange treatment [114]. Contamination with Cl^- ions is also a problem and to avoid this oxidative routes to RuO_4 employing either Ru metal [114] or RuO_2 [114, 115] are preferred. In a procedure employing RuO_2, $RuCl_3 \cdot nH_2O$ (7.3 g) was heated in 300 cm^3 of 5.0 M NaOH at $60°C$ for 30 minutes and then stirred at room temperature overnight. The fine black precipitate of $RuO_2 \cdot H_2O$ was washed with water until free of Cl^- and dried for the oxidation to the tetraoxide. A typical set-up used for the generation of $[Ru(OH_2)_6]^{2+}$ is shown in Fig. 8.8. The glassware should be scrupulously clean. In this case oxidation of RuO_2 to RuO_4 is carried out using $NaIO_4$ and the supporting acid is H_2SiF_6. Flask A contains 5.0 g of $RuO_2 \cdot H_2O$ and 25 g of $NaIO_4$ in 200 cm^3 of H_2O, B contains 35 g of activated lead grains in 1.0 M H_2SiF_6, C contains 100 cm^3 of ice-cold 50% H_2SO_4 and D contains 5.0 g of activated lead in 70 cm^3 of 1.0 M H_2SiF_6. The RuO_4 is generated in flask A and swept across to flask B using a stream of argon gas, helped by the addition of the H_2SO_4 over a one hour period, where it is reduced to Ru^{2+} (aq) by the metallic lead. Any RuO_4 escaping flask B is reduced by passing through flask D. Flask A and B are continuously stirred and

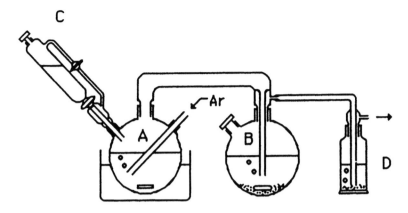

Figure 8.8. Laboratory apparatus for the preparation of $[Ru(OH_2)_6]^{2+}$ from $RuO_2 \cdot H_2O$ [115]

their distance should be as short as possible. The formation of $[Ru(OH_2)_6]^{2+}$ in flask B is indicated by the appearance of a deep purple-red colour. The Pb^{II} ions are removed by decanting the contents of flask B into 100 cm^3 of Ar-saturated 2.0 M H_2SO_4, and filtering (G4 frit) of the $PbSO_4$ precipitate prior to air-free dilution with water (to 4.0 dm^3) and cation-exchange purification (30 × 3 cm column of Dowex 50W X8 resin). Following washing of the loaded column with water (300 cm^3) and 0.1 M Hpts (700 cm^3) the $[Ru(OH_2)_6]^{2+}$ ion is eluted with 1.8 M Hpts. Concentration of the red solution (400 cm^3) by one half and cooling to 0 °C produced a red crystalline mass of $[Ru(OH_2)_6](pts)_2$. Following filtration, washing with ethyl acetate (removal of Hpts) and then diethyl ether, the dried yield was 12.98 g. Samples gave satisfactory elemental analysis. $[Ru(OH_2)_6]^{2+}$ can be assayed by recording the u.v.–visible spectrum in 0.5 M CF_3SO_3H under air-free conditions. The two-band maxima at 534 nm ($\varepsilon = 12\,M^{-1}\,cm^{-1}$) and 394 nm (15 $M^{-1}\,cm^{-1}$) [115] are respectively assigned to the $^1A_{1g}-^1T_{1g}$ and $^1A_{1g}-^1T_{2g}$ transitions of the low-spin t_{2g}^6 ion. The $CF_3SO_3^-$ salt, which is more soluble in alcoholic solvents and THF can be similarly made via cation-exchange elution with CF_3SO_3H, rotary evaporation to constant volume and cooling to −20 °C for two days. The resulting red precipitate is less stable than the pts$^-$ salt and is best used within a few days of preparation. The Tutton-type salts, $M^I_2Ru(SO_4)_2 \cdot 6H_2O$ (M = K, Rb) have been prepared and subjected to normal coordinate analysis of their vibrational spectra [116]. The results are discussed along with those for $CsRu(SO_4)_2 \cdot 12H_2O$ in Section 8.2.2. The X-ray structure of $[Ru(OH_2)_6](pts)_2$ is shown in Fig. 8.9 [114]. The average Ru—O bond length is 212.2 pm, which is significantly longer than the average distance in $[Ru(OH_2)_6]^{3+}$ (202.9 pm) to which it is readily oxidised (see later). $[Ru(OH_2)_6](pts)_2$ is a versatile

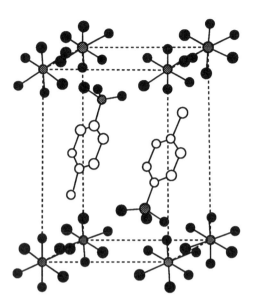

Figure 8.9. X-ray structure of $[Ru(OH_2)_6](p\text{-}CH_3C_6H_4SO_3)_2\cdot H_2O$

lead-in to a wide range of ruthenium coordination and organometallic compounds (Fig. 8.10) [117]. Many others can doubtless be synthesised. The water ligands on $[Ru(OH_2)_6]^{2+}$ are readily replaced by π-acid ligands such as nitrogen heterocycles [117], phosphines [118], CO [119], nitriles [120] and alkenes [121]. The ready formation of a stable N_2 complex [122] in aqueous media reflects the particularly strong π-donor property of Ru^{2+} arising from its low charge and filled t_{2g} orbital set. Formation of the N_2 complex in solution has been followed by ^{17}O and ^{15}N NMR spectroscopy in a sapphire NMR tube supporting pressures of up to 10 MPa. Under a 5 MPa pressure of N_2 the -196.3 ppm resonance of coordinated H_2O on $[Ru(OH_2)_6]^{2+}$ (0.03 m) was replaced by two signals at -165.2 and -86.2 ppm assigned on the basis of their respective integration near to 4:1 to the now distinguishable equatorial and axial H_2O ligands of $[Ru(OH_2)_5N_2]^{2+}$. At the same time, the ^{15}N NMR spectrum was characterised by the growth of two doublets of equal intensity at -82.9 (N adjacent to Ru) and -24.3 ppm (distal N) relative to free N_2 at -71.5 ppm. Use of higher Ru^{2+} concentrations (~ 0.5 m) led to the appearance of a further signal at -98.4 ppm assignable to the axial waters of the dimeric μ-N_2 ion, $[(H_2O)_5Ru(\mu\text{-}N_2)Ru(OH_2)_5]^{4+}$:

$$[Ru(OH_2)_5N_2]^{2+} + [Ru(OH_2)_6]^{2+} \longrightarrow [(H_2O)_5Ru(\mu\text{-}N_2)Ru(OH_2)_5]^{4+} + H_2O$$
$$(8.15)$$

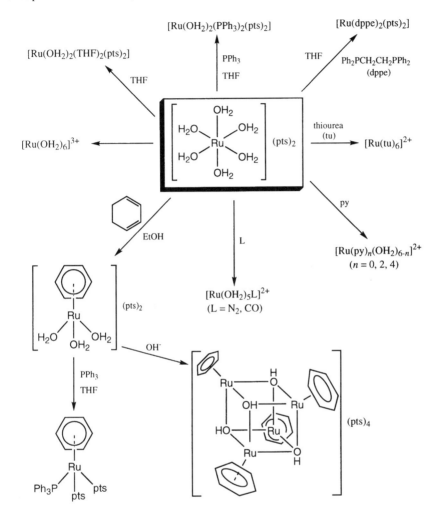

Figure 8.10. Synthetic routes to ruthenium compounds using $[Ru(OH_2)_6]^{2+}$

previously reported by Creutz and Taube [123]. In the ^{15}N NMR the symmetrical μ-N_2 group of the dimer appears as a singlet at -82.4 ppm.

The formation of yellow $[Ru(OH_2)_5(CO)]^{2+}$ was similarly followed under a 5 MPa pressure of CO. Here the axial and equatorial waters occurred at -28.3 and -154.8 ppm, the greater downfield shift in the axial water reflecting stronger π donation to the CO ligand [119].

8.2.1.2   COMPLEX FORMATION AND WATER EXCHANGE ON $[Ru(OH_2)_6]^{2+}$

Merbach and coworkers have determined the rate constant for water exchange on $[Ru(OH_2)_6]^{2+}$ from a ^{17}O NMR study of the $[Ru(OH_2)_6](CF_3SO_3)_2$ salt in solution [124]. The rate was slow enough to permit conventional monitoring using a fast injection method. The growth in the height of the bound H_2O singlet at -196.3 ppm from bulk water (0 ppm) was monitored. A variable temperature and pressure study in $I = 2.0\,m\,(CF_3SO_3{}^-)$ gave rise to the following parameters: $k_{ex}(25\,°C) = (1.8 \pm 0.2) \times 10^{-2}\,s^{-1}, \Delta H^{\ddagger} = 88 \pm 4\,kJ\,mol^{-1}, \Delta S^{\ddagger} = +16 \pm 15\,J\,K^{-1}\,mol^{-1}$ and $\Delta V^{\ddagger} = -0.4 \pm 0.7\,cm^3\,mol^{-1}$. No $[H^+]$ dependence was detected. Rate constants for Cl^-, Br^-, and I^- complexation on $[Ru(OH_2)_6]^{2+}$ were measured by Kallen and Earley in their earlier kinetic study of the catalysis, by $[Ru(OH_2)_6]^{2+}$, of ligand substitution on $[Ru(OH_2)_6]^{3+}$ [113]. Since then a whole range of ligands have been studied regarding the kinetics of 1:1 complex formation reactions on $[Ru(OH_2)_6]^{2+}$ [120]. Selected data are listed in Table 8.8.

Complex formation with Cl^-, Br^-, I^-, $HC_2O_4{}^-$ and N-methylpyrazinium ion (NMP^+) was followed by u.v.–visible spectrophotometry. Complex formation with MeCN, DMSO, 1.4-thioxane, tetrahydrothiophen (THT), maleic and fumaric acids and 2,5-dihydrofuran (DHF) was followed by ^1H NMR. Despite

Table. 8.8.   Kinetic data for monocomplex formation reactions of $[Ru(OH_2)_6]^{2+}$ [14]

Incoming ligand	$10^3 k_f^{298}$ $(M^{-1}s^{-1})$	$\Delta H^{\ddagger}$ $(kJ\,mol^{-1})$	$\Delta S^{\ddagger}$ $(J\,K^{-1}mol^{-1})$	$K_{os}{}^a$ (m^{-1})	$10^3 k_I^{298\,b}$ (s^{-1})	I (m)	Ref.
Cl^-	8.5^c	84.4	-2	1	9	0.3	[113]
Br^-	10.2^c	82.8	-5	1	10	0.3	[113]
I^-	9.8^c	81.5	-10	1	10	0.3	[113]
$HC_2O_4{}^-$	26.0^c	101.1	$+65$	2	13	1.0	[125]
H_2O		88	$+9$		18^d	1.0	[124]
$MeCN^e$	2.07	81.1	-24	0.16	13	0.2–0.4	[120]
$DMSO^e$	1.31	87.3	-7	0.16	8	0.4	[120]
1,4-thioxane^e	2.2	82.4	-2	0.16	14	0.4	[120]
THT^e	2.4^f			0.16	15	0.4	[120]
Maleic acid^e	2.18	84	-15	0.16	11	0.8	[126]
Fumaric acid^e	1.72	128	$+133$	0.16	11	0.8	[126]
2,5-DHF^e	1.06	126	$+120$	0.16	7	1.0	[126]
NMP^+	0.73	77.7	-77.9	0.02	36	0.1–0.2	[120]

^{a} Calculated using the Eigen–Fuoss relationship for the reaction of $[Ni(OH_2)_6]^{2+}$ with the same ligands (see Chapter 10).
^{b} Calculated with the relationship $k_f^{298} = K_{os}k_I^{298}$.
^{c} $M^{-1}\,s^{-1}$.
^{d} Rate constant for exchange of a particular water molecule as obtained from ^{17}O isotopic labelling NMR kinetic experiments.
^{e} In D_2O.
^{f} Estimation from the value at 302 K, $k_f = 3.79\,m^{-1}s^{-1}$ with $\Delta H_f^{\ddagger} = 82.4$ (cf. 1,4-thioxane).

the value of $\Delta V^{\ddagger}_{ex}$ close to zero the kinetic data provide compelling clear evidence for a dissociative mechanism. Thus the variation in observed k_f values can be adequately accounted for by assuming the Eigen–Wilkins ion pair model, $k_f = K_{os}k_I$, with values for K_{os} estimated from the Eigen–Fuoss relationship as used on $[Ni(OH_2)_6]^{2+}$. Ni^{2+} and Ru^{2+} have similar radii and it is felt that these K_{os} values are probably reliable within a factor of 2–4. This leads to a virtually constant value of the interchange constant, k_I, within a factor of 2 of the rate constant for water exchange on $[Ru(OH_2)_6]^{2+}$. There is also a strong correlation of $\Delta H^{\ddagger}$ with $\Delta S^{\ddagger}$ for a wide range of incoming ligands (Fig. 8.11). This is another example of the limitations apparent in the use of the activation volume for a direct diagnosis of the mechanism. The molar volume contraction between six- and five-coordinate complexes on forming the transition state can become quite significant, as has proved to be the case in estimates of the molar volume difference $V_c([Co(NH_3)_6]^{3+}) - V_c([Co(NH_3)_5]^{3+})$ (17–20 cm³ mol⁻¹) [127, 128]. Since the ion pair I_d mechanism is probably relevant for reactions on $[Ru(OH_2)_6]^{2+}$ a molar volume contraction of presumably ~ 15 cm³ mol⁻¹ for the process $\{[Ru(OH_2)_6]^{2+},L\}-\{[Ru(OH_2)_5]^{2+},L\}$ is balanced exactly by the gain in partial molar volume ~ 15 cm³ mol⁻¹ of the partially excluded electros-

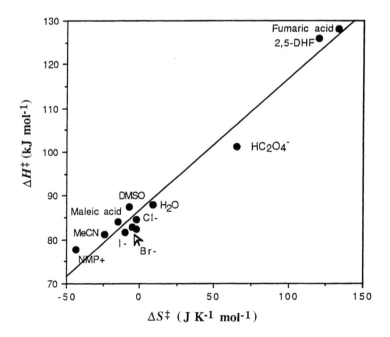

Figure 8.11. Correlation of $\Delta H^{\ddagger}$ with $\Delta S^{\ddagger}$ for 1:1 complexation reactions on $[Ru(OH_2)_6]^{2+}$

tricted H_2O molecule in the transition state, leading to the observed $\Delta V_{ex}^{\ddagger}$ of ~ 0. A similar situation may be relevant for interchange reactions on the t_{2g}^6 ion $[Rh(OH_2)_6]^{3+}$ (see Chapter 9).

$[Ru(OH_2)_6]^{2+}$ is an active catalyst for a number of alkene isomerisations, the active intermediates probably being π-alkene and π-allyl complexes. Certain allyl ethers and alcohols are quantitatively converted to the corresponding vinyl compounds at room temperature after 12 hours in the presence of aqueous ethanolic $[Ru(OH_2)_6]^{2+}$ [129]. Importantly, the major products are selectively the E isomers. Experiments conducted on the isomerisation of hexenes indicate the stepwise process hexe-1-ene $\rightarrow (E)$-hexe-2-ene $\rightarrow (E)$-hexe-3-ene. In experiments conducted with 4-allyl-2-methoxyphenol (giving 2-methoxy-4-propenyl-phenol as product) the derived kinetic parameters ($k = (3.8 \pm 0.1) \times 10^{-5} \, s^{-1}$, $\Delta H^{\ddagger} = 44 \, kJ \, mol^{-1}$, $\Delta S^{\ddagger} = J \, K^{-1} \, mol^{-1}$) are noted as being significantly removed from those characterising water exchange and here the monitoring of a different process is probably relevant. A possible candidate is the formation of an intermediate hydrido-allyl complex following the relatively rapid initial formation of an alkene complex. $[Ru(OH_2)_6]^{2+}$ will also catalyse certain alkene polymerisation reactions [130]. In the catalytic dimerisation of ethene, the reactive intermediates $[Ru(CH_2{=}CH_2)(OH_2)_5](pts)_2$ and $[Ru(CH_2{=}CH_2)_2(OH_2)_4](pts)_2$ (Fig. 8.12) have been individually characterised in solution by ^{17}O NMR spectroscopy [121]. Both but-1-ene and but-2-ene products are formed with once again a selectivity shown towards the E isomer.

The incorporation of organo π-acid ligands into the hexaaqua Ru^{II} coordination sphere is found to markedly labilise the remaining water ligands. Structural studies on a number of these systems are furthermore providing insights into the electronic origin of the apparent labilisation. For example, reaction of $[Ru(OH_2)_6]^{2+}$ with 1,4-cyclohexadiene (or the 1,3 isomer) in ethanol gives the π-arene 'piano stool' complex $[Ru(\pi\text{-}C_6H_6)(OH_2)_3]^{2+}$ (Fig. 8.13) [131]. The reaction stoichiometry:

$$3[Ru(OH_2)_6]^{2+} + 5\langle\!\!\bigcirc\!\!\rangle \longrightarrow 3[Ru(\pi\text{-}C_6H_6)(OH_2)_3]^{2+} + \bigcirc + \bigcirc \qquad (8.16)$$

seems to be involved, consisting of an Ru^{2+}-promoted disproportionation of the cyclohexadiene. No reaction of $[Ru(OH_2)_6]^{2+}$ is observed directly with benzene. The first order rate constant for the appearance of the single aromatic proton of the $\pi\text{-}C_6H_6$ ring using a twenty fold excess of cyclohexadiene was $4.6 \times 10^{-4} \, s^{-1}$ for reaction in THF at 303 K.

Use of sulphate as the counter-ion gave crystals suitable for X-ray investigation. The Ru—O distance to the three H_2O ligands was 211.7 pm, close to that observed in $[Ru(OH_2)_6]^{2+}$. However, the presence of the $\pi\text{-}C_6H_6$ ligand markedly labilises the water ligands vs the hexaaqua ion such that ^{17}O line broadening

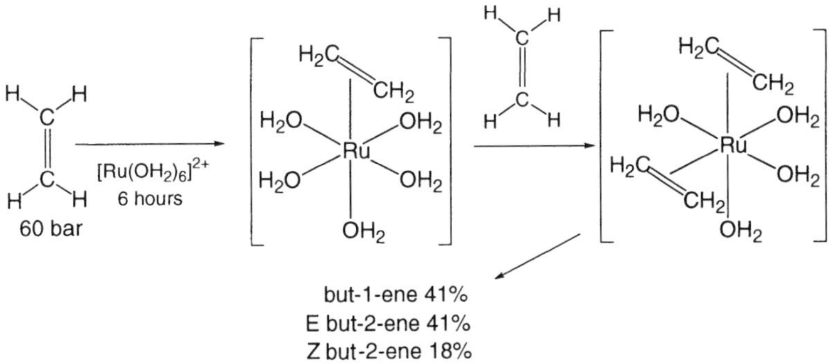

but-1-ene 41%
E but-2-ene 41%
Z but-2-ene 18%

Figure 8.12. Reactive intermediates in the $[Ru(OH_2)_6]^{2+}$ catalysis of the dimerisation of ethene

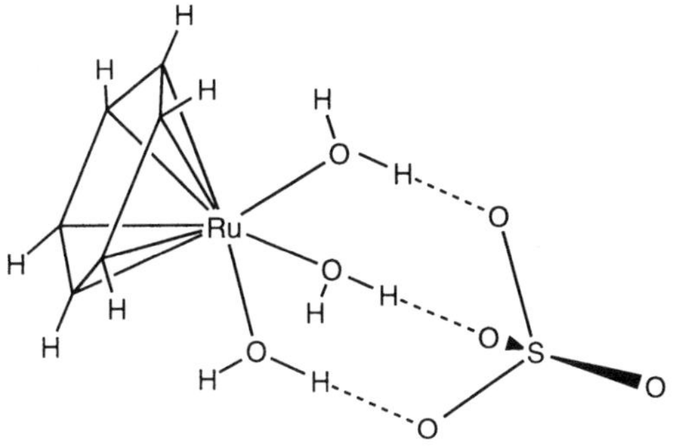

Figure 8.13. Structure of $[Ru[\pi-C_6H_6)(OH_2)_3](SO_4)$ showing the H-bonding to the SO_2^{2-} ion

NMR measurements are now required. At $25°C$, $k_{ex} = 11.5 \pm 3.1$ s^{-1}, $\Delta H^{\ddagger} = 75.9 \pm 3.8$ kJ mol^{-1}, $\Delta S^{\ddagger} = +29.9 \pm 10.6$ J K^{-1} mol^{-1} and $\Delta V^{\ddagger} = +1.5 \pm 0.4$ cm^3 mol^{-1}. In this case, therefore, the strong kinetic *trans* labilising effect (factor of $\sim 10^3$) of the π-benzene moiety is not manifest in a significant lengthening of the Ru—O bonds. This is not the case, however, within the π-1,5-cyclooctadiene complex $[Ru^{II}(\pi-COD)(OH_2)_4]^{2+}$. The structure of $[Ru^{II}(\pi-COD)(OH_2)_4]$ (pts)$_2$ (Fig. 8.14) reveals a significant lengthening in the Ru—OH$_2$ bonds *trans* to the double bonds, which shows up in a significant labilising at these water sites towards water exchange and ligand substitution reactions [132]. Thus, the reversible formation of $[Ru(\pi-COD)(OH_2)_3(CO)]^{2+}$ involves CO substitution at one of the labile equatorial waters. Also of interest is that the axial

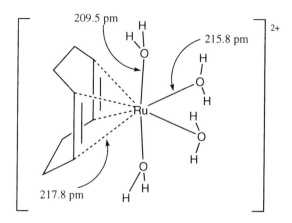

Figure 8.14. Structure of $[Ru^{II}(\pi\text{-COD})(OH_2)_4](pts)_2$

waters appear to be orientated so as to facilitate some degree of π-donation from O to Ru^{II} as opposed to the labile equatorial waters which are more tetrahedral, reflecting only σ donation to Ru^{II}. Thus it appears that removal of the electron density on Ru^{II} towards the π-acid COD in this complex results in more π-acidity at Ru^{II} in the *cis* direction as opposed to the *trans* direction. Thus both *cis* and *trans* compensating effects may be involved in the *trans* labilisation resulting from π-bonded alkenes. The presence of these compensating effects within the three-fold symmetric $[Ru^{II}(\pi\text{-}C_6H_6)(OH_2)_3]^{2+}$ may be the reason why the observed labilisation at the water ligands here is not reflected by changes in the bond lengths.

8.2.1.3 $[Ru(NO)(OH_2)_5]^{3+}$

Conceivably this compound could be considered as a Ru^{III} complex of NO or a Ru^{II} complex of NO^+. It can be made via either route although it was first prepared via reaction of $[Ru(OH_2)_6]^{3+}$ with NO in 1.0 M CF_3SO_3H. The real situation is probably one of considerable M—L electron delocalisation. Two relatively low intensity electronic transitions are seen at 345 nm ($\varepsilon = 57 \, M^{-1} \, cm^{-1}$) and 469 nm ($23 \, M^{-1} \, cm^{-1}$). The energy and intensity of the two bands is suggestive of an assignment to a low-spin t_{2g}^6 Ru^{II} complex ($^1A_{1g}$–$^1T_{1g}$ and $^1A_{1g}$–$^1T_{2g}$), with the former having a higher intensity due to proximity to the t_{2g}–$\pi^*(NO^+)$ MLCT transition. However, alternative assignments to a Ru^{III} complex, with the lower energy band assignable to a d–d/MLCT transition combination, have been suggested.

8.2.1.4 REDOX REACTIONS INVOLVING $[Ru(OH_2)_6]^{2+}$

Solutions of $[Ru(OH_2)_6]^{2+}$ must be protected from the air as the ion is readily air oxidised to yellow $[Ru(OH_2)_6]^{3+}$. Likely intermediates in the O_2 reduction process are $[Ru^{III}(OH_2)_5(O_2^-)]^{2+}$ and $[(H_2O)_5Ru^{III}(\mu\text{-}O_2^{2-})Ru^{III}(OH_2)_5]^{4+}$. With other oxidising reagents both inner- and outer-sphere reductions involving $[Ru(OH_2)_6]^{2+}$ are implicated. The oxidation of $[Ru(OH_2)_6]^{2+}$ by ClO_4^- appears to be substitution controlled [133]. The rate law, $d[Ru^{III}]/dt = 2k_1[ClO_4^-][Ru^{2+}]/(1 + K[Cl^-])$, shows evidence of strong inhibition by Cl^-. At 25 °C, $k_1 = 3.2 \pm 0.1 \times 10^{-3} M^{-1} s^{-1}$ $\Delta H^{\ddagger} = 19.4 \pm 0.3\,kJ\,mol^{-1}$, $\Delta S^{\ddagger} = -5 \pm 1\,J\,K^{-1}mol^{-1}$ (similar to the values in Table 8.8) and $K = 1.0 \pm 0.2\,M^{-1}$, identical to that obtained for formation of $[Ru(OH_2)_5Cl]^+$ in separate studies. The overall stoichiometry and reaction scheme is shown below:

$$2Ru^{2+}(aq) + ClO_4^- + 2H^+ \longrightarrow 2Ru^{3+}(aq) + ClO_3^- + H_2O \quad (8.17)$$

$$Ru^{2+}(aq) + ClO_4^- \underset{k_{-1}}{\overset{k_1}{\rightleftharpoons}} RuOClO_3^+(aq) \quad (8.18)$$

$$Ru^{2+}(aq) + Cl^- \overset{K}{\rightleftharpoons} RuCl^+(aq) \quad (8.19)$$

$$Ru^{2+}(aq) + RuOClO_3^+(aq) + 2H^+ \overset{fast}{\longrightarrow} 2Ru^{3+}(aq) + ClO_3^- + H_2O \quad (8.20)$$

$[Ru(OH_2)_6]^{2+}$ has the fastest rate for reduction of ClO_4^- of all the reducing metal ions studied. This is believed to reflect the highly polarised nature of the t_{2g} orbitals on Ru^{2+}, leading to a favourable π interaction with the empty 3d derived MOs of the Cl—O bonding system. A number of outer-sphere redox reactions involving both the $[Ru(OH_2)_6]^{3+/2+}$ and $[Ru(OH_2)_5Cl]^{2+/+}$ couples have been investigated and are discussed in Section 8.2.2.7.

8.2.2 Ruthenium(III) (d^5): Chemistry of the $[Ru(OH_2)_6]^{3+}$ Ion

8.2.2.1 PREPARATION AND PROPERTIES

Solutions containing the yellow trivalent hexaaqua ion $[Ru(OH_2)_6]^{3+}$ are conveniently prepared by simple air oxidation of purified solutions of $[Ru(OH_2)_6]^{2+}$. Crystals of the air-stable lemon yellow pts$^-$ salt $[Ru(OH_2)_6](pts)_3 \cdot 3H_2O$ can be isolated from concentrated solutions of $[Ru(OH_2)_6]^{3+}$ in Hpts by precipitation, warming to dissolve and then slow cooling (to -2 °C), as described above for $[Ru(OH_2)_6]^{2+}$ [114]. Crystals of $[Ru(OH_2)_6](CF_3SO_3)_3$ can be isolated similarly but are extremely hygroscopic and so for the purpose of handling and long-term storage the pts$^-$ salt is preferred. The $[Ru(OH_2)_6]^{3+}$ ion is also characterised in $CsRu(SO_4)_2 \cdot 12H_2O$ which conforms to the β-alum modification [134]. The alum

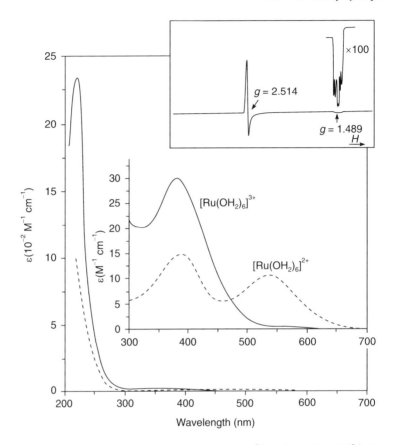

Figure 8.15. Solution electronic spectra for $[Ru(OH_2)_6]^{3+}$ and $[Ru(OH_2)_6]^{2+}$. (Inset: X-band EPR spectrum from a polycrystalline sample of $[Ru(OH_2)_6]^{3+}$ (3% ^{101}Ru, $I = 5/2$) doped into $CsGa(SO_4)_2 \cdot 6H_2O$ at 3 K)

is easily isolated from a concentrated solution of $[Ru(OH_2)_6]$ (pts)$_3$ following the addition of Cs_2SO_4. An X-ray structural investigation on $[Ru(OH_2)_6]$ (pts)$_3 \cdot 3H_2O$ at 22 °C showed evidence of considerable disorder about the Ru—O bonds. However, the average Ru—O distance (202.9 pm) is significantly shorter (by 9 pm) than that present in $[Ru(OH_2)_6]$ (pts)$_2$, a feature that had been predicted some 3 years earlier on the basis of estimates for the $[Ru(OH_2)_6]^{3+/2+}$ self-exchange rate (see below). This can be rationalised as due to a significant degree of π bonding to $t_{2g}^5 Ru^{III}$ from the O atom of OH_2, a feature confirmed in subsequent vibrational spectral studies on both ions.*

---

* The faster self-exchange oberved for the $[Ru(NH_3)_6]^{3+/2+}$ couple is reflected in only a 40 pm difference in Ru—N distance between the two complexes consistent with a smaller preorganisational energy requirement. With NH_3 there is no possibility of π bonding to Ru^{III}.

The electronic and EPR spectra for $[Ru(OH_2)_6]^{3+}$ are shown in Fig. 8.15 [135]. The electronic spectrum consists of a weak band at ~ 550 nm ($\varepsilon \sim 1$ M^{-1} cm^{-1}) (spin-forbidden $^2T_{2g}-^4T_{1g}$ transition) and a stronger band at 390 nm ($\varepsilon = 30$ M^{-1} cm^{-1}) (spin-allowed $^2T_{2g}-^2T_{1g}$, $^2A_{2g}$ transition). An intense oxygen—Ru charge transfer band at ~ 220 nm dominates the remainder of the u.v. portion of the spectrum and masks any remaining higher energy d–d transitions. The EPR spectrum shows evidence of pronounced anisotropy, indicative of a significant electronic distortion to the octahedral geometry. Part of this distortion was viewed as reflecting the trigonal site symmetry of the $[Ru(OH_2)_6]^{3+}$ octahedron within the alum lattice. The paramagnetic susceptibility of the alum obeyed the Curie law from 2.6 to 200K. The temperature-independent magnetic moment (1.92 BM) is unusual for low-spin d^5 ions. Its origin was traced to a combination of spin–orbit coupling and low symmetry field splitting. In addition, the ^{101}Ru hyperfine splitting constants for $[Ru(OH_2)_6]^{3+}$ are found to be a factor of 2 smaller than in the case of $[Ru(NH_3)_6]^{3+}$, rationalised as due to a spreading out of the spin density on to the O atoms of the H_2O ligands as a result of the significant $p_\pi-d_\pi$ O–Ru interaction. A pronounced anisotropy is also observed in the A tansor, viewed as further evidence of a significant orbital contribution to the hyperfine interaction. Thus, as was highlighted in Chapter 1, there is now considerable evidence in support of significant covalency within the M—O bonds of aqua ions of the second and third row transition elements.

8.2.2.2 HYDROLYSIS OF $[Ru(OH_2)_6]^{3+}$

Values of the first hydrolysis constant for $[Ru(OH_2)_6]^{3+}$ ($-\log Q_{11}$) have been estimated from spectrophotometric measurements and from measurements of the shift in the E^θ value for the $[Ru(OH_2)_6]^{3+/2+}$ couple as afunction of $[H^+]$. The spectrophotometric measurements have made use of the intense absorption peak for $[Ru(OH_2)_6]^{3+}$ occuring at 290 nm ($\varepsilon_{RuOH^{2+}} \sim 1300$ M^{-1} cm^{-1}). In 1980 Harzion and Navon reported a value of 2.4 ± 0.1 at $25\,°C$ [136]. Sutin and coworkers studied the E^θ value for $[Ru(OH_2)_6]^{3+/2+}$ as a function of pH [137]. Below a pH of ~ 3 the E^θ value ($+0.217$ V vs NHE) is independent of pH. Above pH 3 the E^θ decreases linearly with pH. The intersection of the two curves gives a measure of the hydrolysis constant ($-\log Q_{11}$, $25\,°C$) as 2.9. Merbach and coworkers reinvestigated the hydrolysis as part of their study of the water-exchange process on $[Ru(OH_2)_6]^{3+}$ (see below) [124]. The dependence of the absorbance at 290 nm (A) was fitted to

$$A = b[Ru^{III}][\varepsilon_{Ru^{3+}} + \varepsilon_{RuOH^{2+}} 10^{pH-pK_{11}}]/(1 + 10^{pH-pK_{11}}) \qquad (8.21)$$

as a function of acidity, temperature and pressure. The following values were obtained $(25\,°C)$: $-\log Q_{11} = 2.73 \pm 0.02$, $\varepsilon_{Ru^{3+}} = 143$ $m^{-1}cm^{-1}$, $\varepsilon_{RuOH^{2+}} =$

$1307 \, \text{m}^{-1} \text{cm}^{-1}$, $\Delta H^{\circ}_{11} = 41.1 \pm 1.9 \, \text{kJ mol}^{-1}$, $\Delta S^{\circ}_{11} = 85.6 \pm 6.4 \, \text{J K}^{-1} \text{mol}^{-1}$ and $\Delta V^{\circ}_{11} = -3.0 \pm 0.6 \, \text{cm}^3 \, \text{mol}^{-1}$. Few data are available on hydrolytic polymerisation products of $\text{Ru}^{\text{III}}(\text{aq})$ although a dimeric form has been identified as the final product of reduction of $\text{Ru}^{\text{IV}}(\text{aq})$ to Ru^{III} (see Section 8.2.3).

8.2.2.3   WATER EXCHANGE AND SUBSTITUTION ON $[\text{Ru}(\text{OH}_2)_6]^{3+}$

The kinetics of water exchange on $[\text{Ru}(\text{OH}_2)_6]^{3+}$ were monitored using variable temperature and pressure ^{17}O NMR on solutions of $[\text{Ru}(\text{OH}_2)_6]$ (pts)$_3 \cdot 3\text{H}_2\text{O}$ at $I = 2.0 \, \text{m}$, $\text{CF}_3\text{SO}_3\text{H}/\text{NaCF}_3\text{SO}_3$ [124]. Rate constants are much smaller by a (factor of 10^4) than those for $[\text{Ru}(\text{OH}_2)_6]^{2+}$ under the same conditions, allowing easy conventional monitoring by isotopic dilution. The coordinated waters on $[\text{Ru}(\text{OH}_2)_6]^{3+}$ occur at $+34.7$ ppm from bulk water, the peak easily being observed despite the paramagnetism. The decrease in the height of the ^{17}O NMR signal from enriched solutions of $[\text{Ru}(\text{OH}_2)_6]$ (pts)$_3$ (0.3 m, 9 atom %) was monitored together with the corresponding increase in the height of the bulk water signal. Both were fitted to a standard exponential function. $\text{Mn}^{2+}(\text{aq})$ (0.1 m) was added as before to relax out the bulk water signal. The relatively narrow linewidth observed for coordinated water on $[\text{Ru}(\text{OH}_2)_6]^{3+}$ ($1/T_2 = 2130 \, \text{s}^{-1}$ at 301.9 K) is only slightly larger than that for free H_2O ($1430 \, \text{s}^{-1}$) and comparable to linewidths obtained for diamagnetic ions under similar conditions, e.g. $\text{Al}^{3+}(\text{aq})$ ($1/T_{2\text{Q}^b} = 2297 \, \text{s}^{-1}$) and $\text{Ga}^{3+}(\text{aq})$ ($1567 \, \text{s}^{-1}$), indicating that quadrupolar relaxation is the dominant process. A marked acidity dependence of the type $a + b[\text{H}^+]^{-1}$ was found, indicating a significant contribution from the monohydroxo ion $[\text{Ru}(\text{OH}_2)_5\text{OH}]^{2+}$. Using values for the hydrolysis constant as a function of temperature and pressure (ranges 2–40 °C, 0.1–190 MPa) the following kinetic results (25 °C) were obtained $k_{\text{ex}}(\text{Ru}^{3+}) = (3.5 \pm 0.3) \times 10^{-6} \, \text{s}^{-1}$, $\Delta H^{\ddagger} = 90 \pm 4 \, \text{kJ mol}^{-1}$, $\Delta S^{\ddagger} = -48 \pm 14 \, \text{kJ K}^{-1} \text{mol}^{-1}$, $\Delta V^{\ddagger} = -8.3 \pm 2.1 \, \text{cm}^3 \text{mol}^{-1}$, $k_{\text{ex}}(\text{RuOH}^{2+}) = 5.9 \times 10^{-4} \, \text{s}^{-1}$, $\Delta H^{\ddagger}_{\text{OH}} = 96 \, \text{kJ mol}^{-1}$, $\Delta S^{\ddagger}_{\text{OH}} = +15 \, \text{J K}^{-1} \text{mol}^{-1}$ and $\Delta V^{\ddagger}_{\text{OH}} = +0.9 \, \text{cm}^3 \, \text{mol}^{-1}$. If one assumes a similar volume contraction on going from the hexaaquated to the pentaaquated complex for both Ru^{3+} and RuOH^{2+}, as in the case above for Ru^{2+} (say $\sim 15 \, \text{cm}^3 \text{mol}^{-1}$), then the exchange mechanism can be assigned by the effective excluded volume of the leaving water molecule (~ 15–$16 \, \text{cm}^3 \text{mol}^{-1}$, I_{d}, Ru^{2+}, RuOH^{2+}; $\sim 7 \, \text{cm}^3 \text{mol}^{-1}$, I, Ru^{3+}). Similar arguments are used for the case of substitution on $[\text{Rh}(\text{OH}_2)_6]^{3+}$ and $[\text{Rh}(\text{OH}_2)_5\text{OH}]^{2+}$ (Section 9.2.2). Assignment of an actual associative mechanism to $[\text{Ru}(\text{OH}_2)_6]^{3+}$ on the basis of the negative $\Delta V^{\ddagger}_{\text{ex}}$ [18] may be premature in the absence of further supportive kinetic data. Unfortunately, there is a scarcity of data regarding substitution reactions on $[\text{Ru}(\text{OH}_2)_6]^{3+}$ with different ligands. One can probably only conclude from the $\Delta V^{\ddagger}_{\text{ex}}$ value that the transition state probably contains a greater contribution from bond-making processes than is the case for $[\text{Ru}(\text{OH}_2)_6]^{2+}$ or $[\text{Ru}(\text{OH}_2)_5\text{OH}]^{2+}$. The 10^4 times smaller rate constants for $[\text{Ru}(\text{OH}_2)_6]^{3+}$ vs $[\text{Ru}(\text{OH}_2)_6]^{2+}$ shows the effect of an increase in one unit of charge as overcoming the loss in CFSE ($0.2\Delta_O$). The charge effect is even more apparent in the 10^7-fold decrease in k_{ex} on going from $[Ru(OH_2)_6]^{2+}$ to $[Rh(OH_2)_6]^{3+}$ (Section 9.2.2) (both t_{2g}^6 ions).

8.2.2.4 The Ru^{III}–chloroaqua system

Adamson has reported that the rates of ligand exchange for Cl$^-$ bound to Ru follows the order Ru^{II} < Ru^{III} < Ru^{IV} [138]. Substitution rates also increase with the number of Cl$^-$ ligands attached. Fine studied the 'ageing' of an aqueous solution of $K_2[RuCl_5(OH_2)]$ and observed initial rapid (minutes) equilibra involving conversions to $[RuCl_4OH_2)_2]^-$ and $[RuCl_3(OH_2)_3]$ with much slower equilibra (days), giving $[RuCl_2(OH_2)_4]^+$ and eventually $[RuCl(OH_2)_5]^{2+}$ [139]. Heating was observed to speed up the attainment of equilibrium markedly [140]. The later stages required several weeks to reach full thermodynamic equilibrium. Connick and Fine [141, 142] and Ohyoshi, Ohyoshi and Shinagawa [143] reported the earliest estimates of the equilibrium constants for the various $[RuCl_n(OH_2)_{6-n}]^{(3-n)+}$ complexes in Cl$^-$ media, the values having been extrapolated to infinite dilution at 298 K by Rard [112]. A preliminary investigation of the kinetics of 1:1 Cl$^-$ substitution on $[Ru(OH_2)_6]^{3+}$ in aqueous NaCF$_3$SO$_3$ media showed compications stemming from a range of products, ill-defined electronic spectra and evidence of autocatalysis [144]. Extensive hydrolysis and the formation of dinuclear species may also contribute to the observed electronic spectra over longer time periods. Ready oxidation to Ru^{IV} and mixed Ru^{III}–Ru^{IV} species can cause further complications. The equilibrium constant of the 1:1 Cl$^-$ anation reaction has been estimated via catalysis by $[Ru(OH_2)_6]^{2+}$ [113].

A series of papers by Taqui–Khan and coworkers has attempted to unravel the complexity of the Ru^{III}–chloroaqua system by reporting the kinetics and thermodynamics of aquation and Cl$^-$ anation on various $[RuCl_n(OH_2)_{6-n}]^{(3-n)+}$ complexes ($n = 1$–4) following spectrophotometric and polarographic monitoring in KCl/HCl media [145, 146]. It was found that in 1.0 M HCl only one principal stable species $[RuCl_3(OH_2)_3]$ was present that could be used as the starting species for equilibration studies which were carried out at $\mu = 0.1$ M KCl/HCl. The precise composition of the Ru^{III}–Cl$^-$ system was found to depend both on $[H^+]$ and $[Cl^-]$. At pH < 0.4 the principal species is $[RuCl_4(OH_2)_2]^-$ whereas at pH 2 it is $[RuCl_2(OH_2)_4]^+$. Both $[RuCl(OH_2)_5]^{2+}$ and $[RuCl_4(OH_2)_2]^-$ convert to $[RuCl_2(OH_2)_4]^+$ over a period of ~ 1 hour at 25 °C at pH 2.0. In 1.0 M HCl the half-lives for conversion of $[RuCl_4(OH_2)_2]^-$, $[RuCl_2(OH_2)_4]^+$ and $[RuCl(OH_2)_5]^{2+}$ into stable $[RuCl_3(OH_2)_3]$ at 25 °C are respectively 1.3 days, 0.24 day and 22.5 days. A protonated form $H[RuCl_4(OH_2)_2]$, $pK_a \sim 1.1$, plays a role in the aquation to give $[RuCl_3(OH_2)_3]$. The kinetic results may be summarised as in Table 8.9.

The somewhat lower $\Delta H^{\ddagger}$ value observed for the aquation pah is in line with the weaker Ru^{III}—Cl bond in these complexes when compared to the Ru^{III}—OH$_2$

Table 8.9. Kinetic parameters for aquation and Cl⁻ anation reactions on various $[RuCl_n(OH_2)_{6-n}]^{(3-n)+}$ complexes in 0.1 M KCl/HCl [145, 146]

Equilibrium	$10^5 k_f$ (M⁻¹s⁻¹) anation	$10^5 k_b$ (s⁻¹) aquation	K_f (M⁻¹)	$\Delta H_f^{\ddagger}$ (kJ mol⁻¹)	$\Delta H_b^{\ddagger}$ (kJ mol⁻¹)	$\Delta S_f^{\ddagger}$ (J K⁻¹ mol⁻¹)	$\Delta S_b^{\ddagger}$ (J K⁻¹ mol⁻¹)
Ru³⁺(aq)/RuCl²⁺(aq)	—	—	148 ± 6[g]	54.8	56.0	−150	−150
RuCl²⁺(aq)/RuCl₂⁺(aq)	1.8	0.5	3.6	29.7	25.1	−234	−247
RuCl₂⁺(aq)/RuCl₃(aq)	2.9	3.4	0.85	27.2	18.4	−224	−263
RuCl₃(aq)/RuCl₄⁻(aq)	4.3	13.0	0.33	65.2	53.1	−129	−163
RuCl₃(aq).H⁺/H[RuCl₄](aq)	0.3	1.2	0.25				
RuCl₄⁻(aq)/RuCl₅²⁻(aq)	—	—	~0.14[b]				
RuCl₅²⁻(aq)/RuCl₆³⁻	—	—	~0.1[b]				

[a] Ref. [113].
[b] Ref. [141].

bond. The highly negative $\Delta S^{\ddagger}$ values might be interpreted as reflecting the importance of a bond-making activation mode to the transition state in line with the water-exchange data [124]. Using cation-exchange chromatography, Cady and Connick claimed successful separation and characterisation of $[RuCl_n(OH_2)_{6-n}]^{(3-n)+}$ ($n = 0, 1$ and 2) in introducing their method for the determination of the charge/M [147]. However, they also found that electronic spectra were irreproducible over extended time periods consistent with further equilibration/hydrolysis in the varying pH and Cl^- media. Suffice it to say the $Ru^{III}-Cl$ system is a highly complex mixture of species. Electronic spectra for the individual monomeric $[[RuCl_n(OH_2)_{6-n}]^{(3-n)+}$ complexes have been determined and these are shown in Fig. 8.16 [142].

$[RuCl_2(OH_2)_4]$ can be handelled conveniently in water/dioxane mixtures at a 'pH' of $\geqslant 2.0$ and has been shown to exhibit catalytic activity towards the O_2 oxidation of a number of organic compounds [148]. $[RuCl_2(OH_2)_4]^+$ is also a mediator in the electrooxidation of certain alkenes [149]. The active species is believed to be monomeric aqua $Ru^V = O$ cation $[Ru(O)Cl_2(OH_2)_3]^+$.

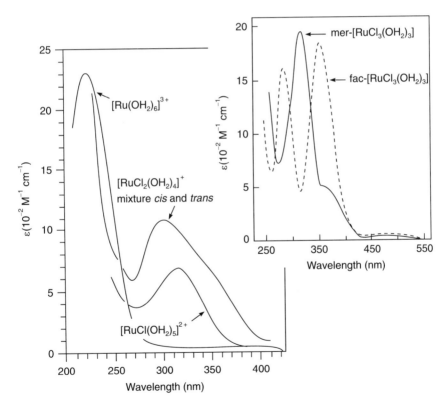

Figure 8.16. Electronic spectra for $[RuCl_n(OH_2)_{6-n}]^{(3-n)+}$ complexes

8.2.2.5 $[Ru(EDTA)(OH_2)]^-$: A REMARKABLY LABILE Ru^{III} AQUA COMPLEX

Although EDTA is potentially a hexadentate ligand it is seen with a range of trivalent metal ions to adopt 'penta-coordination', leaving a one-coordinate H_2O ligand on the metal centre accompanied by a pendent carboxylate arm. The remaining water ligand is more often than not highly labile and $[Ru(EDTA)(OH_2)]^-$ is probably the most labile Ru^{III} aqua complex. For example, substitution reactions occur on the stopped-flow timescale (second-order rate constants (25 °C) for formation of $[Ru(EDTA)L]^{n-}$ are ~ 30–20000 $M^{-1} s^{-1}$; Table 8.10) which can be compared with the half-life of ~ 11 days for water exchange on $[Ru(OH_2)_6]^{3+}$. The 6000-fold range of rate constants and markedly negative activation volumes, in particular the increasingly negative volume along the series thiourea–dimethylthiourea–tetramethylthiourea, have been interpreted in terms of a highly associatively activated transition state [150, 151] (cf. the behaviour of the Ru^{III}–chloroaqua system). A specific interaction (perhaps hydrogen bonded) between the coordinated water ligand and the free carboxylate arm was suggested as leading to an exposed site for attack by the incoming ligand [151, 152]. However, this is not reflected by two independent X-ray structural studies on $[Ru(EDTA)(OH_2)]^-$, which do show a significant lengthening of the Ru—OH_2 bond (213 pm) (comparing with 203 pm for $[Ru(OH_2)_6]^{3+}$) but no interaction with the carboxylate arm (Fig. 8.17) [153].

Table. 8.10. Kinetic parameters for the substitution reaction [a]: $[Ru^{III}(EDTA)(OH_2)]^- + L \rightarrow [Ru^{III}(EDTA)L]^{n-}$

L	$10^{-3}k(25\,°C)$ $(M^{-1} s^{-1})$	$\Delta H^{\ddagger}$ $(kJ\,mol^{-1})$	$\Delta S^{\ddagger}$ $(J K^{-1} mol^{-1})$	$\Delta V^{\ddagger}$ $(cm^3 mol^{-1})$	Ref.
CH_3CN	0.03 ± 0.007	34.7 ± 2.1	-100 ± 16.7	—	[150]
SCN^-	0.27 ± 0.02	37.2 ± 2.1	-75.0 ± 5.5	-9.6 ± 0.3	[150, 151]
N_3^-	1.89 ± 0.08	24.8 ± 1.4	-99 ± 5	-9.0 ± 0.6	
	2.07 ± 0.1^b	26.4 ± 3.0	-94 ± 10	-9.9 ± 0.5	[151]
Me_4tu^c	0.154 ± 0.005	28.9 ± 3.3	-107 ± 11	-12.2 ± 0.5	[151]
Me_2tu^c	1.450 ± 0.025	25.3 ± 1.3	-107 ± 4	-8.8 ± 0.2	[151]
tu^c	2.970 ± 0.050	22.3 ± 1.4	-105 ± 5	-6.8 ± 0.6	[151]
Imidazole^d	1.86 ± 0.10				[152]
Pyridine^e	6.3 ± 0.5				[152]
Isonic^f	8.3 ± 0.6	27.6 ± 2.1	-79 ± 13		[150]
Pyrazine	20 ± 1	23.8 ± 2.1	-83 ± 13		[150]

[a] Acetate buffer pH 5.1–5.5, $\mu = 0.2$ M.
[b] pH 6.0.
[c] Me_4tu = tetramethylthiourea, Me_2tu = dimethylthiourea, tu = thiourea.
[d] Imidazole buffer pH 7.2.
[e] Pyridine buffer pH 5.2.
[f] Isonicotinamide.

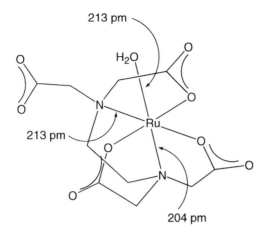

213 pm

H₂O

213 pm

204 pm

Figure 8.17. Structure of $[Ru^{III}(EDTA)(OH_2)]^-$

Nonetheless, the lengthening of the $Ru—OH_2$ bond could allow an accessible site for the incoming ligand to attack. The single H_2O is opposite one of the tertiary nitrogen and correspondingly the $Ru—N$ distance here (203.8 pm) is considerably shorter than that (213 pm) found opposite an Ru–carboxylate bond. Similar bond length variations can be seen in the corresponding $[Fe^{III}(EDTA)(OH_2)]^-$ complex. Related work has shown that the corresponding $[Ru^{II}(EDTA)(OH_2)]^{2-}$ complex is more inert by up to a factor of 10^6 and substitution can indeed be catalysed by the presence of the Ru^{III} complex, a rare if not unique example of 'inverted' redox catalysis, i.e. catalysis of substitution on the lower oxidation state by a higher one [150, 152]. Normally the reverse is true, cf, Cr^{III} vs Cr^{II} and Co^{III} vs Co^{II}. It has thus been suggested that the anomalously high lability at the remaining water of trivalent $[M(EDTA)(OH_2)]^-$ complexes (M = Cr, Fe, Ru, Rh) vs the $[M(OH_2)_6]^{3+}$ is due to facilitation of the seven-coordinate associatively activated transition state and not mere reduction in the overall charge on the complex [151, 152].

$[Ru^{III}(EDTA)(OH_2)]^-$ has been found to be an active catalyst for the activation of oxygen [154], hydrogen [155], carbon monoxide [156], nitric oxide [157] and nitrogen [158]. In the presence of ascorbate it will catalyse the H_2O_2 or O_2 oxidation of saturated ring hydrocarbons such as adamantane and cyclohexane [159]. $[Ru(EDTA)(OH_2)]^-$ has been independently shown to catalyse the oxidation of ascorbate by H_2O_2, involving what appears to be an (EDTA)Ru^{IV}—O—O—Ru^{IV}(EDTA) intermediate [160]. In the absence of reducing agents $[Ru(EDTA)(OH_2)]^-$ will catalyse the O_2-dependent epoxidation of alkenes such as styrene. The active compound here is believed to be the oxo Ru^{IV} complex $[Ru(O)(EDTA)]^-$.

8.2.2.6 OXO-CENTERED TRIAQUATRIRUTHENIUM CARBOXYLATES

The Ru analogue of the Fe_3^{III} complex $[Fe_3(\mu_3\text{-O})(\mu\text{-O}_2CCH_3)_6(OH_2)_3]^+$ was first reported by Wilkinson and coworkers in 1970 as the acetate salt via treatment of $RuCl_3 \cdot nH_2O$ with a mixture of acetic acid and acetic anhydride [161]. Since then the unit has been crystallographically characterised as its BF_4^- [162] and ClO_4^- [163] salts. Interest in the latter stems from the fact that both valence forms $[Ru_3(\mu_3\text{-O})(\mu\text{-O}_2CCH_3)_6(OH_2)_3]^+$ and $[Ru_3(\mu\text{-O})(\mu\text{-O}_2CCH_3)_6(OH_2)_3]^{2+}$ can be isolated depending upon the conditions [163, 164]. Triruthenium hexacarboxylates have a rich redox chemistry with up to five valence forms observable [165, 166]. Solutions containing the Ru_3^{III} complex are best prepared by exchanging the acetate salt on to a Dowex 50W X2 column in the H^+ form, washing out the acetic acid then eluting with $HClO_4$ (0.1–1.0 M). If elution with $<1.0\,M$ $HClO_4$ is carried out and the solutions evaporated in a desiccator, purple-blue crystals of $[Ru_3(\mu_3\text{-O})(\mu\text{-O}_2CCH_3)_6(OH_2)_3](ClO_4)_2 \cdot H_2O$ are obtained [162, 163]. However, if more concentrated $HClO_4$ (2.0 M) eluates are evaporated the bluish-green crystals deposited are of $[Ru_3(\mu_3\text{-O})(\mu\text{-O}_2CCH_3)_6(OH_2)_3](ClO_4) \cdot 2H_2O$ [164]. The crystal structure of this complex is shown in Fig. 8.18.

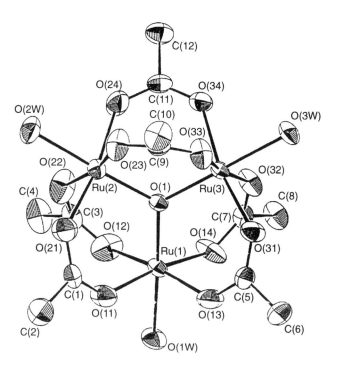

Figure 8.18. Structure of the cationic unit in $[Ru_3(\mu_3\text{-O})(\mu\text{-O}_2CCH_3)_6(OH_2)_3]$ $(ClO_4) \cdot 2H_2O$

In both the cases the filtrate remaining is purple, reflecting the presence of the dicationic form in solution. This is due to oxidation by ClO_4^- as its concentration builds up. If 2.0 M $HClO_4$ solutions are evaporated apparently the Ru^{III}_3 complex can crystallise quickly out of the high ionic strength medium before oxidation occurs. The use of more dilute ClO_4^- solutions result in the Ru^{III}_3 complex being in solution for a time sufficient to allow oxidation by ClO_4^- up to the dicationic form which then crystallises. Further interest has surrounded the structural parameters for the two complexes. Virtually identical $Ru-OH_2$, $Ru-O(\mu_3)$ and $Ru-O$(carboxylate) are found for each, implying that removal of one electron from the Ru^{III}_3 complex has a minimal effect on the structural parameters. This is in agreement with the simplified MO treatment for the M_3O π system, which predicts that the HOMO is essentially non-bonding [167]. There is also no detectable difference between the three Ru atoms in the mixed valence form, implying that the structure is valence delocalised at least on the crystallographic timescale.

The water-exchange rate constant at 25 °C is $(1.08 \pm 0.07) \times 10^{-3}$ s^{-1} [163] which is three orders of magnitude larger than that for $[Ru(OH_2)_6]^{3+}$. The averaged $Ru-OH_2$ distance (212 pm) is longer than in $[Ru(OH_2)_6]^{3+}$ (203 pm). Both of these observations are believed to reflect a significant *trans* influence from the planar π-bonded μ_3-O group, as is apparent in other members of this class of compound [168].

Triruthenium carboxylates are of interest as homogeneous catalysts for a number of conversions, including the hydrogenation and isomerisation of alkenes [169], the oxidation of primary alcohols [170] and the air oxidation of cyclic aliphatic [171, 172] hydrocarbons.

8.2.2.7 ELECTRON TRANSFER REACTIONS INVOLVING THE $[Ru(OH_2)_6]^{3+/2+}$ COUPLE

Redox potentials (E^θ of Ru^{III}/Ru^{II} vs NHE) determined for the various $[RuCl_n(OH_2)_{6-n}]^{(3-n)+}$ complexes (Table 8.11) show a steady trend to more negative values as the number of Cl^- ligands attached is increased. Kinetic studies of the reduction of Ru^{3+}(aq) by $[Cr(OH_2)_6]^{2+}$ indicate, however, an inner-sphere process. A discernable intermediate is seen absorbing between 320 and 400 nm ($\varepsilon_{330} = 420 \pm 30$ M^{-1} cm^{-1}). The rate of appearance of the intermediate obeys the rate law, $d[I]/dt = k[Cr^{2+}(aq)][Ru^{3+}(aq)][H^+]^{-1}$, indicating that the reactive Ru^{III} species is $[Ru(OH_2)_5OH]^{2+}$ and that the intermediate is probably the OH-bridged dimer, $[(H_2O)_5Ru(OH)Cr(OH_2)_5]^{4+}$ [137]. At 25 °C, $k = (9.5 \pm 0.9) \times 10^{-3}$s^{-1}. Corresponding reduction of $[Ru^{III}Cl(OH_2)_5]^{2+}$ by Cr^{2+}(aq), giving $[Cr^{III}Cl(OH_2)_5]^{2+}$ as product, proceeds through a similar Cl^--bridged intermediate [173] and suggests that in the breakdown of the above OH-bridged complex it is a $Ru-O$ bond rather than a $Cr-O$ bond that is cleaved.

Table. 8.11. Redox potentials for several $[RuCl_n(OH_2)_{6-n}]^{(3-n)+}$ complexes

Ru^{III}/Ru^{II} process			E^θ (V) vs NHE
$[Ru(OH_2)_6]^{3+}$	$+e^-$ →	$[Ru(OH_2)_6]^{2+}$	$+0.217$
$[RuCl(OH_2)_5]^{2+}$	$+e^-$ →	$[RuCl(OH_2)_5]^+$	$+0.077^a$
$[RuCl_2(OH_2)_4]^+$	$+e^-$ →	$[RuCl_2(OH_2)_4]$	-0.023^a
$[RuCl_3(OH_2)_3]$	$+e^-$ →	$[RuCl_3(OH_2)_3]^-$	-0.113^a

^{a} Measured at 0 °C vs SCE (0.237 V vs NHE).

The remaining redox reactions involving the $[Ru(OH_2)_6]^{3+/2+}$ couple with metal complex reagents appear to be outer sphere in nature and the results from kinetic measurements have been used to estimate the self-exchange rate constant. Selected data are listed in Table 8.12. The rate constants in Table 8.11 are larger than those for water exchange on either $[Ru(OH_2)_6]^{2+}$ ($k_{ex} \sim 10^{-2}$ s^{-1}) or $[Ru(OH_2)_6]^{3+}$ ($k_{ex} \sim 10^{-6}$ s^{-1}) and for the most part are larger than water exchange or substitution on the oxidising or reducing agents excluding the

Table. 8.12. Rate constant (25 °C) for redox reactions involving the $[Ru(OH_2)_6]^{3+/2+}$ couple [137]

Reagent	ΔE^θ (V)	k_{12} ($M^{-1}s^{-1}$)	k_{22} ($M^{-1}s^{-1}$)	k_{11} ($M^{-1}s^{-1}$)
(a) Oxidation of $[Ru(OH_2)_6]^{2+}$				
$[Ru(NH_3)_5py]^{3+}$	0.082	1.1×10^4	4.7×10^5	15
$[Co(phen)_3]^{3+}$	0.15	5.3×10^1	40	0.24
$[Ru(NH_3)_5isn]^{3+}$	0.167	5.5×10^4	4.7×10^5	14
$[Fe(OH_2)_6]^{3+}$	0.52	2.3×10^3	4.2	0.012
$[Os(bipy)_3]^{3+}$	0.6	2.9×10^8	2×10^9	0.23
$[Ru(bipy)_3]^{3+}$	1.04	1.9×10^9	2×10^9	0.025
(b) Reduction of $[Ru(OH_2)_6]^{3+}$				
$[RuCl(OH_2)_5]^+$	0.131	3.0×10^3		
$[Ru(NH_3)_6]^{2+}$	0.15	1.4×10^4	2×10^4	32
$[V(OH_2)_6]^{2+}$	0.472	2.8×10^2	1×10^{-2}	0.37
(c) Reduction of $[RuCl(OH_2)_5]^{2+}$				
$[Ru(NH_3)_6]^{2+}$	0.064	$\sim 2 \times 10^5$	2×10^4	9.2×10^5
$[V(OH_2)_6]^{2+}$	0.341	1.7×10^3	1×10^{-2}	1.2×10^3

possibility of an inner-sphere pathway. Values of k_{11} calculated from the Marcus approach are seldom too high and significantly lower values frequently result from the very exothermic reactions. An estimate of the true k_{11} was therefore evaluated from a plot of log k_{11} (calculated) vs ΔE^θ for the cross-reaction [137]. This approach gives $(6 \pm 4) \times 10^{-1}$ M^{-1}s^{-1} for the $[Ru(OH_2)_6]^{3+/2+}$ self-exchange in 1.0 M CF_3SO_3H at 25 °C. Substitution of a Cl$^-$ in place of H_2O is found to significantly increase the rate of the Ru$^{3+/2+}$ self-exchange.

The relatively slow water exchange on both $[Ru(OH_2)_6]^{2+}$ and $[Ru(OH_2)_6]^{3+}$ has allowed a direct measurement of the self-exchange process using both ^{17}O and ^{99}Ru NMR spectroscopy [174]. In one set of experiments a sample of $[Ru(OH_2)_6]^{2+}$ is rapidly injected into a solution containing enriched $[Ru(^{17}OH_2)_6]$ (pts)$_3 \cdot 3H_2^{17}O$ in aqueous CF_3SO_3H and the increase and decrease in intensity respectively of the bound water resonance on $[Ru(OH_2)_6]^{2+}$ and $[Ru(OH_2)_6]^{3+}$ were followed. Rapid electronic equilibration $(t_{1/2} \sim 1$ min) is followed by the slow loss of both the signals over a 60 minute period due to the water exchange process on $[Ru(OH_2)_6]^{2+}$. $[Ru(OH_2)_6]^{3+/2+}$ is the only hexaaqua redox couple known where the water exchange on both complexes is slower than the electronic exchange.

In a second set of experiments the linewidth of the bound water signals on $[Ru(OH_2)_6]^{2+}$ was studied as a function of temperature (range of 252–366 K) in the presence of amounts of $[Ru(OH_2)_6]^{3+}$. At temperatures below 315 K the linewidth is governed by quadrupolar relaxation only. At higher temperatures line broadening is observed due to the electronic exchange process. The theory behind the line-broadening method is essentially the same as that described for water exchange on labile diamagnetic systems (see Section 1.2.3). Independent measurement of the linewidth of the ^{99}Ru NMR signal of $[Ru(OH_2)_6]^{2+}$ ($+16050$ ppm from $[Ru(CN)_6]^{4-}$) as a function of temperature was required to allow an independent estimate of the contribution from ^{99}Ru to the overall $1/T_{2Q}$ values. At 25 °C, $k_{11} = 20 \pm 4$ M^{-1}s^{-1}, $\Delta H_{11}^\ddagger = 46.0 \pm 0.8$ kJ mol^{-1} and $\Delta S^\ddagger = -65.7 \pm 2.7$ JK^{-1}mol^{-1}, $[H^+] = 2.5$ M, $I = 5.0$ M, $CF_3SO_3^-$. The values of k_{11} agrees well with the value estimated by Sutin and coworkers [137] in being surprisingly low, only 10 times greater than that for the $[Fe(OH_2)_6]^{3+/2+}$ couple and some 500 times smaller than the value for the $[Ru(NH_3)_6]^{3+/2+}$ couple. It has been concluded that this is due to a non-negligible inner-sphere reorganisation barrier, namely the significant difference in Ru—O bond length (~ 9 pm) between the two hexaaqua ions. The $[Ru(OH_2)_6]^{3+/2+}$ couple is nonetheless one of the best behaved hexaaqua redox couples in electrochemistry. It exhibits essentially diffusion-controlled reversible behaviour on a range of surfaces in cyclic voltammetric experiments (Fig. 8.19). This is believed to be a further reflection of the highly polarised and largely accessible t_{2g} orbitals.

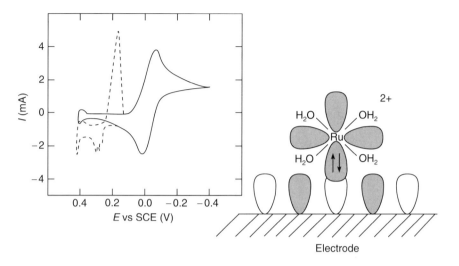

Figure 8.19. Cyclic voltammogram from a solution of $[Ru(OH_2)_6]^{2+}$ (3×10^{-3} M) in 1.0 M Hpts at 25 °C, scan rate 34 mV s^{-1}, Hg drop electrode, area 4.4 mm^2 (Inset: illustration of favourable t_{2g} orbital overlap from $[Ru(OH_2)_6]^{2+}$ with the electrode surface)

8.2.3   Ruthenium(IV) (d^4): Chemistry of Tetrameric Ru^{IV}(aq)

8.2.3.1   PREPARATION AND PROPERTIES

The existence of a red-brown Ru^{IV}(aq) ion has been documented since the early 1950s in papers by Wehner and Hindman [175], Neidrach and Tevebaugh [176] and Atwood and DeVries [177]. Arguments have been presented in favour of monomeric RuO^{2+}(aq) (ruthenyl), or Ru(OH)$_2^{2+}$(aq), [178], and dimeric Ru^{IV}(aq) [177, 179] species. However, further work has since pointed towards the existence of a tetranuclear oxo/hydroxo-bridged structure [180–186]. Studies on the solubility of RuO$_2 \cdot n$H$_2$O in various media were initially interpreted in terms of the formation of monomeric Ru^{IV}(aq) [178]. Subsequent [Ru] dependences were found to be more in line with formation of a polynuclear species and, assuming a tetramer, a solubility product in the range $\sim (0.5–5) \times 10^{-44}$ M^5 was estimated for RuO$_2 \cdot n$H$_2$O in two independent studies [185, 186]. Rard has reviewed much of the literature on Ru(aq) species and attempted a compilation of solution thermodynamic data for both monomeric and tetrameric Ru^{IV}(aq) [112]. However, the evidence for putative monomeric Ru^{IV}(aq) is far from convincing. The detection of fractional oxidation states Ru$^{3.25+}$ and Ru$^{3.75+}$

[180, 183] and even $Ru^{4.25+}$ [184] in subsequent electrochemical studies in conjunction with charge/metal (1.0) and charge/species (4.0) determinations has provided compelling evidence for the tetranuclear formulation.

There are three methods for the preparation of solutions of Ru^{IV}(aq): (a) reduction from RuO_4 (chemically or electrochemically) [176, 178, 180], (b) electrochemical oxidation of either $[Ru(OH_2)_6]^{3+}$ or $[Ru(OH_2)_6]^{2+}$ [182] and (c) reaction of $[RuBr_6]^{2-}$ with BrO_3^- [187]. Treatment of RuO_4 with either Sn or Pb metal results in reduction all the way to Ru^{II}. However, if a milder reducing agent is used, e.g. H_2O_2, the reduction stops at Ru^{IV}. Solutions of Ru^{IV}(aq) are thus conveniently prepared by extracting a solution of RuO_4 in CCl_4 with an aqueous solution of H_2O_2 in 2.0 M $HClO_4$ [182]. The Ru passes into the aqueous phase as it is reduced to Ru^{IV}. Use of lower $[H^+]$ can lead to much precipitation of $RuO_2 \cdot nH_2O$ and thus the use of $[H^+] \geqslant 2.0$ M is recommended. After standing overnight to allow any excess H_2O_2 to decompose, the red-brown Ru^{IV}(aq) solution is subjected to cation-exchange purification on columns of Dowex 50W X2 resin. Ru^{IV}(aq) is strongly retained on the resin such that elution is only achieved by displacement methods, e.g. with a stable trivalent metal ion such as La^{3+} (0.25 M) in $HClO_4$ (0.5 M) [180, 182]. The presence of La^{3+} ions can be easily tested by adding a few drops of 0.5 M $H_2C_2O_4$ (a white precipitate of insoluble $La_2(C_2O_4)_3$). The fact that higher concentrations of $[H^+]$ alone cannot be used as eluent suggests extensive protonation of an oxo-bridged polynuclear structure. Concentrations of Ru^{IV}(aq) (~ 0.2 M) can easily be prepared by displacement from a saturated Dowex and fresh solutions have the electronic spectrum shown in Fig. 8.20 with an absorption maximum at 487 nm ($\varepsilon = 709$ M^{-1} cm^{-1} per Ru) [178] and a shoulder at ~ 300 nm ($\varepsilon \sim 2200 \, M^{-1} \, cm^{-1}$). A broad absorbance is also seen tailing into the near i.r. region above 900 nm. Much polymeric Ru^{IV} material remains on the column and the recovery of elutable Ru^{IV} is never higher than 60%. The polymeric Ru^{IV} can be recovered either via oxidation to RuO_4 or via conversion to elutable anionic chloro complexes by soaking the column for a week in concentrated HCl.

Several workers have investigated the effect of $[H^+]$ on the electronic spectrum in view of evidence for protonation of presumed μ-oxo bridges. The spectrum is only marginally altered in the range 0.01–6.0 M, although above 7.0 M $HClO_4$ more noticeable and moreover irreversible changes occur and the solutions lose their 'reddish' tint. Wallace and Propst determined the charge/Ru at $[H^+]$ 0.5 M to be 1.00 [180] using the method of Cady and Connick [147]. This coupled with a charge/species value of 4.0, estimated from a Donnan membrane equilibrium study [181], indicated formulation under these conditions as $Ru_4O_6^{4+}$(aq) or $Ru_4(OH)_{12}^{4+}$(aq). Most workers have tended to favour the OH-bridged formulation although the oxo-bridged structure would seem more able to explain the protonation behaviour. There are as yet no representative crystal structures for either Ru^{IV}(aq) itself or of derivative complexes. Further work in this area should be encouraged. Tetrameric Ru^{IV}(aq) is diamagnetic.

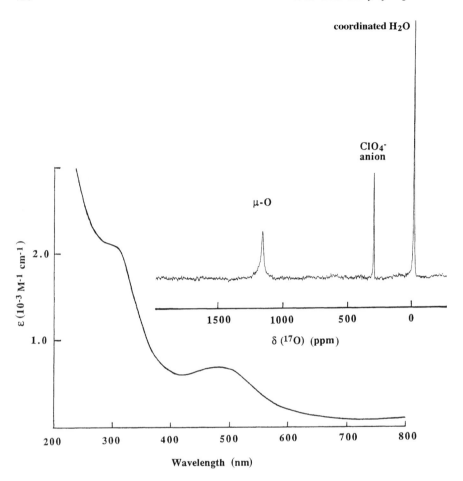

Figure 8.20. Electronic spectrum of $Ru^{IV}(aq)$ in 0.01 M $HClO_4$. (Inset: 54.24 MHz ^{17}O NMR spectrum of a fully enriched sample)

8.2.3.2 OXYGEN-17 NMR STUDIES

O^{17} NMR studies have been carried out in order to provide a further insight into the structure of tetrameric $Ru^{IV}(aq)$ [182]. Addition of an equal volume of $H_2^{17}O$ to a solution of $Ru^{IV}(aq)$ (0.2 M) in 0.02 M $HClO_4$ (containing 0.1 M Mn^{2+}) at $\mu = 1.0$ M, $NaClO_4$, results in the appearance of a single resonance at -24 ppm assignable to coordinated water. This was found to broaden at temperatures above $40\,^\circ$C due to chemical exchange with bulk water. A variable temperature linewidth study led to the following kinetics parameters, $k_{ex}(25\,^\circ C) \sim 30\,s^{-1}$, $\Delta H^{\ddagger} = 85 \pm 16\ kJ\ mol^{-1}$ and $\Delta S^{\ddagger} = +69 \pm 47\ J\,K^{-1}\,mol^{-1}$. Thus ~ 7 orders of magnitude cover the water-exchange rates on the various ruthenium aqua ions. The particular lability of the coordinated water ligands in Ru^{IV}(aq) probably stems from the presence of the μ-bridging oxo/hydroxo groups, although a full $[H^+]$ dependent study has not been carried out. A further observation during the linewidth study was the reversible appearence of new resonances at $+20$, $+120$, $+500$ and $+1150$ ppm. All of these resonances disappeared upon cooling to RT except for the resonance at $+1150$ ppm which is retained. Such behaviour is indicative of some fluctionally (perhaps partial opening up) of the structure at higher temperatures. The preparation of a fully core-enriched sample studiable at RT was thus sought. Preparative routes involving reduction of RuO_4 (e.g. with H_2O_2) or treatment of $[RuBr_6]^{2-}$ with BrO_3^- were viewed as undesirable since full enrichment of all oxygen sites during the assembly process could not be guaranteed. Finally electrochemical oxidation of solutions of fully exchanged $[Ru(^{17}OH_2)_6]^{2+}$ in HBF_4 or CF_3SO_3H solution ($t_{1/2} \sim 100s$) proved the method of choice. The oxidation was carried out on a 0.1 M solution of $[Ru(OH_2)_6]^{2+}$ in 0.5 M CF_3SO_3H at a constant current of 0.5 A with a simple galavanostatic set-up employing a cylinderical Pt gauze working electrode (area $= 6\ cm^2$) and a Pt wire counter electrode in a two compartment cell (volume $\sim 20\ cm^3$). A colour change from reddish-pink (Ru^{2+}) to yellow (Ru^{3+}) and finally to intense red-brown (Ru^{IV}) was observed. The electrolysis ceased when the peak for Ru^{IV}(aq) at 710 nm was at maximum intensity. The overall scheme to prepare a fully enriched Ru^{IV} sample is shown in Fig. 8.21.

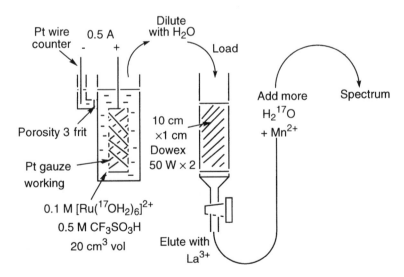

Figure 8.21. Apparatus for the preparation of fully ^{17}O-enriched Ru^{IV}(aq) in acidic aqueous solution

Figure 8.20 also shows the resulting ^{17}O NMR spectrum which, as expected, showed the presence of the additional resonance at 1157 ppm due to the bridging oxo groups.* This shows that the new resonances observed reversibly at elevated temperatures may be due to new terminal H_2O or OH ligands. The ^{17}O spectrum of $Ru^{IV}(aq)$ was, in view of the cation-exchange behaviour, surprisingly insensitive to changes in the $[H^+]$ and no shift in either resonance was observed in the range $0.1-3.0$ M $HClO_4$.

The appearance of only one resonance for bridging oxo (hydoxo) did not in itself allow a definitive assignment of the structure for $Ru^{IV}(aq)$ other than to verify the presence of an oxo-bridged oligomeric core. Attempts to prepare derivative complexes of the core, however, have had mixed success. Rapid ligand substitution occurs consistent with the fast water exchange accompanied by marked changes to the u.v.–visible spectrum. Such studies have also verified further that the polynuclear core is somewhat labile and readily disrupted. For example, a sequence of colour changes occurs upon treatment of $Ru^{IV}(aq)$ with Cl^- from red-brown through yellow, violet and then finally yellow [188]. The composition of the final yellow solution has not been ascertained and conceivably reduction to Ru^{III} chloro-aqua complexes cannot be ruled out. Treatment of $Ru^{IV}(aq)$ with NCS^- and $C_2O_4^{2-}$ leads to reduction and formation of mononuclear Ru^{III} complexes. Using NCS^- the complex $[Ru(SCN)_6]^{3-}$ has been isolated from the final reaction solution [189]. The observation of Cl^--induced depolymerisation indicates a liability of the $Ru—O(OH)$ bonds in the presence of certain anions [185]. Treatment of RuO_4 (in CCl_4) with an aqueous solution of the tripodal anionic ligand L (Fig. 8.22) in a two-phase system gives rise to the dimeric complex $[LRu^{IV}(OH_2)(\mu-O)_2Ru^{IV}(OH_2)L]^{2+}$ [190]. It is notable that this complex lacks the characteristic peak for $Ru^{IV}(aq)$ at 480 nm. A number of poorly characterised complexes with the redox inactive ligands EDTA and MIDA are believed to retain the tetrameric Ru^{IV} core [189]. However, treatment of $Ru^{IV}(aq)$ at pH 2 with a solution of potassium hydridotris(1-pyrazolyl)borate, $K[HBpz_3]$, resulted in the precipitation of a 'purple-brown' complex which gives satisfactory elemental analyses and a strong M^+ peak in the FAB MS in the range $m/e = 1356-1360$ for $\{H_nRu_4O_6(HBpz_3)_4\}^+$ ($n = 0-4$, depending upon the counteranion, ClO_4^-, $CF_3SO_3^-$, NO_3^- or BPh_4^-) (Fig. 8.23) [182]. The mass peak at 1423 corresponds to the additional coordination of a dissociated free pyrazole. The presence of M^{2+} ions at $m/2e$ values is a further indication of the presence of a highly charged cationic fragment. The presence of a ^{17}O NMR resonance at ~ 1180 ppm from a ^{17}O-enriched sample of the complex dissolved in CD_3CN

---

* Related work indicates that, when compared to, for example, the Group 6 oxo-bridged clusters of Mo^{IV} and W^{IV}, the bridging oxo groups on Ru^{IV} (aq) species should occur more significantly downfield. Thus the assignment of the 1157 ppm resonance to a bridging oxo group is not unreasonable and provides a further example of the strong metal dependence of the oxo resonances on different diamagnetic oxo-bridged cations. This probably reflects the stronger covalent bonding and thus delocalisation of metal d-electron density in $M—O—M$ bonds as opposed to $M—OH_2$.

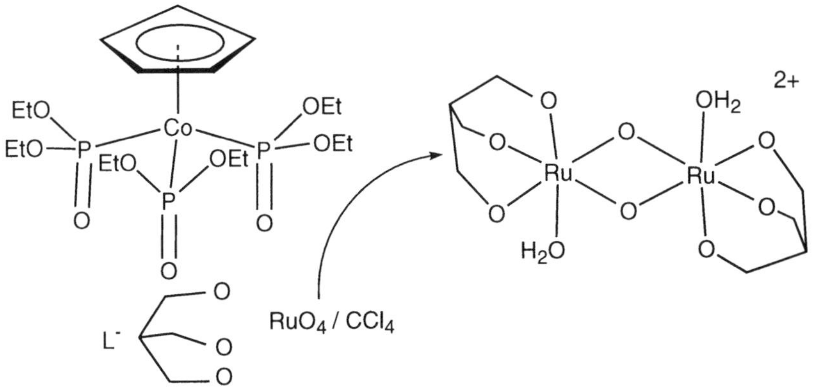

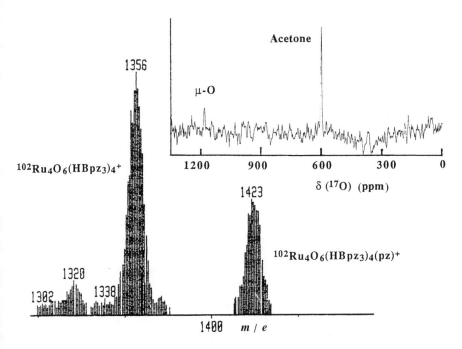

Figure 8.22. Formation of $[LRu^{IV}(OH_2)(\mu\text{-}O)_2Ru^{IV}(OH_2)L]^{2+}$ from RuO_4 ($L = \eta_5\text{-}C_5H_5)Co\{(EtO)_2P=O\}_3]^-)$

Acetone

μ-O

δ (^{17}O) (ppm)

1200   900   600   300   0

1356

$^{102}Ru_4O_6(HBpz_3)_4^+$

1423

$^{102}Ru_4O_6(HBpz_3)_4(pz)^+$

1328

1302   1338

1400   *m / e*

Figure 8.23. Positive ion fast atom bombardment mass spectrum of $H_4[Ru_4O_6(HBpz_3)_4](NO_3)_4$·(Inset: 40.56 MHz oxygen-17 NMR spectrum from a 5 atom % enriched sample (saturated) in CD_3CN (10% acetone added by volume))

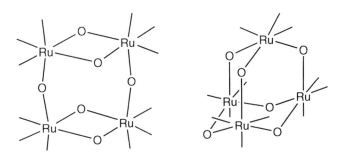

Figure 8.24. Possible structures for the $Ru_4O_6^{4+}$ unit of Ru^{IV} (aq)

(acetone reference at 600 ppm) (Fig. 8.23) confirms that it derives from the same $Ru_4O_6^{4+}$ (aq) core. Peak integration of the coordinated water resonance of Ru^{IV}(aq) indicated the presence of three H_2O ligands on each Ru. Thus Ru^{IV}(aq) was concluded to be the tetrameric ion $H_n[Ru_4O_6(OH_2)_{12}]^{(4+n)+}$, the formulation as written implying highly 'mobile' protons. One can eliminate a linear $\{Ru_4O_6\}$ core structure as having no such reason to form specifically. This leaves two likely 'cyclic' $\{Ru_4O_6\}$ structures (Fig. 8.24). Since Ru^{IV}(aq) can be eluted using La^{3+} in 0.5 M $HClO_4$ but not $\geqslant$ 2.0 M $HClO_4$ alone one assume that protonation of the oxo bridges occurs above $[H^+] = 0.5$ M.

Further support for the $Ru_4O_6^{4+}$ formulation, as predicted by Bremard, Nowogrocki and Tridot [185], may be found in the extensive Russian literature with reports of green and brown Ru^{IV}–sulfato complexes obtained from Ru^{IV}– chloro species in aqueous H_2SO_4 or from reduction of RuO_4 in the same medium [191]. Ginzburg *et al.* [192] have formulated one of the brown complexes as $K_4H_6[Ru_4O_6(SO_4)_6]$ ($Ru^{+3.5}$ state) on the basis of thermal decomposition studies and from i.r. data which appeared to rule out OH-bridge formation. Thus Ginsburg also formulates the protons in the mixed salt as being highly mobile.

8.2.3.3 REDOX PROPERTIES OF TETRAMERIC Ru^{IV} (AQ)

Wallace and Propst [180] and D'Olieslager and coworkers [183, 184] have reported electrochemical studies on tetrameric Ru^{IV}(aq) in ClO_4^- media as a function of $[H^+]$. Cyclic voltammograms of Ru^{IV}(aq) in 1.0 M $HClO_4$ show two discernible quasi-reversible reduction processes that result in the formation of Ru^{III} (Fig. 8.25). Closer analysis of the two processes shows that each corresponds to two one-electron processes, making a total of four discernible steps to Ru^{III} consistent with the tetranuclear formulation. Each one-electron reduction of the tetramer is accompanied by the addition of two protons. If one assumes that $Ru_4O_6^{4+}$ is unprotonated in 1.0 M $HClO_4$ then the final product can be formulated as '$Ru_4^{III}(OH)_4(OH_2)_{12}^{8+}$'. This Ru^{III} species is metastable and decays by

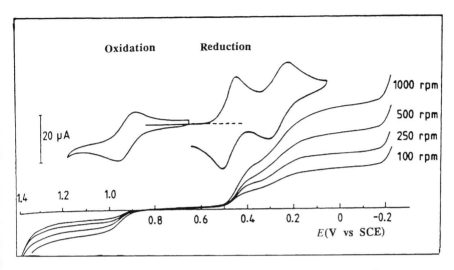

Figure 8.25. Cyclic and rotating disc voltammogram of 1×10^{-2} M Ru^{IV}(aq) in 1.0 M $HClO_4$ at a Pt electrode

a first-order process (no $[H^+]$ dependence) into a more stable form of Ru^{III} [193] from which dimeric complexes such as $[(bipy)_2(H_2O)Ru(\mu\text{-}O)Ru(OH_2)(bipy)_2]^{4+}$ (or the di-μ-OH-bridged derivative, no structural confirmation) have been isolated. Conceivably the stable form of Ru^{III}(aq) (peak at 290 nm ($\varepsilon = 2360$ M^{-1} cm^{-1} per Ru) is the aqua dimer $[Ru_2(\mu\text{-}OH)_2(OH_2)_8]^{4+}$ (cf. Cr^{3+} and Mo^{3+} species, Chapter 6). Attempts to follow the formation of these species from Ru^{IV}(aq) by ^{17}O NMR to the $Ru_2(OH)_2^{4+}$(aq) formulation have not proved definitive. Broad peak maxima around 290 nm are reported to result from Ru^{III}–chloroaqua species at pH 2 after 1–2 hours. These have been tentatively assigned to di-μ-OH-bridged species such as $\{ClRu(\mu\text{-}OH)_2RuCl\}^{2+}$(aq). The depolymerisation reaction from tetrameric Ru^{III}(aq) to dimeric Ru^{III}(aq) is most easily explained in terms of the stacked dimer arrangement for the tetrameric unit Fig. 8.24, with the breakdown of tetrameric Ru^{III}(aq) stemming from the ready cleavage of a bridging H_2O molecule formed in the final protonation step (Fig. 8.26). The stacked di-μ-O-bridged structure also readily explains how such isolated dimer structures are readily formed in the presence of certain chelating ligands (e.g. see Fig. 8.22). However Ru K edge EXAFS measurements on solutions of Ru^{IV}(aq) in 2.0 M $HClO_4$ give support for a highly symmetric structure involving sets of three nearest neighbour backscatterers; Ru—$O(OH_2)$ 2.168 pm, Ru—$O(\mu)$ 1.837 pm and Ru—Ru 3.403 pm. These parameters fit well to the alternative adamantanoid arrangement with Ru—$\hat{O}$—Ru = 13*.

---

*Osman, J. R., Richens, D. T. and Crayston, J. A. unpublished results (1996) (from p. 415).

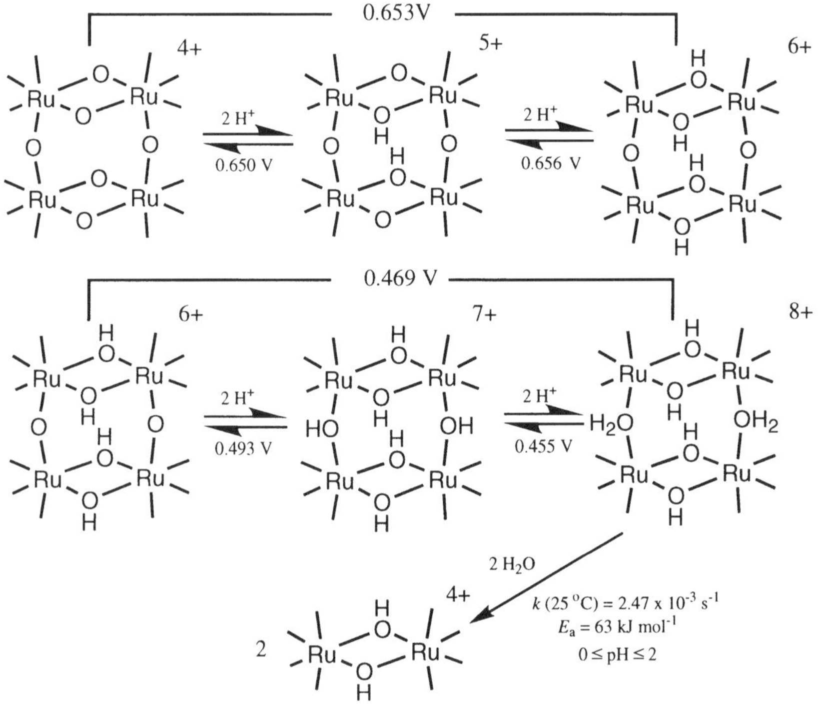

Figure 8.26. Possible species involved in the stepwise reduction of tetrameric Ru^{IV}(aq) to Ru^{III}(aq). ($E°$ values quoted are vs NHE in 1.0 M $HClO_4$)

As Fig. 8.25 shows, Ru^{IV}(aq) also exhibits a single oxidation process. Coulometric studies indicate the removal of 0.25 electron per Ru consistent with formation of a tetrameric $Ru^{+4.25}$(aq) species [184]. The standard potential for this process is quoted as $\sim +1.16$ V vs NHE. Changes to the electronic spectrum, notably loss of the peak maximum at 487 nm and the shoulder at 300 nm are observed on forming $Ru^{+4.25}$(aq). There had previously been a number of reports documenting Ru(aq) species with the formal oxidation state between $+4.2$ and $+4.3$ stemming from electrolytic reduction of RuO_4 in $HClO_4$ [175] and from a number of other chemical and electrochemical oxidative titrations on Ru^{IV}(aq) [179, 194]. It is not clear why the $Ru^{+4.25}$ state should be so stable. It should be ESR active but no studies have yet been reported.

Cyclic voltammograms from 'stable' aqua Ru_2^{III} in 1.0 M $HClO_4$ show an irreversible $2e^-$ reduction process which is followed on the return scan by the development of characteristic reversible waves for the $[Ru(OH_2)_6]^{2+/3+}$ couple [193]. The ratio of peak heights as a function of scan rate supports a ECE mechanism (Fig. 8.27). Further characterisation of the aqua Ru^{III} dimer and

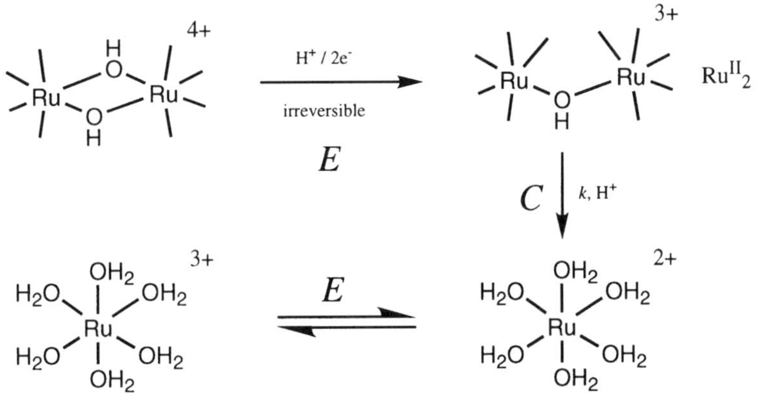

Figure 8.27. Formation of $[Ru(OH_2)_6]^{2+}$ following electrochemical reduction of stable $Ru^{III}_2(aq)$ in 1.0 M $HClO_4$ [193]

related species is required. The tendency of the bridging oxo groups in these species to readily protonate, even in the case of the Ru^{IV} (compare the Group 6 oxo-bridged species), probably stems from the reasonably high electron density maintained at both the O and Ru centres due to the absence of M—M bonding and the lack of an appreciable O—Ru π-bonding component.

Reduction of RuO_4 with insufficient $[H^+]$ to support formation of soluble $Ru^{IV}(aq)$ leads to black colloidal suspensions of the hydrated dioxide $RuO_2 \cdot nH_2O$. Raising the pH of solutions of $Ru^{IV}(aq)$ above 3.0 leads to similar precipitated species. Varying n values have been claimed. Connick and Hurley [195] and Gortsema and Cobble [178] claimed $n = 2$ whereas Fletcher *et al.* reported values of 1.0–1.3 from H_2O_2 reduction of RuO_4 [196]. Reduction with H_2 is, however, reported to give $n = 1$ [197]. Values of n between 1.0 and 2.0 seem generally accepted. Thin films of $RuO_2 \cdot nH_2O$ are of interest for their potential electrocatalytic properties. Most are characterised by strong and continuous i.r. bands from the visible region all the way to ~ 45 cm^{-1} which mask all the vibrational bands due to OH or H_2O. This property has been attributed to the presence of electronic conduction bands [196]. Mixed RuO_2–IrO_2 oxide composites have also been the subject of thin-film electrocatalytic studies [198].

8.2.4 Ruthenium(VI), (VII) and (VIII)(d²–d⁰): The Chemistry of the Tetraoxo Species $[RuO_4]^{-,0}$ and Trigonal Bipyramidal $[RuO_3(OH)_2]^{2-}$

There are no true aqua ions for Ru^V although the species $[Ru^V(O)Cl_2(OH_2)_3]^+$, detected as an intermediate in $[RuCl_2(OH_2)_4]^+$-mediated oxidations, could be said to derive from putative $[Ru(O)(OH_2)_5]^{3+}$. Ru^{VI}, Ru^{VII} and Ru^{VIII} states are, however, represented in the form of a series of oxo-anionic species. For many

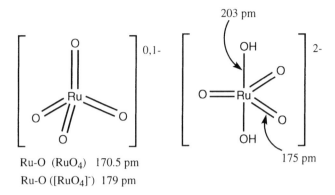

Ru-O (RuO₄) 170.5 pm   175 pm
Ru-O ([RuO₄]⁻) 179 pm

Figure 8.28. Structures of Ru^{VI}, Ru^{VII} and Ru^{VIII} oxo species [199–201]

years it was believed that tetraoxo $[RuO_4]^{2-,-,0}$ anionic structures were relevant for all three states [195]. While this is true for Ru^{VII} and Ru^{VIII} [199, 200], Ru^{VI}(aq) is now known to be present as the trigonal bipyramidal anion $[RuO_3(OH)_2]^{2-}$ [201] (Fig. 8.28).

RuO_4 itself is easily made from lower valent Ru compounds by oxidation of neutral or acidic aqueous solutions with hypochlorite, MnO_4^-, BiO_3^- or O_3. The early preparations involved distillation of RuO_4 from solutions of Ru^{III} sulfate in H_2SO_4 in the presence of MnO_4^- and collection in a liquid nitrogen trap. As an alternative to handling the highly hazardous gaseous oxide more recent procedures have made use of the high solubility of RuO_4 in CCl_4 for extraction and transfer purposes [181, 202]. RuO_4 melts at $25\,^{\circ}C$ and has an appreciable solubility in water (0.13 mol dm^{-3}) at $25\,^{\circ}C$. Aqueous solutions have weakly acidic properties. The formulation as $RuO_4 \cdot H_2O$ may actually be $RuO_3(OH)_2$. Martin [202] measured the first acid dissociation constant for 'H_2RuO_5' as $(6.8 \pm 0.3) \times 10^{-12}$ M which compares well with the value $(1.3 \pm 0.2) \times 10^{-12}$ M determined by Silverman and Levy [200]. Protonation to give $HRuO_4^+$ is reported to occur with an equilibrium constant of $(5.7 \pm 0.8) \times 10^{-15}$ M^{-1} [202], although the existence of this species has been questioned [203]. RuO_4, when generated *in situ* from '$RuCl_3$' in the presence of a suitable cooxidant, is a powerful oxidising reagent in organic chemistry. However, it is much less selective than the less potent OsO_4 and thus has found fewer useful applications (see later).

Reduction of RuO_4 in aqueous alkaline solution with mild reagents such as H_2O_2 gives rise to solutions containing Ru^{VI}(aq) or Ru^{VII}(aq) depending upon the conditions. At high enough concentrations OH^- itself is a sufficient reductant. $[RuO_4]^-$ forms first and eventually converts to trigonal bipyramidal $[RuO_3(OH)_2]^{2-}$. The rate of formation of $[RuO_3(OH)_2]^{2-}$ is also found to increase with the $[OH^-]$. Addition of hypochlorite, which reacts rapidly with any $[RuO_3(OH)_2]^{2-}$, has allowed the selective formation of $[RuO_4]^-$ under certain conditions [195]. The half-life for $Ru^{VII}(aq)/Ru^{VI}(aq)$ self-exchange is quite short (< 5 S) [204] for a process involving a change in coordination number. Rather this implies the existence of a similar structural form in solution for both, perhaps relating to a rapid equilibrium between $[MnO_4]^{n-}$ and its 'hydrated' $[MO_3(OH)_2]^{n-}$ form. Further work is needed to truly substantiate the solution structures of $Ru^{VI}(aq)$ and $Ru^{VII}(aq)$. The E^θ value for the Ru^{VIII}/Ru^{VII} couple has been quoted as $+0.996 \pm 0.005$ V vs NHE (pH 9–12). That for the corresponding Ru^{VII}/Ru^{VI} couple is estimated to be $+0.586 \pm 0.003$ V vs NHE [6]. The quoted potentials are in line with equilibrium constants found for reactions of the oxy anions of ruthenium with the corresponding species of manganese. During electrochemical reduction of RuO_4 in H_2SO_4 a metastable green intermediate is observed on the way to formation of red-brown Ru^{IV}. This intermediate has been suggested to contain complexed $Ru^{VI}O_2^{2+}$ ions [205]. The stabilisation of $[RuO_3(OH)_2]^{2-}$ versus $[RuO_4]^-$ as the $[OH^-]$ increases stems from an oxidation of OH^- ions by $[RuO_4]^-$. The overall reaction may be represented by:

$$2[RuO_4]^- + 2OH^- + H_2O \longrightarrow 2[RuO_3(OH)_2]^{2-} + \tfrac{1}{2}O_2 \qquad (8.22)$$

The rate law indicates a complex mechanism by containing a squared term in $[RuO_4^-]$ and a cube term in $[OH^-]$ [206], perhaps reflective also of the change in coordination number. Electronic spectra for the various high valent Ru(aq) species are shown in Fig. 8.29. The band maxima observed are at 310 nm

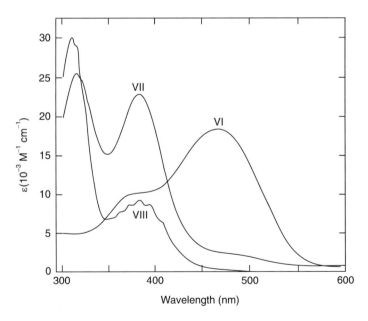

Figure 8.29. Electronic spectra for $Ru^{VI, VII, VIII}(aq)$ species

($\varepsilon = 2960 \text{ M}^{-1} \text{ cm}^{-1}$) and 385 nm (930 M^{-1} cm^{-1}) (RuO$_4$); 310 (2445), 385 (2275) and 460 nm (283 M^{-1} cm^{-1}) (Ru^{VII}); and 385 (1030) and 460 nm (1820 M^{-1} cm^{-1}) (Ru^{VI}). The lack of a band in the 310 nm region for Ru^{VI}(aq) may be reflective of the change to the 'RuO$_5$' trigonal bipyramidal coordination. The above wavelengths are also those used for analysis.

As the pH is lowered, Ru^{VI}(aq) becomes unstable with respect to disproportionation to give Ru^{VII}(aq) and Ru^{IV} as RuO$_2 \cdot n$H$_2$O (n probably 2). Measurement of the equilibrium constant for the disproportionation reaction

$$3[\text{RuO}_3(\text{OH})_2]^{2-} + (2+n)\text{H}_2\text{O} = 2[\text{RuO}_4]^- + \text{RuO}_2 \cdot n\text{H}_2\text{O} + 4\text{OH}^- \quad (8.23)$$

$$[\text{RuO}_3(\text{OH})_2]^{2-} + 2\text{e}^- + (2+n)\text{H}_2\text{O} = \text{RuO}_2 \cdot n\text{H}_2\text{O} + 4\text{OH}^- \quad (8.24)$$

has allowed an estimation for the E^θ value for the Ru^{VI}/Ru^{IV} couple in alkaline solution as $+0.35$ V [205, 207]. RuO$_4$ is a very powerful oxidising reagent if largely unselective. However, because it can be prepared in a catalytic *in situ* form with relative ease it has proved attractive for a number of conversions, in particular the oxidation of alkenes, alkynes, alcohols, ethers and aromatic rings [208]. It is perhaps most commonly used for the oxidation of secondary OH groups in carbohydrates and steroids to the corresponding ketones [209]. It can oxidatively cleave aromatic rings as in the oxidation of napthalene to give phthalic acid [210]. In particular, the ability of RuO$_4$/NaOCl to destroy dioxins [211] and polychlorinated biphenyls (PCBs) [212] may yet be crucial importance in environmental pollution control. RuO$_4$ has found few applications other than those mentioned above. On the other hand, perruthenate, [RuO$_4$]$^-$, has attracted much more attention as a more delicate and selective oxidising reagent for use in the fine chemicals industry. As TBAP (tetrabutylammonium perruthenate, (Bu$_4$N)[RuO$_4$]) or TPAP (the tetrapropyl ammonium salt), [RuO$_4$]$^-$ can be dissolved in a variety of organic solvents such as CH$_2$Cl$_2$ or CH$_3$CN to carry out highly selective oxidations [213–215]. The salts are easily prepared by passing a stream of RuO$_4$ into an aqueous basic solution of Bu$_4$NOH or Pr$_4$NOH wherein the deep green TBAP or TPAP precipitates [213]. TPAP can be obtained in higher yield. The salts will function catalytically if N-methylmorpholine-N-oxide is used as the cooxidant. Thus TPAP will selectively oxidise primary alcohols to aldehydes and secondary alcohols to ketones without attacking double or allylic bonds [216]. Such selective oxidations can also be carried out in the presence of a whole range of functional groups such as epoxides, lactones, indoles, silyl ethers, acetals and tetrahydropyranyl functions. A good example is provided by a step in the total synthesis of the antiparasitic reagent Milbemycin b [215].

Aqueous alkaline ruthenate, [RuO$_3$(OH)$_2$]$^{2-}$, can be used as a useful two-electron oxidant, e.g. for converting primary alcohols to carboxylic acids and secondary alcohols to ketones [216]. It can be made catalytic in the presence of excess S$_2$O$_8^{2-}$ which oxidises any RuO$_2 \cdot n$H$_2$O back to Ru^{VI}.

8.3 OSMIUM

Osmium comes as a complete contrast to the above two elements of Group 8 in having no characterised cationic aqua ions. Despite the existence of well-charac-terised $[M(OH_2)_6]^{3+/2+}$ ions for both iron and ruthenium neither species has been confirmed in the case of osmium. This is also somewhat surprising given that its neighbouring group 9 counterpart iridium forms well-characterised $[Ir(OH_2)_6]^{3+}$. There is no evidence either for an Os^{IV} analogue of tetrameric $[Ru_4O_{6-n}(OH)_n(OH_2)_{12}]^{(4+n)+}$. Water-derived osmium species seem to be largely limited to neutral or anionic oxo derivatives such as *trans*-$[OsO_2(OH)_4]^{2-}$ (Fig. 8.30) and the Os^{VIII} species, *cis*-$[OsO_4(OH)_2]^{2-}$ and tetrahedral OsO_4 (Fig. 8.32).

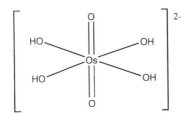

Figure 8.30. Structure of *trans*-$[OsO_2(OH)_4]^{2-}$

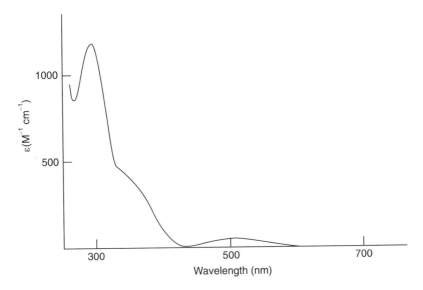

Figure 8.31. Electronic spectrum of *trans*-$[OsO_2(OH)_4]^{2-}$ in aqueous KOH

Indeed, osmium seems more to resemble rhenium in its chemistry than its lighter partner ruthenium or its next neighbour iridium. Since iridium is the first third row transition metal to form a stable hexaaqua ion (in the $+3$ state, t_{2g}^6), the preferred formation of high valent oxo species for osmium and rhenium could arise from a particularly high electronegativity of the low-spin forms of Os^{VI} (Re^V) (t_{2g}^2) and Os^{IV} (Re^{III}) (t_{2g}^4), promoting water deprotonation with strong effective π bonding from the resulting 'oxo' groups into the available empty t_{2g} orbitals. A good example is provided by the low pK_{11} value (0.3) for the neutral Os^{IV} complex $[OsCl_4(OH_2)_2]$ (see below).

8.3.1 $[Os(OH_2)_6]^{3+/2+}$: Do They Exist or Could They Exist?

To answer this question we can examine the available evidence from likely synthetic routes and from the properties of related compounds. Firstly a reasonable estimate for the reduction potential for the '$[Os(OH_2)_6]^{3+/2+}$' couple can be based upon comparisions between other $Ru^{3+/2+}$ and $Os^{3+/2+}$ complex couples and then extrapolation from the known potential for '$[Ru(OH_2)_6]^{3+/2+}$'($+0.21$ V). Spin–orbit coupling stabilises the ground state low-spin t_{2g}^5 configurations of both Os^{III} and Ru^{III}. The corresponding M^{II} states (t_{2g}^6) are not likewise stabilised and the combined effect is to lower the $M^{III/II}$ couple for these heavier Group 8 elements. Spin–orbit coupling constants are, however, much larger for Os^{III} than Ru^{III} and this makes the Os^{III}/Os^{II} couples significantly more negative than their ruthenium counterparts. For example, the $[Os(NH_3)_6]^{3+/2+}$ couple lies 0.83 V more negative than for $[Ru(NH_3)_6]^{3+/2+}$ [217]. The reduction potential for the aquapentaammine couple for Os is shifted slightly positive vs the hexaammine couple [218] paralleling the behaviour for ruthenium. Thus one could infer a value of ~ -0.6 V for '$[Os(OH_2)_6]^{3+/2+}$' which suggests that if '$[Os(OH_2)_6]^{3+}$' exists then the corresponding divalent ion should also.

8.3.1.1 POSSIBLE SYNTHETIC ROUTES TO PUTATIVE $[Os(OH_2)_6]^{3+}$

Two obvious routes might be envisaged: (a) reduction chemically or electrochemically from higher valent compounds and (b) aquation of a suitable lead-in precursor.

Reduction

Extensive studies of the products of electrochemical reduction of *trans*-$[Os^{VI}O_2Cl_4]^{2-}$ in aqueous HCl solution have shown that both Os^{IV} and Os^{III} chloroaqua complexes can be generated [219]. The reduction of *trans*-$[OsO_2Cl_4]^{2-}$ occurs in a single irreversible step to give the Os^{III} complex $[OsCl_4(OH_2)_2]^-$. The potential of the corresponding $[OsCl_n(OH_2)_{6-n}]^{(4-n)+/(3-n)+}$ couples ($n = 4$, 5 or 6) are, however, of interest, the Os^{IV}/Os^{III} potential increasing as water replaces Cl^-:

$$[OsO_2Cl_4]^{2-} + 4H^+ + 3e^- \xrightarrow[\text{irreversible}]{-0.1\,V} [Os(OH_2)_2Cl_4]^- \tag{8.25}$$

$$[Os(OH_2)_2Cl_4] + e^- \xrightleftharpoons{+0.19\,V} [Os(OH_2)_2Cl_4]^- \tag{8.26}$$

$$\Big\Updownarrow Cl^-$$

$$[Os(OH_2)Cl_5]^- + e^- \xrightleftharpoons{+0.10\,V} [Os(OH_2)Cl_5]^{2-} \tag{8.27}$$

$$\Big\Updownarrow Cl^-$$

$$[OsCl_6]^{2-} + e^- \xrightleftharpoons{+0.01\,V} [OsCl_6]^{3-} \tag{8.28}$$

all potentials are quoted vs SCE for 6.0 M HCl

A positive 100 mV shift is seen for every water ligand replacing Cl^-, which, if one makes a linear extrapolation, would imply a value for the reduction potential for the hypothetical $[Os(OH_2)_6]^{4+/3+}$ couple of $\sim +1.6$ V vs SCE under these conditions. Unfortunately, these ideas cannot be tested since at lower Cl^- ion concentration Os^{IV}(aq) precipitates as the hydrous dioxide $OsO_2 \cdot 2H_2O$ from acidic solution such that $[Os(OH_2)_6]^{4+}$, or a related form, has never been isolated. The high 'acidity' of t_{2g}^4 aqua Os^{IV} is indicated by the low pK_{11} (0.3) for formation of $[OsCl_4(OH)(OH_2)]^-$ from neutral $[OsCl_4(OH_2)_2]$ [219]. Here the hydroxo ligand is stabilised by π bonding into the empty t_{2g} orbital of the spin-paired configuration. The standard aqueous potentials for osmium species (Table 8.13) indicate that Os^{VI} is unstable in acidic solution with respect to disproportionation to give Os^{VIII} and $OsO_2 \cdot 2H_2O$ [220].

There have been a number of studies concerning the polarographic and voltammetric reduction of aqueous solutions of OsO_4 under a variety of conditions. In alkaline media, voltammetric reduction of OsO_4, present as the 'perosmate' anion, *cis*-$[OsO_4(OH)_2]^{2-}$, at a Pt cathode is said to involve three steps,

Table. 8.13. Standard reduction potentials for osmium species in acid solution [220]

Couple	E° (V) vs NHE
$OsO_4/[OsO_2(OH)_4]^{2-}$	$+0.46$ V
$[OsO_2(OH)_4]^{2-}/OsO_2 \cdot 2H_2O$	$+1.61$ V
$OsO_4/OsO_2 \cdot 2H_2O$	$+0.96$ V[a]
$OsO_4/[OsCl_6]^{2-}$	$+1.0$ V
$[OsCl_6]^{2-}/[OsCl_6]^{3-}$ (hydrolysed)	$+0.45$ V

[a] In 1.2×10^{-3} M K_2SO_4 or 10^{-2} M Na_3PO_4.

$Os^{VIII}-Os^{VI}$ (positive potential) (presumably as $trans$-$[OsO_2(OH)_4]^{2-}$), $Os^{VI}-Os^{IV}$ (-0.41 V) and $Os^{IV}-Os^{III}$ (-1.02 V) [221]. Meites has studied the polarographic reduction of OsO_4 in both alkaline, neutral and acidic pH [222]. It is reported that controlled potential reduction of OsO_4 at -0.11 V in $\geqslant 0.5$ M NaOH solution consumes two electron to generate Os^{VI}, presumably $trans$-$[OsO_2(OH)_4]^{2-}$. Two further reductions are then seen at -0.37 V (to Os^{IV}) and at -1.27 V (to Os^{III}). In 1.0 M $HClO_4$, 1.0 M HCl, 0.5 M H_2SO_4 and 7.3 M H_3PO_4 the current rises immediately at the dropping mercury electrode from zero applied potential ($\sim +0.2$ V) to consume $5e^-$ per osmium, giving Os^{III}. Concentrations of osmium are around 4×10^{-5} to $\sim 10^{-3}$ M. Under these conditions the Os^{III} product is reported as a brown precipitate. However, the positive reduction potential for the generation of Os^{III} under the acidic conditions is in keeping with the predictions from the redox studies on the chloroaqua complexes (8.25) to (8.28). When reduction is carried out in 4.0 M HCl the Os^{III} product is soluble and assumes a light straw colour. The same colour results when the above brown precipitate is dissolved in 4.0 M HCl, assigning it to a form of hydrous Os_2O_3, a compound still lacking definitive characterisation. A similar $3e^-$ reduction all the way to Os^{III}, also characterised by a spontaneous rising current at zero applied potential, is seen in polarograms from Os^{VI} solutions adjusted into the acid range $0.21-9.1$ M H^+ [117]. The polarographic behaviour of OsO_4 is thus highly reminiscent of ReO_4^- and TcO_4^- (see Section 7.2) rather than to its lighter partner ruthenium.

In solutions of weak acids (pH values of $2-4$) the polarograms consist of two waves, $Os^{VIII}-Os^{IV}$ and $Os^{IV}-Os^{III}$ [222, 223]. At pH values > 5 the characteristic alkaline polarograms appear with appearance of three waves, $Os^{VIII}-Os^{VI}$, $Os^{VI}-Os^{IV}$ and $Os^{IV}-Os^{III}$. This presumably arises because of the higher pH values necessary for the generation and stabilisation of $trans$-$[Os^{VI}O_2(OH)_4]^{2-}$. One can conclude from this that in acid solution both Os^{VI}(aq) and Os^{IV}(aq) species appear to be unstable towards disproportionation and that reduction of the Os^{VIII} occurs spontaneously at the mercury electrode surface. Lay has, however, intimated that a soluble Os^{III}(aq) product ought to be achievable by reduction from high valent osmium complexes in non-complexing acidic solution under the appropriate conditions, given that Os^{III} is formed as a product in both acid and alkaline media. However, as yet the putative Os^{3+}(aq) species has not been characterised [224]. It is conceivable, in view of the polarographic results, that the choice of pH could be crucial. In acid media the production of hydrous Os_2O_3 may be as a result of the $5e^-$ reduction straight from aqueous OsO_4. In weakly acidic solution soluble Os^{IV} and Os^{III} products appear to be formed possibly as hydroxoaqua complexes [222]. Thus the appropriate choice of pH and non-complexing acidic medium may yet permit the isolation of a soluble Os^{III} aqua or hydroxy aqua complex. It has also been suggested that putative 'Os^{3+}(aq)' might be a mixed hydridoaqua species [224].

Aquation

Aquation or hydrolysis of aqueous $[OsCl_6]^{3-}$ might seem another viable route. The electrochemical reduction of *trans*-$[OsO_2Cl_4]^{2-}$ above gives $[Os^{III}Cl_4(OH_2)_2]^-$, but further aquation does not seem straightforward. The complexes $[OsCl_4(OH_2)_2]$ [225], $[OsCl_5(OH_2)]^-$ [225] and $[OsCl_5(OH_2)]^{2-}$ [226] seem nonetheless well characterised. The known synthetic route to dark red Os(acac)₃ via reaction of acacH with $[OsBr_6]^{3-}$ [227] has prompted attempts at the preparation of $Os(CF_3SO_3)_3$. A number of routes to isolable $[OsX_6]^{3-}$ complexes have now been reported involving chemical (Ag metal) [227] and electrochemical [228] reduction of solutions of the commercially available compounds $K_2[OsX_6]$ or $H_2[OsX_6]$ (X = Cl or Br). Dark brown '$OsCl_3 \cdot nH_2O$' is also marketed commercially and in one attempt this was treated with an excess of deoxygenated neat CF_3SO_3H. However, this approach has so far proved unsuccessful; the addition of deoxygenated water to the resulting dark brown solution, after stirring overnight under argon, spontaneously gave rise to the violet colour characteristic of a dioxo Os^{VI} species, presumably *trans*-$H_m[OsO_2Cl_n(OH)_{4-n}]^{(2-m)-}$ [229]. The accumulated evidence would seem to point against the direct use of halide-containing osmium precursors for the synthesis of Os^{3+}(aq). Further efforts aimed at synthesising labile O-donor complexes such as $Os(CF_3SO_3)_3$ should, however, be encouraged.

Finally, little seems known of the nature of alkaline solutions of Os^{III}. Claus and Jacoby first reported that addition of alkali to aqueous solutions of salts of $[OsCl_6]^{3-}$ produces a reddish-brown precipitate of hydrous Os_2O_3 [230]. There is little information on the alkaline solubility of the oxide, likely to require rigorous air-free conditions in view of the polarographic potentials. The oxide was, however, reported to be insoluble in acids but there are few details. There is also little indication of how ageing of the oxide might affect the dissolution process. Further studies in this area are also required.

The fact that $[Os(NH_3)_6]^{3+}$ and $[Os(NH_3)_6]^{2+}$ are, in contrast, both well characterised [224] points to a particular preference for good σ-donor ligands by the low-spin t_{2g}^5 Os^{III} and t_{2g}^6 Os^{II} centres. Water is a weaker σ-donor at the expense of being a potential π-donor, especially upon deprotonation to oxo, which would tend to favour the higher oxidation states, as is observed. Ammonia is also less readily reduced and is less acidic.

8.3.2 Osmium(VI) (d²): The *trans*-$[OsO_2(OH)_4]^{2-}$ Ion

This ion is very conveniently made via treatment of OsO_4 (see below) with ethanolic KOH and isolation as the K^+ salt [231]:

$$OsO_4 + 2\,KOH \xrightarrow{\text{ethanol}} K_2[\textit{trans-}OsO_2(OH)_4] \qquad (8.29)$$

The X-ray structure reported for the K^+ salt confirmed the *trans*-dioxo geometry [232] (Fig. 8.30) analogous to other d^2 compounds, e.g. Re^V, Mo^{IV} and W^{IV}. Solutions containing the pink *trans*-$[OsO_2(OH)_4]^{2-}$ ion can also be obtained by controlled electrochemical reduction of alkaline solutions of OsO_4 [233] or by chemical reduction with CO_3^{2-} or by gentle warming of the alkaline OsO_4 solutions [233]. The electronic spectrum is shown in Fig. 8.31 [231].

The preference for the *trans*-dioxo geometry arises from it being the only one allowing the favourable pairing-up of the two d electrons within the perpendicular $5d_{xy}$ orbital. This then allows for strong π bonding from the oxo groups into the two empty remaining $5d_{xy,yz}$ orbitals, causing a noticeable tetragonal compression. The Os—O distances in the ion are typically; Os—O(oxo) 177 pm and Os—O(OH) 203 pm [232]. The v_{as} and v_s stretching frequencies of the *trans*-OsO_2 group typically occur at 800 and ~ 850 cm^{-1} respectively. $[OsO_2(OH)_4]^{2-}$ is the anion of the weak dibasic acid $H_2[OsO_2(OH)_4]$. The pK values for the acid are p$K_{11} = 8.5$ and p$K_{12} = 10.4$ [234].

The formally 'heptavalent' osmium compound, $Sr_3[OsO_4(OH)_2]_2$, has been found to contain discrete Os^{VIII} and Os^{VI} units, and is thus far the only example of the $[Os^{VI}O_4(OH)_2]^{4-}$ moiety [235].

8.3.3 Osmium(VIII) (d^0): Tetrahedral OsO_4 and *cis*-$[OsO_4(OH)_2]^{2-}$

Tetrahedral OsO_4 is by far the most well-known and well-studied osmium-containing species and it was the form within which the element was first seperated and purified [236]. Gas phase OsO_4 (Fig. 8.32) is a regular tetrahedron (Os—O = 171.1 pm) but a slight distortion is seen in the crystalline state (Os—O = 168.4(7), 171.0(7) pm) [237]. The pale yellow crystals have a solubility in water (25 °C) of 72.4 g dm^{-3} [238] and the resulting solutions are weakly acidic as a result of the following equilibrium giving small amounts of $H_2[cis\text{-}OsO_4(OH)_2]$ [234]:

$$OsO_4 + 2\,H_2O \underset{}{\overset{K = 8 \times 10^{-13}}{\rightleftharpoons}} H_2[OsO_4(OH)_2] \tag{8.30}$$

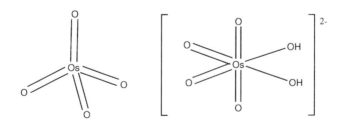

Figure 8.32. Structures of the Os^{VIII} species OsO_4 and *cis*-$[OsO_4(OH)_2]^{2-}$

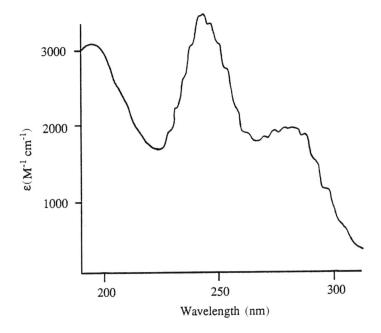

Figure 8.33. Ultraviolet–visible spectrum of OsO_4

As a result solutions of OsO_4 have a small residual conductance. Here Os^{VIII} differs from Ru^{VIII} in being able to expand its coordination number beyond four. The *cis* arrangement of hydroxide ligands allows the maximum number of orbitals to be available for π bonding from the hydroxo groups since Os^{VIII} is $5d^0$. Salts of the deep red *cis*-$[OsO_4(OH)_2]^{2-}$ ion (Fig. 8.32) are known with Li^+ [239], Na^+ [240], Sr^{2+} and Ba^{2+} [241].

Typical distances are; Os—O(oxo) 175 pm and Os—O(OH) 216 pm. The electronic spectrum in water is shown in Fig. 8.33 [206, 242]. The properties of OsO_4 in base have been studied and the change in the electronic spectrum with pH has been used to estimate the pK_{11} value of $H_2[OsO_4(OH)_2]$ as 12.2 ($I = 0.1$ M) [234]. The hydroxo-bridged dimeric ion $[O_4Os—OH—OsO_4]^-$ (Fig. 8.34) is also known in Rb^+ and Cs^+ salts [243]. The dimer is probably formed via reaction of monohydroxo $[OsO_4(OH)]^-$ with OsO_4 in the presence of the M^+ cation.

Os^{VIII} is also present in the solid compound $Sr[OsO_5(OH_2)]$ [244], later formulated as $Sr[OsO_3(OH)_2(OH_2)](OH)_2 \cdot H_2O$ [245]. The Os—OH_2 distance is quoted as 212 pm.

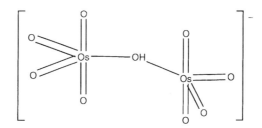

Figure 8.34. Structure of the hydroxo-bridged Os^{VIII} ion in $Rb[O_4Os(OH)OsO_4]$

8.3.4   Water Ligand Replacement Reactions on Os(aq) Species

The only water-exchange process carried out thus far on a cationic Os^{n+}(aq) species has been that on $[Os^{II}(\eta_6\text{-}C_6H_6)(OH_2)_3]^{2+}$ along with its ruthenium counterpart [25]. This might be considered a chance to compare behaviour on the two metals. The kinetic parameters, obtained using the variable temperature ^{17}O NMR line-broadening method, were k_{ex} (25 °C) = 11.8 ± 2.0 s^{-1}, $\Delta H^{\ddagger}$ = 65.5 ± 2.2 kJ mol^{-1}, $\Delta S^{\ddagger}$ = −4.8 ± 6.1 J K^{-1} mol^{-1} and $\Delta V^{\ddagger}$ = +2.9 ± 0.6 cm^3 mol^{-1}. These parameters are virtually identical to those found for the ruthenium analogue (p. 393), meaning that if there is a difference in lability between the two metals then this is not manifest in this class of compound. Here it appears that the rate of water exchange is metal independent and entirely controlled by electronic effects stemming from the π-benzene ligand. The small positive activation volume is suggestive of an interchange process towards the centre of the continuum.

A qualitative picture of the reactivity of $[OsCl_4(OH_2)_2]$ has emerged from the electrochemical studies involving reduction from *trans*-$[OsO_2Cl_4]^{2-}$ [219]. Anation of $[OsCl_4(OH_2)_2]$ to give $[OsCl_5(OH_2)]^-$ is complete within one day in 6.0 M HCl whereas in 0.4 M HCl, 7.0 M LiCl it takes only 2 hours. Hydrolysis and equlibration reactions involving $[OsCl_6]^{2-}$ as well as various $[Os^{IV}Cl_n(OH_2)_{6-n}]^{(4-n)}$ complexes have been reported in a number of other studies [246].

Few other detailed studies have been reported, mostly concerning aquation of the pentaammine complexes, $[Os(NH_3)_5X]^{2+/1+}$ (X = N_2, Cl$^-$ and CF$_3$SO$_3^-$) [224]. A significant difference between the two metals is apparent. Both Os^{II} and Os^{III} are more inert than their ruthenium counterparts, the biggest difference being with the trivalent complexes. For example, aquation of $[M(NH_3)_5(O_3SCF_3)]^{2+}$ to give $[M(NH_3)_5(OH_2)]^{3+}$ at 25 °C is 10^5 times slower on Os^{III} ($k = 8.8 \times 10^{-4}$ s^{-1}) [217] than on Ru^{III}. Os^{III} is however, significantly more labile than Os^{II} (by a factor of ∼ 10^6) the latter believed to react more dissociatively.

REFERENCES

[1] (a) Ohtaki, H., Yamaguchi, T., and Maeda, M., *Bull. Chem. Soc. Japan*, **49**, 701 (1976).
    (b) Kalman, E., Radnai, T., Palinkas, G., Hajdu, F., and Vertes, A., *Electrochim. Acta*, **33**, 1223 (1988).
[2] Herdman, G. J., and Neilson, G. W., *J. Phys. Condens. Matter*, **4**, 649 (1992).
[3] Sham, T. J., Hastings, B. K., and Perlman, M. L., *Chem. Phys. Lett.*, **83**, 391 (1981).
[4] Chmelnik, A. M., and Fiat, D. N., *J. Am. Chem. Soc.*, **93**, 2875 (1971).
[5] Fratiello, A., Kubo, V., Peak, S., Sanchez, B., and Schuster, R. E., *Inorg. Chem.*, **10**, 2552 (1971).
[6] Nicholls, D., in *Comprehensive Inorganic Chemistry* (eds. J. C. Bailar Jr, H. J. Emeleus, R. S. Nyholm and A. F. Trotman- Dickenson), Vol. 3, Pergamon, London, 1973, 1007.
[7] (a) Figgis, B. N., Kucharski, B. N., Reynolds, P. A., and Tasset, F., *Acta Cryst.*, **C45**, 942 (1989).
    (b) Figgis, B. N., Forsyth, J. B., Kucharski, B. N., Reynolds, P. A., and Tasset, F., *Proc. Roy. Soc. Lond.*, **428**, 113 (1990).
[8] (a) Figgis, B. N., Kepert, C. J., Kucharski, B. N., and Reynolds, P. A., *Acta Cryst.*, **B48**, 753 (1992).
    (b) Chandler, G. S., Christos, G. A., Figgis, B. N., and Reynolds, P. A., *J. Chem. Soc. Faraday Trans. I*, **88**, 1961 (1992).
[9] Cotton, F. A., and Meyers, M. D., *J. Am. Chem. Soc.*, **82**, 5023 (1960).
[10] Baes, C. F., and Mesmer, R. E., *The Hydrolysis of Cations*, Krieger, Florida, 1986, pp. 226–8.
[11] Morozumi, T., and Posey, F. A., *Denki Kagaku*, **35**, 633 (1967).
[12] (a) Baes, C. F., and Mesmer, R. E., *The Hydrolysis of Cations*, Krieger, Florida, 1986, p. 229.
     (b) Sweeton, F. H., and Baes, C. F., *J. Chem. Thermodyn.*, **2**, 479 (1970).
[13] Swift, T. J., and Connick, R. E., *J. Chem. Phys.*, **37**, 307 (1962).
[14] (a) Eigen, M., and Tamm, K., *Z. Elektrochem.*, **66**, 93, 107 (1962).
     (b) Bechtler, A., Breitschwerdt, K. G., and Tamm, K., *J. Chem. Phys.*, **52**, 2975 (1970).
[15] Ducummon, Y., Newman, K. E., and Merbach, A. E., *Inorg. Chem.*, **19**, 3696 (1980).
[16] Kustin, K., Taub, I. A., and Weinstock, E., *J. Chem. Phys.*, **40**, 877 (1964).
[17] Eisenstadt, M., *J. Chem. Phys.*, **51**, 4421 (1969).
[18] Holyer, R. H., Hubbard, C. D., Kettle, S. F. A., and Wilkins, R. G., *Inorg. Chem.*, **4**, 929 (1965) and **5**, 622 (1966).
[19] Pagenkopf, G. K., and Margerum, D. W., *Inorg. Chem.*, **7**, 2514 (1968).
[20] (a) Taube, H., *Electron Transfer Reactions of Complex Ions*, Academic Press, New York, 1970, p. 51.
     (b) Stritar. J., and Taube, H., *Inorg. Chem.*, **8**, 2284 (1969).
     (c) Espenson, J., *Inorg. Chem.*, **4**, 121 (1965).
[21] Silverman, J., and Dodson, R. W., *J. Phys. Chem.*, **66**, 846 (1952).
[22] Hupp, J. T., and Weaver, M. J., *Inorg. Chem.*, **22**, 2557 (1983).
[23] Bernhard, P., Helm, L., Ludi, A., and Merbach, A. E., *J. Am. Chem. Soc.*, **107**, 312 (1985).
[24] Brunschwig, B. S., Creutz, C., Macartney, D. H., Sham, T. K., and Sutin, N., *Faraday Discuss. Chem. Soc.*, **74**, 113 (1982).

[25] Jolley, W. H., Stranks, D R., and Swaddle, T. W., *Inorg. Chem.*, **29**, 1948 (1990).
[26] Furholz, U., and Haim, A., *Inorg. Chem.*, **24**, 3091 (1985).
[27] Brown, G., M., Krentzien, H. J., Abe, M., and Taube, H., *Inorg. Chem.*, **18**, 3374 (1979).
[28] (a) Chou, M., Creutz, C., and Sutin, N., *J. Am. Chem. Soc.*, **99**, 5615 (1977).
      (b) Macartney, D. H., and Sutin, N., *Inorg. Chem.*, **22**, 3530 (1983).
[29] (a) Beattie, J. K., Best, S. P., Skelton, B. W., and White, A. H., *J. Chem. Soc. Dalton Trans.*, 2105 (1981).
      (b) Haire, N. J., and Beattie, J. K., *Inorg. Chem.*, **16**, 245 (1977).
[30] Armstrong, R. S., Beattie, J. K., Best, S. P., Braithwaite, G. P., Delfavero, P., Skelton, B. W., and White, A. H., *Aust. J. Chem.*, **43**, 393 (1990).
[31] Best, S. P., and Forsyth, J. B., *J. Chem. Soc. Dalton Trans.*, 395 (1990).
[32] Magini, M., *J. Inorg. Nucl. Chem.*, **40**, 43 (1978).
[33] Herdman, G. J., and Neilson, G. W., *J. Phys. Condens. Matter*, **4**, 627 (1992).
[34] (a) Jorgensen, C. K., *Absorption Spectra and Chemical Bonding*, Pergamon, Oxford, 1962.
      (b) Schlafer, H. L., *Z. Phys. Chem.*, *NF*, **4**, 116 (1955).
[35] Turner, R. C., and Miles, K. E., *Can. J. Chem.*, **35**, 1002 (1975).
[36] Dvir, M., and Low, W., *Phys. Rev.*, **119**, 1587 (1966).
[37] Sylva, R. N., *Pure. Appl. Chem.*, **22**, 115 (1972).
[38] Flynn, C. M., 'The hydrolysis of inorganic iron (III) salts', *Chem. Rev.*, **84**, 31 (1984).
[39] (a) Cornell, R. M., Giovanoli, R., and Schneider, W., 'Review of the hydrolysis of Iron(III) and the crystallisation of amorphous iron(III) hydroxide hydrate', *J. Chem. Tech. Biotechnol.*, **46**, 115 (1989).
      (b) Schneider, W., 'Hydrolysis of iron (III)-chaotic olation versus nucleation', *Comm. Inorg. Chem.*, **3**, 205 (1984).
[40] (a) Schneider, W., *Chimia*, **42**, 9 (1988).
      (b) Schneider, W., and Schwyn, B., 'The hydrolysis of iron in synthetic, biological and aquatic media, in *Aquatic Surface Chemistry*, (ed. W. Stumm), Wiley, New York, 1987, p. 167.
[41] Baes, C. F., and Mesmer, R. E., *The Hydrolysis of Cations*, Krieger, Florida, 1986, pp. 235–237.
[42] Biedermann, G., unpublished work quoted in Schytler, K., *Trans. Roy. Inst. Technol. Stockholm*, 196 (1962).
[43] Khoe, G. H., Brown, P. L., Sylva, R. N., and Robins, R. G., *J. Chem. Soc. Dalton Trans.*, 1901 (1986).
[44] (a) Ciavatti, L., and Grimaldi, M., *J. Inorg. Nucl. Chem.*, **37**, 163 (1975).
      (b) Dousma, J., and DeBruyn, P. L., *J. Colloid. Interface. Sci.*, **56**, 527 (1976).
[45] Daniele, P. G., Rigano, C., Sammartano, S., and Zelano, V., *Talanta*, **41**, 1577 (1964).
[46] Data of Biedermann, G., quoted in the following:
      (a) Hedstrom, B. O., *Ark. Kemi.*, **6**, 1 (1953).
      (b) Milburn, R. M., *J. Am. Chem. Soc.*, **79**, 537 (1957).
      (c) Perrin, D. D., *J. Chem. Soc.*, 1710 (1959).
      (d) Richards, D. H., and Sykes, K. W., *J. Chem. Soc.*, 3626 (1960).
      (e) Behar, B., and Stein, G., *Isr. J. Chem.*, **7**, 827 (1969).
      (f) Fordham, A. W., *Aust. J. Chem.*, **22**, 111 (1969).
[47] Wilkinson, E. C., Dong, Y. H., and Que, L., *J. Am. Chem. Soc.*, **116**, 8394 (1994).
[48] Martinez, P., van-Eldik, R., and Kelm, H., *Ber. Bunsenges. Phys. Chem.*, **89**, 81 (1985).

[49] Martinez, P., and van-Eldik, R., *Ber. Bunsenges. Phys. Chem.*, **89**, 728 (1985).
[50] Murray, K. S., *Coord. Chem. Revs.*, **12**, 1 (1974).
[51] Thich, J. A., Ou, C.-C., Powers, D., Vasilious, B., Mastropaolo, D., Potenza, J. A., and Schugar, H. J., *J. Am. Chem. Soc.*, **98**, 1425 (1976).
[52] Ou, C.-C., Wollman, R. G., Hendrickson, D. N., Potenza, J. A., and Schugar, H. J., *J. Am. Chem. Soc.*, **100**, 4717 (1978).
[53] Schugar, H. J., Walling, C., Jones, R. B., and Gray, H. B., *J. Am. Chem. Soc.*, **89**, 3712 (1967).
[54] Morrison, T. I., Reis Jr, A. H., Knapp, G. S., Fradin, F. Y., Chen, H., and Klippert, T. E., *J. Am. Chem. Soc.*, **100**, 3262 (1978).
[55] (a) Magini, M., Saltelli, A., and Caminiti, R., *Inorg. Chem.*, **20**, 3564 (1981).
      (b) Morrison, T. I., Shenoy, G. K., and Nielsen, L., *Inorg. Chem.*, **20**, 3565 (1981).
[56] Knudsen, J. M., Larsen, E., Moreira, J. E., and Furskov-Neilson, O., *Acta Chem. Scand.*, **A29**, 833 (1975).
[57] Meagher, A., *Inorg. Chim. Acta*, **146**, 19 (1988).
[58] Hazell, A., Jensen, K. B., McKenzie, C. J., and Toftlund, H., *Inorg. Chem.*, **33**, 3127 (1994).
[59] See, for example, Carrano, C., and Spartalian, K., *Inorg. Chem.*, **23**, 1993 (1984).
[60] Knight, R. J., and Sylva, R. N., *J. Inorg. Nucl. Chem.*, **37**, 779 (1975).
[61] Milburn, R. M., and Vosburgh, W. C., *J. Am. Chem. Soc.*, **77**, 1352 (1955).
[62] Biedermann, G., and Schindler, P. W., *Acta Chem. Scand.*, **11**, 731 (1957).
[63] Feitknecht, W., and Michaelis, W., *Helv. Chim. Acta*, **45**, 212 (1962).
[64] Sommer, B. A., and Margerum, D. W., *Inorg. Chem.*, **9**, 2517 (1970).
[65] See, for example, Lutz, B., and Wendt, H., *Ber. Bunsenges. Phys. Chem.*, **74**, 372 (1970).
[66] See, for example, Music, S., Orehovec, Z., Popovic, S., and Czako-Nagy, I., *J. Mater. Sci.*, **29**, 1991 (1994).
[67] Bradley, S. M., and Kydd, R. M., *J. Chem. Soc. Dalton Trans.*, 2407 (1993).
[68] Rustand, J. R., Hay, B., P., and Halley, J. W., *J. Chem. Phys.*, **102**, 427 (1995).
[69] (a) Gotic, M., Popovic, S., Ljubesic, N., and Music, S., *J. Mater. Sci.*, **29**, 2474 (1994).
      (b) Murphy, P. J., Posner, A. M., and Quirk, J. P., *J. Colloid and Interface Sci.*, **56**, 270 (1976).
[70] Atkinson, R. J., Posner, A. M., and Quirk, J. P., *Clays and Clay Minerals*, **25**, 49 (1977).
[71] (a) See, for example, Johnson, J. H., and Lewis, D. G., *Geochim. Cosmochim. Acta*, **47**, 1823 (1983).
      (b) Fischer, W. R., and Schwertmann, U., *Clays and Clay Minerals*, **23**, 33 (1975).
[72] Jolivet, J. P., and Tronc, E., *J. Colloid Interface Sci.*, **125**, 688 (1988).
[73] Henry, M., Jolivet, J. P., and Livage, J., *Structure and Bonding*, **77**, 153 (1992).
[74] Melikov, I. V., Kozlovskaya, L. B., Berliner, L. B., and Prokofiev, M. A., *J. Colloid Interface Sci.*, **117**, 1 (1987).
[75] Murphy, P. J., Posner, A. M., and Quirk, J. P., *J. Colloid Interface Sci.*, **56**, 312 (1976).
[76] (a) Winkelmann, G., Van der Helm, F., and Neilands, J. B. (eds.), *Iron transport in microbes, plants and animals*, VCH, Weinheim 1987.
      (b) Fairbanks, V. F., 'Iron metabolism', in *Hematology*, (eds.) W. J. Williams, E. Butler, A. J. Erslev and M. A. Lichtmann, McGraw-Hill, New York, 1983, p. 300.

[77] Harrison, P. M., Clegg, G. A., and May, K., 'Ferritin structure and function', in
      Iron Biochemistry and medicine II (eds. A. Jacobs, and M. Worwood), Academic
      Press, London, 1980, p. 131.
[78] (a) Powell, A. K., Heath, S. L., Gatteschi, D., Pardi, L., Sessoli, R. Spina, G.,
      Delgiallo, F., and Pierralli, F., *J. Am. Chem. Soc.*, **117**, 2491 (1995).
      (b) Harding, C. J., Henderson, R. K., and Powell A. K., *Angew. Chem. Intl. Ed.*
      (Engl.), **32**, 570 (1993).
[79] Grant, M. W., and Jordan, R. B., *Inorg. Chem.*, **20**, 55 (1981).
[80] Swaddle, T. W., and Merbach, A. E., *Inorg. Chem.*, **20**, 4212 (1981).
[81] Jordan, R. B., *Reaction Mechanisms of Inorganic and Organometallic Systems*,
      Oxford University Press, 1991, pp. 77–80.
[82] Hasinoff, B. B., *Can. J. Chem.*, **54**, 1820 (1976).
[83] Seewald, D., and Sutin, N., *Inorg. Chem.*, **2**, 643 (1963).
[84] Hasinoff, B. B., *Can. J. Chem.*, **57**, 77 (1979).
[85] Funahashi, S., Adachi, S., and Tanaka, M., *Bull. Chem. Soc. Japan*, **46**, 479 (1973).
[86] Grace, M. R., and Swaddle, T. W., *Inorg. Chem.*, **31**, 4674 (1992).
[87] Permutter-Hayman, B., and Tapushi, E., *J. Coord. Chem.*, **6**, 31 (1976).
[88] Pardey, R. N., and Smith, W., *Can. J. Chem.*, **50**, 194 (1972).
[89] Cavasino, F., and Didio, E., *J. Chem. Soc. A*, 1151 (1970).
[90] Ishihara, K., Funahashi, S., and Tanaka, M., *Inorg. Chem.*, **22**, 194 (1983).
[91] Birus, M., and van -Eldik, R., *Inorg. Chem.*, **30**, 4559 (1991).
[92] Funahashi, S., Isahihara, K., and Tanaka, M., *Inorg. Chem.*, **22**, 2070 (1983).
[93] (a) Birus, M., Bradic, Z., Krznaric, G., Kujundzic, N., Pribanic, M., Wilkins, P. C.,
      and Wilkins, R. G., *Inorg. Chem.*, **26**, 100 (1987).
      (b) Birus, M., Krznaric, G., Kujundzic, N., and Pribanic, M., *Croat. Chem. Acta*,
      61, 33 (1988).
[94] Martinez, P., Mohr, R., and van-Eldik, R., *Ber. Bunsenges. Phys. Chem.*, **90**, 609
      (1986).
[95] Shibahara, T., Akashi, H., and Kuroya, H., *J. Am. Chem. Soc.*, **108**, 1342 (1986).
[96] Dimmock, P. W., Dickson, D. P. E., and Sykes, A. G., *Inorg. Chem.*, **29**, 5920
      (1990).
[97] Chrichton, R. R., and Charloteaux-Wauters, M., 'Iron transport and storage', *Eur.*
      J. Biochem., **164**, 485 (1987).
[98] deJong, G., van Dijk, J. P., and van Eijk, H. G., *Clin. Chim. Acta*, **90**, 17 (1990).
[99] (a) Jameson, R. F., Linert, W., Tschinkowitz, A., and Gutmann, V., *J. Chem. Soc.*
      Dalton Trans., 943 (1988).
      (b) Jameson, R. F., Linert, W., and Tschinkowitz, A., *J. Chem. Soc. Dalton Trans*,
      2109 (1988).
[100] Bansch, B., Martinez, P., Uribe, D., Zulunga, J., and van Eldik, R., *Inorg. Chem.*,
      30, 4555 (1991).
[101] (a) Tousek, J., *Coll. Czech. Chem. Commun.*, **27**, 908, 914 (1962).
      (b) Grube, G., and Gmelin, H., *Z. Elektrochem.*, **26**, 153, 459 (1920).
[102] (a) Scholder, R., *Bull. Soc. Chim. Fr.*, 1112 (1965).
      (b) Sudette, R. J., and Quail, J. W., *Inorg. Chem.*, **11**, 1904 (1972).
[103] (a) Carrington, A., Schonland, D., and Symons, M. C. R., *J. Chem. Soc.*, 659
      (1957).
[104] (a) Helferich, B., and Lang, K., *Z. Anorg. Allg. Chem.*, **263**, 169 (1950).
      (b) Krebs, H., *Z. Anorg. Allg. Chem.*, **263**, 175 (1950).
[105] Hrostowski, H. J., and Scott, A.z B., *J. Chem. Phys.*, **18**, 105 (1950).
[106] Murmann, R. K., *J. Am. Chem. Soc.*, **93**, 6058 (1971).
[107] (a) Crowell, C. R., and Yost, D. M., *J. Am. Chem. Soc.*, **50**, 374 (1928).

(b) Grube, G., and Fromm, G., *Z. Elektrochem.*, **46**, 661 (1940).
(c) Belikin, A. V., Borisov, V. V., Sinitsyn, N. M., and Solomonova, A. S., *Russ. J. Inorg. Chem. Engl. Transl.*, **24**, 1693 (1979).
[108] Sawyer, D. T., George, R. S., and Bagger, J. B., *J. Am. Chem. Soc.*, **81**, 5893 (1959).
[109] Seddon, K. R., *Coord. Chem. Revs.*, **41**, 79 (1982).
[110] Evans, I. P., Spencer, A., and Wilkinson, G., *J. Chem. Soc. Dalton Trans.*, 204 (1973).
[111] Seddon, K. R., and Seddon, E. A., *The Chemistry of Ruthenium*, Elsevier, New York, 1985.
[112] Rard, J. A., *Chem. Revs.*, **83**, 1 (1985).
[113] Kallen, T. W., and Earley, J. E., *Inorg. Chem.*, **10**, 1149 (1971).
[114] Bernhard, P., Burgi, H.-B., Hauser, J., Lehmann, H., and Ludi, A., *Inorg. Chem.*, **21**, 3936 (1982).
[115] Bernhard, P., Biner, M., and Ludi, A., *Polyhedron*, **8**, 1095 (1990).
[116] Bernhard, P., and Ludi, A., *Inorg. Chem.*, **23**, 870 (1984).
[117] Bernhard, P., Lehmann, H., and Ludi, A., *Comm. Inorg. Chem.*, **2**, 145 (1983) and *J. Chem. Soc., Chem. Commun.*, 1216 (1981).
[118] Bailey, O. H., and Ludi, A., *Inorg. Chem.*, **24**, 2582 (1985).
[119] Laurenczy, G., Helm, L., Ludi, A., and Merbach, A. E., *Helv. Chim. Acta*, **74**, 1236 (1991).
[120] Aebischer, N., Laurenczy, G., Ludi, A., and Merbach, A. E., *Inorg. Chem.*, **32**, 2810 (1993).
[121] Laurenczy, G., and Merbach, A. E., *J. Chem. Soc., Chem. Commun.*, 187 (1993).
[122] Laurenczy, G., Helm, L., Merbach, A. E., and Ludi, A., *Inorg. Chim. Acta*, **189**, 131 (1991).
[123] Creutz, C., and Taube, H., *Inorg. Chem.*, **10**, 2664 (1971).
[124] Rapaport, I., Helm, L., Merbach, A. E., Bernhard, P., and Ludi, A., *Inorg. Chem.*, **27**, 873 (1988).
[125] Patel, A., Leitch, P., and Richens, D. T., *J. Chem. Soc. Dalton Trans.*, 1029 (1991).
[126] Karlen, T., personal communication in reference 14.
[127] Sisley, M. J., and Swaddle, T. W., *Inorg. Chem.*, **20**, 2799 (1981).
[128] (a) Swaddle, T. W., and Mak, M. K. S., *Can. J. Chem.*, **61**, 473 (1983).
(b) Swaddle, T. W., in *Adv. Inorg. Bioinorg. Mech.* (ed. A. G. Sykes), **2**, 95 (1983) and *Inorg. Chem.*, **22**, 2663 (1983).
[129] Karlen, T., and Ludi, A., *Helv. Chim. Acta*, **75**, 1604 (1992).
[130] Novak, B. M., and Grubbs, R. H., *J. Am. Chem. Soc.*, **110**, 960, 7542 (1988).
[131] Stebler-Rothlisberger, M., Hummel, W., Pittet, P.-A., Burgi, H.-B., Ludi, A., and Merbach, A. E., *Inorg. Chem.*, **27**, 1358 (1988).
[132] (a) Kolle, U., Flunkert, G., Gorissen, R., Schmidt, M. U., and Englert, U., *Angew, Chem. Intl. Ed. (Engl.)*, **31**, 440 (1992).
(b) Dadci, L., Elias, H., Frey, U., Hornig, A., Kolle, U., Merbach, A. E., Paulus, H., and Scheider, J. S., *Inorg. Chem.*, **35**, 306 (1995).
[133] Kallen, T. W., and Earley, J. E., *Inorg. Chem.*, **10**, 1152 (1971).
[134] Beattie, J. K., Best, S. P., Skelton, B. W., and White, A. H., *J. Chem. Soc. Dalton Trans.*, 2105 (1981).
[135] Bernhard, P., Stebler, A., and Ludi, A., *Inorg. Chem.*, **23**, 2151 (1984).
[136] Harzion, Z., and Navon, G., *Inorg. Chem.*, **19**, 2236 (1980).
[137] Bottcher, W., Brown, G. M., and Sutin, N., *Inorg. Chem.*, **20**, 4212 (1981).
[138] Adamson, M. G., *J. Chem. Soc.*, 1370 (1968).
[139] Fine, D. A., Ph.D. Thesis, University of California, Berkeley, California, 1960.

[140] Hrabikova, J., Dolezal, J., and Zyka, J., *Anal. Lett.*, **7**, 819 (1974).
[141] Connick, R. E., and Fine, D. A., *J. Am. Chem. Soc.*, **83**, 3414 (1961).
[142] Connick. R. E., in *Advances in the Chemistry of Coordination Compounds* (ed. S. Kirschner), Macmillan, New York, 1961, pp. 15–20.
[143] Ohyoshi, E., Ohyoshi, A., and Shinagawa, M., *Radiochim. Acta*, **13**, 10 (1970).
[144] Gardiner, A. M., and Richens, D. T., unpublished results, 1991.
[145] Taqui-Khan, M. M., Ramachandraiah, G., and Prakash-Rao, A., *Inorg. Chem.*, **25**, 665 (1986).
[146] Taqui-Khan, M. M., Ramachandraiah, G., and Shukla, R. S., *Polyhedron*, **11**, 23 (1992) and *Inorg. Chem.*, **27**, 3274 (1988).
[147] Cady, H. H., and Connick, R. E., *J. Am. Chem. Soc.*, **80**, 2647 (1958).
[148] (a) Taqui-Khan, M. M., Prakash-Rao, A., and Bhatt, S. D., *J. Mol. Catal.*, **75**, 129 (1992).
       (b) Taqui-Khan, M. M., and Prakash-Rao, A., *J. Mol. Catal.*, **44**, 95 (1988).
[149] Taqui-Khan, M. M., Prakash-Rao, A., and Mehta, S. H., *J. Mol. Catal.*, **78**, 263 (1993).
[150] Mastsubara, T., and Creutz, C., *J. Am. Chem. Soc.*, **100**, 6255 (1978).
[151] Bajaj, H. C., and van Eldik, R., *Inorg. Chem.*, **27**, 4052 (1988).
[152] Matsubara, T., and Creutz, C., *Inorg. Chem.*, **18**, 1956 (1979).
[153] (a) Okamoto, K.-I., Hidaka, J., Iida, I., Higashino, K., and Kanamori, K., *Acta Cryst.*, **C46**, 2327 (1990).
       (b) Taqui-Khan, M. M., Venkatasubramanian, K., Bajaj, H. C., and Shirin, Z., *Ind. J. Chem.*, **A31**, 303 (1992).
[154] (a) Taqui-Khan, M. M., *Pure Appl. Chem.*, **55**, 159 (1983).
       (b) Taqui-Khan, M. M., Hussain, A., Venkatasubramanian, K., and Moiz, M. A., *Inorg. Chem.*, **25**, 3023 (1986).
[155] (a) Taqui-Khan, M. M., Samad, S. A., and Siddiqui, M. R. H., *J. Mol. Catal.*, **53**, 23 (1989).
       (b) Taqui-Khan, M. M., Samad, S. A., Shirin, Z., and Siddiqui, M. R. H., *J. Mol. Catal.*, **54**, 81 (1989).
[156] (a) Taqui-Khan, M. M., Halligudi, S. B., and Shukla, S., *Angew. Chem. Int. Ed. (Engl.)*, **27**, 734 (1988).
       (b) Taqui-Khan, M. M., *Platinum Metals Rev.*, **35**, 70 (1992) (and refs. therein).
[157] Taqui-Khan, M. M., Shirin, Z., Siddiqui, M. R. H., and Venkatasubramanian, K. *J. Mol. Catal.*, **72**, 271 (1992).
[158] Taqui-Khan, M. M., Bhardwaj, R. C., and Bhardwaj, C., *Angew. Chem. Int. Ed. (Engl.)*, **27**, 923 (1988).
[159] Taqui-Khan, M. M., and Shukla, R. S., *J. Mol. Catal.*, **71**, 157 (1992), **72**, 361 (1992), **77**, 221 (1992) and **44**, 73 (1988).
[160] Taqui-Khan, M. M., and Shukla, R. S., *Polyhedron*, **10**, 2711 (1991).
[161] (a) Legzdins, P., Mitchell, R. W., Rempel, G. L., Ruddick, J. D., and Wilkinson, G., *J. Chem. Soc. A*, 3322 (1970).
       (b) Spencer, A., and Wilkinson, G., *J. Chem. Soc. Dalton Trans.*, 1570 (1992).
[162] Almog, O., Bino, A., and Garfinkel-Schweky, D., *Inorg. Chim. Acta*, **213**, 99 (1993).
[163] Powell, G., Richens, D. T., and Powell, A. K., *Inorg. Chim. Acta*, **213**, 147 (1993).
[164] Powell, G., Richens, D. T., and Bino, A., *Inorg. Chim. Acta*, **232**, 167 (1995).
[165] Baumann, J. A., Salmon, D. J., Wilson, S. T., Meyer, T. J., and Hatfield, W. E., *Inorg. Chem.*, **17**, 3342 (1978).
[166] Toma, H. C., and Cunha, C. J., *Can. J. Chem.*, **67**, 1632 (1989).
[167] Cotton, F. A., and Norman, J. G., *Inorg. Chim. Acta*, **6**, 411 (1972).

[168] (a) Anson, C. E., Chai-Sa'ard, N., Bourke, J. P., Cannon, R. D., Jayasooriya, U. A., and Powell, A. K., *Inorg. Chem.*, **32**, 1502 (1993).
(b) Cannon, R. D., and White, R. P., *Prog. Inorg. Chem.*, **36**, 195 (1988).
[169] (a) Fouda, S. A., and Rempel, G. W., *Inorg. Chem.*, **18**, 1 (1979).
(b) Sasson, Y., and Rempel, G. W., *Tetrahedron Lett.*, 4133 (1974) and *Can. J. Chem.*, **52**, 3825 (1974).
[170] Bilgrien, C., Davis, S., and Drago, R. S., *J. Am. Chem. Soc.*, **109**, 3786 (1987).
[171] (a) Barton, D. H. R., Boivin, J., Gastiger, M., Morzyski, J., Hay-Motherwell, R. S., Motherwell, W. B., Ozbalik, N., and Schwartzentruber, K. M., *J. Chem. Soc. Perkin 1*, 947 (1986).
(b) Fish, R. H., Fong, R. H., Vincent, J. B., and Christou, G., *J. Chem. Soc., Chem. Commun.*, 1504 (1988).
[172] Powell, G., Richens, D. T., and Khan, L., *J. Chem. Res. S*, 506 (1994).
[173] Seewald, D., Sutin, N., and Watkins, K. O., *J. Am. Chem. Soc.*, **91**, 7307 (1969).
[174] Bernhard, P., Helm, L., Ludi, A., and Merbach, A. E., *J. Am. Chem. Soc.*, **107**, 312 (1985).
[175] Wehner, P., and Hindman, J. C., *J. Am. Chem. Soc.*, **72**, 3911 (1950).
[176] Niedrach, L. W., and Tevebaugh, A. D., *J. Am. Chem. Soc.*, **73**, 2385 (1951).
[177] Atwood, D. K., and De Vries, T., *J. Am. Chem. Soc.*, **84**, 2659 (1962).
[178] Gortsema, F. P., and Cobble, J. W., *J. Am. Chem. Soc.*, **83**, 4317 (1961).
[179] Cady, H. H., USAEC Report UCRL 3757, 1957.
[180] Wallace, R. M., and Propst, R. C., *J. Am. Chem. Soc.*, **91**, 3779 (1969).
[181] Wallace, R. M., *J. Phys. Chem.*, **68**, 2418 (1964).
[182] Patel, A., and Richens, D. T., *Inorg. Chem.*, **30**, 3789 (1991).
[183] Schauwers, J., Meuris, F., Heerman, L., and D'Olieslager, W., *Electrochim. Acta*, **26**, 1065 (1981).
[184] Heerman, L., Van Nijen, H., and D'Olieslager, W., *Inorg. Chem.*, **27**, 4320 (1988).
[185] Bremard, C., Nowogrocki, G., and Tridot, G., *Bull. Soc. Chim. Fr.*, 392 (1968).
[186] Starik, I. E., and Kositskyn, A. V., *Russ. J. Inorg. Chem., Engl. Transl.*, **2**, 332 (1957).
[187] Deloume, J. P., Duc, G., ad Thomas-David, G., *Polyhedron*, **4**, 875 (1985).
[188] Wehner, P., and Hindman, J. C. *J. Phys. Chem.*, **56**, 10 (1952).
[189] Patel, A., Ph.D Thesis, The University of Stirling, Scotland, 1988.
[190] Power, J. M., Evertz, K., Henling, L., Marsh, R., Schaefer, W. P., Labinger, J. A., and Bercaw, J. E., *Inorg. Chem.*, **29**, 5058 (1990).
[191] (a) Orlov, A. M., Shirokov, Yu. S., Kunaeva, I. V., Chalisova, N. N., and Fomina, T. A., *Russ. J. Inorg. Chem. Engl. Transl.*, **24**, 1373 (1979).
(b) Vdovenko, V. M., Lazarev, L. N., and Khvoristin, Ya. S., *Sov. Radiochem., Engl. Transl.*, **7**, 228 (1965) and **9**, 449 (1967).
[192] Ginzburg, S. I., Yuz'ko, M. I., Fomina, T. A., and Evstaf'eva, O. N., *Russ. J. Inorg. Chem. Engl. Transl.*, **20**, 1082 (1975).
[193] D'Olieslager, W., Heerman, L., and Clarysse, M., *Polyhedron*, **2**, 1107 (1983).
[194] Maya, L., *J. Inorg. Nucl. Chem.*, **41**, 67 (1979).
[195] Connick, R. E., and Hurley, C. R., *J. Am. Chem. Soc.*, **74**, 5012 (1952).
[196] Eletcher, J. M., Gardner, W. E., Greenfield, B. F., Holdoway, M. J., and Rand, M. H., *J. Chem. Soc. A.*, 533 (1968).
[197] Keattch, C. J. and Redfearn, J. P., *J. Less Common Metals*, **4**, 460 (1962).
[198] See, for example, Chen, H., and Transatti, S., *J. Appl. Electrochem.*, **23**, 559 (1993) (and refs. therein).
[199] Schafer, L., and Seip, H. M., *Acta Chem. Scand.*, **21**, 737 (1967).
[200] Silverman, M. D., and Levy, H. A., *J. Am. Chem. Soc.*, **76**, 3317 (1964).
[201] Elout, M. O., Haije, W. G., and Maaskant, W. J. A., *Inorg. Chem.*, **27**, 610 (1988).

[202] Martin, F. S., *J. Chem. Soc.*, 2564 (1954).
[203] Shiloh, M., Givon, M., and Spiegler, K. S., *J. Inorg. Nucl. Chem.*, **25**, 103 (1963).
[204] Luoma, E. V., and Brubaker Jr, C. H., *Inorg. Chem.*, **5**, 1618 (1966).
[205] (a) Martin, F. S., *J. Chem. Soc.*, 3055 (1952).
       (b) Khain, V. S., and Volkov, A. A., *Russ. J. Inorg. Chem., Engl. Transl.*, **23**, 1631 (1978).
       (c) Ardeev, D. K., Seregin, V. I., and Tekster, E. N., *Russ. J. Inorg. Chem., Engl. Transl.*, **16**, 399 (1971).
[206] Carrington, A., and Symons, M. C. R., *J. Chem. Soc.*, 284 (1960).
[207] Eichner, P., *Bull. Soc. Chim. Fr.*, 2051 (1967).
[208] Griffith, W. P., *Platinum Metal Revs.*, **33**, 118 (1989).
[209] Lee, D. G., and van der Engh, M., in *Oxidation in Organic Chemistry*, Vol. B. (ed. W. S. Trahanowski), Academic Press, New York, p. 177 (1973).
[210] Courtney, J. L., in *Organic Syntheses by Oxidation with Metal Complexes* (eds. W. J. Meijs and C. R. H. I. de Jonge, Plenum, New York, 1986, p. 445.
[211] Ayres, D. C., *Nature, Lond.*, **290**, 323 (1981).
[212] (a) Creaser, C. S., Fernandes, A. R., and Ayres, D. C., *Chem. Ind.*, **15**, 499(1988).
       (b) Ayres, D., C., Creaser, C. S., and Levy, D. P., in *Chlorinated Dioxins and Dibenzofurans in the Total Environment, II*, (eds. L. K. Keith, C. Rappe and C. Choudhary), Butterworths, London, 1985, p. 37.
[213] (a) Griffith, W. P., Ley, S. V., Whitcombe, G. P., and White, A. D., *J. Chem. Soc., Chem. Commun.*, 1623 (1987).
       (b) Dengel, A. C., El-Hendaway, A. M., and Griffith, W. P., *Trans. Met. Chem.*, **14**, 230 (1989).
[214] Dengel, A. C., Hudson, R. A., and Griffith, W. P., *Trans. Met. Chem.*, **10**, 98 (1985).
[215] Anthony, N. J., Armstrong, A., Ley, S. V., and Madin, A., *Tetrahedron Lett.*, **30**, 3209 (1989).
[216] Schroder, M., and Griffith, W. P., *J. Chem. Soc., Chem. Commun.*, 58 (1979).
[217] Lay, P. A., Magnuson, R. H., and Taube, H., *Inorg. Chem.*, **28**, 3001 (1986).
[218] Lay, P. A., Magnuson, R. H., and Taube, H., *Inorg. Synth.*, **24**, 269 (1986).
[219] Bremard, C., and Mouchel, C., *Inorg. Chem.*, **21**, 1810 (1982).
[220] (a) Bard, A. J. (ed), *Encyclopaedia of the Electrochemistry of the Elements*, Vol. VI, Marcel Dekker, New York, 1976.
       (b) Goldberg, R. N., and Hepler, L. G., *Chem. Revs.*, **68**, 229 (1968).
       (c) Latimer, W. M., *Oxidation States of the Elements and the Potentials in Aqueous Solution*, Prentice-Hall, Englewood Clipps, New Jersey, 1952, p. 202.
[221] (a) Perichan, J., Palous, S., and Buvet, R., *Bull. Soc. Chim. Fr.*, 982 (1963).
       (b) Lal, S., and Christian, G. D., *Rev. Polarogr.*, **16**, 109 (1970).
[222] Meites, L., *J. Am. Chem. Soc.*, **79**, 4631 (1957).
[223] Cover, R. E., and Meites, L., *J. Am. Chem. Soc.*, **83**, 4706 (1961).
[224] Lay, P. A. and Harman, W. D., *Adv. Inorg. Chem.*, **37**, 219 (1991).
[225] Moskvin, L. N., and Shmatko, A. G., *Russ. J. Inorg. Chem., Engl. Transl.*, **33**, 695 (1988) and *Zh. Neorg. Khim.*, **33**, 1229 (1988).
[226] Bardin, M. B., Goncharenko, V. P., and Ketrush, P. M., *Zh. Anal. Khim.*, **42**, 2013 (1987).
[227] Dwyer, F. P., and Sargeson, A. M., *J. Am. Chem. Soc.*, **77**, 1285 (1955).
[228] (a) Emerson, K. E., and Fergusson, J. E., *Proc. Mont. Acad. Sci.*, **42**, 101 (1983).
       (b) Heath, G. A., Moocvk, K. A., Sharp, D. W. A., and Yellowlees, L. J., *J. Chem. Soc., Chem. Commun.*, 1503 (1985).
       (c) Sun, I. W., Ward, E. W., and Hussey, C. L., *J. Electrochem. Soc.*, **135**, 3035 (1988).
[229] Minhas, S., and Richens, D. T., unpublished results (1993).

[230] Claus, C., and Jacoby, E., *Bull. Acad. St. Petersburg*, **6**, 145 (1863).
[231] (a) Lott, K. A. K., and Symons, M. C. R., *J. Chem. Soc.*, 973 (1960).
(b) Malin, J. S., *Inorg. Synth.*, **20**, 61 (1980).
[232] (a) Porai-Koshits, M. A., Atovmyan, L. O., and Adrianov, V. G., *J. Struct. Chem. (Engl. Transl.)*, **2**, 686 (1961).
(b) Atovmyan, L. O., Adrianov, V. G., and Porai-Koshits, M. A., *J. Struct. Chem. (Engl. Transl.)*, **3**, 660 (1962).
[233] (a) Sartori, C., and Preetz, W., *Z. Anorg. Allg. Chem.*, **572**, 151 (1989).
(b) Singh, A. K., Tewari, A., and Sisodia, A. K., *Natl Acad. Sci. Lett. (India)*, **8**, 209 (1985).
[234] Galbacs, Z. M., Zsednai, A., and Csanyi, L. J., *Trans. Met. Chem.*, **8**, 328 (1983).
[235] Tomaszewska, A., and Mullerbuschbaum, H., *J. Alloys Comp.*, **194**, 163 (1993).
[236] Griffith, W. P., *The Chemistry of the Rarer Platinum Metals*, Interscience, London, 1967.
[237] Krebs, B., and Hasse, K. D., *Acta Cryst.*, **32B**, 1334 (1976).
[238] Anderson, L. H., and Yost, D. M., *J. Am. Chem. Soc.*, **60**, 1823 (1938).
[239] Nevskii, N. N., Ivanov-Emin, B. N., and Nevskaya, N. A., *Dokl. Akad. Nauk, SSSR*, **266**, 628 (1982).
[240] Nevskii, N. N., Ivanov-Emin, B. N., and Nevskaya, N. A., and Belov. N. V, *Dokl. Akad. Nauk, SSSR*, **266**, 1138 (1982).
[241] Ivanov-Emin, N. N., Nevskaya, N. A., Nevskii, N. N., and Izmailovich, A. S., *Russ. J. Inorg. Chem., Engl. Transl.*, **29**, 710 (1984).
[242] (a) Wells, E. J., Jordan, A. D. Alderdice, D. S., and Ross, I. G., *Aust. J. Chem.*, **20**, 2315 (1967).
(b) Foster, S., Felps, S., Johnson, L. W., Larson, D. B., and McGlynn, S. P., *J. Am. Chem. Soc.*, **95**, 6578 (1973).
(c) Roebber, J. L., Weiner, R. N., and Russell, C. A., *J. Chem. Phys.*, **60**, 3166 (1974).
[243] Nevskii, N. N., and Porai-Koshits, M. A., *Dokl. Akad. Nauk, SSSR (Cryst.)*, **270**, 1392 (1983).
[244] Nevskii, N. N., Ivanov-Emin, B. N., and Nevskaya, N. A., *Dokl. Akad. Nauk, SSSR*, **266**, 347(1982).
[245] Jewiss, H. C., Levason, W., Tajik, M., Webster, M., and Walker, N. P. C., *J. Chem. Soc. Dalton Trans.*, 199 (1985).
[246] Preetz, W., and Schatzel, G., *Z. Anorg. Allg. Chem.*, **423**, 117 (1976). Muller, H., Scheible, H., and Martin, S., *Z. Anorg. Allg. Chem.*, **462**, 18 (1980).

Chapter 9

Group 9 Elements: Cobalt, Rhodium and Iridium

Group 9 marks the end of the ability of any of the elements within a group to exhibit the maximum valency and at the same time the oxo anion chemistry that has characterised elements of all the preceding groups. These two phenomena are of course related; the higher the stable valence the more likely oxo anions are to be formed in aqueous solution as stable species. Thus there is a tendency towards low and medium valency compounds down the entire group as a consequence of the steadily increasing Z_{eff} along the period. The dominating valency throughout is $+3$ which, with a few rare exceptions for cobalt, is highly stabilised in octahedral coordination geometry by the maximum crystal field stabilisation energy (2.4Δ) arising from the low-spin t_{2g}^6 configuration. For Co^{II} the lower charge leads to most compounds with simple ligands being high spin. As a consequence, steric and electrostatic factors can quickly favour the tetrahedral geometry, e.g. $[CoCl_4]^{2-}$ and $[Co(OH)_4]^{2-}$, because of the small difference in CFSE between the octahedral and tetrahedral geometry. Rh^{II} is well defined but Rh—Rh dimers are formed in contrast to the monomeric nature of most Co^{II} compounds. As seen in the preceding groups this reflects the tendency of the heavier members of a group to form M—M bonded species. Interestingly, though, the trend is not continued to Ir wherein dimeric Ir^{II} compounds are unknown. Indeed, Ir^{II} is an extremely rare oxidation state. The so-called 'high valent' Rh compounds generated from alkaline solutions of Rh^{III} are now known to consist of superoxo-bridged Rh^{III} aqua dimers. Truly authentic Rh^{IV} compounds are few, but in contrast both Ir^{IV} and Ir^{V} are established in oligomeric aqua species. Much of the aqueous and hydrolytic chemistry of Rh^{III}, and to a certain extent Ir^{III}, parallels that of Cr^{III} (Chapter 6).

9.1 COBALT

Co^{II} and Co^{III} are the only oxidation states relevant to aqueous solutions. Both form simple hexaaqua ions. The relative stability of aqueous Co^{II} vs Co^{III} is, however, markedly dependent upon the nature of the coordinated ligands, this being well illustrated by a consideration of the redox potentials in Table 9.1 [1]. Thus while $[Co(OH_2)_6]^{3+}$ spontaneously oxidises water evolving dioxygen, $[Co(CN)_5]^{3-}$ reduces water evolving hydrogen. The chemistry of Co^{III} is dominated by the formation of a wide variety of stable, kinetically inert, low-spin, t_{2g}^6 octahedral complexes with nitrogen donor ligands, whereas the coordination

Table 9.1. Redox Potentials

$Co^{III/II}$ redox couple	Ligand geometry	E^{θ} (vs NHE) [1]
$[Co(OH_2)_6]^{3+/2+}$	(O_6)	+ 1.92 V
$[Co(C_2O_4)_3]^{3-/4-}$	(O_6)	+ 0.55 V
$[Co(EDTA)]^{-/2-}$	(N_2O_4)	+ 0.38 V
$[Co(NH_3)_6]^{3+/2+}$	(N_6)	+ 0.06 V
$[Co(en)_3]^{3+/2+}$	(N_6)	− 0.18 V
$[Co(CN)_5]^{2-/3-}$	(C_5)	− 0.60 V

chemistry of air-stable Co^{II} is severely limited outside water as ligand. Biological interest in cobalt has stemmed primarily from its known involvement within the class of B_{12} co-enzymes, the only known naturally occurring organometallic system. However, just as interesting and important has been the use of the metal both as a powerful *in vivo* spectroscopic probe (as Co^{II}) for other divalent ions in biological systems, e.g. Fe^{II} in hemoglobin [2] and Zn^{II} in, for example, carbonic anhydrase and carboxypeptidase (see Chapter 12), and in providing active *in vitro* models for dioxygen binding and activation. For example, the synthesis of active Co^{II}-hemoglobin (Coboglobin) [2] led to two important dicoveries relevant to the natural Fe^{II} protein, namely that the protein (a) does not appreciably affect the electronic structure of the heme group and (b) cooperativity does not depend upon the spin state of the metal. Another interesting finding was the higher activity (1.8 times) shown by Co^{II^-} carboxypeptidase when compared to that of the natural Zn^{II} enzyme [3].

9.1.1 Cobalt(II) (d^7): Chemistry of the $[Co(OH_2)_6]^{2+}$ Ion

The reddish-pink $[Co(OH_2)_6]^{2+}$ ion is present in hydrated salts of Co^{II} such as $Co(ClO_4)_2 \cdot 6H_2O$, $Co(NO_3)_2 \cdot 6H_2O$, $CoSO_3 \cdot 6H_2O$, $CoSO_4 \cdot 7H_2O$ and the Tutton salts $M^I_2Co(SO_4)_2 \cdot 6H_2O$ [4]. The synthesis, properties and crystal structures of two sulfonate salts, $[Co(OH_2)_6]$ $(PhSO_3)_2$ [5] and $[Co(OH_2)_6]$ $(pts)_2$ [6], have also been reported. The regular octahedral hexaaqua ion is immediately liberated upon dissolution of these salts in water. Typical Co—OH_2 distances in the octahedral ion are around 208–210 pm, verified in solution by EXAFS [7] and XRD [8] measurements. The electronic spectrum is highly temperature dependent. From both solution and solid state samples of hydrated $CoCl_2$ the expected three spin-allowed transitions are observable from the $^4T_{1g}(F)$ $(t^5_{2g}e^2_g)$ ground state to $^4T_{2g}$ (~ 1500 nm),$^4A_{2g}$ (~ 770 nm) and $^4T_{1g}(P)$ (570 nm) [9] (Fig. 9.1). Aqueous solutions of $Co(ClO_4)_2$ at RT have the $^4T_{1g}(F)$ to $^4A_{2g}$ and $^4T_{1g}(P)$ transitions housed within a broad envelop around 530 nm. The transitions are better resolved in samples of $CoCl_2$. The fully

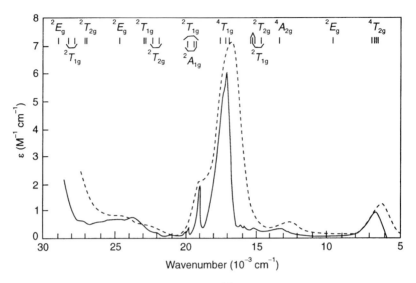

Figure 9.1. Electronic spectrum of $[Co(OH_2)_6]^{2+}$ from hydrated $CoCl_2$ at (---) RT and (—) $-30\,°C$

resolved spectrum has also been obtained from $[Co(OH_2)_6]^{2+}$ doped into $[Zn(OH_2)_6]SeO_4$ [10]. A host of spin-forbidden transitions to doublet states also characterises the electronic spectrum of $[Co(OH_2)_6]^{2+}$. Magnetic moments (μ_{eff}) are also highly temperature dependent and obey

$$\mu_{eff}(BM) = 3.87 \left(\frac{1 - 4\lambda}{\Delta} \right) \qquad (9.1)$$

Values of the spin–orbit coupling constant (λ) are around -180 (negative since $> d^5$). Since Δ is quite small ($\sim 4000\,cm^{-1}$), μ_{eff} values are raised appreciably above the spin-only value occurring in the range 4.4–4.8 BM. Experimental values quoted at room temperature are 4.82 BM ($[Co(OH_2)_6]Cl_2$), 4.91 BM ($Rb_2Co(SO_4)_2 \cdot 6H_2O$) and 4.84 BM ($Co(NO_3)_2 \cdot 6H_2O$) [11].

The ready formation of tetrahedral Co^{II} is exemplified by aqueous $[Co(OH_2)_6]^{2+}$ being found in equilibrium with amounts of tetrahedral $[Co(OH_2)_4]^{2+}$ [12]. $Zn^{2+}(aq)$ (d^{10}) shows evidence of the same equilibrium (Chapter 12). Co^{II} forms the largest number of tetrahedral complexes, their appearance frequently characterised by intense blue colorations. Indeed, the presence of blue $[CoCl_4]^{2-}$ in neutral aqueous solutions of $CoCl_2$ dates back to the early electrophoresis experiments of Donnan and Bassett in 1902 [13]. Here the blue anion was observed migrating towards the anode.

Hydrolysis of $[Co(OH_2)_6]^{2+}$ to give $[Co(OH_2)_5OH]^+$ ($pK_{11} = 9.65$) is well established [14]. At high Co^{II} concentrations (> 0.1 M) significant amounts of

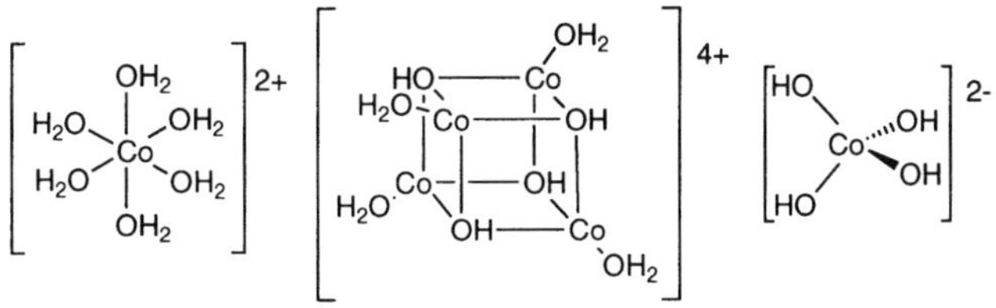

Figure 9.2. The structures of representative aqua hydroxo ions of Co^{II}

tetranuclear $Co_4(OH)_4^{4+}$ (aq) are believed to form ($pK_{44} = 30.5$) prior to precipitation of hydrous $Co(OH)_2$ at around pH 11 [15]. Tetrameric $Co_4(OH)_4^{4+}$(aq) is believed to have a cubane-like structure (Fig. 9.2), similar to that of $Ni_4(OH)_4^{4+}$(aq) (Chapter 10). Tetrahedral coordination at Co as in $[Co(OH)_4]^{2-}$ has been assumed. Interestingly, two forms of hydrous $Co(OH)_2$, pink and blue, have been identified and their solubility products measured [16]. Little else seems known about the nature of these two forms. One could infer that tetrahedral Co^{II} may well be present in the blue form although this may be a dangerous presumption. A form of hydrated octahedral $CoCl_2$ is reported to be 'blue'. However most octahedral Co^{II} complexes are pink in colour such as $Co(py)_4(SCN)_2$. The low water solubility of this latter compound has been successfully used in the gravimetric determination of cobalt.

At higher pH ($ > 12$), the amphoteric hydroxide redissolves to form blue solutions consisting of the tetrahedral hydroxo ions $[Co(OH)_3(OH_2)]^-$ ($pK_{13} = 31.5$) and $[Co(OH)_4]^{2-}$ ($pK_{14} = 46.3$) [15]. The tetrahedral form for the latter (Fig. 9.2) is confirmed by the electronic spectrum showing two bands at 1370 nm ($^4A_2-^4T_1(F)$) and 602 nm ($^4A_2-^4T_1(P)$) [9, 11]. The $^4A_2-^4T_2$ transition (Δ) occurs at very low energy in the normal infra-red range ($\sim 3000\ cm^{-1}$). A more recent article reports the visible band maximum for $[Co(OH)_4]^{2-}$ in 5–10 M NaOH solution occurring at 585 nm [17]. The fact that both octahedral and tetrahedral Co^{II} with O donors can have bands in the ~ 600 nm region reinforces the premise that 'colour' cannot be used as the sole basis for assigning the geometry around Co^{II}. The formation of tetrahedral $[Co(OH)_4]^{2-}$ rather than octahedral $[Co(OH)_4(OH_2)_2]^{2-}$ is believed to be responsible for the stabilisation of Co^{II} vs Co^{III} in alkaline solution and provides an interesting contrast with behaviour of the preceding first row elements manganese and iron.

9.1.1.1 COMPLEX FORMATION AND WATER EXCHANGE ON $[Co(OH_2)_6]^{2+}$

Estimates for the water-exchange rate constant for $[Co(OH_2)_6]^{2+}$ were initially made using ultrasound absorption from complex-forming reactions such as with SO_4^{2-} ions. Typical values reported up to 1970 ranged from $(0.2–1.8) \times 10^6\ s^{-1}$

[18]. Later the Swift–Connick line-broadening ^{17}O NMR method was successfully used by several groups to produce k_{ex} values ranging from 1.1 to $2.4 \times 10^6 \text{ s}^{-1}$ [19]. The rate constant redetermined by the Merbach group using the Swift–Connick approach [20], $(3.2 \pm 0.2) \times 10^6 \text{ s}^{-1}$, is close to the early estimates and reflects the continuing high kinetic lability of first row high-spin M^{2+} ions, the t_{2g}^3 ion, V^{2+} seemingly being the one exception. Activation parameters for the water-exchange process are $\Delta H^{\ddagger} = (46.9 \pm 1.2) \text{kJ mol}^{-1}$, $\Delta S^{\ddagger} = (+37.2 \pm 3.7) \text{J K}^{-1} \text{mol}^{-1}$ and $\Delta V^{\ddagger} = (+6.1 \pm 0.2) \text{cm}^3 \text{mol}^{-1}$ [20]. The activation volume is similar to the positive values obtained from complex formation reactions on $[Co(OH_2)_6]^{2+}$ by NH_3 ($+4.8 \pm 0.7$) [21], PADA ($+7.2 \pm 0.2$) [21] and glycinate$^-$ ($+5 \pm 2$) [22], suggestive of a general dissociative mechanism. The dissociative behaviour is further exemplified by the observation of positive activation volumes for solvent exchange on other $[CoL_6]^{2+}$ complexes with neutral ligands, L = MeOH ($+8.9 \text{cm}^3 \text{mol}^{-1}$), MeCN ($+8.1 \text{cm}^3 \text{mol}^{-1}$) and DMF ($+6.7 \text{cm}^3 \text{mol}^{-1}$) [23]. Thus $[Co(OH_2)_6]^{2+}$ continues the trend towards gradually increasing dissociative character as one moves from left to right along the first row period, a feature that will be further exemplified in the behaviour of $[Ni(OH_2)_6]^{2+}$ (next chapter). Second-order rate constants for complex-forming reactions on $[Co(OH_2)_6]^{2+}$ are fairly constant around $\sim 10^5 \text{M}^{-1}\text{s}^{-1}$ (Table 9.2) in support of the dissociative process. Reactions of $[Co(OH_2)_6]^{2+}$ with amino acids and a number of peptides have been reported

Table 9.2. Rate constants for complexation reactions on $[Co(OH_2)_6]^{2+}$ ($T = 298$ K)

Incoming ligands	$k_f(\text{M}^{-1}\text{s}^{-1})$	Method	Ref.
H_2O	3.2×10^{6a}	^{17}O NMR	[20]
NH_3	9.5×10^5	T jump	[28]
Imidazole	1.3×10^5	T jump	[29]
HF	5.5×10^5	^{19}F NMR	[30]
F^-	1.8×10^5	^{19}F NMR	[30]
PADA[b]	0.4×10^5	T jump	[31]
phen	1.2×10^5	Stopped-flow	[32]
bipy	0.63×10^5	Stopped-flow	[32]
terpy	0.24×10^5	Stopped-flow	[32]
Murexide	1.5×10^5	T jump	[33]
β-Alanine	0.75×10^5	T jump	[34]
Glycinate$^-$	4.6×10^5		[29]
Glycinate$^-$	3.0×10^6		[22]
HEDTA^{3-}	4.0×10^6		[35]
$HP_2O_7^{3-}$	9.3×10^7		[36]
ATP^{4-}	9.2×10^7	T jump	[37]

[a]units of s^{-1}
[b]PADA = pyridine-2-azo-p-dimethylamine.

[24], interest stemming from further reaction of the resulting complexes with dioxygen. The highest formation constants are found for *l*-histidine (log $\beta_1 = 6.9$; log $\beta_2 = 12.34$) and aspartic acid (log $\beta_1 = 5.95$; log $\beta_2 = 10.23$). The 2:1 complex with *l*-histidine shows reversible dioxygen binding properties [25], in contrast to the irreversibility of the corresponding Fe^{II} system. The rate constant (25°C) for dioxygenation was measured as $3.5 \times 10^3 \, M^{-1} s^{-1}$ [26]. The reaction of a number of Co^{II} dipeptide complexes with O_2 in alkaline solution has been reported [27]. The formation of dioxygenated products was found to depend upon the dipeptide and on the stoichiometries of the O_2–Co(dipeptide)$_2$ complex. The studies have shown that a minimum of three N donors is necessary for formation of a dioxygenated complex.

9.1.2   Cobalt(III) (d⁶): Chemistry of the $[Co(OH_2)_6]^{3+}$ Ion

The E^{θ} value in Table 9.1 emphasises Co^{II} as the favoured thermodynamic oxidation state in a field consisting solely of water ligands. The highly oxidising blue $[Co(OH_2)_6]^{3+}$ ion is, however, well characterised in alums [11, 38] and in the hydrated sulphate $Co_2(SO_4)_3 \cdot 18H_2O$ [39], but in solution it is only meta-stable, undergoing spontaneous reduction by water to pink $[Co(OH_2)_6]^{2+}$. Possible reasons for the large E^{θ} of the hexaaqua couple vs, for example the hexaammine couple were discussed in Chapter 1. Generally speaking, O donor ligands favour Co^{II} vs Co^{III} whereas the reverse is true for N donors. As the ligand field strength increases the low-spin t_{2g}^6 configuration of octahedral Co^{III} (maximum gain in CFSE of 2.4Δ) quickly becomes favoured and the E^{θ} (Co^{III}/Co^{II}) drops sharply. The presence of π-acceptor ligands further enhances the CFSE and formation of low-spin Co^{III} by lowering of the t_{2g} orbital energy. On the other hand, π donors, unless of sufficiently weak ligand field strength (e.g. F^-) to force the high-spin $t_{2g}^4 e_g^2$ configuration of Co^{III}, will rather favour the high-spin $t_{2g}^5 e_g^2$ configuration of Co^{II}. This is the case for a field of six water ligands—hence the high E^{θ}. With OH^- as ligand, the extra negative charge on the O atoms results in increased electrostatic repulsion, favouring tetrahedral geometry (there is little difference in CFSE vs octahedral).

Stock solutions of $[Co(OH_2)_6]^{3+}$ are best prepared for subsequent studies in strongly acidic solution ($\geqslant 4.0 \, M$). This is because the rate of the spontaneous reduction to $[Co(OH_2)_6]^{2+}$ by H_2O is suppressed as the acidity is increased. The normal preparative method is via anodic electrolysis of solutions of $[Co(OH_2)_6]^{2+}$ in $\geqslant 4.0 \, M \, H^+$ (usually $HClO_4$, HNO_3 or H_2SO_4) at a Pt electrode at 0°C [40, 41]. In H_2SO_4 solution crystals of the anhydrous cobaltic sulfate are conveniently isolable and stored at 0°C in a desiccator over concentrated H_2SO_4 as a source of stable Co^{3+}(aq). Solutions of the ClO_4^- salt for study can be prepared by addition of a slight excess of $Ba(ClO_4)_2$ in $HClO_4$ to solutions of the anhydrous sulfate [42]. Solutions $\sim 0.02 \, M$ in Co^{3+}(aq) are easily prepared by this method. Alternatively, solutions of Co^{3+}(aq) in ClO_4^- can be prepared by direct anodic electrolysis of $Co(ClO_4)_2$ in $HClO_4$ solution. Concentrations of $Co^{3+}(aq)$ can be assayed by treatment with an excess of $Fe^{2+}(aq)$ followed by back titration of the excess Fe^{2+} with $Ce^{IV}(aq)$. Direct measurement of the $Fe^{3+}(aq)$ generated has also been utilised. Total cobalt is most conveniently determined by allowing spontaneous reduction to $Co^{2+}(aq)$ to occur followed by analysis either directly as $[Co(OH_2)_6]^{2+}$ or as $[CoCl_4]^{2-}$ (by spectrophotometry) or by precipitation as $Co(py)_4(SCN)_2$.

9.1.2.1   ELECTRONIC STRUCTURE AND MAGNETISM OF $[Co(OH_2)_6]^{3+}$

The electronic spectrum of $Co^{3+}(aq)$ in various aqueous $HClO_4$ solutions at ionic stength 1.82 M is shown in Fig. 9.3 [43]. The true hexaaqua ion is present only at the very highest acidities. The appearance of two bands at 401 nm $(\varepsilon = 39.3\,M^{-1}\,cm^{-1})$ and 605 nm $(32.7\,M^{-1}\,cm^{-1})$ for $H^+ = 1.82\,M$ is suggestive of the expected spin-allowed transitions from a low-spin $^1A_{1g}(t_{2g}^6)$ ground state to the $^1T_{1g}$ and $^1T_{2g}$ excited states (both $t_{2g}^5 e_g^1$). The two-electron transition to the ^1E_g state $(t_{2g}^4 e_g^2)$ occurs well into the u.v. region and is not normally observed. The low-spin nature of $[Co(OH_2)_6]^{3+}$ is also reflected by the properties of the ion within the alum lattice. In the series of caesium alum salts the $Co-OH_2$ distance

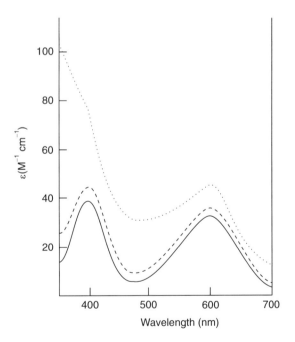

Figure 9.3. Electronic spectrum of $[Co(OH_2)_6]^{3+}$ as a function of $[H^+]$ with $\mu = 1.82\,M$ for ClO_4^- of (—) 1.82 M, (---) 0.5 M and (·····) 0.1 M [43]

in the Co^{3+} salt (187.3 pm) falls right on an extrapolated line from the alum salts of $Ti^{3+}(t_{2g}^1)$ (202.8 pm), $V^{3+}(t_{2g}^2)$ (199.2 pm) and $Cr^{3+}(t_{2g}^3)$ (195.9 pm) in a plot of rM—O vs the number of d electrons, correlating with filling only the t_{2g} orbitals (see Fig. 1.7) [38]. The observed Co—O distance is also too short to be reflective of a high-spin configuration and moreover the decrease in Co—OH_2 distance observed upon comparing hexaaquated Co^{2+} and Co^{3+} structures (21 pm) is sufficiently large to reflect a change from high-spin Co^{2+} to low-spin Co^{3+}. Magnetic measurements also support the low-spin configuration for $[Co(OH_2)_6]^{3+}$. A value < 1 BM has been reported for the rubidium cobalt alum [44] whereas an earlier reported value of 2.1 BM is probably due to Co^{2+} contamination [45]. These values may nonetheless be compared with RT values of around 4.8 BM typically observed for $[Co(OH_2)_6]^{2+}$ and 4.47 BM for $[CoF_3(OH_2)_3]$, a rare example of high-spin Co^{III}. Hunt and Taube and coworkers finally concluded in their 1951 paper that $[Co(OH_2)_6]^{3+}$ was low spin and diamagnetic [46].

9.1.2.2 REACTIONS OF $[Co(OH_2)_6]^{3+}$

Below 0.5 M H^+, the spectrum of $[Co(OH_2)_6]^{3+}$ shows a marked increase in absorbance below 600 nm (Fig. 9.3) due to the generation of hydrolysis products [43]. The kinetics of oxidation reactions by $[Co(OH_2)_6]^{3+}$ on a wide range of substrates is dominated by the involvement of the hydrolysed forms. In the spontaneous reduction of Co^{3+}(aq) by water, involvement of both $[Co(OH_2)_5OH]^{2+}$ and dimeric Co^{III}_2(aq), as either $Co(\mu\text{-}O)Co^{4+}$(aq) or $Co(\mu\text{-}OH)_2Co^{4+}$(aq), have been implicated. In an early study Bawn and White [47] presented evidence for a rate law involving terms in both $[Co^{III}]$ and $[Co^{III}]^2$ implying parallel involvement of both monomer and dimer species in the water oxidation process. In dilute solution only the monomeric form is involved. In each case the proposed rate-determining step for both $[Co(OH_2)_6]^{3+}$ and $[Co(OH_2)_5OH]^{2+}$ involved oxidation of OH^- ions to $OH^{\cdot}$ followed by coupling to form H_2O_2, which then reacts further with the Co^{3+}(aq). Baxendale and Wells [43] subsequently found evidence for a $[Co^{III}]^{3/2}$ term, suggesting that the monomer and dimer react together in the water oxidation step. Here $CoOH^{2+}$ was the dominant reactant according to

$$[Co(OH_2)_6]^{3+} \underset{}{\overset{K_a}{\rightleftharpoons}} [Co(OH_2)_5OH]^{2+} + H^+ \tag{9.2}$$

$$2\,CoOH^{2+}(aq) \underset{}{\overset{K_{22}}{\rightleftharpoons}} Co—O—Co^{4+}(aq) + H_2O \tag{9.3}$$

$$CoOH^{2+}(aq) + Co—O—Co^{4+}(aq) \overset{k}{\longrightarrow} 3\,Co^{2+}(aq) + HO_2^{\cdot} \tag{9.4}$$

$$CoOH^{2+}(aq) + HO_2^{\cdot} \overset{fast}{\longrightarrow} Co^{2+} + H_2O + O_2 \tag{9.5}$$

In each case Co_2^{III}(aq) was formulated as the μ-oxo dimer. A wide range of oxidations promoted by $[Co(OH_2)_6]^{3+}$ appear to proceed through $CoOH^{2+}$(aq). Despite, this, widely differing values for the hydrolysis constant for formation of $CoOH^{2+}$(aq) have been claimed from various studies and the precise value remains somewhat uncertain.

The rate of many oxidation reactions appears to be controlled by substitution into the Co^{3+} coordination sphere. For example, 1:1 complex formation with Cl^- is easily observed prior to slower Cl^- oxidation. Again $CoOH^{2+}$(aq) was found to be the dominant reactant. Sutin and coworkers claimed a K_a value of 0.22 M (25 °C) from a kinetic study of the 1:1 complexation reaction with Cl^- at $I = 3.0$ M (ClO_4^-) [48]. McAuley et al., in later work on the same reaction, but at $I = 0.5$ M [42], proposed an upper limit for the K_a (25 °C) of 1×10^{-2} M, which agrees with values quoted from independent spectrophotometric studies, 1.7×10^{-2} M (Sutcliffe and Weber) [49] and 1.2×10^{-2} M, $I = 1.0$ M (Martinez and coworkers) [50], the latter in conjunction with a kinetic study of the anation reaction with $H_2PO_4^-$. Some of the available data are summarised in Table 9.3.

The variation in K_a values is not merely a function of the $[Co^{III}]$ range studied (in view of the evidence for dimer formation). Sutin and coworkers used a $[Co^{III}]$ range from 3.0 to 5.0×10^{-4} M for their value of 0.22 M [48], which was similar to the range used by Davies and Watkins [51] in obtaining a much smaller estimate. In the water oxidation process Baxendale and Wells, working at $\leqslant 5 \times 10^{-3}$ M $[Co^{III}]$, estimated that K_a must be $\ll 5 \times 10^{-3}$ M on the basis of lack of curvature in plots of ε_{obs} (at 401, 370 and 350 nm) vs $[H^+]^{-1}$ down to $[H^+]$ 0.04 M (ε_2 and ε_1 being respectively the ε values for $[Co(OH_2)_6]^{3+}$ and $[Co(OH_2)_5OH]^{2+}$):

$$(\varepsilon_{obs} - \varepsilon_1)(K_a + [H^+]) = K_a(\varepsilon_2 - \varepsilon_1) \tag{9.6}$$

The lack of a $[Co^{III}]$ dependence on ε_{obs} (605 nm) at 0.1 M $[H^+]$ was surprisingly used as evidence against polymer formation despite $[Co^{III}]$ dependences on ε_{obs} at

Table 9.3. Kinetic data (25 °C) for complex formation and hydrolysis on $[Co(OH_2)_6]^{3+}$

Incoming ligand	$k_1 (M^{-1}s^{-1})^a$	$k_{OH}(M^{-1}s^{-1})^b$	Estimated $K_a(M)$	I(M)	Ref.
Cl^-	$\leqslant 2$	$(2 \pm 1) \times 10^2$	(0.22 ± 0.05)	3.0	[48]
Cl^-	(2.5 ± 1^c)	$(2.0 \pm 0.8) \times 10^2$	1×10^{-2}	0.5	[42]
NCS^-	68.5	$> 1.6 \times 10^4$	$< 5 \times 10^{-3}$	3.0	[51]
$H_2PO_4^-$	—	$(1.45 \pm 0.03) \times 10^{2d}$	1.5×10^{-2}	3.0	[50]

$^a[Co(OH_2)_6]^{3+}$ pathway.
$^b[Co(OH_2)_5OH]^{2+}$ pathway.
^{c}Value at 8 °C.
^{d}Proceeds solely via $CoOH^{2+}$(aq).

lower wavelengths. Martinez and coworkers worked at much higher $[Co^{III}]$ (0.05–0.4 M) for their spectrophotometric study [50] in reporting a K_a value lower than that quoted by Sutin. Sutin had earlier argued against the lower value obtained by Sutcliffe and Weber [49] as due to a failure to take account of hydrolytic polymerization.

The spontaneous water oxidation reaction has thus far precluded estimates of the water-exchange rate constant on Co^{3+}(aq). The complex formation kinetic data are probably reflective of a k_{ex} value (25 °C) on $[Co(OH_2)_6]^{3+}$ around 1–$10\,s^{-1}$ with $[Co(OH_2)_5OH]^{2+}$ reacting some 10^2 times faster.

Davies [52] and McAuley [53] have tabulated kinetic data for oxidations of a wide range of substrates by Co^{3+}(aq). In all of the reactions $CoOH^{2+}$(aq) is the dominant reactant. Tables 9.4 and 9.5 summarise some of the available kinetic data.

Davies and Watkins [51] described values of $k_{OH}K_a < 10^2\,s^{-1}$ as indicating that the reaction is inner sphere (substitution controlled) whereas $k_{OH}K_a$ values $> 10^3\,s^{-1}$ are indicative of outer-sphere reactions. A number of outer-sphere reductions of $[Co(OH_2)_6]^{3+}$ by macrocyclic Co and Ni complexes have been studied by Endicott, Durham and Kumar employing the Marcus approach [41]. The findings are somewhat interesting. Despite the reductions giving a good Marcus correlation of the reaction free energy change, the value of the self-exchange rate constant required for correlation with outer-sphere cross-reaction rate constants is found to be 12 orders of magnitude smaller ($\sim 1 \times 10^{-12}\,M^{-1}s^{-1}$) than the experimentally measured value. Around ~ 7 orders of magnitude of the difference can be traced to a consideration of Franck–Condon factors, the remainder presumed to be electronic. It is concluded that the self-exchange reaction itself must be occurring by a different

Table 9.4. Kinetic data for reduction of $[Co(OH_2)_6]^{3+}$ by various reagents at 25 °C [52, 53]

Reductant	k_1 $(M^{-1}s^{-1})$	$\Delta H^{\ddagger}$ $(kJ\,mol^{-1})$	$\Delta S^{\ddagger}$ $(J\,K^{-1}\,mol^{-1})$	Ref.
Malic acid	4.6	—	—	[54]
Thiomalic acid	$\leqslant 6$	—	—	[54]
Cl^-	2.5^a	89.5	$+50 \pm 30$	[48]
H_2O_2	$\leqslant 2$	—	—	[51]
HNO_2	18	76.6	$+38$	[51]
HN_3	$\leqslant 2$	—	—	[55]
ClO_2	$\leqslant 1$	—	—	[56]
Br^-	$\leqslant 5$	—	—	[51]
SCN^-	86.5	86.2	$+84$	[51]
I^-	8000	81.2	$+105$	[51]

^{a}Value at 8 °C.

Table 9.5. Kinetic data for reduction of $[Co(OH_2)_5OH]^{2+}$ by various reagents [52, 53]

Reductant	$k_{OH}K_a$ (s^{-1})	$\Delta H^{\ddagger}$ $(kJ\,mol^{-1})$	$\Delta S^{\ddagger}$ $(J\,K^{-1}\,mol^{-1})$	Ref.
NH_3OH^+	3.3	94.1	84	[57]
$N_2H_5^+$	1.1	99.6	92	[57]
H_2O_2	23.0	98.3	113	[51]
HN_3	35	96.7	113	[58]
Br^-	30	109.2	155	[51]
SCN^-	80	107.1	155	[51]
I^-	2680	90.0	126	[51]
p-Hydroquinone	1280	77.8	71	[51]
Ascorbic acid	1480	50.6	-8	[59]
Ni^{II}(cyclam)	935	69.0	50	[60]
$[Fe(OH_2)_6]^{2+}$	275	57.7	4.6	[53]
$[Mn(OH_2)_6]^{2+}$	46.5	51.5	-32	[53]

mechanism to that relevant to cross-redox reactions. In this regard it is clear that $[Co(OH_2)_6]^{3+}$ is somewhat unique. A mechanism for the self-exchange process involving a bridging water ligand has been proposed [41].

Efforts to elucidate the precise mechanism of substitution on $[Co(OH_2)_6]^{3+}$ are further confounded by catalysis from amounts of the more highly labile $[Co(OH_2)_6]^{2+}$ ion unavoidably present as a result of the spontaneous water oxidation process. However, in most cases it has proved possible to extract information on the intrinsic rate of ligand substitution on $[Co(OH_2)_6]^{3+}$, which remains some 10^6 times faster than those typically observed on $[Co(NH_3)_6]^{3+}$. In the latter the presence of the low-spin $^1A_{1g}$ ground state is not in doubt and the slow rates of substitution here in part related to the considerable thermodynamic and kinetic stability arising in the case of NH_3 from effective σ donation into the available empty e_g orbitals. However, of interest is the energy gap to the more labile high-spin $^5T_{2g}$ state. In the case of $[Co(NH_3)_6]^{3+}$ the gap is probably too high to be of relevance but conceivably might be low enough in the case of $[Co(OH_2)_6]^{3+}$ to allow appreciable thermal population at $\sim$ RT. Cobalt-59 NMR has been used in order to investigate the degree of participation from the high-spin form. Both ^{59}Co chemical shifts and spin-lattice (T_2) relaxation times were monitored as a function of temperature. The surprising result was that no deviation in either was found, stemming from possible exchange with the paramagnetic high-spin form [61]. What changes there were could be adequately explained by increased formation of $CoOH^{2+}$(aq) at the higher temperatures. Estimates for the lower limit of the spin change energy gap and fraction of the Co^{3+}(aq) in the high-spin form were given as $22.6\,kJ\,mol^{-1}$ and $< 10^{-4}$ respectively. Thus it appears that participation from the high-spin form is not a feature

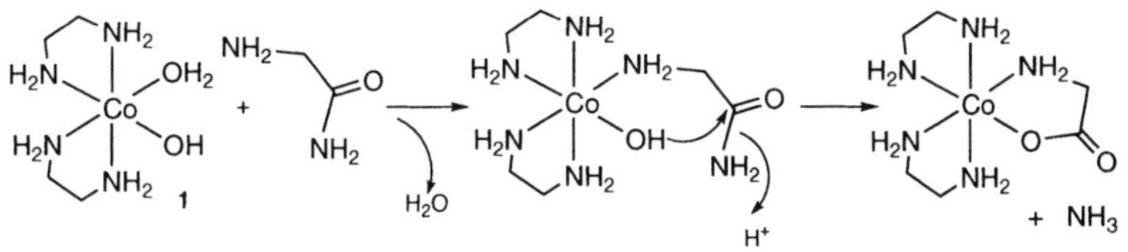

Figure 9.4. Hydrolysis of glycine amide promoted by a *cis*-hydroxo group on Co^{III}

sufficient to explain the high lability of $[Co(OH_2)_6]^{3+}$ vs $[Co(NH_3)_6]^{3+}$. On the other hand, the low-spin configuration of $[Co(OH_2)_6]^{3+}$ is clearly responible for the significant decrease in substitution rate ($\sim 10^2$ times) vs the previous trivalent ion of the first row, the high-spin $t_{2g}^3 e_g^2$ $[Fe(OH_2)_6]^{3+}$. $[Co(OH_2)_6]^{3+}$ also promotes the polymerisation of acrylonitrile and methylmethacrylate and, as either $Co_2(SO_4)_3$ or $Co(ClO_4)_3$ in the corresponding acid, it has found use as an oxidant for a range of organic reactions [62].

Certain *cis*-aqua hydroxy complexes of Co^{III} (usually with N donors in the other positions) show high activity in metal-promoted hydrolysis reactions which, coupled with their well-defined octahedral geometry and substitution inertness, have led to useful mechanistic inights into the reactions of more labile but biologically relevant Zn^{II} and Cu^{II} systems. The bound OH ligand is a powerful nucleophile towards attack at a proximal *cis*-bound substrate, a particularly pertinent example being shown in Fig. 9.4. The intramolecular hydrolysis of glycine amide using structure **1** is some 10^7 times faster than the spontaneous hydrolysis of the uncoordinated amide at the same pH. A number of excellent review articles on this general subject have appeared [63].

9.2 RHODIUM

9.2.1 Rhodium(II) (d^7): Chemistry of the $[Rh_2(OH_2)_{10}]^{4+}$ Ion

9.2.1.1 PREPARATION AND PROPERTIES

The preparation of the Rh_2^{4+} aqua dimer was first reported by Maspero and Taube in 1968 [64]. Apart from a few slight modifications subsequent preparations have varied little from the original Taube preparation [65–67] consisting of reduction of $[Rh(OH_2)_6]^{3+}$ (see later) by $[Cr(OH_2)_6]^{2+}$ in a non-complexing acidic solution followed by Dowex 50W column chromatography. Subsequently Hills, Moszner and Sykes showed that Cr^{2+}(aq) reduction of $[Rh(OH_2)_5X]^{2+}$ (X = Cl, Br) also gives Rh_2^{4+}(aq) after column chromatography [66]. In a typical preparation [65, 66] stoichiometric amounts of Cr^{2+}(aq) (a few cm^3 of a concentrated ~ 0.5 M solution) are added under rigorously air-free conditions to a 0.03 M solution of $[Rh(OH_2)_6]^{3+}$ in 2.0 M $HClO_4$. The solution is then diluted with air-free water to a concentration of ~ 1.0 M H^+ and then loaded on to a column (15 cm × 1 cm) of Dowex 50W X2 400 mesh resin. Following washing with a 1.0 M solution of the desired acid, usually $HClO_4$ or CF_3SO_3H (200 cm^3), the green aqua dimer is eluted with 3 M solutions of the desired final acid, giving *ca.* 0.01 M solutions which can be standardised by the peak maxima observed at 402 nm ($\varepsilon = 63$ M^{-1} cm^{-1}) and 648 nm (46.5 M^{-1} cm^{-1}) [68]. Solutions stored air free can be kept at 0 °C for three weeks without significant deterioration.

The dimeric Rh_2^{4+} structure is relevant from observations of only a weak paramagnetism, contrasting with $[Co(OH_2)_6]^{2+}$, and cation-exchange elution behaviour characteristic of a 4+ cation. As yet there are no reported X-ray structures of the aqua dimer due to an inherent instability of the ion towards oxidation and disproportionation. However, an X-ray structure of the corresponding acetonitrile solvate $[Rh_2(CH_3CN)_{10}]$ $(BF_4)_4$ has been reported by Dunbar which shows the expected 'staggered' structure and longer bonds to the two axial ligands, (Fig. 9.5) [69]. A similar structure for the aqua dimer is probably relevant and indeed ^{17}O NMR studies by Merbach and coworkers

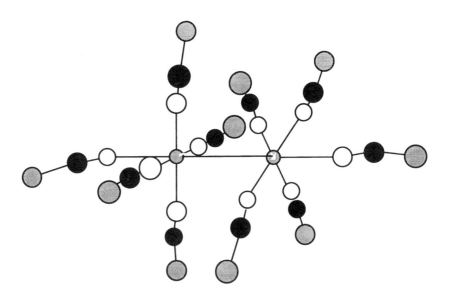

Figure 9.5. Structure of the cationic unit in $[Rh_2(CH_3CN)_{10}](BF_4)_4$ [69]

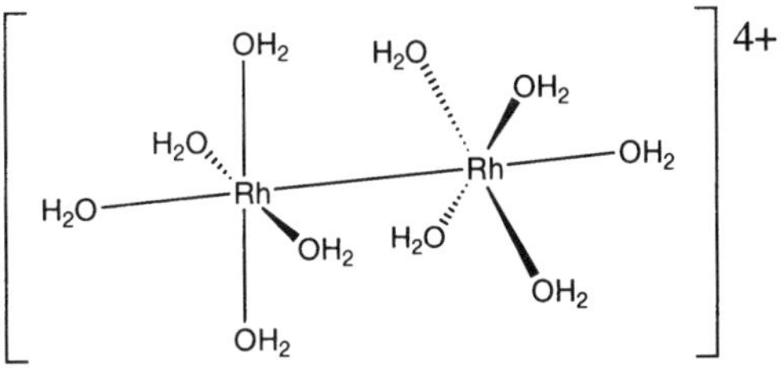

Figure 9.6. Likely structure for $[Rh_2(OH_2)_{10}]^{4+}$. (The staggered conformation is assumed to be the minimum energy solution structure)

have revealed evidence for kinetically distinct equatorial and axial water ligands [65]. A single Rh—Rh bond is believed relevant with presumably free rotation in solution. The Rh—Rh distance in the acetonitrile solvate, 262.1 pm, is longer than that in numerous Rh_2^{II}(bridging anion)$_4L_2$ complexes, e.g. $[Rh_2(O_2CCH_3)_4(OH_2)_2]$ (238.6 pm). The shorter Rh—Rh distance in the bridged dimers is believed to be a direct result of the 'bite' of the bridging ligand pulling the Rh atoms closer. Therefore a Rh—Rh distance of ~ 260 pm, as in the acetonitrile solvate, is presumed for the aqua dimer (Fig. 9.6).

In the electronic structure of dimeric Rh^{II} complexes the single Rh—Rh bond arises from sufficient d electrons being present (14) to populate both the δ^* and the two π^* levels in the M—M bonded framework (cf. the treatment for Mo_2^{4+} in Chapter 6, leaving behind just a net σ bond. Theoretical studies of Rh_2^{4+} complexes based upon the electronic and molecular structure of the tetraacetate suggest that the two lowest energy transitions of Rh_2^{4+} arise from $\delta^*(Rh—Rh)-\sigma^*(Rh—Rh)$ and $\delta^*(Rh—Rh)-\sigma^*(Rh—O)$ excitations. The electronic spectrum of $[Rh_2(OH_2)_{10}]^{4+}$ in 3.0 M $HClO_4$ is shown later in Fig. 9.8. Rh_2^{4+} is one of three known M—M bonded dimeric aqua ions; Mo_2^{4+}(aq) (quadruple bond, see Chapter 6) and Hg_2^{2+}(aq) (single bond, see Chapter 12) are the others.

9.2.1.2 REDOX REACTIONS INVOLVING THE GENERATION OF $[RH_2(OH_2)_{10}]^{4+}$

Hills, Moszner and Sykes have studied the kinetics of the Cr^{2+}(aq) reduction of $[Rh(OH_2)_6]^{3+}$ to give Rh_2^{4+}(aq) [66]. The rate law, $k = a[H^+]^{-1}$, indicates that the reaction proceeds through $[Rh(OH_2)_5OH]^{2+}$. At 25 °C, $a = 93 \pm 3\,s^{-1}$. Therefore it seems that Cr^{2+}(aq) has a tendency to react with pentaaqua $[Rh(OH_2)_5X]^{2+}$ species (X = OH, Cl and Br). The formation of both $CrCl^{2+}$(aq) and $CrBr^{2+}$(aq) as products when $X = Cl$ and Br respectively suggests an inner-sphere mechanism:

$$[Cr(OH_2)_6]^{2+} + [Rh(OH_2)_5]X^{2+} \xrightarrow{k_{Cr}} [(H_2O)_5Cr-X-Rh(OH_2)_5]^{4+} + H_2O$$

(9.7)

$$[(H_2O)_5Cr-X-Rh(OH_2)_5]^{4+} \xrightarrow{fast} [Cr(OH_2)_5X]^{2+} + [Rh(OH_2)_5]^{2+}$$ (9.8)

$$2[Rh(OH_2)_5]^{2+} \xrightarrow{fast} [Rh_2(OH_2)_{10}]^{4+}$$ (Rh—Rh bond   (9.9)
formation)

Interestingly, when the less reducing $[V(OH_2)_6]^{2+}$ ion is used the result is reduction all the way to Rh metal. This was eventually traced to a disproportionation reaction of Rh_2^{4+}(aq) occurring in the presence of free Cl^- and Br^- ions, confirmed in separate experiments. The presence of free halide ions confirmed the outer-sphere nature of the V^{2+}(aq) reduction:

$$V^{2+} + RhX^{2+} \xrightarrow{k_V} V^{3+} + X^- + Rh^{2+}$$ (9.10)

$$3\,Rh^{2+} + 2\,X^- \xrightarrow{fast} Rh(s) + 2\,RhX^{2+}$$ (9.11)

These observations provide a particularly good example of the mechanistic pathway, and not merely the thermodynamic driving force, as leading to the desired final product.

9.2.1.3 SUBSTITUTION REACTIONS ON $[RH_2(OH_2)_{10}]^{4+}$ AND RELATED DERIVATIVES

Complexation of Cl^- and Br^- on $[Rh_2(OH_2)_{10}]^{4+}$ as mentioned above, leads to disproportionation of Rh^{II} to give Rh^{III} complexes and Rh metal. This seems to be a reoccurring phenomenon with anionic monodentate ligands tending to prefer forming Rh^{III} complexes. Potential bridging bidentate ligands will, however, preserve the Rh—Rh dimer unit and lead to stable complexes such as in the case of $[Rh_2(O_2CCH_3)_4(OH_2)_2]$.

Exchange of bulk water on $[Rh_2(OH_2)_{10}]^{4+}$ has been studied using ^{17}O NMR [65]. The ^{17}O NMR spectrum of $[Rh_2(OH_2)_{10}]^{4+}$ is consistent with the solution structure in Fig. 9.6 showing the presence of distinguishable equatorial and axial water ligands. The eight equatorial waters appear as a single broad resonance at -162 ppm from bulk water at 298 K. The signal attributable to the two axial waters is not observable within the range $-400 - +500$ ppm at the lowest temperature studied (272.3 K) and thus the rate constant for exchange (k_{ex}) was estimated to be $> 10^4 s^{-1}$ at 298 K. The rate constant for the equatorial waters was estimated by the temperature dependence of the linewidth of the -162 ppm resonance using the methods described earlier (see, for example, Section 1.2.3). The thermal instability of $[Rh_2(OH_2)_{10}]^{4+}$ only permits an estimate of the rate constant as lying between 10 and $50 s^{-1}$. This is because of the high temperatures (~ 330 K) needed for observation of a significant contribution from k_{ex} ($\geqslant 10\%$) to the observed value of $1/T_2$.

More reliable information was gained from a ^{1}H NMR study of the acetonitrile exchange on the related, but more thermally stable, $[Rh_2(CH_3CN)_{10}]^{4+}$ cation [65]. The ^{1}H NMR spectrum of $[Rh_2(CH_3CN)_{10}]^{4+}$ in CD_3NO_2 was consistent with the presence of the structure in Fig. 9.6. In this case both types of acetonitrile solvent ligand are observable in the ^{1}H NMR at $+2.72$ ppm (equatorial) and $+2.39$ ppm (axial) (versus the signal from incompletely deuterated nitromethane at $+4.3$ ppm). Exchange at the equatorial acetonitriles is slow and is easily studied by isotopic dilution. At 298 K, $k_{ex} = 3.1 \times 10^{-5}$, $\Delta H^{\ddagger} = 65.6$ kJ mol^{-1}, $\Delta S^{\ddagger} = -111.0$ J K^{-1} mol^{-1} and $\Delta V^{\ddagger} = -4.9$ cm^3 mol^{-1}. Exchange at the two axial acetonitriles was too fast to be observable by ^{1}H NMR. A mechanism was proposed for both water and acetonitrile exchange on the respective $[Rh_2(solv)_{10}]^{4+}$ complexes wherein a concerted migration of an axial solvent ligand to the equatorial plane occurs with release of an equatorial solvent ligand to the bulk taking place through a contracted transition state. A similar mechanism is believed to operate between the axial and equatorial ligand sites on $[VO(OH_2)_5]^{2+}$ (see Chapter 5) and on $[TiO(DMSO)_5]^{2+}$. Structural similarities are apparent from Fig. 9.7. A similar mechanism is also believed to operate for water exchange on the aqua Mo_2^{4+} dimer (Chapter 6). Direct exchange at the labile axial site with bulk solvent is assumed to proceed through a dissociatively activated mechanism. The k_{ex} value of $\geqslant 10^4 s^{-1}$ for the aqua complex is consistent with estimated values ($1-9 \times 10^5 s^{-1}$) for axial solvent exchange on the acetato dimer $[Rh_2(O_2CCH_3)_4L_2]$ for both $L = H_2O$ and acetonitrile.

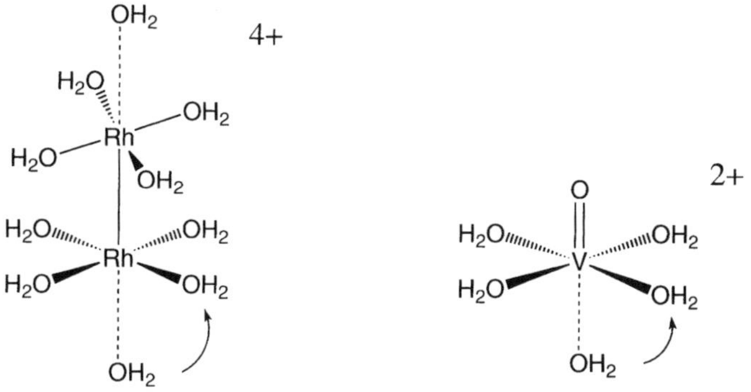

Figure 9.7. Structural similarity between $[Rh_2(OH_2)_{10}]^{4+}$ and $[VO(OH_2)_5]^{2+}$

9.2.1.4   OXIDATION OF $[Rh_2(OH_2)_{10}]^{4+}$

Rh_2^{4+}(aq) can be readily reoxidised to Rh^{III} compounds using a number of chemical oxidants. Treatment of solutions of $[Rh_2(OH_2)_{10}]^{4+}$ with X_2 ($X = Cl$, Br) is a one-step route to the halopentaaqua Rh^{III} ions $[Rh(OH_2)_5X]^{2+}$ following column chromatography [66]. This is a useful synthetic route since $[RhX(OH_2)_5]^{2+}$ complexes are difficult to isolate in good yield via direct reaction of X^- on Rh^{3+}(aq) owing to the invariable presence of bis and tris halo complexes. This is due to labilisation of the remaining water ligands once one X^- has substituted (see Section 9.2.2). The reaction of $[Rh_2(OH_2)_{10}]^{4+}$ with O_2 [70] is also discussed in Section 9.2.2 in the context of leading to the formation of superoxo-bridged aqua Rh_2^{III} species such as $[Rh_2(\mu-O_2^-)(OH_2)_8(OH)_2]^{3+}$.

9.2.2   Rhodium(III) (d^6): Chemistry of the $[Rh(OH_2)_6]^{3+}$ Ion

9.2.2.1   PREPARATION AND PROPERTIES

The preparation of $[Rh(OH_2)_6](ClO_4)_3$ dates back to the work of Ayres and Forrester in the late 1950s. In the original preparation [71] 2.0 g $RhCl_3 \cdot xH_2O$ is dissolved in water (25 cm^3) to which 30 cm^3 of 72% $HClO_4$ solution is added. The resulting mixture is then boiled to concentrate the solution to ~ 25 cm^3. A red-brown precipitate appears but redissolves upon further boiling of the solution. The mixture is cooled and a further 10 cm^3 72% $HClO_4$ is added and fuming continued to a final volume of between 10 and 15 cm^3. The fact that Rh^{III} is not oxidised during this process is an indication of its redox stability and provides a contrast with the behaviour of Ir^{III} (see later). On cooling, needles of $[Rh(OH_2)_6](ClO_4)_3$ separate which can be isolated and washed with ice-cold 70% $HClO_4$ until a constant yellow colour of the washings is obtained. The crystals can then be dried at 110°C at 10^{-3} mm Hg as light yellow fluffy needles. In the presence of moisture the crystals are found to darken in colour due to hydrolytic polymerisation. Freshly prepared samples of the perchlorate salt have been shown to give satisfactory elemental analyses.

Modifications to the preparation have been reported. Hills, Moszner and Sykes boiled $RhCl_3$ hydrate in 72% $HClO_4$ under a slight water pump vacuum to help reduce the volume during the fuming process. Solutions could then be diluted with water to 0.05 M $[H^+]$ and loaded on to a column of Dowex 50W X8 resin (10 cm $\times$ 1.6 cm). The column is washed with 0.5 M and then 1.0 M $HClO_4$ which results in the removal of chloro complexes having a charge < 2.0. Elution with $\geqslant 1.0$ M $HClO_4$ gives $[Rh(OH_2)_6]^{3+}$ as a bright yellow band [66]. Concentrated (> 0.1 M) solutions of Rh^{3+}(aq) can be obtained with the use of a short saturated column of Dowex 50W X2 200–400 mesh resin and elution with 3.0 M $HClO_4$.

Solutions of $[Rh(OH_2)_6]^{3+}$ are conveniently standardised for solution studies by using the electronic spectrum (Fig. 9.8). The two-band maxima expected for

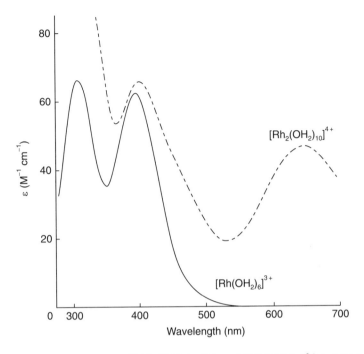

Figure 9.8. Electronic spectra of $[Rh_2(OH_2)_{10}]^{4+}$ and $[Rh(OH_2)_6]^{3+}$ in 3.0 M $HClO_4$

the low-spin octahedral t_{2g}^6 configuration are observed at 396 nm ($\varepsilon = 62$ M^{-1} cm^{-1}) ($^1A_{1g}$–$^1T_{1g}$) and 311 nm (67.4 M^{-1} cm^{-1}) ($^1A_{1g}$–T_{2g}). The two peaks are well defined and of almost equal height [66, 72, 73], contrasting with some earlier measurements [74] from solutions that may have been contaminated with hydrolysis products.

A low-resolution X-ray structure of $[Rh(OH_2)_6](ClO_4)_3 \cdot 3H_2O$ has been reported [75]. The positions of the H atoms were not ascertained and there were some disorder problems. The somewhat long Rh—O distances (~ 213 pm) obtained compared with those obtained from the alum, $Cs[Rh(OH_2)_6](SO_4)_2 \cdot 6H_2O$ (Rh—O = 201.6 pm) [76] and from EXAFS and large-angle X-ray scattering (LAXS) studies on Rh^{3+}(aq) in solution (Rh—O = 203 ± 2 pm) [77] were attributed to an influence of specific ClO_4^-–H_2O interactions in the crystalline perchlorate salt. However, this could also be as a result of the inherent disorder. The latter studies have confirmed retention of the regular octahedral geometry in solution. The EXAFS and LAXS studies confirmed retention of the octahedral geometry in the primary shell of water molecules in solution while at the same time finding conclusive evidence for a second hydration shell of 13 ± 1 water molecules at a Rh—O distance of 402 ± 2 pm [77]. This supported earlier infra-red findings [78]. Indeed, $[Rh(OH_2)_6]^{3+}$ and $[Cr(OH_2)_6]^{3+}$, in comparative studies, are found to be remarkably similar regarding the structure of the primary and secondary hydration shells. Other similarities between Rh^{3+}(aq) and Cr^{3+}(aq) will become apparent later. A rigid well-defined H-bonding network between primary and secondary hydration shells of water molecules is believed to be responsible for the slow proton exchange kinetics in aqueous solutions of $[Rh(OH_2)_6]^{3+}$ [79]. A primary hydration number of 5.9 ± 0.4 in solution has also been confirmed by ^{18}O isotopic dilution [80]. $Cs[Rh(OH_2)_6](SO_4)_2 \cdot 6H_2O$ can be prepared by treatment of freshly precipitated $Rh(OH)_3$ with 1.0 M H_2SO_4 followed by the addition of Cs_2SO_4 (roughly stoichiometric). A similar procedure has been successfully used to prepare the corresponding Ir^{3+} alum. The Rh^{3+} alum, like its Co^{3+} analogue and also that of Ir^{3+}, belongs to the α class of alums (Section 1.2.1.2) in possessing a significant tilt of the H—O—H plane from lying along the M—O bond vector [76]. This implies that, as with the Co^{3+} alum, an important contributory factor to the orientation of the water molecule is the accommodating of sp^3 hybridisation at oxygen to permit strong σ-bonding into the empty e_g orbitals of the low-spin t_{2g}^6 M^{III} ion. The result for the α alums is six waters forming a trigonal antiprismatic arrangement along the threefold axis, which could be described as a puckered ring perpendicular to the three fold axis. In the β alums this ring is almost planar.

9.2.2.2 HYDROLYSIS OF $[RH(OH_2)_6]^{3+}$

The dominant involvement of $[Rh(OH_2)_5OH]^{2+}$ in the mechanism of reduction of Rh^{3+}(aq) to Rh_2^{4+}(aq) has already been highlighted [66]. Early estimates of pK_{11} for $[Rh(OH_2)_6]^{3+}$ were based upon kinetic measurements of ligand substitution reactions which were shown to involve a contribution from $RhOH^{2+}$(aq) from $1/[H^+]$ terms apparent in the rate expressions [74, 81]. Many of the early measurements were hampered by slow attainment of equilibria which were eventually traced to hydrolytic polymerisation under the conditions used. However Ayres and Forrester [71], by conducting titrations rapidly, were able to study the formation of $RhOH^{2+}$(aq) and estimate a pK_{11} value (298 K) of 3.2 for a dilute solution which was essentially the same as an earlier value obtained by Plumb and Harris (3.4) in their water-exchange study in 2.5 M $NaClO_4$ [80]. As part of their reinvestigation of the water-exchange process Laurenczy, Merbach and coworkers [82] performed potentiometric titrations on Rh^{3+}(aq) in $I = 5.0$ m $(NaClO_4)$ solution at variable temperature and pressure with appropriate calibration and corrections. The experimental value for pK_{11} (3.86 ± 0.05), determined at 278.2 K, is close to the value (4.09) obtained at the same temperature but at $I = 1.0$ M $(NaClO_4)$ [83]. The value for pK_{11} at 298 K (3.45), computed using $\Delta H_{11}^\circ = 32.8$ kJ mol^{-1} [84], is in good agreement with the two earlier estima-

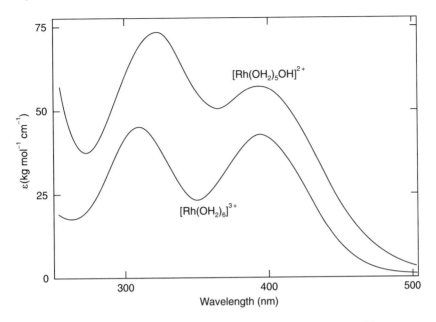

Figure 9.9. Electronic spectrum for $[Rh(OH_2)_6]^{3+}$ and $[Rh(OH_2)_5OH]^{2+}$ at 278.2 K, $I = 5.0\,m\,(NaClO_4)$

tions [80, 85]. Figure 9.9 shows electronic spectra for $[Rh(OH_2)_6]^{3+}$ (experimental) and for $[Rh(OH_2)_5OH]^{2+}$ (calculated) at 278.2 K in $I = 5.0\,m\,(NaClO_4)$ [82].

9.2.2.3 CHEMISTRY OF THE AQUA DIMER $[RH_2(\mu\text{-OH})_2(OH_2)_8]^{4+}$
AND RELATED SPECIES

Spiccia and coworkers have studied the products obtained upon the deliberate addition of excess base to solutions of $[Rh(OH_2)_6]^{3+}$. A similar strategy to that used in the study of the hydrolysis products of $Cr^{3+}(aq)$ was employed, the slow attainment of equilibria allowing the various oligomeric $Rh^{3+}(aq)$ species to be separated by cation-exchange chromatography. Several polynuclear species were obtained, of which one has been characterised as the aqua dimer $[Rh_2(\mu\text{-OH})_2(OH_2)_8]^{4+}$ [72]. The dimer elutes well after $[Rh(OH_2)_6]^{3+}$ and alongside the corresponding $[Cr_2(\mu\text{-OH})_2(OH_2)_8]^{4+}$ ion (Chapter 6) on columns of Dowex 50W X2 cation-exchange resin, indicating a similar $4+$ charge. The charge/Rh atom (2.1 ± 0.1) was determined as described for $Cr_2(\mu\text{-OH})_2^{4+}(aq)$ in Chapter 6. $Rh_2(\mu\text{-OH})_2^{4+}(aq)$ is characterised by an intense peak at 242 nm $(\varepsilon \sim 1750\,M^{-1}\,cm^{-1})$ assigned to an allowed CT from OH^- to Rh^{III} which masks the higher energy $^1A_{1g}$–$^1T_{2g}$ transition around 310 nm (Fig. 9.10b). The lower

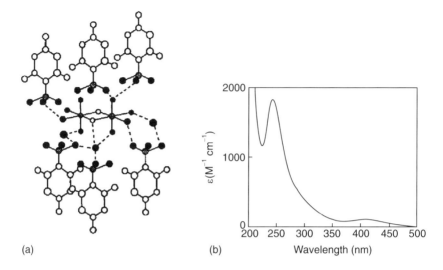

(a)   (b)   Wavelength (nm)

Figure 9.10. (a) X-ray structure of $[(H_2O)_4Rh(\mu\text{-OH})_2Rh(OH_2)_4](dmtos)_4 \cdot 8H_2O$. (b) Electronic spectrum of $[(H_2O)_4Rh(\mu\text{-OH})_2Rh(OH_2)_4]^{4+}$ in 1.0 M $NaClO_4$/0.02 M $HClO_4$

energy $^1A_{1g}$–$^1T_{1g}$ transition is still observable at ~ 400 nm. Thus the presence of hydrolysis products profoundly increases the extinction coefficient of Rh^{3+}(aq) solutions below 350 nm. The X-ray structure of the $Rh_2(\mu\text{-OH})_2^{4+}$(aq) ion as its mesitylenesulfonate (dmtos$^-$) salt is shown in Fig. 9.10(a) [72].

The crystals were obtained by a similar procedure to that used to crystallise $Cr_2(\mu\text{-OH})_2^{4+}$(aq) via initial precipitation of the yellow active 'dimer hydroxide' followed by dissolution in aqueous Hdmtos and crystallisation from a slowly evaporating solution. The non-bonded Rh—Rh distance is 303 pm. The quality of the X-ray data did not allow an assessment of whether the bridging OH groups influenced the Rh—OH_2 distances directly *trans* to them. Structural similarities between the aqua dimers of Cr^{III} and Rh^{III} are apparent to the extent that the mixed $\{Cr^{III}Rh^{III}\}$ aqua dimer can be crystallised by addition of $[Cr(OH_2)_6]^{3+}$ (at pH 2) to an ice-cold alkaline solution of Rh^{III}(aq) at pH 13 [86]. Acidification to pH 2, cation-exchange purification from monomers and higher polymers, precipitation of the 'active hydroxide' with imidazole followed by crystallisation from Hdmtos leads to isolation of the mixed metal dimer $[Cr^{III}Rh^{III}(\mu\text{-OH})_2(OH_2)_8](dmtos)_4 \cdot 4H_2O$, the structure of which is shown in Fig. 9.11.

Spiccia and coworkers have also studied the changing composition of aged Rh^{3+}(aq) solutions (0.08 M) at 0 °C after various times following treatment with base (up to a pH of 13) before quenching via reacidification [72]. Even after an ageing time of 2 minutes 80% of the Rh could be recovered as $[Rh(OH_2)_6]^{3+}$,

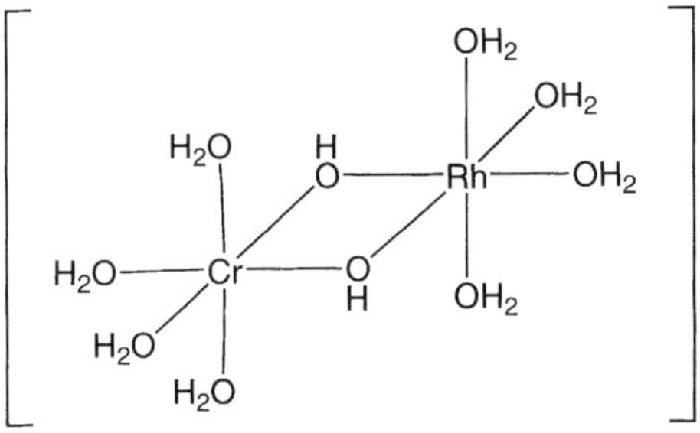

Figure 9.11. Structure of the cationic unit in $[(H_2O)_4Rh(\mu\text{-}OH)_2Cr(OH_2)_4]$ $(dmtos)_4 \cdot 4H_2O$ [86]

reflecting the greater inertness of Rh^{III} vs Cr^{III} (see Section 6.1.2). An ageing time of 90 minutes, however, produced six Rh-containing fractions obtainable following cation-exchange chromatography on Sephadex SP C25. After an ageing time of 1 day none of the products could be eluted from the Sephadex column, even with 4.0 M $NaClO_4$ and 2.0 M $Ba(ClO_4)_2$.

Read *et al.* subsequently carried out some elegant ^{17}O and ^{103}Rh NMR studies on the products of hydrolysis of $Rh^{3+}(aq)$ solutions [87]. In a typical procedure the solid acid-free salt $[Rh(OH_2)_6](ClO_4)_3$ (3.0 g) was dissolved in 120 cm^3 of 1.0 M NaOH and left to age for 45 minutes at 25° C before quenching via the addition of 1.0 M $HClO_4$ to a final pH of 1.7. The solution was then absorbed on to a column of Sephadex SP C25 resin and eluted with increasing concentrations of $NaClO_4$ in 0.02 M $HClO_4$. The first two fractions contained $[Rh(OH_2)_6]^{3+}$ and the aqua dimer. However, a third fraction was obtained using 2.0 M $NaClO_4$. Of interest were the results of ^{103}Rh NMR investigations on these fractions. $[Rh(OH_2)_6]^{3+}$ and the aqua dimer gave rise under these conditions to single ^{103}Rh signals at 9915.8 and 9997.7 ppm respectively. However, fraction 3 gave rise to two ^{103}Rh NMR signals at 9966 ppm and 10003 ppm in a 2:1 ratio consistent with a trimeric $Rh^{III}(aq)$ species. A close similarity between Rh and Cr in their hydrolysis chemistry is indeed clearly apparent. Closer analysis, however, of the two trimer resonances revealed that they are split into a triplet (central Rh) and a doublet (terminal Rh). This immediately rules out the 'incomplete cuboidal' $[Rh_3(\mu_3\text{-}OH)(\mu\text{-}OH)_3(OH_2)_9]^{5+}$ structures that have been favoured for trimeric $Cr^{III}(aq)$. This leaves three alternative structures, shown in Fig. 9.12. All would seem equally plausible on the basis of the NMR findings. One might have a slight preference for structure **III** on the basis that its single unique Rh is most like that of the Rh centres present in the dimer to which it closely resonates. Moreover, only $[Rh(OH_2)_6]^{3+}$ has the facial arrangement of waters with a three-fold symmetrical arrangement as present in the two other identical Rh atoms in

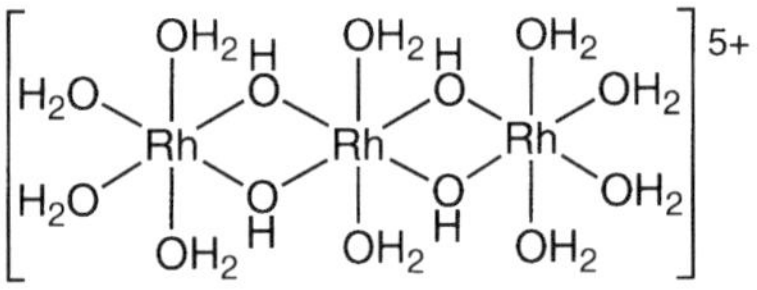

(I)

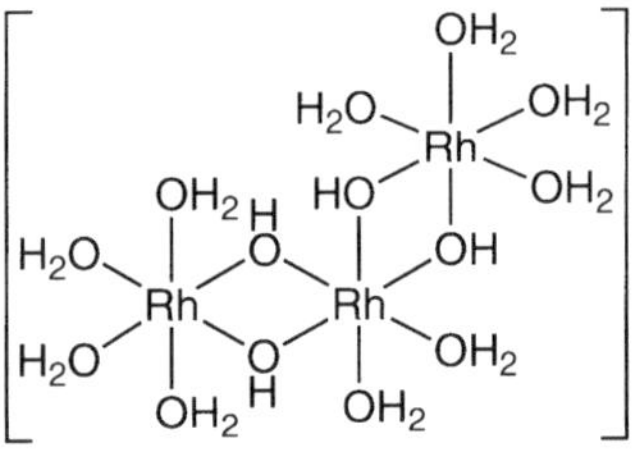

(II) 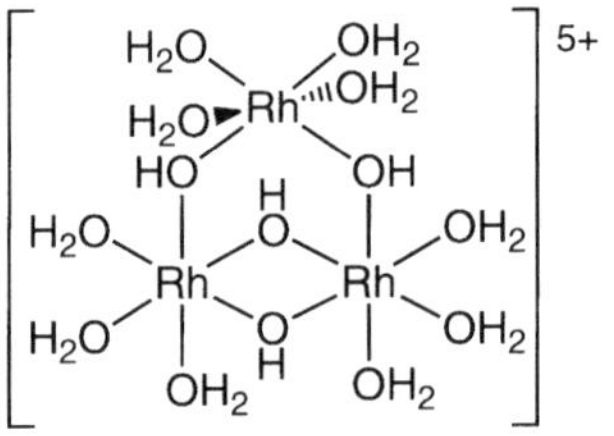   (III)

Figure 9.12. Possible structures for trimeric Rh^{III}(aq)

structure **III**. However, it would be dangerous to rule out either of the two other possibilities on this basis.

^{17}O NMR spectra for the hydrolysis products were obtained from hydrolysing ^{17}O-enriched solutions of $[Rh(OH_2)_6]^{3+}$ [87]. The aqua dimer shows two resonances at -122.2 and -130.6 ppm (from bulk $H_2O = 0$ ppm) assigned to H_2O opposite μ-OH and H_2O opposite H_2O respectively. The μ-OH resonance appeared at -320.1 ppm. Relative peak heights were in agreement with the di-μ-OH formulation. For the trimer a broad envelope of resonances was observed around -120 ppm with the μ-OH groups appearing at -303 ppm. The water ligands on $[Rh(OH_2)_6]^{3+}$ are represented by a single peak at -142 ppm. In these studies a number of higher oligomers were observed eluting very slowly from the Sephadex column, but these have not yet been studied in detail.

^{103}Rh NMR chemical shifts have also been measured successfully for all ten isomers in the series $[RhX_n(OH_2)_{6-n}]^{(3-n)}$ (X = Cl or Br, $n = 0$–6) [88, 89]. Assignment of shifts to *trans/cis* and *mer/fac* isomer pairs was based on measurements of the isolated complexes. A Nephelauxetic dependence (decrease in δ with increasing number of halide ligands with $\delta_{Br} > \delta_{Cl}$, $\delta_{trans} > \delta_{cis}$ and $\delta_{mer} > \delta_{fac}$) was demonstrated. ^{103}Rh chemical shifts were found to correlate with the ligand field parameter $\beta/\Delta E$, where β is the Nephelauxetic ratio and ΔE is the energy of the $^1A_{1g}$–$^1T_{1g}$ transition. Table 9.6 shows selected ^{103}Rh NMR data for various aqua species.

Table 9.6. ^{103}Rh NMR chemical shifts for various $[RhX_n(OH_2)_{6-n}]^{(3-n)}$ species [88, 89]

Complex	n	δ(chloro)	δ(bromo)
$[Rh(OH_2)_6]^{3+}$	0	9880	9880
$[RhX(OH_2)_5]^{2+}$	1	9503	9327
trans-$[RhX_2(OH_2)_4]^+$	2	9208	8928
cis-$[RhX_2(OH_2)_4]^+$	2	9142	8846
mer-$[RhX_3(OH_2)_3]$	3	8870	8447
fac-$[RhX_3(OH_2)_3]$	3	8753[a]	8347
trans-$[RhX_4(OH_2)_2]^-$	4	8620	8056
cis-$[RhX_4(OH_2)_2]^-$	4	8541	7946
$[RhX_5(OH_2)]^{2-}$	5	8298	7493
$[RhX_6]^{3-}$	6	8075	7007

[a]Recorded at 3 °C.

9.2.2.4 WATER EXCHANGE AND LIGAND SUBSTITUTION ON $[Rh(OH_2)_6]^{3+}$

The first measurements of the water-exchange process were carried out by Plumb and Harris in 1964 [80]. They used an ^{18}O isotopic labelling technique coupled with mass spectrometry detection at a very high ionic strength (12.0 M). It was not fully explained why such a high ionic strength was used and in later comparisons with the rate constants for Cl^- and Br^- anation, extrapolation to ionic strength ~ 0.2 M was required. Rate laws are of the form $k_1 + k_{OH}[H^+]^{-1}$, with k_1 referring to reaction via $[Rh(OH_2)_6]^{3+}$ and k_{OH} to reaction via $[Rh(OH_2)_5OH]^{2+}$. Rate constant obtained at 50 °C ($k_1(H_2O) = 3 \times 10^{-7} s^{-1}$, $I = 2.0$ M, $k_1(Cl^-) = 1.6 \times 10^{-7} M^{-1} s^{-1}$, $I = 2.5$ M and $k_1(Br^-) = 1.3 \times 10^{-7} M^{-1} s^{-1}$, $I = 2.0$ M) were interpreted in terms of a D mechanism (no incoming ligand dependence was observed), which was further promoted by the monohydroxy form $[Rh(OH_2)_5OH]^{2+}$, $k_{OH}(Cl^-)$ (50° C) $= 5.9 \times 10^{-4} M^{-1} s^{-1}$ and $k_{OH}(Br^-) = 4.4 \times 10^{-4} M^{-1} s^{-1}$, with pK_{11} for formation of $RhOH^{2+}$(aq) estimated as 3.2. For some time these conclusions were accepted but with some reservations and eventually the Rh^{3+}(aq) exchange process was reinvestigated by Merbach and coworkers using variable temperature and pressure ^{17}O NMR [82]. Here it was found desirable to use solid samples of the perchlorate salt for the purpose of enriching $[Rh(OH_2)_6]^{3+}$ with the ^{17}O isotope. The technique for conventional monitoring of the slow exchange of water on diamagnetic aqua ions using isotope dilution ^{17}O NMR is described in Section 1.2.3.2. Here the exchange was conveniently followed by observing the slow loss of enriched bound signal coinciding with an increase in the bulk water signal at 0 ppm. For the $[H^+]$ dependent study four acidities 1.0, 1.5, 3.0 and 5.0 m were used with $[Rh^{3+}]$ (~ 0.1 m) at $I = 5.6$ m ($NaClO_4$). An $[H^+]$ dependence of the form $k_1 + k_{OH}k_{11}[H^+]^{-1}$, reflecting $K_{11} \ll [H^+]$ ($pK_{11} = 3.45$) and an involvement from

Table 9.7. Kinetic data for ligand substitution reactions on $[Rh(OH_2)_6]^{3+}$ and $[Rh(OH_2)_5OH]^{2+}$

Incoming ligand	Ionic strength	k_{298} (s^{-1})	$\Delta H^\ddagger$ $(kJ\,mol^{-1})$	$\Delta S^\ddagger$ $(J\,K^{-1}\,mol^{-1})$	Ref.
$[Rh(OH_2)_6]^{3+}$					
H_2O	5.6^c	2.2×10^{-9}	131.2	$+29.3$	[82]
Cl^-	2.5	2.99×10^{-9a}	125.0	$+11.2$	[74]
Br^-	2.0	2.75×10^{-9a}	120.2	-5.5	[81]
$H_2C_2O_4$	3.0	1.21×10^{-5b}	76.7	-81.7	[90]
Hacac	0.1^d	7.36×10^{-3b}	64.3	-70.0	[91]
$[Rh(OH_2)_5OH]^{2+}$					
H_2O	5.6^c	4.2×10^{-5}	103	—	[82]
Cl^-	2.5	1.01×10^{-5a}	127.8	$+88.2$	[74]
Br^-	2.0	0.62×10^{-5a}	131.1	$+95.0$	[81]
Thiourea	0.1^e	0.17×10^{-5}	113.8	$+26.6$	[92]

^{a}Values refer to interchange within an ion pair ($K_{IP} = 1.0\,M^{-1}$).
^{b}Units of $M^{-1}s^{-1}$.
^{c}Units of molality (mol kg^{-1}).
^{d}At pH 2.0.
^{e}In ethanol (30%) v/v in water at 'pH' 4.0.

$[Rh(OH_2)_5OH]^{2+}$. At 298 K the findings were $k_1 = 2.2 \times 10^{-9}\,s^{-1}$, $\Delta H_1^\ddagger = 131 \pm 23\,kJ\,mol^{-1}$, $\Delta S_1^\ddagger = +29 \pm 69\,J\,K^{-1}\,mol^{-1}$, $\Delta V_1^\ddagger = -4.2 \pm 0.6\,cm^3\,mol^{-1}$, $k_{OH} = 4.2 \times 10^{-5}\,s^{-1}$, $\Delta H_{OH}^\ddagger = 103\,kJ\,mol^{-1}$ and $\Delta V_{OH}^\ddagger = +1.5\,cm^3\,mol^{-1}$, the k_{OH}/k_1 ratio of 19 100 illustrating the dominant participation from $[Rh(OH_2)_5OH]^{2+}$ which had been underestimated in the earlier studies. Kinetic data for substitution reactions on $[Rh(OH_2)_6]^{3+}$ are presented in Table 9.7.

The extremely high reactivity of $[Rh(OH_2)_5OH]^{2+}$ is believed to be a result of particularly strong σ- and π-donation from the OH^- ligand labilising the remaining water ligands via stabilisation of the trigonal bipyramidal transition state. The k_{OH}/k_1 value for the Rh species (19 100) compares with the next highest value (750) for the corresponding $FeOH^{2+}(aq)/Fe^{3+}(aq)$ ratio (see the previous chapter). A plausible explanation for the very high k_{OH}/k_1 ratio for Rh could be the existence of the D_{CB} (conjugate base) mechanism occurring on $[Rh(OH_2)_5OH]^{2+}$ despite the value of $\Delta V_{OH}^\ddagger$ close to zero ($+1.5\,cm^3\,mol^{-1}$). The limitations of using $\Delta V^\ddagger$ values as a guide to mechanism were eluded to in Section 1.4. Even a small $\Delta V^\ddagger$ of 1.5 could reflect extreme dissociative character if offsetting negative volume contributions are apparent, such as from a significant Rh—O bond contraction in forming the trigonal transition state. Such a situation may also be true of the negative $\Delta V^\ddagger$ value (-4.2) for $[Rh(OH_2)_6]^{3+}$, given the lack of evidence for bipyramidal incoming ligand involvement. Indeed, the k_{obs} value for anation of $Rh^{3+}(aq)$ by NCS^- in the pH range 1.0–3.0 ($[NCS^-]$ in fourfold excess over $[Rh^{3+}]$) is found to be independent of the $[NCS^-]$ concentration [93]. This is found again to be the case for complexation by thiourea in the concentration range 0.005–0.015 M ($Rh = 0.0005$ M) at $I = 0.1$ M ($NaClO_4$) [92]. Here it appears that saturation kinetics, as a result of ion pairing, cannot be responsible for the [ligand] independence so a D mechanism has been proposed:

$$[Rh(OH_2)_6]^{3+} \underset{k_{-1}}{\overset{k_1}{\rightleftharpoons}} [Rh(OH_2)_5]^{3+} + H_2O \tag{9.12}$$

$$[Rh(OH_2)_5]^{3+} + L \xrightarrow{k_2} [Rh(OH_2)_5L]^{3+} \tag{9.13}$$

$$k_{obs} = \frac{k_1 k_2 [L]}{k_{-1} + k_2 [L]} \tag{9.14}$$

when $\qquad k_2[L] \gg k_{-1}$, k_{obs} becomes independent of [L]

which is in line with the original conclusions of Harris and coworkers [74, 81] involving rate-determining formation of five-coordinated $[Rh(OH_2)_5]^{3+}$. Such a process, however, would be expected to show a large positive $\Delta V_{ex}^{\ddagger}$ instead of the negative value ($-4.2 \, cm^3 \, mol^{-1}$) observed. The conclusion is that either the D mechanism is particularly facilitated within 'reactant-pair' complexes $\{[Rh(OH_2)_6,L]\}$ or there is a significant contraction to the Rh coordination sphere on forming the five-coordinate Rh^{III} transition state, this being sufficient to overcome the increase in partial molar volume due to the excluded electrostricted water molecule giving a much lower $\Delta V_{obs}^{\ddagger}$. Similar arguments were used in Chapter 8 to explain the near-zero $\Delta V_{ex}^{\ddagger}$ value for $[Ru(OH_2)_6]^{2+}$ amidst considerable evidence for an I_d mechanism. In the case of Rh^{3+} the higher positive charge could promote further volume contraction in forming the pentaaqua ion transition state sufficient in this case to change the sign of $\Delta V^{\ddagger}$ to a negative value. As noted in Chapter 1, Rh^{3+} may be considered a fairly 'soft' metal centre despite the higher charge [94].

Studies carried out on various $[RhCl_n(OH_2)_{6-n}]^{(3-n)}$ complexes give further weight to the arguments in favour of the D mechanism. Table 9.8 summarises the findings. Large positive $\Delta V^{\ddagger}$ values characterise aquation and anation processes on the chloroaqua ions [95]. In each case the observed $\Delta V^{\ddagger}$ is close to V° for the

Table 9.8. Activation volumes for aquation and anation reactions on chloroaqua Rh^{III} complexes [95]

Complex	Incoming ligand	Outgoing ligand	$\Delta V^{\ddagger}$
$[RhCl_4(OH_2)_2]^-$	Cl^-	H_2O	$+14.7 \pm 1.6$
$[RhCl_5(OH_2)]^{2-}$	Cl^-	H_2O	$+15.7 \pm 6.5$
$[RhCl_5(OH_2)]^{2-}$	H_2O	Cl^-	$+14.4 \pm 0.5$
$[RhCl_6]^{3-}$	H_2O	Cl^-	$+21.5 \pm 0.6$

electrostricted leaving ligand [96]. Alternatively, it can be said that the leaving groups are fully solvated. The larger size of Cl (vs H_2O) and its stronger *trans* effect are factors promoting the D process. Furthermore, the lability of the remaining water ligands increases with the number of halide ligands attached. The corresponding decrease in [103]Rh NMR chemical shift shows that this stems from increased σ donation into the Rh e_g orbitals. Since halides are better σ donors than the O atoms of H_2O this facilitates dissociative loss preferentially of a water ligand.

Finally, particular interest stems from the kinetic data obtained for complexation by the O-donor chelates $H_2C_2O_4$ [90] and Hacac [91] on $[Rh(OH_2)_6]^{3+}$ wherein the rate constants and activation parameters obtained are not in keeping with the conclusions reached above for monodentate ligands. Rate constants for $H_2C_2O_4$ and Hacac complexation are respectively four and six orders of magnitude faster than the water-exchange process. It is clear that these rate constants relate to a different mechanism in operation for these O-donor ligands. Moreover, the corresponding anation reaction of $H_2C_2O_4$ on $[Ir(OH_2)_6]^{3+}$ gives rise to virtually identical rate and activation parameters to those for $[Rh(OH_2)_6]^{3+}$ [97]. It will become apparent that substitution rates involving dissociatively activated ligand substitution on Ir[III] complexes are normally between 10 and 100 times slower than corresponding reactions on Rh[III]. The conclusion is that the rate-determining step for complexation by these C—O-donor ligands involves C—O and not M—O(OH_2) bond breakage (Fig. 9.13), followed by rapid chelation. Ghosh and De have studied complexation on $[Rh(OH_2)_6]^{3+}$ by a range of β-diketones [91] and found that k_{obs} decreases as the methyl groups of Hacac are successively replaced by phenyl, correlating with an increase in $\Delta H^{\ddagger}$ and a positive shift in $\Delta S^{\ddagger}$. Surprisingly these workers did not recognise the possibility of rate determining C—O bond breakage in explaining these observations and therefore did not try to correlate the C—O bond strength with the rate constant obtained for each ligand. However, the concerted (associative) nature of these reactions was indicated by subsequent studies conducted in media of varying dielectric constant (water/dioxane mixtures at a 'pH' of 2) wherein a direct correlation of increasing rate constant with decreasing dielectric constant was observed in line with the Laidler–Eyring equation [98]. Rate constants, activation parameters ($\Delta H^{\ddagger} = 54.4 \, kJ \, mol^{-1}$, $\Delta S^{\ddagger} = -77 \, J \, K^{-1} \, mol^{-1}$) and solvent dielectric effects for the bimolecular reaction of $[Rh(OH_2)_6]^{3+}$ with salicylaldoxime ($salH_2$), giving

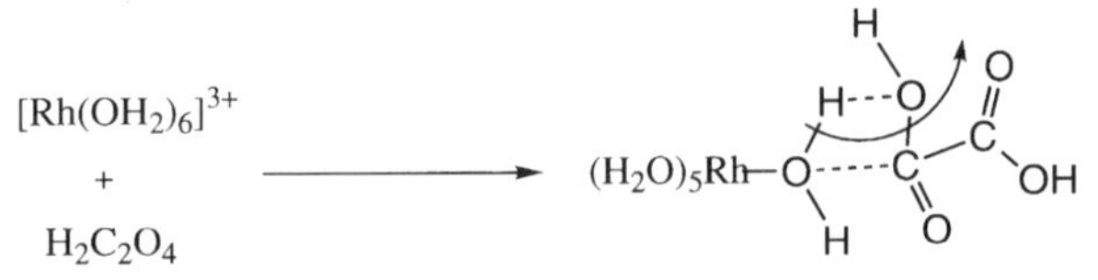

$[Rh(OH_2)_6]^{3+}$
+
$H_2C_2O_4$

$\longrightarrow$

$(H_2O)_5Rh$...

Figure 9.13. Concerted mechanism for complexation (or addition) of substrate (in this case $H_2C_2O_4$) to $[Rh(OH_2)_6]^{3+}$

$[Rh(OH_2)_4(salH)]^{2+}$, suggest a similar operation of a concerted C—O bond-breaking mechanism [99].

With regard to the mechanism of ligand substitution on $[Rh(OH_2)_6]^{3+}$ it is pertinent at this stage to compare corresponding studies conducted on $[Rh(NH_3)_5(OH_2)]^{3+}$. Comparisons between the behaviour of corresponding hexaaqua and aquapentammine complexes are often useful in providing mechanistic insights. $\Delta V^{\ddagger}$ values for water exchange on $[Rh(NH_3)_5(OH_2)]^{3+}$ (-4.1) and $[Rh(OH_2)_6]^{3+}$ (-4.2) are identical. However, the presence of five NH_3 ligands labilises the remaining water ligand site by a factor of ~ 4000 vs having five H_2O ligands present ($k_{ex}(25°C) = 8.4 \times 10^{-6}\,s^{-1}$ vs $2.2 \times 10^{-9}\,s^{-1}$) [100]. Rate constants and activation parameters for Cl^-, Br^-, I^- and NCS^- anation on $[Rh(NH_3)_5(OH_2)]^{3+}$ are essentially constant and identical to those for water exchange (Table 9.9).

A dissociative mechanism is indicated despite the negative value for $\Delta V^{\ddagger}_{H_2O}$ for the same reasons as argued above for the $[Rh(OH_2)_6]^{3+}$. However, reaction of $[Rh(NH_3)_5(OH_2)]^{3+}$ with CO_2 and SO_2 must proceed, in view of the much higher rate constants, via the simple 'concerted' addition reaction (Fig. 9.13), without Rh—O bond cleavage. A similar deduction for the reaction of $[Rh(NH_3)_5(OH_2)]^{3+}$ with each of $H_2C_2O_4$, $HC_2O_4^-$ and $C_2O_4^{2-}$ [102] is more difficult since the rate constants obtained are close to the value for water exchange itself at this temperature.

In conclusion, the nature of substitution reactions on Rh^{III} poses a dilemma regarding the interpretation of the substitution mechanism on the basis of the

Table 9.9. Kinetic data for substitution reactions on $[Rh(NH_3)_5(OH_2)]^{3+}$

Incoming ligand	$T(°C)$	k $(M^{-1}s^{-1})$	$\Delta H^{\ddagger}$ $(kJ\,mol^{-1})$	$\Delta S^{\ddagger}$ $(JK^{-1}mol^{-1})$	$\Delta V^{\ddagger}$ $(cm^3\,mol^{-1})$	Ref.
H_2O	25	8.41×10^{-6a}	102.9 ± 1.2	$+3.4 \pm 4.6$	-4.1 ± 0.4	[100]
	50	2.22×10^{-4a}				
Cl^-	50	2.01×10^{-4c}	107 ± 2.1	$+13.0 \pm 4.0$	$+3.0 \pm 0.7^b$	[101]
Br^-	50	1.44×10^{-4c}	106 ± 1.3	$+8.4 \pm 4.2$		[101]
I^-	50	1.15×10^{-4c}	102 ± 0.8	-4.2		[101]
NCS^-	50	1.19×10^{-4c}	99.6	-13.0 ± 4.2		[101]
$H_2C_2O_4$	60	1.5×10^{-4d}	35.0 ± 2.1	-213.0 ± 4.2		[102]
$HC_2O_4^-$	60	1.4×10^{-4d}	56.1 ± 2.9	-142.0 ± 8.4		[102]
$C_2O_4^{2-}$	60	1.2×10^{-4d}	$54.4 = 2.1$	-138.0 ± 4.2		[102]
CO_2	25	4.9×10^2	71.0 ± 4.2	$+50.0 \pm 12.6$	-4.7 ± 0.8^e	[103]
SO_2	25	1.8×10^8				[104]

^{a}Units of s^{-1}.
$^b I = 4.0\,M\,(NaClO_4)$.
$^c I = 0.2\,M\,(NaClO_4)$.
$^d I = 1.0\,M\,(NaClO_4)$.
$^e 7 < pH < 9.9$.

sign of $\Delta V^{\ddagger}$. Despite negative $\Delta V^{\ddagger}_{H_2O}$ values ($\sim -4\,\mathrm{cm^3\,mol^{-1}}$) for both $[\mathrm{Rh(OH_2)_6}]^{3+}$ and $[\mathrm{Rh(NH_3)_5(OH_2)}]^{3+}$ rate constants for substitution reactions are found to be largely insensitive to the nature of the incoming ligand in support of a $\mathrm{Rh-OH_2}$ bond-breaking activation process. Thus a clear interpretation of the mechanism from $\Delta V^{\ddagger}$ measurements may not always prove possible in the cases where relatively small values of $\Delta V^{\ddagger}$ ($\pm 4\,\mathrm{cm^3\,mol^{-1}}$) are relevant.

9.2.3 'Oxidants' Containing Rhodium

This title has been given to a series of papers by Ellison and Gillard [105–108] concerning investigations into alleged 'high oxidation state' complexes of Rh that can be generated from oxidation of certain aqueous solutions of $\mathrm{Rh^{III}}$. Claus in 1860 first reported highly coloured solutions resulting from treatment of alkaline $\mathrm{Rh^{III}}$ solutions with oxidants such as $\mathrm{Cl_2}$, $\mathrm{Br_2}$ or via hypochlorite or hypobromite [109]. Blue or violet solutions are also said to result from electrochemical oxidation [110] or chemical oxidation (e.g. with $\mathrm{BiO_3}^-$ [111] or ozone [112]) of acidic or basic solutions of $\mathrm{Rh^{III}}$. These blue or violet products were generally formulated as containing $[\mathrm{RhO_4}]^{n-}$ anions ($n = 2, 3,$ or 4). The original Claus' blue solid compound was made via precipitation with $\mathrm{Ba^{2+}}$, this being viewed as evidence for oxo anion formation. It has now been shown that these coloured

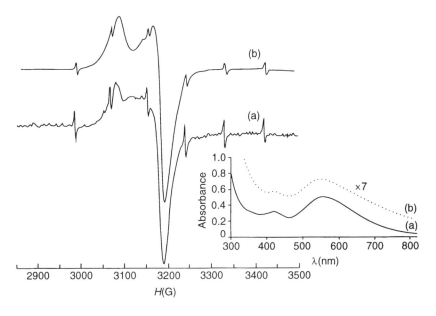

Figure 9.14. EPR spectrum from solutions of (a) Claus' blue and (b) the μ-superoxo aqua dimer of $\mathrm{Rh^{III}}$ [105]. (Inset: electronic spectra obtained from the same solutions.) (Reproduced by permission of The Royal Society of Chemistry)

solutions contain superoxo-bridged dimeric complexes of Rh^{III}. Indeed, EPR and electronic spectra (Fig. 9.14) taken from a sample of Claus' blue eluted from a Dowex 50W X2 column and that of an authentic sample of the product of O_2 oxidation of Rh_2^{4+}(aq), formulated by Moszner and Ziolkowski as '$[Rh_2(\mu\text{-}O_2^{-})$ $(OH_2)_8(OH)_2]^{3+}$' [70], are essentially identical [105]. It has been further established that the blue superoxo-bridged anions generated in base give insoluble Ba^{II} products. Like the behaviour of known $Rh^{III}(\mu\text{-}O_2^{-})$(aq) species, Claus' blue is stable in acid (violet colour) but decomposes slowly in base (blue colour). Varying analyses of solid compounds obtained via the addition of Ba^{2+}(aq) to these purple or violet solutions suggest a range of formulations of the type $Ba_x[Rh_y(\mu\text{-}O_2^{-})_a(\mu\text{-}O_2^{2-})_b(\mu\text{-}O)_c(\mu\text{-}OH)_d(OH_2)_e]\cdot f\,H_2O$. Claus' blue itself is probably a mixture of closely related aqua and chloro μ-superoxo Rh_2^{III} complexes. Analysis conducted on a number of μ-superoxo Rh^{III} carboxylates, however, suggest the presence of tetranuclear structures [113]. Table 9.10 shows selected data for a range of μ-superoxo dimeric Rh^{III} compounds.

The intriguing aspects of the Rh^{III} 'oxidations' is the origin of the $\mu\text{-}O_2^{-}$ bridge since the reactions are carried out in the absence of O_2. The likely mechanism (Fig. 9.15) therefore probably involves generation of Rh^{III}—O$^{\bullet}$ radicals, presumably via H atom abstraction from Rh^{III}—OH, which ultimately dimerise to form a μ-peroxo dimer which becomes further oxidised. Thus we have the unusual case of apparent oxidation of OH^{-} to O_2^{-} mediated by Rh^{III}. This transformation is

Table 9.10. Properties of μ-superoxo dirhodium(III) compounds

Complex	EPR parameters ($T = 77$ K)				$\lambda_{max}(\varepsilon)$	Ref.
	g_1	g_2	g_3	g_{av}		
$[Rh_2(\mu\text{-}O_2^{-})(\mu\text{-}O)(OH_2)_8]^{3+}$	2.082	2.020	2.040	2.041	560 (217)[a]	[70]
					420 (140)	
					235 (3800)	
Claus' blue			(blue)	2.044	628 (1122)[b]	[105]
			(violet)	2.044	526 (525)[c]	
Ba_n[Claus' blue] (green)	2.030	2.012	1.999			
$[Rh_2(\mu\text{-}O_2^{-})(py)_8Cl_2]^{3+}$				2.019	602 (2820)	[114]
cis-$[Rh_2(\mu\text{-}O_2^{-})(en)_4Cl_2]^{3+}$	2.086	2.022	1.992	2.034	493 (8710)	
$trans$-$[Rh_2(\mu\text{-}O_2^{-})(en)_4Cl_2]^{3+}$	2.076	2.022	1.999	2.033	n/a	[115]
$[Rh_2(\mu\text{-}O_2^{-})(en)_4(NO_2)_2]^{3+}$	2.097	2.030	1.989	2.039	540 (9120)	[114]
$[Rh_2(\mu\text{-}O_2^{-})(py)_8(OH_2)_2]^{5+}$	2.093	2.029	1.998			[116]
$[Rh_2(\mu\text{-}O_2^{-})(bipy)_4Cl_2]^{3+}$	2.088	2.020	2.004	2.038		[117]
$[Rh_2(\mu\text{-}O_2^{-})(bipy)_8(OH_2)_2]^{5+}$	2.089	2.021	1.992	2.034		[117]
'Rhodium(IV) sulfate' (violet)				2.036	n/a	[112]
'Rhodium(IV) sulfate' (blue)				2.040	570 (n/a)	[112]
					410–420 (n/a)	

[a] In 3.0 M $HClO_4$.
[b] Acid solution.
[c] Basic solution.
n/a = not applicable.

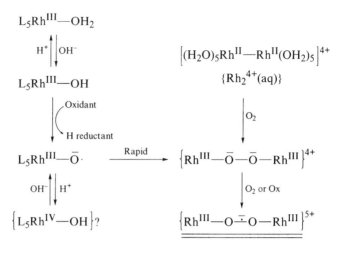

Figure 9.15. Mechanistic routes to μ-superoxo dirhodium(III) compounds

normally very difficult to achieve selectively. The presumed generation of Rh^{III}—O, rather than a high oxidation state Rh compound, of which it may be in equilibrium, is a reflection of the high redox stability of Rh^{III} and the strength of the Rh—O bond which remains intact throughout the transformation. This behaviour will be shown to be shown to be in contrast to the behaviour of Ir^{III} (aq) wherein authentic high oxidation state Ir^{n+} (aq) species have been identified (see Section 9.3.1.3).

The progress of the reaction of Rh_2^{4+} (aq) with O_2 is found to depend on the flux of O_2 [70]. If air is used as the source of O_2, yellow solutions containing Rh^{III} (aq) compounds are produced of which $Rh_2(\mu\text{-}OH)_2^{4+}$ (aq) (peaks at 400 and 228 nm) is a major component. These solutions can be reduced by Cr^{2+} (aq) to regenerate Rh_2^{4+} (aq). However, if the green Rh_2^{4+} solutions are stirred rapidly in a stream of pure O_2 a violet colour is produced characteristic of μ-superoxo Rh_2^{III} species in acid solution. The $[Rh_2(\mu\text{-}O_2^-)(OH)_2(OH_2)_8]^{3+}$ structure was concluded on the basis of cation-exchange behaviour typical of a $3+$ ion and experimental measurement of the charge per Rh as $1.5+$ using the Cady and Connick method [118]. In the absence thus far of an X-ray structure, however, the presence of the two terminal OH ligands is questioned. The alternative formulation containing a μ-oxo bridge, $[(H_2O)_4Rh(\mu\text{-}O_2^-)(\mu\text{-}O)Rh(OH_2)_4]^{3+}$ (Fig. 8.16) might be preferred. The structure of the analogous octaammine complex, $[Rh_2(\mu\text{-}O_2^-)(OH)(NH_3)_8]^{4+}$ [119], indeed shows a dibridged Rh_2^{III} unit but here with a μ-OH bridge identical to that present in the extensive range of corresponding Co^{III} complexes. The redox potential for the couple $[Rh_2(\mu\text{-}$

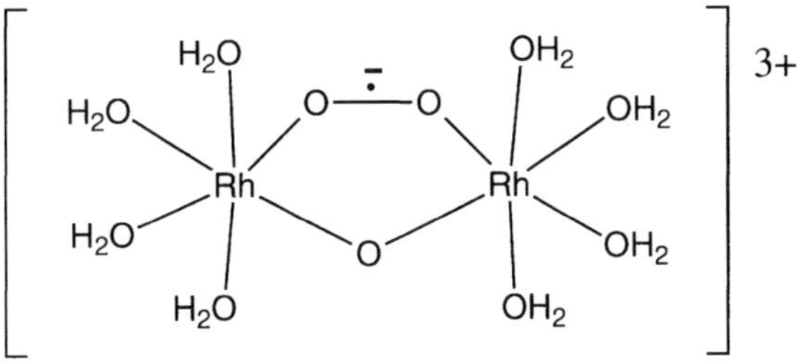

Figure 9.16. Proposed structure for the μ-superoxo aqua dimer of Rh^{III}

$O_2^-)(\mu\text{-}O)(OH_2)_8]^{3+}/[Rh_2(\mu\text{-}O_2^{2-})(\mu\text{-}O)(OH_2)_8]^{2+}$ has been measured as $+1.032\,V$ vs NHE in $3.0\,M$ $HClO_4$ and is fully reversible from polarographic studies [120]. The potential for the corresponding μ-superoxo octaammine couple is $+0.732\,V$, which is some $300\,mV$ more negative than the potential for the corresponding complexes of Co^{III} [121].

9.2.4 Existence of $Rh^{IV}(aq)$?

Authentic high oxidation state Rh compounds are rare. The Rh^{IV} anions $[RhF_6]^{2-}$ and $[RhCl_6]^{2-}$ exist in solid compounds. A yellow intermediate species, observed during the anodic electrolysis of alkaline Rh^{III} solutions, has been attributed to the formation of $[Rh^{IV}(OH)_6]^{2-}$ [122]. However, with time these solutions turn green and the resulting final products are μ-superoxo Rh^{III} dimers. Conceivably such 'Rh^{IV}—OH' species might actually be the protonated form of Rh^{III}—O^-, the peroxo dimer precursor compound (Fig. 9.15).

9.3 IRIDIUM

There are no aqua ions known for iridium below the $+3$ state. Much of the aqueous chemistry is restricted to the $+3$ state. However, unlike rhodium, and in keeping with the general increase in accessibility of higher oxidation states down a group, there is an authentic solution chemistry of oligomeric Ir^{IV} and even Ir^{V} aqua species.

9.3.1 Iridium(III) (d^6): Chemistry of the $[Ir(OH_2)_6]^{3+}$ Ion

9.3.1.1 PREPARATION AND CHARACTERISATION

Gamsjager and Beutler first characterised the $[Ir(OH_2)_6]^{3+}$ ion in 1976 [123] and reported on its hydrolysis behaviour in 1979 [124]. While it will become apparent that much of Ir^{III} chemistry is reminiscent of Rh^{III} there are significant differences, not least the much greater inertness and ready oxidation of Ir^{III} species. Thus while $[Rh(OH_2)_6]^{3+}$ can readily be prepared from $[RhCl_6]^{3-}$ by boiling in concentrated $HClO_4$ solution initial attempts to perform the same process on $[IrCl_6]^{3-}$ resulted in failure. This is now recognised as due to the ready oxidation of Ir^{III} to Ir^{IV} under such conditions, resulting in a mixture of monomeric and oligomeric aqua chloro complexes of Ir^{IV} and perhaps even mixed-valence Ir^{III}–Ir^{IV} species. Gamsjager and Beutler's success lay in their recognition of the fact that the Ir—Cl bonds of $[IrCl_6]^{3-}$ are labilised in alkaline solution. In fact $[IrCl_6]^{2-}$ salts can also be used as the lead-in (often available at higher purity) since rapid reduction to $[IrCl_6]^{2-}$ occurs in the presence of OH^- ions:

$$2[IrCl_6]^{2-} \text{ (red-brown)} + 2OH^- = 2[IrCl_6]^{3-} \text{ (green)} + 0.5O_2 + H_2O \quad (9.15)$$

This occurs within the first 5–10 minutes. In a typical preparation 1.0 g of $Na_2[(IrCl_6]$ is treated with $600\,cm^3$ of 0.1 M NaOH. The solution is kept between 35 and 40 °C for 3–4 hours wherein the gradual replacement of Cl^- by OH^- is indicated by a gradual colour change from green ($[IrCl_6]^{3-}$) to light yellow ($[Ir(OH)_6]^{3-}$). At the first hint of a blue colour (Ir^{IV}), usually a good indication that the hydrolysis is complete, an excess of ascorbic acid ($\sim 200\,mg$) is added to protect the $[Ir(OH)_6]^{3-}$ solution from air oxidation. The solution is cooled to 5 °C, maintained for 30 minutes and then the pH is lowered to ~ 8 with 0.1 M $HClO_4$ solution wherein a light yellow precipitate of $[Ir(OH)_3(OH_2)_3]$ is formed. After filtration and washing thoroughly with distilled water the hydroxide is dissolved in $40\,cm^3$ of 0.1 M $HClO_4$, producing a yellow solution. This solution is then diluted further with $50\,cm^3$ of water and chromatographed on a column of Dowex 50W X8 (50/100 mesh) in the H^+ form. Chloroaqua iridium complexes can be removed by discarding the first 5–10 cm^3 fractions upon elution with 2.0 M $HClO_4$. A number of polynuclear Ir^{III}(aq) species are also formed in this procedure but remain strongly held to the resin, allowing pure solutions of $[Ir(OH_2)_6]^{3+}$ to be eluted. Use of a second column of Dowex 50W X2 resin can allow a change of acid to, for example, CF_3SO_3H, and give more concentrated solutions as high as 0.37 M in Ir^{3+} as well as allowing elution of both $[Ir(OH_2)_6]^{3+}$ and oligomeric species. The electronic spectrum of 0.37 M $[Ir(OH_2)_6]^{3+}$ in 3.89 M CF_3SO_3H is shown in Fig. 9.17 [125]. The two bands in the u.v. region are due to the $^1A_{1g}$–$^1T_{2g}$ (313 nm, $\varepsilon = 32.5\,M^{-1}\,cm^{-1}$) and $^1A_{1g}$–$^1T_{2g}$ (267 nm, 36.4 $M^{-1}\,cm^{-1}$) transition respectively of the low-spin t_{2g}^6 configuration. The shoulder at $\sim 400\,nm$ is due to a spin-forbidden transition. Iridium can be conveniently determined by the method of Zinser and Page [126] involving the fuming of samples in concentrated $HClO_4$ resulting in oxidation to purple Ir^{IV}. The Ir^{IV} is then conveniently standardised by reductive titration back to Ir^{III} with either standard Fe^{II}(aq) (chemical) or by use of a Pt metal cathode (electrochemical).

The increase in d–d transition energy down the group, reflecting the increase in the strength of the ligand field 3d–4d–5d, is easily seen in this the *only* transition

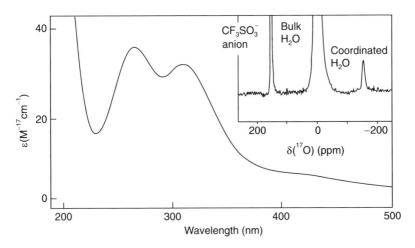

Figure 9.17. Electronic spectrum of 0.37 M $[Ir(OH_2)_6]^{3+}$ in 3.89 M CF_3SO_3H. (Inset: 54.21 MHz oxygen-17 NMR spectrum of the same solution at natural abundance)

element group allowing a comparison of hexaaqua ions for all three group members. The natural abundance ^{17}O NMR spectrum of $[Ir(OH_2)_6]^{3+}$ is also shown in Fig. 9.17. The water ligands are found upfield of free water at -154 ppm. $[Ir(OH_2)_6]^{3+}$ is also characterised in the caesium alum [127, 128]. The preparation involves dissolution of 'Ir(OH)$_3$' above in 2.0 M H_2SO_4 followed by addition of Cs_2SO_4. Upon standing initially at room temperature, and then at 5 °C, pale yellow crystals of the alum are obtained which can be filtered off, washed with ice-cold water and air dried, with a typical yield ~ 59–60%. The absorption spectrum of $Cs[Ir(OH_2)_6](SO_4)_2 \cdot 6H_2O$ in 1.0 M CF_3SO_3H has peak maxima at 271 nm ($\varepsilon = 44 \, M^{-1} \, cm^{-1}$) and 314 nm (40 $M^{-1} \, cm^{-1}$). The ε values, here determined using the alum stoichiometry, may be compared with those obtained above [125]. The ratio of the absorbance at the 271 nm maximum to the 233 nm minimum (3.14) obtained from the alum [127] is probably indicative of a purer product free of hydrolytic polymeric species.

Gamsjager and Beutler have measured the hydrolysis constant of $[Ir(OH_2)_6]^{3+}$ at various temperatures at a constant ionic strength of 1.05 mol kg^{-1} (NaClO$_4$) [124]. Values determined at 25 °C (K values in molality units) are $pK_{11} = 4.37 \pm 0.02$, $pK_{12} = 9.57 \pm 0.05$ and $pK_{13} = 10.22 \pm 0.05$. Thus Ir^{3+}(aq) is somewhat less acidic than Rh^{3+}(aq). This seems to correlate with slightly longer Ir—OH$_2$ bonds vs Rh—OH$_2$ in their respective caesium alums, 204.1(3) pm (Ir) vs 201.6(3) pm (Rh) [128]. The Ir—OH$_2$ distances in the alum are typical; the average Ir—OH$_2$ distance in the X-ray structure of the cation *trans*-$[Ir(OH_2)_4Cl_2]^+$, occurring in the double salt $(H_3tacn) [Ir(OH_2)_4Cl_2](SO_4)_2$, being 204.4(13) pm [129]. The symmetric $\nu_1(MO_6)$ stretching mode for $[Ir(OH_2)_6]^{3+}$ has been located at $536\,cm^{-1}$ from low-temperature polarised single-crystal Raman spectra (PSCR) of the alum [130]. As mentioned in Chapter 1, the combination of single-crystal XRD and PSCR measurements have demonstrated, where comparisons have been possible, a different relationship of $\nu_1(MO_6)$ with $M\!-\!OH_2$ distance occurring along the first row transition metal series than occurs down a typical transition metal group. This behaviour has been argued as evidence of a higher degree of covalency in the $M\!-\!OH_2$ bonds for the heavier transition metals. All of the alums of Group 9 belong to the α alum type.

9.3.1.2 WATER EXCHANGE AND LIGAND SUBSTITUTION ON $[Ir(OH_2)_6]^{3+}$

Indications are that $[Ir(OH_2)_6]^{3+}$ is the most inert hexaaqua metal ion known [131]. An early attempt at measuring the water-exchange process on solutions of $[Ir(OH_2)_6]^{3+}$ in $ClO_4{}^-$ media by ^{18}O enrichment at $120\,^\circ C$ in sealed tubes was compounded by oxidation to purple $Ir^{IV}(aq)$ (see below); thus the use of non-oxidising strong acids is essential. A study of the water-exchange rate constant under conditions of high temperature $(80\text{--}130\,^\circ C)$ and pressure $(0.1\text{--}210\,MPa)$ in non-oxidising $CF_3SO_3{}^-$ media has recently been completed [132]. This has been achieved using the same isotopic dilution method as performed on $[Rh(OH_2)_6]^{3+}$. The preparation of oxygen-17 enriched samples of $[Ir(OH_2)_6]^{3+}$ has made use of the greater lability in OH^- media. In a typical procedure active 'Ir(OH)$_3$' is dissolved in a 10 M NaOH solution made up with 5 atom % $H_2{}^{17}O$ sufficient to ensure a 40-fold excess of OH^- over Ir^{III}. Ascorbic acid is then added and the resulting $[Ir(OH)_6]^{3-}$ solution kept at $0\,^\circ C$ for 16 hours to ensure complete equilibration. The mixture is then neutralised carefully with ice-cold pure CF_3SO_3H to reprecipitate the active 'Ir(^{17}OH)$_3$' which is then rapidly dissolved in $50\,cm^3$ of 1.0 M CF_3SO_3H. Following dilution and loading on to a 10 cm column of Dowex 50W X2 (200–400 mesh) cation-exchange resin, the $[Ir(^{17}OH_2)_6]^{3+}$ is eluted with 4.0 M CF_3SO_3H to give stock solutions suitable for the isotope dilution studies. During the column procedure, performed as quick as possible, negligible loss of enrichment can be assumed. Samples were then adjusted to the required acidity and ionic strength with $NaCF_3SO_3$. Water exchange was monitored by observation of the collapse in the intensity of the enriched bound water signal at $-154\,ppm$ as it equilibrates with bulk water (0 ppm) at natural abundance. As a check the rise in the intensity of the free water signal can also be monitored. High temperatures ($\geqslant 100\,^\circ C$) are required to observe the exchange on a monitorable timescale and at temperatures of $\sim 140\,^\circ C$ can allow re-enrichment of samples of the aqua ion. The need to employ high temperature for the measurements is exemplified in the half-life of 7 days estimated from a run conducted at $80\,^\circ C$. The observed rate law carries an $[H^+]$

dependence of the following type:

$$\text{Rate}/[\text{Ir}^{3+}(\text{aq})] = k_{\text{obs}} = \underset{\text{aqua ion}}{k_1} + \underset{\text{monohydroxy ion}}{k_{\text{OH}} K_{11} [\text{H}^+]^{-1}} \qquad (9.16)$$

implying an involvement from the monohydroxy ion, $[\text{Ir}(\text{OH}_2)_5\text{OH}]^{2+}$. Similar rate expressions characterise water exchange on the trivalent hexaaqua ions of Fe^{3+}, Cr^{3+}, Ru^{3+}, Rh^{3+} and Ga^{3+}. Table 9.11 shows the resulting kinetic parameters for the process (conducted at $I = 5.1\,\text{m}$, $\text{NaCF}_3\text{SO}_3^-$, $[\text{H}^+] = 0.5\text{–}5.0\,\text{m}$) along with corresponding kinetic and thermodynamic data for other trivalent transition metal hexaaqua ions. The $\text{p}K_{11}$ value for Ir^{3+} determined by potentiometric titration (4.45) is in good agreement with the earlier value (4.37) obtained by Beutler and Gamsjager.

The extrapolated water-exchange rate constant at $25\,°\text{C}$ for $[\text{Ir}(\text{OH}_2)_6]^{3+}$, $(1.1 \pm 0.1) \times 10^{-10}\,\text{s}^{-1}$, is only 20 times smaller than that for $[\text{Rh}(\text{OH}_2)_6]^{3+}$ $(2.2 \times 10^{-9}\,\text{s}^{-1})$. The marked inertness of $\text{Ir}^{3+}(\text{aq})$, however, stems from a much smaller contribution from the labilising monohydroxy ion, $k_{\text{OH}} = 4.0 \times 10^{-7}\,\text{s}^{-1}$, compared with the value for $\text{Rh}^{3+}(\text{aq})$, $k_{\text{OH}} = 4.2 \times 10^{-5}\,\text{s}^{-1}$. Clearly the OH^- ligand is much less effective at labilising the remaining coordinated waters on $[\text{Ir}(\text{OH}_2)_5\text{OH}]^{2+}$ than in the case of $[\text{Rh}(\text{OH}_2)_5\text{OH}]^{2+}$. Since both have filled t_{2g}^6 configurations the OH^- ligand cannot be considered to labilise via effective π donation. The much higher electron density on the Ir^{3+} may then be a feature suppressing effective σ donation from OH^-, as might be the high energy of the σ-accepting e_g orbitals in view of larger ligand field splitting energy. Values of $\Delta V_{\text{ex}}^{\ddagger}$ on the hexaaqua ions are comparable for each, perhaps reflective, in the case of $[\text{Ir}(\text{OH}_2)_6]^{3+}$, of a slightly larger volume contraction to the transition state and stemming from its slightly larger effective radius, 68 pm (vs 66.5 pm for Rh^{3+}).

The extreme inertness of $\text{Ir}^{3+}(\text{aq})$ in acidic solution contrasts with the relative lability of $\text{Ir}^{3+}(\text{aq})$ in strongly alkaline solution as the $[\text{Ir}(\text{OH})_6]^{3-}$ ion ($t_{1/2}$ for exchange with OH^- is 125 min at $0°\text{C}$ in $0.5\,\text{M}$ $[\text{OH}^-]$ from an ^{18}O isotope dilution study [133]). The observed rate constant shows an unusual reciprocal dependence on $[\text{OH}^-]$:

$$\text{Rate}/[\text{Ir}(\text{OH})_6^{3-}] = k_{\text{obs}} = k_{\text{OH}} [\text{OH}^-]^{-1} \qquad (9.17)$$

suggesting that the exchange actually goes via the highly labilised aqua complex $[\text{Ir}(\text{OH})_5(\text{OH}_2)]^{2-}$. One might speculate on the possible facilitation here (via powerful electron donation from the five OH^- groups) of an extreme D process (stabilisation of the five-coordinate intermediate) similar to that proposed for both $[\text{RhCl}_5(\text{OH}_2)]^{2-}$ and $[\text{IrCl}_5(\text{OH}_2)]^{2-}$ (see below). The lability at $\text{Ir}^{3+}(\text{aq})$ clearly increases with the number of OH^- groups attached. Solutions of $[\text{Ir}(\text{OH})_6]^{3-}$ in alkaline solution are very oxygen sensitive, and catalysis from trace amounts of the readily formed, and likely more labile, Ir^{IV} complexes, $[\text{Ir}(\text{OH})_6]^{2-}$ or $[\text{Ir}(\text{OH})_5(\text{OH}_2)]^-$, is also a possibility.

Table 9.11. Rate constants and activation parameters for water exchange on hexaaqua and monohydroxypentaaqua trivalent metal ions $k_{obs} = k_1 + k_2[H^+]^{-1}$, $k_2 = k_{OH}K_{11}$

Species	Parameter	Fe^{III}	Cr^{III}	Ru^{III}	Rh^{III}	Ir^{III}	Ga^{III}
$[M(OH_2)_6]^{3+}$	k_1^{298} (s^{-1})	1.6×10^2	2.4×10^{-6}	3.5×10^{-6}	$(2.2 \pm 2.6) \times 10^{-9}$	$(1.1 \pm 0.1) \times 10^{-10}$	4.0×10^2
	$\Delta H_1^{\ddagger}$ (kJ mol^{-1})	64.6	108	89.8	131.2 ± 22.4	130.5 ± 0.6	67.1
	$\Delta S_1^{\ddagger}$ (J K mol^{-1})	+12.1	+11.6	-48.2	29.3 ± 65.8	$+2.1 \pm 0.5$	+30.1
	$\Delta V_1^{\ddagger}$ (cm^3 mol^{-1})	-5.4	-9.6	-8.3	-4.1 ± 0.5	-5.7 ± 0.5	+5.0
	$k_{1,O}^{358}$ (s^{-1})					$(8.5 \pm 0.2) \times 10^{-7}$	
$[M(OH_2)_5OH]^{2+}$	k_2^{298} (m s^{-1})			1.1×10^{-6}	$(1.5 \pm 0.6) \times 10^{-8}$	$(1.4 \pm 0.6) \times 10^{-11}$	
	$\Delta H_2^{\ddagger}$ (kJ mol^{-1})			136.9	135.8 ± 6.7	138.5 ± 4.5	
	$\Delta S_2^{\ddagger}$ (J K^{-1} mol^{-1})			+100.5	$+60.8 \pm 19.6$	$+11.5 \pm 11.6$	
	$\Delta V_2^{\ddagger}$ (cm^3 mol^{-1})			-2.1	$+1.2 \pm 0.3$	-0.2 ± 0.8	
	pK$_{11}^{298}$		4.1	2.7	3.5	4.45 ± 0.03	3.9
	k_{OH}^{298} (s^{-1})	1.2×10^5	1.8×10^{-4}	5.9×10^{-5}	4.2×10^{-5}	4.0×10^{-7}	
	k_{OH}/k_1	750	75	170	19100	3600	275
	ΔV_{11}° (cm^3 mol^{-1})	+1.1	-3.8	-3.0	-0.2	-1.5 ± 0.3	+1.5
	$\Delta V_{OH}^{\ddagger}$ (cm^3 mol^{-1})	+7.0	+2.7	+0.9	+1.5	+1.3	+6.2

The water-exchange rate constant for $[Ir(OH_2)_6]^{3+}$ appears to correlate well with estimated upper limits for the rate constant of Cl^- anation ($< 10^{-9}\,M^{-1}\,s^{-1}$) based upon there being little or no detectable spectral change occurring to $[Ir(OH_2)_6]^{3+}$ solutions at 40 °C in 3.7 M Cl^- at pH 2 over a period of 10–15 days [131]. This observation is not now surprising in view of the estimated half-life for exchange on $[Ir(OH_2)_6]^{3+}$ at 40 °C (15 years) and moreover the similarity in the spectrum of the final product (likely to be *trans*-$[Ir(OH_2)_4Cl_2]^+$ due to *trans* labilisation by the first Cl^- ligand) to that of $[Ir(OH_2)_6]^{3+}$ (see Fig. 9.18). Studies using NCS^- are compounded by background absorbance changes due to products formed from the reaction of NCS^- with $[H^+]$ over the long reaction times and high temperatures required.

As is the case on Rh^{III}, the rate of substitution (aquation) increases with the number of Cl^- ligands attached to Ir^{III} [134]. A kinetic study of Cl^- anation carried out on $[IrCl_5(OH_2)]^{2-}$ shows saturation behaviour in $[Cl^-]$ which, in view of the anion–anion interaction, cannot be due to ion pairing. Thus the same D mechanism has been proposed as in the case of corresponding reactions on Rh^{III} (9.12) to (9.14). In the case of $[IrCl_5(OH_2)]^{2-}$ the values of k_1 in (9.14)) for the process $[IrCl_5(OH_2)]^{2-} = [IrCl_5]^{2-} + H_2O$ and k_{-1}/k_2 (representing the competition between Cl^- and H_2O for the five-coordinate intermediate) are respectively $3.82 \times 10^{-6}\,s^{-1}$ and 1.85 M. The D mechanism also applies to aquation on $[IrCl_6]^{3-}$ and $[IrCl_5(OH_2)]^{2-}$ wherein k_{obs} (25 °C) respectively are $(3.40 \pm 0.05) \times 10^{-5}\,s^{-1}$ and $(1.19 \pm 0.03) \times 10^{-6}\,s^{-1}$. These rate constants are some 100 times smaller than for the corresponding reactions on Rh^{III} [95]. The

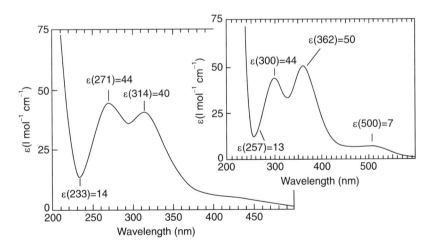

Figure 9.18. Electronic spectrum of $[Ir(OH_2)_6]^{3+}$ (from $Cs[Ir(OH_2)_6](SO_4)_2$) and *trans*-$[Ir(OH_2)_4Cl_2]^+$ in 1.0 M CF_3SO_3H

greater inertness of the Ir^{III} complexes within the D mechanism is believed to arise from stronger $Ir-OH_2$ bonds stemming from a further increase in the already large LFSE of the low-spin t_{2g}^6 configuration due to more extensive 5d orbital overlap.

Only oxalic acid, together with its anions, have lent themselves to the monitoring of complexation on $[Ir(OH_2)_6]^{3+}$ [97]. Here reaction rate constants are significantly several orders of magnitude in excess of that for the water-exchange process. This is a result of operation of the same mechanism as relevant for oxalate anation on $[Rh(OH_2)_6]^{3+}$ wherein the $M-OH_2$ bond is retained and a $C-O$ bond breakage process is relevant. The mechanism (Fig. 9.13) consists of the electrophilic $C=O$ carbon of the oxalic acid being effectively attacked by the bound H_2O ligand on the Ir^{III} acting as a nucleophile. Thus kinetic data (Table 9.12) for oxalate anation on both $[Ir(OH_2)_6]^{3+}$ and $[Rh(OH_2)_6]^{3+}$ are virtually identical in support of a metal-independent $C-O$ bond-breaking activation step. These may also be contrasted with the kinetic data characterising water exchange on the respective hexaaqua ions. This kind of mechanism is facilitated when (a) $M-OH_2$ bond breakage occurs with a much slower rate than $C-O$ bond breakage, as it does here, and (b) the H_2O ligand is an effective enough nucleophile. For Group 9 aqua complexes the filled t_{2g}^6 shell does not allow effective π-bonding from H_2O to the metal which would otherwise 'tie up' the available lone pair. Thus the necessitated sp^3 hybridisation at oxygen, coupled with the high intrinsic electron density on the metal, results in the bound H_2O ligand on Ir^{3+} being an effective nucleophile. The higher rate constants (factor of 10^3) for reactions of the monohydroxo forms reflects the much higher nucleophilicity of bound OH^- vs H_2O, the rate constant, for Rh^{III}, still some 10^3 times higher than that for water exchange on the monohydroxo form.

Table 9.12. Kinetic data for oxalic acid complexation on inert trivalent metal aqua ions

	$k^{298}(M^{-1}s^{-1})$	$\Delta H^{\ddagger}(kJ\,mol^{-1})$	$\Delta S^{\ddagger}(JK\,mol^{-1})$
$[Co(NH_3)_5(OH_2)]^{3+}$	2.0×10^{-4a}	57.3	-150.6
$[Co(en)_2(NH_3)(OH_2)]^{3+}$	5.4×10^{-5a}	95.8	-37.6
$[Rh(OH_2)_6]^{3+}$	1.2×10^{-5a}	76.7 ± 3.7	-81.7 ± 10.9
$[Ir(OH_2)_6]^{3+}$	1.3×10^{-5a}	73.1 ± 4.6	-94.1 ± 13.9
$[Rh(OH_2)_5OH]^{2+}$	1.7×10^{-2b}	57.9 ± 1.9	-85.7 ± 5.8
$[Ir(OH_2)_5OH]^{2+}$	1.4×10^{-2b}	n/a	n/a

^{a}Values at $40\,°C$.
^{b}Closer fit to the data results from an assumption of MOH^{2+} reacting with $H_2C_2O_4$ rather than M^{3+} reacting with $HC_2O_4^-$. The limitations of the 'proton ambiguity' problem are, however, recognised. Values of the hydrolysis constant (K_{11}) for $[Rh(OH_2)_6]^{3+}$ and $[Ir(OH_2)_6]^{3+}$ are taken to have values of $3.2 \times 10^{-4}\,M$ [82] and $4.2 \times 10^{-5}\,M$ [124] respectively.
n/a = not available.

Correspondingly, the higher rate constants for reaction with $[Co(NH_3)_5(OH_2)]^{3+}$ reflect a further build-up of electron density on Co^{III} as a result of effective σ-donation from the five NH_3 ligands [135].

9.3.1.3 OXIDATION OF $[Ir(OH_2)_6]^{3+}$: CHEMISTRY OF DIMERIC $Ir^{III}(AQ)$, $Ir^{IV}(AQ)$ AND $Ir^V(AQ)$ IONS

There are few examples of authentic compounds of Ir^{IV} or Ir^V. Purple solutions have been reported to result [126, 131] when Ir^{III} solutions are heated in $HClO_4$ solution. Purple solutions also result when $[IrCl_6]^{2-}$ solutions are similarly treated. Careful redox titrations have confirmed the presence of Ir^{IV}. Numerous papers, many in the Russian literature, have described the formation of blue or purple colours during various chemical oxidations of $Ir^{III}(aq)$ solutions, most of them describing attempts to develop colorimetric methods for the estimation of the metal [136]. $[Ir(OH_2)_6]^{3+}$ is reported to be mildly air sensitive. However, cyclic voltammetric studies on freshly columned pure solutions of $[Ir(OH_2)_6]^{3+}$ in 2.0 M $HClO_4$ show no evidence of an oxidation process up to $\sim +1.6$ V (vs NHE) at a platinum electrode. It is now clear that the apparent 'air sensitivity' results from the build-up of 'hydrolytic polymers' (oxidisable at more negative potentials) during 'ageing' of the $[Ir(OH_2)_6]^{3+}$ solutions.

Only powerful reagents, such as $Ce^{IV}(aq)$, O_3, etc., with redox potentials in excess of $+1.6$ V will oxidise freshly prepared solutions of $[Ir(OH_2)_6]^{3+}$ [125]. If $[Ir^{III}]$ is kept below 7×10^{-3} M in 2.0 M $HClO_4$ the product of the oxidation (normally carried out at $0\,^\circ$C) is a brownish-green solution containing Ir^V (from reduction back to Ir^{III} with either Fe^{2+}, I^- or Eu^{2+}). Electrolysis of an iridium wire anode in 2.0 M $HClO_4$ can also be used to generate the same brownish-green solutions. During the reduction process a deep purple coloration results (Ir^{IV}) prior to the generation of yellow solutions of Ir^{III}. The whole process can also be carried out in a two-compartment electrochemical cell of the kind illustrated in Fig. 9.19 [137]. A constant current of 1.0 A at $0\,^\circ$C is usually applied in each

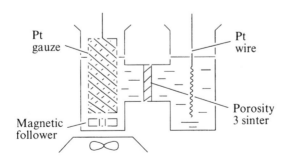

Figure 9.19. Electrochemical cell used to generate $Ir^V(aq)$ and its reduction products

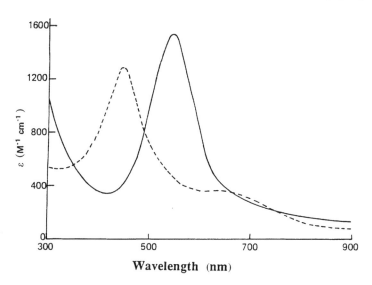

Wavelength (nm)

Figure 9.20. Ultraviolet–visible spectra (ε values per Ir) for solutions of $Ir^V(aq)$ (----) and $Ir^{IV}(aq)$ (—) generated via electrochemical oxidation of $[Ir(OH_2)_6]^{3+}$. (Ir products are a mixture of dimeric and polymeric forms)

direction for the oxidative/reductive processes. The spectra of the Ir^V and Ir^{IV} products are illustrated in Fig. 9.20. Redox titrations at the maximisation of the purple and yellow colours confirm the presence of Ir^{IV} and Ir^{III} in these solutions. However, subsequent analysis of the yellow Ir^{III} solutions revealed that the solutions did not contain $[Ir(OH_2)_6]^{3+}$. Cyclic voltammetric studies revealed evidence for a number of oxidation processes in the region $+0.4$ to $+1.5$ V with the yellow solution much more strongly held on a Dowex 50W X2 column indicating a charge $> 3+$. Two yellow species could be subsequently eluted with solutions of 2.0 M $HClO_4$. The spectra of these two Ir^{III} species, $Ir^{IIIA}(aq)$ (eluted first) and $Ir^{IIIB}(aq)$ (eluted second), are shown in Fig. 9.21 (Ir analyses carried out by the Zinser and Page method [126]). The elution behaviour suggests an overall charge of $4+$ for $Ir^{IIIA}(aq)$ and $5+$ for $Ir^{IIIB}(aq)$. A further yellow band (polymeric $Ir^{III}(aq)$) is strongly held and not elutable from the column, even with concentrated 10 M $HClO_4$.

Solution structures for $Ir^{IIIA}(aq)$ and $Ir^{IIIB}(aq)$ have been deduced from ^{17}O NMR studies. During the enrichment procedure for $[Ir(OH_2)_6]^{3+}$, via $[Ir(OH)_6]^{3-}$, and subsequent acidification, an additional yellow species of charge $> 3+$ was elutable from the column with 2.0 M $HClO_4$ after $[Ir(OH_2)_6]^{3+}$ with an identical ultraviolet–visible spectrum and electrochemistry to that of Ir^{IIIA}. The subsequent ^{17}O NMR solution of this species revealed two closely spaced peaks of equal height in the bound water region at -137 and -144 ppm and

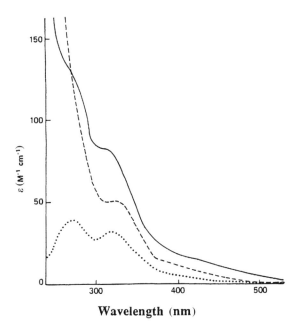

Wavelength (nm)

Figure 9.21. Ultraviolet–visible spectra (ε values per Ir) for Ir^{IIIA}(aq) (---), Ir^{IIIB}(aq) (—) and [Ir(OH$_2$)$_6$]$^{3+}$ (........)

a further broad feature around -220 ppm (Fig. 9.22). These features are very similar to those obtained for $[(H_2O)_4Rh(\mu\text{-OH})_2Rh(OH_2)_4]^{4+}$ [87] and thus Ir^{IIIA}, consistent with the $4+$ charge, is similarly assigned to $[(H_2O)_4Ir(\mu\text{-OH})_2Ir(OH_2)_4]^{4+}$. The features at -220 ppm is assigned to μ-OH. A ^{17}O-enriched sample of Ir^{IIIB}(aq) was prepared in a different manner. It was found that both Ir^{IIIA}(aq) and Ir^{IIIB}(aq) gave rise upon oxidation to respectively blue (λ_{max} 584 nm, ε 1190) and purple (λ_{max} 547, ε 1700) Ir^{IV} solutions. The blue form of Ir^{IV}(aq) was subsequently found to be metastable, spontaneously converting to the purple form with an inverse [H$^+$] dependence.

Reduction of the purple form generates Ir^{IIIB}. Thus oxidation of an enriched Ir^{IIIA} solution (in H$_2$^{17}O) allows a preparative route through to enriched Ir^{IIIB}. This was achieved and the resulting Ir^{IIIB} spectrum is shown in Fig. 9.22 for a 5 atom % enriched solution. Here the addition of Mn^{2+}(aq) was necessary in order to relax the intense resonance of enriched bulk water. The feature around -145 ppm is assigned to coordinated water with the μ-OH group appearing at ~ -300 ppm. The individual water ligand *trans* to μ-OH was only resolvable by deconvolution (some broadening by the Mn^{2+}). However, the integration, H$_2$O/OH ($\sim$ 10:1), and cation-exchange behaviour of a 5$+$ ion suggests an assign-

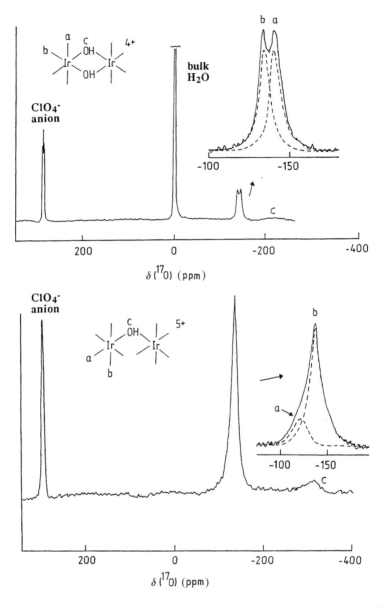

Figure 9.22 Oxygen-17 NMR spectra in 2.0 M $HClO_4$ for enriched solutions of Ir^{IIIA} (in natural abundance H_2O) and Ir^{IIIB} (in 5% enriched $H_2{}^{17}O$ with 0.1 M Mn^{2+} present). (Inset: deconvolution of the bound water region)

ment of Ir^{IIIB} to the mono-ol dimer ($[(H_2O)_5Ir(\mu\text{-}OH)Ir(OH_2)_5]^{5+}$. Consistent with these assignments the u.v.–visible spectral features of both aqua dimers are very similar to those obtained from the respective diol and mono-ol complexes $[(H_3N)_4Ir(\mu\text{-}OH)_2Ir(NH_3)_4]^{4+}$ and $[(H_2O)(H_3N)_4Ir(\mu\text{-}OH)Ir(NH_3)_4(OH_2)]^{5+}$, characterised by Galsbol, Simonsen and Springborg [138]. Here acid dissociation of the water ligands on the mono-ol form ($pK_a = 1.9$) is promoted via stabilisation of the terminal OH$^-$ ligand in the form of a bridging $H_3O_2^-$ ligand (Fig. 9.23).

Spectral changes observed for the mono-ol aqua dimer as a function of $[H^+]$ in the range 0.01–2.0 M ($pK_a = 0.7$) indicates a similar acid dissociation process. The presence of a 4+ species for Ir^{IIIB} at a pH of > 1.0 has been confirmed by charge/Ir studies (Cady and Connick method [118]) at pH 1.6. Acid dissociation of Ir^{IIIA} before a pH of 2 also seems relevant of the basis of similar charge/Ir studies.

Oxygen-17 NMR spectra for the blue and purple Ir^{IV} (aq) species are observable, the blue species possessing two peaks at $+43$ and -46 ppm, assigned to coordinated water, which collapse to a single peak for the purple form at -9 ppm [137]. Similar assignments to dioxo(diol) (blue) and monooxo (purple) forms seem relevant, in keeping with observations of quasi-reversible CV waves linking

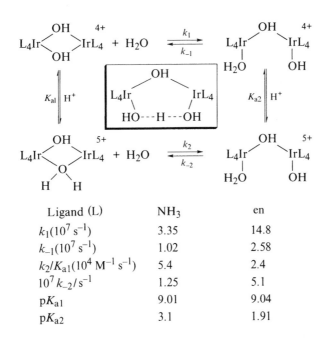

Ligand (L)	NH$_3$	en
$k_1(10^7\,s^{-1})$	3.35	14.8
$k_{-1}(10^7\,s^{-1})$	1.02	2.58
$k_2/K_{a1}(10^4\,M^{-1}\,s^{-1})$	5.4	2.4
$10^7\,k_{-2}/s^{-1}$	1.25	5.1
pK_{a1}	9.01	9.04
pK_{a2}	3.1	1.91

Figure 9.23. Solution chemistry of diol and mono-ol bridged dimers of Ir^{III}

the respective Ir^{III} and Ir^{IV} species. The Ir^{IVA}/Ir^{IIIA} and Ir^{IVB}/Ir^{IIIB} couples appears at $+1.26$ and $+1.23$ V (vs NHE) respectively in 2.0 M $HClO_4$. Oxidation to Ir^V(aq) is poorly resolved but is seen prominently in the CVs of Ir^{IIIA}(aq) occurring at more positive potentials in the sequence Ir^{IIIA}(aq) $(+1.49$ V), Ir^{IIIB}(aq) $(\sim +1.6$ V) and $[Ir(OH_2)_6]^{3+}$ $(> +1.6$ V). Little is known further of the nature of the metastable brownish-green Ir^V(aq) species. Only pure solutions of $[Ir(OH_2)_6]^{3+}$ give brown-green Ir^V as the first detectable oxidation product. The irreversible nature of the electrochemical process suggests that any monomeric Ir^V produced (conceivably IrO^{3+}(aq) or IrO_2^+(aq)) rapidly oligomerises. Indications are that a mixture of various oxo-bridged dimeric and polymeric forms are formed. The Ir^V(aq) species is metastable, decaying via oxidation of water to O_2 to give brown-purple polymeric form of Ir^{IV} which slowly deposits black hydrous IrO_2. Figure 9.24 summarises the interconversions of Ir^{III}, Ir^{IV} and Ir^V(aq) species in $HClO_4/NaClO_4$ solutions.

The $Ir = O^{3+}$(aq) formulation for putative metastable monomeric Ir^V is attractive on the basis of the structural characterisation of bright green $Ir^VO(\eta_1$-

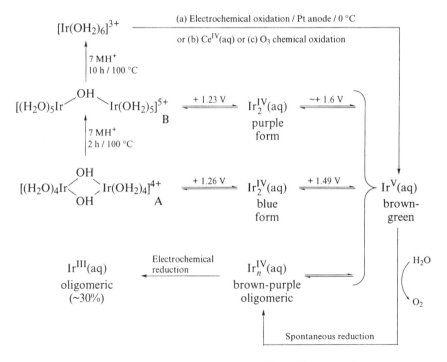

Figure 9.24. Redox interconversions involving Ir^{III}, Ir^{IV} and Ir^V(aq) species in $HClO_4/NaClO_4$ solution

mesityl)$_3$, a rare example of a terminal Ir=O compound (Ir—O = 172.5 pm) [139]. The longer Ir—O bond length vs that in the corresponding Re^V—O compound ($\sim$165 pm) suggests that Ir=O bonds are relatively weak. No terminal IrO$_2^+$ compounds seem to have been isolated.

The redox route to the [(H$_2$O)$_5$Ir(μ-OH)Ir(OH$_2$)$_5$]$^{5+}$ ion is of interest in that, in the absence of such similar redox chemistry for Rh^{III}, the corresponding Rh^{III} species has yet to be characterised. However, it might ultimately be preparable directly from the diol form since heating solutions of [(H$_2$O)$_4$Ir(μ-OH)$_2$Ir(OH$_2$)$_4$]$^{4+}$ in 7.0 M CF$_3$SO$_3$H at $\sim$80 °C under air-free conditions for several days converts it to the mono-ol dimer (purple Ir^{IV} product) and eventually the monomer [Ir(OH$_2$)$_6$]$^{3+}$ (brown-green oxidation product).

9.3.2 Trimeric Sulfato-Bridged Ir Aqua Species

The existence of sulfato-bridged trinuclear Ir complexes dates back to work of Delepine [140] and Lecoq [141] at the turn of the century. Lecoq's salt was first isolated by heating Na$_2$[IrCl$_6$]·6H$_2$O in boiling concentrated H$_2$SO$_4$. The bright green product was originally formulated as Ir$_2$(SO$_4$)$_3$·3K$_2$SO$_4$ but subsequent careful analytical work showed the correct formula to be K$_{10}$[Ir$_3$(μ-O)(μ_3-SO$_4$)$_6$(SO$_4$)$_3$]·3H$_2$O consisting of a planar triangular arrangement of Ir atoms around the central μ_3-O group, each bridged by two SO$_4^{2-}$ ions (Fig. 9.25) [136j, 141]. Lecoq's salt exists in all four oxidation states from Ir$_3^{III}$ to Ir$_3^{IV}$. Delepines's salt is formed when the NH$_4^+$ salt is heated in a similar manner, but in this case a rare example of a planar μ_3-nitrido complex is formed with formula K$_4$[Ir$_3$(μ_3-N)(μ-SO$_4$)$_6$(OH$_2$)$_3$] [141, 142] (Fig. 9.25). Here only one stable redox state

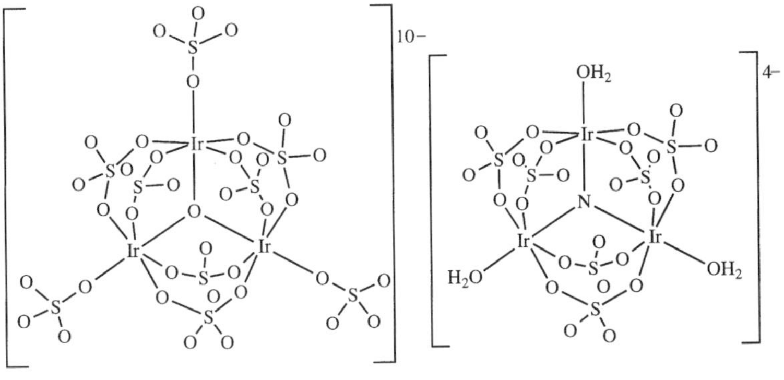

Figure 9.25. Structures of trimeric [Ir$_3$O(SO$_4$)$_9$]$^{10-}$ and [Ir$_3$N(SO$_4$)$_6$(OH$_2$)$_3$]$^{4-}$

$Ir^{III,IV,IV}$ is relevant, in contrast to the rich redox chemistry of the planar μ_3-oxo compound, the difference presumably stemming from the presence of the μ_3-N ligand.

Kinetic studies of substitution at the terminal water ligands of Delepine's salt have been carried out at $I = 1.0$ M, $LiClO_4$ [143]. Observable rate constants for equilibration of Cl^- ($k_f = 3.1 \times 10^{-4}$ $M^{-1}s^{-1}$, $k_b = 1.6 \times 10^{-4} s^{-1}$) and Br^- ($k_f = 16.0 \times 10^{-4}$ $M^{-1}s^{-1}$, $k_b = 2.4 \times 10^{-4} s^{-1}$) and N_3^- ($k_f \sim 1.8 \times 10^{-4}$ $M^{-1}s^{-1}$, $k_b \sim 3.5 \times 10^{-8} s^{-1}$) at 50 °C indicate significant *trans* labilisation of the water ligands from the planar μ_3-N ligand.The mixed-valence complex appears to be valence delocalised on the timescale of the ligand substitution process.

REFERENCES

[1]   Farina, R., and Wilkins, R. G. *Inorg. Chem.*, **7**, 514 (1968).
[2]   Jones, R. D., Summerville, D. A., and Basolo, F., *Chem. Rev.*, **79**, 139 (1979).
[3]   Coleman, J. E., and Vallee, B. L., *J. Biol. Chem.*, **236**, 2244 (1961).
[4]   Cotton, F. A., Daniels, L. M., Murillo, C. A., and Quesada, J. F., *Inorg. Chem.*, **32**, 4861 (1993).
[5]   Varela, A. J., Guerrero, A., Santos, A., Ramirez, A., Sabater, F. F., Carrera, S. M., and Blanco, S. G., *An Quim.*, **84B**, 194 (1988).
[6]   Martinezzaporta, M. A., Guerrero, A., Amil, A. R., Martinezcarrera, S., and Garciablanco, S., *An. Quim.*, **84B**, 201 (1988).
[7]   (a) Morrison, T. I., Reis Jr, A. H., Gebert, E., Iton, L. E., Stucky, G. D., and Suib, S. L., *J. Chem. Phys.*, **72**, 6276 (1980).
     (b) Sandstrom, D. R., Stults, B. R., and Greegor, R. B., in *EXAFS Spectroscopy*, (eds. B. K. Teo and D. C. Joy), Plenum, New York, 1981.
[8]   (a) Bol, W., Gerrits, G. J. A., and van Panthaleon van Eck., C. L., *J. Appl. Crystallogr.*, **3**, 486 (1970).
     (b) Ohtaki, H., Yamaguchi, T., and Maeda, M., *Bull. Chem. Soc. Japan*, **49**, 701 (1976).
     (c) Wertz, D. L., and Kruh, R. F., *Inorg. Chem.*, **9**, 595 (1970).
     (d) Magini, M., and Giubileo, G., *Gazz. Chim. Ital.*, **111**, 449 (1981).
[9]   (a) Holmes, O. G., and McClure, D. G., *J. Chem. Phys.*, **26**, 1686 (1957).
     (b) Jorgensen, C. K., *Adv. Chem. Ser.*, **8**, 1495 (1954) and *Adv. Chem. Phys.*, **5**, 33 (1963).
     (c) Lever, A. B. P., *Inorganic Electronic Spectroscopy*, 2nd edn, Elsevier, New York, 1983.
[10] Gailey, K. D., and Palmer, R. A., *Chem. Phys. Lett.*, **13**, 176 (1972).
[11] Figgis, B. N., and Lewis, J., *Prog. Inorg. Chem.*, **6**, 37 (1964).
[12] Swift, T. J., *Inorg. Chem.*, **3**, 526 (1964).
[13] Donnan, F. G., and Bassett, H., *J. Chem. Soc.*, 939 (1902).
[14] (a) Collados, M. P., Brito, F., and Cadavieco, R. D., *An. Fiz. Quim.*, **63B**, 843 (1967).
     (b) Bolzan, J. A., and Arvia, A. J., *Electrochim. Acta*, **7**, 589 (1962).
     (c) Shanker, J., and DeSousa, B. C., *Aust. J. Chem.*, **16**, 1119 (1963).
[15] Gayer, K. H., and Garrett, A. B., *J. Am. Chem. Soc.*, **72**, 3921 (1950).
[16] Feitknecht, W., and Hartmann, L., *Chimia*, **8**, 95 (1954).
[17] Ezhov, B. B., Kamnev, A. A., and Malandin, O. G., *Koord. Khim.*, **14**, 30–5 (1988).

[18] (a) Bechter, A., Breitschwerdt, K. G., and Tamm, K., *J. Chem. Phys.*, **52**, 2975 (1970).
(b) Eigen, M., and Tamm, K., *Z. Elektrochem.*, **66**, 93, 107 (1962).
[19] (a) Swift, T. J., and Connick, R. E., *J. Chem. Phys.*, **37**, 307 (1962).
(b) Chmelnick, A. M., and Fiat, D., *J. Chem. Phys.*, **47**, 3986 (1967).
(c) Hoggard, P. E., Dodgen, H. W., and Hunt, J. P., *Inorg. Chem.*, **10**, 959 (1971).
[20] Docummun, Y., Newman, K. E., and Merbach, A. E., *Inorg. Chem.*, **19**, 3696 (1980).
[21] Caldin, E. F., Grant, M. W., and Hasinoff, B. B., *J. Chem. Soc. Faraday Trans. I*, **68**, 2247 (1972).
[22] Grant, M. W., *J. Chem. Soc. Faraday Trans. I*, **69**, 560 (1973).
[23] Cossy, C., Helm, L., and Merbach, A. E., *Helv. Chim. Acta*, **70**, 1516 (1987).
[24] (a) Perrin, D., in *Stability Constants of Metal Ion Complexes, Part B, Organic Ligands*, IUPAC Chemical Data Series, Pergamon, Oxford, 1982.
(b) Martell, A. E., and Smith, R. M., *Critical Stability Constants*, Vol. 1, *Amino Acids*, Plenum, New York, 1985.
[25] Vasil'ev, V. P., and Zimina, I. D., *Zh. Neorg. Khim.*, **32**, 1429 (1987).
[26] Walters, K. L., and Wilkins, R. G., *Inorg. Chem.*, **13**, 752 (1974).
[27] Gillard, R. D., and Spencer, A., *J. Chem. Soc. A*, 2718 (1969).
[28] Rorabacher, D. B., *Inorg. Chem.*, **5**, 1891 (1966).
[29] Hammes, G. G., and Steinfield, J. I., *J. Am. Chem. Soc.*, **84**, 4639 (1962).
[30] Eisenstadt, M., *J. Chem. Phys.*, **51**, 4421 (1969).
[31] Wilkins, R. G., *Inorg. Chem.*, **3**, 520 (1964).
[32] Holyer, R. H., Hubbard, C. D., Kettle, S. F. A., and Wilkins, R. G., *Inorg. Chem.*, **4**, 929 (1965) and **5**, 622 (1966).
[33] Geier, G., *Helv. Chim. Acta*, **51**, 94 (1968).
[34] Kustin, K., Pasternack, R. F., and Weinstock, E. M., *J. Am. Chem. Soc.*, **88**, 4610 (1966).
[35] Wilkins, R. G., and Eigen, M., 'Mechanisms of inorganic reactions', *Adv. Chem. Ser.*, **49**, 55 (1963).
[36] Hammes, G. G., and Morell, M. L., *J. Am. Chem. Soc.*, **86**, 1497 (1964).
[37] Hammes, G. G., and Levison, S. A., *Biochemistry*, **3**, 1504 (1964).
[38] Best, S. P., Armstrong, R. D., and Beattie, J. K., *J. Chem. Soc. Dalton. Trans.*, 2105 (1981).
[39] Swann, S., and Xanthakos, T. S., *J. Am. Chem. Soc.*, **53**, 400 (1931).
[40] McAuley, A., and Whitecombe, T. W., *Inorg. Chem.*, **27**, 3090 (1988) (and refs. therein).
[41] Endicott, J., Durham, W., and Kumar, K., *Inorg. Chem.*, **21**, 2437 (1982).
[42] McAuley, A., Malik, M. N., and Hill J., *J. Chem. Soc. A*, 2461 (1970).
[43] Baxendale, J. H., and Wells, C. F., *Trans. Faraday Soc.*, **53**, 800 (1957).
[44] Bommer, H., *Z. Anorg. Allg. Chem.*, **246**, 275 (1941).
[45] Ray, P., and Sen, D., *J. Ind. Chem. Soc.*, **12**, 190 (1935).
[46] Friedman, H. L., Hunt, J. P., Plane, R. A., and Taube, H., *J. Am. Chem. Soc.*, **75**, 4028 (1951).
[47] Bawn, C. E. H., and White, A. G., *J. Chem. Soc.*, 332 (1951).
[48] Conocchioli, T. J., Nancollas, G., and Sutin, N., *Inorg. Chem.*, **5**, 1 (1966).
[49] Sutcliffe, L. H., and Weber, J. R., *Trans. Faraday Soc.*, **52**, 1225 (1956).
[50] Ferrer, M., Llorea, J., and Martinez, M., *J. Chem. Soc. Dalton. Trans.*, 229 (1992).
[51] Davies, G., and Watkins, K. O., *J. Phys. Chem.*, **74**, 3388 (1970).
[52] Davies, G., and Warnqvist, B., *Coord. Chem. Revs.*, **5**, 349 (1970).
[53] Brodovich, J. C., and McAuley, A., *Inorg. Chem.*, **20**, 1667 (1981).
[54] Hill, J., and McAuley, A., *J. Chem. Soc. A*, 1169 2405 (1968).

[55] Murmann, R. K., Sullivan, J. C., and Thompson, R. C., *Inorg. Chem.*, **7**, 1876 (1968).
[56] Thompson, R. C., *J. Phys. Chem.*, **72**, 2642 (1968).
[57] Jijie, K., and Santappa, M., *Proc. Ind. Acad. Sci. Sec. A*, **69**, 117 (1969).
[58] Thompson, R. C., and Sullivan, J. C., *Inorg. Chem.*, **9**, 1590 (1970).
[59] Curtis, N. F., *J. Chem. Soc.*, 2644 (1964).
[60] Ferraudi, G., *Inorg. Chem.*, **18**, 3230 (1979).
[61] Navon, G., *J. Phys. Chem.*, **85**, 3547 (1081).
[62] (a) Pelizzi, E., and Mentasti, E., *J. Chem. Soc. Dalton Trans.*, 2222 (1976).
     (b) Hanotier, J., Cameraman, P., Hanotier-Bridoux, M., and DeRaditsky, P., *J. Chem. Soc. Perkin Trans. II*, 2247 (1972).
[63] See for example, Hay, R. W., 'Lewis acid catalysis and reactions of coordinated ligands', in *Comprehensive Coordination Chemistry* (eds. J. A. McCleverty, G., Wilkinson and R. D. Gillard), Vol. 6, Pergamon, New York, 1987, p. 411.
[64] Maspero, F., and Taube, H., *J. Am. Chem. Soc.*, **90**, 7361 (1968).
[65] Pittet, P.-A. Dadci, L., Zbinden, P., Abou-Hamdan, A., and Merbach, A. E., *Inorg. Chim. Acta*, **206**, 135 (1993).
[66] Hills, E. F., Moszner, M., and Sykes, A. G., *Inorg. Chem.*, **25**, 339 (1986).
[67] Moszner, M., and Ziolkowski, J. J., *Inorg. Chim. Acta*, **145**, 299 (1988).
[68] Wilson, C. R., and Taube, H., *Inorg. Chem.*, **14**, 405 (1975).
[69] Dunbar, K. R., *J. Am. Chem. Soc.*, **110**, 8247 (1988).
[70] Moszner, M., and Ziolkowski, J. J., *Inorg. Chim. Acta*, **145**, 299 (1988).
[71] Ayres, G. H., and Forrester, J. S., *J. Inorg. Nucl. Chem.*, **3**, 365 (1957).
[72] Cervini, R., Fallon, G. D., and Spiccia, L., *Inorg. Chem.* **30**, 831 (1991).
[73] Wolsey, W. C., Reynolds, C. A., and Kleinberg, J., *Inorg. Chem.*, **2**, 463 (1963).
[74] Swaminathan, K., and Harris, G. M., *J. Am. Chem. Soc.*, **88**, 4411 (1966).
[75] Fallon, G. D., and Spiccia, L., *Aust. J. Chem.*, **42**, 2051 (1989).
[76] Beattie, J. K., Best, S. P., Moore, F. W., Skelton, B. W., and White, A. H., *Aust. J. Chem.*, **46**, 1337 (1993).
[77] Read, M. C., and Sandstrom, M., *Acta Chem. Scand.*, **46**, 1177 (1992).
[78] Bergstrom, P.-A., Lindgren, J., Read, M. C., and Sandstrom, M., *J. Phys. Chem.*, **95**, 7650 (1991).
[79] Read, M. C., Private communication, 1992.
[80] Plumb, W., and Harris, G. M., *Inorg. Chem.*, **3**, 542 (1964).
[81] Buchacek, R. J., and Harris, G. M., *Inorg. Chem.*, **15**, 926 (1976).
[82] Laurenczy, G., Rapaport, I., Zbinden, D., and Merbach, A. E., *Mag. Res. Chem.*, **29**, S45 (1991).
[83] Beutler, P., and Gamsjager, H., *Chimia*, **29**, 525 (1975).
[84] Beutler, P. PhD Thesis, University of Berne, 1976.
[85] Forrester, J. S., and Ayres, G. H., *J. Phys. Chem.*, **63**, 1979 (1959).
[86] Crimp, S. J., Fallon, G. D., and Spiccia, L., *J. Chem. Soc., Chem. Commun.*, 197 (1992).
[87] Read, M. C., Glaser, J., Sandstrom, M., and Toth, I., *Inorg. Chem.*, **31**, 4155 (1992).
[88] Read, M. C., Glaser, J., and Sandstrom, M., *J. Chem. Soc. Dalton Trans.*, 233 (1992).
[89] Carr, C., Glaser, J., and Sandstrom, M., *Inorg. Chim. Acta*, **131**, L53 (1987).
[90] Patel, A., Leitch, P., and Richens, D. T., *J. Chem. Soc. Dalton Trans.*, 1029 (1991).
[91] Ghosh, A. K., and De, G. S., *Ind. J. Chem.*, **33**, 247 (1994).
[92] Ghosh, A. K., and De, G. S., *Transition Met. Chem.*, **17**, 260 (1992).
[93] Ghosh, A. K., and De, G. S., *Transition Met. Chem.*, **17**, 435 (1992).
[94] Pearson, R. G., *J. Am. Chem. Soc.*, **85**, 3533 (1963) and *J. Chem. Educ.*, **45**, 581, 643 (1968).

[95] Hyde, K. E., Kelm, H., and Palmer, D. A., *Inorg. Chem.*, **17**, 1647 (1978).
[96] Swaddle, T. W., in *Adv. Inorg. Bioinorg. Mech.* (ed. A. G. Sykes), **2**, 95 (1983) and *Inorg. Chem.*, **22**, 2663 (1983).
[97] McMahon, M. R., McKenzie, A., and Richens, D. T., *J. Chem. Soc. Dalton Trans.*, 711 (1988).
[98] Laidler, K. J., and Eyring, H., *Ann. N.Y. Acad. Sci.*, **39**, 303 (1940).
[99] Ghosh, A. K., and De, G. S., *Ind. J. Chem.*, **33**, 929 (1994).
[100] Swaddle, T. W., and Stranks, D. R., *J. Am. Chem. Soc.*, **94**, 8357 (1972).
[101] Poe, A. J., Shaw, K., and Wendit, M. J., *Inorg. Chim. Acta*, **1**, 371 (1967).
[102] van Eldik, R., *Z. Anorg. Allg. Chem.*, **416**, 88 (1975).
[103] (a) Palmer, D. A., and Harris, G. M., *Inorg. Chem.*, **13**, 565 (1974).
(b) Spitzer, U., van Eldik, R., and Kelm, H., *Inorg. Chem.*, **21**, 2821 (1982).
[104] van Eldik, R., *Inorg. Chim. Acta*, **42**, (1980).
[105] Ellison, I. J., and Gillard, R. D., *J. Chem. Soc., Chem. Commun.*, 851 (1992).
[106] Buckley, A. N., Busby, J. A., Ellison, I. J., and Gillard, R. D., *Polyhedron*, **12**, 247 (1993).
[107] Edwards, N. S. A., Ellison, I. J., Gillard, R. D., and Mile, B., *Polyhedron*, **12**, 371 (1993).
[108] Ellison, I. J., Gillard, R. D., Mozsner, M., Wilogocki, M., and Ziolkowski, J. J., *Polyhedron*, **13**, 1351 (1994).
[109] Claus, C., *Petersb. Akad. Bull.*, **2**, 177 (1860).
[110] (a) Grube, G., and Autenreith, H., *Z. Elektrochem.*, **44**, 296 (1938).
(b) Kiseleva, I. N., Ezerskaya, N. A., Afanas'eva, M. V., and Shubochkin, L. K., *Zh. Anal. Khim.*, **43**, 2235 (1988).
[111] Syrokomsky, V. S., and Proshenkova, N. N., *Zh. Anal. Khim.*, **2**, 247 (1947).
[112] Sidorov, A. A., Komozin, P. N., Pichkov, V. N., Miroschnickenko, I. V., Golovin, K. A., and Sinitsyn, N. M., *Russ. J. Inorg. Chem.*, **30**, 1477 (1985).
[113] Moszner, M., and Ziolkowski, J. J., *J. Coord. Chem.*, **25**, 255 (1992).
[114] Gillard, R. D., and Pedrosa de Jesus, J. D., *J. Chem. Soc. Dalton Trans.*, 1895 (1984).
[115] Raynor, J. B., Gillard, R. D., and Pedrosa de Jesus, J. D., *J. Chem. Soc. Dalton Trans.*, 1165 (1982).
[116] Baranowskii, J. B., Zilaev, A. N., Dikareva, L. M., and Rotov, A. W., *Zh. Nieorg. Khimii*, **11**, 2892 (1986).
[117] Caldemaru, H., deArmond, K., and Henck, K., *Inorg. Chem.*, **17**, 2030 (1978).
[118] Cady, H. H., and Connick, R. E., *J. Am. Chem. Soc.*, **80**, 2646 (1958).
[119] Springborg, J., and Zender, M., *Acta, Chem. Scand.*, **41**, 484 (1987).
[120] Moszner, M., Wilgocki, M., and Ziolkowski, J. J., *Coord. Chem.*, **20**, 219 (1989).
[121] Richens, D. T., and Sykes, A. G., *J. Chem. Soc. Dalton Trans.*, 1621 (1982).
[122] Kiseleva, I. N., Ezerskaya, N. A., and Shubochkin, L. K., *Sov. J. Anal. Chem.* (*Engl. Trans.*), **41**, 1132 (1986).
[123] Beutler, P., and Gamsjager, H., *J. Chem. Soc. Chem. Commun.*, 554 (1976).
[124] Gamsjager, H., and Beutler, P., *J. Chem. Soc. Dalton Trans.*, 1415 (1979).
[125] Castillo-Blum, S. E., Richens, D. T., and Sykes, A. G., *J. Chem. Soc., Chem. Commun.*, 1120 (1986).
[126] Zinser, E. J., and Page, J. A., *Anal. Chem.*, **42**, 787 (1970).
[127] Gajhede, M., Siminsen, K., and Skov, L. K., *Acta Chem. Scand.*, **47**, 271 (1993).
[128] Armstrong, R. S., Beattie, J. K., Best, S. P., Skelton, B. W., and White, A. H., *J. Chem. Soc. Dalton Trans.*, 1973 (1983).
[129] Flenburg, C., Simonsen, K., and Skov, L. K., *Acta Chem. Scand.*, **47**, 862 (1993).
[130] Best, S. P., Armstrong, R. S., and Beattie, J. K., *J. Chem. Soc. Dalton Trans.*, 299 (1992).

[131] Castillo-Blum, S. E., Sykes, A. G., and Gamsjager, H., *Polyhedron*, **6**, 101 (1987).
[132] Cusanelli, A., Frey, U., Richens, D. T., and Merbach, A. E., *J. Am. Chem. Soc.*, **118**, 5265 (1996).
[133] Gamsjager, H., and Murmann, R. K., *Inorg. Chem.*, **28**, 379 (1989).
[134] Domingos, A. J. P., Domingos, A. M. T. S., and Peixoto Cabral, J. M., *J. Inorg. Nucl. Chem.*, **31**, 2563 (1969).
[135] van Eldik, R., and Harris, G. M., *Inorg. Chem.*, **14**, 10 (1975).
[136] (a) Bardin, M. B., and Ketrush, P. M., *Russ. J. Inorg. Chem. (Engl. Transl.)*, **18**, 693 (1973).
   (b) Tikhonova, L. P., and Zayats, V. A., *Russ. J. Inorg. Chem. (Engl. Transl.)*, **21**, 1184 (1976).
   (c) Pilipenko, A. T., Falendysh, N. F., and Parkhomenko, E. P., *Russ. J. Inorg. Chem. (Engl. Transl.)*, **20**, 1683 (1975).
   (d) Tikhonova, L. P., Zayats V. Ya., and Svarkozskaya, I. P., *Russ. J. Inorg. Chem. (Engl. Transl.)*, **24**, 1502 (1979).
   (e) Pshenitsyn, N. K., Ginszburg, S. I., and Sal'skaya, L. G., *Russ. J. Inorg. Chem. (Engl. Transl.)*, **4**, 130 (1959).
   (f) Bardin, M. B., Houng, Th. T., and Kelrush, P. M., *Zh. Anal. Khim.*, 1282 (1976).
   (g) Rose, D., Liver, F. M. Powell, A. R., and Wilkinson, G., *J. Chem. Soc. A*, 1690 (1969).
   (h) Fine, D. A., *J. Inorg. Nucl. Chem.*, **12**, 2731 (1970).
   (i) Ginsburg, S. I., and Yuz'ko, M. I., *Russ. J. Inorg. Chem. (Engl. Transl.)*, **10**, 444 (1965).
   (j) Ginsburg, S. I., Yuz'ko, M. I., and Sal'skaya, L. G., *Russ. J. Inorg. Chem. (Engl. Transl.)*, **8**, 429 (1963).
[137] Castillo-Blum, S. E., Richens, D. T., and Sykes, A. G., *Inorg. Chem.*, **28**, 954 (1989).
[138] Galsbol, F., Simonsen, K., and Springborg, J., *Acta Chem. Scand.*, **46**, 915 (1992).
[139] Hay-Motherwell, R. S., Wilkinson, G., Hussain-Bates, B., and Hursthouse, M. B., *Polyhedron*, **12**, 2009 (1993).
[140] Delepine, M., *Nouveau Traite de Chemie Minerale*, Vol. XIX (ed. P. Pascal), Masson et Cie, Paris, 1958 and *C. R. Acad. Sci.*, **251**, 2633 (1960) (and refs. therein).
[141] Lecoq de Boisbaudran, *C. R. Acad. Sci.*, **96**, 1336, 1406 (1883).
[142] Brown, D. B., Robin, M. B., McIntyre, J. D. E., and Peck, W. F., *Inorg. Chem.*, **9**, 2315 (1970).
[143] Hills, E. F., Richens, D. T., and Sykes, A. G., *Inorg. Chem.*, **25**, 3144 (1986).

Chapter 10

Group 10 Elements: Nickel, Palladium and Platinum

Nickel was found to be an essential component of the enzyme jack bean urease in 1975 [1] and its biochemical importance has been further enhanced by the recent discovery of its involvement in a number of bacterial hydrogenases [2], carbon monoxide dehydrogenase [3] and in the F430 cofactor of methyl coenzyme M reductase [4]. It is a puzzle why nickel-dependent enzymes and proteins are not more widespread in nature when compared to, for example, copper or zinc, given the high redox stability of the Ni^{II} state. A likely explanation is the relative substitution inertness and weaker Lewis acidity of $Ni^{2+}(aq)$. However, the nickel centre is almost certainly playing a role as a Lewis acid in urease and this, coupled with the recognition of a growing number of Ni-dependent proteins and cofactors, may mean that these conclusions could be somewhat premature [5].

The pale green paramagnetic octahedral $[Ni^{II}(OH_2)_6]^{2+}$ ion and derivatives obtained from it dominate the aqueous chemistry of nickel. This contrasts with the yellow diamagnetic square planar ions, $[M^{II}(OH_2)_4]^{2+}$, formed by the heavier members palladium and platinum. There have been few, if any, opportunities to compare rates and mechanisms of comparative substitution reactions on, for example, Ni and Pd species. No simple aqua ions representing Ni^{III} or Ni^{IV} exist, although coordination to specific arrangements of donor atoms is known to readily stabilise these oxidation states [6]. For Ni^{III} the most biologically relevant of these are N-deprotonated peptides. Pd^{IV} forms few representative stable complexes but a monomeric aqua cation of Pt^{IV}, probably $[Pt(OH_2)_5(OH)]^{3+}$, may exist in concentrated acids under high dilution. The study of the aqueous chemistry, more specifically the hydrolysis chemistry, of a range of square planar N-donor ligand chloro (aqua) complexes of Pt^{II} under physiological conditions is central to an understanding of the *in vivo* fate of antitumour active complexes such as *cis*-platin, *cis*-$[PtCl_2(NH_3)_2]$ and related second and third generation analogues. Surprisingly, despite much research into DNA binding by *cis*-platin and analogues and possible catalysis of, polyphosphate hydrolysis, only recently has the hydrolysis aqueous chemistry of *cis*-platin itself been fully studied and understood [7]. On the other hand, free $[Ni(OH_2)_6]^{2+}$ has been implicated in mutagenesis [8].

The tetrahedral nickel site found in the aqua cluster ion $[Mo_3NiS_4(OH_2)_{10}]^{4+}$ (see Section 10.1.2) is remarkable for its relative substitution inertness compared with simple tetrahedral or octahedral nickel complexes, a feature believed to reflect the considerable electron delocalisation present in these sulfur-bridged clusters. Such studies may ultimately have relevance towards an understanding of the role played by nickel in certain bacterial hydrogenases and in carbon monoxide dehydrogenase, recent work having indicated that S-bonding to Ni is important in these systems with the nickel closely related to, of possibly even part of, an Fe—S cluster [3, 9].

10.1 NICKEL

10.1.1 Nickel(II) (d^8): Chemistry of the $[Ni(OH_2)_6]^{2+}$ Ion

This ion is well established in simple salts such as $Ni(NO_3)_2 \cdot 6H_2O$ and $[Ni(OH_2)_6](C_6H_5SO_3)_2$, the latter being the subject of a recent X-ray structural

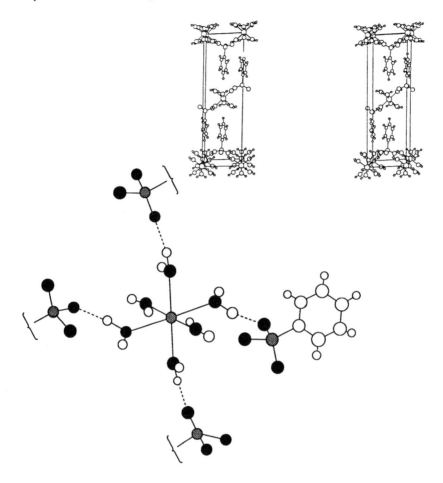

Figure 10.1. X-ray structure of the salt $[Ni(OH_2)_6](C_6H_5SO_3)_2$. (Inset: stereo-view of the crystal packing

study [10] (Fig. 10.1). Solutions of $[Ni(OH_2)_6]^{2+}$ are immediately generated upon dissolution of simple Ni^{2+} salts in water containing non- or weakly coordinating counter-anions such as SO_4^{2-}, ClO_4^-, pts$^-$ or $CF_3SO_3^-$. In the X-ray structural study on $[Ni(OH_2)_6](C_6H_5SO_3)_2$ the complex was obtained as single crystals following aerobic evaporation of an acetonitrile solution of the tetrahedral thiolato complex $(NEt_4)_2[Ni(SC_6H_4)_4]$. Here aerobic oxidation of $C_6H_5S^-$ to $C_6H_5SO_3^-$ accompanies atmospheric hydration of Ni^{2+} to form $[Ni(OH_2)_6]^{2+}$. In the X-ray structure the $C_6H_5SO_3^-$ ions are hydrogen bonded

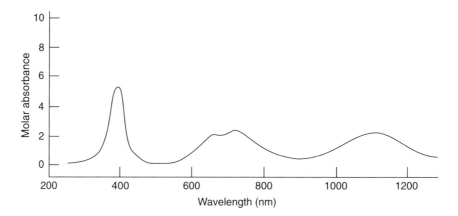

Figure 10.2. Electronic spectrum of $[\text{Ni(OH}_2)_6]^{2+}$

to the six water molecules while the packing diagram (Fig. 10.1) shows a stacked array of benzene rings. The paramagnetism, $\mu_{\text{eff}}(\text{RT}) \sim 3.2$ BM, and characteristic electronic spectrum of $\text{Ni}^{2+}(\text{aq})$ (Fig. 10.2) (three spin-allowed transitions observed from the $^3A_{2g}$ ground state to higher $^3T_{2g}$, $^3T_{1g}(\text{F})$ and $^3T_{1g}(\text{P})$ states at 1113 nm ($\varepsilon = 2.1$ M^{-1} cm^{-1}), 714 nm (2.3 M^{-1} cm^{-1}) and 397 nm (5.2 M^{-1} cm^{-1})) both indicate retention of the octahedral d^8 hexaaqua ion in solution. The coordination number of six has also been independently verified by both neutron scattering and by NMR peak area measurement from concentrated aqueous solutions of Ni^{2+} salts [11].

10.1.1.1 WATER EXCHANGE AND COMPLEX FORMATION ON $[\text{Ni(OH}_2)_6]^{2+}$

Water exchange on $[\text{Ni(OH}_2)_6]^{2+}$ occurs with a rate some 10^2 times slower than on the corresponding octahedral Fe^{2+} or Co^{2+} aqua ions and significantly slower than that on $\text{Cu}^{2+}(\text{aq})$ (see Chapter 11), reflecting the relatively large LFSE resulting from $t_{2g}^6 e_g^2$ configuration. The rate constant (25 °C) has been measured by several groups. Parameters determined by Merbach and coworkers [12, 13] are respectively $k_{\text{ex}} = (3.15 \pm 0.18) \times 10^4$ s^{-1}, $\Delta H_{\text{ex}}^{\ddagger} = +56.9$ kJ mol^{-1}, $\Delta S_{\text{ex}}^{\ddagger} = +32.0$ J K^{-1} mol^{-1} and $\Delta V_{\text{ex}}^{\ddagger} = (+7.2 \pm 0.3)$ cm^3 mol^{-1}. Values for $\Delta V_{\text{ex}}^{\ddagger}$ (cm^3 mol^{-1}) determined for $[\text{M(OH}_2)_6]^{2+}$ ions along the first row are respectively V^{2+} (-4.2), Mn^{2+} (-5.4), Fe^{2+} ($+3.8$), Co^{2+} ($+6.1$) and Ni^{2+} ($+7.2$ cm^3 mol^{-1}), indicative of a definite trend on moving across from left to right. The expression of an increasingly more positive activation volume is explained as reflecting the gradual increase in occupancy of the

Table 10.1. Selected rate constants and activation parameters for substitution reactions on $[Ni(OH_2)_6]^{2+}$ [12–21]

Incoming ligands	$10^{-3}k_f$ $(M^{-1}s^{-1})$	K_{os}^a (M^{-1})	$10^{-4}k_I$ (s^{-1})	$\Delta V^{\ddagger}$ $(cm^3 mol^{-1})$	Ref.
H_2O			3.15^b	$+7.2 \pm 0.3$	[12, 13]
F^-	8.4				[14]
HF	3.1				[14]
NH_3	4.5	0.15	3.0	$+6.0 \pm 0.3$	[15]
Bipyridine	5.1			$+5.5 \pm 0.3$	[16]
Terpyridine	3.0			$+6.7 \pm 0.2$	[16]
Pyridine	3.6	0.15	2.0		[17]
$CH_3CO_2^-$	300	20	1.5		[17]
$CH_3PO_4^{2-}$	280	40	0.7		[17]
$PADA^c$	4.0			$+7.7 \pm 0.3$	[18]
Imidazole	5.0			$+11.0 \pm 1.6$	[19]
Isoquinoline				$+7.4 \pm 1.3$	[20]
Glycinate				$+7 \pm 1$	[21]

^{a} Estimated from the Fuoss equation [22].
^{b} k_{ex}.
^{c} PADA = (pyridine)-2-azo-p-dimethylaniline.

t_{2g} orbitals,* which disfavours the approach of the entering ligand, as well as a correspondingly smaller volume change to the coordination sphere (lower compressibility) on forming the five-coordinate transition state. Interchange rate constants follow the usual trend of being largely insensitive to the nature of the entering ligand in support of a mechanism largely controlled by water dissociation from the hexaaqua ion. Rate constants for $Ni^{2+}(aq)$ are found to fall in a particularly narrow range of ~ 1–$4 \times 10^4 s^{-1}$ (Table 10.1) and, as expected, these reactions are similarly characterised by significantly positive activation volumes [12–21]. As argued in Section 1.4, it is possible that substitution on $Ni^{2+}(aq)$ is approaching the extreme limit for a dissociatively activated process.

Substitution of H_2O by NH_3 on $[Ni(OH_2)_6]^{2+}$ results in a higher water-exchange rate constant for the remaining water ligands [17, 24]. For example, exchange of the remaining water on $[Ni(NH_3)_5(OH_2)]^{2+}$ occurs with a rate constant of > 100 times that for each water on the hexaaqua complex (allowing for statistical factors). NH_3 is a better σ donor than H_2O and the effect may stem from a build-up of σ^* density at the Ni^{2+} centre, leading to promotion of the dissociative process. Activation volumes $(cm^3 mol^{-1})$ for solvent exchange on

---

*The shift to gradually more positive activation volume is also seen for exchange on other hexasolvates in the series Mn–Fe–Co–Ni [23], showing that the effect is a general one in support of the ideas presented in Section 1.4.

$[NiL_6]^{2+}$ (L = MeOH (+ 11.4) and CH_3CN (+ 9.6)) [23] also support the existence of a mechanism on Ni^{2+} towards, and possibly approaching, the dissociative limit. With bulkier leaving ligands such as DMF, replacement of DMF by NCS^- is observed to occur with a rate in excess of that of the DMF exchange itself. As a result a truly dissociative (D) process, involving a discrete five-coordinate intermediate, has been proposed to account for this observation [23a, 25].

10.1.1.2 COMPLEXATION BY AMINO ACIDS AND PEPTIDES ON $[Ni(OH_2)_6]^{2+}$

Complex formation with almost all common amino acids on Ni^{2+}(aq) has been reported [26] and their formation constants measured [27]. The highest values for 1:1 and 2:1 complexes are with aspartic acid (log β_1 = 7.16, log β_2 = 12.4), cysteine (log β_1 = 9.82, log β_2 = 20.07) and histidine (log β_1 = 8.67, log β_2 = 15.54). Binding to cysteine has been implicated in the case of jack bean urease. X-ray structures of a number of amino acid complexes of stoichiometry $NiL_2(OH_2)_2$ (L = *l*-serine, β-alanine, *l*-tyrosine, *l*-aspartic acid and *l*-histidine) have been reported [28]. The ready formation of stable amino acid complexes has led to an appreciation of the preferred conformation around the metal. Thus the 2:1 complex with *l*-(+)-histidine was found to be slightly more stable than that with the *d*-(−) enantiomer. With methionene, the 2:1 complex containing both enantiomers was found to be the most stable [28]. Complexation of Ni^{2+} to a number of dipeptides has been reported [29], the 1:1 complex with glycyl-*l*-histidine having the highest formation constant of those reported [27]. Significantly high formation constants are found for the 1:1 complex of Ni^{2+} with the oligopeptides glyglygly-*l*-his (log β_1 = 6.08) and glygly-*l*-his (log β_1 = 8.38). These compounds are believed to be good models for the strong and specific binding of Ni^{2+} and Cu^{2+} to the N terminus of human serum albumin. The extensive complex chemistry of both amino acids and peptides with Ni^{2+} in addition to Fe^{2+}, Co^{2+}, Cu^{2+} and Zn^{2+} has been the subject of extensive reviews [29].

$[Ni(OH_2)_6]^{2+}$ has recently been implicated as being responsible for the observed mutagenetic responses at the guanine phosphoribosyl transferase gene locus [8]. Binding of Ni^{2+} ions to DNA is believed to be involved in the mutagenesis and clearly further work is needed in this area.

10.1.1.3 HYDROLYSIS OF $[Ni(OH_2)_6]^{2+}$: THE CUBOIDAL $NI_4(OH)_4^{4+}$ (aq) ION

As in the case of Fe^{2+}(aq) and Co^{2+}(aq), hydrolysis of $[Ni(OH_2)_6]^{2+}$ is not extensive and alkaline solutions are required to generate $[Ni(OH_2)_5OH]^+$ (pK_{11} = 9.86) [30] prior to precipitation of green hydrous $Ni(OH)_2$ at around pH 10. The hydrolysis behaviour also parallels Co^{2+}(aq) to the extent that at high concentrations (> 0.1 M) the tetranuclear species $Ni_4(OH)_4^{4+}$(aq) (pK_{44} = 27.7) is formed above pH 8, in addition to formation of the soluble tetrahedral anions

$[Ni(OH)_3(OH_2)]^-$ ($pK_{13} = 30$) and $[Ni(OH)_4]^{2-}$ ($pK_{14} > 44$) above pH 12. Much more is known of $Ni_4(OH)_4^{4+}(aq)$ than of the corresponding Co^{II} species. The kinetics of reaction of $[Ni(OH_2)_6]^{2+}$ (~ 0.2 M) with OH^- ions has been followed at 230 nm by the stopped-flow technique [31]. Here $[Ni(OH_2)_6]^{2+}$ and $Ni_4(OH)_4^{4+}(aq)$ differ markedly in absorbance. A two-stage process is observed involving rapid (< 2 ms) formation of a range of hydroxo-nickel oligomers which condense to larger oligomers with increasing $[OH^-]$ and decreasing $[Ni^{2+}]$, finally reacting with the excess $Ni^{2+}(aq)$ to form $Ni_4(OH)_4^{4+}(aq)$. The kinetics of acid decomposition of $Ni_4(OH)_4^{4+}$ (aq) to give $[Ni(OH_2)_6]^{2+}$ has been likewise followed at 235 nm and the following rate law established [32]:

$$\frac{-d(Ni_4(OH)_4^{4+})}{dt} = \frac{k_1 k_2 [Ni_4(OH)_4^{4+}][H^+]}{k_{-1} + k_2 [H^+]} \tag{10.1}$$

The activation energy (71 kJ mol^{-1}) was rationalised by a distortion of a cuboidal structure (Fig. 10.3) similar to that established in $Pb_4(OH)_4^{4+}(aq)$, in order to provide the site for $[H^+]$ attack. Cuboidal Ni_4^{III} units are also established in compounds such as $[Ni_4(OMe)_4(acac)_4(ROH)_4]$ (R = Me or Et) and $[Ni_4(OMe)_4(O_2CCH_3)_2L_4]$ (L = $Me_2C(NC)CH_2CH_2C(NC)Me_2$).

10.1.2 Cuboidal $[Mo_3NiS_4(OH_2)_{10}]^{4+}$

Reaction of nickel metal with $[Mo_3S_4(OH_2)_9]^{4+}$ (see Section 6.2.3.4) in 2.0 M HCl solution gives rise to the blue-green mixed-metal cuboidal ion $[Mo_3NiS_4(OH_2)_{10}]^{4+}$ [33] analogous in structure to that of the Fe com-

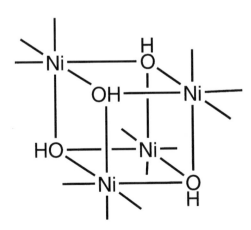

Figure 10.3. Proposed cuboidal structure for $Ni_4(OH)_4^{4+}(aq)$

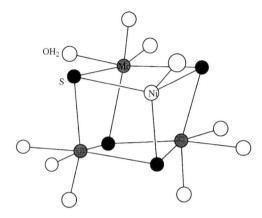

Figure 10.4. X-ray crystal structure of $[Mo_3NiS_4(OH_2)_{10}]^{4+}$

pound (Section 8.1.2.5):

$$Mo_3S_4^{4+} + Ni \longrightarrow Mo_3NiS_4^{4+} \qquad (10.2)$$

The same mixed cube can also be isolated following the reaction of $[Mo_3S_4(OH_2)_9]^{4+}$ with $NiCl_2 \cdot 6H_2O$ in 0.5 M HCl in the presence of a 100-fold excess of $NaBH_4$ [34]:

$$Mo_3S_4^{4+} + Ni^{2+} + 2e^- \longrightarrow Mo_3NiS_4^{4+} \qquad (10.3)$$

In each case the cube is purified by Dowex 50W X2 cation-exchange chromatography (elution with 2.0 M HCl or 2.0 M $HClO_4$). The X-ray structure (Fig. 10.4) reveals a fairly regular tetrahedral geometry around the Ni atom. The 3:1 stoichiometry was confirmed by oxidative titration to yield Ni^{2+} and $Mo_3S_4^{4+}$(aq). Substitution of the single water ligand on the Ni site is unusually slow for a tetrahedral Ni^{II} site (k_f, 25 °C, $9.4 M^{-1}s^{-1}$(Cl^-) and $45 M^{-1}s^{-1}$ (NCS^-)) and significantly slower than on $[Ni(OH_2)_6]^{2+}$ (Table 10.1). There are few if any studies of substitution on tetrahedral Ni^{II} for comparison. More importantly, however, it may be premature to assign an oxidation state of $+2$ to the Ni site in view of evidence in support of considerable electron delocalisation within these sulfur-bridged cuboidal cluster ions. Cu^{2+} (aq) readily displaces the Ni to give the Mo—Cu cluster $[Mo_3CuS_4(OH_2)_{10}]^{4+}$ [35]. An electron transfer mechanism, involving the generation of Cu^I, has been proposed for this reaction, a feature which may explain why Ni^{2+} (no possibility of an electron transfer pathway) is unable to displace Fe from $[Mo_3FeS_4(OH_2)_{10}]^{4+}$ to give the Ni cube. These studies also imply, as for the other mixed cubes, that the nickel atom could be in a formal oxidation state below Ni^{II}.

10.1.3   Nickel(III) (d⁷): Evidence for a Transient Ni³⁺(aq) Ion

There is a growing awareness of the importance of this oxidation state in the nickel-dependent hydrogenases [2]. In water, however, no stable aqua ion of Ni^{III} appears to exist. Two forms of hydrous 'Ni_2O_3', referred to as α- and β-NiO(OH), are established, which dissolve in acids with the formation of a 'transient' Ni^{3+}(aq) species before evolution of dioxygen and reduction to $[Ni(OH_2)_6]^{2+}$ occurs [6]. Ni^{III} can, however, be stabilised by certain arrangements of donor ligands, the best examples being tetraazamacrocycles and oligopeptides able to provide strong planar ligand fields to promote formation of a low-spin d^7 configuration at the metal. The complexes with peptides [36], in which hard-base N-deprotonated donors are frequently involved in coordination, are important biologically and a role for Ni^{III} cannot be ruled out. In the hydrogenases S coordination to Ni^{III} is believed to be involved [2]. There is no aqua ion representative of Ni^{IV}.

10.2   PALLADIUM AND PLATINUM

10.2.1   Palladium(II) and Platinum(II) (d⁸): Chemistry of Square Planar [Pd(OH₂)₄]²⁺ and [Pt(OH₂)₄]²⁺

The highly favoured formation of square planar ML_4 complexes for both d^8 Pd^{II} and Pt^{II} contrasts with Ni^{II} and results in authentic four-coordinate square planar aqua ions for both of the heavier members. The synthesis and characterisation of $[Pd(OH_2)_4]^{2+}$ and $[Pt(OH_2)_4]^{2+}$ (Fig. 10.5) in the mid 1970s owes much to the pioneering work of Elding and coworkers in Sweden who, along with others [37–50], have made substantial contributions towards an understanding of the mechanisms of substitution processes at four-coordinate square-planar complexes, the studies on the tetraaqua complexes being of a fundamental nature.

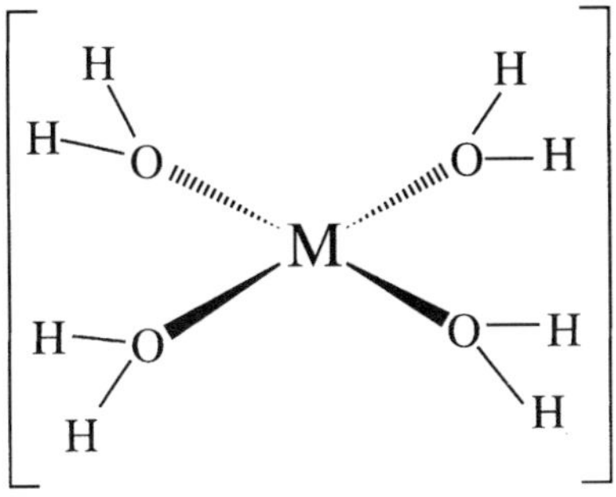

Figure 10.5. Square planar structure for the tetraaqua ions of Pd^{II} and Pt^{II}

10.2.1.1 PREPARATION AND PROPERTIES OF $[Pd(OH_2)_4]^{2+}$ AND $[Pt(OH_2)_4]^{2+}$

The 'lead-ins' to Pd^{II} and Pt^{II} aqueous chemistry have been principally the $[PdCl_4]^{2-}$ and $[PtCl_4]^{2-}$ ions usually supplied as their K^+ salts. Rapid aquation of $[PdCl_4]^{2-}$ occurs extensively in non-complexing acidic solution [37] and ultimately $[Pd(OH_2)_4]^{2+}$ is formed. An alternative method to ensure a chloride-free solution involves precipitation firstly of 'Pd(OH)$_2$', usually from chloride or nitrate salts, followed by copious washing of the hydroxide and dissolution in non-complexing strong acids [51]. An additional problem with this method is the likely presence of polynuclear hydrolysis products or colloidal species. Cation-exchange chromatography can be used for subsequent purification; however, a superior preparative method, developed by Jorgensen and Rasmussen [52], has involved initial dissolution of Pd metal in fuming nitric acid. In a typical preparation portions of palladium sponge (4.5 g) are dissolved in hot fuming 100% nitric acid (300 cm^3) contained in a 400 cm^3 beaker. Following cautious boiling to reduce the volume to 150 cm^3 the mixture is transferred to an evaporating dish and the HNO$_3$ removed by repeated evaporations with several 50 cm^3 portions of 70% perchloric acid to a final volume of 80 cm^3 [37]. Concentrations of Pd^{2+}(aq) can be calculated from the absorbance at the 380 nm band maximum ($\varepsilon = 82.8$ M^{-1}cm^{-1}) (Fig. 10.6). This absorption represents the d_{xy}–$d_{x^2-y^2}$ transition (Δ). Stable solutions containing up to 0.5 M Pd^{2+}(aq) have been reported and used in subsequent water-exchange kinetic studies [40].

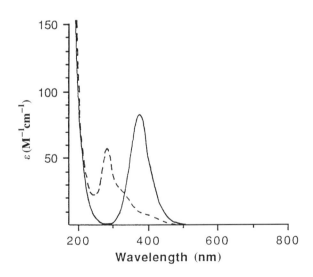

Figure 10.6. Electronic spectra for $[Pd(OH_2)_4]^{2+}$ (—) and $[Pt(OH_2)_4]^{2+}$ (---)

Similar acid-catalysed aquation of $[PtCl_4]^{2-}$ only produces micromolar quantities of $[Pt(OH_2)_4]^{2+}$ owing to the high kinetic and thermodynamic stability of $[PtCl(OH_2)_3]^+$ [41]. As a result aquation of the remaining chloride requires metal ion catalysis, the use of Hg^{2+} and Ag^+ being well documented, the latter being preferred since the excess Ag^+ is easily removed. In a typical preparation [38] $K_2[PtCl_4]$ (3.5 g, 8.5 mmol) is added to deoxygenated 1.0 M perchloric acid (400 cm^3) in a glass-stoppered 1000 cm^3 flask. Heating at 70 °C for 60–90 minutes causes the Pt salt to completely dissolve. A solution of $AgClO_4$ (0.2 M, 400 cm^3) is then added slowly over a period of 6–7 hours. The slow addition avoids precipitation of sparingly soluble $Ag_2[PtCl_4]$. The excess of Ag^+ ensures quantitative production of $[Pt(OH_2)_4]^{2+}$. The excess Ag^+ (60 mM) is then conveniently removed by electrolysis to Ag metal at a Pt cathode. Solutions containing ~ 8 mM Pt^{2+}(aq) are normally generated at this point. Higher concentrations can be produced by simple evaporation of the solutions. Solutions of $[Pt(OH_2)_4]^{2+}$ can be standardised using the absorbance of the d_{xy}–$d_{x^2-y^2}$ transition at 273 nm ($\varepsilon = 56.5$ M^{-1} cm^{-1}). In addition, two shoulders at lower energy, probably due to spin-forbidden transitions, are observed in the spectrum at 390 nm (10) and 320 nm (15) (Fig. 10.6). These are not observed in the spectrum of $[Pd(OH_2)_4]^{2+}$.

10.2.1.2   HYDROLYSIS OF $[Pd(OH_2)_4]^{2+}$ AND $[Pt(OH_2)_4]^{2+}$

A pK_a of 2.3 ± 0.1 has been measured for $[Pd(OH_2)_4]^{2+}$ [53]. Estimates for pK_a of $[Pt(OH_2)_4]^{2+}$ put the value $> \sim 2.5$ [38]. Solutions of both $[Pd(OH_2)_4]^{2+}$ and $[Pt(OH_2)_4]^{2+}$ held at a pH of > 1 for several hours show a slow general rise in absorbance at all wavelengths, indicative of the formation of polynuclear hydrolysis products. These seem not to have been fully characterised. Dinuclear, cyclic trinuclear and tetranuclear OH-bridged Pt^{II} complexes have been characterised with N-donor ligands such as $[Pt_2(\mu\text{-}OH)_2(NH_3)_4](NO_3)_2$, $[Pt_3(\mu\text{-}OH)_3(NH_3)_6](NO_3)_3$ and $[Pt_4(\mu\text{-}OH)_4(en)_4](NO_3)_4$ (Fig. 10.7) [54]. The yellow dimer and cyclic trimer are isolated as hydrolysis products of *cis*-$[Pt(NH_3)_2(OH_2)_2]^{2+}$ when kept at a pH of between 6 and 7 for 4 days. The tetramer is similarly isolated following hydrolysis of the corresponding diaqua(1,2-diaminoethane) complex. Such oligomeric species are therefore of possible relevance to the ultimate fate of the anticancer drug '*cis*-platin' and related species *in vivo* at physiological pH [7] and in particular to the possible aggregation and concentration of the metal. A direct Pt—Pt bond in these oligomeric species is not suggested from the bond lengths.

Such core structures may be representative of the hydrolysis products generated in the pH range > 2.5 in solutions of Pd^{2+}(aq) and Pt^{2+}(aq) prior to precipitation of the hydroxide. White amphoteric 'Pt(OH)$_2$' precipitates around a pH of 4 but redissolves above a pH of 10. On standing, the precipitate darkens at room temperature as it converts to a black Pt^{II} oxide. Once formed this black

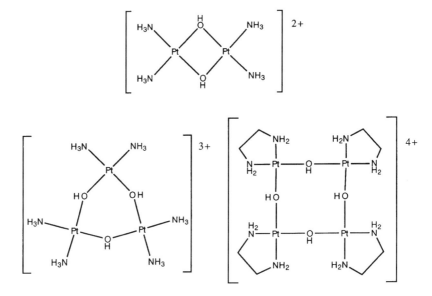

Figure 10.7. Structures of some oligomeric OH-bridged Pt^{II} cations

oxide will dissolve in 5 M HCl but remains completely insoluble in concentrated (70%) perchloric acid. The freshly prepared white hydroxide dissolves readily in 1 M perchloric acid to produce mainly $[Pt(OH_2)_4]^{2+}$, but with small amounts of hydrolysis products. Similar hydrolysis behaviour is observed in the case of $[Pd(OH_2)_4]^{2+}$.

10.2.1.3   WATER LIGAND SUBSTITUTION REACTIONS

The fundamental process of water exchange has been studied as a function of both temperature and pressure for both $[Pd(OH_2)_4]^{2+}$ [40] and $[Pt(OH_2)_4]^{2+}$ [42, 43]. The results are summarised in Table 10.2.

Table 10.2. Kinetic parameters for water exchange on $[Pd(OH_2)_4]^{2+}$ and $[Pt(OH_2)_4]^{2+}$ at 25 °C [40, 42, 43]

Complex	k_{ex} (s^{-1})	$\Delta H^{\ddagger}_{ex}$ (kJ mol^{-1})	$\Delta S^{\ddagger}_{ex}$ (JK^{-1} mol^{-1})	$\Delta V^{\ddagger}_{ex}$ (cm^3 mol^{-1})
$[Pd(OH_2)_4]^{2+}$	$(5.6 \pm 0.4) \times 10^{2\,a}$	49.5 ± 1.9	-26 ± 6	-2.2 ± 0.2
$[Pt(OH_2)_4]^{2+}$	$(3.9 \pm 0.3) \times 10^{-4\,a}$	89.7 ± 2.4	-9 ± 8	-4.6 ± 0.2
	$(4.8 \pm 0.1) \times 10^{-4\,b}$	92 ± 2	0 ± 5	

^{a} Studied by ^{17}O NMR [42].
^{b} Studied using ^{18}O with an oxidative quenching technique [43].

Table 10.3. Rate constants for 1:1 complex formation on $[Pd(OH_2)_4]^{2+}$ and $[Pt(OH_2)_4]^{2+}$ [40–50, 53–55]

Incoming ligand	$k_{Pd}(M^{-1}s^{-1})$	$k_{Pt}(M^{-1}s^{-1})$	k_{Pd}/k_{Pt}	Ref.
H_2O	41^a	2.8×10^{-5a}	1.5×10^6	[40, 42]
Cl^-	1.8×10^4	2.66×10^{-2}	6.8×10^5	[45, 46]
Br^-	9.2×10^4	2.11×10^{-1}	4.4×10^5	[45, 46]
SCN^-	4.4×10^5	1.33	3.3×10^5	[46, 55]
I^-	1.14×10^6	7.7	1.4×10^5	[47]
$(CH_3)_2SO$	2.45	8.4×10^{-5}	2.9×10^4	[48, 49]
CH_3CN	3.09×10^2			[54]
Me_2S	1.51×10^5	3.6	4.2×10^4	[48]
$(NH_2)_2CS$	9.6×10^5	13.9	6.9×10^4	[55]

^{a} Rate constant converted to second-order units ($4k_{ex}/55.56$).

These findings support the general concensus that an associative mechanism is relevant for substitution on these square planar ions. Further support comes from significant dependence of rate on the incoming ligand. A recent paper by Elding and coworkers [44] has summarised the available kinetic data for complexation on both square planar aqua ions. These are listed in Table 10.3.

Rates for substitution of $[Pt(OH_2)_4]^{2+}$ are significantly lower than those on $[Pd(OH_2)_4]^{2+}$ and are believed to stem from essentially an enthalpic effect (typical $\Delta H^{\ddagger} \sim 56 \pm 6 \, kJ \, mol^{-1}$ for Pt^{II} vs $\sim 28 \pm 6 \, kJ \, mol^{-1}$ for Pd^{II}). The largest rates are found for incoming soft ligands with the ratio k_{Pd}/k_{Pt} increasing with increasing hardness, showing that Pt^{II} is more sensitive than Pd^{II} to the nature of the entering ligand reflective of its softer nature. These trends could also reflect an increasing covalence in the Pt—L bonds. Steric effects of the entering ligands seem harder to rationalise [44]. A further observation is the apparent independence of $\Delta V^{\ddagger}$ on the leaving ligand L for the reverse aquation reaction on a number of $[PdL(OH_2)_3]^{(2-n)+}$ derivatives studied (L = H_2O, $(CH_3)_2SO$ and CH_3CN). Furthermore, the value of $\Delta V^{\ddagger}$ is only slightly negative, suggesting that the leaving group L is still tightly bound in the transition state, its release expected to cause a large expansion. This has been used as evidence in favour of an A mechanism rather than I_a [50]. On the other hand, the small values for $\Delta V^{\ddagger}_{ex}$ (Table 9.2) would appear to suggest an I_a process. It has been pointed out, however, that small $\Delta V^{\ddagger}$ values could result within an A process if a substantial lengthening of two of the M—OH$_2$ bonds, e.g. those 'axial', occurs within the trigonal bipyramidal intermediate, thus offsetting somewhat the expectedly larger $-\Delta V^{\ddagger}$ [40, 42]. The associative nature of the substitution reaction is further exemplified by the well-established two-term rate law [39]:

$$\text{Rate} = k_1 [ML_3X] + k_2 [ML_3X][Y] \tag{10.4}$$

wherein k_1 defines the solvent coordination path and k_2 the path involving the incoming ligand Y. Leaving ligand effects on rate have also been detected in Pt^{II} substitution reactions and are attributed to the greater covalency in Pt—L bonds. Square planar complexes containing good π-acceptors or σ-donors exhibit well-defined *trans* labilising effects arising from these ligands. These effects have been rationalised within the general associative mechanism [39]. Strong field chelating lligands, frequently macrocycles, are required to stabilise square planar Ni^{II} and as a result no meaningful comparisons between simple Ni^{II} and Pd^{II} square complexes have been possible in order to establish the trend in rate and mechanism. The rapid water exchange rate observed for $[Ni^{II}(cyc-lam)(OH_2)_2]^{2+}$, $k_{ex} = (2.1 \pm 0.4) \times 10^7 \, s^{-1}$ [56], is rather reflective of the weak axial coordination of water in the 'octahedral' diaqua form and of the equilibrium which strongly favours the square planar species. There has been much recent research into the catalytic activity of *cis*-diaqua Pt^{II} complexes towards the hydrolysis of certain polyphosphates and phosphate esters.

10.2.2 Palladium(III) and Platinum(III) (d⁷): Dimeric Pt–Pt Species

The aqueous chemistry of this oxidation state appears to be entirely that of a number of M—M bonded dinuclear complexes of Pt(III) with μ-bridging oxo anionic ligands, good examples being $K_2[Pt_2(SO_4)_4(OH_2)_2]$ [57] and $Na_2[Pt_2(HPO_4)_4(OH_2)_2]$ [58]. Typical Pt—Pt distances are around 250 pm in the range for a single bond. Conversion of $[Pt_2(HPO_4)_4(OH_2)_2]^{2-}$ to the corresponding μ-sulfato-bridged anion can be achieved upon heating a 50% aqueous H_2SO_4 solution of the μ-$HPO_4{}^{2-}$ complex at 110 °C for 2 hours. Addition of excess $(NH_4)_2SO_4$ precipitates the ammonium salt at 4 °C [59]. Hetrocyclic bases such as pyridine readily replace the weakly bonded axial water ligands. There are no true aqua cations of this oxidation state for either metal.

10.2.3 Palladium(IV) and Platinum(IV) (d⁶)

There are very few well-characterised Pd^{IV} aqua complexes. The ^{195}Pt NMR chemical shift of a number of chloro aqua Pt^{IV} complexes have been recorded as products obtained following the oxidation of a number of Pt^{II} chloro aqua complexes, including $[Pt(OH_2)_4]^{2+}$, with dry chlorine gas [43]. Here the coordination number change required in the Pt^{II}–Pt^{IV} redox process is facilitated by an oxidative addition reaction. In the absence of such a process oxidation to Pt^{IV} can be kinetically very slow. The reaction with Cl_2 is complete within a few milliseconds and has been used to quench the water-exchange process on various Pt^{II} complexes allowing rate constant measurement. With $[Pt(OH_2)_4]^{2+}$ the major product of the oxidation is $[Pt^{IV}Cl(OH_2)_5]^{3+}$ ($\delta(^{195}Pt) = 2640$ ppm from $[PtCl_6]^{2-}$). As expected, all Pt^{IV} aqua complexes are highly inert low-spin octahedral species. Over several hours, however, a range of chloroaqua complexes of Pt^{IV} can be observed in the ^{195}Pt NMR spectrum. These have been assigned to slow equilibration reactions giving *trans*-$[PtCl_2(OH_2)_4]^{2+}$ (2248 ppm), *cis*-$[PtCl_2(OH_2)_4]^{2+}$ (2173 ppm), *mer*-$[PtCl_3(OH_2)_3]^+$ (1646 ppm), *fac*-$[PtCl_3(OH_2)_3]^+$ (1548 ppm) and *trans*-$[PtCl_4(OH_2)_2]$ (1029 ppm). For $[PtCl(OH_2)_5]^{3+}$ coupling to the ^{35}Cl and ^{37}Cl nuclei was observed under high resolution, as were a range of $^{18}O/^{16}O$ isotopomers following the reactions carried out in labelled $H_2^{18}O$.

10.2.3.1 EXISTENCE OF PUTATIVE $[Pt(OH_2)_6]^{4+}$?

Formation of the hexaaqua ion $[Pt(OH_2)_6]^{4+}$ was not observed in the above studies and this species presently remains unknown. This may seem initially surprising in view of the fact that $[PtCl(OH_2)_5]^{3+}$ is well characterised in 1.0 M $HClO_4$. The use of higher concentrations of $HClO_4$ and/or metal ion catalysis (e.g. with Ag^+) might have been expected to assist aquation of the remaining chloride. This has not been achieved. Extensive hydrolysis of putative $[Pt(OH_2)_6]^{4+}$, even in concentrated $HClO_4$, is, however, indicated from the work of several groups [60, 61]. Indeed, the possibility of a 4+ hexaaqua ion complex existing for Pt^{IV} in aqueous media has been recently discounted on the basis of related experimental and thermodynamic evidence [61]. Thus $[Pt(NH_3)_5OH]Cl_3$ and not the aqua complex is found to crystallise from concentrated HCl, suggestive of an extremely low pK_{11} (< -1) for the water ligand [62]. Even under conditions of high dilution in concentrated $HClO_4$ solution it is likely therefore that the species present would be $[Pt(OH_2)_5(OH)]^{3+}$ and not $[Pt(OH_2)_6]^{4+}$. Hydrolysis of $[Me_3Pt(OH_2)_3]^+$ leads to formation of tetrameric $[(Me_3PtOH)_4]$, the cuboidal structure ensuring retention of an octahedral geometry around each Pt^{IV} centre. Although not sustantiated, similar cuboidal oligomeric hydroxoaqua structures could conceivably be present in hydrolysed Pt^{IV}(aq) solutions [60]. Should the putative $[Pt(OH_2)_6]^{4+}$ be characterised it would be expected to be the most inert metal ion known, more inert than even $[Ir(OH_2)_6]^{3+}$.

REFERENCES

[1]   (a) Nixon, N. E., Gazzola, C., Blakekey, R. L., and Zerner, B., *J. Am. Chem. Soc.*, **97**, 4131 (1975).
      (b) Andrews, R. K., Blakeley, R. L., and Zerner, B., in *Advances in Inorganic Biochemistry* (eds. G. L. Eichhorn and L. G. Marzilli), Vol. 5, Elsevier, New York, 1984, p. 245.
      (c) Andrews, R. K., Blakeley, R. L., and Zerner, B., in *The Bioinorganic Chemistry of Nickel* (ed. J. H. Lancaster), VCH, New York, 1988, pp. 141–66.
[2]   Scott, R. A., Wallin, S. A., Czechowski, M., Dervartanian, D. V., LeGall, J., Peck, H. D., and Moura, I., *J. Am. Chem. Soc.*, **106**, 6864 (1984).
[3]   Bastian, N. R., Diekert, G., Niederhoffe, B.-K., Teo, B.-K., Walsh, C. T., and Orme-Johnson, W. H., *J. Am. Chem. Soc.*, **110**, 5581 (1988).

[4]   Diekert, G., Klee, B., and Thaeur, R. K., *Arch. Microbiol.*, **124**, 103 (1980).
[5]   Kolodziej, A. F., *Prog. Inorg. Chem.*, **41**, 493 (1994).
[6]   (a) Haines, R. L., and McAuley, A., *Coord. Chem. Rev.*, **39**, 77 (1981).
       (b) Bhattacharga, S., Mukherjee, R., and Chakravorty, A., *Inorg. Chem.*, **25**, 3448
           (1986).
       (c) Munn, S. F., Lannon, A. M., Laranjeira, M. C. M., and Lappin, A. J., *J. Chem.
           Soc. Dalton Trans.*, 1371 (1984).
[7]   (a) Miller, S. E., and House, D. A., *Inorg. Chim. Acta*, **166**, 189 (1989), **161**, 131
           (1989), **173**, 53 (1990) and **187**, 125 (1991).
       (b) Wen, H., Miller, S. E., House, D. A., and Robinson, W. T., *Inorg. Chim. Acta*,
           184, 111 (1991).
       (c) Miller, S. E., Gerard, K. J., and House, D. A., *Inorg. Chim. Acta*, **190**, 135(1991).
[8]   Pappenhagen, T. L., Kennedy, W. R., Bowers, C. P., and Margerum, D. W., *Inorg.
       Chem.*, **24**, 4356 (1985).
[9]   (a) Maroney, M. J., Colpas, G. J., Bagyinka, C., Baidya, N., and Mascharak, P. K.,
           J. Am. Chem. Soc., **113**, 3962 (1991).
       (b) Yamamura T., Nakakura, N., Yasui, A., Sasaki, K., and Arai, H., *Chem. Lett.*, **5**,
           875 (1991).
[10]   Groh, S. E., Riggs, P. J., Baldacchini, C. J., and Rheingold, A. L., *Inorg. Chim. Acta*,
       174, 17 (1990).
[11]   (a) Bechtold, D. B., Liu, G., Dodgen, H. W., and Hunt, J. P., *J. Phys. Chem.*, **82**, 333
           (1978).
       (b) Soper, A. K., Neilson, G. W., Enderby, J. E., and Howe, R. A., *J. Phys. Chem.*,
           10, 1793 (1977).
[12]   Ducummon, Y., Earl, W. L., and Merbach, A. E., *Inorg. Chem.*, **18**, 2754 (1979).
[13]   Ducummon, Y., Newman, K. E., and Merbach, A. E., *Inorg. Chem.*, **19**, 3696 (1980).
[14]   Eisenstadt, M., *J. Chem. Phys.*, **51**, 4421 (1969).
[15]   Rorabacher, D. B., *Inorg. Chem.*, **5**, 1891 (1966).
[16]   (a) Holyer, R. H., Hubbard, C. D., Kettle, S. F. A., and Wilkins, R. G., *Inorg.
           Chem.*, **4**, 929 (1965) and **5**, 622 (1966).
       (b) Mohr, R., and van Eldik, R., *Inorg. Chem.*, **24**, 3396 (1985).
[17]   Wilkins, R. G., *Acc. Chem. Res.*, **3**, 408 (1970) and *Comments Inorg. Chem.*, **2**, 187
       (1983).
[18]   Caldin, E. F., Grant, M. W., and Hasinoff, B. B., *J. Chem. Soc. Faraday Trans. I*, **68**,
       2247 (1972).
[19]   (a) Hammes, G. G., and Steinfield, J. I., *J. Am. Chem. Soc.*, **84**, 4639 (1962).
       (b) Yu, A. D., Waissbuth, M. D., and Grieger, R. A., *Rev. Sci. Instrum.*, **44**, 1390
           (1973).
[20]   Ishihara, K., Funahashi, S., and Tanaka, M., *Inorg. Chem.*, **22**, 2564, (1983).
[21]   Grant, M. W., *J. Chem. Soc. Faraday Trans.*, **69**, 560 (1963).
[22]   (a) Fuoss, R. M., *J. Am. Chem. Soc.*, **80**, 5059 (1958).
       (b) Eigen, M., *Z. Phys. Chem.*, **1**, 176 (1954).
[23]   (a) Cossy, C., and Merbach, A. E., *Helv. Chim. Acta*, **70**, 1516 (1987).
       (b) Helm, L., Lincoln, S. F., Merbach, A. E., and Zbinden, D., *Inorg. Chem.*, **25**,
           2550 (1985).
       (c) Sisley, M. J., Yano, Y., and Swaddle, T. W., *Inorg. Chem.*, **21**, 1141 (1982).
[24]   Moore, P., *Pure Appl. Chem.*, **57**, 347 (1985).
[25]   Fielding, L., and Moore, P., *J. Chem. Soc., Chem. Commun.*, 49 (1988).
[26]   Sacconi, L., Mani, F., and Bencini, A., in *Comprehensive Coordination Chemistry*
       (eds G. Wilkinson, R. D. Gillard, and J. A. McCleverty), Vol. 5, Pergamon, London,
       1987, p. 219 (and refs. therein).

[27] (a) Perrin, D., *Stability Constants of Metal Ion Complexes, Part B, Organic Ligands*, IUPAC Chemical Data Series, Pergamon, Oxford, 1982.
 (b) Martell, A. E., Smith, R. M., *Critical Stability Constants*, Vol. 1, *Amino Acids*, Plenum, New York, 1985).
[28] Freeman, H. C., Guss, J. M., and Sinclair, R. L., *Acta Cryst.*, *B*, **34**, 2459 (1978).
[29] (a) Hay, R. W., and Nolan, K. B., Amino Acids and Peptides, Royal Society of Chemistry Specialist Periodical Reports, **15**–22, 1991.
 (b) Burger, K. (ed.), *Biocoordination Chemistry*, Ellis Horwood, Chichester, 1990.
[30] Baes, C. F., and Mesmer, R. E., *The Hydrolysis of Cations*, Krieger, Florida, 1986, p. 241 (and refs therein).
[31] Clare, B. W., and Kepert, D. L., *Aust. J. Chem.*, **28**, 1489 (1975).
[32] Kolski, G. B., Kildahl, N. K., and Margerum, D. W., *Inorg. Chem.*, **8**, 1211 (1969).
[33] Shibahara, T., Yamasaki, M., Akashi, H., and Katayama, T., *Inorg. Chem.*, **30**, 2693 (1991).
[34] Dimmock, P. W., Lamprecht, G. J., and Sykes, A. G., *J. Chem. Soc. Dalton Trans.*, 955 (1991).
[35] Shibahara, T., Asano, T., and Sakane, G., *Polyhedron*, **10**, 2351 (1991).
[36] Cho, S. T., and McAuliffe, C. A., *Prog. Inorg. Chem.*, **19**, 51 (1975).
[37] Elding, L. I., *Inorg. Chim. Acta*, **6**, 647 (1972).
[38] Elding, L. I., *Inorg. Chim. Acta*, **20**, 65 (1976).
[39] (a) Cross, R. J., *Chem. Soc. Rev.*, **14**, 197 (1985) and *Adv. Inorg. Chem.*, **34**, 219 (1989).
 (b) van Eldik, R., *Inorganic High Pressure Chemistry, Kinetics and Mechanisms*, Elsevier, Amsterdam, 1986, Ch. 4.
 (c) Cattalini, L., 'The mechanism of square planar substitution', in *Inorganic Reaction Mechanisms* (ed. J. O. Edwards), Wiley, New York, 1972, p. 266.
 (d) Lanza, S., Minnitti, D., Moore, P., Sachinidis, J., Romeo, R., and Tobe, M.L., *Inorg. Chem.*, **23**, 4428 (1984).
[40] Helm, L., Elding, L. I., and Merbach, A. E., *Helv. Chim. Acta*, **67**, 1453 (1984).
[41] Elding, L. I., *Acta. Chem. Scand.*, **24**, 1331 (1970).
[42] Helm, L., Elding, L. I., and Merbach, A.E., *Inorg. Chem.*, **24**, 1719 (1985).
[43] Groning, O., and Elding, L. I., *Inorg. Chem.*, **28**, 3366 (1989).
[44] Elmroth, S., Bugarcic, Z., and Elding, L. I., *Inorg. Chem.*, **31**, 3551 (1992).
[45] Elding, L. I., *Inorg. Chim. Acta*, **6**, 683 (1972).
[46] Elding, L. I., *Inorg. Chim. Acta*, **28**, 255 (1978).
[47] Elding, L. I., and Olsson, L.F., *Inorg. Chim. Acta*, **117**, 9 (1986).
[48] Ducummon, Y., Merbach, A. E., Hellquist, B., and Elding, L. I., *Inorg. Chem.*, **26**, 1759 (1987).
[49] Elding, L. I., and Groning, A. -B, *Inorg. Chim. Acta*, **31**, 243 91978).
[50] Hellquist, B., Elding, L. I., and Ducummon, Y., *Inorg. Chem.*, **27**, 3620 (1988).
[51] (a) Templeton, D. H., Watt, G. W., and Garner, C. C., *J. Am. Chem. Soc.*, **65**, 1608 (1943).
 (b) Droll, H. A., Block, B. P., and Fernelius, W. C., *J. Phys. Chem.*, 1000 (1957).
 (c) Burger, K., and Dryssen, D., *Acta Chem. Scand*, **17**, 1489 (1963).
 (d) Schukarev, S. A., Lobaneva, O. A., Ivanova, M. A., and Kovanova, M. A. *Zh. Neogr. Khim.*, **9**, 2791 (1964).
[52] Jorgensen, C. K., and Rasmussen, L. *Acta Chem Scand.*, **22**, 2313 (1968).
[53] Nabivanets, B. I., and Kalabina, L. V., *Russ. J. Inorg. Chem.*, **15**, 818 (1970).
[54] Faggiani, R., Lippert, B., Lock, C. J. L., and Rosenberg, B., *J. Am. Chem. Soc.*, **99**, 777 (1977) and *Inorg. Chem.*, **16**, 1192 (1977).
[55] Bugarcic, Z, PhD Thesis, Kragujevac University, 1989.
[56] Pell, R. G., Dodgen, H. W., and Hunt, J. P., *Inorg. Chem.*, **22**, 529 (1983).

[57] Muraveiskaya, G. S., Kukina, G. A., Orlova, V. S., Evstaf'eva, O. N., and Porai-Koshits, M. A., *Dokl. Akad. Nauk SSSR*, **226**, 76 (1976).

[58] Cotton, F. A., Falvello, L. R., and Han, S., *Inorg. Chem.*, **21**, 1709 (1982).

[59] Conder, H. L., Cotton, F. A., Falvello, L. R., Han, S., and Walton, R. A., *Inorg. Chem.*, **22**, 1887 (1983).

[60] (a) Nabivanets, B. I., Kalabina, L. V., and Kudritskaya, L. N., *Russ. J. Inorg. Chem.*, **16**, 1736 (1971).

      (b) Simanova, S. A., Nabivanets, B. I., Kutsyi, V. G., Kalabina, L. V., and Bash makov, V.I., *Appl. Chem. USSR*, **63**, 419 (1990).

[61] Heck, L., *Proceedings of the Fifth International Conference on the Platinum Group Metals*, St. Andrews, UK, 1993, p. A109.

[62] Tschugajeff, L., *Z. Anorg. Allg. Chem.*, **137**, 1 (1924).

Chapter 11

Group 11 Elements: Copper, Silver and Gold

11.1   COPPER

The elements of Group 11 have been collectively termed the 'coinage metals' reflecting their use in such materials. Although silver and gold have long since been superseded copper, in the form of durable corrosion-resistant alloys, still finds wide use. Copper is still extensively used in piping for household gas and water supplies and for electrical wiring. Similarly to nickel, also a coinage metal, it is quite unreactive to water even at fairly low pH, requiring high concentrations of CO_2 at low pH for significant corrosion to set in. This primarily stems from the metal's highly positive redox potential ($E^\theta(Cu^{2+}(aq)/Cu(s)) = +0.33$ V, 1 M H^+). Reduction of $Cu^{2+}(aq)$ to $Cu^+(aq)$ occurs at a more negative potential ($+0.15$ V), leading to an acute instability of the $Cu^+(aq)$ ion with respect to disproportionation (see below). Copper is also the third most abundant metal in the human body (80–120 mg) after iron and zinc, the metal essential for a range of enzymes and proteins involved in such diverse functions as O_2 transport and activation (tyrosinase, laccase, hemocyanin), electron transfer (plastocyanin, azurin), iron metabolism (cytochrome c oxidase) and the removal of superoxide ion (bovine superoxide dismutase). Maintaining the delicate balance of copper in the body is of crucial importance and diseases are known to be attributable to either a deficiency or an excess of the metal, the most well known of these being Wilson's disease involving deposition of the metal in the liver, brain and kidneys. Molybdenum is a known antagonist of copper uptake in ruminant animals, as mentioned earlier. The crucial job of transporting copper within the human blood is believed to be performed by serum albumin, which contains the same specific and strong binding site for Cu^{2+} as mentioned above for Ni^{2+}. Thus the serum albumin model tripeptide glygly-*l*-his has been successfully used as a treatment for Wilson's disease. The importance of tying up the copper during mobilisation points towards the acute toxicity of free Cu^{2+}, the only biologically relevant aqua species in water. The $Cu^+(aq)$ ion has only a transient existence in water, whereas free $Cu^{3+}(aq)$ is unknown in the absence of specific stabilising ligands, the most biologically important of these being N-deprotonated peptides, cf. Ni^{3+}. The rapid complexation of Cu^{2+} upon entry to the cell is probably of crucial importance and, as will be seen, this is indeed a feature of its solution chemistry. Biological aspects of the chemistry of copper and copper biochemistry have been extensively reviewed [1].

11.1.1   Copper(I) (d^{10}): The Cu^+ (aq) Ion

$Cu^+(aq)$ has only a fleeting existence in aqueous solution due to the highly favourable bimolecular disproportionation reaction, $K = \sim 10^6$ M^{-1}, to give $Cu^{2+}(aq)$ and Cu metal as well as the low solubility of the brick-red oxide Cu_2O:

$$2\,Cu^+(aq) \underset{}{\overset{K = 10^6 M^{-1}}{\rightleftharpoons}} Cu^{2+}(aq) + Cu(s) \qquad (11.1)$$

Despite the thermodynamic instability of $Cu^+(aq)$, arising from the much more effective solvation of the d^9 Cu^{2+} ion by water, the disproportionation reaction itself is quite slow in the absence if catalysis. Solutions containing 0.01 M $Cu^+(aq)$, stable for up to 10 hours, have been prepared in 0.1 M $HClO_4$ at 0 °C by the reaction of $[Cu^I(1,3\text{-cyclooctadiene})](ClO_4)$ in MeOH with $HClO_4$ followed by extraction of the diene [2]. A yellow Cu^I hydroxide can be precipitated from these solutions at high pH and if rapidly redissolved in $HClO_4$ can regenerate the metastable solutions of $Cu^+(aq)$ provided the hydroxide is rapidly dissolved and not allowed to contact the glass surface of the vessel which leads to efficient catalysis of the disproportionation. Attempts to similarly dissolve Cu_2O in $HClO_4$ lead to spontaneous disproportionation. $Cu^+(aq)$ is similar reported to persist for several hours following reduction of $Cu^{2+}(aq)$ under O_2-free conditions with $V^{2+}(aq)$ or $Cr^{2+}(aq)$ [3] $(E^\theta(Cu^{2+}(aq)/Cu^+(aq)) = 0.153$ V). Remarkably the value of K in equation (11.1) decreases to 10^{-21} M^{-1} [4] if the medium is changed to acetonitrile which effectively solvates Cu^I by forming the stable tetrahedral $[Cu(NCCH_3)_4]^+$ ion [5]. The ability of acetonitrile to stabilise Cu^I is further illustrated by the fact that only 2 CH_3CN molecules/Cu are required to suppress the disproportionation in water and moreover $[Cu(NCCH_3)_4]^+$ itself persists in water-acetonitrile mixtures provided that at least four CH_3CN molecules/Cu are present in the aqueous mixture (Fig. 11.1) [6, 7]. In studies with both Cu^I and Cu^{II} present it has been further shown that CH_3CN preferentially solvates Cu^I at the complete exclusion of water while the complete reverse is true for Cu^{II} [8]. Acetonitrile also uniquely stabilises Au^I in solution (see Section 11.3.1).

The properties of Cu^I–water–CH_3CN systems have been the subject of detailed study, not least because of the interest in stabilising Cu^I for hydrometallurgical applications due to the lower energy requirement for subsequent reduction to the metal. It has not been established whether there is a change in the coordination geometry at Cu^I when H_2O replaces CH_3CN in the coordination sphere. $[Cu(NCCH_3)_4]^+$ itself is well established as a useful lead-in reagent for the preparation of many other Cu^I compounds. Comproportionation to Cu^I is also favoured at high pH when the insolubility of the brick red Cu_2O promotes the reaction:

$$Cu(s) + Cu^{2+}(aq) + 2\,OH^- \underset{}{\overset{K = 10^{23} M^{-2}}{\rightleftharpoons}} Cu_2O + H_2O \qquad (11.2)$$

This has the consequence of limiting the concentration of $Cu^{2+}(aq)$ that can co exist with the metal as the pH is raised and limits the usefulness of the Cu^{2+}/Cu^0 electrode.

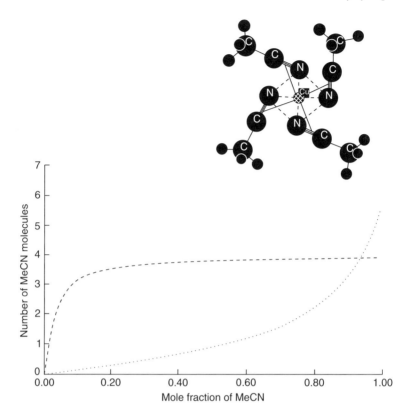

Figure 11.1. Preferential solvation of acetonitrile around Cu^+ (-----) and of water around Cu^{II} (············) as a function of mole fraction of acetonitrile in acetonitrile–water mixtures [8]. (Inset: solution structure of the $[Cu(NCCH_3)_4]^+$ ion deduced from $^{63,65}Cu$ NMR spin-lattice relaxation)

Redoxing between Cu^{II} and Cu^I is central to the function of many copper-dependent enzymes and proteins, e.g. rapid Cu^{II}/Cu^I electron transfer is established in the type 1 'blue copper' proteins stellacyanin, plastocyanin and azurin [1,9]. The different geometry preferences for Cu^I (tetrahedral) and Cu^{II} (tetragonally distorted octahedral or square), responsible for the rather slow rates of electron transfer found for many simple Cu^{II}/Cu^I couples, is overcome in the blue copper proteins by provision of an 'entatic' highly distorted tetrahedral geometry (almost trigonal) which favours neither oxidation state and as a result permits rapid electron transfer between each [9]. In addition Cu^I is involved in the reduced forms of hemocyanin and the monooxygenase tyrosinase and may be responsible for the EPR silent copper observed in several of the other multi-copper oxidases [1].

11.1.2 Copper(II) (d⁹): The Chemistry of the $[Cu(OH_2)_6]^{2+}$ Ion

11.1.2.1 PROPERTIES

The aqueous chemistry of copper is dominated by the blue $[Cu(OH_2)_6]^{2+}$ ion. It is well established that the very rapid rates of ligand complexation reactions on this species arises from a significant tetragonal distortion from regular octahedral as a result of the Jahn–Teller effect operating on the $t_{2g}^6 e_g^3$ configuration [10]. The characteristic electronic spectrum of $[Cu(OH_2)_6]^{2+}$ (Fig. 11.2) shows three overlapping transitions from the $^2B_{1g}$ ground state arising from the tetragonal distortion to the regular octahedral geometry. The extremely fast rate of water exchange observed ($k_{ex}(25\,°C) \sim 10^9\,s^{-1}$) [11, 12] is a direct consequence of the dynamic properties of the distortion, resulting in the presence of the two weakly bonded water ligands in axial positions at any time. A similar effect responsible

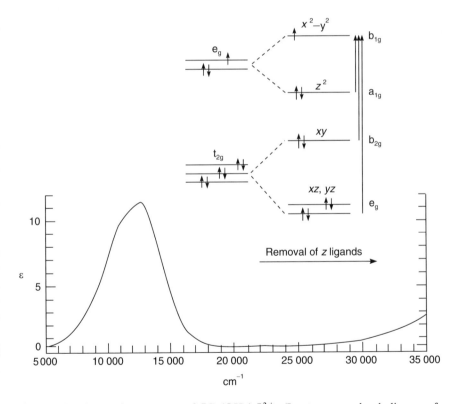

Figure 11.2. Electronic spectrum of $[Cu(OH_2)_6]^{2+}$. (Inset: energy level diagram for a tetragonally distorted octahedral field showing the observed transitions)

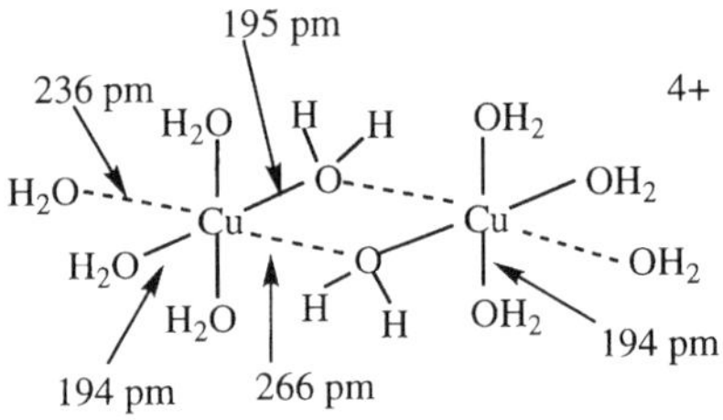

Figure 11.3. The dimeric $[Cu_2(OH_2)_{10}]^{4+}$ cation in the salt $[Cu_2(OH_2)_{10}][Cu(OH_2)_6]$-$(ZrF_7)_2$

for the fast exchange/substitution rates on $[Cr(OH_2)_6]^{2+}$ $(t_{2g}^3 e_g^1)$ is recalled (Section 6.1.1). Structural studies on $[Cu(OH_2)_6]^{2+}$, carried out in solution using X-ray techniques [13] and neutron diffraction [14], show the tetragonal distortion very clearly. The four equatorial waters are at a mean Cu—O distance of between 196 and 200 pm. However, the mean Cu—O distance to the two more distant axial waters has proved more difficult to ascertain due in part to interferences within the radial distribution function from the secondary solvation shell of water molecules and, in some cases, from the counter-ions. The novel dimeric Cu_2^{II} cation $[Cu_2(OH_2)_{10}]^{4+}$ is established in the salt $[Cu_2(OH_2)_{10}][Cu(OH_2)_6](ZrF_7)_2$. The crystal structure (Fig. 11.3) reveals four equatorial waters at a mean Cu—O distance of ~ 194 pm and two 'axial' waters at a mean Cu—O distance of ~ 250 pm [15]. The structure of $CuSO_4 \cdot 5H_2O$ has also been the subject of a recent reinvestigation (Fig. 11.4) and shows clearly the presence of the four short Cu—O bonds to the equatorial waters at 197 pm with

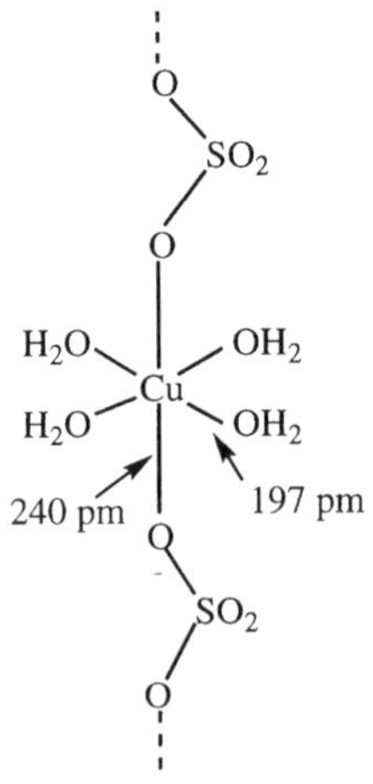

Figure 11.4. The coordination sphere around the Cu^{2+} centre in $CuSO_4 \cdot 5H_2O$

two elongated Cu—O bonds at 240 pm to SO_4^{2-} ions [16]. Finally, a recent crystal structure of $[Cu(OH_2)_4](SiF_6)_2$, isolated from the reaction of $Cu^{II}CO_3$ with hexafluorosillicic acid, shows evidence of a pronounced tetragonal distortion with an average Cu—$O(OH_2)$ (equatorial) distance of 195 pm and two elongated Cu—$F(SiF_6^{2-})$ (axial) interactions at 234 pm [17].

11.1.2.2 WATER EXCHANGE AND THE DYNAMIC JAHN–TELLER EFFECT ON $[Cu(OH)_2)_6]^{2+}$

The precise nature of the Jahn–Teller effect itself has been the subject of much discussion and it appears that two distinct situations are relevant from solid state studies: a fluxional elongated rhombic octahedral distortion (temperature variable) and a static elongated rhombic octahedral distortion (non-temperature variable). In either case the result is an elongated along two of the bonds with respect to the other four. $[Cu(OH_2)_6]^{2+}$ belongs to the former category and, in both solid state and solution samples, the ion has been the subject of a number of detailed studies [18, 19]. Potential energy surfaces and the energy barrier to fluxionality in a number of the Tutton-type salts, $M_2[Cu(OH_2)_6]$-$(SO_4)_2(M = NH_4^+, Cs^+, Na^+)$, have been determined [19, 20]. A recent ^{17}O NMR study of the dynamics of the water molecules on aqueous $[Cu(OH_2)_6]^{2+}$ has been carried out using a theory developed for the interpretation of the scalar relaxation in the presence of the Jahn–Teller inversion and the water exchange between the aqua ion and bulk water [12]. A model wherein the inversion was several orders of magnitude faster than water exchange proved to be that which was most consistent with the experimental results. Thus all water ligand molecules on $[Cu(OH_2)_6]^{2+}$ are effectively equivalent on the timescale of the water exchange. Other models ignoring the influence of the inversion were inconsistent with the observed T_1 values [12]. The inversion (Fig. 11.5) is a picosecond process

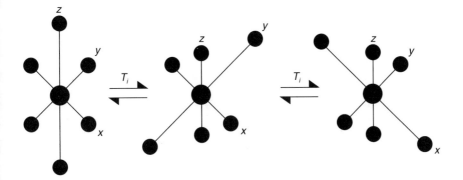

Figure 11.5. The Jahn–Teller inversion process on Cu^{2+} leading to an equivalence of all ligand positions on the picosecond timescale

characterised by a lifetime of $(5.1 \pm 0.6) \times 10^{-12}$ s and an activation energy of (3.5 ± 1.5) kJ Mol^{-1}. The water-exchange process is characterised by a rate constant, $k_{ex}(25\,°C)$, of $(4.4 \pm 0.1) \times 10^9$ s^{-1} and activation parameters, $\Delta H^{\ddagger}_{ex} = (11.5 \pm 0.3)$ kJ mol^{-1}, $\Delta S^{\ddagger}_{ex} = (-21.8 \pm 0.9)$ J K^{-1} mol^{-1} and $\Delta V^{\ddagger}_{ex} = (+2.0 \pm 1.5)$ cm^3 mol^{-1} [12], the value of $\Delta V^{\ddagger}_{ex}$ for $[Cu(OH_2)_6]^{2+}$, is somewhat less than found for methanol solvent exchange on $[Cu(CH_3OH)_6]^{2+}$ $(+8.3$ cm^3 mol$^{-1})$, for which a negative $\Delta S^{\ddagger}_{ex}$ value $(-44.0$ J K^{-1} mol$^{-1})$ is also obtained [21]. These results have been interpreted in terms of a dissociatively activated process, thus continuing the trend established along the first row. The smaller $\Delta V^{\ddagger}_{ex}$ for $[Cu(OH_2)_6]^{2+}$ vs that found for $[Ni(OH_2)_6]^{2+}$ $(+7.2$ cm^3 mol$^{-1})$ is believed to reflect the lengthening of the axial Cu—OH$_2$ bonds within the tetragonally distorted structure; i.e. the ground state structure for $[Cu(OH_2)_6]^{2+}$ now resembles a transition state already well along the reaction coordinate towards dissociative exchange of a water molecule. Rates of formation of complexes on Cu^{2+}(aq) are technically difficult to measure. Temperature jump methods have been widely employed. A fairly constant $k_{Cu^{2+}}/k_{Ni^{2+}}$ ratio for reactions with a wide variety of ligands is supportive of a common dissociative mechanism [22].

11.1.2.3 [Cu(TREN)(OH$_2$)]$^{2+}$

The rate constant for water exchange on this complex is 2.5×10^5 s^{-1}, some four orders of magnitude slower than that for the tetragonally distorted octahedral $[Cu(OH_2)_6]^{2+}$. The reason may well lie in the change in geometry to trigonal bipyramidal. This geometry is suggested from the X-ray structure of [Cu(tren)(NCS)](NCS). Interestingly, significantly negative activation volumes are relevant for both water exchange and anation/aquation reactions involving pyridine and substituted pyridines on [Cu(tren)(OH$_2$)]$^{2+}$ (range -4.7 to -10 cm^3 mol^{-1}). Here the trigonal bipyramidal arrangement may be responsible for a greater ligand field activation barrier towards operation of the normal dissociative mechanism, allowing an alternative more associatively activated path to compete successfully [23].

11.1.2.4 COMPLEXATION BY AMINO ACIDS AND PEPTIDES ON [Cu(OH$_2$)$_6$]$^{2+}$

Amongst the numerous complexes formed by Cu^{2+} are well defined and stable complexes with almost all known amino acids and many oligopeptides. As with Co^{2+} and Ni^{2+}, formation constants for many of these complexes have been measured and a number of reviews have appeared [1, 23]. The highest formation constants for 1:1 and 2:1 complexes are with histidine ($\log \beta_1 = 10.16$, $\log \beta_2 = 18.1$), 2,3-diaminopropionic acid (DAPA) ($\log \beta_1 = 10.62$, $\log \beta_2 = 19.81$) and 2,4-diaminobutyric acid (DABA) ($\log \beta_1 = 10.62$, $\log \beta_2 = 18.61$), paralleling the behaviour observed for Co^{2+} and Ni^{2+}. The higher formation constants observed for Cu^{2+} reflect the strong in-plane bonding within the tetragonally distorted octahedral geometry. As a result, chelated 3:1 complexes with, for example, glycine and alanine are unknown. Cysteine is oxidised to the disulphide in the presence of Cu^{2+}. Complexes with N donor ligands are generally more stable than those with O donors. As observed with Pt^{II}, particularly stable Cu^{II} oligopeptide complexes are characterised by strong in-plane bonding to deprotonated peptide N atoms. For Cu^{II} the strongest tend to be those containing histidine residues, reflecting strong bonding to the imidazole N atom. Peptide and other ligand complexes containing deprotonated thiol groups have provided potential models for 'blue copper' sites. Unfortunately, oxidation of the thiol group to the disulfide is a frequent problem. However, several stable complexes of Cu^{II} with the peptides α-mercaptopropionylglycine and α-mercaptopropionyl-*l*-cysteine have been reported, with the former reproducing many of the blue copper spectroscopic parameters [24]. It is possible that the presence of S-donor ligands on Cu is just as important as the distorted 'entatic' tetrahedral geometry for promoting rapid electron transfer between Cu^{II} and Cu^{I} within the blue copper sites.

11.1.2.5 HYDROLYSIS OF $[Cu(OH_2)_6]^{2+}$

Hydrolysis of $[Cu(OH_2)_6]^{2+}$ is not extensive prior to precipitation of hydrous $Cu(OH)_2$, thus paralleling the behaviour observed for both $[Co(OH_2)_6]^{2+}$ and $[Ni(OH_2)_6]^{2+}$. Indeed, there has been some controversy as to the precise nature of the species involved. Pedersen's experimental results were interpreted as involving three species, $CuOH^+$ ($pK_{11} = 7.97$), $Cu_2(OH)^{3+}$ ($pK_{21} = 6.82$) and $Cu_2(OH)_2^{2+}$ ($pK_{22} = 10.89$) [25]. The formation of dimeric species is generally agreed [25–27] to occur above a pH of 5, reaching a maximum at a pH of 9 prior to precipitation of $Cu(OH)_2$ at around a pH of 10. However, in a later potentiometric study formation only of the dimer $Cu_2(OH)^{3+}$ ($pK_{21} = 5.75 \pm 0.1$) was claimed [26]. At a pH above 10, formation of $[Cu(OH)_4]^{2-}$ ($pK_{14} = 39.6$) and possibly $[Cu(OH)_3]^-$ ($pK_{13} < 27.8$) also parallels the behaviour of the preceding two metals of the first row. The stability of $[Cu(OH_2)_6]^{2+}$ to hydrolysis and hence mobility in the physiological pH range, coupled with its tendency to undergo rapid complexation reactions with a wide range of biological donor sites, are probably responsible for the toxicity caused by appreciable levels of free copper within the body.

11.1.2.6 COMPLEXATION REACTIONS ON THE HYDROXYCUPRATE IONS $[Cu(OH)_3]^-$ AND $[Cu(OH)_4]^{2-}$

Reaction rates involving complex formation on both $[Cu(OH)_3]^-$ and $[Cu(OH)_4]^{2-}$ are slower than those occurring on $[Cu(OH_2)_6]^{2+}$ and as such have proved amenable to study by stopped-flow methods (typical k_f ($25°$ C) $= 10^3$–

$10^7 \, M^{-1} s^{-1}$). Furthermore, complexation reactions with, for example, poly-amines are simplified by the lack of competing ligand protonation processes in the strongly basic media employed (up to 0.5 M in OH^-). Of particular interest has been the study of reactions involving complexation with a number of tetraaza macrocyclic ligands which have given insights into the stepwise processes involved. The original studies were reported by Rorabacher and Margerum and coworkers [28] who proposed a mechanism (Scheme 11.1) involving initial interaction of an N donor at an axial position (k_1/k_{-1}) followed by Jahn–Teller inversion to the basal plane. Further substitution of N donors for OH(H) then occurs in the basal plane to form the final planar Cu^{II} complex. The rate-determining step for simple unsubstituted macrocycles appears to be involved with the inversion step (k_2/k_{-2}) prior to the second Cu—N bond formation. Steric effects (substituents at both N and C on the ring) are found to shift the rate-determining step to the second bond formation (k_3/k_{-3}), particularly for reactions on $[Cu(OH)_4]^{2-}$ [29]. Further work has appeared to show that the rate-determining step is shifted to the second bond formation for both hydroxycuprate complexes upon the introduction of alkyl substituents at both N and C [30]. These effects were also seen for open-chain ligands. In all cases complexation on $[Cu(OH)_3]^-$ is faster than on $[Cu(OH)_4]^{2-}$ by in some cases > 300 times. In a more recent study the rate of complexation is found to increase with increasing ring size of the tetraaza macrocycle [31].

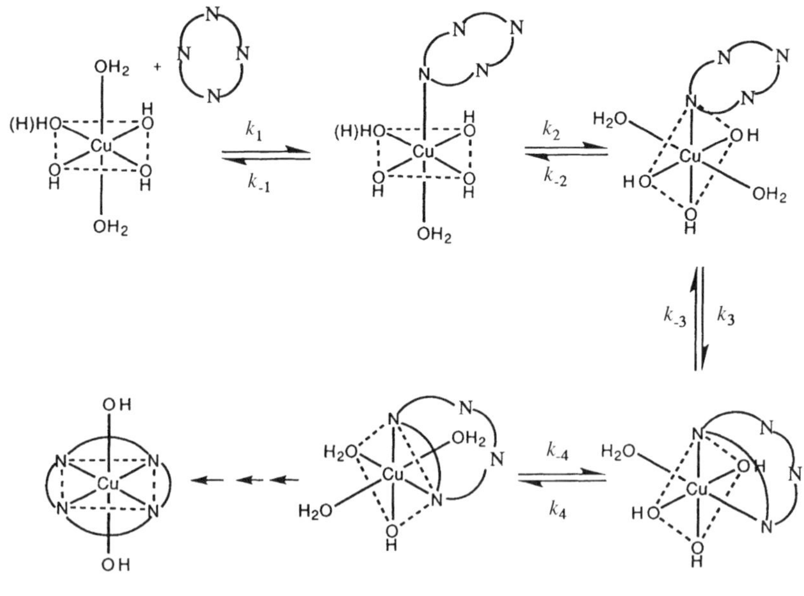

Scheme 11.1. Mechanism for the complexation of tetrazamacrocycles on hydroxycuprate (II) ion in strongly basic media.

11.1.3   The Cuboidal $[CuMo_3S_4(OH_2)_{10}]^{4+/5+}$ Aqua Ions

In contrast to the $NiMo_3S_4$ cube ion (Chapter 10), the corresponding $CuMo_3S_4$ cube has now been characterised in two oxidation states. The $4+$ cube crystallises from Hpts solutions as the edge-shared double-cube aqua ion $[(H_2O)_9Mo_3S_4CuCuS_4Mo_3(OH_2)_9](pts)_8 \cdot 2H_2O$ (Chapter 6) [32] but exists as a monomeric species in solution in equilibrium with the dimer [33]. The monomeric $5+$ cube is observed as an intermediate in the stepwise $1e^-$ oxidation of the $4+$ cube to $Cu^{2+}(aq)$ and $[Mo_3S_4(OH_2)_9]^{4+}$. Both inner- and outer-sphere pathways operate for this oxidation process. The $4+$ cube is prepared by treatment of $[Mo_3S_4(OH_2)_9]^{4+}$ with either Cu metal [32] or with $Cu^{2+}(aq)$ in the presence of BH_4^- [33]. In each case the product is purified using Dowex 50 W X2 cation-exchange column chromatography under rigorous O_2-free conditions. The $5+$ cube can, however, be prepared directly by treatment of $[Mo_3S_4(OH_2)_9]^{4+}$ with $Cu^+(aq)$, $k_1(25\,°C) = 980\,M^{-1}s^{-1}$. The $Cu^+(aq)$ here is prepared by $[Cr(OH_2)_6]^{2+}$ reduction of $[Cu(OH_2)_6]^{2+}$ in $HClO_4$ and standardised via the amount of Cr^{2+} consumed. Alternatively, a suspension of CuCl can be used for preparations in Hpts solution. The Cu^+ reaction occurs in two stages (Scheme 11.2). Initial attachment of the Cu^+ to one μ-S group is followed by reorientation to establish contact with the remaining two μ-S groups in a Cu^+-independent process ($k_2(25\,°C) = 15.2\,s^{-1}$). Both $4+$ and $5+$ cubes are O_2 sensitive and can conveniently be assayed by deliberate O_2 oxidation to a 1:1 mixture of $[Mo_3S_4(OH_2)_9]^{4+}$ and $Cu^{2+}(aq)$. The $5+$ cube has a characteristic absorption band at 472 nm ($\varepsilon = 2200\,M^{-1}\,cm^{-1}$) whereas the $4+$ cube absorbs with additional peaks at 325 nm ($3700\,M^{-1}\,cm^{-1}$) and in the NIR at 975 nm ($473\,M^{-1}\,cm^{-1}$). The 975 nm band is only observed from pts$^-$ solutions and, in the absence of a possible assignment to amounts of the edge-shared double-cube in equilibrium, has been assigned to pts$^-$ complexation. It is recalled, however, that the edge-shared double cube only crystallises from pts$^-$ solutions [32].

Rapid 1:1 complexation of Cl^- ($t_{1/2} < 3\,ms$) is observed at the tetrahedral Cu site ($K = 3500\,M^{-1}$) which may be compared with a K of $500\,M^{-1}$ for 1:1 complexation at $Cu^+(aq)$ [34]. Assignment of a definitive oxidation state to the

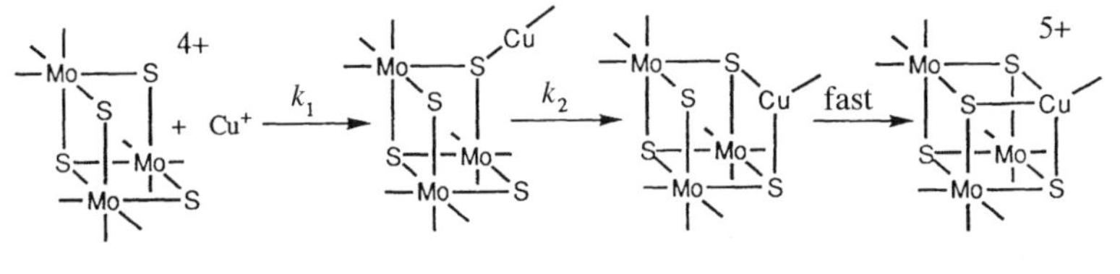

Scheme 11.2. Reaction of $Cu^+(aq)$ with $[Mo_3S_4(OH_2)_9]^{4+}$ to give $[CuMo_3S_4(OH_2)_{10}]^{5+}$

Cu centre in these compounds is difficult, although the similarity of spectra would seem to suggest a constant average oxidation number for the Mo_3S_4 moiety and a change therefore from effectively Cu^{II} to Cu^I. As a result of these findings the ready displacement of Ni and Fe from their hetero metal $[Mo_3MS_4(OH_2)_{10}]^{4+}$ ions by Cu^{2+}, reported to give the $4+$ Cu cube [35], has been assigned to an electron transfer mechanism involving firstly generation of $Cu^+(aq)$ and $[Mo_3S_4(OH_2)_9]^{4+}$ and then their subsequent reaction to generate the $5+$ cube. Further careful experiments have indeed established that the $5+$ cube is the final Cu-containing product of these reactions [33].

11.1.4 Copper(III) (d^8): The Existence of a Cu^{3+}(aq) Ion

A transient $Cu^{III}(aq)$ species can be generated by pulse radiolysis of solutions of $[Cu(OH_2)_6]^{2+}$ [36]. In neutral solutions the species is characterised by an absorption maximum at 290 nm ($\varepsilon \sim 5700\ M^{-1}cm^{-1}$) and a second-order decomposition pathway giving $[Cu(OH_2)_6]^{2+}$ and H_2O_2. In acid solution the decomposition is first order and may involve the regeneration of OH radicals. The kinetic parameters for the relevant reactions are given by

$$Cu^{2+}(aq) + OH \underset{k_{-1} = (4.2 + 1.4)\times 10^4\ s^{-1}\ (pH\ 3.5)}{\overset{k_1 = (3.1 + 0.3)\times 10^8\ M^{-1}s^{-1}}{\rightleftharpoons}} Cu^{III}(aq) \qquad (11.3)$$

$$Cu^{III}(aq) + OH \xrightarrow[\text{fast}]{} Cu^{2+} + H_2O_2 \qquad (11.4)$$

$$2\,OH \xrightarrow[\text{fast}]{} H_2O_2 \qquad (11.5)$$

$$Cu^{III}(aq) + Cu^{III}(aq) \xrightarrow[k_2 = (2.3 + 0.3)\times 10^7\ M^{-1}s^{-1}\ (neutral)]{} 2\,Cu^{2+} + H_2O_2 \qquad (11.6)$$

In a later study, $Cu(OH)_3$ was suggested as the predominant species at a pH of $\geqslant 4$ with a half-life of ~ 2 ms at a pH of 4.4 [37]. $Cu(OH)_3(aq)$ is reported to have an absorption maximum at 280 nm ($\varepsilon = (7.4 \pm 0.7) \times 10^3\ M^{-1}cm^{-1}$) with $Cu(OH)_2^+(aq)$ quoted as having a maximum at 290 nm ($\varepsilon = 3.5 \pm 0.3) \times 10^3\ M^{-1}cm^{-1}$) [37]. A yellow dimeric Cu^{III} oxo-hydroxo species has been proposed as an intermediate in the $Cu(OH)_2$ catalysed decomposition of OCl^- and OBr^- [38]. Yellow soluble species have long been identified as resulting from solutions of $Cu^{2+}(aq)$ following treatment with OCl^- [39]. Many workers have formulated the species as $[Cu(OH)_4]^-$ but there appears little definitive proof [40]. A square planar donor set providing a strong ligand field would be expected to stabilise Cu^{III} and this is indeed the case. Of importance biologically are N-deprotonated peptides and as such a role for Cu^{III} in a number of the copper enzymes is a distinct possibility. The strongest indication is provided from EPR measurements on galactose oxidase [41]. Certain Cu^{II}–peptide complexes, like those on Ni^{II} undergo ready decarboxylation reactions almost certainly promoted by Cu^{III} intermediates. The solution chemistry and possible biological importance of Cu^{III} has been considered in an excellent review by Margerum and Owens [42].

11.2 SILVER

The aqueous chemistry of silver encompasses three oxidation states. Interest in detailing the solution/hydrolysis behaviour of soluble silver aqua species has stemmed from possible spin-offs for hydrometallurgical applications. However, there is only one stable oxidation state in water (Ag^{I}) with Ag^{II} (d^9) and Ag^{III} (d^8) both powerful oxidants, the oxidising power of Ag^{II} contrasting sharply with Cu^{II}(aq). The stable Ag^+(aq) ion is well established. Similarities to the alkali metal cations have been noted. A cationic Ag^{2+}(aq) ion exists as a highly reactive species in acidic solution whereas Ag^{III}, as the yellow square planar tetrahydroxo ion $[Ag(OH)_4]^-$, exists in strongly alkaline solutions. The oxidising properties of both Ag^{II}(aq) and Ag^{III}(aq) have been the subject of a number of detailed kinetic studies.

11.2.1 Silver(I) (d^{10}): Chemistry of the $[Ag(OH_2)_4]^+$ Ion

The well-known insolubility in water of many Ag^{I} compounds points to rather poor hydration of the Ag^+ ion and indeed many insoluble Ag^{I} salts, e.g. AgCl, AgBr, etc., are anhydrous. However, highly water-soluble Ag^{I} salts are found with the anions F^-, NO_3^- and ClO_4^-, with the latter showing hygroscopicity. The Ag^+(aq) ion itself has been the subject of much debate as to its precise hydration number. Early reports from X-ray diffraction studies favour a value of two [43]. More recently values ranging from 0.7 (NMR) [44] to between 3 and 4 (conductivity [45] and compressibility [46] studies) have been claimed. Later reports favour the value of 4 [47]. Finally, in 1984 the results of an EXAFS study conducted on solutions of Ag^+(aq) in nitrate and perchlorate solutions, up to 9.0 M in anion, favour a hydration number of 4 in a presumed tetrahedral arrangement (Fig. 11.6) [48]. The mean Ag—O(OH_2) distance was ~ 240 pm. In the nitrate solutions evidence for inner-sphere coordination of one NO_3^- ion within the primary hydration sphere was found over all concentrations of the anion investigated (Fig. 11.7). Similar coordination of one ClO_4^- was also found but only at high concentrations ~ 9.0 M.

Such strong evidence for anion coordination was presumed a likely explanation for the variable and sometimes lower hydration numbers found in the earlier studies. The ionic radius of Ag^+ (114 pm) is similar to that of Na^+ (113 pm) [49] and obvious similarities to the alkali metal cations have been noted. As might be expected, complexation reactions on Ag^+ are extremely facile, with typical rate

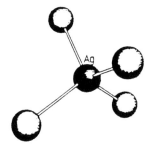

Figure 11.6. Tetrahedral four-coordinated structure for $Ag^+(aq)$

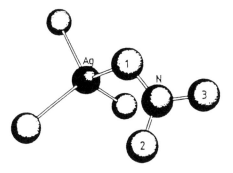

Figure 11.7. Inner-sphere coordination of nitrate to $Ag^+(aq)$

constants $\sim 10^6\,M^{-1}\,s^{-1}$ [50]. Nonetheless, these rates are significantly slower than those occurring on comparable alkali metal cations, the somewhat slower metal ion dissociation rates being responsible for the greater stability of many Ag^I cryptates vs their alkali metal counterparts [51]. As observed with Cu^I, preferential solubility of Ag^I by CH_3CN in water–CH_3CN mixtures is observed, reflecting the low affinity of Ag^I for H_2O [8]. However, the rise in the number of coordinated CH_3CN molecules with mole fraction of CH_3CN is more gradual than in the case of Cu^I, suggesting that H_2O competes more effectively at Ag^I, reflective of the existence of a stable $M^+(aq)$ ion for this element. Tetrahedral $[Ag[NCCH_3)_4]^+$ appears to be the relevant species in solution at high mole fractions of CH_3CN.

The hydrolysis of $[Ag(OH_2)_4]^+$ has been studied with formation of $AgOH$ characterised by a pK_{11} value of (12.0 ± 0.3) [52]. $AgOH(s)$ is amphoteric and in more strongly alkaline solution $[Ag(OH)_2]^-$ ($pK_{12} = 24.85$) [53] is formed. The pK_{12} value was based upon measurements of the solubility of Ag_2O in both neutral and alkaline solution. The equilibrium between Ag^+ and OH^- to give $\frac{1}{2}Ag_2O$ and $\frac{1}{2}H_2O$ is characterised by a log K value of 7.7 [54]. No evidence has been found for polynuclear Ag^I species in these solutions.

11.2.2   Silver(II): Chemistry of the Ag^{2+}(aq) Ion

$[Ag(OH_2)_4]^+$ is a well-established catalyst of a range of aqueous substrate oxidations by thermodynamically strong but kinetically sluggish reactants such as peroxodisulfate, $S_2O_8{}^{2-}$. It is now known that these reactions involve the generation of the highly reactive oxidant Ag^{2+}(aq). The nature of the Ag^{2+}(aq) solutions and the mechanisms involved in these catalytic oxidation reactions remain somewhat uncertain, despite a number of detailed investigations [55].

Early descriptions of attempts to prepare solutions of Ag^{2+}(aq) were given by Noyes, Hoard and Pitzer [56]. Some of the earliest methods involved dissolution in nitric acid of the black anodic deposits of AgO, obtained during the electrolysis of neutral $AgNO_3$ solutions and the oxidation of Ag^I compounds by $S_2O_8{}^{2-}$ [57], O_3 [56, 58], PbO_2, $BiO_3{}^-$ or F_2 [57]. The most popular method has proved to be the dissolution of freshly precipitated AgO, via treatment of a concentrated aqueous solution of $AgNO_3$ with $K_2S_2O_8$, in the appropriate non-complexing acid of choice [57]. The reaction of OH radicals, generated by pulse radiolysis, with Ag^+(aq) solutions has also been used [37, 59]. The generally agreed mechanism for generation of Ag^{2+}(aq) from Ag^+(aq) and $S_2O_8{}^{2-}$ is given by

$$Ag^+ + S_2O_8{}^{2-} \xrightarrow{\text{slow}} Ag^{2+} + SO_4{}^{2-} + SO_4{}^{-\cdot} \qquad (11.7)$$

$$Ag^+ + SO_4{}^{-\cdot} \xrightarrow{\text{fast}} Ag^{2+} + SO_4{}^{2-} \qquad (11.8)$$

Solutions of Ag^{2+}(aq) in 1.5–6.0 M $HClO_4$ show an absorption maximum around 475 nm ($\varepsilon = 140\,M^{-1}cm^{-1}$) and a shoulder around 575 nm [58] (Fig. 11.8). In more concentrated $HClO_4$ solution (6.0–11.5 M) an additional maximum at 275 nm is found which is sensitive to the $[H^+]$ [60]. The 275 nm maximum ($\varepsilon = 5400\,M^{-1}cm^{-1}$) is also observed from solutions of Ag^{2+}(aq) generated by pulse radiolysis [37]. In 1.5–6.0 M nitric acid solution different features are apparent (maximum around 380–400 nm, $\varepsilon = 2300\,M^{-1}cm^{-1}$) which are sensitive to the nitric acid concentration suggesting complexation of Ag^{2+}(aq) by $NO_3{}^-$ [58] (Fig. 11.9). Some of the findings in $HClO_4$ were similarly interpreted as suggesting $ClO_4{}^-$ complexation [57].

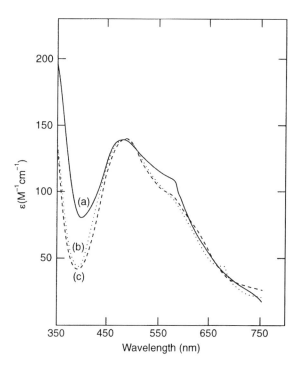

Figure 11.8. Electronic spectrum of Ag^{2+}(aq) in varying perchloric acid solutions: (a) 6.0 M, (b) 3.0 M, (c) 1.5 M

11.2.2.1 HYDROLYSIS OF AG^{2+}(AQ)

Highly conflicting results seem to be apparent from attempts to measure the hydrolysis constants for Ag^{2+}(aq). Honig and Kustin reported a hydrolysis constant for the generation of $AgOH^+$ from Ag^{2+} of 0.69 M at 22.5 °C [61]. Values between 0.1 and 0.7 M, at $I = 5.6$ M, have been claimed by other groups [62]. A much smaller K_{11} value, $10^{-5.35}$ M, has been claimed by Asmus *et al.* [37] from the solutions generated via pulse radiolysis. They further claimed the existence of both $AgOH^+$ and $Ag(OH)_2$ in solutions at a pH of > 4. It was subsequently found that oxo anion complexation on Ag^{2+}(aq) is significant and moreover can control the rate and mechanism of reduction of the ion by H_2O [58, 63, 64]. It remains likely therefore that acute anion complexation effects are responsible for the conflicting hydrolysis data. Both Ag^{2+} and $AgOH^+$ appear to be relevant to many oxidation reactions carried out by Ag^{II}(aq), but none of the studies appears to have shed further light on the precise value of the hydrolysis constant.

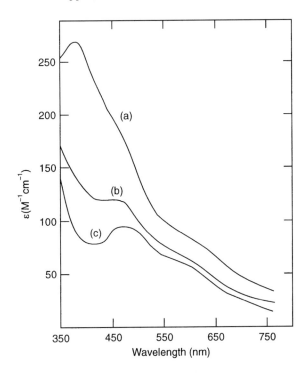

Figure 11.9. The effect of added nitric acid on the spectrum of Ag^{2+}(aq) in 3.0 M perchloric acid: (a) 0.3 M, (b) 0.12 M, (c) 0.06 M

The apparent 'splitting' observed in the low-energy electronic bands of Ag^{2+}(aq) (Fig. 11.6) has been argued as evidence in favour of a Jahn–Teller distorted octahedral geometry [58] (Fig. 11.10) similar to that for Cu^{2+}(aq). However, in view of the uncertainty as to the hydrolysis constant and the degree of anion coordination in Ag^{2+}(aq) solutions these remain premature deductions. Axial coordination of the anions (Fig. 11.10) is a distinct possibility to explain the spectral changes observed in the various media. Many square planar complexes of Ag^{II} are also known, e.g. $[Ag(py)_4]^{2+}$, $[Ag(phen)_2]^{2+}$ and $[Ag(bipy)_2]^{2+}$. Especially pertinent examples are those formed by tetraazamacrocyclic ligands which can stabilise Ag^{II} to the extent of disproportionating Ag^{I}, giving rise to metallic Ag (mirror) and the Ag^{II} complex.

Reduction potentials for the Ag^{2+}/Ag^+ couple vary in the different media as expected (Table 11.1). Higher E^θ values are found to be relevant to the solutions of Ag^{2+} in $HClO_4$, where anion complexation may be less extensive. The potential obtained in 4 M $HClO_4$ solutions at 25 °C (2.00 V) tends to be the one most quoted. The values reflect the strongly oxidising nature of Ag^{2+}(aq) solutions.

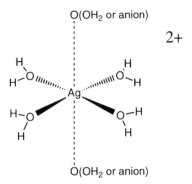

Figure 11.10. Possible coordination geometry for the Ag^{2+} aqua ion in perchloric acid solution

Table 11.1. Reduction potential for the couple Ag^{2+}/Ag^+

E^θ (V vs NHE)	Temperature ($^\circ$C)	Medium	Ref.
1.914 ($\pm$ 0.002)	0	1.0–4.0 M HNO_3	[65]
1.932	25	4.0 M HNO_3	[66]
2.000	25	4.0 M $HClO_4$	[66]
1.978	−5	6.5 M $HClO_4$	[67]
1.453	25	H_2SO_4	[68]

11.2.2.2 OXIDATIONS BY AG^{2+}(AQ)

A convenient assay developed for determining Ag^{2+}(aq) solutions has made use of the oxidation of Ce^{III}(aq) solutions to Ce^{IV}(aq) [58]. Solutions of Ag^{2+}(aq) are, however, metastable owing to spontaneous reduction by H_2O. For a study carried out in ClO_4^- media, the following mechanism, involving disproportionation to generate Ag^{III}(aq), was proposed to account for the observed rate law [58]:

$$2\,Ag^{2+} \underset{}{\overset{K}{\rightleftharpoons}} Ag^+ + Ag^{3+} \tag{11.9}$$

$$Ag^{3+} + H_2O \underset{}{\overset{K_1}{\rightleftharpoons}} AgO^+ + 2\,H^+ \tag{11.10}$$

$$AgO^+ \underset{RDS}{\overset{k}{\longrightarrow}} Ag^+ + 1/2\,O_2 \tag{11.11}$$

$$\frac{-d[Ag^{2+}]}{dt} = \frac{kKK_1[Ag^{2+}]^2}{[Ag^+][H^+]^2} \tag{11.12}$$

In addition, an apparent $[ClO_4^-]^2$ was noted. Wells [69] argued subsequently that a mechanism involving the generation of dimeric Ag^{2+}(aq) species could also explain the $[Ag^{2+}]^2$ dependence. The involvement of $AgOH^+$ was also considered [69]. However, other evidence [56, 70] points towards the relevance of the disproportionation process. A more complicated mechanism seems relevant to the reduction of Ag^{2+}(aq) by water in HNO_3 solutions [63, 64]. Overall, considerable evidence has been gathered pointing to anion complexes in solutions of Ag^{2+}(aq), even in the case of ClO_4^-, and these are involved to differing degrees in the mechanisms for water oxidation. By working at high $[H^+]$ and by having excess Ag^I present, however, the reduction of Ag^{2+}(aq) by H_2O can be suppressed sufficiently so as to allow detailed study of the oxidation of other substrates.

For a wide range of Ag^I catalysed $S_2O_8^{2-}$ oxidations [71] a constant value of k (25 °C) ($(6 \pm 3) \times 10^3\ M^{-1}\ s^{-1}$) is found in support of the common rate-determining process (11.7), the generation of the active Ag^{2+}(aq) species. In many cases this has been verified by observation of faster rates for oxidation by independently generated Ag^{2+}(aq) [72, 73]. Under the strongly acidic reaction conditions typically employed, 1.0–6.0 M $HClO_4$, fully protonated Ag^{2+} was expected to be the main constituent species [37, 61, 62]. For the oxidation of Fe^{2+}(aq), Co^{2+}(aq) and VO^{2+}(aq) the rate was independent of $[H^+]$, suggesting Ag^{2+}(aq) as the dominant reactive species. In many other oxidations by Ag^{2+}(aq), however, the observed $[H^+]$ dependence is suggestive of an involvement from $AgOH^+$(aq). In their review [55] Mentasti, Baiocchi and Coe considered the data from a number of $[H^+]$-dependent reactions. In the absence of a seemingly reliable value for K_{11} (Ag^{2+}(aq)) few definitive conclusions could be reached since many of the reactions were with substrates also exhibiting acid dissociation within the same pH range, resulting in a proton transfer ambiguity.

11.2.3 Silver(III) (d^8): Chemistry of the $[Ag(OH)_4]^-$ Ion

11.2.3.1 SYNTHESIS AND CHARACTERISATION

Ag^{III}(aq) has only a fleeting existence in acidic aqueous solution, as mentioned above in the context of its involvement in the autoreduction of Ag^{2+}(aq) by H_2O. However, in alkaline solution d^8 Ag^{III}(aq) is well characterised as the square planar tetrahydroxo ion $[Ag(OH)_4]^-$. Early reports describe stable square planar Ag^{III} complexes such as $[Ag(IO_4(OH)_2)_2]^{3-}$ and $[Ag(TeO_2(OH)_4)_2]^-$ isolated following electrolytic [74] or chemical ($S_2O_8^{2-}$) [75] oxidation of alkaline Ag^I solutions in the presence of periodate and tellurate. Similar square planar Cu^{III} complexes are known. A saturated alkaline solution of AgO can also be used [76]. These findings finally led to recognition of the disproportionation reaction, (11.9), which was promoted in alkaline solution, and this led to AgO being itself reformulated as $Ag^I Ag^{III} O_2$. Finally, in 1968 Cohen and Atkinson [77]

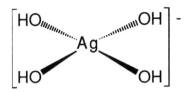

Figure 11.11. Square-planar structure for the tetrahydroxoargenate(III) anion

characterised the alkaline Ag^{III} species as the $[Ag(OH)_4]^-$ ion (Fig. 11.11). The method of choice for its preparation has proved to be the electrolytic oxidation in aqueous alkali of a silver foil anode usually in conjunction with a Pt cathode [78, 79]. In a typical procedure [78] approximately $200\,cm^3$ of a $1.2\,M$ NaOH solution is placed in a covered $250\,cm^3$ polyethylene beaker, the cover being

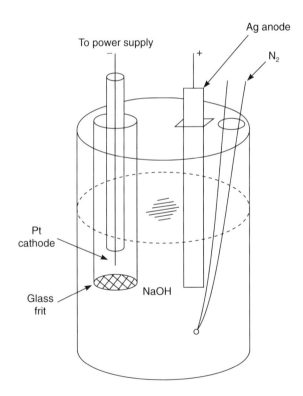

Figure 11.12. Typical electrolysis cell for the preparation of $[Ag(OH)_4]^-$

equipped with appropriate ports to accommodate the silver anode, a 1 cm wide × 5 cm long strip of silver foil, a fritted Pt cathode compartment and a glass capillary for purging the solution with N_2 gas. A typical cell design is illustrated in Fig. 11.12. Electrolysis is carried out at a constant current of 0.7 A for 35–60 min. At the end of this period the solution is a deep yellow colour and, following filtration, can be adjusted as required with amounts of NaOH and/or $NaClO_4$. The single broad absorption maximum at 267 nm ($\varepsilon = 1.17 \times 10^4\,M^{-1}\,cm^{-1}$) of $[Ag(OH)_4]^-$ (Fig. 11.13) [78, 79] serves for standardisation purposes. Solutions of $[Ag(OH)_4]^-$ are metastable and decompose at 25 °C with a half-life of ~ 240 min in 12 M NaOH, decreasing to ~ 100 min in 1.2 M NaOH and to < 30 min in 0.1 M NaOH. Solutions prepared in 1.2 M NaOH and stored at 0 °C have proved normally sufficient for the study of the rapid complexation/oxidation reactions that are observed. Kinetic experiments have typically been carried out at $I = 1.2$ M in the $[OH^-]$ range 0.12–1.2 M with substrate concentrations in 10–100 times excess over $[Ag(OH)_4]^-$.

11.2.3.2 REACTIONS INVOLVING $[Ag(OH)_4]^-$

Kirschenbaum and coworkers [78–92] have studied a wide range of substitution reactions on, and oxidations carried out by, $[Ag(OH)_4]^-$ in aqueous alkaline media and have commented upon the mechanisms involved. Substitution reactions are observed with non-oxidisable O-donor ligands such as periodate, tellurate [77], phosphate, borate, pyrophosphate, carbonate and arsenate [80]. Reactions studied involving reduction of $[Ag(OH)_4]^-$ include those with SO_3^{2-} [79], $S_2O_3^{2-}$ [81], I^- [82], SCN^- [83], CN^- [84], $H_2PO_2^-$ [85], HPO_3^{2-}

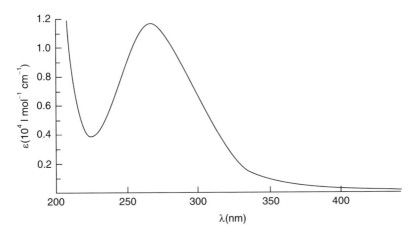

Figure 11.13. Electronic spectrum of $[Ag(OH)_4]^-$ in 1.2 M NaOH

Table 11.2. Kinetic data for inner-sphere reduction of $[Ag(OH)_4]^-$ by various substrates

Incoming ligand	k $(M^{-1} s^{-1})$	$\Delta H^{\ddagger}$ $(kJ\,mol^{-1})$	$\Delta S^{\ddagger}$ $(J\,K^{-1}\,mol^{-1})$	Ref.
SO_3^{2-}	$(9.1 \pm 0.1) \times 10^2$	27	-153	[79]
HPO_3^{2-}	9.5×10^{-2}			[79]
I^-	76 ± 1	32.5 ± 0.6	-100 ± 2	[82]
SCN^-	21.7 ± 0.3	28.3 ± 1.2	-129 ± 7	[83]
Thiourea	1.46×10^3			[86]
$H_2AsO_3^-$	$(6.2 \pm 1.0) \times 10^4$	9.6 ± 4.2	-140 ± 15	[87]
$HAsO_3^{2-}$	$(3.7 \pm 0.3) \times 10^4$	10.5 ± 3.3	-123 ± 11	[87]
AsO_3^{3-}	$(1.8 \pm 0.5) \times 10^4$	46 ± 4	-8.4 ± 33	[87]
N_3^-	7.3 ± 1.5			[88]
HO_2^-	$(4.2 \pm 0.6) \times 10^{5\,a}$	25 ± 5	-113 ± 5	[89]
H_2O	$(2 \pm 1)^b$			[81]

^{a} Units of $M^{-2} s^{-1}$ for third-order term involving $[OH^-]$.
^{b} Aquation process, units of s^{-1}.

[79], thiourea [86], arsenite [87], N_3^- [88], H_2O_2 [89] as well as a number of organic molecules and ligands [90]. Reactions can involve one-electron (e.g. with HO_2^- [89], $[MnO_4]^{2-}$ [91] and $[Fe(CN)_6]^{3-}$ [91]) or two-electron (e.g. with I^-, NCS^-, N_3^- and CN^-) reduction of $[Ag(OH)_4]^-$. The latter, in the case of certain oxo anions, can be accompanied by O-atom transfer [79, 85, 87]. In

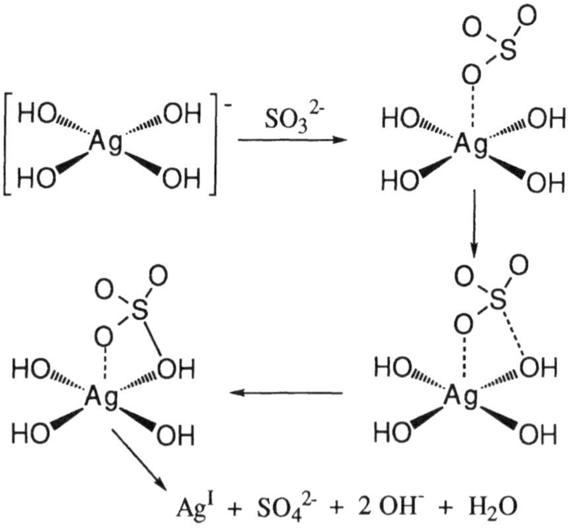

Figure 11.14. Mechanism for two-electron reduction of $[Ag(OH)_4]^-$ by SO_3^{2-} involving O atom transfer

a number of cases the aqua species $[Ag(OH)_3(OH_2)]$ is involved [81, 86] and its formation can be rate determining $(k \ (25\,^\circ C) = (2 \pm 1)\,s^{-1})$. The activation parameters, low $\Delta H^{\ddagger}$ and appreciably negative $\Delta S^{\ddagger}$, observed for many of the reactions (Table 11.2), support an associative mechanism for ligand substitution involving initial attack of the substrate at the axial position to give a five-coordinate intermediate followed by entry into the equatorial positions. In the O-atom transfer mechanism the axially coordinated substrate interacts with the equatorial OH groups followed by dissociation of the oxidised substrate and generation of Ag^I. This mechanism is illustrated for the reduction with $SO_3{}^{2-}$ in Fig. 11.14 [79]. In this reaction an alternative mechanism was also proposed involving Ag—OH attack on a non-coordinated S^{IV} centre while at the same time allowing rapid donation of the electron pair back into the empty $d_{x^2-y^2}$ orbital of Ag^{III}(aq) [79]. The 10^4 times faster rate for reduction by $SO_3{}^{2-}$ vs by $HPO_3{}^{2-}$ may be related to the presence of a pair of electrons on the S atom, but not on the P atom of phosphite, for the electron donation to Ag^{III}. Moreover, the non-reduction of $[Ag(OH)_4]^-$ by $NO_2{}^-$ [79] is an observation believed to be consistent with the inability of N to expand its octet and accept a pair of electrons from one of the OH groups while bonded to the Ag^{III} centre. These findings have been cited as further evidence in support of the O atom transfer mechanism.

The presence of third-order terms in the rate laws for the reductions involving 1,2-diaminoethane [92], $HO_2{}^-$ [88], thiourea [86] and $S_2O_3{}^{2-}$ [81] is cited as further evidence for the involvement of five-coordinate intermediates. In the case of $N_3{}^-$ [88] two $N_3{}^-$ ions are incorporated forming cis-$[Ag(OH)_2(N_3)_2]^-$ before electron transfer occurs to liberate N_2 and Ag^I. This concerted process is seemingly able to overcome the large activation barrier inherent for direct one- or two-electron oxidation of a single azide. In the reduction by $S_2O_3{}^{2-}$ the coordination of a second $S_2O_3{}^{2-}$ molecule on Ag^{III} leads to a pathway for reductive elimination of the product $S_4O_6{}^{2-}$ [81]. The studies have shown that $[Ag(OH)_4]^-$ is a versatile oxidant capable of employing different pathways depending upon the requirements of the substrate. In almost all reactions, however, formation of a five-coordinate intermediate appears to be involved in the initial step.

11.3 GOLD

Gold was the first pure metal to be recognised by man and has been prized ever since. It is one of the few metals to be found naturally in its elemental state, a fact reflecting its unreactivity and indeed reluctance to exhibit an extensive chemistry in an oxygen atom environment. The name 'aqua regia' was eventually given to the concentrated acidic mixture ($cHCl/cHNO_3$, 3:1) found uniquely to dissolve the metal by stabilising Au^{III} in the form of the chloro complex $[AuCl_4]^-$ (see below). A limited aqueous chemistry exists for gold in oxidation states I to III. Truly authentic Au^{II} complexes are, however, in contrast to Cu^{II}, extremely rare useless stabilised by special ligands, a fact that correlates with predictions from atomic ionisation potentials. Most apparent 'Au^{II} compounds' in fact consist of an equimolar mixture of Au^I and Au^{III} often easily distinguishable in structures by their respective preferred geometries, linear and square planar respectively, Au$_4$Cl$_8$ being good example [93]. The instability of Au^{II}, and in turn Ag^{II}, towards disproportionation can also be explained on crystal field grounds.

11.3.1 Existence of a Stable Au(aq) Ion

There are no stable aqua complexes characterised for Au^I (contrasting with Cu^I and Ag^I) and indeed few stable aqueous complexes of this oxidation state. The reason is due to a highly favourable disproportionation reaction to give Au metal and Au^{III}, Au^{II} as we have seen being similarly unstable. Partially aquated Au$^+$ ($\sim 10^{-4}$ M) can be stabilised to a limited extent, however, in acetonitrile–water mixtures (cf. Cu^I) and estimates of the Au$^+$/Au^0 reduction potential have been made by measuring the shift in the potential of solutions of the stable solvate [Au(NCCH$_3$)$_2$]$^+$ [94] upon dilution in aqueous HClO$_4$ [95]. The value extrapolated ($+ 1.695$ V) shows that Au^I is a powerful oxidising agent, more so than Au^{III}, which explains the favourable disproportionation reaction in aqueous media:

$$3\,Au^I \longrightarrow 2\,Au + Au^{III} \tag{11.13}$$

The stability of solvated Au$^+$ was estimated to be around 5 min in mixtures containing only 0.4 % acetonitrile in aqueous 0.1 M HClO$_4$, where the major constituent species might be the partially hydrated ion [Au(NCCH$_3$)(OH$_2$)]$^+$ [95].

A range of stable square planar d^8 Au^{III} complexes are, however, well established, some of the simplest examples being [AuCl$_4$]$^-$ [96] and [Au(NH$_3$)$_4$]$^{3+}$ [97]. Chloroauric acid (H$_3$O[AuCl$_4$]·3H$_2$O) is obtained as yellow crystals upon evaporation of solutions of gold dissolved in aqua regia. Salts of [AuCl$_4$]$^-$ are the usual lead-in to aqueous Ag^{III} chemistry. There is in fact, an extensive coordination chemistry of ammine, amine and haloammine complexes of Au^{III} [98,99] which have, like those of Pd^{II} and Pt^{II}, lent themselves to a further study of the intimate mechanisms surrounding ligand substitution processes on d^8 square planar complexes [100]. The putative [Au(OH$_2$)$_4$]$^{3+}$ ion has not been characterised and it appears that gold has a remarkably low affinity for water as ligand. For example, in substitution reactions on Au^{III}, unlike those on Pd^{II} and Pt^{II}, rate laws show terms only dependent upon the concentration of the incoming ligand (no spontaneous (aquation) path is seen), showing that water competes extremely poorly [100]. Substitution rates on Au^{III} are in fact acutely sensitive to the nature of the entering ligand and the classical associative mechanism is implied. Rates typically observed are higher than those on comparable Pd^{II} complexes. As with Pt^{II}, leaving ligand effects are also apparent but less so. Where apparent, like on Pt^{II}, the rates are reflective of the leaving ligand's affinity for the soft nature of Au^{III} and also to a degree of covalency in the Au^{III}—L bonds.

Table 11.3. pK_{11} values for coordinated protic ligands on Au^{III} complexes

Complex	pK_{11}	Assignment	Ref.
$[Au(NH_3)_3(OH_2)]^{3+}$	-0.7	Ionisation of OH_2	[99]
$[AuCl_3(OH_2)]$	2.72	Ionisation of OH_2	[101]
$[Au(dien)(OH)]^{2+}$	5.8	Ionisation of NH	[102]
$[Au(en)_2]^{3+}$	6.3	Ionisation of NH	[103]
$[Au(NH_3)_4]^{3+}$	7.5	Ionisation of NH	[99]

Other evidence points to an intrinsic instability of Au^{III} aqua species. Firstly, when water replaces other ligands on Au^{III} there is a marked shift in the potential for Au^{III} reduction into the region which is suggestive of spontaneous reduction of Au^{III} by water. Indeed, many Au^{III} complexes are already quite strong oxidants, e.g.

$$[AuCl_4]^- + 3e^- \longrightarrow Au^0 + 4Cl^- \qquad (E^\circ = 1.00\,V) \qquad (11.14)$$

Secondly, although Au^{III} is soft and prefers bonding to soft ligands (good π donor) it is also an efficient σ-acid-forming stable complexes with good σ donors as in the case of the ammine, amine and Cl^- complexes. As a result, in the few cationic Au^{III}–OH_2 complexes known, the pK_a of the aqua ligand is extremely low. This is illustrated by the pK_{11} values listed in Table 11.3. A remarkable example is provided by the hydroxy complex $[Au(dien)(OH)]^{2+}$ wherein the pK_a for ionisation of the next proton (on dien) is as low as 5.8 [102]. It is clear from these data that $[Au(OH_2)_4]^{3+}$ itself would be expected to have little chance of existence with $Au^{3+}(aq)$ extensively hydrolysed even in concentrated acid. Indeed, the only stable binary Au^{III}–O(OH) compounds known are Au_2O_3 and $Au(OH)_3$ [98, 104].

11.3.2 Gold(III) (d^8). Studies on 'Auric Hydroxide' $Au(OH)_3$

Freshly precipitated $Au(OH)_3$ is notable for its remarkable stability in water over a wide range of pH (Fig. 11.15). Baes and Mesmer [105] have reviewed available data on the solution behaviour of $Au(OH)_3$. Much of it is old data which have not been substantiated in recent work. For example, early claims for the formation of $Au(OH)^{2+}(aq)$ and $Au(OH)_2^+(aq)$ from dissolution of $Au(OH)_3$ in sulfuric acid should be tempered since the likely formation of sulfato complexes of Au^{III} was not taken into account and there were also uncertainties about the degree of bisulfate dissociation and activity coefficients [107]. Thus the hydrolysis constants and speciation behaviour quoted in ref. [105] for $Au(OH)^{2+}(aq)$ and $Au(OH)_2^+(aq)$ should be treated with extreme caution in view of the expected high acidity of aqua ligands on cationic Au^{III} species. The solubility (pK_{s10}) of

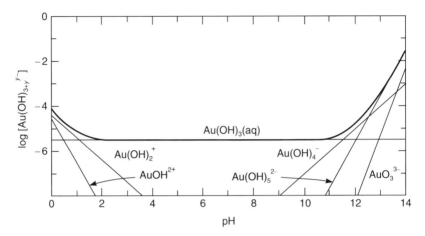

Figure 11.15. Concentrations of Au^{III}(aq) species in equilibrium with $Au(OH)_3$, $I = 1$ m, 25 °C

$Au(OH)_3$ in water (5.51 $\pm$ 0.07) is remarkably high for a neutral hydroxide [106]. $Au(OH)_3$ is also unusually stable in solution over a wide range of [Cl$^-$], up to $\sim 10^{-2}$ M in Cl$^-$ in the pH range 5–11. The solubility of $Au(OH)_3$ in base increases above pH 10 until in 0.42 M NaOH a solid phase appears consisting of $Na_2[Au(OH)_5]$ (sometimes given as $Na_2[HAuO_3]$). Thereafter the solubility in base increases again. The data up to 1.0 M in NaOH solution have been interpreted by the formation of both $[Au(OH)_4]^-$ [108] and $[Au(OH)_5]^{2-}$, the latter in equilibrium with the solid phase. The pK_{s11} value for formation of $[Au(OH)_4]^-$ from $Au(OH)_3$ and OH$^-$ is given as 3.28 $\pm$ 0.2 [106]. Above 1 M NaOH the linear rise in solubility has been attributed to the formation of AuO_3^{3-}. No evidence has been provided for polynuclear Au^{III} hydrolysis products [109] although colloids are reported from solutions of $Au(OH)_3$ in < 0.07 M NaOH [106]. Structural studies conducted on Au_2O_3 show the expected square planar arrangement of $Au^{III}O_4$ units linked by oxo bridges [104]. A square planar Au(—O)$_4$ array (Au—O$_{av}$ = 203 pm) has been found in the salts $M[Au(OH)_4]_2$ (M = Sr^{II}, Ba^{II}) [110]. Finally, the tetrameric OH-bridged structure $(Au$—$O_{av} = 223$ pm) found in $[Me_2Au(OH)]_4$ [111] emphasises the extensively hydrolysed nature of Au^{III}(aq) species and the ready formation of hydroxy complexes for this oxidation state.

REFERENCES

[1]   (a) Karlin, K. D. and Zubieta, J. (eds.), *Copper Coordination Chemistry, Biochemical and Inorganic Perspectives*, Adenine Press, New York, 1983 and 1986.
      (b) Nriagu, J. O., *Copper in the Environment*, Wiley, New York, 1980.

(c) Lontie, R., *Copper Proteins and Copper Enzymes*, Vols. 1–3, C.R.C. Press, Boca Raton, Florida, 1984.

(d) Kitajima, N., *Adv. Inorg. Chem.*, **39**, 1 (1992).

(e) Walters, K. L., and Wilkins, R. G., *Inorg. Chem.*, **13**, 752 (1974).

(f) Gillard, R. D., and Spencer, A., *J. Chem. Soc. A*, 2718 (1969).

[2] Altermatt, J. A., and Manahan, S. E., *Inorg. Nucl. Chem. Lett.*, **4**, 1 (1968).

[3] Parker, O. J., and Espenson, J. H., *Inorg. Chem.*,**8**, 185, 1523 (1969) and *J. Am. Chem. Soc.*, **91**, 1968 (1969).

[4] Ahrland, S., Nielsson, K., and Tagesson, B., *Acta Chem. Scand.*, **A40**, 418 (1983).

[5] (a) Csoregh, I., Kierkegaard, P., and Norrestam, R., *Acta Cryst.*, **B31**, 314 (1975).

(b) Parker, A. J., *Aust. J. Chem.*, **30**, 1423 (1977).

[6] Gill, D. S., Rodeshauser, L., and Delpeuch, J. -J., *J. Chem. Soc. Faraday Trans. I*, **86**, 2847 (1990).

[7] Gill, D. S., Arora, K. S., Tewari, J., and Singh, B., *J. Chem. Soc. Faraday Trans. I*, **84**, 1729 (1988).

[8] Marcus, Y., *J. Chem. Soc. Dalton Trans.*, 2265 (1991).

[9] (a) Chapman, S. K., in *Perspectives in Bioinorganic Chemistry*, (eds. R. W. Hay, J. R. Dilworth, and K. B. Nolan, Jai Press, London, Vol. **1**, 1992.

(b) Sykes, A. G., *Adv. Inorg. Chem.*, **36**, 377 (1991).

[10] See, for example, Magini, M., *Inorg. Chem.*, **21**, 1535 (1983) (and refs. therein).

[11] (a) Swift, T. J., and Connick, R. E., *J. Phys. Chem.*, **37**, 307 (1962) and **41**, 2553 (1964).

(b) Hunt, T. J., and Friedman, H. L., *Prog. Inorg. Chem.*, **30**, 327 (1983).

[12] (a) Powell, D. H., Helm, L., and Merbach, A. E., *J. Chem. Phys.*, **95**, 9258 (1991).

(b) Powell, D. H., Furrer, P., Pittet, P.-A., and Merbach, A. E., *J. Phys. Chem.*, **99**, 16622 (1995).

[13] (a) Licheri, G., Musini, A., Paschina, G., Piccaluga, G., Pinna, G., and Sedda, A. F., *J. Chem. Phys.*, **80**, 5308 (1984).

(b) Tajiri, Y., and Wakita, H., *Bull. Chem. Soc. Japan*, **59**, 2285 (1986).

(c) Onori, G., Santucci, A., Scafati, A., Belli, M., and Della Lonza, S., *Chem. Phys. Lett.*, **149**, 289 1988).

[14] (a) Salmon, P. S., and Neilson, G. W., *J. Phys. Condensed Matter*, **1**, 5291 (1989).

(b) Salmon, P. S., Neilson, G. W., and Enderby, J. E., *J. Physics C*, **21**, 1335 (1988).

[15] Fisher, J., and Weiss, R., *Acta Cryst. Sec. B.*, **29**, 1963 (1973).

[16] Varghese, J. N., and Maslen, E. N., *Acta Cryst. Sec. B.*, **41**, 184 (1985).

[17] Cotton, F. A., Daniels, L. M., and Murillo, C. A., *Inorg. Chem.*, **32**, 4868 (1993).

[18] (a) Hathaway, B. J., *Structure and Bonding*, Vol. 57, Springer, Berlin, 1984, p. 55.

(b) Hathaway, B. J., Duggan, M., Murphy, A., Mullane, J., Power, C., Walsh, A., and Walsh, B., *Coord. Chem. Revs.*, **36**, 267 (1981) (and refs. therein).

[19] (a) Silver, B. L., and Getz, D., *J. Chem. Phys.*, **61**, 638 (1974).

(b) Hathaway, B. J., and Hodgson, P. G., *J. Inorg. Nucl. Chem.* **35**, 4071 (1973).

(c) Melnik, M., *Coord. Chem. Revs.*, **47**, 239 (1982).

(d) Alcock, N. W., Duggan, M., Murray, A., Tyagi, S., Hathaway, B. J., and Hewat, A., *J. Chem. Soc. Dalton Trans.*, 7 (1984).

[20] (a) Shields, K. G., and Kennard, C. H. L., *Acta Cryst.*, **C1**, 189 (1972).

(b) Brown, G. M., and Chidambaram, R., *Acta Cryst.*, **B25**, 676 (1969).

(c) Mani, N. V., and Ramaseshan, S., *Z. Kristallogr.*, **115**, 97 (1961).

[21] Helm, L., Lincoln, S. F., Merbach, A. E., and Zbinden, D., *Inorg. Chem.*, **25** 2250 (1986).

[22] Wilkins, R. G., *Pure Appl. Chem.*, **33**, 583 (1973).

[23] Powell, D. H., Merbach, A. E., Fabian, I., Schindler, S., and van Eldik, R., *Inorg. Chem.*, **33**, 4468 (1994).
[24] Sugiura, Y., and Hirayama, X., *J. Am. Chem. Soc.*, **99**, 1581 (1977).
[25] Pederson, K. J., *K. Dan. Vidensk. Selsk., Mat.-Fys. Medd.*, **7**, 20 (1943).
[26] Neher-Neumann, E., *Acta Chem. Scand.*, **38A**, 517 (1984).
[27] (a) Baes, C. F., and Mesmer, R. E., *The Hydrolysis of Cations*, Krieger, Florida, 1986, pp. 267–74 (and refs. therein).
     (b) Kalihana, H., Amaya, T., and Maeda, M., *Bull. Chem. Soc. Japan*, **43**, 3155 (1970).
[28] Lin, C.-T., Rorabacher, D. B., Cayley, G. R., and Margerum, D. W., *Inorg. Chem.*, **14**, 919 (1975).
[29] (a) Liang, B.-F., and Chung, C.-S., *Inorg. Chem.*, **20**, 2152 (1981).
     (b) Liang, B.-F., Margerum, D. W., and Chung, C.-S., *Inorg. Chem.*, **18**, 2001 (1979).
     (c) Liang, B.-F., and Chung, C.-S., *Inorg. Chem.*, **19**, 1867 (1981).
     (d) Chen, F.-T., Lee, C.-S., and Chung, C.-S., *Polyhedron*, **2**, 1301 (1983).
[30] Drumhiller, J. A., Montavon, F., Lehn, J.-M., and Taylor, R. W., *Inorg. Chem.*, **25**, 3751 (1986).
[31] Hay, R. W., and Hassan, M. M., private communication, 1994.
[32] Shibahara, T., Akashi, H., and Kuroya, H., *J. Am. Chem. Soc.*, **110**, 3314 (1988).
[33] Nasreldin, M., Li, Y.-J., Mabbs, F. E., and Sykes, A. G., *Inorg. Chem.*, **33**, 4283 (1994).
[34] Ahrland, S., and Rawsthorne, J., *Acta Chem. Scand*, **24**, 157 (1970).
[35] Shibahara, T., Asano, T., and Sakane, G., *Polyhedron*, **10**, 2351 (1991).
[36] (a) Meyerstein, D., *Inorg. Chem.*, **10**, 638 (1971).
     (b) Baxendale, J. H., Fielden, E. M., and Keene, J. P., in *Pulse Radiolysis*, (eds. M. Ebert, J. P. Keene, A. J. Swallow, and J. H. Baxendale), Academic Press, New York, 1965, p. 217.
[37] Asmus, K. D., Boniface, M., Toffel, P., O'Neill, P., Schulte-Frohlinde, D., and Steeden, S., *J. Chem. Soc. Faraday Trans. I*, 1820 (1978).
[38] Gray, Jr, J. S., Taylor, R. W., and Margerum, D. W., *Inorg. Chem.*, **16**, 3047 (1977).
[39] Thenard, L., *Ann. Chim. Phys.*, **9**, 51 (1818).
[40] (a) Magee, Jr, J. S., and Wood, R. H., *Can. J. Chem.*, **45**, 1234 (1965).
     (b) Lister, M. W., *Can. J. Chem.*, **31**, 638 (1953).
[41] Hamilton, G. A., Adolf, P. K., deJersey, J., duBois, G. C., Dyrkasc, G. R., and Libby, R. D., *J. Am. Chem. Soc.*, **100**, 1899 (1978).
[42] Margerum, D. W., and Owens, G. D., in *Metal Ions in Biological Systems* (ed. H. Sigel), Vol. 13, Marcel Dekker, New York, 1981.
[43] Maeda, M., Maegava, Y., Yamaguchi, T., and Ohtaki, H., *Bull. Chem. Soc. Japan*, **52**, 2545 (1979).
[44] Akitt, J. W., *J. Chem. Soc., Dalton Trans.*, 175 (1974).
[45] Gusev, N. I., *Russ. J. Phys. Chem.*, **47**, 1309 (1973).
[46] Allam, D. S., and Lee, W. H., *J. Chem. Soc.*, 5 (1966).
[47] Brown, R. D., and Symons, M. C. R., *J. Chem. Soc., Dalton Trans.*, 426 (1976).
[48] Yamaguchi, T., Johansson, G., Holmberg, B., Maeda, M., and Ohtaki, H., *Acta Chem. Scand.*, **A38**, 437 (1984).
[49] Shannon, R. D., *Acta Cryst.*, **A32**, 751 (1976).
[50] Farrow, M. M., Purdie, N., and Eyring, E. M., *Inorg. Chem.*, **14**, 1584 (1975).
[51] Cox, B. G., Garcia,-Rosas, J., Schneider, H., and van Truong, N., *Inorg. Chem.*, **25**, 1165 (1986).
[52] Biedermann, G., and Sillen, L. G., *Acta Chem. Scand.*, **14**, 717 (1960).

[53] Antikainen, P. J., and Dryssen, D., *Acta Chem. Scand.*, **14**, 86 (1960).
[54] Nasanen, R., and Merilainen, P., *Suomen. Kem.*, **33B**, 197 (1960).
[55] Mentasti, E., Baiocchi, C., and Coe, J. S., *Coord. Chem. Revs.*, **84**, 131 (1984).
[56] Noyes, A. A., Hoard, J. L., and Pitzer, K. S., *J. Am. Chem. Soc.*, **57**, 1221 (1935).
[57] (a) Hammer, R. N., and Kleinberg, J., *Inorg. Syn*, **4**, 12 (1953).
     (b) Miler, J. D., *J. Chem. Soc. A*, 1778 (1968).
     (c) Anderson, J. M., and Kochi, J. K., *J. Org. Chem.*, **35**, 986 (1970).
     (d) Indrayan, A. K., Mishra, S. K., and Gupta, Y. K., *Inorg. Chem.*, **20**, 450 (1981).
[58] Kirwin, J. B., Peat, F. D., Proll, P. J., and Sutcliffe, L. H., *J. Phys. Chem.*, **67**, 1617 (1963).
[59] (a) Bonifacic, M., and Asmus, K.-D., *J. Phys. Chem.*, **80**, 2426 (1976).
     (b) Kumar, A., and Neta, P., *J. Am. Chem. Soc.*, **102**, 7284 (1980).
[60] Rechnitz, G. A., and Zamochnik, S. B., *Talanta*, **11**, 713 (1964).
[61] Honig, D. S., and Kustin, K., *J. Inorg. Nucl. Chem.*, **32**, 1599 (1970).
[62] (a) Hammes, P., Rich, L. D., Cole, D. L., and Eyring, E. M., *J. Phys. Chem.*, **75**, 929 (1971).
     (b) Eigen, M., Kruse, W., Maass, G., and DeMaeyer, L., *Prog. React. Kinet.*, **2**, 287 (1964).
[63] Noyes, A. A., Coryell, C. D., Stitt, F., and Kossiakoff, A., *J. Am. Chem. Soc.*, **59**, 1316 (1937).
[64] Po, H. N., Swinehart, J. N., and Allen, T. L., *Inorg. Chem.*, **7**, 244 (1968).
[65] Noyes, A. A., and Kossiakoff, A., *J. Am. Chem. Soc.*, **57**, 1238 (1935).
[66] Noyes, A. A., De Vault, D., Coryell, C. D., and Deahl, T. S., *J. Am. Chem. Soc.*, **59**, 1326 (1937).
[67] Biedermann, G., Maggio, F., Ramano, V., and Zingales, R., *Acta Chem. Scand.*, **35A**, 287 (1981).
[68] Scrocco, E., Marmani, G., and Mirone, P., *Bull. Soc. Sci. Fac. Chim. Ind. Bologna*, **8**, 119 (1950).
[69] Wells, C. F., *J. Inorg. Nucl. Chem.*, **36**, 3856 (1974).
[70] Kumar, A., and Neta, P., *J. Phys. Chem.*, **83**, 3091 (1979).
[71] Yost, D. M., and Russell Jr, H., *Systemmatic Inorganic Chemistry*, Prentice Hall, New York, 1946.
[72] Arselli, P., Baiocci, C., Mentasti, E., and Coe, J. S., *J. Chem. Soc. Dalton Trans.*, 475 (1984).
[73] Huchital, D. N., Sutin, N., and Warnquist, B., *Inorg. Chem.*, **6**, 838 (1967).
[74] Jensovsky, L., and Skala, M., *Z. Anorg. Allg. Chem.*, **312**, 26 (1961).
[75] Malatesta, L., *Gazz, Chim. Ital.*, **71**, 467 (1964).
[76] Cohen, G. L., and Atkinson, G., *Inorg. Chem.*, **3**, 1711 (1964).
[77] Cohen, G. L., and Atkinson, G., *J. Electrochem. Soc.*, **115**, 1236 (1968).
[78] Kirschenbaum, L. J., Ambrus, J. H., and Atkinson, G., *Inorg. Chem.*, **12**, 2832 (1973).
[79] Kirschenbaum, L. J., Kouadio, I., and Mentasti, E., *Polyhedron*, **8**, 1299 (1989).
[80] Rush, J. D., and Kirschenbaum, L. J., *Polyhedron*, **5**, 1573 (1985).
[81] Rush, J. D., and Kirschenbaum, L. J., *Inorg. Chem.*, **24**, 744 (1985).
[82] Kouadio, I., Kirschenbaum, L. J., and Mehrotra, R. N., *J. Chem. Soc. Dalton Trans.*, 1929 (1990).
[83] Kirschenbaum, L. J., and Sun, Y. F., *Inorg. Chem.*, **30**, 2360 (1990).
[84] Sun, Y. F., and Kirschenbaum, L. J., *J. Coord. Chem.*, **26**, 127 (1992).
[85] Mehrotra, R. N., and Kirschenbaum, L. J., *Inorg. Chem.*, **28**, 4327 (1989).
[86] Kirschenbaum, L. J., and Panda, R. K., *Polyhedron*, **7**, 2753 (1988).
[87] Kirschenbaum, L. J., and Rush, J. D., *Inorg. Chem.*, **22**, 3304 (1983).

[88] Borish, E. T., and Kirschenbaum, L. J., *Inorg. Chem.*, **23**, 2355 (1984).
[89] Borish, E. T., and Kirschenbaum, L. J., *J. Chem. Soc. Dalton Trans.*, 749 (1983).
[90] (a) Kumar, A., and Panwar, A., *Bull. Chem. Soc. Japan*, **67**, 1207 (1994).
      (b) Satchell, D. P. N., Satchell, R. S., and Bhavnani, S., *J. Chem. Soc. Perkin Trans II*, 1543 (1993).
      (c) Sun, Y. F., Kirschenbaum, L. J., and Kouadio, I., *J. Chem. Soc. Dalton Trans.*, 2311 (1991).
      (d) Kirschenbaum, L. J., and Rush, J. D., *J. Am. Chem. Soc.*, **106**, 1003 (1984).
[91] Kirschenbaum, L. J., Borish, E. T., and Rush, J. D., *Israel. J. Chem.*, **25**, 159 (1985).
[92] Kirschenbaum, L. J., *J. Inorg. Nucl. Chem.*, **38**, 881 (1976).
[93] Dell'Amico, D. B., Calderazzo, F., Marchetti, F., Merlino, S., and Perego, G., *J. Chem. Soc., Chem. Commun.*, 31 (1977).
[94] (a) Bergerhoff, G., *Z. Anorg. Allg. Chem.*, **327**, 139 (1964).
      (b) Goolsby, A. D., and Sawyer, D. T., *Anal. Chem.*, **40**, 1978 (1968).
[95] (a) Kissner, R., Latal, P., and Geier, G., *J. Chem. Soc., Chem. Commun.*, 136 (1993).
      (b) Johnson, P. R., Pratt, J. M., and Tilley, R. I., *J. Chem. Soc., Chem. Commun.*, 606 (1978).
[96] Bonamico, M., Dessy, G., and Vaciago, A., *Atti. Acad. Nazl. Lincei. Rend. Classe Sci. Fis. Mat. Nat.*, **39**, 504 (1965).
[97] Mason, W. R., and Gray, H. B., *J. Am. Chem. Soc.*, **90**, 5721 (1968).
[98] Puddephatt, R., in *Comprehensive Coordination Chemistry*, (eds. G. Wilkinson, R. D. Gillard, and J. A. McCleverty), Vol. 7, Pergamon, New York, 1987, pp. 862–91.
[99] (a) Skibsted, L. H., and Bjerrum, J., *Acta Chem. Scand.*, **28A**, 764 (1974).
      (b) Weishaupt, M., and Strachle, J., *Z. Naturforsch. Teil B*, **31**, 554 (1976).
[100] Skibsted, L. H., *Adv. Inorg. Bioinorg. Mech.*, **4**, 137 (1986).
[101] Carlsson., L., and Lundgren, G., *Acta Chem. Scand*, **21**, 819 (1967).
[102] Baddley, W. H., Basolo, F., Gray, H. B., Nolting, C., and Poe, A. J., *Inorg. Chem.*, **2**, 921 (1963).
[103] Block, B. P., and Bailar, J. C., *J. Am. Chem. Soc.*, **73**, 4722 (1951).
[104] Hydes, P. C., and Middleton, H., *Gold. Bull.*, **12**, 90 (1979).
[105] Baes, C. F., and Mesmer, R. E., *The Hydrolysis of Cations*, Krieger, Florida, 1986, pp. 279–86.
[106] Johnson, H. L., and Leland, H. L., *J. Am. Chem. Soc.*, **60**, 1439 (1938).
[107] Jirsa, F., and Jelinek, H., *J. Electrochem. Soc.*, **30**, 286, 534 (1924).
[108] Bjerrum, N., *Bull. Soc. Chim. Belges.*, **57**, 432 (1948).
[109] Jander, G., and Krien, G., *Z. Anorg. Allg. Chem.*, **304**, 154 (1960).
[110] Jones, P. G., and Sheldrick, G. M., *Acta Cryst.*, **C40**, 1776 (1984).
[111] Glass, G. E., Konnert, J. H., Miles, M. G., Britton, D., and Tobias, R. S., *J. Am. Chem. Soc.*, **90**, 1131 (1968).

Chapter 12

Group 12 Elements: Zinc, Cadmium and Mercury

Zinc is an extremely important biological element [1]. It is the second most abundant metal in the human body after iron. An adult human body contains 2–3 g of zinc with all of it involved in an active or structure-forming role in some 200 enzymes and proteins. The primary role for zinc (as Zn^{II}) is in the form of a Lewis acid promoting hydrolytic processes [2], the most well-known examples being the enzymes carboxypeptidase (CP), carbonic anhydrase (CA) and the alcohol dehydrogenases (AD). Zn^{II} is also found in one form of purple acid phosphatase from red kidney beans, associated with Fe^{III}, in bovine superoxide dismutase, associated with Cu^{II} and in alkaline phosphatase as a homodimeric unit. Many other proteins also have Zn^{II} present in a tertiary structure defining capacity. Like Cu^{II} and Ni^{II}, Zn^{II} favours binding to a whole range of O-, N- and S-donor sites, probably the most important biologically of these being the imidazole group of histidine. Because of the absence of LFSE for the d^{10} configuration, Zn^{II} has the added advantage of a flexible and variable coordination geometry with four, five and six coordination all commonly observed, good examples being $[Zn(OH_2)_6]^{2+}$, $[ZnCl(OH_2)_5]^+$ and *trans*-$[ZnCl_2(OH_2)_4]$ (octahedral), $[ZnCl_3(OH_2)_2]^-$ (trigonal bipyramidal, axial H_2O) and $[ZnCl_4]^{2-}$ (tetrahedral) [3]. Distortions from regular geometry are also commonly encountered situations. The flexible geometry requirement, coupled with quite fast, but not overly so, rates of ligand substitution processes and an absence of redox chemistry, is almost certainly crucial to its biological role as a Lewis acid catalyst. Moreover, this ability is further exemplified by the reluctance of $Zn^{2+}(aq)$ to hydrolyse. In both CP and CA the coordinating amino acid ligands play a crucial part in tuning the acidity of the Zn^{II} centre to cause lowering of the pK_a for acid dissociation of an H_2O ligand into the physiological pH range (~ 7.0). The result is the generation of a potent bound hydroxide nucleophile which is crucial to the enzyme mechanism (see below) [2]. Similar lowering of the pK_a for bound H_2O has now been achieved in a number of simple $[LZn^{II}(OH_2)]$ complexes which have allowed ligand control (by L) of the coordination number and geometry around the Zn^{II} [4]. Amalgamated zinc (Zn/Hg) is a useful aqueous solution heterogeneous reductant (E^θ $(Zn^{2+}(aq)/Zn^0) = -0.76$ V) known in organic chemistry as the Jones reductor. The amalgamation has the effect of reducing the overpotential for H^+ reduction allowing efficient reduction of substrates under acidic conditions, its use for the generation of reducing aqua ions such as $[Cr^{II}(OH_2)_6]^{2+}$ and $[Mo^{III}_2(\mu\text{-}OH)_2(OH_2)_8]^{4+}$ having been well documented in both Chapters 1 and 6. The Zn^{2+}/Zn^0 couple forms the cathode half-reaction in a number of well-established battery systems such as the Daniel cell and subsequent 'dry' cell modifications, several of which remain in active commercial use today.

Cadmium and mercury are highly toxic and environmentally hazardous elements. Both metals as M^{II} can substitute for biological Zn^{II} in its enzymes and proteins [5], particularly where Zn—S bonding is involved, since both Cd^{II} and Hg^{II} form stronger bonds to sulfur. The result is invariably deactivation. It is believed that this deactivation, in the case of carbonic anhydrase, may relate in part to the generally lower tendency of Cd^{II} (aq) species to hydrolyse compared with Zn^{II} (see below). In some cases substitution by Cd^{II} has allowed the preparation of apo (metal-free) enzyme samples which can then be used to substitute other metals acting as spectroscopic probes for the Zn^{II} site, e.g. Co^{II} (u.v–visible and EPR) [6] as well as the use of Cd^{II} itself (^{113}Cd NMR) [5]. The ionic radius of Cd^{2+} (109 pm) is also similar to that of Ca^{2+} (114 pm) (for six coordination) so that accumulation of Cd^{II} in, for example, muscle and bone tissue is a further potential hazard. This aspect has, however, been put to beneficial use in a process developed for the removal of trace Cd^{II} levels in natural water. The method uses treatment with a slurry of limestone ($CaCO_3$) to which the Cd^{II} readily binds, displacing some of the Ca^{II}. The solubility of $CdCO_3$ is also somewhat less than that of $CaCO_3$ and it does not form a bicarbonate. The result is the effective removal of the Cd^{II} from the water to be replaced by Ca^{II}. Despite these premises, however, it remains likely that the principal cause of acute cadmium toxicity to man stems from the high affinity of Cd^{II} for sulfur and in particular for the thiol group of the amino acid cysteine. Indeed, it is believed that the metallothioneins, the largely cysteine-containing proteins that bind large quantities of zinc and also copper, play a detoxifying role by efficiently 'mopping' up free Cd^{II} and Hg^{II}. The metallothioneins are the only biological materials known to accumulate both of these elements [7]. A role for the metallothioneins in the aggregation and transport of zinc (and copper) is also a possibility.

Hg^{II}, in the form of the organoaqua ion $[CH_3Hg(OH_2)]^+$ (methyl mercury) was responsible for an ecological disaster in Minamata, Japan, in 1952. Hg^{II} leaked into a bay from a local on-shore chemical works wherein it was converted to the highly mobile and lipid soluble CH_3Hg^+ (aq) upon contact with certain marine microorganisms. This allowed it to pass readily through the cell walls of gills to contaminate large quantities of the fish which were the stable diet of the local community. As a result 52 local people died and many others received acute mercury poisoning before the leak was traced. Like cadmium, mercury's acute toxicity largely stems from its high affinity for sulfur and for the sulfur-containing amino acids cysteine and methionene. Indeed, the old name for a thiol, mercaptan, literally means 'mercury captor'. Trace amounts of mercury tend to accumulate in human hair and nail tissue which can result, in acute cases, in premature hair loss and nail brittleness. Acute mercury poisoning also causes headaches, tremors, inflammation of the bladder and eventually memory loss. It also affects the nervous system. The expression 'made as a hatter' relates to the nervous disorder 'hatter's shakes' suffered by workers once involved in the use of mercury in poorly ventilated workshops for the manufacture of felt for hats.

All three Group 12 elements have a stable M^{II}(aq) ion (d^{10}). However, for mercury, in contrast to the other two, the Hg^I state ($s^1 d^{10}$) is also well established in aqueous solution and there have even been claims for the existence of Hg^{III}. Zn^+ (aq) and Cd^+ (aq), on the other hand, have only a fleeting existence owing to the higher solvation energy of the M^{2+} (aq) species and the relatively low second ionisation potential promoting favourable disproportionation. For Hg^+(aq) the second ionisation potential is anomalously high and this, coupled with the formation of a reasonably strong Hg^I—Hg^I bond, is effectively responsible for the suppression of the disproportionation reaction. In addition elemental liquid mercury itself has an appreciable solubility in water (50 ppb) and plays an active role in equilibration processes with other soluble Hg species. ^{1}H NMR studies of molar shifts in the OH protons induced by the Group 12 M^{2+}(aq) cations in solution have been interpreted as indicating much higher covalency in the M–$O(OH_2)$ interactions for Cd^{2+} and Hg^{2+} vs Zn^{2+}. For mercury, despite the stronger covalency and high formation constants for soft donors such as sulfur, the species remain extremely kinetically labile, providing a good example of kinetics overriding thermodynamics.

12.1   ZINC AND CADMIUM

12.1.1   Zinc(I) and Cadmium(I) ($s^1 d^{10}$): The M^+(aq) Ions

Solutions containing Zn^+(aq) and Cd^+(aq) can be obtainable by reduction of corresponding Zn^{2+}(aq) and Cd^{2+}(aq) solutions with hydrated electrons [8]. The rate constant is rather slow ($k < 10^5$ M^{-1} s^{-1}) and for zinc in alkaline solution the reaction is reported to be reversible [9]. The spectra of Zn^+(aq) and Cd^+(aq) have been estimated by pulse radiolysis at different wavelengths. Both ions have intense absorptions centred around 300 nm (Fig. 12.1) assigned to

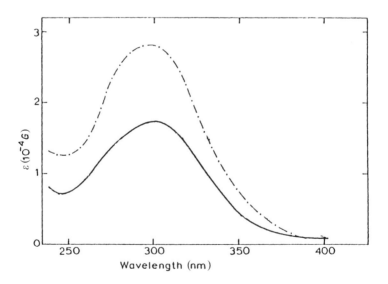

Figure 12.1. Electronic spectra of Zn^+(aq) (—) and Cd^+(aq) (—·—)

charge transfer [10], which have proved useful for following subsequent reactions. Stanbury has reported reduction potetials for the M^{2+}(aq)/M^+(aq) couples ($M = Zn$ and Cd) in a comprehensive review [11]. Cd^+(aq) is the weaker reductant, E^θ (Cd^{2+}/Cd^+) = -1.4 ± 0.4 V, compared to E^θ (Zn^{2+}/Zn^+) = -1.6 ± 0.4 V. Both estimations followed refinements of a Marcus treatment for the reduction of $[Co(NH_3)_6]^{3+}$ by both ions [12]. The potential for the Cd^{2+}/Cd^+ couple was consistent with the slow reduction of Cd^{2+}(aq), but not Zn^{2+}(aq), by CO_2^- [10] and the rapid reduction of $[Ru(bipy)_3]^{2+}$ by Cd^+(aq) [13]. Estimates of the free energy of hydration of both atomic zinc and cadmium as 10 kJ mol^{-1} have allowed estimates for the E^θ values for the Zn^+/Zn^0 and Cd^+/Cd^0 couples as respectively -1.01 and -0.31 V [11]. Thus both Zn^+(aq) and Cd^+(aq) are prone to disproportionation. For Zn^+(aq) this has been experimentally observed ($k = 3.5 \times 10^8$ M^{-1} s^{-1}) [14]. On the other hand, Cd^+(aq) behaves somewhat like Hg^+(aq) in showing evidence of rapid dimerisation to give Cd_2^{2+}(aq) [15]. This somewhat surpresses the disproportionation tendency although, unlike Hg_2^{2+}(aq). Cd_2^{2+}(aq) is highly reducing and rather unstable. Solutions of Zn^+(aq) have been claimed to reduce Cd^{2+}(aq), consistent with the relative potentials [16], but this has been disputed [17]. Both Zn^+(aq) and Cd^+(aq) react with a range of inorganic oxidants, including O_2, H_2O_2, Cu^{2+} and various oxo anions, with second-order rate constants $\sim 2 \times 10^9$ M^{-1} s^{-1} [18]. Both ions are reported to react more slowly with N_2O however, ($k_{Zn} \sim 2 \times 10^7$ M^{-1}s^{-1}; $k_{Cd} \sim 3 \times 10^6$ M^{-1}s^{-1}), to generate solutions which rapidly oxidise Br$^-$ to Br$_2$ and I$^-$ to I$_2$ [19]. Cl$^-$ is untouched. These solutions have been claimed to contain the d^9 M^{III} species ZnO^+(aq) and CdO^+(aq). The redox behaviour would be consistent with a reduction potential (to M^{2+}(aq)) of $\sim +1.2$ V.

12.1.2 Zinc(II) and Cadmium(II) (d^{10}): Chemistry of the $[Zn(OH_2)_6]^{2+}$ and $[Cd(OH_2)_6]^{2+}$ Ions

12.1.2.1 STRUCTURE AND PROPERTIES

This is the stable oxidation state for all three group members. $[Zn(OH_2)_6]^{2+}$ is formed spontaneously upon dissolution of simple Zn^{II} salts in water. Its octahedral structure has been verified from X-ray crystal structures of a number of $[Zn(OH_2)_6]X_2 \cdot nH_2O$ salts (X = hydrogen acetylenedicarboxylate, $n = 2$ [20]; X = benzenesulfonate, $n = 0$ [21]; X = p-toluenesulfonate, $n = 0$ [22]; X = 2,3-dihydroxybenzoate, $n = 2$ [23]; X = isonicotinate N-oxide [24]; X = salicylatoborate, $n = 4$ [25] and X = ethylenediaminedisuccinatonickel(II), $n = 2$ [26]. The crystal packing arrangement in the p-toluenesulfonate salt (Fig. 12.2) is typical of many such salts. Typical Zn—OH$_2$ distances are around 209–210 pm. The observed Zn—OH$_2$ distance is found to be sensitive to the presence of hydrogen bonding in the crystal. Solutions of both Zn^{2+}(aq) and Cd^{2+}(aq) have been the subject of an extensive number of X-ray diffraction and

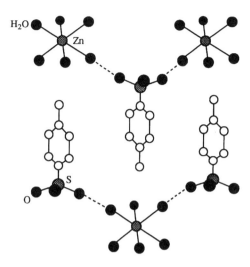

Figure 12.2. Unit cell structure of $[Zn(OH_2)_6](p\text{-}CH_3C_6H_4SO_3)_2$ showing the crystal packing arrangement

vibrational spectroscopic studies in aqueous solution. $[Zn(OH_2)_6]^{2+}$ is reported to persist at most concentrations of zinc salts with non-coordinating anions [27–29], although a change to a hydration number of four (tetrahedral) is said to result from solutions of $Zn(ClO_4)_2$ in aqueous $HClO_4$ as the acid concentration is increased [30]. Such a change to tetrahedral coordination is believed to occur as the mole fraction of water is decreased. Early temperature jump studies on aqueous $Zn(NO_3)_2$ solutions coupled with pH detection suggested a conversion to tetrahedral $[Zn(OH_2)_4]^{2+}$ as the temperature is raised [31] (cf. Co^{2+}(aq), Chapter 9). In more recent studies, however, predominantly hexaaqua coordination is said to persist over all salt concentrations studied [27]. Thus the extent of the octahedral–tetrahedral equilibrium (Fig. 12.3) under different conditions

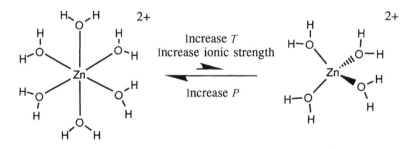

Figure 12.3. Octahedral-tetrahedral equilibrium for Zn^{2+}(aq)

remains a subject of contentious debate. The presence of coordinating anions such as SO_4^{2-} [32] and Cl^- [3, 33] also leads to a change in coordination (hydration) number as the concentration of anion is increased. A large-angle X-ray scattering study on aqueous $ZnCl_2$ is described in terms of octahedral geometry for $[ZnCl(OH_2)_5]^+$, trigonal bipyramidal geometry (planar $ZnCl_3$) for $[ZnCl_3(OH_2)_2]^-$ and tetrahedral coordination for $[ZnCl_4]^{2-}$. Typical Zn—Cl and Zn—OH_2 distances were 228 and 210 pm [33]. Similar findings are apparent from studies on the $ZnBr_2$ and ZnI_2 aqueous systems [34]. In a ^{67}Zn NMR study it was found that the oxy anions ClO_4^-, NO_3^- and SO_4^{2-} did not induce a detectable shift in the ^{67}Zn NMR resonance at concentrations of up to 0.08 m. Aqueous 2.0 m $Zn(ClO_4)_2$ solutions are reported as having a ^{67}Zn NMR linewidth of 50 Hz assigned to the presence of octahedral $[Zn(OH_2)_6]^{2+}$ [35]. Solution Zn—$O(OH_2)$ distances vary from 208 pm (ClO_4^- salt), 208 pm [29] and 215 pm [36] (SO_4^{2-} salt) to 209 pm [37] and 217 pm [27a] (NO_3^- salt). The mean value of ~ 210 pm compares well with the values from single-crystal X-ray diffraction studies. A Raman study conducted on $[Zn(OH_2)_6]MF_6$ (M = Si or Ti) has allowed assignment of the symmetric Zn—$O(OH_2)$ stretching frequency as 384 cm^{-1} [38]. The asymmetric Zn—O vibration occurs at 278 cm^{-1}.

Octahedral $[Cd(OH_2)_6]^{2+}$ is established in the solid state X-ray structures of $Cd(ClO_4)_2 \cdot 6H_2O$ (Cd—$O(OH_2)$ = 231 pm) [39] and the Tutton salt, $(NH_4)_2$-$[Cd(OH_2)_6](SO_4)_2$ (Cd—$O(OH_2)$ = 228 pm) [40], as well as in aqueous solution (Cd—$O(OH_2)$ = 229 pm) [37, 41]. Typical Cd—O distances of ~ 230 pm are established in other aqua complexes [39]. Aqueous 0.1 M $Cd(ClO_4)_2$ is the standard reference solution for ^{113}Cd NMR studies [42, 43]. Typical ^{113}Cd NMR chemical shifts are in the range 800 to -100 ppm from aqueous Cd^{2+}. Substitution by N donors shifts the resonances downfield although the largest downfield shifts are found for sulfur ligands, Cd—S compounds covering the range from 100 to 800 ppm.

12.1.2.2 HYDROLYSIS OF $[Zn(OH_2)_6]^{2+}$ AND $[Cd(OH_2)_6]^{2+}$

As with the preceding three first row M^{2+} ions, hydrolysis of $[Zn(OH_2)_6]^{2+}$ requires alkaline conditions before hydroxy species are formed. Hydrolysis constants and speciation, as in the case of other metal ions, has been largely deduced on the basis of the pH dependence of the solubility of $Zn(OH)_2$. The available data have been interpreted as involving hydrolysis of $[Zn(OH_2)_6]^{2+}$ to give $ZnOH^+$(aq) ($pK_{11} = 8.96$), $Zn_2(OH)^{3+}$(aq) ($pK_{21} = 9.0$) and eventually hydrous $Zn(OH)_2$ ($pK_{12} = 16.9$) at around pH 10 [44]. $Zn(OH)_2$ is amphoteric and at higher pH the anions $Zn(OH)_3^-$ (aq) ($pK_{12} = 28.4$), $[Zn(OH)_4]^{2-}$ ($pK_{14} = 41.2$) and $[Zn_2(OH)_6]^{2-}$ ($pK_{26} = 57.8$) are formed, the latter only in the presence of Cl^- [45]. The hydrolysis of $[Zn(OH_2)_6]^{2+}$ is known to be promoted by Cl^- ions [44]. A different value for pK_{11} (7.84) was reported in a later study [46] although there was close agreement in the pK values for the other monomeric hydrolysis products. The formation of ZnOH$^+$(aq) has been questioned in other work [47] wherein the data were satisfactorily accounted for by the presence only of dimeric Zn$_2$(OH)$^{3+}$ in addition to Zn(OH)$_2$. It appears that the value for the first hydrolysis constant for Zn^{2+}(aq) remains somewhat debatable. Most workers quote values for pK_{11} of between 9 and 10. Hydrolytic polymerisation to give Zn$_4$(OH)$_4^{4+}$(aq) (cf. Co^{2+} and Ni^{2+}), in addition to Zn$_2$(OH)$^{3+}$, has also been claimed [48]. The Zn^{2+}(aq)–CO$_2$ system above pH 6 around a log P_{CO_2} of -4 shows a dominant phase consisting of Zn$_5$(OH)$_6$(CO$_3$)$_2$ [49]. A similar 'pen-tazinc' moiety is established in the 'basic' nitrate Zn$_5$(OH)$_8$(NO$_3$)$_2$·2H$_2$O [50].

[Cd(OH$_2$)$_6$]$^{2+}$ is reported as having a slightly less tendency to hydrolyse than [Zn(OH$_2$)$_6$]$^{2+}$ [44] and, like Zn^{2+}(aq), requires alkaline conditions. Hydrolysis eventually occurs to give CdOH$^+$(aq) (pK_{11} = 10.08), Cd$_2$(OH)$^{3+}$(aq) (pK_{21} = 9.39) and Cd(OH)$_2$(aq) (pK_{12} = 20.35). In addition, the formation of Cd$_4$(OH)$_4^{4+}$(aq) (pK_{44} = 32.85) seems well established for solutions 0.1 M in Cd^{II}(aq). In more strongly alkaline solution the mononuclear anion Cd(OH)$_4^{2-}$(aq) (pK_{14} = 47.35) is formed. The additional formation of Cd(OH)$_3^-$(aq) (pK_{13} > 33) is somewhat more uncertain.

12.1.2.3   SUBSTITUTION REACTIONS ON [Zn(OH$_2$)$_6$]$^{2+}$ AND [Cd(OH$_2$)$_6$]$^{2+}$

Only a few detailed kinetic studies of complexation reactions on [Zn(OH$_2$)$_6$]$^{2+}$ and [Cd(OH$_2$)$_6$]$^{2+}$ have been described. Both ions are highly labile with rates on Zn^{2+} being somewhat lower. This is largely believed to be a reflection of the smaller size of the Zn^{2+} ion (74 vs 95 pm for Cd^{2+}) although mechanistic differences may also be apparent (see below). Reaction rates on both ions are fast enough to require the use of sophisticated fast reaction monitoring techniques. Ultrasonic absorption spectroscopy has been successfully applied in a number of cases. Few reactions have lent themselves to optical spectroscopic monitoring reflective of the absence of d–d bands. Certain reactions have allowed stopped-flow and temperature-jump monitoring of optical absorption changes. Ligands such as bipyridine and phenathroline give rise to changes within the u.v. region sufficient to allow kinetic monitoring of 1:1 equilibration reactions with both Zn^{2+}(aq) and Cd^{2+}(aq) using stopped-flow spectrophotometry. Moreover, monitoring under variable high pressure has been possible [51]. These reactions have provided a rare opportunity to compare in detail the complexation behaviour of the two elements. Table 12.1 lists kinetic data for these reactions along with selected data obtained using ultrasonic absorption and temperature-jump methods [52].

From the early ultrasonic relaxation studies, reactions on Zn^{2+}(aq) have long been interpreted as involving rate-determining dissociative loss of a coordinated water. The activation volumes for the 1:1 complexation reactions with bipy, $+7.1$ cm^3 mol^{-1} (Zn^{2+}) and -5.5 cm^3 mol^{-1} (Cd^{2+}), are of interest in indicating a changeover in mechanism down the group from I$_d$ (Zn) to I$_a$ (Cd)

Table 12.1 Kinetic data for selected complexation reactions on $[Zn(OH_2)_6]^{2+}$ and $[Cd(OH_2)_6]^{2+}$

Incoming	ligand	$k(25\,°C)$ $(M^{-1}s^{-1})$	$\Delta H^{\ddagger}$ $(kJ\,mol^{-1})$	$\Delta S^{\ddagger}$ $(J\,K^{-1}\,mol^{-1})$	$\Delta V^{\ddagger}$ $(cm^3\,mol^{-1})$	Ref.
$[Zn(OH_2)_6]^{2+}$						
NH_3[a]		0.8×10^7	—	—	—	[53]
Glycinate[a]		15.0×10^7	—	—	$+4$	[54]
bipy[b]	k_f	$(2.3 \pm 0.2) \times 10^6$	35.4 ± 0.6	-4 ± 2	$+7.1 \pm 0.4$	[51]
	k_b	13 ± 5	62.8 ± 0.9	-12 ± 3	$+3.6$	[51]
phen[b]	k_f	0.2×10^7	33.4	—	—	[55]
2-Cl-phen[b]	k_f	1.1×10^6	37.9	-2	$+5.0$	[56]
	k_b	8.9×10^2	57.3	$+4 \pm 12$	$+4.1 \pm 0.4$	[56]
terpy[b]		1.3×10^6	33.4	-17	—	[55]
Murexide[c]		0.8×10^7	25.1	-25.1	—	[57]
Acetate[a]	k_f	3×10^7	—	—	—	[58]
Cl^{-}[a]	k_f	1.3×10^7	—	—	—	[59]
Br^{-}[a]	k_f	1.5×10^7	—	—	—	[59]
SO_4^{2-}[a]	k_f	3.2×10^7	—	—	—	[60]
$[Cd(OH_2)_6]^{2+}$						
bipy[b]	k_f	$(3.3 \pm 0.3) \times 10^6$	24.0 ± 2.3	-39.7 ± 8.2	-5.5 ± 1.0	[51]
	k_b	160 ± 20	46.1 ± 2.8	-48.4 ± 9.7	-6.9 ± 1.2	[51]
terpy[b]		3.2×10^6	25.1	-29.3	—	[55]
Murexide[c]		1.26×10^8	—	—	—	[61]
NCS^{-}[a]	k_f	4.9×10^8	—	—	—	[62]
	k_b	2.1×10^7	—	—	—	[62]
Acetate[a]	k_f	2.5×10^8	—	—	—	[58]
Cl^{-}[a]	k_f	4.0×10^8	—	—	—	[63]
SO_4^{2-}[a]	k_f	$> 10^8$	—	—	—	[63]
$EDTA^{4-}$[d]		1×10^9	—	—	—	[64]

[a] By ultrasonic absorption.
[b] By stopped-flow spectrophotometry.
[c] By temperature jump.
[d] By ^1H NMR.

[51]. Such a changeover has been attributed to the increase in ionic radius but may also be simply related to changes in partial molar volume [65]. On the basis of Swaddle's correlation of $V°$ with $\Delta V^{\ddagger}$ for water exchange on a range of aqua cations, values for $\Delta V_{ex}^{\ddagger}$ $(cm^3\,mol^{-1})$ were predicted as $+6$ (Zn^{2+}) and -8 (Cd^{2+}).

The slightly larger spread of rate constants for complexation with different ligands on Cd^{2+} (aq) vs Zn^{2+} (aq) (Table 12.1) [52, 63, 66] would also seem to be consistent with more associative character for Cd^{2+}. In this regard it is pertinent to note the formation of several seven-coordinate Cd^{II} complexes, namely $Cd(BrO_3)_2 \cdot 2H_2O$ [67] and $Cd(O_3SCH_2SO_3) \cdot 3H_2O$ (distorted pentagonal bipyramid) [68]. Zn^{II} complexes invariably show a maximum of six coordination. Values for $k_{ex}(H_2O)$ are difficult to estimate exactly. Ultrasonic relaxation data obtained from substitution by various ligands, if interpreted in an Eigen–Wilkins manner, would be suggestive of $k_{ex}(H_2O)$ values of $\sim 3 \times 10^7$ s^{-1} and $\sim (0.2$–$4) \times 10^8$ s^{-1} respectively for Zn^{2+}(aq) and Cd^{2+}(aq) [63, 65]. From theoretical estimates of the K_{os} values for a number of complexation reations, Merbach and coworkers [51] found reason to extend the upper limit for $k_{ex}(H_2O)$ on Zn^{2+} (aq) to $\sim 6 \times 10^8$ s^{-1}. Higher rate constants and a wider range of values are generally found for complexation reactions on Cd^{2+}(aq) [52, 55, 57–59, 61, 63, 69] vs on Zn^{2+}(aq). The reason may be mechanistic [51, 65], but probably also stems from the greater covalency in the bonds to Cd^{2+}. For this reason the upper limit for $k_{ex}(H_2O)$ set for Zn^{2+}(aq) [51] may be somewhat high. The predicted/experimental values for $\Delta V_{ex}^{\ddagger}$ for Zn^{2+}(aq) show that the gradual changeover in mechanism from associative (early members) to dissociative (later ones) along the first row is not manifested in the limiting positive value for Zn^{2+} predicted by Swaddle and Mak to be $+9$ cm^3 mol^{-1} [65]. The reason may be the overriding decrease in ionic radius, promoted by poor shielding from the filled shell of d orbitals. Also contributions from the small amount of tetrahedral $[Zn(OH_2)_4]^{2+}$, in equilibrium with the hexaaqua from [31], have yet to be fully evaluated.

Carbon-13 NMR has been used with some success in the study of aqueous complexation reactions of Zn^{2+}(aq) by a number of polyamines [70]. Some insights into possible solution conformations of the complexes were gained. A growing number of studies are using the diamagnetism of d^{10} Zn^{II} and Cd^{II} complexes to employ NMR in the study of solution structures in complexes.

The most important Zn^{II}–OH_2 moiety overall is probably that present at the active site of a whole range of hydrolytic enzymes [4a] and as such requires some discussion. Three distinct mechanisms have been proposed [2] for the Lewis acid catalysis: (a) hydrogen bonding via a coordinated water molecule (OH group) on the Zn^{2+}, leading to promoted nuclephilic hydrolysis by a free water molecule at a carbonyl group on the substrate, (b) direct coordination of the carbonyl group at the Zn^{2+}, leading to enhanced hydrolysis or (c) as in (b) with hydrolysis occurring via a water (OH$^-$) ligand bound to the Zn^{2+} (Fig. 12.4). In the so–called hydration–dehydration mechanism for CA all three aspects may be involved [2].

The usefulness of Co^{2+} as a spectroscopic substitute for the Zn^{2+} has been highlighted [1a, 2, 6]. As a result it is generally agreed that the geometry at the Zn^{2+} centre in both CP and CA is a distorted tetrahedron with coordination of substrate possible in a fifth position. Coordination of Zn^{2+} to three histidine imidazoles and a water molecule is relevant for CA, whereas for CPA coordination to two histidines and a glutamate carboxyl is involved. As mentioned above, a crucial role of the amino acid ligands might be to promote a lowering of the pK_a value for the Zn^{II}–OH_2 group into the physiological range of pH (~ 7), cf. pK_a for Zn^{2+}(aq) $= \sim 9$. This has also been achieved in a number of simple model Zn^{II}

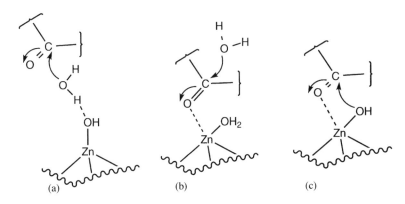

Figure 12.4. Mechanisms for Zn^{II}-promoted hydrolysis of carbonyl substrates

complexes such as that with the macrocyclic ligand 1, 5, 9-triazacyclododecane (Fig. 12.5) and derivatives thereof [4a, 71]. In further work by Kimura's group a coordinating pendent hydroxyethyl group placed on the macrocycle was found to readily deprotonate with a pK_a of 7.4 [72]. The resulting pendent $CH_2CH_2O^-$ group (Fig. 12.6) was found to be an even more potent nucleophile than coordinated hydroxide for the catalysis of *p*-nitrophenylacetate hydrolysis [72]. These findings are of interest in pointing towards a possible nucleophilic role for certain proximal serine residues as well as illustrating the nucleophilicity of Zn^{II}–OH. These model systems also mimick AD in some of their reactions

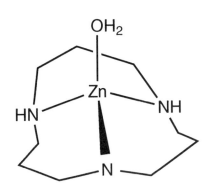

Figure 12.5. Structure of the aqua (1,5,9-triazacyclododecane)zinc(II) cation

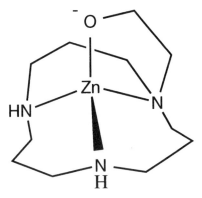

Figure 12.6. Structure of N-2'-oxyethyl-1,5,9-triazacyclododecanezinc(II)

(hydride transfer) [73]. The presence of coordinated cysteine in AD illustrates the diversity of binding sites for Zn^{2+} in its enzymes. The rate of uptake of a number of divalent metal ions including Zn^{2+} by apo-CPA has been studied and compared with the rates of water exchange and of complexation with simple neutral ligands such as 1,10-phenanthroline. The rate constant for Zn^{2+} was found to be approximately 10 times that for Co^{2+} uptake, exactly paralleling the difference between the formation rate constants for the two metal ions with 1,10-phen [74]. Thus it was concluded that the metal ion binding site in CPA is flexible and reminiscent of the properties of a simple ligand. Moreover, relative formation constants for binding of M^{2+} to apo-CPA are in the order Mn(CPA) < Co(CPA) < Ni(CPA) < Cu(CPA) > Zn(CPA), consistent with the Irving–Williams series. As a result of this flexibility in coordination geometry, despite the relatively high levels of free Zn^{2+} in the body, the element is readily displaced from its enzyme binding sites by Cd^{2+} and Hg^{2+} [5], this being responsible in part for the acute toxicity of these metal ions. The biochemistry and biologically relevant coordination chemistry of Zn^{2+} has been reviewed in an excellent series of articles on zinc enzymes edited by Spiro [1a] as well as in a number of other texts [1b, 2, 4a].

12.1.2.4   COMPLEXATION OF Zn^{2+} (AQ) BY AMINO ACIDS AND PEPTIDES

The highest formation constants for 1:1 and 1:2 amino acid complexes of Zn^{2+}(aq) are not surprisingly with cysteine (log $\beta_1 = 9.17$, log $\beta_2 = 18.18$), penicillamine (log $\beta_1 = 9.5$, log $\beta_2 = 19.4$), histidine (log $\beta_1 = 6.51$, log $\beta_2 = 12.04$), DAPA (log $\beta_1 = 6.31$, log $\beta_2 = 11.66$) and DABA (log $\beta_1 = 6.7$, log $\beta_2 = 12.1$), reflecting favourable tridentate coordination [75, 76]. Penicillamine is unusual in that the stepwise formation constant for the 1:2 complex is actually greater than that of the 1:1, believed to be due to a change from tridentate to bidentate coordination. Zn^{2+} complexes with a number of simple peptides have been studied [76]. Interestingly, it appears that even in alkaline solution Zn^{2+} does not promote amide deprotonation in contrast to the behaviour of Ni^{2+} and Cu^{2+}. The lack of LFSE and greater affinity for OH^- as a ligand in basic solution in the case of Zn^{2+} is believed responsible. Zn^{2+} complexes with a number of cysteinyl peptides have been recently studied, their enhanced stability arising from favourable bonding to the deprotonated thiol group [77].

12.2   MERCURY

Much has been written of the reasons for the extremely low melting point $(-38.7\,^{\circ}\text{C})$ and heat of vaporization (59.1 kJ mol^{-1}) of elemental mercury as stemming from relativistic effects [78]. These relate to the anomalously low energy and thus stability of the outer 6s valence electrons. This arises from a 'knock-on' effect resulting from the inner core 1s electrons being accelerated to radial velocities that approach the speed of light, $\sim c/2$ in the case of mercury, resulting in relativistic mass increase and subsequent Bohr radius contraction and lowering of energy. This energy lowering is transmitted successively to all s electrons to include the valence 6s electrons of gold and mercury. A further consequence to the valence shell is the resulting large 6s–6p energy gap which limits effective hybrid orbital formation to linear sp, explaining the preponderance of linear coordination in compounds of Au^I and both Hg^I and Hg^{II}. Indeed, almost pure 6s character has been envoked. The liquid property of mercury at room temperature and in turn the inertness and unreactivity of metallic gold, in particular the very high Au^+/Au^o potential, have been cited as indicative of an inaccessible valence shell. It is for these reasons that mercury differs somewhat in its chemistry from that of zinc and cadmium and requires separate consideration.

Despite the presence of two stable oxidation states for mercury there is very little redox chemistry to speak of, primarily as a result of the ready formation of compounds with Hg—Hg bonds for Hg^I but not for Hg^{II}. This gives some degree of kinetic control to the Hg^I/Hg^{II} redox change while in turn stabilising Hg^I towards disproportionation, cf. Zn^I and Cd^I. The filled d^{10} shell is also unavailable, as it is for both zinc and cadmium, and likewise stable compounds of Hg^{III} have yet to be isolated.

12.2.1 Mercury(I) ($s^1 d^{10}$): Chemistry of the $[Hg_2(OH_2)_2]^{2+}$ Ion

12.2.1.1 STABILITY AND PROPERTIES

The cation $Hg_2{}^{2+}$(aq) is the last of the series of M—M bonded dimeric aqua ions following $Mo_2{}^{4+}$(aq) and $Rh_2{}^{4+}$(aq) and can be regarded as the *simplest* binuclear compound that contains a single metal–metal bond. The linear diaqua structure (Fig. 12.7) found in a number of hydrated Hg^I salts, such as $Hg_2(NO_3)_2 \cdot 2H_2O$ [79], $Hg_2SiF_6 \cdot 2H_2O$ [80] and $Hg_2(ClO_4)_2 \cdot nH_2O$ [81] ($n = 2$ or 4), persists in aqueous solutions [82]. Typical Hg—Hg and Hg—OH_2 distances of ~ 250 and ~ 214 pm are found in the solid hydrates. The Hg—Hg—OH_2 group is almost linear, the angle ranging from 160 to 180° in

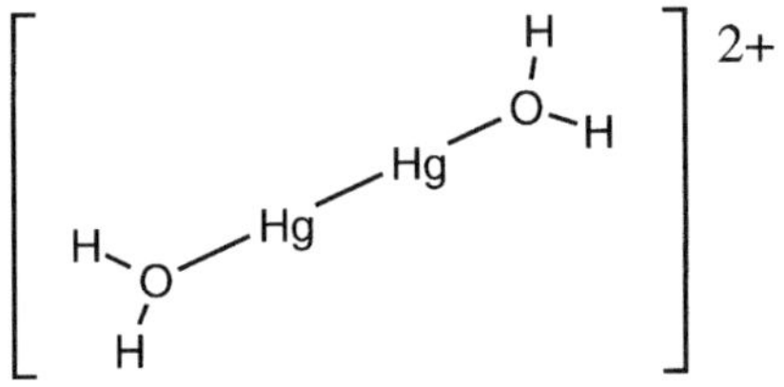

Figure 12.7. Structure of the $[Hg_2(OH_2)_2]^{2+}$ cation in hydrated Hg^I_2 salts

different compounds. This can be attributed to H bonding and/or weak coordination of anions perpendicular to the Hg—Hg bond direction. The $[Hg_2(OH_2)_2]^{2+}$ cation is easily preparable via dissolution of soluble mercurous salts in water or acids such as $HClO_4$. Earlier procedures also involved treatment of mercury with concentrated HNO_3 and precipitation of HgO via sodium carbonate. Disso-lution of the HgO in 2.0 M $HClO_4$ generated an Hg^{2+}(aq) solution which was converted to Hg_2^{2+}(aq) via treatment with excess mercury (stirring) for 24 hours [83]. Alternatively, commercial HgO or $Hg(ClO_4)_2 \cdot 6H_2O$ can be used as the Hg^{II} lead-ins [84]. Concentrations of Hg^I can be estimated by titration against NaCl using bromophenol blue as indicator. Alternatively, the intense band maximum at 237 nm ($\varepsilon = 2.67 \times 10^4$ M^{-1} cm^{-1}) of Hg_2^{2+}(aq) (Fig. 12.8) can be used [83]. This band has been assigned to the $\sigma(6s)–\sigma^*(6s)$ transition within the Hg—Hg bond [83]. This assignment has been confirmed in photolysis studies at the 237 nm peak which generates Hg^{2+}(aq) (Fig. 12.8) [85]. Photolytic disproportionation of Hg_2^{2+}(aq) has also been reported [86]. In an earlier report Cartledge [87] calculated the dissociation constant for the Hg—Hg bond to form monomeric Hg^+(aq) as 10^{-31}M by estimating the hydration energy of Hg^+(aq). In pulse radiolysis studies reduction of Hg_2^{2+}(aq) by H atoms is said to produce Hg_2^+(aq) ($\lambda_{max} = 285$ nm ($\varepsilon = 9000$ M^{-1} cm^{-1}) which decays by a second-order reaction attributed to disproportionation to give Hg_2^{2+}(aq) and Hg^0 [88]. Hg^+(aq) can be similarly produced by reduction of Hg^{2+}(aq). The

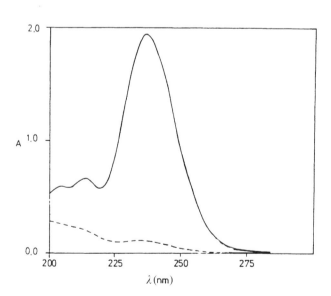

Figure 12.8. Electronic spectra of $[Hg_2(OH_2)_2]^{2+}$ (—) and $[Hg(OH_2)_6]^{2+}$ (----) in 0.01 M $HClO_4$

transient Hg^+(aq) ion generated is reported to have $\lambda_{max} = 272$ nm ($\varepsilon = 7800$ M^{-1} cm^{-1}). Subsequent decay processes following the pulse have been attributed firstly to rapid disproportionation giving Hg^0 and Hg^{2+}(aq) ($k = 2.5 \times 10^9 M^{-1}$ s^{-1}) which then recombine slowly to form Hg_2^{2+}(aq) [89]. It was presumed that water ligand exchange (dissociation) was occurring too slowly on Hg^+(aq) for the dimerisation reaction to compete with the disproportionation [90]. The decay of $Hg^I Cl$ (from reduction of aqueous $Hg^{II}Cl_2$), on the other hand, is attributable to dimerisation ($k = 4 \times 10^9$ M^{-1} s^{-1}) [91]. In his review Stanbury [11] reported potentials calculated for a number of aqueous Hg couples, Hg^{2+}/Hg^+ (0.0 V), $Hg^+, Hg^{2+}/Hg_2^{2+}$ (+1.83 V), Hg^+/Hg^0 (+1.33 V) and Hg_2^{2+}/Hg^0, Hg^+ (−0.5 V). The calomel reference electrode, used widely in potentiometry and voltammetry, is based upon the potential of the Hg_2Cl_2/Hg^0 couple in saturated aqueous KCl. The potential is stable and its temperature dependence is known to a high degree of accuracy. At 25 °C the E^θ value is +0.244V.

12.2.1.2 HYDROLYSIS OF $[Hg_2(OH_2)_2]^{2+}$

The following reaction has an equilibrium quotient of 132 (25 °C) [92] which shows that Hg_2^{2+}(aq) is marginally stable towards disproportionation:

$$Hg^0 + Hg^{2+} \rightleftharpoons Hg_2^{2+} \tag{12.1}$$

A consequence of this equilibrium, however, is that stable mercurous hydroxide cannot form since, on addition of OH^-, Hg^{2+} is more rapidly hydrolysed, resulting in precipitation of HgO which promotes the disproportionation process. In one study, the effect of Hg^{2+}(aq) production was taken into account and a pK_{11} value of 5.0 was estimated for the formation of $Hg_2(OH)^+$ in 0.5 M $NaClO_4$ [93]. Further information of Hg^I hydrolysis products has relied upon pulse radiolysis studies on HgO in water. Reduction of HgO by e^-(aq) reportedly generates the monomeric hydroxy species HgOH in a rapid process $k = 2.3 \times 10^{10}$ M^{-1} s^{-1}):

$$HgO + e^-(aq) \xrightarrow{H_2O} HgOH + OH^- \tag{12.2}$$

'HgOH' is characterised by a peak at 233 nm and shoulder at ~ 265 nm ($\varepsilon = 5.3 \times 10^3$ M^{-1} cm^{-1}) and decays by a second-order process ($k = 2.2 \times 10^9$ M^{-1} s^{-1}) independent of ionic strength, consistent with dimerisation, to give $Hg_2(OH)_2$ [94]. From the pH dependence of the absorption spectrum a pK_{11} value for the reaction

$$Hg^+ + H_2O \rightleftharpoons HgOH + H^+ \tag{12.3}$$

was estimated to be 5.1. These studies also allowed subsequent monitoring of the slower disproportionation of $Hg_2(OH)_2$ ($k = 1.2 \times 10^4$ s^{-1}) [94]:

$$Hg_2(OH)_2 \text{ (or } Hg_2O) \longrightarrow Hg^0 + Hg(OH)_2 \text{ (or } HgO) \tag{12.4}$$

The potential for the $Hg(OH)^+/HgOH$ couple was estimated to be -0.09 V [11, 94]. Vibrational spectra on the crystalline hydrolysis products from mercurous nitrate have been reported [95]. A crystal structure of one of these phases exhibited infinite chains consisting of —Hg—Hg(OH)— units (Hg—Hg = 250 pm) with additional Hg_2^{2+} ions (Hg—Hg = 249.8 pm) coordinated to the OH groups in the chains and nitrate ions (Hg—O = 221 pm).

Kinetic studies of the oxidation of Hg_2^{2+}(aq) in perchlorate solution appear to be consistent with different processes for both one- and two-electron reagents [84]. Rate-determining Hg—Hg bond breaking accompanies oxidation of Hg_2^{2+}(aq) in the case of one-electron oxidants (e.g. Ce^{IV}, Mn^{III}, $[Fe(phen)_3]^{3+}$ and $[Ru(bipy)_3]^{3+}$) whereas rapid disproportionation, followed by rate-determining oxidation of Hg^0(aq) to Hg^{2+}(aq), appears to be relevant for two-electron oxidants, e.g. BrO_3^-, Tl^{III}:

$$\text{1e}^-\text{ oxidants} \begin{cases} Hg_2^{2+} - e^- \xrightarrow{k} Hg^+ + Hg^{2+} & (12.5) \\[2em] Hg^+ - e^- \xrightarrow{\text{fast}} Hg^{2+} & (12.6) \end{cases}$$

$$\text{2e}^-\text{ oxidants} \begin{cases} Hg_2^{2+} \underset{}{\overset{\text{fast, }K}{\rightleftharpoons}} Hg^0 + Hg^{2+} & (12.7) \\[2em] Hg^0 - 2e^- \xrightarrow{k} Hg^{2+} & (12.8) \end{cases}$$

Consistent with the two-electron oxidation mechanism the rate law here carries an $[Hg^{2+}]^{-1}$ term. Reactions rates for one-electron oxidants tend to be slower than those for two-electron oxidants [84]. The sluggish reactivity (Ce^{IV}, $+1.73$ V; Mn^{III}, $+1.51$ V) or non-reactivity (Np^{VI}, $+1.14$ V) of certain one-electron oxidants would appear to be consistent with the high potential for the Hg^{2+}, Hg^+/Hg_2^{2+} couple estimated independently as $+1.83$ V [11, 96]. However, the oxidants $[Ru(bipy)_3]^{3+}$ ($+1.27$ V) and $[Fe(phen)_3]^{3+}$ ($+1.11$ V) are found to react quite readily with Hg_2^{2+}(aq) and certainly much faster than either Ce^{IV} or Mn^{III}, despite their lower reduction potentials. The Fe and Ru oxidants are highly favoured outer-sphere reactants possessing empty π^*-acceptor orbitals on the ligands which might facilitate e^- transfer from the somewhat shielded 6s valence orbitals of Hg_2^{2+}(aq). The irreversibility of the oxidation process, breakage of dimeric Hg_2^{2+} to form monomeric products, presumably drives the reaction to completion. Preference for an outer-sphere mechanism would also appear to be likely given the reluctance of Hg_2^{2+}(aq) to form further coordinate bonds other than those to the two H_2O ligands. Furthermore, the catalysis of the Ce^{IV} oxidation of Hg_2^{2+}(aq) by $IrCl_6^{3-}$ is strongly indicative of a favourable outer-sphere process [97].

The formation of one-dimensional linear chain structures for many Hg_2^{2+} compounds illustrates further the reluctance of the Hg_2^{2+} ion to form coordinate bonds other than along the Hg—Hg bond direction and likely linear sp or purely s character of the metal valence orbitals. At the same time complex formation tends to promote disproportionation of Hg_2^{2+} to generate the more extensively complexed Hg^{2+} ion. As a result the coordination chemistry of Hg_2^{2+}(aq) is somewhat limited [98].

12.2.2 Mercury(II) (d^{10}): Chemistry of the $[Hg(OH_2)_6]^{2+}$ Ion

12.2.2.1 STRUCTURE AND PROPERTIES

Hg^{2+}(aq) is normally described as being hexacoordinated by water [98]. A degree of uncertainty has, however, stemmed from the ready hydrolysis of Hg^{2+}(aq) and the variety of coordination geometries that mercury adopts in complexes. As with Zn^{2+}(aq) and Cd^{2+}(aq) the d^{10} Hg^{2+}(aq) ion has no electronic spectroscopic handle. However, X-ray scattering studies of acidified perchlorate solutions of Hg^{2+}(aq) are in support of retention of the same octahedral $[Hg(OH_2)_6]^{2+}$ structure [99] as present in the X-ray structure of crystalline $Hg(ClO_4)_2 \cdot 6H_2O$ (Fig. 12.9)[100].

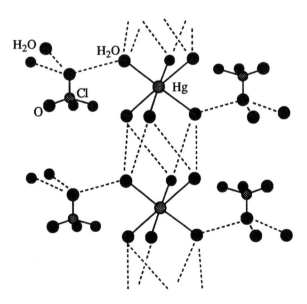

Figure 12.9. View of the structure of $[Hg(OH_2)_6](ClO_4)_2$ showing a string of $Hg(OH_2)_6$ octahedra along the z axis

Most other salts are either insoluble or sparingly so. Solutions are usually standardised by precipitative titrimetry with, for example, Cl^- or NCS^- (cf. Ag^+) since there is no spectroscopic handle. In other crystalline hydrated oxoanion salts, octahedral $Hg^{II}O_6$ units are present involving coordination to both water and the oxo anions. In certain anhydrous salts such as Hg_3TeO_6, both distorted octahedral HgO_6 and tetrahedral HgO_4 units are present. Anhydrous $HgSO_4$ contains distorted HgO_4 tetrahedra whereas in the monohydrate $HgSO_4 \cdot H_2O$ highly distorted HgO_6 octahedra (two short and four long Hg—O bonds, see below) are present. These structures illustrate the flexible coordination geometry around Hg^{II}. Extensive complexation by oxo anions is a feature of the coordination chemistry of aqueous Hg^{II}, in contrast to that of Hg^I [98].

Solutions containing $[Hg(OH_2)_6]^{2+}$ are readily produced by dissolution of HgO in non-complexing or weakly coordinating acids such as $HClO_4$ or HNO_3 or by simple dissolution of the highly soluble commercial perchlorate and nitrate salts. An extensive number of basic salts result if an excess of HgO is employed.

12.2.2.2   HYDROLYSIS OF $[HG(OH_2)_6]^{2+}$

HgO is only weakly basic and hydrolysis of Hg^{2+}(aq) readily occurs in weakly acidic solution characterised by the formation of $HgOH^+$(aq) ($pK_{11} = 3.4$), $Hg_2(OH)^{3+}$ ($pK_{21} = 3.33$), $Hg(OH)_2$(aq) ($pK_{12} = 6.17$) and eventually $Hg(OH)_3^-$ (aq) ($pK_{13} = 21.1$) [101, 102]. In addition the formation of polynuclear species such as formally $Hg_3(OH)_3^{3+}$ (aq) ($pK_{33} = 6.42$) is required in order to explain the hydrolysis data at $[Hg^{II}] \geqslant 0.1$ m prior to precipitation of $Hg(OH)_2$ [101]. The hydrolysis products of Hg^{2+}(aq) have been the subject of a number of detailed structural studies. Results obtained from X-ray scattering studies of concentrated $Hg(ClO_4)_2$ solutions ($\geqslant 3.5$ M) have led to the assignment of structures for the various Hg^{II}(aq) species, as shown in Fig. 12.10 [99]. A notable

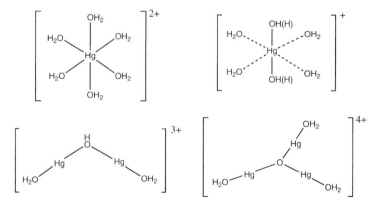

Figure 12.10. Solution structures for hydrolysis products of Hg^{2+}(aq)

observation is the tendency for Hg^{II} to move towards linear two coordination as the extent of hydrolysis increases. In $HgOH^+$(aq) two of the Hg—O bonds are shortened to distances between 190 and 210 pm at the expense of a lengthening of the other four Hg—O bonds (to ~ 250 pm). This may be compared with an average Hg—O distance of 234–241 pm found in the regular octahedral $[Hg(OH_2)_6]^{2+}$ ion from both solid state and solution studies. The structure change from distorted octahedral to two-coordinate linear has been rationalised by Barnum using an empirical approach based upon comparisons of predicted and experimental ΔG_f^θ of various hydroxy aqua species vs the number of coordinated hydroxides [103]. The results of an infra-red study on solutions of $[M(OH_2)_6](ClO_4)_2$ salts for all three Group 12 elements, coupled with earlier SCF calculations, were interpreted as indicating a significantly low 5d–6s excitation energy for the $[Hg(OH_2)_6]^{2+}$ ion, presumably as a result of the relativistic contraction in the 6s orbital energy. The apparent broad range of Hg—O bond lengths in Hg^{2+}(aq) species, stemming from the distortion from regular octahedral towards two-coordinate linear, was rationalised by a second-order Jahn–Teller effect resulting from a vibronically coupled excitation from the ground state A_g ($d^{10}s^0$) to the lowest excited E_g state (formally $d^9 s^1$) [104]. A similar change in structure upon hydrolysis, in this case probably from tetrahedral to linear, was suggested for d^{10} Ag^+(aq) (cf. the conflicting reports of the hydration number of Ag^+(aq) in Chapter 11).

The triangular μ_3-O centred structure $[Hg_3O(OH_2)_3]^{4+}$ (Fig. 12.10), or rather its conjugate base $[Hg_3O(OH)OH_2)_2]^{3+}$, was suggested for the 3+ trinuclear hydrolysis product on the basis of crystal structures obtained from a number of basic Hg^{II} perchlorates, each of which showed a planar μ_3-oxo structure and an approach to linear two coordination at Hg, the more distant Hg—O contacts now occurring at > 270 pm [105]. Three different crystalline types of basic perchlorate salt have been identified; triclinic $Hg_5O_2(OH)_2(ClO_4)_2(OH_2)_2$, orthorhombic $Hg_7O_4(OH)_2(ClO_4)_4$ and monoclinic $Hg_2O(OH)(ClO_4)$ [99, 105]. In the aqua dimer the angular $[(H_2O)Hg(OH)Hg(OH_2)]^{3+}$ structure (Fig. 12.10) can be considered as a section of the 'zig-zag' ($—HgOH^+—)_n$ chain network structurally characterised in a number of other basic salts such as $Hg(OH)NO_3$ [106], $Hg_3(OH)_2(SO_4)_2 \cdot H_2O$ [107] and $Hg(OH)(BrO_3)$ [108]. Typical Hg—Hg distances in the polynuclear species range from 338 to 362 pm, similar to those in other oxo centred basic trinuclear metal complexes ruling out direct Hg—Hg bonding. The results appear to suggest that linear O—Hg—O units are the basic building blocks for the polynuclear hydrolysis products of Hg^{2+}(aq) and its basic salts.

For more dilute Hg^{II} solutions ($< 10^{-5}$ m) treatment of the hydrolysis data only requires consideration of monomeric $HgOH^+$(aq) in addition to $Hg(OH)_2$(aq) [101, 102]. $Hg(OH)_2$(aq) is itself unstable with the respect to the formation of hydrated HgO.

12.2.2.3 COMPLEX FORMATION AND WATER EXCHANGE ON $[Hg(OH_2)_6]^{2+}$

Substitution reactions at Hg^{2+}(aq) occur extremely rapidly and study by relaxation techniques is required, some of which date back to the earliest studies of rapid equilibration reactions. Eigen and Eyring studied the rate of formation of $HgCl_2$ from Hg^{2+}(aq) using an electric field dispersion method. A rate constant of $(5-10) \times 10^9$ $M^{-1}s^{-1}$ (24–25 °C) was deduced [109]. Tamura *et al.* have reported rate constants for equilibration of acetate with Hg^{2+}(aq) using ultrasonic relaxation, $k_f = (7.8 \pm 0.9) \times 10^9$ $M^{-1}s^{-1}$, $k_b = (1.6 \pm 0.2) \times 10^6$ s^{-1} (30 °C) [110]. Equilibration of Hg^{2+}(aq) with phen, bipy and OH^- have been examined by the temperature-jump method [111]. Rate constants for phen and bipy are close to the diffusion controlled limit (log $k = 10.5$, phen; ~ 9.5, bipy) whereas those for phenH$^+$ and bipyH$^+$ were some 10^2–10^3 times slower. The invariance of rate constants for Hg^{2+}(aq) close to the diffusion-controlled limit (10^{10} s^{-1}) has been interpreted as suggesting the same Eigen mechanism as that occurring for reactions on other divalent ions, namely the rate-determining dissociation of a water ligand. Eigen estimated the rate constant for this process on Hg^{2+}(aq) as 2×10^9 s^{-1} [63]. This places the water-exchange rate on Hg^{2+}(aq) as having the same magnitude as that occurring on Cr^{2+}, Cu^{2+}, the lighter lanthanides and the heavier alkali and alkaline earth metals. Most aqueous solution studies on Hg^{2+}(aq) have concentrated on measurements of stability constants which have illustrated the high affinity of Hg^{II} for soft ligand donors such as sulfur and iodide. This tendency is amplified in the studies conducted on the methylmercury cation (Section 12.2.4).

The affinity of Hg^{II} for halides and in particular sulfur has been utilised in the catalysis of aquation and metal ion dissociation processes, the latter particularly when sulfur donors are involved . Satchell and co-workers have studied the Hg^{2+}(aq)-promoted desulfurisation of thiocarbamates as well as the promoted hydrolysis of a number of sulfur-containing compounds including disulfides and 2-phenyl-1,3-oxathiolan [112]. Rorabacher and coworkers have reported efficient metal ion dissociation from macrocyclic tetrathiaether complexes promoted by Hg^{2+}(aq) [113]. The affinity for sulfur donors is moreover exemplified in the development of an Hg^{II} selective electrode based upon the use of a neutral crown thiaether ligand [114].

12.2.3 The Double-Cubane Ion $[(H_2O)_9Mo_3S_4HgS_4Mo_3(OH_2)_9]^{8+}$

A method for the spectrophotometric determination of mercury has made use of the observation of incorporation of the metal into the trinuclear aqua ion $[Mo_3S_4(OH_2)_9]^{4+}$ to form the edge-shared double-cubane aqua ion $[(H_2O)_9-Mo_3S_4HgS_4Mo_3(OH_2)_9]^{8+}$ [115].

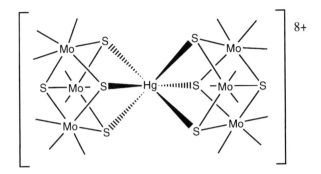

Figure 12.11. Structure of the cationic unit in $[(H_2O)_9Mo_3S_4HgS_4Mo_3(OH_2)_9(pts)_8 \cdot 20H_2O$

The structure of $[(H_2O)_9Mo_3S_4HgS_4Mo_3(OH_2)_9](pts)_8 \cdot 20H_2O$ [116] is shown in Fig. 12.11. As in the case of other such edge-shared mixed Mo—S cubane ions (see Section 6.2.3.4), the Hg atom can be considered as being in an octahedral HgS_6 environment arising from complexation to what is in effect two electron-rich $[Mo_3S_4(OH_2)_9]^{4+}$ 'ligands'. The octahedral HgS_6 arrangement is actually quite rare. The preference for Hg^{II}, when surrounded by sulfurs, to be tetrahedrally four-coordinated or linearly two-coordinated is exemplified in the structure of the $Hg_3S_4{}^{2-}$ ion, present in $Na_2[Hg_3S_4] \cdot 2H_2O$, wherein both the latter two situations are found [117].

12.2.4 The Methylmercury Cation: $[CH_3Hg(OH_2)]^+$

The methylmercury cation (Fig. 12.12) [118] is one of the most simplest and most harmful cationic species existing in the environment [119]. This stems from its acute lipophilicity, mobility, kinetic lability and high affinity for the soft donor

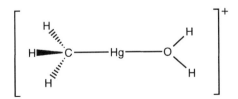

Figure 12.12. The methylmercury cation

sulfur [120–122]. Degradation of mercury compounds, particularly if contact with the marine environment is involved, invariably results in the generation of $CH_3Hg(aq)^+$ and related species thereof (Fig. 12.13). As a result there has been extensive studies of the hydrolysis and complexation behaviour of the methylmercury cation in solution [120]. The hydrolysis of $[CH_3Hg(OH_2)]^+$ has been studied by a number of groups [118, 121–123]. Hydrolysis constants have been reported for the formation of $CH_3HgOH(aq)$ ($pK_{11} = 4.65$) and $(CH_3Hg)_2OH^+(aq)$ ($pK_{21} = 1.73$) [121]. The formation of the oligonuclear species only becomes significant for concentrations of methylmercury in excess of 10^{-3} M. In addition, the formation of the μ_3-oxo-centred trinuclear cation $(CH_3Hg)_3O^+(aq)$ has been proposed [121, 122], its existence having been confirmed in a number of isolated crystalline salts [124]. The formation of neutral $CH_3HgOH(aq)$ would be expected to further enhance the lipophilicity of methyl mercury.

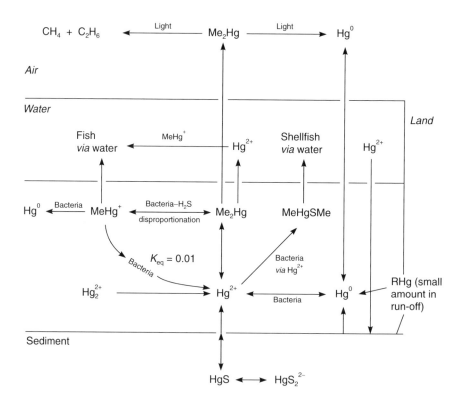

Figure 12.13. The environmental mercury cycle

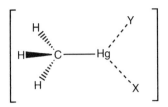

Figure 12.14. Associative transition state in the replacement of ligand X in CH_3HgX by Y

Complexation rates on $[CH_3Hg(OH_2)]^+$ are often too fast to follow directly as they are close to, if not at, the diffusion-controlled limit and studies have largely concentrated on ligand exchange processes of the type

$$CH_3HgX + Y \underset{k_b}{\overset{k_f}{\rightleftharpoons}} CH_3HgY + X \tag{12.9}$$

Studies by Geier and coworkers [125] show a large range of exchange rate constants for different X ligands and a transition state of the form shown in Fig. 12.14 is indicated. An associative I_a mechanism has also been deduced on the basis of rate-equilibria correlations for a variety of incoming ligands on $CH_3Hg^+(aq)$ [125]. Stability constant data for CH_3Hg^+ is extensive and has been comprehensively reviewed [120]. The largest values are with S^- ligands such as mercaptalbumin, glutathione and cysteine, S^{2-}, CH_3HgS^- and $S_2O_3^{2-}$. CH_3HgSR species generally have log β_1 values $= 14–18$. i.e. some 8 log units higher than those for, for example, harder N donors. The softer halides such as I^- are also characterised by high values of log β_1 (~ 12). Despite the high stability constants, the extreme kinetic lability and in particular the facile associative ligand exchange mechanism makes $CH_3Hg^+(aq)$ highly mobile, which is probably the prime reason for its acute toxicity at levels well below the normal thiol level.

Hg—C bonds are thermodynamically unstable (typical Hg—C bond energies are in the range 80–100 kJ mol^{-1}). Despite this $CH_3Hg^+(aq)$ persists in an aqueous environment due to its extreme mobility and an even lower affinity for the oxygen atom of water. Moreover, a steady state flux of $CH_3Hg^+(aq)$ in the aqueous environment is readily maintained by efficient methylating bacteria acting on 'inorganic' mercury. The half-life of methylmercury in the human body has been measured as around 70 days which may be compared with 4 or 5 days for inorganic Hg^{II}.

As with inorganic mercury, accumulation in the brain and irreversible damage to the nervous system remains the main hazard associated with methyl mercury intake. The CH_3Hg^+ cation has also been implicated in DNA cleavage and chromosomal damage. As a result the interaction of $CH_3Hg^+(aq)$ with nucleosides and nucleotide bases has received extensive study [126]. Aqueous reactions of potential importance in the environmental and atmospheric chemistry of mercury have been recently reviewed [127].

12.2.5 ^{199}Hg NMR Studies on Hg(aq) Species

Growing use is being made of the ^{199}Hg NMR nucleus ($I = \frac{1}{2}$) in the study of the hydrolysis/coordination/organometallic chemistry of mercury compounds [42] as high-field FT-NMR machines become routine instruments in most chemical laboratories. Chemical shifts are normally reported relative to $(CH_3)_2Hg$ (0.0 ppm). Solutions of soluble inorganic Hg^{II} salts, i.e. containing $Hg^{2+}(aq)$, are found at the highest field between -2200 and -2400 ppm whereas dimeric $Hg_2^{2+}(aq)$ is found further downfield at ~ -1450 ppm. It is unusual to find the chemical shift of a lower oxidation state of a metal at a lower field. Here the strong downfield influence ($\sim 800-900$ ppm) stems from the presence of the Hg—Hg bond [128]. Another noticeable observation is the unusual breadth of the resonances for $Hg_2^{2+}(aq)$, a feature which is believed to stem from an exchange reaction between $Hg_2^{2+}(aq)$ and some $Hg^{2+}(aq)$ which is presumed present as a result of disproportionation and/or oxidation [129]. Consistent with this presumption the ^{199}Hg NMR linewidths of both $Hg^{2+}(aq)$ and $Hg_2^{2+}(aq)$ broaden upon addition of salts of the other species. ^{199}Hg NMR has also been used widely to study complex equilibria involving $CH_3Hg^+(aq)$ [130]. Large downfield shifts are found with sulfur donor ligands such as glutathione (400 ppm). ^{199}Hg chemical shifts cover the range from $+350$ (a few HgR_2 compounds) to -2600 ppm (HgO_6 compounds) [42].

REFERENCES

[1] (a) Spiro, T. G. (ed.), *Zinc Enzymes*, Wiley Interscience, New York, 1983.
   (b) Kimura, E., *Pure Appl. Chem.*, **65**, 355 (1993).
[2] (a) Suh, J. H., *Acc. Chem. Res.* **25**, 273 (1992).
   (b) Silverman, D. N., and Lindskog, S., *Acc. Chem. Res.*, **21**, 30 (1988).
   (c) Christianson, D. W., and Lipscomb, W. N., *Acc. Chem. Res.*, **22**, 62 (1989).
   (d) Matthews, B. W., *Acc. Chem. Res.*, **21**, 333 (1988).
[3] (a) Kaatze, U., Lonnecke, V., and Pottel, R., *J. Phys. Chem.*, **91**, 2206 (1987).
   (b) Morris, D. F. C., Short, E. L., and Water. D. N., *J. Inorg. Nucl. Chem.*, **25**, 975 (1963).
   (c) Quicksall, C. O., and Spiro, T. G., *Inorg. Chem.*, **5**, 2232 (1966).
[4] (a) Kimura, E., *Prog. Inorg. Chem.*, **41**, 443 (1994).
   (b) Zhang, X. P., van Eldik, R., Koike, T., and Kimura, E., *Inorg. Chem.*, **32**, 5749 (1993).
   (c) Koike, T., and Kimura, E., *J. Am. Chem. Soc.*, **113**, 8935 (1991).

[5]   Summers, M. F., van Rijn, J., Reedijk, J., and Marzilli, L. G., *J. Am. Chem. Soc.*, **108**, 4254 (1986) (and refs. therein).
[6]   Coleman, J. E., and Vallee, B. L., *J. Biol. Chem.*, **236**, 2244 (1961).
[7]   Dalgarno, D. C., and Armitage, I. M., *Adv. Inorg. Biochem.*, **6**, 113 (1984).
[8]   (a) Buxton, G. V., and Sellers, R. M., *Coord. Chem. Revs.*, **22**, 195 (1977).
      (b) Adams, G. E., Baxendale, J. H., and Boag, J. W., *Proc. Chem. Soc.*, 241 (1963).
      (c) Baxendale, J. H., and Dixon, R. H., *Proc. Chem. Soc.*, 23 (1963).
[9]   Gogolev, A. V., Makarov, I. E., and Pikaev, A. V., *High Energy Chem., Engl. Trans.*, **15**, 85 (1985).
[10] Buxton, G. V., and Sellers, R. M., *J. Chem. Soc. Faraday Trans. I*, 558 (1975).
[11] Stanbury, D. M., *Adv. Inorg. Chem.*, **33**, 69 (1989).
[12] Navon, G., and Meyerstein, D., *J. Phys. Chem.*, **74**, 4067 (1970).
[13] Meisel, D., Matheson, M. S., Mulac, W. A., and Rabini, J., *J. Phys. Chem.*, **81**, 1449 (1977).
[14] Rabani. J., Mulac, W. A., and Matheson, M S., *J. Phys. Chem.*, **81**, 99 (1977).
[15] Kelm, M., Lilie, J., and Henglein, A., *J. Chem. Soc. Faraday Trans. I*, 1132 (1975).
[16] Baxendale, J. H., Keene, J. P., and Stott, D. A., *J. Chem. Soc., Chem. Commun.*, 715 (1966).
[17] Meyerstein, D., and Mulac, W. A. *J. Phys. Chem.*, **72**, 784 (1968).
[18] Sellers, R. M., *J. Chem. Educ.*, **58**, 114 (1981).
[19] Buxton, G. V., Sellers, R. M., and McCracken, D. R., *J. Chem. Soc. Faraday Trans. I*, 1464 (1976).
[20] Gupta, M P., and Agrawal, J. L., *Acta Cryst.*, **C6** 103 (1977).
[21] Broomhead, J, M., and Nicol, A. D. I., *Acta Cryst.*, **1**, 88 (1948).
[22] Hargreaves, A., *Acta Cryst.*, **10**, 191 (1957).
[23] Cariati, F., Erre, L., Micera, G., Panzanelli, A., Ciani, G., and Sironi, A., *Inorg. Chim. Acta*, **80**, 57 (1983).
[24] Knuuttila, P., *Polyhedron*, **3**, 303 (1984).
[25] Zviedre, I. I., Bel'skii, V. K. and Mardanenko, V. K., *Latv. Psv. Zin. Akad. Vest Kim.*, 387 (1985).
[26] Bezrukavnikova, I. M., Polynova, T. N., Kovaleva, I. B., Mitrofavanova, N. D., and Povai-Koshits, M. A., *Koord. Khim.*, **18**, 150 (1992).
[27] (a) Dagnall, S. P., Hague, D. N., and Towl, A. D. C. *J. Chem. Soc. Faraday Trans. II*, 2161 (1982).
      (b) Kuznetsov, V. V., Trostin, V. N., and Krestov, G. A. *Izv. Vyssh. Uchebn. Zaved. Khim. Khim. Teknol.*, **24**, 709 (1981).
[28] (a) Licheri, G., Paschina, G., Piccaluga, G., and Pinna, G., *Z. Naturforsch, Teil A*, **37**, 1205 (1982).
      (b) Radnai, T., Palinkas, G., and Camaniti, R., *Z. Naturforsch, Teil A*, **37**, 1247 (1982).
[29] (a) Ohtaki, H., Yamaguchi, T., and Maeda, M., *Bull. Chem. Soc. Japan*, **49**, 701 (1976).
      (b) Ohtaki, H., and Johansson, G., *Pure. Appl. Chem.* **53**, 1357 (1981).
[30] Andreev, P., Myund, L. A., and Lilich, L. S., *Chem. Abstr.*, **83** 66398 (1975).
[31] Swift, T. J., *Inorg. Chem.* **3**, 526 (1964).
[32] Audinos, R., and Zana, R., *J. Chim. Phys. Phys.-Chim. Biol.*, **78**, 183 (1981).
[33] Maeda, M., Ito, T., Hori, M., and Johansson, G., *Abstracts of the 30th International Conference on Coordination Chemistry*, Kyoto, Japan, 322 (1994).
[34] (a) Goggin, P. L., Johansson, G., Maeda, M., and Wakita, H., *Acta Chem. Scand.*, **A38**, 625 (1984).

(b) Wakita, H., Johansson, G., Sandstrom, M., Goggin, P. L., and Ohtaki, H., *J. Solution Chem.*, **20**, 643 (1991).

[35] (a) Epperlein, B. W., Kruger, H., Lutz, O., and Nole, A., *Z. Naturforsch, Teil A.* **30**, 1237 (1975).

(b) Epperlein, B. W., Kruger, H., Lutz, O., and Schwenk, A., *Z. Naturforsch, Teil A*, **29** 1553 (1974).

[36] Shapovalov, I. M., and Radchenko, I. V., *Russ. J. Struct. Chem.* **12**, 705 (1971).

[37] Bol, W., Gerrits, G. J. A., and Panthaleon van Eck, C. L., *J. Appl. Crystallog.*, **3**, 486 (1970).

[38] Jenkins, T. J., and Lewis, J., *Spectrochim. Acta*, **37A**, 47 (1981).

[39] (a) Prince, R. H., in *Comprehensive Coordination Chemistry* (eds. G. Wilkinson, R. D. Gillard and J. A. McCleverty), Vol. 7, Pergamon, New York, 1987, pp. 926–1004.

(b) Richens, D. T., 'Zinc, cadmium and mercury', *Royal Society of Chemistry Annual Reports*, **86A**, 61–75 (1989).

[40] Montgomery, H., and Lingfelter, E. C., *Acta Cryst.*, **20** 728 (1966).

[41] Ohtaki, H., Maeda, M., and Ito, S., *Bull. Chem. Soc. Japan*, **47**, 2217 (1974).

[42] Dechter, J. J., *Prog. Inorg. Chem.*, **33** 472–485 (1985).

[43] Summers, M. F., *Coord. Chem. Revs.*, **86**, 43 (1988).

[44] Baes, C. F., and Mesmer, R. E., *The Hydrolysis of Cations*, Krieger, Florida, 1986, pp. 287–94 (and refs. therein).

[45] Schorsch, G., *Bull. Soc. Chim. Fr.* 1449 (1964).

[46] Reichle, R. A., McCurdy, K. G., and Helper, L. G., *Can. J. Chem.*, **53**, 3841 (1975).

[47] Ferri, D., and Salvatore, F., *Ann. Chim. (Rome)*, **78**, 83 (1988).

[48] Zinevich, N. I., and Garmash, L. A., *Zh. Neorg. Khim.*, **20**, 2838 (1975).

[49] Schindler, P., Reinert, M., and Gamsjager, H., *Helv. Chim. Acta*, **52**, 2327 (1969).

[50] Mannoorettonnil, M., and Gilbert, J., *Bull. Soc. Chim. Belg.*, **83**, 9 (1974).

[51] Ducummon, Y., Laurenczy, G., and Merbach, A. E., *Inorg. Chem.*, **27**, 1148 (1988).

[52] Margerum, D. W., Cayley, G. R., Weatherburn, D. C., and Pagenkopf, G. K., in *Coordination Chemistry* (ed. A. E. Martell) ACS Monograph 174, Washington DC, 1978, Ch. 1.

[53] Rorabacher, D. B., *Inorg. Chem.*, **5**, 1891 (1966).

[54] (a) Miceli, M., and Steuhr, J. E., *Inorg. Chem.*, **11**, 2763 (1972).

(b) Grant, M. W., *J. Chem. Soc. Faraday Trans. I*, 560 (1973).

[55] Holyer, R. H., Hubbard, C. D., Kettle, S. F. A., and Wilkins, R. G., *Inorg. Chem..*, **4**, 929 (1965).

[56] Laurenczy, G., Ducummon, Y., and Merbach, A. E., *Inorg. Chem.*, **28**, 3024 (1989).

[57] Berwick, A., and Robertson, P. M., *Trans. Faraday. Soc.*, **63**, 678 (1967).

[58] Maass, G., *Z. Phys. Chem.*, **20**, 138 (1968).

[59] Tamura, K., *J. Phys. Chem.*, **81**, 820 (1977).

[60] Fittipaldi, F., and Petrucci, S., *J. Phys. Chem.*,**71**, 3414 (1967).

[61] Geier, G., *Ber. Bunsenges.*, **69**, 617 (1965) and *Helv. Chim. Acta*, **51**, 97 (1968).

[62] Tamura, K., *J. Phys. Chem.*, **91**, 4596 (1987).

[63] Eigen, M., *Pure Appl. Chem.*, **6**, 105 (1963).

[64] Sudmeier, J. L., and Reilley, C. N., *Inorg. Chem.*, **5**, 1047 (1966).

[65] Swaddle, T. W., and Mak, M. K. S., *Can. J. Chem.*, **61**, 473 (1983).

[66] Hewkin, D. J., and Prince, R. H., *Coord. Chem. Revs.*, **5**, 45 (1970).

[67] Murty, V. V. S., Seshasayee, M., and Murty, B. V. R., *Indian J. Phys.*, **55A**, 310 (1981).

[68] Charbonnier, F., Faure, R., and Loiseleur, H., *Rev. Chim. Miner.*, **16**, 555 (1979).

[69] Sharps, J. A., Brown, G. E., and Stebbins, J. F., *Geochim. Cosmochim. Acta*, **57**, 721 (1993).

[70] See, for example, Hague, D. N., and Moreton, A. D., *J. Chem. Soc. Dalton Trans.*, 1171 (1989).

[71] Koike, T., and Kimura, E., *J. Am. Chem. Soc.*, **113**, 8395 (1991).

[72] Koike, T., and Kimura, E., *Abstracts of the 30th International Conference on Coordination Chemistry*, Kyoto, Japan, 1994, p. 289.

[73] Kimura, E., Shionoya, M., Hoshino, A., Ikeda, T., and Yamada, Y., *J. Am. Chem. Soc.*, **114**, 10134 (1992).

[74] Billo, E. J., Brito, K. K., and Wilkins, R. G., *Bioinorg. Chem.*, **8**, 461 (1978).

[75] (a) Perrin, D., *Stability Constants of Metal Ion Complexes Part B, Organic Ligands*, IUPAC Chemical Data Series, Pergamon, Oxford, 1982.
(b) Martell, A. E., and Smith, R. M., *Critical Stability Constants*, Vol. 1, *Amino Acids*, Plenum, New York, 1985.

[76] (a) Hay, R. W., and Nolan, K. B, in *Amino Acids and Peptides*, Royal Society of Chemistry Specialist Periodical Reports **15–22**, 1991.
(b) Burger, K. (ed.), *Biocoordination Chemistry*, Ellis Horwood, Chichester, 1990.

[77] Cherifi, K., Decock-le Reverend, B., Varnagy, K., Kiss, T., Sovago, I., and Kozlowski, H., *J. Inorg. Biochem.*, **38**, 69 (1990).

[78] (a) Pitzer, K. S., *Acc. Chem. Res.*, **12**, 271 (1979).
(b) Pyykko, P. and Desclaux, J.-P., *Acc. Chem. Res.*, **12**, 276 (1979).
(c) McKelvey, D. R., *J.Chem., Educ.*, **60**, 112 (1983).
(d) Pyykko, P., *Chem. Revs.*, **88**, 563 (1988).

[79] (a) Grdenic, D., *J. Chem. Soc.*, 1312 (1956).
(b) Grdenic, D., Sikirica, M., and Vichovic, I., *Acta. Cryst.*, **31B**, 2174 (1975).

[80] Dorm, E., *Acta Chem. Scand.*, **25**, 1655 (1971).

[81] Johansson, G., *Acta Chem. Scand.*, **20**, 553 (1966).

[82] Mason, W. R., *Inorg. Chem.*, **21**, 147 (1983).

[83] Pugh, W., *J. Chem. Soc.*, 1824 (1937).

[84] Davies, R., Kipling, B., and Sykes, A. G., *J. Am. Chem. Soc.*, **95**, 7250 (1973).

[85] Vogler, A., and Kunkely, H., *Inorg. Chim. Acta*, **162**, 169 (1989).

[86] Horwath, O., Ford, P. C., and Vogler, A., *Inorg. Chem.*, **32**, 2614 (1993).

[87] Cartledge, G. R., *J. Am. Chem. Soc.*, **93**, 906 (1941).

[88] Faraggi, M., and Amozig, A., *Int. J. Radiat. Phys Chem.*, **4**, 353 (1972).

[89] Fujita, S., Horii, H., and Taniguchi, S., *J. Phys. Chem.*, **77**, 2868 (1973).

[90] (a) Waltz, W., Akhtar, S. S., and Eager, R. L., *Can. J. Chem.*, **51**, 2525 (1973).
(b) Sharma, B. K., *Can. J. Chem.*, **46**, 2757 (1968).

[91] (a) Nazhat, N. B., and Asmus, K.-D., *J. Phys. Chem.*, **77**, 614 (1973).
(b) Baxendale, J. H., Garner, C. D., Senior, R. G., and Sharpe, P., *J. Am. Chem. Soc.* **98**, 637 (1976).

[92] Sillen, L. G., *Svensk Kem, Tidskr.*, **58**, 52 (1946).

[93] Forsling, W., Hietanen, S., and Sillen, L. G., *Acta Chem. Scand.*, **6**, 901 (1952).

[94] Fujita, S., Horii, H., and Taniguchi, S., *J. Phys. Chem.*, **79**, 969 (1975).

[95] Tan, K. H., and Taylor, M. J., *Aust. J. Chem.*, **31**, 2601 (1978).

[96] Ershov, B. G., *Russ. Chem. Rev. Engl. Trans.*, **50** 1119 (1981).

[97] Yatsimirskii, K. B., Tikhonova, L. P., and Svarkovskaya, I. P., *Russ. J. Inorg. Chem.*, **14**, 1572 (1969).

[98] Brodersen, K., and Hummel, H.-U., in *Comprehensive Coordination Chemistry* (eds., G. Wilkinson, R. D. Gillard and J. A. McCleverty), Vol. 7, Pergamon, New York, 1987, pp. 1048–96.

[99] (a) Johansson, G., *Acta Chem. Scand.*, **25**, 2787, 2799 (1971).

(b) Sandstrom, M., Persson, I., and Ahrland, S., *Acta Chem. Scand.*, **A32**, 607 (1978).

[100]   Johansson, G., and Sandstrom, M. *Acta Chem. Scand.*, **A32**, 109 (1978).

[101]   (a) Ahlberg, I., *Acta. Chem. Scand.*, **16**, 887 (1962).

(b) Baes, C. F., and Mesmer, R. E., *The Hydrolysis of Cations*, Krieger, Florida, 1986, pp. 302–12.

[102]   Hietanen, S., and Sillen, L. G., *Acta Chem. Scand.*, **6**, 147 (1952).

[103]   Barnum, D. W., *Inorg. Chem..*, **22**, 2297 (1983).

[104]   Bergstrom, P.-A., Lindgren, J., Sandstrom, M., and Zhou, Y., *Inorg. Chem.*, **31**, 150 (1992).

[105]   Johansson, G., *Acta Chem. Scand.*, **25**, 1905 (1971).

[106]   Matkovik, B., Ribar, B., Prelensik, B., and Herak, R., *Inorg. Chem.*, **13**, 2006 (1974).

[107]   Bjornlund, G., *Acta Chem. Scand.*, **28A**, 169 (1974).

[108]   Bjornlund, G., *Acta Chem. Scand.*, **25**, 1645 (1971).

[109]   Eigen, M., and Eyring, E. M., *Inorg Chem.*, **2**, 636 (1963).

[110]   Tamura, K., Harada, S., Hiraissh, M., and Yasunaga, T., *Bull. Chem. Soc. Japan*, **51**, 2928 (1978).

[111]   Gross, H., and Geier, G., *Inorg. Chem.*, **26**, 3044 (1987).

[112]   (a) See, for example, Satchell, D. P. N., Satchell, R. S., and Wassef. W. N., *J. Chem. Soc. Perkin Trans II*, 1091 (1992).

(b) Satchell, D. P. N., Satchell, R. S., *J. Chem. Res. S*, 262 (1988).

(c) Penn, D., and Satchell, D. P. N., *J. Chem. Soc. Perkin Trans II*, 813 (1982).

[113]   See, for example, Diaddario, L. L., Ochrymowycz, L. A., and Rorabacher, D. B., *Inorg. Chem.*, **31**, 2347 (1992).

[114]   Masuda, Y., and Sekido, E., *Bunseki Kagaku, E.*, **39**, 683 (1990).

[115]   Aikoh, H., and Shibahara, T., *Physiol. Chem. Phys. Med. NMR*, **22**, 187 (1992).

[116]   Shibahara, T., Akashi, H., Yamasaki, M., and Hashimoto, K., *Chem. Lett.*, **4**, 689 (1991).

[117]   Herath Banda, R. M., Craig, D., Dance, I. G., and Scudder, M., *Polyhedron*, **10**, 41 (1991).

[118]   Goggin, P. L., and Woodward, L. A., *Trans. Farad. Soc.*, **56**, 1591 (1960) and **58**, 1495 (1962).

[119]   Craig, P. J., in *Comprehensive Organometallic Chemistry* (eds G. Wilkinson, F. G. A. Stone and E. W. Abel), Vol. 2, Pergamon, New York, 1982, pp. 979–1015.

[120]   Wardell, J. L., in *Comprehensive Organometallic Chemistry* (eds. G. Wilkinson, F. G. A. Stone and E. W. Abel), Vol. 2, Pergamon, New York, 1982, pp. 863–978.

[121]   Jawaid, M., Ingman, F., and Liem, D. H., *Acta Chem. Scand.*, **32A**, 333 (1978).

[122]   Schwarzenbach, G., and Schellenberg, M., *Helv. Chim. Acta*, **48**, 28 (1965).

[123]   Clarke, J. H. R., and Woodward, L. A., *Trans. Farad. Soc.*, **62**, 3032 (1966).

[124]   (a) Libich, S., and Rabenstein, D. L., *Anal. Chem.*, **45**, 118 (1973).

(b) Rabenstein, D. L., Evans, C. A., Tourangeau, M. C., and Fairhurst, M. T., *Anal Chem.*, **47**, 338 (1975).

[125]   (a) Erni, I. W., and Geier, G., *Helv. Chim. Acta*, **62**, 1007 (1979).

(b) Geier, G., and Gross, H., *Inorg. Chim.Acta*, **156**, 91 (1989).

(c) Raychera, J. M. T., and Geier, G., *Inorg. Chem.*, **18**, 2486 (1979).

[126]   See, for example,

(a) Prizant, L., Olivier, M. J., Rivest, R., and Beauchamp, A. L., *Can. J. Chem.*, **59**, 1311 (1981).

(b) Buncel, E., Norris, A. R., Racz, W. J., and Taylor, S. E., *Inorg. Chem.*, **20**, 98 (1981).

(c) Buncel, E., Hunter, B. K., Kumar, R., and Norris, A. R., *J. Inorg. Biochem.*, **20**, 171 (1984).

(d) Norris, A. R., Kumar, R., and Buncel, E., *J. Inorg. Biochem.*, **22**, 11 (1984).

[127] Munthe, J., and McElroy, W. J., *Atmos, Env. A, Gen. Topics*, **26**, 553 (1992).

[128] Kruger, H., Lutz, O., Nolle, A., and Schwenk, A., *Z. Physik A*, **273**, 325 (1975).

[129] Peringer, P., *J. Chem. Res. S*, 194 (1980).

[130] Sudmeier, J. L., Birge, R. R., and Perkins, T. G., *J. Mag. Reson.*, **30**, 491 (1978).

Appendix 1

Transition Metal Ion Electronic Spectra: Electron Configurations and Term Symbols

As m_l and m_s apply to individual electrons so M_L and M_S apply to multielectron systems:

$$M_L = \Sigma m_l \qquad M_S = \Sigma m_s$$
$$M_L = L, (L-1), \ldots, 0, \ldots, -L$$
$$M_S = S, (S-1), \ldots, \ 0, \ldots, -S$$

Likewise as the d subshell is orbitally five fold degenerate so the D term also has five M_L values, $2, 1, 0, -1$ and -2 and

$$
\begin{array}{ll}
S & \text{corresponds to } M_L = 0 \\
P & \text{corresponds to } M_L = 1, 0, -1 \\
D & \text{corresponds to } M_L = 2, 1, 0, -1, -2 \\
F & \text{corresponds to } M_L = 3, 2, 1, 0, -1, -2, -3 \\
G & \text{corresponds to } M_L = 4, 3, 2, 1, 0, -1, -2, -3, -4, \text{etc.}
\end{array}
$$

Example: the d^2 configuration of V^{3+}
There are 45 different electron configurations allowed within the Pauli exclusion principal. These are resolvable into a set of degenerate energy states defined by their M_L and M_S values and each is given a label known as the *term symbol*. The term symbol is represented by ^{2S+1}L. The expression $2S+1$ known as the spin multiplicity and L (M_L) relates to the orbital multiplicity. Therefore the term symbol is an expression of the total spin and total orbital multiplicity.

To illustrate the rules for finding the respective terms and their symbols for a given electron configuration, we can first examine the situation for a simpler case: the p^2 configuration of gaseous carbon:

m_l values				
1	0	-1	M_S	M_L
/	/		1	1
X			0	2
/	\		0	1
\	/		0	1
\	\		-1	1
/		/	1	0
/		\	0	0
\		/	0	0
\		\	-1	0
	X		0	0
	/	/	1	-1
	/	\	0	-1
	\	/	0	-1
	\	\	-1	-1
		X	0	-2

Rules for Extracting Terms

1. Look for the highest M_L value $(=L)$. Here $M_L(=L)=2$ so have a D term present.
2. What value(s) of M_S are associated with it?
   Only $M_S = 0$ so $S = 0$. Therefore the term symbol is ^1D.
3. Now remove all configurations corresponding to it (five in all), i.e.

$$M_S = 0, \; M_L = 2, 1, 0, -1, -2$$

   This leaves ten configurations to sort.
4. Look for the remaining highest M_L value $(=L)$.
   Now $M_L(=L)=1$ so have a P term present.
5. Look for the corresponding M_S values again; now $M_S = 1, 0$ and -1 so the term symbol is ^3P.
6. Remove all the configurations corresponding to ^3P (9 altogether, 3×3).
7. This leaves one configuration left over, $M_L = 0$, $M_S = 0$, so there is a ^1S term.

Therefore the 15 PEP allowed configurations making up the threefold degenerate p^2 situation transform into *three* energy terms ^1D, ^3P and ^1S. Which is the lowest in energy, i.e. the ground state? This is the one with the highest S (electrons having the same m_s value and lying in different orbitals) and if more than one such state is relevant then take the highest L value. Therefore here it is ^3P.

Thus the electronic configuration of gaseous carbon atoms corresponds to a ground state ^3P term with higher energy terms ^1D and ^1S. Note the transitions to these would be spin-forbidden (change in S).

d" Configurations

Applying the same rules the following sets of terms arise from the stated free ion high-spin d^n configurations:

d^1, d^9	^2D
d^2, d^8	$^3F, ^3P, ^1G, ^1D, ^1S$
d^3, d^7	$^4F, ^4P, ^2H, ^2G, ^2F, ^2D(2), ^2P$
d^4, d^6	$^5D, ^3H, ^3G, ^3F(2), ^3D, ^3P(2), ^1I, ^1G(2), ^1F, ^1D(2), ^1S(2)$
d^5	$^6S, ^4G, ^4F, ^4D, ^4P, ^2I, ^2H. ^2G, ^2F(2), ^2D(3), ^2P, ^2S$

Note that the numbers in parentheses indicate how many times a given term occurs. The same set of terms applies to d^n as to d^{10-n} since the 'hole' concept applies.

Crystal Field Splitting of Terms

Just as d orbitals are split by the presence of a crystal field, terms are as well. The labels are the same as for the splitting of the orbitals:

Orbitals of free ion	Orbitals in a cubic octahedral or tetrahedral field	Free ion terms	Terms in a cubic octahedral or tetrahedral field
s	a_1	S	A_1 (orbital singlet)
p	t_1	P	T_1 (triplet)
d	$e + t_2$	D	E (doublet) $+ T_2$ (triplet)
f	$a_2 + t_1 + t_2$	F	A_2 (singlet) $+ T_1 + T_2$
g	$a_1 + e + t_1 + t_2$	G	$A_1 + E + T_1 + T_2$

Electronic Transitions for d" Configurations–Orgel Diagrams

Consider the d^2 configuration of V^{3+}, ground state free ion term $= ^3F. \; ^3F$ splits in a cubic field to give $^3A_2, \; ^3T_1$ and $^3T_2$ terms. Which one corresponds to the ground state term for the $[V(OH_2)_6]^{3+}$ ion? In an *octahedral* field the orbital configuration is t_{2g}^2. There are three possible ways of expressing t_{2g}^2 about the three degenerate t_{2g} orbitals:

$$d_{xy}^1, d_{xz}^1, d_{yz}^0 \quad d_{xy}^1, d_{xz}^0, d_{yz}^1 \quad d_{xy}^0, d_{xz}^1, d_{yz}^1$$

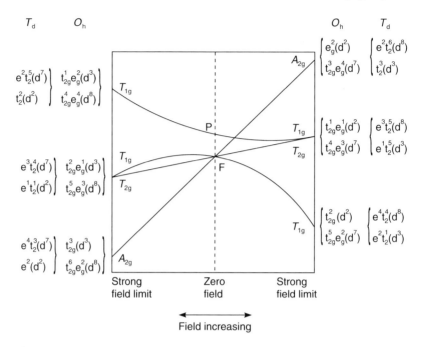

Figure A1.1. Orgel diagram for configurations having free ion F ground states

Therefore it cannot be the singlet state $^3A_2$. It is one of the ^3T states, for which we refer to the Orgel diagram (Fig. A1.1). It is the $^3T_1$ state. Since the higher ^3P term also gives a $^3T_1$ state, differentiated as $^3T_1(P)$, we can predict *three* spin-allowed transitions for $[V(OH_2)_6]^{3+}$ from the ground state $^3T_1(F)$ to higher energy spin multiplicity 3 states. (Note that since the field is octahedral, having a centre of inversion, we add the symmetry label g for gerade.) So we have for $[V(OH_2)_6]^{3+}$:

$$^3T_{1g}-^3T_{2g} \quad (580\,nm)$$
$$^3T_{1g}-^3T_{1g}(P) \quad (400\,nm)$$
$$^3T_{1g}-^3A_{2g} \quad (\text{obscured}) \text{ (see Section 5.1.2.2)}$$

Similarly for $[V(OH_2)_6]^{2+}$ (d^3) we can show that there is now the singlet ground term $^4A_{2g}$ (corresponding to t_{2g}^3) and the spin-allowed transitions:

$$^4A_{2g}-^4T_{2g} \quad (843\,nm)$$
$$^4A_{2g}-^4T_{1g}(F) \quad (556\,nm)$$
$$^4A_{2g}-^4T_{1g}(P) \quad (\sim 390\,nm, sh) \text{ (see Section 5.1.1)}$$

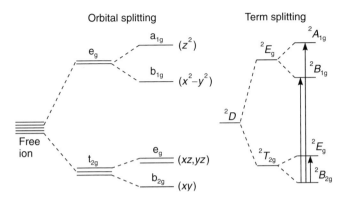

Figure A1.2. Tetragonal field for a d^1 ion

Note:

1. The A_2–T_2 separation is 10 Dq (or Δ), the crystal field splitting parameter.
2. The Orgel diagram is the same for d^n and d^{n+5} configurations but is inverted for d^n and d^{10-n}, i.e. d^8 is the same as d^3 but inverted to d^2.
3. A tetrahedral field has the diagram inverted with respect to an octahedral field with the same d^n configuration.

Crystal Field Splitting in Non-cubic fields

Within a square planar or tetragonally distorted octahedral crystal field a further splitting of the terms can result (Fig. A1.2), the term labels now reflecting individual properties of the d orbitals with respect to symmetry operations. For example:

Term in octahedral field	Tetragonal or square planar field
A_1	A_1
A_2	B_1
E	A_1 (z^2) + $B_1(x^2 - y^2)$
T_1	A_2 + E
T_2	B_2 (xy) + E (xz, yz)

Example: d^1 ion $[Ti(OH_2)_6]^{3+}$ (Section 4.1.2) or $[VO(OH_2)_5]^{2+}$ (Section 5.1.3).

For $[VO(OH_2)_5]^{2+}$ the presence of the $V{=}O^{2+}$ group causes a strong tetragonal distortion resulting in an appreciable $^2B_{2g}$–^2E_g splitting comparable

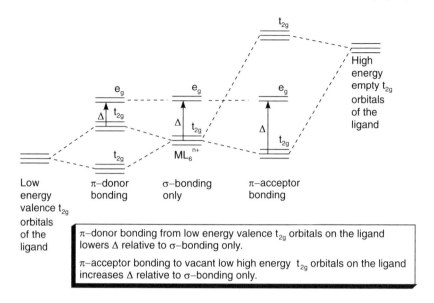

Figure A1.3. The influence of π-bonding on the splitting of the d-orbitals within an octahedral transition metal complex

in energy to the $^2B_{2g}-^2B_{1g}$ transition resulting in a broad envelope of bands around 760 nm. The $^2B_{2g}-^2A_{1g}$ transition lies in the u.v. region below 350 nm. In the case of $[Ti(OH_2)_6]^{3+}$ the splitting due to the tetragonal field is less with both the $^2B_{2g}-^2A_{1g}$ and $^2B_{2g}-^2B_{1g}$ transitions now occurring within a broad band envelope around 500 nm (Fig. 4.1). The $^2B_{2g}-^2E_g$ transition occurs in the infra-red region.

The d^9 ion $[Cu(OH_2)_6]^{2+}$, also the free ion ground state term ^2D (Section 11.1.2), is the classic example of a tetragonal crystal field. Here the crystal field term splitting diagram is inverted and the ground state term is now $^2B_{1g}$, corresponding to the 'hole' in the partially filled x^2-y^2 orbital. The three transitions (adopting the hole concept) now become:

$$d^9[Cu(OH_2)_6]^{2+} \quad \begin{array}{l} ^2B_{1g}-^2A_{1g} \text{ (lowest energy)} \\ ^2B_{1g}-^2B_{2g} \\ ^2B_{1g}-^2E_g \end{array}$$

Ligand Field Theory and the Introduction of π-Bonding

Ligand field theory describes metal–ligand bonding in terms of molecular orbital theory and describes the consequences of the introduction of π-bonding. This can occur in two ways depending upon whether the acceptor π-symmetry t_{2g} (t_2) orbitals are available on the metal ion or on the ligand. It can be easily appreciated that the metal ion will best behave as a π acceptor of electron density *from* the ligand if it has a low t_{2g} (t_2) orbital population (i.e. a transition metal ion to the far left of the period or a metal ion in a high oxidation state). Conversely π-bonding *to* the ligand is facilitated when the metal ion is in a low oxidation state (plenty of filled t_{2g} (t_2) orbitals and a high electron density) and the ligand has available π-acceptor t_{2g} (t_2) symmetry orbitals.

Water cannot behave as a π-acceptor but has the ability to act both as a σ-donor and a π-donor in bonding situations to a metal ion. Deprotonation of the water ligand increases the π-donating ability, i.e. O^{2-} better than OH^- better than H_2O. This is exemplified in the low coordination numbers and short $M-O$ bond lengths $(M{=}O)(\pi)$ bonding) found in many oxo compounds such as ClO_4^-, RuO_4 and VO_4^{3-}, etc.; π-bonding from OH^- ligands may be involved in the so-called conjugate-base labilising effect responsible for the enhanced lability of water ligands found in many hydroxy aqua compounds (see Section 1.4). Figure A1.3 shows the effect of the introduction of π-bonding on the splitting of the d orbitals within an octahedral field of ligands.

Appendix 2

Effective Ionic Radii (ER, pm) as Used Throughout the Book

(Taken from Shannon, R. D., *Acta Cryst.*, **A32**, 751 (1976))

Ion	CN	ER	Ion	CN	ER
Group 1 elements			Group 3 elements, lanthanides, actinides		
Li^+	4	59	Sc^{3+}	6	74.5
	6	76		(7)	(80)
	8	92		8	87
Na^+	4	99	Y^{3+}	6	90
	6	102		8	102
	8	118	La^{3+}	6	103
K^+	4	137		8	116
	6	138	Ce^{3+}	6	101
	8	151		8	114
Rb^+	6	152	Pr^{3+}	6	99
	8	161		8	113
	10	166	Nd^{3+}	6	98
	12	172		8	111
Cs^+	6	167	Pm^{3+}	6	97
	8	174		8	109
	10	181	Sm^{3+}	6	96
	12	188		8	108
			Eu^{3+}	6	95
				8	107

contd. overleaf

Appendix 2 *(Contd.)*

Ion	CN	ER
Group 2 elements		
Be^{2+}	4	27
	6	45
Mg^{2+}	4	57
	6	72
	8	89
Ca^{2+}	6	100
	8	112
	10	123
Sr^{2+}	6	118
	8	126
	10	136
Ba^{2+}	6	135
	8	142
	10	152
Ra^{2+}	8	148
Lanthanides, actinides		
Ce^{4+}	6	87
	8	97
	10	107
	12	114
Eu^{2+}	6	117
	8	125
	10	135
Yb^{2+}	6	102
	8	114
Th^{4+}	6	94
	8	105
	10	113
	12	121
Pa^{4+}	6	90
	8	101
U^{3+}	6	103
U^{4+}	6	89
	8	100
	12	117
U^{6+}	4	52

Ion	CN	ER
Gd^{3+}	6	94
	8	105
Tb^{3+}	6	92
	8	104
Dy^{3+}	6	91
	8	103
Ho^{3+}	6	90
	8	102
Er^{3+}	6	89
	8	100
Tm^{3+}	6	88
	8	99
Yb^{3+}	6	87
	8	99
Lu^{3+}	6	86
	8	98
Ac^{3+}	6	112
Group 5 elements		
V^{2+}	6	79
V^{3+}	6	64
V^{4+}	6	58
	8	72
V^{5+}	4	36
	6	54
Nb^{3+}	6	72
Nb^{4+}	6	68
	8	79
Nb^{5+}	4	48
	6	64
	8	74
Ta^{4+}	6	68
Ta^{5+}	6	64
	8	74
Group 6 elements		
Cr^{2+}	6	80

Appendix 2 (*Contd.*)

Ion	CN	ER
	6	73
	8	86
Np^{4+}	6	87
	8	98
Pu^{4+}	6	86
	8	96
Am^{3+}	6	98
	8	109
Cm^{3+}	6	97
Bk^{3+}	6	96
Cf^{3+}	6	95
Cf^{4+}	6	82
	8	92
Group 4 elements		
Ti^{3+}	6	67
Ti^{4+}	4	42
	6	61
	8	74
Zr^{4+}	4	59
	6	72
	8	84
Hf^{4+}	4	58
	6	71
Group 7 elements		
Tc^{4+}	6	65
Tc^{5+}	6	60
Tc^{7+}	4	37
	6	56
Re^{4+}	6	63
Re^{5+}	6	58
Re^{6+}	6	55
Re^{7+}	4	38
Group 8 elements		
Fe^{2+}	4	63

Ion	CN	ER
Cr^{3+}	6	62
Cr^{4+}	4	41
	6	55
Cr^{5+}	4	35
	6	49
Cr^{6+}	4	26
	6	44
Mo^{3+}	6	69
Mo^{4+}	6	65
Mo^{5+}	4	46
	6	61
Mo^{6+}	4	41
	6	59
W^{4+}	6	66
W^{5+}	6	62
W^{6+}	4	42
	6	60
Group 7 elements		
Mn^{2+}	4	66
	6	67
	8	96
Mn^{3+}	6	65
Mn^{4+}	4	39
Group 10 elements		
Ni^{2+} T_d	4	55
sq	4	49
	6	69
Ni^{3+}	6	60
Ni^{4+} ls	6	48
Pd^{2+} sq	4	64
Pd^{3+}	6	76
Pd^{4+}	6	62
Pt^{2+} sq	4	60
Pt^{4+}	6	63

contd. overleaf

Appendix 2 (*Contd.*)

Ion		CN	ER
	hs	6	78
	ls	6	61
		8	92
Fe^{3+}		4	49
	hs	6	65
	ls	6	55
		8	78
Fe^{4+}		6	59
Fe^{6+}		4	25
Ru^{3+}		6	68
Ru^{4+}		6	62
Ru^{5+}		6	57
Ru^{7+}		4	38
Ru^{8+}		4	36
Os^{4+}		6	63
Os^{5+}		6	58
Os^{6+}		6	55
Os^{7+}		6	53
Os^{8+}		4	39

Group 9 elements

Ion		CN	ER
Co^{2+}		4	58
		6	75
Co^{3+}	hs	6	61
	ls	6	55
Co^{4+}		4	40
		6	53
Rh^{3+}		6	67
Rh^{4+}		6	60
Rh^{5+}		6	55
Ir^{3+}		6	68
Ir^{4+}		6	63
Ir^{5+}		6	57

Group 13 elements

Ion	CN	ER
Al^{3+}	4	39
	6	54
Ga^{3+}	4	47
	6	62

Group 11 elements

Ion		CN	ER
Cu^+		2	46
		4	60
		6	77
Cu^{2+}	T_d	4	57
	sq	4	57
		6	73
Cu^{3+}	ls	6	54
Ag^+		2	67
	T_d	4	100
	sq	4	102
		6	115
Ag^{2+}	sq	4	79
		6	94
Ag^{3+}	sq	4	67
		6	75
Au^+		6	137
Au^{3+}	sq	4	68
		6	85
Au^{5+}		6	57

Group 12 elements

Ion	CN	ER
Zn^{2+}	4	60
	6	74
	8	90
Cd^{2+}	4	78
	6	95
	8	110
Hg^+	3	97
	6	119
Hg^{2+}	2	69
	4	96
	6	102

Group 16 elements

Ion	CN	ER
Se^{4+}	4	50
Se^{6+}	4	28
	6	42
Te^{4+}	4	66

Appendix 2 (*Contd.*)

Ion	CN	ER
In^{3+}	4	62
	6	80
	8	92
Tl^+	6	150
	8	159
	12	170
Tl^{3+}	4	75
	6	89
	8	98

Group 14 elements

Ion	CN	ER
Si^{4+}	4	26
	6	40
Ge^{4+}	4	39
	6	53
Sn^{4+}	4	55
	6	69
	8	81
Pb^{2+} py	4	98
	6	119
	8	129
	10	140
	12	149
Pb^{4+}	4	65
	6	78
	8	94

Group 15 elements

Ion	CN	ER
As^{3+}	6	58
As^{5+}	4	34
	6	46
Sb^{3+} py	4	76
	6	76
Sb^{5+}	6	60
Bi^{3+}	6	103
	8	117

Ion	CN	ER
Te^{4+}	6	97
Te^{6+}	4	43
	6	56
Po^{4+}	6	94
	8	108
Po^{6+}	6	67

Group 17 elements

Ion	CN	ER
Cl^-	6	181
Cl^{5+} py	3	12
Cl^{7+}	4	8
	6	27
Br^-	6	196
Br^{3+} sq	4	59
Br^{5+} py	3	31
Br^{7+}	4	25
	6	39
I^-	6	220
I^{5+} py	3	44
	6	95
I^{7+}	4	42
	6	53

Others

Ion	CN	ER
H^+	1	-38
	2	-18
O^{2+}	2	135
	4	138
	6	140
	8	142
OH^-	2	132
	4	135
	6	137

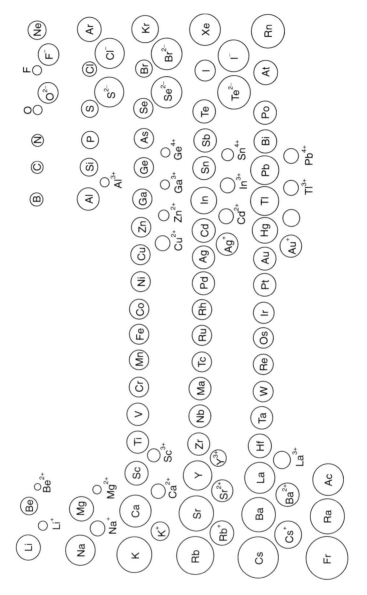

Figure A2.1 Diagrammatic representation of the trends in atomic and ionic radii with atomic number across the periodic table

Appendix 3

Single Ion Hydration Enthalpies (kJ mol^{-1})

(From Phillips, C. G., and Williams, R. J. P., *Inorganic Chemistry*, Vol. 1, Clarendon Press, New York, 1965, p. 161)

H^+	-1089.7	Ce^{3+}	-3368.5
		Pr^{3+}	-3411.2
Li^+	-514.1	Nd^{3+}	-3440.1
Na^+	-405.4	Pm^{3+}	-3476.1
K^+	-320.9	Sm^{3+}	-3512.9
Rb^+	-296.2	Eu^{3+}	-3545.5
Cs^+	-263.2	Gd^{3+}	-3569.4
		Tb^{3+}	-3603.7
Tl^+	-325.9	Dy^{3+}	-3635.1
		Ho^{3+}	-3665.6
Be^{2+}	-2487	Er^{3+}	-3689.4
Mg^{2+}	-1922.1	Tm^{3+}	-3715
Ca^{2+}	-1592.4	Yb^{3+}	-3737.1
Sr^{2+}	-1444.7	Lu^{3+}	-3758.5
Ba^{2+}	-1303.7		
Ra^{2+}	-1260	Ce^{4+}	-6488.5
Zn^{2+}	-2044.3	Cr^{2+}	-1849.7
Cd^{2+}	-2384.9	Cr^{3+}	-4401.6
		Mn^{2+}	-1845.6
Sn^{2+}	-1554.4	Fe^{2+}	-1920.0
Pb^{2+}	-1479.9	Fe^{3+}	-4376.5
		Co^{2+}	-2054.3
Al^{3+}	-4659.7	Ni^{2+}	-2105.8
Sc^{3+}	-3960.2	Cu^+	-594.1

Contd.

Appendix 3 (*Contd.*)

Y^{3+}	-3620	Cu^{2+}	-2100.4
La^{3+}	-3282.8	Ag$^+$	-475.3
Ga^{3+}	-4684.8		
In^{3+}	-4108.7	Hg^{2+}	-1853.5
Tl^{3+}	-4180		

Index